PRODUCTION MANAGEMENT AND ENGINEERING SCIENCES

SCIENTIFIC PUBLICATION OF THE INTERNATIONAL CONFERENCE ON ENGINEERING SCIENCE AND PRODUCTION MANAGEMENT (ESPM 2015), TATRANSKÁ ŠTRBA, HIGH TATRAS MOUNTAINS, SLOVAK REPUBLIC, 16–17 APRIL 2015

Production Management and Engineering Sciences

Editors

Milan Majerník, Naqib Daneshjo & Martin Bosák
Faculty of Business Economics with Seat in Kosice,
University of Economics in Bratislava, Slovak Republic

CRC Press
Taylor & Francis Group
Boca Raton London New York

CRC Press is an imprint of the
Taylor & Francis Group, an **informa** business

A BALKEMA BOOK

Published by:
CRC Press/Balkema
P.O. Box 447, 2300 AK Leiden, The Netherlands
e-mail: Pub.NL@taylorandfrancis.com
www.crcpress.com – www.taylorandfrancis.com

First issued in paperback 2020

© 2016 by Taylor & Francis Group, LLC
CRC Press/Balkema is an imprint of the Taylor & Francis Group, an informa business

No claim to original U.S. Government works

Typeset by MPS Limited, Chennai, India

ISBN 13: 978-0-367-73762-7 (pbk)
ISBN 13: 978-1-138-02856-2 (hbk)

Visit the Taylor & Francis Web site at
http://www.taylorandfrancis.com

and the CRC Press Web site at
http://www.crcpress.com

Contents

Scientific and technical support of sustainable production

Experimental research of technologies for remediation of tailing ponds

Production Management and Engineering Sciences – Majerník, Daneshjo & Bosák (Eds)
© 2016 Taylor & Francis Group, London, ISBN: 978-1-138-02856-2

Introduction

Sustainable and competitive socio-economic development, either in terms of global or regional and local point of view, is ever more associated with optimizing its material and environmental dimension in the form of "green growth." International calls from this area and the derived strategic development documents of European Union and Slovak republic are focussing on the process of balanced fulfilment of technical, economic, social and environmental objectives. Production management, related technical support sciences and economic practice must respond to these developing trends.

Monography of Production Management and Engineering Sciences composed of monothematic scientific submissions by experts from several countries, describe the status quo and presents the progressive development trends in industrial engineering and integrated quality (technical, environmental, security, economic, social, energetic, ...), as well as in management of sustainable and competitive enterprise production, supported by technical fields such as computer simulation and modelling processes, metrology and testing, authorization and standardization, with focus on implementation of the best techniques and technologies available, Innovation Union and Horizon 2020 included.

Researchers present a case study of experimental research and development of new technology supporting qualitative environmental-safety growth and sustainable development of production management in the field of energetics in Slovakia and European Union.

Editors

Progressive trends in production management

Production Management and Engineering Sciences – Majerník, Daneshjo & Bosák (Eds)
© 2016 Taylor & Francis Group, London, ISBN: 978-1-138-02856-2

Assessment of ergonomic risk for firefighters

M. Balážiková, M. Tomašková & M. Dulebová
Faculty of Mechanical Engineering, Technical University in Košice, Košice, Slovak Republic

ABSTRACT: Firefighters have a legal obligation to rescue people, animals and property during fires, accidents and other emergencies, where the incident stands for any case in which life or health of a person, animal or property is threatened. The paper analyses the ergonomic load on a fireman while using hydraulic rescue equipment. The chosen method RULA (Rapid Upper Limb Assessment) aims to quickly assess stress in the spine and upper limbs. In this method, individual body parts are observed. Each body part is assigned a value, the amount of which is a deviation from the neutral position. Continuous value is calculated for each group separately (A-arm, B-torso, neck and legs), and eventually the final result is calculated by combination and transformation of the individual values of given groups.

1 ERGONOMIC AND SAFETY

Ergonomics is defined as a multi-disciplinary branch, which deals with human activity (within system of work) and their bonds (man and machine in the working process) to the work equipment and working environment (chemical, physical, biological, social and organizational). The main objective is to optimize those aspects affecting the humans at the workplace with regard to the workload. Ergonomics as a whole plans working and natural environment, proposes work tools, construction of machines and health protection with aim to alleviate load on humans while improving the performance and efficiency of work (Karwowski 1999).

A traffic accident is an event in traffic involving killing or injury to persons or damage to property directly related to motor and non-motor vehicles.

Types of traffic accidents:

1. Simple traffic accident: an accident, in which one person was injured lightly, or one person was injured seriously, alternatively one person was killed.
2. Mass traffic accident: an accident in which five people were injured lightly or 3 people were injured seriously, at least two persons were killed, or there was a crash of 5 or more vehicles.
3. Traffic accident involving dangerous substance: a traffic accident involving a vehicle transporting a dangerous substance, with the risk of leakage of the substance into the environment with subsequent risks to the population (Occupational safety to the recovery facility, 2015).

The individual processes performed during an action on site of traffic accident involving passenger cars can be divided into the following sections, Figure 1:

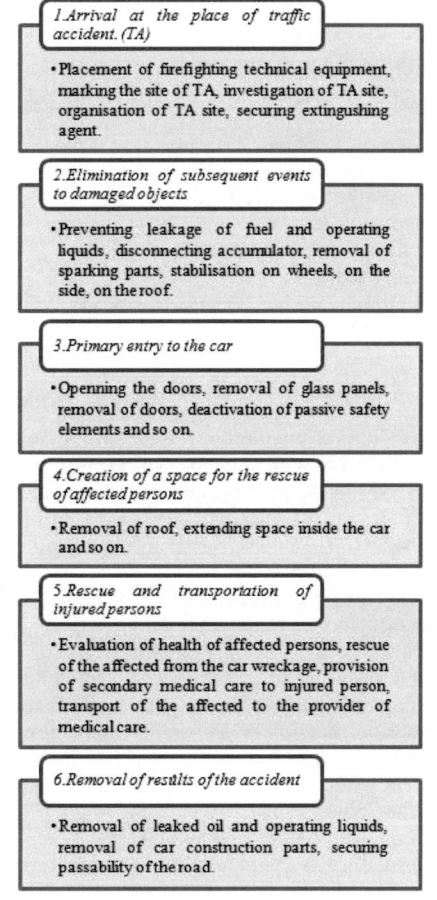

Figure 1. Activities performed in the individual stages of intervention.

1. Arrival at the place of the accident.
2. Elimination of subsequent events to damaged objects.
3. Primary entry to the car.
4. Creation of a space for the rescue of affected persons.
5. Rescue and transportation of injured persons.
6. Removal of results of the accident (Tomašková 2014).

2 TECHNICAL EQUIPMENT USED IN RESCUE WORKS IN TA

Firefighters use technical equipment for special activities during the intervention in traffic accidents in order to rescue persons, e.g. from crashed cars (Occupational safety to the recovery facility, 2015).

They are divided into:

1. Hydraulic extrication equipment,
2. Pneumatic lifting and sealing bags,
3. Manual extrication equipment,
4. Other equipment and tools.

Hydraulic extrication equipment includes:

1. Hydraulic shears,
2. Hydraulic expansion pliers,
3. Hydraulic expansion cylinder,
4. Hydraulic cutter,
5. Hydraulic door opener,
6. Mini shears,
7. Hydraulic wedge (Occupational safety to the recovery facility, 2015).

2.1 *Work safety at the site of a traffic accident*

Hydraulic and pneumatic rescue equipment may only be used by a person who has been trained and informed on operation, maintenance and work safety. Range of personal protective equipment according to the extent of risk is determined by the chief of the intervention.

When lifting a passenger car by means of rescue equipment nobody can enter under the car, attention must be paid to limbs. Stability of the car is continuously ensured. The purpose of stabilization is to prevent vibration, displacement, elimination of suspension function of the car. Stabilization must be performed carefully with regard to the injured persons, without jerky movements with the car.

Other objects, which could endanger the firefighter, must be stabilized as well (tree, pole, load). It is necessary to ensure that the high pressure hose couplings are properly connected and secure. It must be observed whether the hydraulic hoses do not form loops and are not placed over sharp edges.

Firefighter working with hydraulic tools, Figure 2, must stand on a stable surface, holding the instrument in both hands in designated places. If an

Figure 2. Placement of extrication equipment in fire engine.

instrument forms a sharp angle with the vehicle, the firefighter does not stand between the vehicle and the instrument, but on the outside of the instrument.

Firefighter holding the separated parts of the construction stands next to the platform and not against sharp edges. Hands must not be placed between the jaws of the instrument. Cutting hard and hardened materials is prohibited. Once the jaws of the scissors are beginning to open to the side, or they begin to curl, cutting must be discontinued.

The entire surface of the jaws is used when working with the instruments (they may be expanding, cutting—a combined instrument). During the rescue it is necessary to pay attention to the placement of generators that serve to activate the airbags. Cutting into the generator can cause an explosion and firing of parts of the package.

3 RULA METHOD (RAPID UPPER LIMB ASSESSMENT)

The aim of ergonomics may not be just the effort to adapt shapes of objects and tools so that their shape is as much as possible consistent with the extent of the human body, but also study of the cumulative exposure to adverse (risk) factors and the subsequent proposal of new measures to help reduce the mental, physical and mental load on the worker. It is therefore necessary to pay sufficient attention to ergonomics and quality of working environment (Bosák 2013), as only then it is possible to achieve the improvement of conditions at the workplace. In this performance it can be noted that in addition to other aspects of safety, taking care of this area contributes to the reduction of occupational diseases and to the reduction of work accidents (Karwowski 1999).

RULA Method, Figure 3, is aimed at rapid assessment of strain to neck and upper limbs. Each body part is assigned a value, the amount of which is a deviation from the neutral position. Continuous value is calculated for each group separately (A-arm, B-torso, neck and legs), and eventually the final result is calculated by combination and trans-

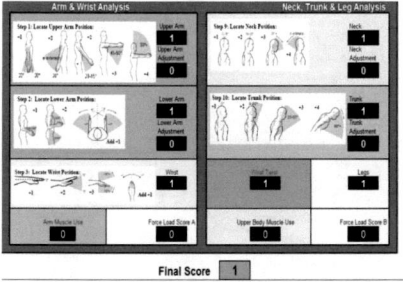

Figure 3.   RULA Method – worksheet..

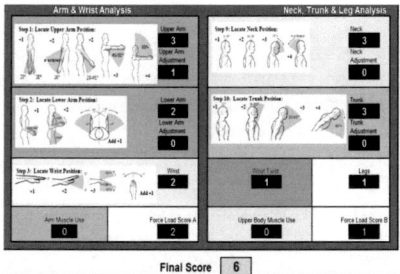

Figure 4.   Assessment of ergonomic risk for firefighters during manipulation with hydraulic tools.

Table 1.   The resulting level of risk in RULA method.

RULA score	Level of risk	Measures
1–2	Negligible risk	No necessary measures
3–4	Low risk	Continuous observation
5–6	Moderate risk	Necessary implementation of measures
7 and more	High risk	Immediate implementation of measures

formation of the individual values of given groups (Jack and Process Simulate Human 2015).

The resulting level of risk in RULA Method is displayed in Table 1.

4   THE ASSESSMENT OF ERGONOMIC HAZARD FOR FIREFIGHTERS

One of the most commonly used ways to improve ergonomic aspects is using Tecnomatix Jack Simulation Software, DHM—Digital Human Modeling software. This software represents great progress enabling the designer to design a suitable working environment. The most significant advantage is the possibility to perform various experiments on the model without incurring additional funds.

By means of modeling and simulation of the working environment and working operations in the digital environment it is possible to identify human factors and resolve ergonomic risks before practical implementation. Basis of examining the interaction of man and his work environment is the physical structure of the body and the possibility of load. The human body is part of the physical world and the same physical laws apply to it as they do to other objects. The aim of ergonomics at this level is to optimize the interaction between the body and the physical laws of its surrounding with the use of knowledge of anthropometry. This requires understanding of the working position in terms of strain to joints and muscles during work.

Currently, the trend on the market moves the industry towards investment in new methods for increasing the flexibility, increasing the productivity. The focus is placed on the interaction between the man and the machine. In this case it is necessary to focus on flexible production system, which will be studied in laboratory conditions. The research will cover the area of ergonomic features in this system. The main idea of ergonomics is to reduce the physical load by designing or modifying job position, working methods and tools. The ambition is to eliminate excessive stress and load, limit unsuitable postures at work and reduce the number of movements needed to complete the work.

The aim of the article is to assess the ergonomic risks (stress on the spine and upper limbs) for firefighters when using extrication tools in a traffic accident, which is placed in fire engines. Such extrication tools support a maximum load of 7–25 kg. Ergonomic risk assessment was carried out by RULA method in two ways. The aim was to compare these two types of assessment. The first form was paper-based RULA method, Fig. 6, and the second was the modeling of work position in Technomatix software and the evaluation by RULA method in that software.

Evaluation of the degree of risk by paper form of RULA method is shown in Figure 4. The resulting level of risk was calculated to 6, which is a Moderate risk, which calls for introduction of measures to minimize it.

The second type of ergonomic risk assessment was modeling of work position of a firefighter during the manipulation with hydraulic tools in the Tecnomatix Jack Simulation software and its subsequent evaluation by RULA method, which is part of the software. In this procedure it was not necessary to enter value deviations from the neutral position, which the software automatically assigned from the modeled (illustrative) position. Evaluation of the level of risk by software RULA method is depicted in Figure 5 (illustrative modeling of work position in Figure 6). The resulting level of risk was calculated to 5, which is also a moderate risk, which calls for introduction of measures to minimize it.

Proposal of measures to prevent health problems associated with motional support system in the case

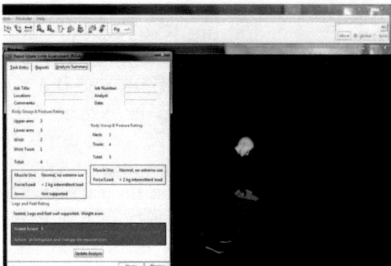

Figure 5. Evaluation of the level of risk for the positions of a firefighter by software RULA method.

Figure 6. Manipulation with extrication equipment.

of a firefighter is more complex. As a firefighter has precisely determined safety rules of behavior on the site of a traffic accident, and the value of the level of risk was calculated to Moderate risk due to the capacity of the extrication tools, which can be 7–25 kg, so the proposal would be alternation of firefighters in the performance of the activity. Manipulation with extrication tools may be necessary at every traffic accident, or every other.

5 CONCLUSION

Ergonomics is gaining importance in the sphere of work. It is important for optimizing working processes and eliminating identified difficulties causing distress and health problems to the worker. The aim of this paper was to assess the risk of damage to musculoskeletal system using modern ergonomic methods. By selecting an appropriate method it is possible to determine and assess the unacceptable position of the body in the performance of work. RULA Method was used for analysis, it is primarily used for classification of load on neck, torso, arms and legs. The result of the analysis was to determine the level of risk and the need for the changes. After the assessment of the results, measures are proposed to reduce the negative effects on human health and improve working conditions.

The aim of this paper is to inform about the possibility of using the tool for RULA method application for rapid detection of current status of the workload of workers in practice. However, it is better to implement the principles of ergonomics at the beginning (prevention) and not retroactively (treatment). Ergonomic requirements should be introduced as soon as possible in working conditions of operations and, where possible, integrated into the project.

ACKNOWLEDGEMENT

The paper has been developed within project APVV-0337-11 Research on new and emerging risks of industrial technologies within the integrated security management as a prerequisite for sustainable development management.

This paper was created thanks to the support of operational program Research and Development, for the project: University Science Park TECHNICOM for innovative applications with support for knowledge technologies, code ITMS: 26220220182, co-financed from the European Regional Development Fund.

REFERENCES

Bosák, M., Krištanová, A., Kravec, M., Lešková, Ľ., Čorba, J. & Čorba, J. 2013. Environmental-economic aspects of management. *Financial management of firms and financial institutions*: Czech Republic Ostrava: VŠB—Technical University of Ostrava.

Jack & Process Simulate Human [online]. [cit. 12.3.2015]. Available on: http://www.plm.automation.siemens.com/en_us/products/tecnomatix/manufacturing-simulation/human-ergonomics/jack.shtml.

Karwowski, W. & Maras, W.S. 1999. *The occupational ergonomics handbook*.

Lueder, R.K. 1996. A proposed RULA for computer users. Proceeding of the Ergonomics Summer Workshop. *San Francisco: UC Berkeley Center for Occupational and Environmental Health*.

Occupational safety to the recovery facility. [online]. [cit.9.3.2015]. Available on: http://elearnhazzsk.webnode.sk/technicke-prostriedky-pouzivane-pri-vyslobodzovacich-zachrannych-pracach/bezpecnost-prace-s vyslobodzovacim-naradim.

Qianxiang, Z., Songtao, D., Zhongqi, L. & Xiaohui, Z. *Geometric dimension Model of virtual human for ergonomic analysis of man-machine system. Advances in Physical ergonomics and safety, United States of America*.

Regulation of the Minister of the Interior no. 26/2002 on the rules of safety and health at work in the Fire and agency Corps as subsequently amended (no. 54/2003 of 27 October 2003).

Tomašková, M. 2014. The performance of firefighters in rescuing persons in traffic accidents using extrication equipment. *Monitoring and Expertise in Safety Engineering*. 4(3):18–24.

Production Management and Engineering Sciences – Majerník, Daneshjo & Bosák (Eds)
© 2016 Taylor & Francis Group, London, ISBN: 978-1-138-02856-2

Creating productive and arranged workplace through 5S method in a company

A. Bašistová & L. Stankovič
Faculty of Business Economics with Seat in Košice, University of Economics, Bratislava, Košice, Slovak Republic

M. Ferencová
Faculty of Public Administration, Pavol Jozef Šafárik University, Košice, Slovak Republic

R. Haluška
NEO Business Software s.r.o., Brno, Czech Republic

ABSTRACT: A group of enterprises that have also been severely affected by the crisis are the manufacturing companies in leather industry in Slovakia. How is it possible that some companies are more successful and efficient than the others? The answer consists in discovery and completion of the mission of its existence and in continuous improvement. Article approaches the insights of various domestic and foreign authors on the issue of downsizing business, methods Kaizen and 5S. In the empirical part, we added supplementary methods: Visual management and Operational standards to provide the most comprehensive view. The article brings partial results of the primary research we made in a private enterprise with exclusively domestic capital, focusing on the production of synthetic and natural rubber latex gloves. They affect both the comparison of the situation before and after the implementation of 5S method, and also the quantification of benefits from the implementation of 5S.

1 INTRODUCTION

For a better understanding of the substance of Lean Enterprise and Lean Manufacturing, there is a need for a better insight into this issue. Therefore, in order to create a complete notion we choose different understandings and definitions by several authors. We begin with a quote by Napoleon Hill that captures the essence of lissomness and, concurrently, it presents a challenge of each top company that wants not only to promote but also to maintain in top positions on the global market: "Whatever the mind can conceive and believe, it can achieve" (Stole 1987). Rother says, "Lean manufacturing is a paradigm and a way of thinking about production. It is a philosophy that shortens the continuous time by eliminating the waste that high quality products at low cost are delivered on time" (Košturiak & Frolík 2006). The hearts of Lean manufacturing are flexible, motivated people constantly solving problems. The essence of Lean manufacturing is to look for unusual solutions, and it concerns not only production but also the company as a whole. According to Košturiak, a Lean company practically means doing only the activities that are needed. Doing them correctly right for the first time, making them faster than others and spending less money when doing so. It is an increase in company performance so that on a given area the company was able to produce more than their competitors that with a given number of people and equipment it manufactures a higher added value than the other, that at the time it handles more purchase orders that various business processes and activities consume less time. "The thinness of the company is that we are doing exactly what our customer wants, with a minimum number of actions that do not add a value to the product or service". To be lean means: to produce more money and to produce them faster with less effort" (Košturiak & Frolík 2006). The method "A productive and organized workplace-5S" is a well-known method which aims to eliminate waste in the workplace. Husáriková describes the 5S method as a method of five steps that create a work environment which is tidy and clean and where things are arranged properly. Today, the use of this method should be mandatory for all manufacturing companies because its absence means waste, mess, lack of self-discipline, poor work ethic, unnecessarily high costs, poor quality of production and failure to fulfill obligations (Husáriková 2011).

Our objective is to facilitate the control of gloves production shop in the company by introducing the method of "Productive and organized workplace-5S", and to quantify the benefits of implementing the "Productive and organized workplace-5S" project. Therefore, it is important to understand and to have a deeper understanding of concepts, as well as their mutual relations and broader ties. One of the goals of a lean workplace according to 5S method is to improve transparency of the manufacturing, and also to remove basic forms of waste, to improve cleanliness in the workplace, to increase productivity of the work shift, to define a standard layout in the workplace, to ensure clear rules in the workplace, to create a standardized, visual and well-arranged workplace 5S, to improve the work environment, to improve the staff motivation, and to improve the working atmosphere (Michalska & Szewieczek 2007).

2 MODERN PHILOSOPHIES—LEAN ENTERPRISE, LEAN MANAGEMENT, KAIZEN

Lean manufacturing cannot function without close link with product development and production engineering, logistics and administration of the business. "It is therefore a mistake that in many businesses, for example, manufacturing processes and product development are physically separated from each other." Thinness is being built yet in pre-production stages and a large part of the Lean Enterprise is heavily influenced by logistic chain or processes in the administration. This effort is of course the process of downsizing which occurs on all organizational levels of business management, including senior management. The idea of lean manufacturing takes many forms. According to Křivánek, "the idea of slenderness" in its present form can probably be found in the last century, whether in philosophy of the company Baťa or carmaker Ford, which were and still are considered to be visionaries of the era. Finally, the standard of a "lean model" was successfully applied in full by Toyota Company. Toyota, however, would have not achieved its target without building a "new" corporate culture founded on teamwork, education, solidarity and humility of people of the enterprise" (Hines & Taylor 2000) in addition to improving the manufacturing technologies (innovations and continuous improvements).

Other authors (Burget et al. 2013; Hines & Taylor 2000, Shah & Peter 2003) speaks about a lean workplace as a workplace where there is only what is necessary, and in places that are designated for that purpose. This is particularly the elimination of unnecessary items from the workplace, maintenance of order on the workplace and creation of an organized, standardized and orderly workplace.

It is important that the workplace be well organized according to the needs and requirements of production workers. Matsushita, founder of Matsushita Electric Industries, responded the question whether he worries about copying the know-how by the competitors: "Technology can be copied, information can be obtained, capital can be bought...but the organization's ability to operate effectively and give the highest priority to the most important things cannot be bought or installed. The atmosphere of trust and encouragement is always made at home; it is the only competitive advantage which cannot be imitated" (Košturiak & Frolík 2006). And that is Kaizen. Košturiak defined Kaizen as an event. According to him, Kaizen is the life, it begins with self-reflection, self-awareness and, humility which underlies the ability to learn and improve on. It's the only thing in the world that we can improve 24 hours a day. This concerns mainly us—we must improve ourselves, our own life, our time management, our ability to devote their time to important things and only then we can improve relations and cooperation with co-workers, and ultimately, we improve things and processes around. "This is an ongoing process, for the Japanese it is as natural as breathing for a man" (Košturiak 2010). Hines and Taylor briefly said about Kaizen that it is a continuous improvement of quality, technology, processes, corporate culture, productivity, safety and management (Hines & Taylor 2000).

3 PRODUCTIVE AND ORGANIZED WORKPLACE-5S

According to Burieta, specialist in 5S method, "The 5S is a summary of basic steps for the elimination of waste in the workplace, an essential element for improving a part of some other methodologies and concepts (Kaizen, TPM, Lean Enterprise...). Wastage can be an activity, operation, material or element of a process which does not add a value to the product (non-value-added activity) or service while increasing the price that the customer is not willing to accept" (Kormanec 2010). In the hierarchy of lean manufacturing it belongs to the field of processes standardization and lean workplace. "By applying 5S an increased smoothness of material flow, a more productive deployment of equipment, a more efficient placement of materials and stocks can be achieved. By means of 5S method, an improvement in production quality, or an increase in productivity by eliminating the waste, or an increase in safety in the workplace can be achieved (Xie & Zhang 2004). A thorough implementation of five pillars of the 5S method is the beginning of laying the foundations for the development of improvement actions that ensure the company's survival. The company's survival is crucial especially for the preservation of jobs." These five pillars are defined as sorting, setting policy, shine, standardization and preservation." Since those words begin with S in Japanese and English they are also referred to as pillars of the 5S (Hines & Taylor 2000).

Creation of good working conditions is the primary step of each program for enhancing the enterprise. The 5S system is a set of five basic rules:

1. Sort (sort—organization).
2. Set in Order (give it in order—systemization and visual layout).
3. Shine (cleanliness—clean).
4. Standardize (standardization—creation of standards of work).
5. Sustain (maintenance—improvement).

4 THE OBJECTIVE AND METHODOLOGY FOR OPTIMIZING THE 5S METHOD IN THE COMPANY

As noted above, the five pillars are essential for improving the business. When people hear about the five pillars for the first time, it may be difficult for them to understand why precisely these five pillars are so important. In practice, when implementing changes we quite often encounter different attitudes and questions from employees that are positive but also negative. When implementing changes it is always necessary to think of them and to try to understand them, in order to introduce the changes smoothly and without unnecessary obstacles. Therefore, it is also our main objective of this paper, by introducing the method of "Productive and organized workplace-5S" we want to facilitate the control of the gloves production workshop in the company and to quantify the benefits of implementing the project "Productive and organized workplace-5S". When determining the main goal we were coming out of the tree analysis of the causes and consequences of the current problems in the production. We use core methods and other additional methods that interrelate. The particular core methods of Kaizen, Productive and organized workplace-5S, Visual management, SOP-Operational standards are interdependent. By using these methods we measure the visible help in improving and simplifying the management procedures and ensuring a smoother running of the gloves workshop.

For the needs of our research, we treat them as follows:

- Kaizen,
- Productive, arranged workplace,
- Visual management,
- SOP-Operational Standards.

5 SAMPLE, DATA COLLECTION AND RESULTS

In the scope of Kaizen, Productive and arranged workplace-5S, Visual management and Operational Standardization we created and implemented the following standards and improvements within the gloves production workshop:

1. Standard of the shop producing gloves and comprising:
   - A standardized layout of 5S workshop of gloves production.
   - Layout after the change (implementation of 5S).
   - Rules for the workshop.
   - Map of processes in gloves manufacturing.
   - The manufacturing process of leather gloves—graphic illustration.
   - Standard of the product Leather gloves—photos with descriptions.
   - A standardized technological process optimized.
   - Technological process of leather gloves manufacturing:
     - The current technological procedure according to videos + observations.
     - A standardized technological process of leather gloves manufacturing—optimized.
   - Standards of inspection and audit procedures.
   - Manual as waste is measured.

The workplaces were divided into smaller offices and we introduced them as:

1. T1 standard for the workplace of cutting out parts 5S.
2. T2 standard for the workplace of stamping and packaging of cotton gloves.
3. T3 standard for the workplace of reversing, end working and packaging of leather gloves.
4. T4 standard for the workplace of cutting, stamping, packaging of leather products.
5. T5 standard for the workplace of sewing leather gloves.
6. T6 standard for the workplace Adler.
7. T7 standard for the workplace—cotton gloves.
8. Standard for the shop manufacturing gloves.

2. Individual workstations T1–T7 include the following standards:
   - Name of workplace.
   - Standardized inventory of items in the workplace.

- Visualization (photograph) of the standardized clean workplace.
- Standard of workplace cleaning.

3. Other improvements implemented in the workshop:
   - Use of the "Cause & Effect Tree" method
   - Use of tags and labels on shelves, desks, instruments indicating the item name and place of storage.
   - Color Coding—orientation address of products stored.
   - Color coded floor, lines on the walls and on the desktops.
   - Use of checklists for 5S Audit.

Due to the fact that we elected the troubleshooting process by analyzing the cause and effect tree: top issues—result—the real cause of problems—the root causes of problems—measure—a key measure, we resolved at once, through election of this procedure, several problems the production was constantly fighting with.

After we resolve the problems, we defined few steps to implement the project "Productive and organized workplace—5S", to make improvement in workplace:

Phase 1: At the beginning of the pilot project 5S, Robert Haluška was appointed project manager while assuming responsibility for the successful implementation of 5S pilot project in the gloves production shop. The knowledge regarding the successful implementation of 5S method in the company he acquired by completing a two-day workshop entitled "Productive & Organized Workplace—5S". The purpose of the workshop was to understand the essence of the modern method 5S and to master its principles, and furthermore, to learn how to identify and eliminate the waste.

Phase 2: Step by step we gathered the information about the current state of the production we acquired from production workers and company management, for example, through interviews, questionnaires, observations, surveys, processes and manufacturing operations in order to create a real clear picture of the current situation and the problems in the manufacturing. Based on the information we have developed a comprehensive analysis of the causes and effects. We have created a tree diagram of the current problems in the gloves production workshop with the aim to understand the relationships and to identify the root causes and the key measures that must be taken in the workshop. The output of the problem analysis (cause and effect tree) was that we clearly formulated the key top issue on which we focused our attention in the article: "I one glove manufacturing workshop

4 manufacturing processes were running concurrently (leather gloves, cotton gloves, latex gloves and leather products). It was difficult to control, coordinate and inspect the processes in progress of these production processes. There were no control mechanisms and standardization which would facilitate the control of the production."

Gradually we understood the current situation in the workshop and now we had a complex view of the past and current problems, and the relationship between various processes and issues. In the pilot project 5S we focused on the implementation of the method Productive and organized workplace-5S which is intended to pave the way for the introduction of Kaizen culture and the adoption of other key measures. Given that the workshop produces four different kinds of products, it was necessary to determine which of the production processes when shortened in duration will have the biggest impact on improving the productivity of work as a whole. Based on measurements and comparisons (Table 1) we found out that the production process of leather gloves takes at longest, compared to the latex and cotton gloves. Leather products have not been taken into account as their production is rare.

Production of one pair of leather gloves lasts up to 26.4 minutes, production of cotton gloves takes 8.21 minutes and one pair of latex gloves lasts 4.36 minutes. So we decided to streamline the production of "leather" gloves. We needed to know the current status of waste in the workplace. Based on the calculations and analysis of waste we found out that the rate of wastage (useless movement, search, pacing, and waiting) in individual workers in the workplace is around 30% to 45%. We recorded the current state of various forms of waste by using camera and digital camera to be able to compare the changes after the project 5S completion. When calculating the return on investment, however, we took into account the fact that the pure continuous time of manufacturing operations did not include certain activities that do not add value but must have been carried out because, at that time it, the given operation or act could not be replaced or excluded, such as manual cutting of rubber bands which are then stitched on the glove, or hand winding of the yarn bobbin (sewing machines) by means of winding machine. Therefore, in calculating the return on investment we took into account the wastage of 10%.

Phase 3: Piecemeal we carried out individual steps of 5S method. In the first step 1S (sorting), we visually mapped the current state of the work and workplace and we removed all the objects which could be or have to be removed unconditionally. By implementing 1S (sorting) we found out that

Table 1. Comparison of process times and their evaluation.

Sample monitored	Continuous time of manufacturing (random sample per 430 pairs)	Comparison of results by finaltimes	Selection of variant (of sample monitored)
A-Latex gloves	1874 min/430 p (1 pair=4,36 min)	3.	No
B-Leather gloves	*11352 min/430 p (1 pair=26,4 min)*	*1.*	*Yes*
C-Cotton gloves	3530 min/430 p (1 pair=8,21 min)	2.	No

as a result of sorting and removal of unnecessary objects, materials and machinery we obtained a clear savings of 30% in surface area and we made the workplace significantly transparent.

In the second step 2S (layout, visualization and standard setting) we created a new, more efficient layout of individual workplaces designated as T1–T7. We arranged the items needed and we put them in their proper place. We set the standards so that they could be found and then saved by "anyone" with minimum effort and time while we were trying to keep the principle that the proposal of a new workplace meets all the criteria of maximum efficiency and productivity but it also had to meet the requirements of ergonomics and high security. From now on, every needed object is assigned a specific address and each employee works with minimal waste of walking, searching and waiting.

In the third step 3S (cleaning) we put the workplace and machinery into exemplary, the best possible condition and we created a standard of a clean workplace. In the standard of the clean workplace we defined the areas to be cleaned within the territory of the workplace, and we described who would perform the cleaning, when and how often, and what cleaning tools should be used for cleaning.

In the fourth step 4S (establishment of rules and standardization), we created new standards for a productive and arranged workplace—5S where we clearly described the governing rules and we illustrated these standards graphically. We thereby tried to ensure compliance with the changes introduced in the steps 2S and 3S, thus preventing the unwanted return to the state before the implementation of this methodology. These new standards are to serve as guides or templates for the workers and to help them work better and easier.

Implementing the final step 5S (conservation, improvement and audits) we have created and implemented a system of checking compliance with the established standards 5S through regular audits 5S in individual workplaces T1–T7. The purpose of periodic audits is to build the 5S culture, to strengthen discipline of workers and to pave the way for the introduction of Kaizen culture.

In the results of the work we compared the status of the workplace before and after implementation of changes in the gloves workshop and we enumerated particular benefits of the project 5S. The estimated period of implementation was three months. We managed to keep that time period. The expected rate of elimination of waste was 10% (walking, searching, and movements). After completing the 5S project we achieved a reduction in various forms of waste by 40% (walking, searching, and movements). We expected an increase in productivity of the workshop by 5% per year. This result will be assessed after project implementation. The estimated annual increase in productivity of the workshop is 10–15%. Initially we assumed that we improve the utilization of the production area by 15%, now we can say that we have exceeded this expectation and we have achieved an improvement by 20%. Regarding the expectation as to the shortening of the production time by 10% we can say that most likely it will be exceeded and it will be around 10–15%. The expected savings of general and administrative expenses was 10%, and we can assume that the savings will be varying between 10% and 15% per annum. The expected increase in annual handing over of finished products from the workshop at 10% productivity increase of the workshop will be at least 10%, which represents the value of the annual increase of production by more than 23,988 Euros. The only expectation we did not meet was the height of the single investment in the pilot project 5S, which was originally set to €1,500 and at the end of the project it was exceeded by €1,000, which is a negligible amount in relation to the total benefits that the pilot project 5S has brought.

6 DISCUSSION AND CONCLUSION

In this article we clarified the views of various domestic and foreign authors with regard to the issue of company downsizing, the Kaizen method and the method of productive and organized workplace-5S. Through a case study of a particular enterprise we presented the application of this method as well as advantages and disadvantages of 5S. However, any change has its followers and opponents, and so it was with the introduction of 5S method in the company. Therefore, we briefly describe the main types of resistance we had to face. Should we do not deal with this resistance it could jeopardize the efforts of the enterprise for implementing 5S. We therefore often meet the following attitudes:

- What is amazing in respect of sorting and setting right?
- Why to clean when it gets dirty again?
- Sorting and setting right shall not support production.
- We have already introduced sorting and setting right.
- We did 5S years ago and no result.
- We have too much work and cannot address the 5S activities.
- Why should we introduce the five pillars 5S?

If you encounter such attitudes then you should put the question: "Is this tool correctly understood by the people?" "We dare to say that in some companies, this tool is not well understood." To the question of manufacturing operators "What is it 5S?" they often answer "cleaning". Many times the method was badly explained or remained misunderstood. Action 5S is a fundamental building block for further implementation of advanced Kaizen methods and approaches of "downsizing". Where method 5S is not established stably there is no reason to implement the methods of Kaizen and the method FLOW. 5S is part of the basic stability of processes and that of the whole system of the production. "Many managers are still not familiar with this method and they do not realize what are the real benefits and effects of the correct application of the method 5S in practice. Their attitude in respect of 5S method implementation is often very cold, without interest, as if this did not concern them. After all, they have "their" serious problems (hustling contracts, material, completing reports, managing the enterprise, etc.)." Right the correct understanding and good completion of this method which is otherwise called the "5S of good housekeeping" can bring huge benefits for the enterprise in form of newly discovered material, time or financial resources. At times, in the mind of people, this method turns to 3S as "recurrent cleaning" (Bauer et al. 2012). In addition, people use the 5 pillars of 5S in their personal lives, without being aware of it at all. They exercise sorting and setting right when leaving things like trash cans or towels in adequate and well-known locations. Where home

environment fills up and becomes disordered, it usually starts working ineffectively and there are conflicts and problems. Hardly any production can reach such a degree of standardization of the procedures of five pillars as an organized man in his daily life.

6.1 Conclusions and benefits for the enterprise as a result of 5S implementation

Based on our topic, based on our research, there we are summarized the benefits for the company achieved by an implementing 5S method:

- Visualizing the waste.
- Reduces the waste:
  - By indicating minimum and maximum levels.
  - Errors are addressed through a "mistake defiant" equipment and visual management.
  - Unnecessary movements are removed using standardized techniques.
  - Simplifies the search for necessary items.
- Improvement of smoothness of the material flow, or more productive deployment of equipment, or more efficient location of materials and supplies can be achieved:
  - Reduction of stocks to one tenth.
  - Reduction of necessary working space by half.
  - Reduction of necessary investments in machinery, equipment or tools.
- Shortening of the time of product manufacturing.
- Reducing the hassle output by half.
- Reduction of hours of human efforts necessarily expended in the production to half.
- Zero defect deliver higher quality.
- Zero waste means less cost.
- Zero delay delivers reliable supplies.
- No injuries promote safety.
- No fault improving plant availability.
- No complaints generate greater confidence and trust.
- Zero incidences of "red numbers" bring business growth (Peterson & Smith 1998).

It is also possible to achieve an improvement in production quality, or to increase productivity by eliminating the waste and to increase safety in the workplace. Another advantage is to improve the attitudes of workers, or the improvement of corporate culture and working environment (Hirano 1995).

In conclusion we can say that after a thorough assessment of the successful fulfillment of all the set targets, namely the introduction of the method

Productive and well-arranged workplace-5S, we facilitated the control of the gloves production workshop in an enterprise and we also quantified the benefits of 5S project implementation.

ACKNOWLEDGEMENT

This contribution is the result of the project implementation VEGA No. 1/0328/13 Modeling causal relationships of innovations in Small and medium-sized enterprises.

REFERENCES

Bauer, M. et al. 2012. *Kaizen—The path to lean and flexible enterprise.* Brno: Biz Books.

Burget, M., Chodura, D. & Révay, M. 2001. *Evaluation and analysis of manufacturing processes—Lean.* VŠB: TU Ostrava.

Burieta, J. 2012. *Slim seminar: 5S—productivity and arranged workplace.* Žilina: IPA Slovakia.

Hines, P., & Taylor, D. 2000. *Going lean.* Cardiff: Lean Enterprise Research Centre.

Hirano, H. 1995. *5S for operators; 5 pillars of the visual workplace.* Portland: Productivity Press.

Husáriková, I. 2011. *Kaizen and the possibility of its application in the Czech Republic.* Zlín: UTB.

Kormanec, P. 2010. *SMED.* Žilina: *Internal materials* IPA Slovakia.

Košturiak, J., & Frolík, Z. 2006. *Slim and innovative company.* Praha: Alfa Publishing.

Michalska, J., & Szewieczek, D. 2007. The 5S methodology as a tool for improving the organization. *Journal of Achievements in Materials and Manufacturing Engineering.*

Peterson, J., & Smith, R. 1998. 5S Pocket Guide. Portland: Productivity, Inc.

Shah, R., & Peter, T.W. 2003. Lean manufacturing: context, practice bundles, and performance. *Journal of operations management.*

Stole, C.W. 1987. *Success through a positive mental attitude.* New York: Pocket books.

Xie, Y., & Zhang, Y. 2004. *Management of Central Sterile Supply Department (CSSD): 5S Method.* Chinese Journal of Nosoconmiology.

Production Management and Engineering Sciences – Majerník, Daneshjo & Bosák (Eds)
© 2016 Taylor & Francis Group, London, ISBN: 978-1-138-02856-2

Assessment methods of the influence on environment in the context of eco-design process

L. Bednárová
University of Economics in Bratislava, Kosice, Slovak Republic

L. Witek, R. Piętowska-Laska & A. Laska
Rzeszow University of Technology, Rzeszow, Poland

ABSTRACT: This article presents a new approach relating to the existing methodological achievements in the process of eco-design, consisting of the integration of environmental, economic and social aspects throughout the product lifecycle. This approach is associated with an expanded product liability, which enables companies to see the priorities that guided them in the past in a different light. The main aim of this article is to describe the process of eco-design, with particular reference to the use of selected methods for environmental assessment of products. In the works some methods and tools for ecological assessment were characterized and evaluated. The discussed methods create a deeper and more disaggregated analysis of the environmental aspects, facilitate the search for technological solutions aimed at minimizing the amount of resources and waste, and enable the evaluation of the effectiveness of solutions at the design stage of products.

1 INTRODUCTION

Dynamic economic growth combined with rapid population growth causes excessive use and impact on the natural environment. In times of ecological crisis discourse on sustainable development becomes important. The concept of sustainable development has made fundamental redefinition in the way of thinking and acting of companies in industrial goods sector which include issues related to the environment and its protection (Pacana et al. 2014). They are becoming increasingly aware of the need to improve manufacturing processes and manufactured products. The work presents a new approach relating to the existing methodological achievements in the process of eco-design, consisting of the integration of environmental, economic and social aspects throughout the product lifecycle—from raw material extraction through manufacturing processes and product manufacturing, its distribution, to the use and disposal of waste at the end of its usage. This approach is associated with an expanded product liability, which enables companies to see the priorities that guided them in the past in a different light. Due to the current state of the environment and limited natural resources, it appears necessary to make environmental assessment of the manufacturing process, which enables in a clear and concise to control environmental activities and to control the operation of the company in the field of environmental protection (Králiková et al. 2008).

The need to introduce environmental requirements for the design and development stages of new products is confirmed by many studies over the past several years (Bovea & Pérez-Belis 2012, Karlsson & Luttropp 2006, Li et al. 2015). Current practices are focused around the traditional approach involving achieving the highest possible profit, maintaining high quality at the possibly lowest costs (Asiedu & Gu 1996). Environmental requirements are still considered as a compulsory cost factor and restrictions for product design.

2 CONCEPT OF ECO-DESIGN

Eco-design refers to the systematic inclusion of environmental factors in product design at the stage of its conception (Tukker et al. 2000, Burchart-Korol 2010). Eco-design aims at creating products with minimal environmental impact throughout the product lifecycle from the conception phase to its final use—an approach "from the cradle to the grave" (Karlsson & Luttropp 2006). A holistic approach to the design, taking into account environmental criteria, allows connecting customers' needs with environmental liability of the company (Luttropp & Lagerstedt 2006). A number of studies have shown the use of eco-design methodology in different areas of the company (Brones et al. 2014, González-García et al. 2014, Morrison et al. 2013, Pochat et al. 2007). It allows reducing the costs by verification and modification of the product in the early stages of conception. It brings tangible benefits associated with the reduction of material and energy consumption of the products at every stage of the product lifecycle (Lewandowska & Foltynowicz 2006). It implies values for customers and society, giving the possibility to follow consumers' changing requirements and creating their new needs. Although the primary objective of eco-design is the reduction of impact on the environment, another major benefit is the increase of competitiveness by reducing costs and the prospects of entering new markets with new products (Borchardt et al. 2011, Knight & Jenkins 2009). Due to the rare use of eco-design, it is difficult for the companies to know the potential benefits of eco-design in the development process of a new product. The main barriers are the lack of technical knowledge and methods, as well as the lack of tools for the eco-design based on the technology and know-how (Theyel 2000).

3 METHODS AND TOOLS OF ECO-DESIGN

Knowledge of product environmental performance evaluation issues in its lifecycle is becoming increasingly important for the competitive position of the product and the company. A major problem in the evaluation and presentation of the products environmental performance is the selection of methods at the phase of eco-design, their evaluation and the manner of their suitability. Until now, a number of methods, techniques and tools have been established that allow methodical approach to the analysis and evaluation of the impact on the environment. These methods are still being developed by both practitioners and theorists (Table 1). Many studies point to the implementation of eco-design methodology in the activities of companies in the industrial goods sector (González-García et al. 2014, Hernandez et al. 2012, Yang et al. 2011).

4 CRITICAL OVERVIEW AND EVALUATION OF METHODS

One of the most fundamental methods for analysing decision-making processes of industrial goods is *Life Cycle Assessment* (LCA). Studies have proven that its usefulness is also valuable in the design of new production technologies and products which includes the whole lifecycle of the product (Yeang et al. 2008). The advantage of LCA is standardized structure and internationally accepted methodology. LCA principles were defined in the international ISO standards: 14000 series standards and ISO 14040. This complex, time consuming and costly analysis, unfortunately, makes a lot of problems during the implementation. The discrepancy of the analysis results and the roughness of collected data influence the inadequate interpretation and incomplete identification of improvement opportunities (Nowosielski et al. 2003).

Depending on the purpose of the study and the way of using the results, the level of quality requirements for the same analysis and results obtained with the use of it differentiate. The greater the number of stakeholders involved in decision-making process and the greater misalignment of interests, the higher the level of reliability the study should demonstrate. Each stage of the LCA process has specific sources of uncertainty which affect the level of results aggregation. Analysis progress is followed by propagation of uncertainty, which reaches its maximum at the highest level of results. This raises the dilemma of methods to be used in order to estimate it, as different uncertainty range will be presented by measurement uncertainty of the given size, and another will be presented by model parameter selection, which is a simplification of reality. The results of indicators indicate the type of impact on the environment and identify its main sources. Nevertheless, there is no possibility to interpret whether these results remain stable regardless of the products made and the quality of used data (Lewandowska 2006).

MIPS (Material Input per Service Unit) indicator and Ecoindicator 99 fulfil a useful role in the process of eco-design. MIPS indicator investigates the use of natural resources per specified unit. Similarly as in the case of LCA, the entire lifecycle of a product or process may be analysed. This is a method simpler than LCA (Ritthoff et al. 2002). It allows determining the potential impact of the product on the environment, but only in a quantitative manner. Additionally, it allows comparisons between the effects of particular goods and services and the alternative goods and services. Such an activity may be used to identify innovative products and processes. The main drawback is the lack of a qualitative assessment of the used materials. Gathering of large amounts of data is required, which makes this method costly. Ecoindicator 99 is a popular method of evaluating the state of the environment on the basis of a set of 99 indicator species based on an assessment of biodiversity. In order to calculate the value of this indicator, expressing the environmental impact of the analysed technology, we need to develop a tree of life for the discussed operation and technology,

Table 1. Characteristics of selected methods in the process of eco-design.

Method and references	Description
Checklists, (Borchardt et al. 2009, Piech 1999, Kishita et al. 2010)	Checklists represent logically organized and structured set of questions or statements previously prepared for the purpose of actions. The set of questions is used as a kind of procedure (algorithm), involving carrying out recommendations contained in the procedure.
Matrix Materials, Energy, Toxic Emissions (MET), (Knight & Jenkins 2009)	The matrix presents the lifecycle phases of production (supply of materials / components, production of final product, distribution to customers, product use and end of life). A declaration of material cycle, energy consumption and toxic emissions is assigned for each of these phases. Once the environmental aspects during product designing are estimated by this method, it is important to combine this estimation with other essential aspects (social, technical and financial or business and customers' benefits).
Environmentally Responsible Product/ Process Assesment (ERPA), (Hur et al. 2005, Bovea & Pérez-Belis 2012)	The results of this method provide a comprehensive overview of the environmental aspects of the products. It involves the creation of a matrix with dimensions 5×5, where the matrix cell is a stage of lifecycle (5 stages of the lifecycle) and the category associated with the impact on the environment. The environmental impact is assessed in 5 categories: the choice of materials, energy consumption, solid waste, liquid waste and gaseous emissions. Each element of the matrix receives an environmental impact assessment at a scale of 0–4 (the higher the number the better the result). Checklist for each of the environmental factors and the development of guidelines for the scoring for each question is required. It allows further determination in terms of product development.
Materials, Energy, Chemicals, Others Matrix (MECO), (Volínová 2012)	The method allows obtaining a concise overview of environmental problems. It involves assignment of the factors causing environmental problems at every stage of the lifecycle (but does not focus on the actual categories of the environment impact). Four areas are distinguished, according to the cause of the product impact on the environment: materials, energy, chemicals and others. MECO gives a first indications where there is a significant impact on the environment and shows where there is a lack of data. The main difficulty in the work is to obtain information from third parties about the elements of a product environmental impact. For easy data processing it seems reasonable to use a computer program.
Environmental Product Life Cycle Matrix (EPLC), (Bovea & Pérez-Belis 2012)	The matrix similar to ERPA, where there is no distinction between processes or products. It analyses the interactions between the stages of the product lifecycle and the various categories of impact. Each entry in the matrix is marked with a number from 0 to 4. The number 0 is the least impact on the environment, 4 indicates the highest environmental burden.
Product Investigation, Learning and Optimization Tool (PILOT), (Ostad-Ahmad-Ghorabi & Wimmer 2005)	Method using a computer that helps to integrate eco-design process with product development. It consists of three components: the product lifecycle, product development, product improvement. Facilitation is the provision of practical "checklists".
Environmental Benchmarking Matrix (EBM), (Bednarova 2009)	This is a comparative research method used to assess its activities in the scope of environmental aspects in relation to the best practice in its own sector of activity. Benchmarking can be performed once, but often, however, it is treated as a continuous process in which organizations continually attempt to improve their processes and products. The method is based on an assessment of five central areas: energy, materials, packaging, potential toxic substances and recycling.
Environmental Quality Function Deployment (EQFD), (Yang et al. 2011)	QFD method is known and used in the design and improvement of the quality of industrial products. The essence of the QFD method is the use of quality matrix in order to show correlations that occur between the customer's requirements (in consecutive rows of the matrix) and the technical characteristics of the product (in columns). It allows a thorough understanding of the customer's needs, before engineering solutions will be generated. The advantage of this method is a saving at the cost-intensive stage of product testing (elimination of unnecessary prototypes, reduction of testing and verification of solutions in practice) thanks to the earlier finalising of the product concept. The advantage of this method is the focus on generation of added value for the consumer. A limitation of the method can be a prolonged design stage which is time-consuming.
Eco-Design Checklist Method (ECM), (Wimmer 1999)	The method determines the requirements which are formulated in checklists. They are divided into categories regarding the whole product and its parts and functions. 40 parts and 20 product requirements were determined that are included in 13 checklists. The analysis is carried out in three dimensions with respect to the products, parts or functions. A holistic approach to the product is based on three levels of analysis: part-time level, function level and product level. The method clearly presents the weaknesses of the product.

select the appropriate functional unit, develop inventory tables, calculate the value of ecoindicator corresponding to the technology functional unit, analyse the environmental burden for each category i.e. Human Health, Ecosystem Quality and Resources.

The choice of the appropriate method depends on many factors, among others the desired level of analysis precision, the level of its detail, the availability of data, expected outcomes, time-consumption and costs (Baran & Janik 2013). Criteria for evaluation of individual methods are defined in the Table 1, where structural and functional criteria were detailed. Criteria C1–C7 are described in (Kłos & Kasprzak 2007).

The evaluation of methods taking into account the described criteria was presented in Table 2. In some cases, there is the recommendation to use several methods, depending on the stage of eco-design. In the earlier phases the qualitative methods are used, and in later phases the quantitative methods supported by appropriate computer tools are applied. The quantitative methods offer the possibility of their use for multi-criteria analysis and optimization of decision making process (Janicka & Hewelke 2007).

Analysis of the impact on the environment in the context of eco-design cannot be separated from economic and social aspects, which should also be included in the whole lifecycle (Burchart-Korol 2011, Hoogmartens et al. 2014). Hence the importance of LCC *(Life Cycle Cost)* and LSCA *(Life Sustainability Cycle Assessment)*. LCC has been studied by (Islam et al. 2015, Joachimiak-Lechman 2014). Lifecycle cost calculation summarizes all costs associated with the lifecycle of the product that directly relate to one or more entities (producers, consumers, suppliers and other people involved). In contrast to LCA, LCC has impact assessment component. Under this method, the traditional cost calculation developed in the company is extended by the costs and benefits associated with e.g. the pre-production stage or different variants of waste management generated at the end of life. These are complex methods based on developed scenarios, models, assumptions and various methods of valuation (Baran & Janik 2013). In relation with the fact that the social dimension plays an important role in assessing sustainable development, SLCA becomes significant. This is a method for the identification and assessment of the social consequences that occur throughout the lifecycle (Macombea et al. 2013, Martínez-Blanco et al. 2014, Reitinger et al. 2011).

The use of the LCA, LCC and SLCA methods at the eco-design phase gives more analysis capabilities of industrial goods, facilitates the search for technological solutions aimed at minimizing the amount

Table 2. Criteria for evaluation of selected methods and their characteristics.

Criteria	Characteristics
C1	Focus on evaluation object (product, process, activity)—defining environmental consequences of the use of the product, the process implementation or conducted activity as the object of the study;
C2	Complexity—coverage of all spheres of object existence and all interactions;
C3	Measurability—the possibility to express data using numerical values;
C4	Objectivity—reproducibility and reliability of the results,
C5	Allocation mainstreaming—the possibility of allocation of interactions for different: spheres of the examined object lifecycle, impact categories, the areas in which they occur;
C6	Susceptibility to standardization—the applicability in the legislative process as a tool for verification;
C7	Strategy and development planning—applicability as a tool supporting the decision during the planning process;
C8	Cost-absorption of performing the analysis—costs associated with collection of input data, computer software purchase costs, labour costs;
C9	Complexity of the method—whether it requires a lot of data, special software or expert knowledge, interpretation of data is difficult, data from many departments of the company is required;
C10	Method time consuming—time the method takes from the moment of gathering information to obtaining the output data;
C11	Possibility of comparison with competing products;
C12	Taking into account the costs associated with product lifecycle;
C13	Taking into account the social consequences that occur in the lifecycle of the product

of resources and waste and enables the social and economic assessment of functioning solutions.

5 DISCUSSION AND CONCLUSION

Nowadays, companies in the market of industrial goods while creating the concept of a new product pay attention to the assessment of its impact on the environment. To do this, they employ the methods and tools allowing rational use of raw materials, water and energy in all phases of the product lifecycle. This allows designing a product that meets the stakeholders' requirements and contributes to the

Table 3. Evaluation of methods based on selected criteria.

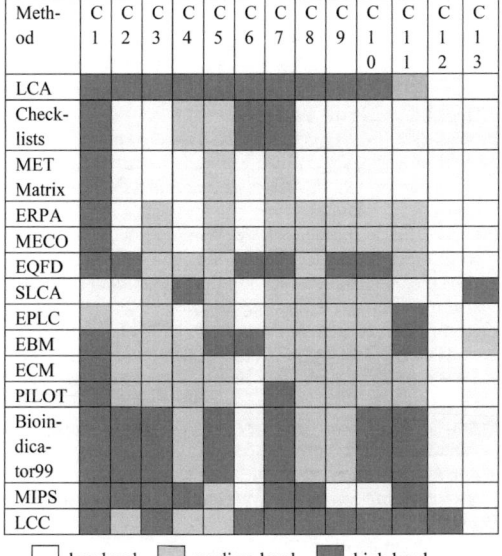

Method	C1	C2	C3	C4	C5	C6	C7	C8	C9	C10	C11	C12	C13
LCA													
Checklists													
MET Matrix													
ERPA													
MECO													
EQFD													
SLCA													
EPLC													
EBM													
ECM													
PILOT													
Bioindicator99													
MIPS													
LCC													

☐ low level ▨ medium level ■ high level

implementation of objectives of sustainable development strategies and improvement of production processes and products. Unfavourable factors limiting the use of advanced methodically analysis are time-consuming, costs and complexity of analysis. This generates too general and distant from reality. Consequently, trials are carried out aimed at detailing methods models. Many of the tools in the use of eco-design require expert knowledge, which makes them difficult to use in small businesses (Pochat et al. 2007). It seems important whether the given method is suitable for the full range environmental impact assessment of overall practices associated with the development, mass adoption, aging, replacement and disposal of a specific product. No less important is the scope of the analysis, whether it covers only the primary effects, or secondary and tertiary, whether it takes into account the synergies effect, cumulative effects, domino, cascading and ricochets effects. In the technology assessment methodology ordering criteria include compatibility with other methods and verifiability of results, versatility in use (whether given method can measure everything you want at a comparable level of results quality, or the results are more reliable when measuring one set of objects, and less certain for other objects) and normativity including evaluative assumptions which must be made, for example to determine the level of significance of certain parameters.

Objectification and quantification of the environmental impact during the product design allows searching for solutions most favourable to the environment in every phase of its life. Such an approach can lead to the development of entirely new products, with significantly less environmental impact than the functional equivalents, or to change the concept of meeting given need.

REFERENCES

Asiedu, Y. & Gu, P. 1996. Product life cycle cost analysis: state of the art review. *International Journal Production Research* 36 (4): 883–908.

Baran, J. & Janik, A. 2013. Zastosowanie wybranych metod analizy i oceny wpływu cyklu życia na środowisko w procesie ekoprojektowania. In R. Knosala (ed.) *Innowacje w zarządzaniu i inżynierii produkcji*: 22–33. Opole: Oficyna Wydawnicza Polskiego Towarzystwa Zarządzania Produkcją.

Bednárová, L., Liberko, I. & Weiss, R. 2009. Benchmarking v riadení podnikov, *Acta Montanistica Slovaca* 14 (1): 86–91.

Borchardt, M.A., Marcos, H., Wendt, B., Giancarlo, M., Pereira, A., Miguel, A. & Sellitto, B. 2011. Redesign of a component based on eco-design practices: environmental impact and cost reduction achievements, *Journal of Cleaner Production* 19: 49–57.

Borchardt, M., Poltosi, L., Sellitto, M. & Pereira, G. 2009. Adopting eco-design practices: case study of a midsized automotive supplier, *Environmental Quality Management* 19: 7–22.

Bovea, M.D. & Pérez-Belis, V. 2012. A taxonomy of eco-design tools for integrating environmental requirements into the product design process, *Journal of Cleaner Production* 20: 61–71.

Brones, F., de Carvalho, M.M. & de Senzi Zancul, E. 2014. Eco-design in project management: a missing link for the integration of sustainability in product development?, *Journal of Cleaner Production* 80: 106–118.

Burchart-Korol, D. 2011. Application of Life Cycle Sustainability Assessment and Socio-Eco-Efficiency Analysis in Comprehensive Evaluation of Sustainable Development, *Journal of Ecology and Health* 15 (3): 107–110.

González-García, S., Salinas-Mañas, L., Agarcia-Lozano, R., & other 2014. Application of eco-design methodology in SMSes run according to lean management: the case of a furniture publishing company, *Environmental Engineering & Management Journal (EEMJ)* 13 (12): 2977–2988.

Hernandez, R.J., Bhamra, T. & Bhamra, R. 2012. Sustainable product service systems in small and medium enterprises (SMEs): Opportunities in the leather manufacturing industry, *Sustainability* 4: 175–192.

Hoogmartens, R., Van Passel S., Van Acker K. & Dubois, M. 2014. Bridging the gap between LCA, LCC and CBA as sustainability assessment tools, *Environmental Impact Assessment Review* 48: 27–33.

Hur, T., Lee, J., Ryu, J. & Kwon, E. 2005. Simplified LCA and matrix methods in identifying the environmental aspects of a product system, *Journal of Environmental Management* 75 (3): 229–237.

Islam, H., Jollands, M. & Setunge, S. 2015. Life cycle assessment and life cycle cost implication of residential buildings-A review, *Renewable and Sustainable Energy Reviews* 42: 129–140.

Janicka, M. & Hewelke, P. 2007. Eco-design as important instrument of environmental protection on example of electric and electronic equipment, *Przegląd Naukowy Inżynieria i Kształtowanie Środowiska* 38: 29–38.

Joachimiak-Lechman, K. 2014. Środowiskowa ocena cyklu życia (LCA) i rachunek kosztów cyklu życia (LCC). Aspekty porównawcze, *Ekonomia i Środowisko* 1 (48): 80–96.

Karlsson, R. & Luttropp, C. 2006. Eco-design: what's happening, An overview of the subject area of Eco-design and of the papers in this special issue, *Journal of Cleaner Production* 14: 1291–1298.

Kishita, Y., Low, B.H., Fukushige, S., Umeda, Y., Suzuki, A. & Kawabe, T. 2010. Checklist-Based Assessment Methodology for Sustainable Design, *Journal of Mechanical Design* 132 (9): 091011–091018.

Kłos, Z. & Kasprzak, J. 2007. Metody oceny ekobilansowej, *Problemy Jakości* 6: 27–31.

Knight, P. & Jenkins, J. 2009. Adopting and applying eco-design techniques: a practioner's perspective, *Journal of Cleaner Production* 17 (5): 549–558.

Králiková, R., Rovňák, M. & Králik, M. 2008. Visualization of environmental data. 8th International Scientific Conference SGEM 2008: Modern Management of Mine Producing, Geology and Environmental Protection, 16–20. June 2008, Albena, Bulgaria., p. 577–584, ISBN 954918181-2

Lewandowska A. 2006. *Środowiskowa ocena cyklu życia produktu na przykładzie wybranych typów pomp przemysłowych.* Poznań: Uniwersytet Ekonomiczny.

Lewandowska, A. & Foltynowicz, Z. 2006. Prośrodowiskowe działania źródłem innowacji w przedsiębiorstwach. In Z. Kłos (ed.), *Rozwój przedsiębiorstw w aspekcie projakościowego doskonalenia i innowacyjności*; Proc. Intern. Symp., Boszkowo, 10–12 May 2006. Poznań: Wydawnictwo Politechniki Poznańskiej.

Li, J., Zeng, X. & Stevels, A. 2015. Eco-design in Consumer Electronics: Past, Present, and Future, *Critical Reviews in Environmental Science & Technology* 45 (8): 840–860.

Luttropp, C. & Lagerstedt, J. 2006. Eco-design and ten golden rules: generic advise for merging environmental aspects into product development, *Journal of Cleaner Production*, 14 (15–16): 1396–1408.

Macombea, C., Leskinenb, P., Feschetc, P. & Antikainend R. 2013. Social life cycle assessment of biodiesel production at three levels: a literature review and development needs, *Journal of Cleaner Production* 52 (1): 205–216.

Majerník, M., Bosák, M., Szaryszová, P., Tarča, A. & Štofová, L. 2014. Model for assessing the environmental performance of product systems within sustainable development, In. *Geo conference on ecology, economics, education and legislation* SGEM 2014, 17–26 June 2014, Albena, ISBN 978-619-7105-19-3, p. 285–292.

Majerník, M. & Štofová, L. 2014. Quality management in the integrated system as a tool for business excellence and sustainability, *Interdisciplinarity in Theory and Practice*. Editura Adoram, No. 1 (2014), p. 59–63.

Majerník, M., Szaryszová, P., Bosák, M., Štofová, L. & Kabdi, K. 2015. Integrated management of environmental—safety and technical risks of plant producing automobiles and automobile components, *Communications: scientific letters of the University of Žilina*. 17 (1): 28–33.

Martínez-Blanco, J., Lehmann, A., Muñoz, P., Anton, A., Traverso M., Rieradevall, J. & Finkbeiner M. 2014. Application challenges for the social Life Cycle Assessment of fertilizers within life cycle sustainability assessment, *Journal of Cleaner Production* 69 (15): 34–48.

Morrison, J.R., Azhar, M., Lee, T. & Suh, H.W. 2013. Axiomatic Design for eco-design: eAD, *Journal of Engineering Design* 24 (10): 711–737.

Nowosielski, R., Gołombek, K.J. & Jaskuła, A. 2003. Wybrane aspekty oceny cyklu życia technologii, *Journal of Achievements in Materials and Manufacturing Engineering* 12: 665–668.

Ostad-Ahmad-Ghorabi, H. & Wimmer, W. 2005. Tools and Approaches for Innovation Through Ecode-sign-Sustainable Product Development, *Journal of Mechanical Engineering Design* 8 (2): 6–13.

Pacana, J., Pacana, A. & Bednárová, L. 2014. Strength calculations of dual-power path gearing with FEM, *Acta Mechanica Slovakia, The Technical University of Košice* 18 (2): 14–21.

Piech, K. 1999. Metody wyznaczania kierunku rozwiązywania problemów. In J. Antoszkiewicz, *Metody rozwiązywania problemów w warunkach małego przedsiębiorstwa*: 63–74. Warszawa: Wydawnictwo SGH.

Pochat, S., Bertoluci, G. & Froelich, D. 2007. Integrating eco-design by conducting changes in SMES, *Journal of Cleaner Production* 15 (5): 671–680.

Reitinger, C., Dumke, M., Barosevcic, M. & Hillerbrand, R. 2011. A conceptual framework for impact, Assessment within SLCA, *International Journal Life Cycle Assessment* 16: 380–388.

Ritthoff, M., Rohn, H. & Liedtke, C. 2002. Calculating MIPS: Resource Productivity of Products and Services, Services, Wuppertal.

Theyel, G. 2000. Management practices for environmental innovation and performance, *International Journal of Operations & Production Management* 20 (2): 249–266.

Tomčíková, M. &, Živčák, P. 2012. *The practical contribution of information systems in times of globalization.* In: Management 2012, Prešov: Bookman, 2012. pp. 253–256, ISBN 978-80-89568-38-3

Tukker, A., Haag, E. & Eder, P. 2000. Eco-design: European state of the art. Part I: Comparative analysis and conclusions. An ESTO project report. Prepared for the European Commission—Joint Research Centre Institute for Prospective Technological Studies. Seville, on line at: http://ftp.jrc.es/EURdoc/eur19583en.pdf.

Volínová, L., 2011. Environmental assessment using MECO matrix—case study, Intensive Programme *Renewable Energy Sources*, Železná Ruda-Špičák, University of West Bohemia on line at: http://home.zcu.cz/~tesarova/IP/Proceedings/Proc_2011/Files/Volinova.pdf.

Wimmer, W. 1999. The eco-design Checklist Method: A Redesign Tool for Environmental Product Improvements, Proceeding eco-design'99 Proceedings of the First International Conference on Environmentally conscious design and inverse manufacturing, Computer Society Washington, DC, USA, *IEEE*, 685–688.

Yang, M., Khan, F.I., Sadiq, R. & Amyotte, P. 2011. A rough set-based quality function deployment (QFD) approach for environmental performance evaluation: a case of offshore oil and gas operations, *Journal of Cleaner Production* 19 (13): 1513–1526.

Yeang, K. & Yeang, D.L. 2008. *Eco-design: a manual for ecological design.* John Wiley & Sons.

The phenomenon of noise in a manufacturing company (automotive industry)

K. Bílecká
Bosal Ltd., Brandýs nad Labem, Czech Republic

R. Petříková
DTO CZ Ltd., Ostrava, Czech Republic

ABSTRACT: The safety of workers is a priority for the Bosal company, which is reflected in our motto: "Safety first!". Bosal is a company engaged in the manufacture of exhaust systems; this production is associated with noise emissions of the manufacturing facility. Noise of the working environment is being addressed and solved legislatively and in many standards, yet the measurement and interpretation of noise parameters is very complex and difficult to apply in practice. Noise is a significant negative factor affecting the health and quality of life of the people burdened. The basis for tackling noise issue is to understand the parameters of noise and noise effects on the human organism. Solution to the issue of the noise reduction is to implement a proven measure that will be useful and beneficial for all Bosal Group plants and that will promote cooperation within the group.

1 INTRODUCTION

The quality of the work environment is an important factor affecting the quality of life. The state of the work environment affects our health and mental condition, thereby the productivity and quality of work performance and consequently extends into the private life. It is therefore in the interest of both employer and employee that the working environment is as beneficial as possible. Only a healthy, well-balanced worker can perform quality work and contribute to the company.

The issue of the quality of the work environment is also an essential part of the Corporate Social Responsibility (CSR), in particular its social pillar. CSR activities contribute significantly to an increase of the level of corporate culture (trust, partnership, loyalty), reputation, reduction in staff turnover which significantly increases the overall competitiveness of the company.

1.1 *About us*

Bosal Group employs over 5,500 people worldwide in about 40 manufacturing plants and 20 distribution centers. Bosal is one of the biggest suppliers of exhaust systems for the automotive market in the world. In addition to conventional systems it also supplies insulated manifolds and catalytic converters. The plant in the Czech Republic in Brandys nad Labem produces original parts of exhaust systems for a variety of major car manufacturing customers (VW Group, GM, Suzuki, Renault).

The value that is of paramount importance for Bosal is the safety (Safety First): Our goal is to work without accident, create an ergonomic and healthy working environment. Uncompromising approach to safety and health of our employees is a matter of course for us. We take great care to ensure that all workers in our factory always use proper personal protective equipment. Occupational safety is a matter of each and every one of being responsible not only for themselves but also for others! (Martínek 2015)

We would not only want to declare our values in policies or visions, but turn them into concrete improvements. Therefore, we were not satisfied with the current level of noise of production, but we would like to limit the impact of noise on workers as much as possible.

Part of the production exhaust system is cutting, sheet metal stamping, sawing, bending, perforating of pipes, welding, labeling etc. So it is not a "quiet" production. These processes are accompanied by noise emissions that worsen the working environment in the production shop.

2 PINCH OF NECESSARY THEORY

2.1 *What turns the sound into noise?*

Sound is vibration of airborne particles creating a sound (or acoustic) waves. The source of oscillation, i.e. the source of noise may be vibrating objects, machines, air currents or shocks. Sound propagates not only in air but also in other elastic media such as water, concrete, steel. The speed of sound propagating through a solid substance is higher than the speed of sound propagating through the air.

Noise is any sound that is unwanted in a given moment. Its perception is largely subjective (loud music does not necessarily have to be perceived as noise, but the sound of a mosquito is perceived as annoying noise. It depends on the concrete situation (we want to concentrate or talk). Subjectively we distinguish loudness, pitch and timbre. By time during divide sound into a steady, variable, intermittent or impulse.

2.2 *How to describe the noise?*

Basic parameters for sound description:

Frequency—the number of cycles of a periodic motion per second (Hz), low frequency sound—(masculine bass, diesel engine, transformer), high frequency sound (a female soprano, mosquito whistling, whistling kettles).

Sound pressure—the pressure change against atmospheric pressure, which spreads through the air as waves (Pa). Since the human ear responds to sound pressure, we can hear sounds.

Sound pressure level—twenty times the logarithm of the ratio of sound pressure for a particular sound to the reference sound pressure (dB).

Acoustic output—the amount of energy emitted from a source of sound for a certain time period, i.e. one second (W), is used to describe a sound source because it is unchanged in relation to a sound source.

The harmful effects of noise on a person are assessed using sound exposure ("noise dose")—the amount of sound energy absorbed by the ear of the person concerned (sound pressure level and duration of exposure).

If the sound during its propagation encounters an obstacle, part of the sound energy is reflected, part is absorbed and part is transmitted by the obstacle. The part that is reflected, absorbed or transmitted depends on the physical properties and dimensions of the obstacle, as well as on the frequency of sound.

Emissions vs. imissions—noisy equipment produces sound, i.e. radiates sound energy = emissions; sound, which comes to the workplace, to the human ear = imission.

3 HOW DOES THE NOISE CAUSE HARM?

The World Health Organization has recognized hearing loss caused by noise as "the most widespread occupational disease with irreversible consequences". It is estimated that the negative effects of noise, resulting in loss of hearing are observed in ten percent of world population.

Hearing has the function of alarming body. A person receives most of the warning impulses from the environment by hearing. The organism can not physiologically exclude hearing from action. The human body can not "turn off" sound, the central nervous system processes all acoustic stimuli even in the course of sleep.

3.1 *So that we could imagine the impact of sound on the human ear, we need to briefly introduce the structure of the ear. How does the human ear look like?*

The ear is divided into three parts: the outer ear, i.e. auricle (intercepts and modifies the sound); auditory canal (transmits sound to the eardrum); ear drum; middle ear i.e. hammer, which pushes on the anvil, which then presses on the stirrup, whereby the vibration from the eardrum transfer into vibrations of fluid in the inner ear, middle ear operates as an acoustic amplifier. Without the middle ear more than 99% of acoustic energy would be reflected and thus would not be available as an audible sound. Another role of the middle ear is to protect acoustic apparatus from the loud sounds. The inner ear (cochlea) represents the actual sound receptor and analyzer. Its structure is so fragile that it gets damaged by noise as the first. Along the cochlear canal snails hair cells are distributed, which are responsible for initiating neural impulses in response vibrations generated by sound. The inner ear behaves as a mechanical-nerve acoustic frequency analyzer.

3.2 *How noise affects the human organism?*

Hearing loss at a certain frequency is related to the damage of hair cells in the specific parts of the cochlea. Partial damage to hearing cells leads to the death of hair cells in various parts of the inner ear. This process is very dangerous because due to inflammation the destroyed cell may cause destruction and death of the neighbouring cells. One of the first signs of hearing damage is ringing in the ears. This ringing (tinnitus) consists of ringing sounds and tones, even when no sound is actually transmitted to the ear. Excessive stimulation of hair cells leads to a "temporary threshold shift" or an increase in hearing threshold due to excessive stimulation; when exposure to noise is over, it slowly disappears. Increase of the threshold is perceived as a feeling of a certain desensitization of hearing and it is the first sign of acoustic system fatigue. After prolonged or repeated exposure to loud noise the threshold increase passes to a "permanent hearing threshold shift." Permanent hearing threshold elevation is factual hearing loss. Complete deafness, which occurs upon disappearance of both the inner and outer hair cells, also causes nerve fiber degeneration. People with problems or hearing loss, which was caused by noise, can hear when someone speaks to them, but may not well understand a part or all of their communication because they do not capture letter sounds with higher frequency in pronouncing such as F, T, K, C or S.

Noise can have side effects even at noise levels and exposures that are harmless to the inner ear. These include the impact on mutual communication (misunderstandings leading to wrong decisions) impact on the fulfillment of tasks (performance degradation), annoyance or irritability, stress, fatigue, restlessness, aggression, difficulty in identifying and recognizing danger and warning signals, and sleep disorders, decreased quality of sleep, motor functions. The amount of adverse effects of noise on individuals varies and depends on subjective conditions. The body tries to defend against noise by accelerating heartbeat and increased blood pressure, releasing hormones such as cortisol and adrenaline into the bloodstream, which are to prepare the body to fight stress. In the long term the human body cannot respond and leads to cardiovascular disorders (high blood pressure, increased heart rate), headache and migraine, heart attack, decreased immunity of the organism, peptic ulcers, thyroid gland dysfunction, irritability, dilatation of pupils of the eye, and others. (European Commission 2009, Jandák 2011, Havránek 1990).

4 OBLIGATIONS OF MACHINERY MANUFACTURERS IN THE EU

Machinery directive 98/37/EC requires machinery manufacturers in the EU to provide information regarding its noise emissions as a part of every offer. Machinery must be designed and constructed so that risks resulting from the emission of airborne noise are reduced to the lowest possible level, taking account the technical progress and the availability of means for noise reduction, in particular at source. Information on noise emissions must be given in the instructions for use and technical documentation for machinery, which allows to choose a quieter machinery with greater transparency in the market with machines. In order to reduce noise in the workplace, the purchasers of machinery should request information on noise emission values from different machine manufacturers and compare these values to be able to choose the quietest machines offered by different suppliers. In order to ensure that the requested noise emission values are comparable,

it is suggested that the machinery purchasers ask manufacturers of machines for issuing noise emission declaration based on European standards. This noise emissions declaration provides reliable technical information on noise emission values as their designation is based on European standards specific to the machinery. This clearly defines the way of measuring the operating and mounting conditions as well as the declaration and verification procedure, for a large number of very diverse machines. Noise machines allow potential purchasers to choose such a machine from a variety of brands that has the lowest noise emissions. At the same time it allows the employer to fulfill the legal obligation of the machine user to source quiet work equipment to the maximum extent possible. (European Commission 2009).

Emission values indicated by manufacturers are not directly comparable with the actual emission levels in the workplace. Under actual conditions at the workplace, sound pressure level, i.e. the sound level measured at the specific workplace, may differ from the declared emission sound pressure levels by more than 10 dB, as determined under free field, due to the noise from other sources, reflections from walls, ceilings, floors or machine surfaces and operating conditions differing from the conditions laid down in standards. Confusion of these fundamentally different values that specify emissions and imissions or even exposure, which includes also the time of exposure, explains many discussions and many misunderstandings arising between machine manufacturers and their customers. In some cases, the employer (purchaser) and machine owner may wish to check whether machine noise emission values exceed the values declared by the manufacturer in the declaration of noise emissions or values given or the contract. This normally occurs when the values of occupational noise exposure are higher than expected after the installation of a new machine. The value of imissions in the workspace can be approximately calculated by applying EN ISO 11690-3 standard, which uses declared noise emission values as the basic input data. EN ISO 4871 standard provides methods of verifying (checking) noise emission information and their declaration by the manufacturer. (European Commission 2009).

4.1 *Results of a study aimed at verifying data concerning noise of machines on the market in the EU in operation manuals as required by machinery directive and noise emission directive (contributions from the 14 EU member states and EFTA)*

Analysis was conducted on 1531 instruction manual for 40 categories of machinery. The analysis of instruction manuals was carried out from all aspects: the traceability of the data and noise, missing quantitative data, the credibility of noise emission values (reference to the measurement procedures applied and working conditions of machines, used standards) information about the residual risk of noise (e.g. the guidelines for use of protective equipment), terminology, language of the instruction manual. The ascertained detected were of varied nature and importance (absence of any data on the machine noise, references to inappropriate standards, specifying of untrusted data, etc.). Of the total number of manuals reviewed, 80% showed some shortcomings. (Havliš 2013).

5 PERSONAL PROTECTIVE EQUIPMENT

Personal Protective Equipment (PPE) are not a solution to the problem of noisy workplaces and are merely a necessary consequence. Personal protective equipment must be proportionate to the risk occurring and must not increase the risk; correspond to the existing conditions at the workplace; take account of ergonomic requirements and the worker's state of health. Employees must be provided with enough information about PPE. Protective equipment is provided by the employer free of charge to ensure its good working order and satisfactory hygienic conditions by means of the necessary maintenance, repairs and replacements. The ear protective devices include ear muffs, ear plugs and custom shaped protectors. If it is necessary to use hearing protectors, it is also necessary to allow employees to choose from several types so as not to impair their comfort. Greatly underestimated phenomenon in the workplace are short periods of time during, at which for some reason the workers cease to protect against noise, even though they are instructed about the dangers of exposure to it. If the hearing protection is removed for 10 minutes of the working day only, the overall effect of protection is reduced very drastically. If protectors are removed for a short time interval, at this point a sharp rise of the sound energy occurs, which must be perceived by our ear. (Jandák 2011).

6 BOSAL MEASURES IN THE CZECH REPUBLIC

As mentioned in the introduction, production processes in our company are causing a noisy work environment of the production shop. We have not set any rules yet that would solve the purchase of new machinery, tools in terms of noise, organizational and technical measures, measures for noise reduction in production etc. Currently, the workers are required to wear personal protective equipment, have regular breaks and undergo medical examinations.

The planned measures to reduce noise and their consequences in the production shop of Bosal Czech Republic:

A new noise measurement in the production shop and measurement targeted to specific risk processes will be conducted.

Requirements for noise emissions of machines will be included in the specifications for ordering of machines.

The machine specification will be compared in terms of noise parameters.

The acceptance of the machine will include a check of the noise emissions of equipment in the operating conditions.

Emphasis on preventive and predictive maintenance of equipment also in terms of the possibility of noise interference.

Workers will be explained the importance of using personal protective equipment and the consequences of failure to use them to their health and mental condition.

Employees will be given a possibility to choose the most suitable hearing protectors.

Workers will be trained in the proper use of hearing protectors, their maintenance and hygiene rules of their use.

The employees will be provided maximum of information aimed at preventing the negative effects of noise.

Employees will be involved in the process of improving their work environment and made familiar with all of the intended actions and the reasons for them.

Assessment of the possibility of individual adaptation of workplaces—e.g. partitioning compartmentation, use of curtains, covers; ban on throwing components from a height; replacement of optical, acoustic signaling; change of technology e.g. In grinding, embossing ("scraping" instead of stamping); using the remote control—greater distance from the noisy equipment. E.g. covering walls and ceiling with absorbing materials may pose a costly and concurrently inefficient solution, especially if the noise source is close to the employee. An inefficient solution can be a cabin for employees, if they have to leave it several times. Clearly the most efficient measure constituting the basis of the noise reduction policy must be the selection of low-noise machinery.

An essential part of the solution will primarily be the collection and transfer of experience and good practice in the context of similar processes and events within the Bosal group.

(Note: Maximum utilization of production sites and the layout of production lines in the spirit of one piece flow does not allow grouping of noisy workplaces in one area and reduce their noise collectively).

We see the biggest contribution in setting the rules for the acquisition of new machinery and assessment of individual workplaces during their creation, transfer. The issue must be addressed by all relevant positions across the Bosal group, since they are involved in the process of buying machines: both workers of a particular plant and headquarters staff. Setting of uniform and transparent rules, responsibilities and competences is therefore the only way to effectively address measures to reduce noise.

7 CONCLUSION

From the introduction of measures we expect a more favourable working environment conducive to improvement of the health and mental condition of employees, increase of productivity and quality of work, reduced staff turnover and increased loyalty.

REFERENCES

European Commission, Luxembourg: Office for Official Publications of the European Commission. 2009. Jak odstranit nebo snížit expozici zaměstnanců hluku při práci. *Nezávazná příručka osvědčených postupů provádění směrnice 2003/10S "Hluk na pracovišti"* (How to eliminate or reduce workers' exposure to noise at work. *Non-binding guide to best practice in implementing Directive 2003/10S "Noise at workplace"*).

Havliš, V. 2013. Seminář Tiché stroje—vyšší konkurenceschopnost a lepší zdraví 2006/45/ES, 2000/14/ES Studie zaměřená na prověření uvádění údajů o hluku strojů na trhu v EU v návodech k obsluze podle požadavků strojní směrnice a směrnice pro emise hluku a zahrnovala příspěvky ze 14 států EU a EFTA (Silent machines Seminar—increased competitiveness and better health 2006/45/EC, 2000/14/EC study focused was on checking particulars concerning noise machines on the market in the EU in the instruction manuals for the requirements of the Machinery Directive and guidelines for noise emissions and included contributions from 14 countries of the EU and EFTA).

Havránek, J. 1990. Hluk a zdraví (Noise and health). Prague: Avicenum.

Jandák, Z. 2011. Problematika hluku na pracovišti (The issue of noise at workplace), State Health Institute in Prague.

Martínek, P. 2015. *Bosal Business Management System.* Brandýs nad Labem: Bosal.

Production Management and Engineering Sciences – Majerník, Daneshjo & Bosák (Eds)
© 2016 Taylor & Francis Group, London, ISBN: 978-1-138-02856-2

Benchmarking of production performance of plastics and rubber producers in Zlin region

R. Bobák, P. Pivodová & J. Filla
Faculty of Management and Economics, Tomas Bata University, Zlin, Czech Republic

ABSTRACT: The paper informs about the research at the Department of Industrial Engineering and Information Systems of Faculty Management and Economics Tomas Bata University in Zlín. Research is focused on manufacturing and logistics performance of plastics and rubber manufacturers in the Zlín Region. The area has the largest concentration of these fields in the Czech Republic. The characteristics of included organizations from the perspectives of legal form, number of employees, turnover, integration in supply networks (automotive, electronics, furniture, construction, food) and the resulting plastics cluster are evaluated in the paper. It carries out export of selected economic indicators, manufacturing and logistics performance of its financial statements in the database Albertina CZ Gold Edition and checks their explanatory power and accuracy of the annual reports organizations. The present results performance are comparing in conclusion with the relative benchmarking metrics averages of cluster organizations and organizations of surveyed fields in Zlín Region.

1 INTRODUCTION

The aim of this paper is to report the results of research of manufacturing and logistics performance industrial producers. These studies are conducted on the Department of Industrial Engineering and Information Systems, Faculty of Management and Economics, Tomas Bata University in Zlin since 1998. Because of the educational and scientific specialization on university near Rubber and Plastics industry to we attend to detail results of industrial producers with that field business. Due to the high concentration of these industries in the Zlin Region a set of researched organizations is territorially limited in this area. Since 2006, some organizations of the plastics industry have banded together in Plastics Cluster. Selected metrics manufacturing and logistics performance of these enterprises is therefore analysed on the benchmark comparison of the average set of categories of industrial producers in the Zlin Region.

2 THEORETICAL SOLUTIONS FOR BENCHMARKING OF PRODUCTION AND LOGISTICS PERFORMANCE METRICS

Nowadays, the production and logistics performance of organizations is the subject of current research in terms of influencing factors, evaluation and measurement methods. Factors, especially the logistics performance of logistics service providers

in the context of 3PL and evaluation methods, are dealt with by (YinanQi 2000) and other authors (Kee-Hung & Lai 2008) empirical findings that electronic integration is positively associated with logistics performance. (Kee-Hun & Lai 2010) surveyed 227 trading firms in Hong Kong and performed a factor analysis of the survey data. (Caplice 1995) addresses this shortcoming by developing a set of evaluation criteria for logistics performance measurement systems and applying it in two case studies. (Halley 2009) with statistical analyses identified four clusters of respondents with regard to their supply-chain management practices. (Schmoltzi 2011) provide a overview of the motives, structure and performance attributes of horizontal cooperation between logistics service providers. (Gotzamani 2010) research proved a positive relationship between quality performance and financial performance for 3PL providers. (Ariff & Khan 2009) presented that collaborative distribution, order commitment, distribution flexibility and inventory management are the key SCM distribution practices and have significant impact on organizational performance. (Wank, 2007) unveils significant relationships between shipper sophistication of logistics function, manufacturing process structures, and the choice of type of 3PL. (Remko 2011) findings indicate that development of measurement systems contributes significantly to the expansion of third party logistics alliances in the supply chain.

Logistics services performance measurement and metrics are mainly dealt with by (De Toni 2001), (Chin-San-Lu 2010) and (Ming-Jen-Schen 2006).

They reveal wide range of methods from the general analytical through the multi-criteria scale. Generally are preferred systems of indicators. There are mostly used systems of indicators based on the Balanced Scorecard. (Lai 2007) indicates that efforts to improve information technology systems are properly aligned with a company's overall strategy. (Gunasekaran 2001) presented a list of key performance metrics. The emphasis is on performance measures dealing with suppliers, delivery performance, customer-service, and inventory and logistics costs in a SCM.

Other authors in their works deal procedures aimed at industrial and logistics clusters performance metrics. (Grando 2006) presents the results of a survey that involves a comparison among three samples of large, small-to-medium and district plants of the mechanical industry. (Liu 2006) put forward a conception of producer logistics service, analyses its positive effect on manufacturing firms in industrial clusters. He discusses logistics outsourcing between manufacturing firm as a logistics outsourcing party and third part logistics enterprise as a logistics service provider. (Groznik 2008) search of higher competitiveness of organisations for innovative business models, in order to foster economic benefits.

In our department, indicators based on the Balanced Scorecard concept listed in Table 1. (Bobák 2004) were also applied as the production

Table 1. Production and logistics performance metrics.

Metrics	Symbol	Indicator
Financial	PHV	Operating profit/loss (thousand CZK)
	CA	Total assets (thousand CZK)
	PHV/CA	Return of operating profit/loss on total assets
Customer	VYK	Outputs (thousand CZK)
Internal processes	NAK = VYK – PHV	Operating costs (thousand CZK)
	ZAS	Stocks (thousand CZK)
	PRAC	Number of employees
	HIM	Tangible fixed assets (thousand CZK)
	ZAS/VYK*360	Stock turnover days
Learning and Growth	VYK/PRAC	Productivity per employee (thousand CZK/employee)
	HIM/PRAC	Availability of tangible fixed assets per employee (thousand CZK/employee)

and logistics performance metrics of the selected organizations. The method used for selecting data from the firms' monitor and subsequent export of selected indicators of performance measures into a spreadsheet application was developed within the research of a faculty project conducted in the period 1999–2004 in (Trnka 2004). Due to the different size of organizations, information capability is more meaningful in case of relative indicators, related to unit characteristics. These indicators are the basis for a detailed evaluation of time trends in the longer term even for benchmark comparisons between individual organizations and the averages of the whole group of producer sand companies included in the cluster. A similar approach is used by other researchers at the faculty (Škodáková 2008) in a research project supported by GACR.

3 OBJECTIVES AND METHODOLOGY

On March 26, 2015, there were found 15 117 organizations in the Czech Republic with prevailing subgroups CZ NACE or complete CZ NACE in the Albertina CZ Gold Edition—Firms' Monitor database collected by Bisnode Czech Republic, a.s. (Albertina 2015). Selecting legal entities with the entry in the Commercial Register located in the Zlín Region, the group was limited to 1 221 organizations, including 561 legal entities. The last binding condition regarding the evaluation of production and logistics performance metrics was available in the database in the form of financial statements for the year 2013. This condition was fulfilled by 165 organizations, 32 of which are considered to be members of the established plastics cluster, 11 other members of the cluster have locations in other regions. Among the cluster members are also three educational organizations whose financial statements are not in the database and, therefore, they were not included in the research (Plastics cluster 2015). Based on available financial statements for the period 1998–2013, selected indicators of production and logistics performance for the specific manufacturing organizations in the Zlín Region and the cluster were exported to a MS Excel table for a statistics and benchmark comparison. The indicators of the number of employees were specified from the collection of documents based on the notes to the financial statements and annual reports in (Commercial register Czech Republic 2015). The research will try to answer two following questions:

a. What are the characteristics of organizations that have become members of the plastics cluster?
b. Has the membership in the cluster influenced the development of their production and logistics performance metrics?

4 RESULTS

4.1 *Basic features of the groups*

Overview of the elemental characteristics of both files in terms of the prevailing group CZ-Nace, district, legal form, number of employees, annual turnover is shown in Table 2.

Table 2. Basic features of the groups.

Prevailing subgroup CZ NACE	Region	Region %	Cluster	Cluster %
22000 Manufacture of rubber and plastic products	1	0.61		
22110 Manufacture of rubber tyres and tubes; retreading of tyres	5	3.03		
22190 Manufacture of other rubber products	10	6.06		
22200 Manufacture of plastic products			1	2,33
22210 Manufacture of plastic plates, sheets, tubes and profiles	12	7.27	7	16.28
22220 Manufacture of plastic packaging	13	7.88	6	13.95
22230 Manufacture of plastic products for construction industry	11	6.67	2	4.65
22290 Manufacture of other plastic products	30	18.19	8	18.60
Other	83	50.30	19	44.19
Total subgroups CZ NACE	165	100	43	100
Kroměř District	24	14.55	5	11.63
Uherské Hradiště District	28	16.97	3	6.98
Vsetín District	40	24.24	4	9.30
Zlín District	73	44.24	20	46.51
Other districts outside the Zlín Region	0	0	11	25.58
Total districts	165	100	43	100
1–9 employees			5	11.63
10–49 employees	73	44.24		18.60
50–249 employees	64	38.79	17	39.53
250 and more employees	28	16.97	12	27.91
The number of employees is not specified			1	2.33
Total employees	165	100	43	100
Cooperative	6	3.63	3	6.98
Joint-stock company	28	16.97	10	23.26
Limited Company	130	78.79	30	69.77

(*Continued*)

Table 2. (*Continued*).

Prevailing subgroup CZ NACE	Region	Region %	Cluster	Cluster %
Other legal form	1	0.61	0	0
Total legal form	165	100	43	100
Annual turnover up to 59 million CZK	53	32.12	13	30.23
Annual turnover 60–299 million CZK	67	40.61	15	34.88
Annual turnover 300–1450 million CZK	36	21.82	11	25.58
Annual turnover above 1450 million CZK	9	5.45	4	9.30
Total annual turnover	165	100	43	100

Table 3. Result ANOVA analysis of subgroup CZ NACE—p-value = 5,741E-06.

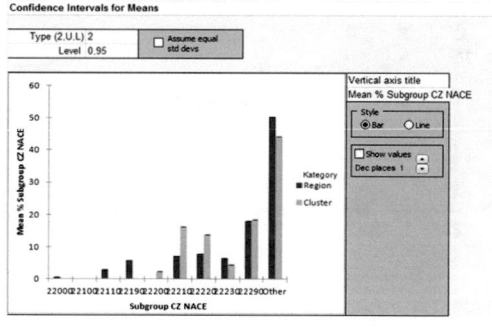

Table 4. Result ANOVA analysis of legal form—p-value = 0,0015684.

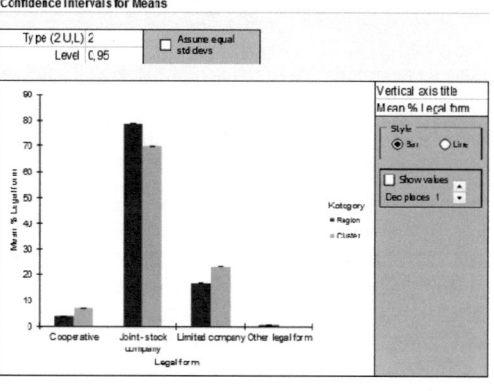

The structure characteristic of both files in terms of business branch, legal form, and turnover is compared using ANOVA statistical analysis methods, in the tables Table 3–5, results by allowing

Table 5. Result ANOVA analysis of turnover—
p-value= 0,0007465.

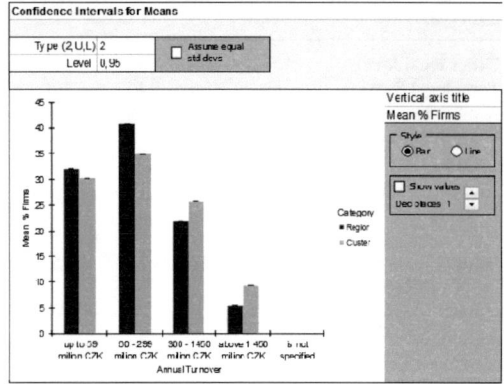

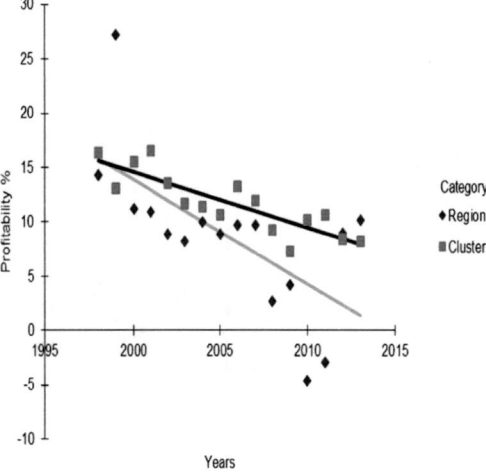

Figure 1. Evolution of the average values of profitability in %.

us to compare the trends in average values of the relative performance metrics of both files.

4.2 Benchmarking of production and logistics performance metrics

Evolution of the average values of production and logistics performance metrics obtained in rubber and plastics organizations in the Zlín Region and plastic cluster members for the years 1998–2013 are expressed in Figures 1–4. The data sets of accounts 43 organizations plastics cluster and 165 manufacturing plants processing rubber and plastics in the Zlín Region of the average values were calculated relative metrics manufacturing and logistics performance and linear trends of their development.

The average value of profitability in % decreases. The decrease in enterprises in the cluster is lower,

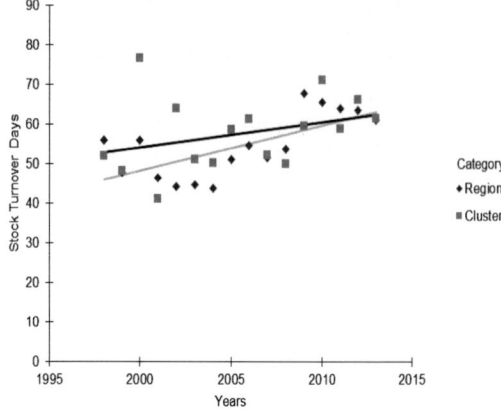

Figure 2. Evolution of the average values of stock turnover in days.

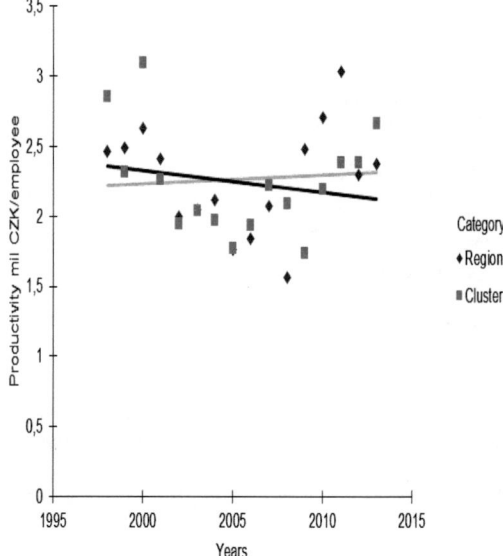

Figure 3. Evolution of the average values of productivity CZK mil per employee.

values are more favorable than the enterprises in the region.

The average value of inventory turnover time in days at businesses in the region increases, a negative trend. Enterprise's clusters exhibit a moderately increasing trend; the trend is favourable.

A larger number of medium and large enterprises in the file region have a positive effect on productivity growth. Firms grouped in a cluster have long been slightly decreasing trend in productivity. In the last five years, they also have a positive increasing trend.

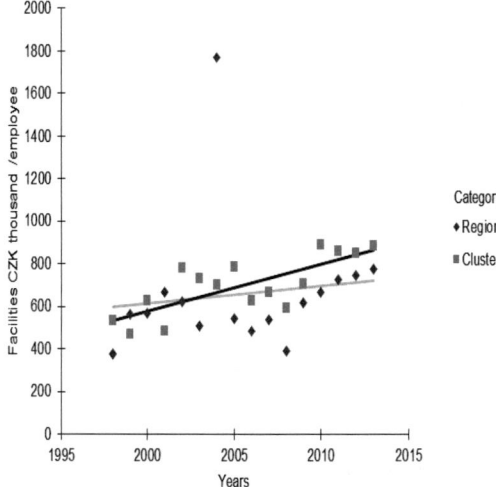

Figure 4. Evolution of the average values of the availability of tangible fixed assets per employee.

Average tangible property of amenities thousand CZK per worker has in the cluster organizations and firms in the regions almost the same trend growth. For firms within the cluster reaches higher values.

5 DISCUSSION

Evolution of the average of plastic and rubber products manufacturers in the Zlín Region in synchronization from 1998 to 2013 shows favourable characteristics only in the indicator Availability of tangible fixed assets per employee and Productivity per employee. Indicators Return of operating profit/loss on total assets and Stock turnover days embody less favourable characteristics of the development; there was a partial reduction in profitability and increase in stock turnover. The positive effect of the plastics cluster established in 2006 is reflected primarily in the comparison of the results of organizations, which favourably acknowledgement the positive characteristics. These are a lower growth in stock turnover a slight decrease in profitability and higher increase in the availability of tangible fixed assets per employee, above the region average. This can be regarded as the seed of synergistic effect of common purchasing policy and capacity utilization.

6 CONCLUSION

In the surveyed producers of plastic and rubber products in the Zlín Region focused in the period

1998–2013 dominated by the number of producers of plastics. The products were intended for supply networks of automotive, electrical, food, furniture and construction industries. From a regional point of view, there prevailed companies operating in the Zlín District. Most companies had the legal form of a limited company. Regarding the size, organizations with more than 50 employees prevailed with a turnover over to 60 million CZK. Approximately 25% of the producers have been incorporated into the plastics cluster. The majority of companies were running their businesses in the Zlín District, with the legal form of a limited company, in terms of size rather medium with 50 to 250 employees and with a turnover above 60 million CZK. These organizations support supply networks of automotive, electrical, food, furniture and construction industries. Cluster enterprises positively influence the average values of production and logistics performance metrics in the whole sector. The development of these characteristics of firms included in the cluster is positive. The impairment is reflected in the profitability indicator, but even that one is at the level above the average of the group of manufacturers of rubber and plastic products in the Zlín Region. Increasing inventory turnover time is to companies in the region also lower positive trend. To the positive turnaround in productivity development comes in the last five years.

ACKNOWLEDGEMENT

The paper was prepared with the support of the Internal Grant TBU in Zlín VaV-IP-RO/2014 Modelling of process parameters and the development of a methodology for optimizing complex systems of company process management in the global economy.

REFERENCES

Albertina CZ Gold Edition 2/2014. Bisnode Czech Republic, a.s, 2014, http://www.bisnode.cz/produkt/albertina/.
Arif Khan, K., Bakkuppa B., Metri Ba. & Sahay, Bs. 2009. Impact of agile supply chain delivery practises on firms' performance: clusters analysis and validation. *Supply Chain Management an International Journal 2009, Vol. 14, Iss. 1*, pp. 41–48, DOI: http://dx.doi.org/10.1108/13598540910927296.
Bobák, R. 2004 Příspěvek logistiky a průmyslového inženýrství ke konkurenceschopnosti výrobního systemu organizace. In. *Kvalita—Inovace-Prosperita (Quality-Innovation—Prosperity) No. 1/2004, Vol. VIII.* Košice: Q-IMPULZ, FEI TU Košice, pp. 9–16, ISSN 1335-1745.
Caplice, Ch. & Yosi, Sch. 1995. A Review and Evaluation of Logistics Performance Measurement Systems.

International Journal of Logistics Management 1995, Vol. 6, Iss. 1, pp. 61–74, DOI (Permanent URL): 10.1108/09574099510805279.

Chin-Shan-Lu & Ching-Chiao-Yang. 2010. Logistics service capabilities and firm performance of international distribution center operators. *The Service Industries Journal 2010, Vol. 30, Iss. 2*, pp. 281–298, ISSN 0264—2069.

Commercial Register and Collections of Documents. Ministry of Justice of the Czech Republic, https://or.justice.cz/ias/ui/rejstrik.

De Toni A. & Toncia, S. 2001. Performance Measurement Systems: Models, Characteristics and Measures. *International Journal of Operations & Production management 2001, Vol. 21, Iss. 1–2*, pp. 46–71, ISSN 0144-3577.

Gotzamani, K. Longidis P. & Vouzas, F.2010. The logistics services outsourcing dilemma: quality management and financial performance perspectives. *Supply Chain Management an International Journal 2010, Vol. 15, Iss. 6*, pp. 438–457, DOI (Permanent URL): 10.1108/13598541011080428.

Grando, A. & Belvedere, V. 2006. District's manufacturing performances: A comparison among small—to medium- sized and district enterprises. *Journal of Production Economics 2006, Vol.104,Iss.1,*pp.85–99,DOI: http://dx.doi.org/10.1016/j.ijpe.2005.01.007.

Groznik, A. 2008. E—logistics: Informatization of Slovenian transport logistics cluster. *Proceedings 2nd WSEAS International Conference on Management Marketing and Financing 24–26 March 2008 Athens. 2008*, pp. 139–143, ISBN 978-960-6766-48-0.

Gunasekaran, A., Pate,l C. & Tirtiagen, E. 2001. Performance measures and metrics in a supply chain environment. *International Journal of Operations & Production management 2001, Vol. 21, Iss. 1–2*, pp. 71–87, DOI: http://dx.doi.org/10.1108/01443570110358468.

Halley, A. & Beaulie, M. 2009. Mastery of Operational Competencies in the Context of Supply Chain Management. *Supply Chain Management an International Journal 2009, Vol. 14, Iss.1,*pp.49–63,DOI: http://dx.doi.org/10.1108/13598540910927304.

Kee-Hung Lai, Ch., Wong, W.Y. & Cheng, T.C.E. 2008. A coordination—theoretic investigation of the impact of electronic integration on logistics performance. *Information & Management 2008, Vol. 45, Iss. 1*, pp. 10–20, DOI: http://dx.doi.org/10.1016/j.im.2007.05.007.

Kee-Hung Lai, Ch., Wong, W.Y. & Cheng. T.C.E. 2010. Bundling digitized logistics activities and its performance implications. *Industrial Marketing Management 2010, Vol. 39, Iss 2*, pp. 273–286, DOI: http://dx.doi.org/10.1016/j.indmarman.2008.08.002.

Lai, F., Zhao, X. & Wang, Q. 2007. Taxonomy of information Technology strategy and its impact on the performance of third-party logistics (3PL) in China. *International Journal of Production Research 2007, Vol. 45, Iss. 10*, pp. 2195–2218, DOI: http://dx.doi.org/10.1080/00207540600693531.

Liu, W. & Cui Apod. 2006. Study in producer logistics service and its outsourcing from manufacturing firms: a perspective of industrial clusters. *Proceedings IEEE International Conference on Service Operations and Logistics and Informatics 21–23 June 2006 Shanghai. 2006*, pp. 699–704, ISBN 1-4244-0317-0.

Ming-Jen Cheng & Yen- Chun Jim Wu. 2006. Key Determinants of Performance Assessment for U.S TPL: DEA and Cluster Analysis. *Proceedings IEEE International Conference on Service Operations and Logistics and Informatics 21–23 June 2006 Shanghai. 2006*, pp. 370–375, ISBN 1-4244-0317-0.

Plastics cluster, an Interest Association of Legal Entities Zlín, www.plastr.cz.

Remko, I. & Van Hoek. 2001. The contribution of Performance measurement to the expansion of third party logistics alliances in the supply chain. *International Journal of Operations & Production management 2001, Vol. 21, Iss. 1–2*, pp. 15–19, DOI: http://dx.doi.org/10.1108/01443570110358431.

Schmoltzi, C. & Wallenburg, Cm. 2011. Horizontal cooperation between logistics service providers, motives, structure, performance. *International Journal of Physical Distribution and Logistics Management 2011, Vol. 41, Iss. 5–6*, pp. 552–576, DOI: http://dx.doi.org/10.1108/09600031111147817.

Skodakova, P., Pavelkova, D. & Vymola, T. 2008. ICT application for benchmarking of financial performance of clusters. In. *Center for Investigations into Information Systems 2008*, ISSN 1214-9489, http://www.cvis.cz/eng.

Trnka, F. 2004. *Výzkum konkurenční schopnosti českých průmyslových výrobců.* Zlín: Tomas Bata University in Zlín, Faculty of Management and Economics, 2004, pp.112–141, ISBN 80-7318-219-X.

Wanke P., Arkader R. & Hijjar, M. F.2007. Logistics sophistication, manufacturing segment and the choice of logistics providers. *International Journal of Operations and Production Management 2007, Vol. 27, Iss. 5*, pp. 542–559, DOI: http://dx.doi.org/10.1108/01443570710742401.

Yinan, Qi. Kenneth, K., Boyer, Xiande & Zhao. 2000. Supply Chain Strategy, Product Characteristics and Performance Impact: Evidence from Chinese Manufactures. *Decision Sciences 2000, Vol. 40, Iss. 4*, pp. 667–695, ISSN 0011- 7315.

Production Management and Engineering Sciences – Majerník, Daneshjo & Bosák (Eds)
© 2016 Taylor & Francis Group, London, ISBN: 978-1-138-02856-2

Methodological aspects of the implementation of energy audit in relation to EnMS and EMS

E. Boďová
University of Central Europe, Skalica, Slovak Republic

K. Šoltésová
Slovak Innovation and Energy Agency, Bratislava, Slovak Republic

ABSTRACT: Within II. climate—energy package in relation to the EU 2020 strategy was approved by the new directive 2012/27/EU on energy efficiency, which priority objective is to regulate the legal framework for energy efficiency in order to achieve the objective of energy efficiency 2020 of 20% primary energy savings. One of measures is to achieve this; it is also an obligation on businesses of EU member states to carry out periodically an energy audit.

The Slovakia has transposed the directive in question by NR SR Nr.321/2014 Coll. on energy efficiency, which is addressed in section 14 requirements regarding performance of an energy audit in the Slovakia and the possibility of carrying out an energy audit as part of an established and certified system of Energy Management (EnMS) according to international standard ISO 50001: 2012 and Environmental Management System (EMS) according to international standard ISO 14001: 2004.

1   INTRODUCTION

One of the initiatives of the Europe 2020 is initiative "Resource efficient Europe" adopted by the Commission on 26 January 2011. This initiative identifies energy efficiency as a major element in ensuring the sustainability of energy sources used. On 8 March 2011 the Commission adopted a Communication entitled Energy Efficiency Plan 2011. The announcement confirmed that the Union is not on track to achieve its energy efficiency targets. This is so despite progress in national energy efficiency policies referred to in the first national Energy Efficiency Action Plans, submitted by Member States on the basis of the requirements of European Parliament and Council Directive 2006/32/EC of 5 April 2006 about energy end-use efficiency and energy services. At the same time based on a study Climate Change 2013- The Physical Science Basis (Working Group IPPC Summary for Policymakers, October 2013) came to confirm the likelihood of further human impact on climate change and the need for substantial and sustained reduction in greenhouse gas emissions (Directive of the European parliament and of the council 2012/27).

To further limit climate change on earth, by the Commission under this activity also adopted a roadmap for moving to a competitive low carbon economy in 2050, stating the need to focus more on energy efficiency in that respect (European commission, 2014).

In the Slovak Republic started to be applied in addition to ISO 14001 additional marketing tool— multirate energy management system according to ISO 50001, which is based on principals deming cycle and is compatible with other uses management systems.

In terms of using the results of the certification audit of EMS EnMS or an organization may apply for the purposes of compulsory energy audit, it needs to consider three important aspects:

– Introduction EnMS or EMS so that organizations properly identify their energy goals, resp. environmental objectives and achieved improvement of energy management with the energy policy.
– Professional competence auditors pursuant to legal requirements and requirements arising from the accreditation schemes, including ISO 50003: 2014.
– To do certification audit in accordance with legal requirements for an energy audit.

2   REQUIREMENTS FOR THE ENERGY EFFICIENCY UNDER THE DIRECTIVE ON ENERGY EFFICIENCY

One measure to address this situation through setting ambitious goals was the adoption of the new EU Directive on Energy Efficiency (EE), which is

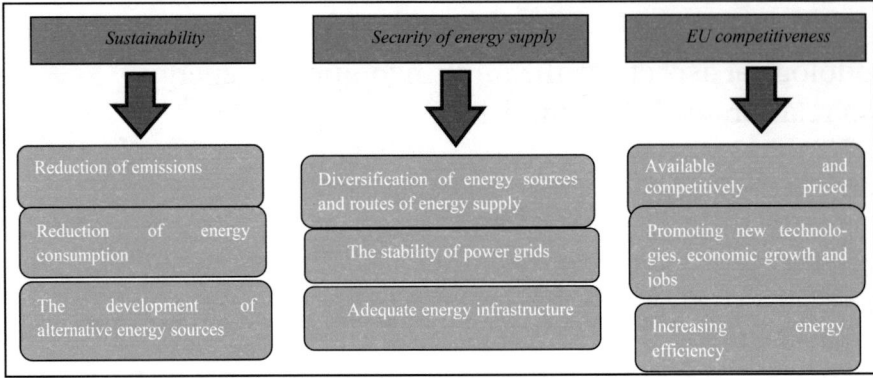

Figure 1.   The EU energy policy objective for 2030.

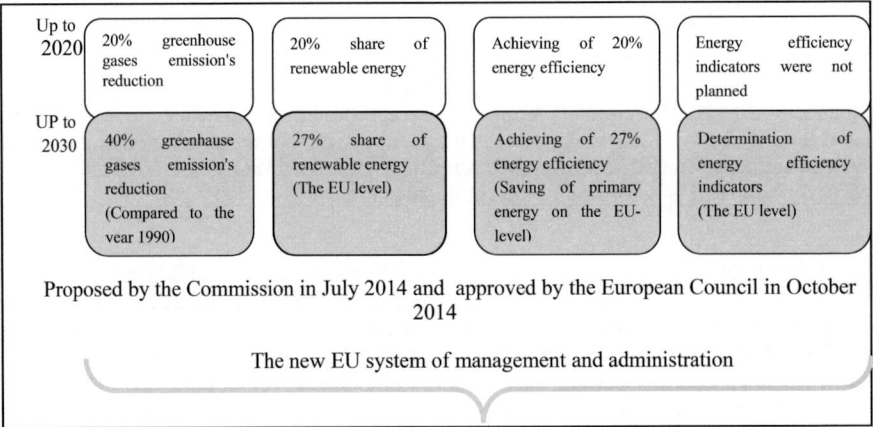

Figure 2.   The main EU energy policy objectives for 2030.

to be proactive approach to their achievement pro-active approach members states for possible use of harmonized conformity assessment tools. The main priority of EE Directive on the adjustment of the legal framework in the field of energy efficiency so in order to achieve the objective of energy efficiency for the period until 2020 of 20% primary energy savings. Directive went into force from 12.05.2012 with effect from 5.6.2014. Among the important Framework Directive in relation to EE are:

- Establishes indicative targets for energy efficiency, Member States—taking the form of expression of the goals laid down by Member States themselves in the form of absolute primary energy consumption and absolute final energy consumption.
- Energy performance of buildings—starting form the need to renew annually 3% of buildings owned by the state, in the first step of 1.1.2014 buildings with a total floor area greater than 500 m² and the second step of 9.7.2015 buildings from which a total floor area greater than

250 m². The level of the renewal meet minimum energy performance requirements under Directive EE Energy performance of Buildings.

- Energy audit, the
  - obligation to perform energy audits for businesses every four years, not to small and medium-sized enterprises according to the European definition of SMEs,
  - first energy audit must be carried out until 12.05.2015,
  - energetic audit may be part of the certification audit under the environmental management systems certification within the energy management system certification according to European or international standards,
  - energetic auditor must comply with the requirements of the EE (Directive of the European parliament and of the council 2012/27).

The performance requirements of an energy audit, including the outcomes of the energy audit are set out in the Directive on EE in the following paragraphs:

In the definitions (Article 2 of the Directive):

– *"energy audit"* a systematic procedure for to have adequate knowledge of the existing energy consumption profile of a building or group of buildings, an industrial or commercial operation or installation or private or public services and to identify and quantify cost-effective energy savings opportunities, which includes a report on the findings;
– *"Europe standards"* means a standard adopted by the European Committee for Standardisation, the European Committee for Electrotechnical Standardisation or the European Tele-communications Standards Institute and made available for public use;
– "international standard" is a standard adopted by an international standardisation organisation and made available to the public.

Energy audits and energy management systems in part (article 8 of the directive)

– Energy audits can be performed alone or as part of a broader environmental audit. Member states may require that part of the energy audit to assess the technical and economic feasibility of connection to the existing or planned network of district heating or cooling;
– Member States are encouraged to ensure that all end users are available high quality energy audits which are cost-effective a) carried out in an independent manner by qualified and/ or accredited experts according to qualification criteria, or b) which are implemented and supervised by independent authorities under national legislation. Energy audits mentioned in the first subparagraph may be carried out by experts or energy auditors provided that the Member State has implemented a system to ensure and check their quality, which includes, if necessary, an annual random selection, at least a statistically significant percentage of all the energy audits carried out. In order to ensure high quality energy audits and energy management systems established by Member States are transparent and non-discriminatory minimum criteria for energy audits based on Annex VI to Directive.

In the part Availability of qualification, accreditation and certification systems (Article 16 of the Directive), while

– If a member state considers that the national level of technical competence, objectivity and reliability under inadequate, to ensure that energy audits for 31 December 2014 made available the certification and/or accreditation schemes and/or equivalent qualification schemes become or be available on that date, including appropriate training programs where necessary.

In Annex VI of the Directive, the energy audits based on the following guidelines:

– they are based on actual, measured, traceable operational data on energy consumption and (for electricity) load profiles,
– contain detailed examination of energy consumption profile of a building or group of buildings, industrial operations or installations, including transportation,
– based, wherever possible, of Cost Analysis based on Life-Cycle (LCCA) instead of Simple Payback Period (SPP) to take account of long-term savings, residual values of long-term investments and discount rates;
– are balanced and sufficiently representative to permit the establishment of a reliable picture of the overall energy efficiency and reliably identify the most significant opportunities for improvement;
– energy audits shall allow detailed and validated calculations for the proposed measures in order to provide clear information on potential savings;
– data used in the energy audit must be a store that has been retrospective analysis in time and research (SK L 315/38 Official Journal of the European Union 14.11.2012).

3 ENERGY MANAGEMENT SYSTEMS (ENMS), ENVIRONMENTAL MANAGEMENT SYSTEMS (EMS), APPLICATION, REQUIREMENTS FOR IMPLEMENTATION IN ORGANIZATIONS, THE REQUIREMENTS FOR ACCREDITATION OF CERTIFICATION BODIES

EnMS in the international arena before 2011 used primarily by national standards (Denmark—DS 2403: 2001, Sweden—SS 627750: 2003, Ireland— IS 393: 2005, Spain—UNE 216301: 2007, China—GB/T 23331: 2009, USA—ANSI/SME 2000—available from 2008) or regional standards as the EU was EN 16001: 2009 standard. In 2011, it enters into force on the international standard EN ISO 50001, which simultaneously replaced the EN 16001: 2009 standard.

According to ISO (International Organization for Standardization) in 2013 was 4826 certified organization, which is 2,236 more than in 2012, so this is an increase of 113%. Certification EnMS is applied in 78 countries and the most certifications are applicable in Germany, U.K. and Italy (EN ISO 50001:2011).

The basic requirements for the implementation of processes in organizations are based on the Deming Cycle: General requirements, Management responsibility, Energy Policy, Energy Planning, Implementation and Operation, Checking,

Management Review, Annex A—Guidance on the use of this International Standard and Annex B—Correspondence Between ISO 50001: 2011, ISO 9001: 2008, ISO 14001: 2004 and ISO 22000: 2005.

EMS started to apply at national level since 1992 through the National Standards BS 7750: 1992 (U.K.), IS310 (Ireland), X30-200 (France), UNE 77-801 (Spain). First International Standard issued by the ISO in 1996 (Starkey, 1998).

Currently applied for the certification standard first revised in October 2004 and is expected to issue a standard as a result of second review (EN ISO 50001:2011). According to ISO in 2013 was certified organizations 301, 647, which is 16,993 more than in 2012, so this is an increase of 6% (EN ISO 50001:2011).

The basic requirements for the implementation of processes in organizations are based on the Deming Cycle: General requirements, Environmental Policy, Planning, Implementation and Operation, Checking, Management Review, Annex A—Guidance on the use of this International Standard and Annex B—Correspondence between ISO 14001:2004 and ISO 9001: 2000. (EN ISO 14001:2004)

To be harmonized certification schemes of booth systems (and other—ISO. 9001), International Accreditation Forum (IAF) in cooperation with the European Organization for Accreditation (EA), ISO and CEN (European Committee for Standardization) standards developed to introduce accreditation schemes for conformity assessment bodies certifying management systems (Certification Bodies—CB) in general.

In relation to international acceptance CB must establish procedures and ensure competence auditors pursuant to the accreditation requirements under the accreditation documents. The basic accreditation standards include ISO/IEC 17021: 2011 Conformity Assessment—Requirements for bodies providing audit and certification of management systems; ISO/IEC TS 17021-2: 2012 Part 2: Requirements on competence for auditing and certification of Environmental Management Systems; ISO 50003:2014 Energy Management Systems. Requirements for bodies providing audit and certification of Energy management systems; ISO 19011:2012 Guidelines for auditing of management systems (IAF 2015; snas 2015).

4 IMPLEMENTATION OF ENERGY AUDIT THROUGH EnMS OR EMS

As can be seen, the most important processes in the construction EnMS or EMS organizations shared planning process. Since its correct setting depends on percentage of achieving the environmental performance of which a special section focuses on energy management energy they may have for the company in addition to the environmental dimension as manifested later in reducing greenhouse gas emissions and the economic effects of earlier energy savings.

Within the EMS planning binds with the correct and complete identification of environmental aspects of all processes, activities and products and services to businesses. The organization shall identify for her significant environmental aspects which form the environmental objectives and targets and programmes through their implementation in operating conditions managed and evaluated by monitoring and measuring environmental indicators. Environmental indicators the organization defines itself and can be used for this purpose, ISO,

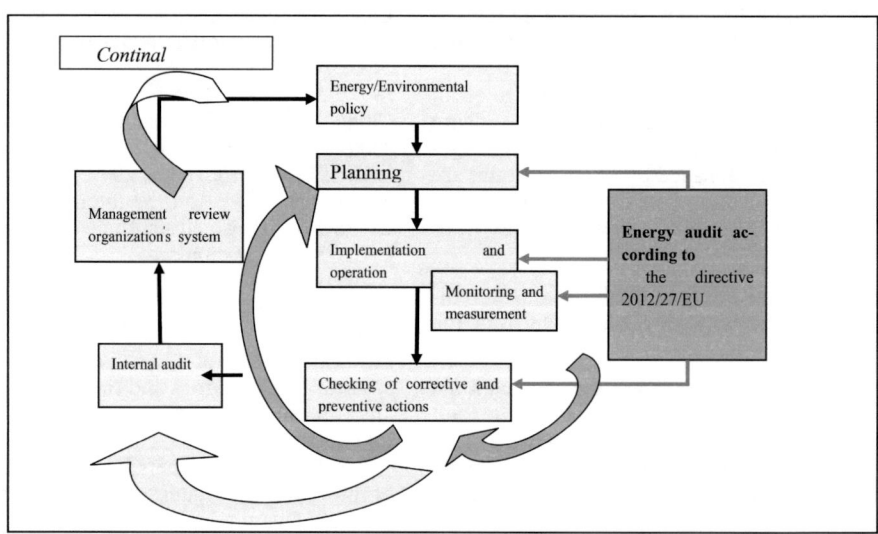

Figure 3. The inclusion of energy audit requirements according to the directive 2012/27/EU into energy management or environmental management systems.

which provides indicators of operational environmental indicators behavior and behavioral indicators management. EMS efficiency mainly depends on these factors not only for the organization itself, but mainly for the social significance of the results of the EMS (EN ISO:).

EnMS the environmental dimension of the whole company focuses on improving the energy performance of the company through better use of energy and its consumption by introducing energy-efficient approaches. In the planning, the organization shall identify all energy inputs: through past and present energy use, and must identify all the variables that affect the significant energy use. An important part of planning is to explore forms of energy use and consumption, identifying areas of significant energy use and consumption and identify opportunities for improving energy management. Within the output section of the planning, the organization shall determine a reference level of energy use and consumption levels based on several years of investigation as the basis for comparison of increasing efficiency through planned objectives and specific energy objectives. In order to assess the achievement of planned efficiency, the organization appropriately set indicators of energy management, and develop and implement action plans and programs to the operating conditions (Starkey, 1998).

Among the important indicators (EnPI) may be included: Contribution of each energy carrier, Efficiency in internal energy conversion, CO_2 emission. Share of energy costs in revenue, Employees specific energy consumption, Area-specific energy consumption, Production quantity specific energy consumption, Value-specific energy consumption, Production quantities specific energy cost and Energy intensity of a process/product (Directive of the European parliament and of the council 2012/27).

These processes are the basis for carrying out an energy audit within the meaning of the EE (see Figure 3).

As with all management systems based on processes Deming Cycle organization if it wishes to join the certification process must be carried out checking the achievement of planned objectives through results of monitoring and measurement and results obtained internal audit and the results of preventive and corrective measures. The organization must take place at least annually or before the certification audit at least one management review at the top level. Thus it is provided a continuous method of execution and evaluation of the commitments made to senior management in accordance with the adopted environmental/energy policy with a view to improving the environmental/energy efficiency. These results, as long as the organization is certified by an accredited certification body audited, and its EnMS or EMS is maintained under the three-year cycle, the review

and presents to the public and government institutions a certain guarantee of responsible conduct proactive organization. Auditors with audit and surveillance audits during the three-year cycle processing audit report, which must include determining whether the organization is in compliance with its policy, the legal requirements and other relevant regulations that they meet the objectives adopted and whether the system is effective management and shows increasing organizational performance.

Based on the evaluation of the individual steps of the system requirements may have to perform an energy audit of the directive on EE incorporated into the EnMS or EMS processes: Planning, Implementation and operation including monitoring and measurement and Checking of corrective and preventive actions.

5 DISCUSSION AND CONCLUSIONS

Options implementing mandatory energy audit which businesses outside the category of small and medium-sized carried out at least once every four years under the Directive on EE are either perform an audit independently or within an established and certified system of energy or environmental management. EE Directive does not impose directly the possibility of using the results of registered organizations in a Eco-Management and Audit Scheme (EMAS), which is governed by European Parliament and Council Regulation (EU) 1221/2009. EMAS is one of the instruments of EU environmental policy, which is the only one in comparison with International and European standards, such as EMS according to EN ISO 14001 and EnMS according to EN ISO 50001, incorporating mandatory management and evaluation of key environmental performance indicators, which include energy efficiency and greenhouse gas emissions. As it is a regulation, all Member States are obliged to this scheme, which is introduced in the EU by 1993, implement and promote its use.

At the same time those organizations are EMAS registered, have a high environmental performance guarantee, as they are under the control of state institutions. Compared to classical certification there is a higher degree of credibility and transparency of the organizations and procedures of information through mandatory disclosure of verified environmental statement registered organizations.

At the same time EMAS has higher potential for environmental technical innovation and economic performance than conventional EMS, which is confirmed by their studies and Rennings, Ziegler (2006). Increasing energy efficiency as one of the EU's strategic objectives should be clearly realized mainly through EMAS, which is in its complexity, the degree of applicability and guarantees one of the most effective instruments.

Reliability competency energy auditors under the national system of training and supervision in the Member States. Each Member State creates its own system requirements for energy auditors and supervision of their activities to be in line with the requirements of the EE. In addition, an energy audit can also be carried out in the company by a person in the enterprise. Here there is a violation of the independence and objectivity—it is important principles that are relevant when considering the accreditation body under the accreditation of conformity assessment bodies—CB (13). Further, when not created additional mechanisms of the Commission on the Directive on EE, it is arguable that national systems of supervision of the execution of energy audits competent authorities for energy efficiency will be harmonized. The success and reliability of performance and competence of energy auditors is appropriate to link the national surveillance systems and national accreditation bodies.

Reliability EnMS and EMS. Based on studies Amores-Salvado (2015), hereinafter Morris Rondinelli (2002) and the percentage performance of the thread approach from senior management and a good set of organizational structures, employee involvement and appropriate adjustment of environmental indicators. Bernardo (2009) reports on a study that 87% of organizations surveyed confirmed integrate EMS into other management systems, the easiest way is possible to integrate policy objectives, resources, documentation, records, internal audit and communication. Specific objectives and specific procedures require specific activities.

Increasing the financial performance of the company. According to the conclusions Pham (2015) it is established that during three years of operations ISO 50001 in the market is the application itself as a competitive tool or as a tool to improve environmental and economic performance of enterprises. Application of ISO 50001 and ISO 14001 compared to the incomparably slower start-up and its use by businesses is minimal interest.

Following a review of the background to be caught following conclusions:

Businesses that are required to perform an energy audit, and they conform to the requirements of ISO 14001 are promising to implement ISO 50001, however it is not clear whether EnMS implemented, as they can rebuild their EMS on more robust in relation to the requirements of the EE, which will for them easier access.

To be a reliable energy audit within the possible approach of the Directive on EE, it is appropriate that national systems licensors energy auditors cooperate with the national accreditation bodies which accredit conformity assessment bodies (CB) with a view to unifying requirements for energy auditors and their work.

REFERENCES

Amores-Salvado, J. 2015. The importance of the complementarity between environmental management systems and environmental innovation capabilities: A firm level approach to environmental and business performance benefits, *Technological Forecasting & Social Change* 96 (2015), pp. 288–297.

Bernardo, M., Casadesus, M., Karapetrovic, S. & Heras, I. 2009. How integrated are environmental, quality and other standardized management systems? An empirical study, *Journal of Cleaner Production* 17, pp. 742–750.

Directive of the European parliament and of the council 2012/27/EU of 25 October 2012 *on energy efficiency*, which amends Directive 2009/125/EC and 2010/30/EU and cancelling Directives 2004/8/EC 2006/32/ES, Official Journal of the EU, L 315/3.

EN ISO 14001:2004 *Environmental management systems—Requirements with guidance for use.*

EN ISO 50001:2011 *Energy management systems—Requirements with guidance for use.*

European commission, 2014. *A policy framework for climate and energy in the period from 2020 to 2030*, Brussels, 22.1.2014, COM 15 final, www. eu.europe. eu/energy.

European commission: Energy 2020. *A strategy for competitive, sustainable and secure energy*, Brussels, 10.11.2010, COM(2010) 639 final, www.eu.europe.eu/energy.

IAF, 2015. www.iaf.nu.

ISO, 2015. www.iso.org.

ISO/IEC 17021:2011 *Conformity assessment—Requirements for bodies providing audit and certification of management systems.*

Javied, T., Rackow, T. & Franke, J. 2015. Implementing energy management system to increase energy efficiency in manufacturing companies, *12th Global Conference on Sustainable Manufacturing*, Procedia CIRP 26, p. 160.

Morrow, D. & Rondinelli, D. 2002. Adopting Corporate Environmental Management Systems: Motivations and Results of ISO 14001 and EMAS Certification, *European Management Journal* Vol. 20, No. 2, (2002), pp. 159–171.

Pham, T.,H. 2015. Energy management systems and market value: Is there a link? *Economic Modelling* 46 (2015), pp. 70–78.

Rennings, K., Ziegler, A., Ankele, K. & Hoffman, E. 2006. The influence of different characteristics of the EU environmental management and audit scheme on technical environmental innovations and economic performance. Analysis. Institute for Ecological economy Research, Berlin, 2006, *Ecological Economist*, Vol. 57, pp. 45–59.

SNAS, 2015. www.snas.sk.

Starkey, R. 1998. The Standardization of Environmental Management Systems, In R. Welford (Ed.), *Corporate Environmental Management* (p.64), London, 1998.

Vašáková, S. 2014. Framework for climate and energy, EU 2030, *Proceedings of the conference Industrial emissions*, ASPEK, Bratislava, 2014.

Production Management and Engineering Sciences – Majerník, Daneshjo & Bosák (Eds)
© *2016 Taylor & Francis Group, London, ISBN: 978-1-138-02856-2*

Mineral resource extraction and its political risks

M. Cehlár, L. Domaracká, I. Šimko & M. Puzder
Faculty BERG, Institute of Earth Resources, Technical University of Kosice, Kosice, Slovak Republic

ABSTRACT: In many cases, external or previously unrecognized political risk of a project may be caused by repulsion of inhabitants after many years of researching the quantity and quality of mineral resource distribution. However, in the European countries or even in the countries of political stability, political risk is perceived as a mere detail. The article focuses on political risks of mineral resources extraction in Slovakia and abroad. It presents few examples of cases and it describes them in the context of extraction risks. In the concluding section, the article describes possible impacts of risks reaching up to the legislation level. However, political commitment may lead to a project termination after a detailed survey, good economic indicators and other positive areas. Public opinion or a protest against mineral resource extraction are closely related to a phase of mineral resource extraction, when all the investments should be refunded.

1 CONNECTION BETWEEN THE VARIABLES

At present, we live in the age when a mining company has no difficulties to pay for any study, such as expert study, prefeasibility study, feasibility study, or even an expertise, to map the actual reserves at a deposit. Moreover, it focuses on the Public Relations that should present a company as an effective contribution to a region (economic, social, etc.) in the first place. On the other hand, the project might lead to the result that it is not possible to proceed to the mineral resource extraction phase (mining permission), after detailed research of good economic indicators and other positive directions, in which all the costs are expected to return. This situation is predominantly connected with public opinion, comments, or even public repulsion to mineral resource extraction. It may be solved by project variation via EIA (Environmental Impact Assessment—by choosing environmentally, economically and socially acceptable option) (Majerník et al. 2014). However, this whole project may also lead to a protest of citizens, which may result in a legislative provision that will prevent the mineral resource extraction. In specific cases, it may be conditioned by a local referendum of concerned municipalities, which might not achieve positive results.

Consequently, the article deals with extraction from a political point of view, being a very sensitive element of the whole project.

2 POLITICAL RISKS OF MINING COMPANIES REGARDING SHALE GAS EXTRACTION IN EUROPE

The need of energy resource diversification in Europe has arisen after extremely successful years of shale gas extraction in the states of North America. This fact influenced the world prices of gas with regard to renewal of resources which consequently multiplied and the price of gas decreased on global markets. However, it could be considered a conspiracy theory to claim that a geopolitical factor of trade relations was able to change the legal standards of Member States, where Europe trades the "cheap" gas with the Russian Federation, which has been offering affordable and relatively stable supplies of natural gas. Natural gas represents very important trade element for the EU Member States, including the Slovak Republic, which is the largest transporter of natural gas, in terms of capacity (Khouri et al. 2009). The above mentioned is underlined by the opinion of the European Economic and Social Committee:

"One of the techniques, which appeared in the last decades, is extraction by means of high volume hydraulic fracturing, so-called fracking that has been improved very quickly in the USA and it has brought undisputable advantages, as the natural gas resources increased. Natural gas is available for economic usage and its increase led to significant price decrease. On the other hand, fracturing raises concerns in relation to public health and environmental impacts, while the public complains about insufficient level of transparency and consultations regarding activities connected with shale gas. The particular communities have to be more informed and it is also necessary to engage them in individual projects, including the procedure of influence evaluation, as required by the applicable legal obligations." (EU committee 2014)

Implementation of a relevant opinion, or transposition of directives or standards, will cause that companies which were carrying out survey and subsequently were counting on extraction may gradually evaluate the damages caused to the company at the time when these facts were not known (Muchova et al. 2013). The circumstances in the mining industry change as a consequence of public requirements, but also company lobbing. In this case, the circumstances change due to international trade with gas at the level of political risk by the European directives, or the European Commission regulations.

3 EXISTING OIL EXTRACTION AND POLITICAL RISK INFLUENCE ON ITS RESERVES AND PRICE ON GLOBAL MARKETS

Generally, political risk does not need to show only legislative parameters. It often represents political unrest, which is primarily caused by dissatisfaction of citizens living in an unstable region. These people express their attitudes via bomb attacks, or even hostage-taking in the strategic areas and societies.

"Today the Brent oil prices have been higher in London after Middle East unrest. Political risk and especially the tension due to possible interruption of oil extraction have brought many bulls to this Black Gold. The market-risk premium of oil had been caused by automobile bomb attacks, where 60 people died during weekend in Iraq. There have been few analysts who had predicted the fact that political risk in Iraq can not influence the increase of commodity prices. However, today events have probably changed their minds. The investors have perceived this situation as a fact that a threat of oil extraction interruption is always possible. Present sudden price increase of oil could not be stopped even if the oil reservoirs in the American Cushing exceed the expectations of analysts for the last three weeks. The second factor that influences the increase of oil prices is the fact that oil refineries process less oil due to their regular repair and maintenance stoppages." (FOREXPO 2014).

One day in Iraq (unrest caused by bomb attacks) caused the increase of oil price on global markets with commodities by 2.7%. It means that the political risk is a factor of mineral uniqueness, which may influence its prices from the short-term, as well as long-term point of view, in case it is required in various economies equally to energetically universal oil. Fortunately, the unrest in question is often successfully eliminated, as it is organized by individuals and not by organized groups, which would be able to damage transfer lines that lead to refineries for a long period of time.

4 NO CONTINUATION OF MINERAL RESOURCE SURVEY IN THE SLOVAK REPUBLIC

Technical, technological, economic, environmental risks, as well as unpredictable risks that may not be known to investors or management, are very often related to investment into shares of mining companies, as there is no sign of resistance in the region where the political-civil risk may appear in the future. As emphasised below, though in connection with investments in silver, at present, discussions are being held regarding the issues of legislative entitlement, or procedure, of mining companies after performance of geological works and surveys, and the subsequent entitlement, the right to extract of potential cost-effective mineral resource extraction (Rybar et al. 2012).

"Investors undergo even greater risks when investing into the shares of individual mining companies than when purchasing a particular metal. Quality management, amount and management of exploration costs in, political risk of a country (threats of nationalization, tax increase, or regulation tightening), risk of strikes, etc., play a significant role in this case. Consequently, Sprott recommends small investors without any expertise in the field of mining companies to accumulate preferably precious metals in their physical form (unlike all the speculations with shares, options, future markets, etc.), in order to eliminate risks of a counterparty (e.g. in case of bankruptcy of a given company, or stock exchange)." (Tonka 2014)

In particular, the Slovak companies which have recently invested in the survey of mineral resources, are the companies previously focused on the survey of gold, other precious metals, and energy resource, as well as technologically valuable uranium that is present in the form of molybdenum-uranium ore.

5 GOLD DEPOSITS IN DETVA

"Prospectors of the EMED Slovakia drilled 44 oblique boreholes of 11 km length at the Detva deposit during 2007–2009. These boreholes were sampled in meter intervals, which led to verification of gold ore element deposits. In 2010, they developed a calculation of reserves and submitted it to the State Commission for approval. The reserves calculation was approved and the certificate of exclusive deposit was issued. At the Detva deposit, there are approximately 30 million tons of recoverable gold ore (1gr of gold (Au) per 1 t of ore), according to this calculation. The total amount of gold is 20 t." (Emed 2012)

Cyanide leaching technology was determined as the BAT for extraction and processing of the mineral on the basis of the data provided by the EMED, which carried out the research of gold deposit with the estimate of more than 20 t of reserves.

"EMED Slovakia, Cypriot EMED Mining subsidiary, plans to extract gold at the Biely Vrch Deposit in the immediate vicinity of Detva engineering plant. However, the Mining Authority has not decided on their request for determination of the mining area yet. Regional authorities, as well as engineering companies concerned, do not agree with this intention. Civil association, Podpoľanie nad zlato, is most intensively engaged in the struggle against it." (SITA 2014)

"Detva has been cooperating with the submitters of the amendment to the Mining Act and was supporting this legislative process since June, 2014. The law amendment is supposed to prohibit extraction and mineral processing by cyanide leaching technology. This type of extraction would only be possible if the citizens support it in the Referendum." (TASR 2014).

Consequently, since September, 2014, the following has been approved in the amendment to the Mining Act: "The cyanide leaching technology is prohibited by law and it will not be necessary to organize any local Referendum, which represents administrative and financial burden. This requirement was submitted by citizens and representatives of regional authorities concerned." (TASR 2014).

Despite the fact that a company invested millions of Euros in the mineral survey, in 2009, the calculation of reserves was made in accordance with the internationally recognized Code of Reserve Calculation of the JORC. In 2010, before the EMED entered the Toronto stock exchange, the calculation was modified according to the Canadian standard of reserve calculation NI 43-101. It required a lot of effort and financial resources; nevertheless, the company encountered a political risk, particularly in a form of revolt by civic association (activists) who achieved, through political pressure, politicizing of this issue, and media presentations, that the politicians who prohibited the extraction and processing of minerals by means of cyanide leaching technology by the Mining Act amendment in the National Council of the Slovak Republic, heard the voice of people.

6 PROPOSAL OF THE METHOD TO BE USED IN THE PREVENTION OF POLITICAL RISKS

Elimination of political risks regarding the extraction could be carried out by unambiguous provision of true information to the public and respective groups. A method which would quantify various opinions on this issue would certainly decrease a possibility of political publicizing of this theme. Mutual transparent environment enables proper decision-making by all parties involved, regardless of whether they are agreements or disagreements. Particularly the politicizing of themes brings uncertainty, anxiety and it rarely results in decisions other than biased and emotional decisions.

Similarly to other analyses, particularly the SWOT analysis, the methodology should include, strengths and weaknesses of a project and positive and negative contributions referred to as opportunities and threats. SWOT analysis represents a conceptual base for the systematic analysis. It focuses on the characteristic of key factors affecting influence a strategic status of a company. It also represents a constant confrontation of internal sources and capabilities of a company—the project, including the changes thereof in its environments. The SWOT analysis uses results of previous analyses of opportunities and risks, which are contained in the external environment. It is an approach towards the synthesis as an output for the strategy formulation. The fundamental contribution of the SWOT analysis lies in a thorough identification of external and internal factors and a consequent evaluation of their mutual influences and interconnections (Hekelova 2008).

It is advisable to demonstrate on a generalised example, what items can be included in this process, at least in the beginning of the awareness. Table 1 presents the variables which most frequently enter into the process.

A desired outcome of the analysis is a strategy that would facilitate proper determination of the procedure of public awareness, while considering all the input factors, and prevent the politicization of this sensitive issue. In the chosen approach, we have considered external factors, particularly defining opportunities and threats:

Definition of opportunities:

- New job opportunities—P1
- Infrastructure improvement—P2
- Support of entrepreneurs who operate in the field of reclamation and regeneration of the environment and country fostering—P3
- Profound environmental monitoring—P4.

Definition of threats:

- Health burden for inhabitants—H1
- Biotopes intrusion (fauna, flora)—H2
- Natural scenery disturbance—H3
- Arbitration—H4
- Tightening the legislation on waste management—H5.

Internal factors are represented in form of strengths and weaknesses.

Strengths:
- New job opportunities—S1
- Infrastructure improvement—S2
- Support of entrepreneurs focused on the environment regeneration and country fostering—S3
- Profound environmental monitoring—S4
- Minimum marks on the surface—S5.

Weaknesses:
- Negative influence on the environment—SS1
- Negative influence on inhabitants' health—SS2
- Insufficient awareness of a company and the extraction methods—SS3
- Negative influence of media on inhabitants—SS4.

We produced a decision matrix on the basis of these defined variables. The individual variables are coded by a letter and a number, as stated with the variables of the individual factors. The evaluation is carried out while comparing the importance of individual signs by means of three degrees.

Table 1. Analysis of external and internal environment.

Weaknesses	Strengths
Negative influence on the environment	Municipal budget increase, GDP formation
Negative influence on inhabitants' health	Employment deficit balancing
Insufficient awareness of a company and extraction methods	Zone visibility enhancement
Negative influence of media on inhabitants	Profound environmental monitoring
Negative influence on a zone	Minimum marks on the surface
Threats	Opportunities
Health burden of inhabitants	New job opportunities
Biotopes intrusion (fauna, flora)	Infrastructure improvement
Violation of natural scenery	Support of entrepreneurs focused on environment regeneration and country fostering
Arbitration	Profound environmental monitoring
Tightening of the legislation on waste management	Minimum marks on the surface

- 1 = a sign is more important than the compared one
- 0 = a sign is less important than the compared one
- 0.5 = importance of signs is the same.

In the Table 2 is an example of calculation of strengths, qualification of internal factors. Where:

CF—Compared factors.
S-Sn—Strengths
I—Importance
R—Rating
WR—Weight rating.

Calculation of the sums in individual Tables provides a weight of each variable. In the next step, we assigned an importance to each variable of the external and internal factors. Similarly, we chose the level of importance in a range from 1–5, while 5 is the highest importance and 1 represents the least important factor. The level of importance is multiplied by a weight and it gives the values, the sum of which equals to a total objectified value of external and internal factors. The strengths and weaknesses are presented as opposites on a single axis and mutually subtracted. Both of them represent a qualitative (in this case also quantitative) demonstration of the same internal environment. A similar procedure will be applied to opportunities and threats. The final strategy is formed as the sum of vectors. This strategy represents the basic recommendation for a strategic procedure that respects a requirement of close harmony between the internal capabilities and the external environment. In our case, the opportunities prevail over the threats and the weaknesses over the strengths. The process is illustrated in Figure 1.

In this case study, we identified a period when a strategy of alliance should be chosen in order to prevent possible political risks at the beginning of extraction. Alliance should probably include a presentation of a potential extraction with specialists, non-profit organizations, and provide a detailed explanation, without any misleading and emotions, of the issues regarding the extraction from the above mentioned points of view.

Table 2. Strengths—quantification of internal factors.

CF	S1	S2	S3	S4	S5	Sum	I	R	WR
S1	x	0.5	0.6	1	0.6	2.7	0.27	5	1.35
S2	0.5	X	0.2	0.7	0.4	1.8	0.18	3	0.54
S3	0.4	0.8	X	0.9	0.5	2.6	0.26	4	1.04
S4	0	0.3	0.1	x	0	0.4	0.04	2	0.08
S5	0.4	0.6	0.5	1	x	2.5	0.25	3	0.75
Total						10	1		3.76

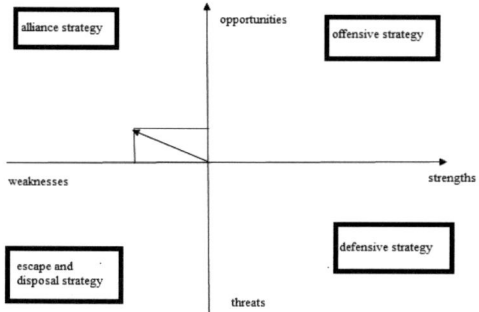

Figure 1.   Graphic display of strategy calculated by vectors.

7   CONCLUSIONS

It is remarkable how many issues are covered by a political risk of the mineral resource extraction. Firstly, a potential extraction of energy mineral resource is influenced by political risk of shale gas extraction via fracturing in the whole European Union. Secondly, implementation of opinions and directives has a significant influence on economy and stable price of natural gas in Europe. Thirdly, findings that disfavour the political situation in politically unstable regions influence the prices of oil on the global markets, risks of extraction, transport, or even refineries. Finally, the local issue of gold survey prohibition in Detva, which led to adopting the amendment to the Mining Act that prohibits extraction and processing of mineral resource by means of cyanide leaching technology. It is not only a phenomenon of militant zones, but also a result of possible concerns, often without any expert ground, regarding potential extraction.

A similar case, presented in mediated after several smaller and also bigger protests, was a potential uranium extraction near Košice. The permission process could have begun after the deposit survey. The Ministry of the Environment of the Slovak Republic adopted the amendment to the Geological Act, whereas the approved amendment stipulated that citizens of Košice as well as urban districts concerned have the right to express their opinions on the potential extraction in the Referendum.

In the conclusion, the above mentioned facts emphasise the unforeseeable and geographically specific political risks that the companies trading with extremely intense capital, without any knowledge of a given region's internal environment, undergo while investing their funds or shareholders' funds, doing so inadvertently or without due diligence. Consequently, it is necessary to carry out a profound analysis of a region and commence the discussion in accordance with the results. In case of a threat of a political risk, it is always more appropriate to

proceed similarly to other projects, i.e. to resign before the project becomes loss-making.

REFERENCES

Európska komisia, 2014. Stanovisko Európskeho hospodárskeho a sociálneho výboru na tému "Oznámenie Komisie Európskemu parlamentu, Rade, Európskemu hospodárskemu a sociálnemu výboru a Výboru regiónov o prieskume a ťažbe uhľovodíkov (ako napríklad bridlicového plynu) v EÚ s použitím vysokoobjemového hydraulického štiepenia" COM(2014) 23 final:NAT/629—EESC-2014-01320-00-00-AC-TRA (EN) 1/8. [08.02.2014]. <http://edz.bib.uni-mannheim.de/edz/doku/wsa/2014/ces-2014-1320-en.pdf>.
EMED, 2014. Účelová publikácia vydaná spoločnosťou Eastern Mediterranean Resources—Slovakia (EMED Slovakia), s.r.o., Ložisko zlata—Detva., Detva 2012 [08.02.2014]. <http://www.envigeo.sk/DATA/LOZ_GEOLOGIA/TE_studia_Detva_BV_4a.pdf>.
Forexpro. 2014: Násilie v Iraku tlačí Brent nahor. Bloomberg. [08.02.2014]. <http://openiazoch.zoznam.sk/cl/139566/Nasilie-v-Iraku-tlaci-Brent-nahor>.
Hekelová, E. 2008. Marketing. Vydavateľstvo FX, Bratislava, Bratislava, 112 s., ISBN 978-80-89313-17-4.
Khouri, S., Al-Zabidi, D. & Al-Zabidi, A. 2009. Prieskum možností financovania projektov v energetike so zameraním na trh so zemným plynom. In: Acta Montanistica Slovaca. 14, (1):75–81. ISSN 1335-1788.
Majerník, M., Bosák, M., Szaryszová, P., Tarča, A. & Štofová, L. 2014. Model for assessing the environmental performance of product systems within sustainable development, In 14th International Multidisciplinary Scientific Geoconference SGEM 2014: Ecology, Economics, Education and Legislation, Conference Proceedings, Volume III, Albena, 285–292 s., ISSN 1314-2704.
Muchova, M., Pavolova, H., Zuzik, J., & Gonos, J. 2013. Risk Management in Mining Company with Medium-Scale Surface Mine—2013. In: Lecture Notes in Management Science: ICMIBI 2013. Internacional Conference on Management Innovation and Business Innovation. Singapore Management and Sports Science Institute, 2013 Vol. 15 (2013), p. 239–244. -ISBN 978-981-07-5034-3 - ISSN 2251-3051.
Rybar, P., Štrba, Ľ., Molokáč, Š. Rybárová, M., Kudelas, D., Domaracký, D., Molokáč, M. & Hvizdák, L. 2012. Model of rock melt penetration into in radial fractures evolved under high pressure and temperature conditions—2012. In: Underground Mining Engineering. Vol. 20, no. 21, p. 143–150. ISSN 0354-2904.
SITA, 2014. Ťažba zlata kyanidom v Kremnici a Detve sa konať nemusí, [08.02.2014]. <http://www.etrend.sk/ekonomika/tazba-zlata-kyanidom-v-kremnici-a-detve-sa-konat-nemusi.html>.
TASR, 2014. Mesto Detva víta novelu zákona, ktorá umožní odmietnuť ťažbu kyanidom, [08.02.2014]. <http://www.teraz.sk/ekonomika/detva-zlato-tazba-kyanid/96605-clanok.html>.
Tonka, J. 2014. Striebro je investícia desaťročia. [online]. [08.02.2014]. <http://finweb.hnonline.sk/svet-akcii-127/striebro-je-investicia-desatrocia-519669>.

Production Management and Engineering Sciences – Majerník, Daneshjo & Bosák (Eds)
© 2016 Taylor & Francis Group, London, ISBN: 978-1-138-02856-2

Online reputation of companies that have implemented the EMAS III scheme

J. Čorba
Faculty of Business Economy, University of Economics in Bratislava, Košice, Slovak Republic

Ľ. Nastišin
Faculty of Management, University of Prešov, Prešov, Slovak Republic

ABSTRACT: Nowadays, company reputation is considered a very valuable asset in the world market. Every company has both a traditional and an online reputation, regardless of whether it's wanted or not and it is the reputation of companies that is integrated in human resources management and in corporate strategies. Similarly, companies try to achieve competitive advantage through the implementation of environmental management systems such as EMAS scheme III respectively ISO 14001. The fact that participation in these systems is not mandatory, involvement of organizations increases the attractiveness and credibility, thus gaining a significant competitive advantage, which in itself includes improving the image, increasing the opportunities in the market and more satisfied customers. The aim of this paper is to evaluate sentiment by analyzing online reputation of companies that implement EMAS III in the motor industry.

1 INTRODUCTION

A general view prevails in the majority of companies that activities related to environmental protection cause unnecessary costs and thereby endanger the competitiveness and development of the company. In recent years we tend to see views that tend to improve environmental performance with the result of a prosperous business and financial performance. Governmental and nongovernmental organizations put pressure on consumers and it forces companies to deal with issues of environmental protection. Companies are a natural part of their surroundings, as the natural environment becomes a natural component of their marketing and managing environment.

Companies must make reasonable use of natural resources on every level, to protect natural wealth, producing as little waste and generally influence their waste in order to be sustainable in nature. These factors currently affect the supplier-customer chain, not only in manufacturing enterprises, therefore, management should consider the company's environmentally oriented approach followed by competitive advantage in the market. One possibility of reducing the negative environmental impact is the introduction of environmental management systems, either ISO 14001 or EMAS III (Piatrik et al. 2003). The companies access these systems voluntarily and at their own expense in order to create conditions for maintaining and improving their economic effectiveness, international competitiveness, meeting the requirements of sustainable development, also in terms of reputation and corporate image.

The aim of the present paper is to evaluate sentiment by analyzing the online reputation of companies that implement EMAS III in the motor industry. This article is one of the partial outputs of the currently solved research grant VEGA no. 1/0145/14 entitled "Online reputation management (ORM) as a tool to increase competitiveness of Slovak SMEs and its utilization in conditions of Central European virtual market" and post on solutions research projects MVP. I-15-109-00 Increasing economic and environmental performance of the company with an emphasis on integrated management systems.

1.1 Environmental management and its implementation in the enterprise

Environmental management is a set of voluntary environmental policy instruments enabling the introduction of a system approach to the solution of environmental protection in the organization. (Kollar, V., 1998). Implementation of this system consists of several steps that the company has to successfully complete. One of the main steps for an effective and successfully functioning system of EMS is the correct identification of environmental aspects and their environmental impacts, which is part of the so-called initial analysis of the environment in the event of the introduction of EMS according to ISO 14001 or evaluation report by environmental verifiers in the event of the introduction of EMS according to EMAS III (Virčíková & Palfy 2001).

An Eco-management and Audit Scheme is a voluntary environmental management tool for organizations that want to recover and improve their environmental performance. It is structured to, through the better use of resources, comply with environmental legislation, management of its significant direct and indirect environmental aspects and the pursuance of its objectives and processes to help increase their competitiveness. EMAS registration gives companies the ability to demonstrate to all parties (customers, public authorities) that they assess, manage and reduce the environmental impact of its activities and products. (Piatrik et al. 2008).

1.2 *Reputation*

It is defined as the often used concept of marketing and the Internet, and usually means the overall web presence. It can also be compared to leaving marks in the digital world. All activities are interrelated and complementary. (Janouch 2011).

Each company has as a standard as well as online reputation, regardless of whether or not they want it. Reputation always exists (Marsden, 2013). The reputation of the company is considered a very valuable asset. Balmer and Greyser (2003) characterize the company's reputation as a state which is formed over a longer period on the basis of what the company does and how it is maintained. Highhouse defines it as a stable global assessment of the companies that is shared by many sectors (Helm et al. 2011) It is pure reaction of customers, investors, employees and shareholders. It is a collective evaluation of individual impressions (Gottschalk 2011).

1.2.1 *From image to reputation*
Companies often invest large amounts of funds to support the development and marketing activities for the purpose of creating communication campaigns that support the company's image and will thus serve as an encouragement to customers to making a purchase (Leboff 2011). This argument is supported Smaizien and Jucevicius (2009), who argue that companies prefer to focus primarily on the image and omit their reputation. Leboff (2011) says that the image is no guarantee of positive references and recommendations. These can only be achieved through positive reputation. In other words, the basis of modern marketing is not only to care about the image, but also to ensure the reputation, which cannot be influenced and is given a long chain of events.

Bennet and Kottasz distinguished image and reputation based on a time horizon as the main characteristic that distinguishes these two concepts. In other words, the company's image can be created in a relatively short time. Reputation is not generated in the long term and thus cannot be altered or redirected as quickly as image (In: Smaiziene, Jucevicius, 2009). This approach is also supported by Jackson (2004) and Cornelissen (2004), who also emphasize time as the main differential element. Fill (2009) perceive reputation as a broader set of image. It considers that reputation is time consuming compared to the image, which may be affected more quickly. On this basis, it can be argued that the image and reputation are not synonymous but they are interrelated and influence each other.

1.2.2 *Creating reputation*
According to Freedom (2009), reputation exists in three forms, namely primary, secondary and cyclical. Building a reputation was associated mainly with marketing and communications. Burke et al. (2011) argues that currently the company reputation is already integrated in human resources management and in corporate strategies. It is generally accepted that there is a reputation within the company and gradually penetrates out. Fombrun and Foss (2011) notes that it is good if the organization cares about its reputation and highlights the following factors:

- Resolution principle—strong reputation is the result of a unique position in the customer's mind.
- The focusing principle—a strong reputation is the result of a focus on communication and activities directed to one main area.
- Consistency principle—a strong reputation arises as a result of the consistency of communication and activities with internal and external environment.
- The principle of identity—a strong reputation arises when a company is acting in a manner that is consistent with the principles of corporate identity.
- The principle of transparency—a strong reputation is the result of transparency in all spheres of society.

1.2.3 *Reputation in the online environment*
Walter (2013) says that the reputation in life and in business is everything. This means that it is a very fragile element and it takes just one mistake for it to produce often irreversible damage. This is true even stronger in the digital environment, in which the radical transparency and customers require the most power. By Chernatony et al. internet is a place where customers have the opportunity to share information about a brand, but equally it

is also a place where these companies can control information about them. The negative comments on the internet can quickly and severely damage the image and reputation of the brand (Siano et al. 2011), reportedly a personal communication (electronic word of mouth) is an important aspect of on-line reputation. According to Henning-Thuraua (2004), this form of communication can be defined as a positive or negative opinion established potential, current or former customers about the product or company through the Internet.

2 METHODS AND MATERIAL

The present study focuses on the 30 production companies with implemented environmental management systems EMAS III active on the world market. As a criterion for ensuring a homogeneous sample, filed of business as well as company size by number of employees was taken into account. We chose companies focusing on motor vehicle production or production of components used in the largest car companies based in the European Union under the NACE code 29.10 (Manufacture of motor vehicles). In addition to these factors on which market, we took into account the size of the organization, which was a company with at least 500 employees.

Sentiment analysis is one of the most widely used rating system s for online reputation in Europe. This area has already been used in the past by Rajzák et al. (2010) in the context of the evaluation of on-line reputation of banks. The assessment is recorded in the first ten results of a search engine. Analyzing each partial result and sentiment was followed by the final assessment of an entity that serves as an evaluation criterion or success or failure of the company in selected segments. In order to minimize the presence of customized search results by user it used a proxy server, which served as an anonymity tool and found only the most relevant results. The final score of the trial and was independently evaluated by three persons, namely to minimize the subjective factor representation of the results of the resulting values represent the average of all the ratings. The business name was chosen as the search phrase for all organizations. The sentiments of the individual results and the score are shown in the following table.

Table 1. Resulting sentiments and their evaluation.

Sentiment / Position of the result	1	2	3	4	5	6	7	8	9	10
+	20	19	18	17	16	15	14	13	12	11
x	10	9	8	7	6	5	4	3	2	1
±	2	2	2	2	2	2	2	2	2	2
-	-20	-19	-18	-17	-16	-15	-14	-13	-12	-11

3 RESULTS

The obtained final score of selected subjects and the position has been transformed into a summary table showing the result of the analysis. The total points indicate the strength of sentiment of all analyzed subjects. This score also serves as a final assessment factor of success or failure.

The following table presents a summary of results.

The final assessment can be read quite clearly, as well as what caused it. In general, the resulting value is regarded as a positive reputation when it is greater than 55 points. A negative reputation is considered with a score under 30 points. This interval serves as a guide for the subjects, how they are doing and therefore any difference between 29 and 30 points cannot be considered significant but rather that these values are at the limit of the normal distribution, when a positive or neutral reputation begins to overlap negative. The highest, but not the maximum score, was acquired by Volkswagen AG (Σtop = 70) versus the lowest acquired

Table 2. The resulting sentiments and evaluation.

pos.	Company name	1	2	3	4	5	6	7	8	9	10	Σ
1.	Volkswagen AG	10	9	2	7	16	5	4	3	12	2	70
2.	MAN Truck & Bus AG	10	2	8	17	6	5	14	3	2	2	69
3.	Audi AG	10	9	2	2	6	5	4	13	2	11	64
4.	BMW AG Werk Regensburg und Innovationspark Wackersdorf	20	2	8	17	2	5	2	3	2	2	63
5.	Volkswagen AG	10	2	2	7	16	5	4	3	2	11	62
6.	MAN Truck & Bus Österreich AG, Standort Steyr	10	9	8	7	2	2	2	2	2	11	55
7.	Mercedes-Benz Manufacturing Hungary Kft.	10	9	8	7	6	5	4	2	2	2	55
8.	Volkswagen Osnabrück GmbH	10	2	8	7	6	2	2	13	2	2	54
9.	MERCEDES-BENZ ESPAÑA, S.A.	10	9	8	7	6	2	4	2	2	2	52
10.	Automobili Lamborghini S.p.A.	10	2	2	7	6	5	4	2	12	2	52
11.	Honda Manufacturing of the UK	10	2	2	7	6	5	4	3	2	11	52
12.	Daimler AG Mercedes-Benz Werk Hamburg	10	9	8	7	6	2	2	2	2	2	50
13.	AUDI Hungaria Motor Kft.	10	9	8	7	2	5	2	3	2	2	50
14.	Hyundai Motor Manufacturing Czech s.r.o.	10	2	2	7	2	2	14	3	2	2	46
15.	KS Aluminium-Technologie GmbH	2	9	2	2	16	5	2	2	2	2	44
16.	Schaeffler Austria GmbH	10	2	8	2	2	2	2	13	2	1	44
17.	Daimler AG Mercedes-Benz Werk Bremen	10	9	8	2	2	5	2	2	2	1	43
18.	BMW Bayerische Motorenwerke AG BMW Werk 4.1	10	2	2	2	16	2	2	2	2	2	42
19.	Volkswagen Sachsen GmbH Motorenwerk Chemnitz	10	9	8	2	2	2	2	3	2	1	41
20.	BMW Bayerische Motoren Werke AG Werk 01.10 und Werk 01.30	10	2	2	7	6	5	2	2	2	2	40
21.	BMW Bayerische Motorenwerke AG Werk Dingolfing	10	2	8	7	2	2	2	2	2	2	39
22.	MAN Trucks sp. z.o.o.	10	2	8	2	6	2	2	3	2	2	39
23.	BMW Group Anlagenbetrieb im Forschungs- und Innovationszentrum (FIZ) Werk 1.50	10	9	2	2	2	2	2	3	2	2	36
24.	Daimler AG Werk Düsseldorf	10	9	8	2	6	5	4	-13	-12	2	21
25.	MAGNA STEYR Fahrzeugtechnik AG & Co KG	10	2	2	2	2	2	2	-13	2	1	12

points of MAGNA STEYR Fahrzeugtechnik AG & Co KG (Σbottom = 12). The average score (Σaverage = 47.8) in the results is clearly seen that any positive mention outside of the results on their own website the company has a significant effect on the improvement of the position. This is apparent. Only two of the examined subjects scored negatively and regardless of the other positions where they scored well, were moved to the very end of the list.

4 CONCLUSION

The European Union is one of the largest car manufacturers in the world and therefore the automotive industry has a crucial place in the prosperity of Europe. It is a valuable employer of skilled workforce and a key point of knowledge and innovation. The automotive industry operates as the largest private investor in research and development in Europe and also contributes significantly to gross domestic product (GDP) in the EU, exports are much larger than imports. Overall, the motor industry in Europe is confronted with a tense situation, because while non-European markets rise quickly, the European demand for cars has stagnated, as we can clearly see by the evolution of registrations of new passenger cars in vehicle registration in Europe. The number of registrations of new cars decreased in January 2014 by 8.7% and thus it's at its lowest level in recent years. This factor should be one of the primary reasons why manufacturers should increase the emphasis in the development of marketing strategies with regard to an environmentally-oriented approach.

REFERENCES

Balmer, J. & Greyser, S. 2003. *Revealing the Corporation: Perspectives on Identity, Image, Reputation, Corporate Branding and Corporate-level Marketing.* Oxford: Routledge. ISBN: 978-0-4152-8421-9.

Burke, J. et al., 2011. *Corporate Reputation: Managing Opportunities and Threats.* UK: Gower Publishing Ltd. ISBN: 978-0-566-09205-3.

Chovancová, J. 2011. Systémy environmentálneho manažérstva. Prešovská univerzita v Prešove, Fakulta manažmentu, 2011. 96 str. ISBN: 978-80-555-0485-8.

Fill, C., 2009. *Marketing Communications: Interactivity, Communities and Content. 5th Edition.* UK: Pearson Education Ltd. ISBN: 978-0-273-71722-5.

Fombrun, C.J. & Foss, C.B. 2001. *The Reputation Quotient,* Part 1: Developing a Reputation Quotient. [on-line]. [cit. 2014-02-09]. Available at: http://www.reputationinstitute.com/frames/press/01_15_14_GUAGE.pdf.

Gottschalk, P., 2011. *Corporate Social Responsibility, Fovernance and Corporate Reputation.* USA: World Scientific Publishing Co. Pte. Ltd. ISBN: 978-981-4335-17-1.

Helm, S. et al., 2011. *Reputation Management. Berlin: Springer-Verlag.* ISBN: 978-3-642-19265-4.

Henning-Thueau, T. et al., 2004. *Electronic Word-of-mouth Via Consumer-opinion Platforms: What Motivates Consumers to Articulate Themselves on the Internet?* [on-line]. [cit. 2014-02-09]. Available at: http://www.gremler.net/personal/research/2004_Electronic_WOM_JIM.pdf.

Jackson, K.T. 2004. *Building Reputational Capital: Strategies for Integrity and Fair Play that Improve the Bottom Line.* USA: Oxford University Press. ISBN: 978-0-1951-6138-0.

Janouch, V. 2011. *333 tipů a triků pro internetový marketing.* Computer Press. ISBN 978 8025134023.

Kollár, V. 1998. Ekológia—kompas environmentálneho riadenia a rozhodovania, EU Bratislava, vydavateľstvo EKONÓM, Bratislava.

Leboff, G., 2011. *Sticky marketing—Jak zaujmout, získat a udržet si zákazníky.* Prague Management Press, ISBN: 978-80-7261-235-2.

Marsden, H. 2013. *Guard Your Reputation On-line.* Birmingham: Smartebookshop. ISBN: 123-0000-194-89-3.

Piatrik, M., Kollár, V., Vincíková, S. & Rusko, M. 2003. *Environmentálny manažment II.* Banská Bystrica: FPV UMB, 127 s. ISBN 80-8055-861-2.

Piatrik, M., Šudý, M. et al. 2008. *Environmentálne manažérske systémy.* Banská Bystrica: Univerzita Mateja Bela, 2008. 253 s. ISBN 978-80-8083-691-7.

Rajzák, P. et al., 2010. *Systém pre hodnotenie on-line reputácie bánk.* Proceedings of the Faculty of Electrical Engineering and Informatics of the Technical University of Košice, pp. 652–657, ISBN: 978-80-553-0460-1.

Siano, A. et al., 2011. *Exploring the Role of On-line Consumer Empowerment in Reputation Building: Research Questions and Hypotheses.* [on-line]. [cit. 2014-02-09]. Available at: http://www.academia.edu/1096337/Exploring_the_role_of_on-line_consumer_empowerment_in_reputation_building_Research_questions_and_hypotheses.

Smaiziene, I. & Jucevicius, R. 2009. *Corporate Reputation: Multidisciplinary Richness and Search for a Relevant* Definition. In: Inzinerine Ekonomika-Engineering Economics. pp. 91–101. ISSN: 1392-2785.

Virčíková, E. & Palfy, P. *Environmentálne manažérstvo—teória a metodika.* Košice: Štroffek,. 267 s. ISBN 80-88896-15-0.

Production Management and Engineering Sciences – Majerník, Daneshjo & Bosák (Eds)
© 2016 Taylor & Francis Group, London, ISBN: 978-1-138-02856-2

Sustainable universities—increased efficiency and reduced consumption

M. Davidová

Institute of Environmental Engineering, Faculty Mining and Geology, VŠB- Technical University of Ostrava, Ostrava, Czech Republic

ABSTRACT: Universities shall start with practicing sustainable development in everyday life of students and teachers, reducing use of resources, recycling waste etc. According to the research, universities in Europe are lagging behind the developed world trends at applying sustainable solutions. As an outcome are better educated, practice oriented graduates expected on one side. On the other side environmental, economic and social benefits shall reduce pollution, cost burden (releasing resources for other purposes like modern equipment, hardware, software, library), increase synergies. Additionally, consumer behavior, lifestyle and quality-of-life expectations of future intellectual elite will start changing. It must be ensure wide dissemination of results to the community, industry, decision-makers, public authorities and other stakeholders by establishing active cooperation within the higher education institutions of EU not only serve the European science. The Lisbon strategy has defined the direction of development in the European Union for the next decades.

1 INTRODUCTION

We are in a race between political tipping points and natural tipping points. Can we cut carbon emissions fast enough to save the Greenland ice sheet and avoid the resulting rise in sea level? Can we close coal-fired power plants fast enough to save the glaciers in the Himalayas and on the Tibetan Plateau, the ice melt of which sustains the major rivers and irrigation systems of Asia during the dry season? Can we stabilize population by reducing fertility before nature takes over and stabilizes our numbers by raising mortality? The thinking that got us into this mess is not likely to get us out. We need a new mindset. Let me paraphrase a comment by environmentalist Paul Hawken in a 2009 college commencement address. In recognizing the enormity of the challenge facing us, he said: First we need to decide what needs to be done. Then we do it. And then we ask if it is possible. (Brown 2009).

The use of services and related products which respond to basic needs and bring a better quality of life while minimizing the use of natural resources and toxic materials as well as the emissions of waste and pollutants over the life cycle of the service or product so as not to jeopardize the needs of future generations (EEA 2005, Cohen & Murphy 2001, Jackson 2006).

Education for Sustainable Development allows every human being to acquire the knowledge, skills, attitudes and values necessary to shape a sustainable future.

Education for Sustainable Development means including key sustainable development issues into teaching and learning; for example, climate change, disaster risk reduction, biodiversity, poverty reduction, and sustainable consumption. It also requires participatory teaching and learning methods that motivate and empower learners to change their behavior and take action for sustainable development. Education for Sustainable Development consequently promotes competencies like critical thinking, imagining future scenarios and making decisions in a collaborative way.

Education for Sustainable Development requires far-reaching changes in the way education is often practiced today. (UNESCO 2014, ESF Education for a sustainable future 2014).

The environment is fundamental to human wellbeing and development, although the two are usually in contradiction. Scientific and technological development has offered many new products, comfortable supply of energy for heating and cooling, individual and just in time transportation abilities, requiring an ever increasing exploitation of natural resources. Also, improved material and energy efficiencies of processes, products, and services are being overtaken by the increasing number of people in the World, and the growing standard of many of them. Therefore, a radical change in thinking and lifestyles is needed to accompany the scientific and technological development, and to make this development sustainable. It needs educating people, especially young ones, in sustainable production and consumption.

The Lisbon strategy has defined the direction of development in the European Union for the next decades. Sustainable Development (SD) is becoming one of the key principles in the strategy: "A European Union Strategy for SD", and "Green Paper on Energy Efficiency or doing more with less", foster research, development and innovations. Universities are very important in this respect: "The role of the universities in the knowledge of Europe", Bergen statement of the European ministers within the Bologna process and other European Union documents are aimed to include sustainability concept in education.

Nowadays, students are not exposed enough to SD approach, and environmental protection is not carried out as a 'learning by doing' activity. Employers' wish to make future graduates more aware of the needs, and better skilled in the practices of SD. Therefore, universities shall start with practicing SD in everyday life of their students and teachers, reducing their use of resources, recycling waste, increasing material and energy efficiencies, optimizing transportation costs, increase human capital etc. Teaching and learning by using research and development should be practiced together with their colleagues in other departments, in all the areas taught at a particular university, solving complex problems of the modern society.

Universities in the United States (USA) foster SD principles by establishing University networks. According to the internet research, universities in Europe are lagging behind the world trends at applying sustainable solutions. Europe also continues to perform poorly than the USA in new, fast-growing fields, in the number of world citations and patents. Europe is better at producing knowledge than at assessing and applying knowledge. Thus, a gap between EU and USA is growing. The action could start filling the gap with the dissemination of new knowledge, sustainable performance, consequently influencing the scientific production. It is important not only to fill the gap between EU and USA, but also to overcome it and become world leaders in the field of the sustainability. European Union will remain competitive only if it fosters creativity (research, development, innovations), diligence, economical behavior (consumption of resources and materials), and ethics (honesty, tolerance, respect, etc.). Sustainable Universities shall combine economic, societal and environmental objectives into a European perspective. The role of universities is inevitable in this process. Idea of a project could be complementary to Technology Platforms, which bring together companies, research institutions and public authorities. Furthermore, its projects could be included

into the framework program, where cooperation, ideas, people, and capacities are engaged to reduce energy and water consumption, and save the environment by sustainable management of resources.

The United Nations Decade on Education for Sustainable Development (ESD) is an international recognition of the key role that Education and Communication can play in enabling and enhancing sustainable development efforts, and processes leading towards these. The recognition that education is a critical agent of transformation in terms of changing lifestyles, attitudes and behaviour, in increasing participation in visioning and realizing a sustainable world, and facilitating the use of Communication, Education, Participation and Awareness (CEPA) to foster the change needs in different sectors, needs to be further strengthened. Reflection, visioning and sharing are the crucial elements of ESD (UNESCO 2014).

It must be ensure wide dissemination of results to the community, industry, decision-makers, public authorities and other stakeholders by establishing active cooperation within the higher education institutions of EU not only serve the European science. It is necessary to contribute and disseminate understanding of interactions between all the three dimensions of SD (economic benefit, environmental protection, social welfare) and will help researchers to bring together their ideas and increase mobility, give an opportunity to build synergies between faculty and local community, and to fully integrate the principles of SD into institution's activities. Furthermore, it may encourage other business and non-business organizations to assume the universities' best practices.

Sustainability of universities will depend on departmental, university, interuniversity and international levels. It will bring together different laboratories of the same department to reduce pollution, increase effectiveness, and raise human capital of the department. It will stimulate cooperation between departments to solve problems, e.g. turn university buildings into passive ones (ecobuilding).

Universities now have the opportunity to reorient the traditional functions of teaching and research by generating alternative ideas and new knowledge. They must also be committed to responding creatively and imaginatively to social problems and in this way educate towards sustainable development.

Engineering education, with the support of the university community as well as the wider engineering and science community, must:

- have an integrated approach to knowledge, attitudes, skills and values in teaching incorporate disciplines of the social sciences and humanities

- promote multidisciplinary teamwork
- stimulate creativity and critical thinking
- foster reflection and self-learning
- strengthen systematic thinking and a holistic approach
- train people, who are motivated to participate and who are able to take responsible decisions

Today's engineers must be able to:

- understand how their work interacts with society and the environment, locally and globally, in order to identify potential challenges, risks and impacts
- understand the contribution of their work in different cultural, social and political contexts and take those differences into account
- work in multidisciplinary teams, in order to adapt current technology to the demands imposed by sustainable lifestyles, resource efficiency, pollution prevention and waste management
- apply a holistic and systemic approach to solving problems and the ability to move beyond the tradition of breaking reality down into disconnected parts
- participate actively in the discussion and definition of economic, social and technological policies, to help redirect society towards more sustainable development
- apply professional knowledge according to universal values and ethics
- listen closely to the demands of citizens and other stakeholders and let them have a say in the development of new technologies and infrastructures.

In order to achieve the above, the following aspects of the educational process must be reviewed:

- the links between all the different levels of the educational system
- the content of courses
- teaching strategies in the classroom
- teaching and learning techniques
- research methods
- training of trainers
- evaluation and assessment techniques
- the participation of external bodies in developing and evaluating the curriculum
- quality control systems.

These aspects cannot be reviewed in isolation. They need to be supported by an institutional commitment and all decision makers, in the form of:

- redefinition of institutions' and universities' missions, so that they are adapted to new requirements in which sustainability is a leading concern

- institutional commitment to quality
- institutional support for changing educational paradigms and objectives research funding.

Students are important actors in the process of Sustainable Development (SD). All the assistants were agree about founding a platform with the main objective of join efforts and work together towards a sustainable universities.

Some vital issues for a successful student conference are the following:

- Set up a frame for information exchange
- Leave as much freedom to the participants in designing solutions, action plans
- Set concrete/reachable goals for the conference to obtain useful results
- Get feedback after the conference
- Socializing
- Create an informal atmosphere
- Create environment in which participants can talk right away about issues they are most interested in (in which participants share as much useful information as possible)

2 TECHNOLOGICAL EDUCATION AND ETHICAL ISSUES IN SD

Often the outcome of the work of engineers has broad implications beyond technological issues, social, economic and environmental implications. Therefore, pressures may be exercised to have an influence on the above referred outcome: designs, recommendations, evaluations, etc.

This conflict is even more important when decisions have to be made under uncertainty, as it usually the case. And this brings ethical issues to the scene.

Education in ethical values should not be implemented as a separate item from technology instruction, on the contrary, ethical values should be incorporated in the technical subjects wherever appropriate.

Topics that should be considered when teaching ethical issues in engineering:

- Understand the strong subjective, judgement-based components that are inevitably present in engineering practice
- Avoid limiting the scope of analysis to short term issues only, include long-term implications in the decision making process
- Avoid unnecessary fragmentation of viewpoints on the projects and engineering evaluations
- Include conflicting views when dealing with complex topics
- Pay attention to social perception and communication aspects and pay special attention to

designing education programs and applications that target improvements for poor people

- Encourage in the university and in the format of teaching and in the relationship student/ professors the following values: respect for other people, respect for the environment, solidarity, honesty, tolerance, team-work, etc
- Incorporate/transmit a global view of sustainable development within each specific industrial area

Technological change in SD:

- to provide students in engineering with sufficient scientific knowledge basic ecology courses are essential to relate sustainable development to politics and science
- training is essential to implement (EU) directives, such training opens wider possibilities towards SME's with respect to sustainable development
- the genesis of technology is a function of social, institutional and cultural conditions which are on their turn are shaped by technology
- engineers need a firm understanding thereof on top of their skills.

Simulation tool and games for teaching SD:

- In future it would be interesting to play games (not only presentations)

3 INTEGRATED SD IN THE CURRICULUM

How should lectures be designed and developed to make clear the complexity of our global society?

- To approach complexity, lectures should be designed to deal with concrete objects and their relationships with the rest of nature and society, to illustrate the way in which an analysis of the problem itself is affected by ecological and environmental services, as well as by other societal activities.
- How should the multidisciplinary approach in educational programs be dealt with?
- Professors participation
- Different study areas interacting strongly
- Basic knowledge in ecology and SD
- Activities like seminars, workshop, etc.
- Research about regional problems
- Developing a disciplinary expertise
- How can the aspects of stakeholder participation be dealt with educational programs?
- Information and social sensitivity
- Strong association with governments, companies, social groups and communities
- Trough research and outreach programs.

4 TOWARDS SUSTAINABLE UNIVERSITIES FOR SD EDUCATION

What are the key success factors and barriers?

- Key success factors include financial resources, commitment from the senior management, adopting a sustainability policy, using the students to "fuel" initiatives, twinning studies with other institutions and celebrating successes
- Key barriers are the participation of students and staff, the externalities of the university's activities and changing the culture of an institution
- Student experience—how is it affected?
- Opportunity to take up new courses (optional and mandatory)
- Putting environmental consciousness into practice (laboratory work, field work, involvement in Environmental Management Systems)
- International partnerships and student ambassador schemes
- Feedback of sustainability initiatives is generally positive

Sustainability literacy—how can it be achieved across the university?

- Integration of sustainable development into all subjects
- Evaluation needs to be addressed as sustainability literacy means something different for different subjects
- Introduce an environmental management system and involve students. Use this as the means, not the goal! This is also an effective way of teaching

5 INTEGRATING SD IN THE CURRICULUM

It is possible to say, that the clue to bring about the integration of SD in the curriculum is to involve all the teachers, and from the oral presentations it seems that the best way to do it is, furthermore having an up-down policy which promotes the integration, working teacher to teacher and guide each teacher to recognise the role he or she can play in SD.

6 ETHICS AND VALUES IN ENGINEERING EDUCATION

Is there a point in teaching values and ethics to engineering students? The students have already developed their values and ethic before the university time? Is there any chance? Is the university the right place? Which strategies can be used to teach ethics and values in the curriculum? Shall the curriculum be redesign?

7 SPECIFIC COURSE ON SD

A specific course on SD is needed in order to guaranty a sustainability basic knowledge, although not everybody thinks it has to be compulsory. Universities must redirect the teaching-learning process in order to become real change of people who are capable of making significant contributions by creating a new model for society.

Responding to change is a fundamental part of a university's role in society. There is evidence that sustainable development has already been incorporated in engineering education in a number of institutions around the world.

Today decade offers a great opportunity to consolidate and replicate this existing good practice across the international higher education community.

8 SCIENTIFIC PROGRAM AND INNOVATION

The program should include environmental, economic and societal part. Vision, mission statement and strategy of member universities will be adapted to this program. Sustainability report should be elaborated and published every year by each university.

a. Environmental program
Energy usage reduction will be carried out by increased insulation of buildings, improved heating and ventilation systems, installing renewable energy sources (combined heat and power systems, boilers using biomass, heat pumps, wind turbines, introducing heating control, solar collectors), more efficient lightning, optimize commuting of students/employees and transportation of goods. Minimization of water use, CO_2 and waste production will be mastered by good housekeeping, rain-water use, wastewater regeneration and reuse, waste separation and recycling, product repair and reuse, cleaner products and processes, pollution prevention and control, water and air analyses, health and safety in laboratories, dangerous waste handling. Food supply and quality, organic waste reuse will be regarded too.

b. Economics program
Environmental management systems (ISO 14000, EMAS), environmental accounting, investment return and cost evaluation of all proposals, green procurement, purchasing and selling of goods and services, cost control and reduction should be aimed.

c. Societal program should include:
- Education of students, employees and general public in energy, waste and consumption minimization, using the Earth Day, Water Day and similar events.
- Organization of energy and water conservation, waste and environmental impact minimization, student employment service.
- Health care and fitness.
- Social policy and services, psychological and employment help to students.
- Environmental law, rules of conduct, ethical codex, and intellectual property rights.
- Life style revolution by zero waste and renewable energy thinking.
- Etc.

9 OBJECTIVES AND EXPECTED SCIENTIFIC IMPACT

For human systems, the discussion of sustainability has focused on the problem of resource dependency (Adger 2000) and the management of place-based ecosystems (Berkes & Folke 1998). There is also some work on the relationships among sustainability, urban design, and resilience.

As an outcome are better educated, practice oriented graduates expected on one side. On the other side environmental, economic and social benefits shall reduce pollution, cost burden (releasing resources for other purposes like modern equipment, hardware, software, library), increase synergies. Additionally, consumer behaviour, lifestyle and quality-of-life expectations of future intellectual elite will start changing.

Networking will produce ideas, case studies, and benchmarks. Individual research and development executed in project and seminar reports, theses at all the three educational levels (bachelor, master, doctoral), will result in investment proposals, retrofits, and best practices. The results will be measured in terms of annual savings in energy, water and materials usage, in waste produced and recycled per individual, and in university rankings regarding sustainability.

Direct scientific impact is expected in the environmental management and governance of universities and in knowledge transfer.

In European policy documents on sustainable consumption, such as "Household Consumption and the Environment," the general approach is to articulate policies that could increase levels of sustainable consumption (EEA 2005). The policies generally include changing the signals for pricing, such as for the consumption of water and electricity; providing attractive alternatives, such as improved public transportation; and making available educational and voluntary approaches, such as product labeling.

The EU actively promotes human rights and democracy and has the most ambitious emission reduction targets for fighting climate change in the world. (Europa 2012a).

Education, research and public finance are stressed as important instruments in facilitating the transition to a more sustainable production and consumption patterns. And because monitoring and follow-up are crucial for effective implementation, the renewed strategy contains a strong governance cycle. Every two years (started in 2007) the Commission is to produce a progress report on the implementation of the strategy. This report is to form the basis for discussion at the European Council, which will give guidance to the next steps in implementation (Europa 2012b).

Creating a sustaining industrial system calls for a new definition of quality in product, process and facility design. Quality is embodied in designs that allow industry to enhance the well being of nature and culture while generating economic value. Pursuing these positive aspirations at every level of commerce anchors intelligent, sustaining design deep within corporate business strategy. And when good design drives the business agenda, the path toward sustainability turns from trying to be "less bad" to identifying new opportunities to generace a wide spectrum of value. (McDonough & Braungart 2002).

REFERENCES

Adger, W.N. 2000. Social and ecological resilince: are they related? Progress in Human Geography 24, no.3: 347–64.

Berkes, F. & Folke, C. 1998. Linking social and Ecological Systems: Management Practices and Social Mechanisms for Building Resilience. Cambridge: Cambridge University Press.

Brown, R.L. 2009. Plan B4 Mobilizing to save civilization: 7–8, W. W. Norton & Company, Inc., 500 Fifth Avenue, New York, N.Y. 10110.

Cohen, M.J. & Murphy, J. 2001. Exploring sustainable consumption: Environmental policy and the social sciences. New York: Pergamon.

EEA (European Environment Agency), 2005, Household Consumption and the Environment, EEA Report, 11/2005, Copenhagen: EEA.

ESF Education for a sustainable future, 2014. Accessible from: http://www.ceeindia.org/esf/esf.asp.

Europa, 2012. Accessible from: http://ec.europa.eu/environment/eussd/.

Europa, 2012. Accessible from: http://europa.eu/about-eu/basic-information/index_en.htm.

Jackson, T., 2006. The earthscan reader in sustainable consumption. Sterling, VA: Earthscan UNESCO, 2014. Accessible from: http://www.unesco.org/new/en/education/themes/leading-the-international-agenda/education-for-sustainable-development/.

McDonough, W. & Braungart, M. 2002. Corporate Environmental Strategy, Elsevier Science Inc.Vol. 9, No. 3, 251.

UNESCO, 2014. Accessible from: http://www.unesco.org/new/en/education/themes/leading-the-international-agenda/education-for-sustainable-development/.

Production Management and Engineering Sciences – Majerník, Daneshjo & Bosák (Eds)
© *2016 Taylor & Francis Group, London, ISBN: 978-1-138-02856-2*

Improvement of health and safety conditions at selected workplaces

M. Dobosz & P. Saja
Podkarpackie Centrum Usług Dydaktycznych, Podkarpackie Teaching Service Centre, Rzeszów, Poland

A. Pacana
Department of Manufacturing and Production Engineering, Rzeszow University of Technology, Politechnika Rzeszowska, Rzeszów, Poland

A. Woźny
Faculty of Management, Rzeszow University of Technology, Politechnika Rzeszowska, Rzeszów, Poland

ABSTRACT: In the paper the process of improving the health and safety conditions at transshipment terminal of logistics company was presented. The specificity of the place of work makes that workers are exposed to constant overload of musculoskeletal system, which is determined by the process of transporting loads of considerable weight. The diverse nature of the work, which largely depends on the size and weight of the cargo, makes that a real estimate of the maximum load per worker becomes extremely difficult. The aim of this paper is to show a positive change in the conditions of safe and hygienic work on the selected positions. Partial mechanization process of transporting goods had a positive effect on the spirit of the people and culture of the organization. Internal responsibility of employees at all levels for the state of safety culture creates the necessary basis for the organization of working conditions.

1   INTRODUCTION

Improvement is a natural phenomenon in developing companies. Also, cultural management is an extremely important aspect of the company functioning, as thanks to the culture of the organization relationship between an employer and employees is formed, and also between a company and its clients. One of the elements of culture is the culture of health and safety. Building and shaping the culture patterns of health and safety at work is an ongoing process to achieve real change because it is extremely difficult and staged.

Quality and efficiency of safety culture change must be made in any space of a company. However, if any culture, especially the culture of health and safety fulfills its role, a system that will draw attention to each element of a change should be developed. Occupational health and safety management systems are a part of a complex system of general business management. The implemented system is a set of effectively interacting elements that shape the policy of health and safety in the organization.

Regardless of profession, an employee is always exposed to hazards that can cause partial or long-term injury. Negative determinants of the labor process form an occupational hazard that must be estimated for each job. Regulation of the Minister of Labor and Social Policy, dated on September 26, 1997 (Journal of Law of 2003, No. 169, item 1650 later amended) § 2 point 7 defines precisely what an occupational hazard is. By this, it is meant the likelihood of adverse events related to performed work, resulting in a loss, especially adverse health effects due to occupational hazards in the working environment or the way of performing work. An important element in an innovative approach to occupational risk is that it should be assessed individually for each employee. Individual external and internal features significantly affect the strength or weak correlation to exposure to the hazard in the workplace. This fact has become a reason for conducting research on the impact of individual characteristics of an employee handling transshipment terminal in a logistics company on work culture and organizational changes.

2   ANALYSIS OF CURRENT CONDITIONS OF HEALTH AND SAFETY AT TRANSSHIPMENT TERMINAL

Determination of health and safety conditions at a transshipment terminal was initiated by the correctly estimated occupational risk. The range of methodical estimation of occupational risk run by a team of professional advisers showed the most significant risks and implemented necessary preventive measures.

Work on the transshipment terminal seems to be repetitive and therefore monotonous. The scope of activities, namely, unloading, loading, reloading goods shows no significant variation of the working environment. However, focusing on the essence of this one can observe that the variability of the load every time shapes other working conditions. Regulation of the Minister of Labor and Social Policy on health and safety on manual handling points in:

§ 21 1. The maximum permissible mass of the moving charge in a truck on the flat and hard surface, plus the mass of the truck, shall not exceed:

1. 350 kg—for a 2-wheel truck,
2. 450 kg—for a 3- or 4-wheel truck.

2. When moving loads in a truck at gradients greater than 5% of the load weight, including the weight of the truck, it must not exceed:

1. 250 kg—for a 2-wheel truck,
2. 350 kg—for a 3- or 4-wheel truck.

Weight variability, as well as the size of the transported goods by a pallet truck make that an employee is constantly exposed to overload of the musculoskeletal system. The differentiation of the load is extremely important because the weight and size influence directly on the tension of individual muscle groups. Organizational culture on the transshipment terminal of a logistics company prohibits the transport of goods with a total weight with a truck of 450 kg (in the plant a 4—wheel pallet trucks are used). However, the variability of the product and its size often make that the goods transported by an employee exceed the legal limit.

During a routine check of safety and health at work on the transshipment terminal one of the employees reported to the safety and health inspector that the continued use of a pallet truck leads to significant discomfort and work overload and influences negatively on the skeletal muscle system. It is important that the employee who reported the problem is characterized by high seniority and old age. According to the risk assessment card of that employee, the age and seniority strongly correlates with the risk resulting from their work, namely the occurrence of the skeletal muscle overloads.

An individual approach to occupational risk assessment cemented the belief of the inspectors that broader perspective to the problem should have a positive impact on the culture of the organization and safety at work. The interviews conducted by a team of advisers of health and safety with randomly selected employees showed that most of them negatively relates to the facts of the working environment and work culture itself.

The interviews showed poor motivation to the work, which created a basis for reducing the productivity and quality of work performed. An analysis of the situation was a basis to take actions to improve the health and safety conditions at the transshipment terminal.

3 IMPROVING SAFETY AT TRANSSHIPMENT TERMINAL

Team of advisers for occupational health and safety at the transshipment terminal paid particular attention to overload of the musculoskeletal system during work. The analysis showed a significant overload and therefore forbade to load, unload or handle of cargo in excess of 450 kg. However, the variability of the working environment does not guarantee that each pallet will have the proper weight There is also a significant element of the risk exposure, which is almost constant.

Combining of work environment variation due to the load and exposure to the hazard caused some of employees (mainly those with more years of service, or age) began to feel the pain in the musculoskeletal system. These were the employees who began the process of changing the culture of health and safety at work.

While the process of change was initiated, namely there was a need for changes in working conditions, a broader analysis of the facts was made. Employees' awareness should be considered as the value of culture change on the transshipment terminal in a logistics company as for the improvement of working conditions and employees' determination to make this change. Training and lectures by inspectors outlined a new policy and explained the existing and future threats in the workplace. Showing of good practices and positive habits is the key to achieve the goals. Trainings prepared staff to change the culture of work, as well as the workers got familiar with the new procedures associated with the use of electric pallet trucks.

The turning point of the organization culture change in the transshipment terminal, and also the culture of health and safety was a change of hand pallet trucks to electric ones. Employees aware of the new procedures can build new habits in the workplace.

The use of electric pallet trucks was a basis to the change of organizational culture on health and safety. The scope of the changes makes that the working environment for transshipment terminal logistics company has been significantly transformed. The change is a practical example of changing the culture of the organization proposed by E. Schein. Natural transition from the foundation, by value and ending on the artifact caused very positive consequences in the consciousness of health and safety of workers and employers.

That change was an impulse to explore the awareness of the work culture, as well as an approach to essence of security policy in the transshipment terminal. Team of advisers for health and safety conducted quantitative research one month after the change. The research concerned all the interested employees.

The first question the employees had to answer concerned subjective feelings of improvement of work comfort. The essence of the change was to improve working conditions by replacing HPT for electric pallet trucks.

The data in Figure 1 show that the employees' feelings about the organization culture change introduced by the use of electric pallet trucks are positive. The vast majority of workers of the terminal, namely 75% (answers "definitely yes" and "rather yes") indicated that the changes improved working conditions. One-fifth of respondents (20%) did not experience a clear improvement in working conditions after the implementation of electric pallet trucks. It is worth noting that only 5% of workers said they did not experience any improvement in working conditions.

An analysis of the data showed that the change improves working conditions. It is important that the working environment evolved. Change in working conditions resulted in the situation that the team of advisers needed to update the risk assessment, since the exposure to overload of the musculoskeletal system decreased significantly. Employees of the terminal are still operating HPT, but the introduction of their electric counterparts caused the employee, through an individual assessment of the load, is free to choose the way the goods are transported.

Another analyzed aspect was a subjective assessment of employees as to the effectiveness of the change in the organization of the work environment. The primary objective of changing the culture of the organization of health and safety at transshipment terminal was to reduce the overload of the skeletal muscle system involved in the loading, unloading and handling of goods.

A month after the introduced change, as many as 75% (response definitely yes and rather yes) of terminal employees felt the reduction of the overload of the skeletal muscle system. Less than 17% of respondents were not able to clearly determine whether the change in safety culture of the organization declined or did not reduce the effects of exposure to overload to the muscular skeletal system. Only 8% of employees of the transshipment terminal said after a month of using electric pallet

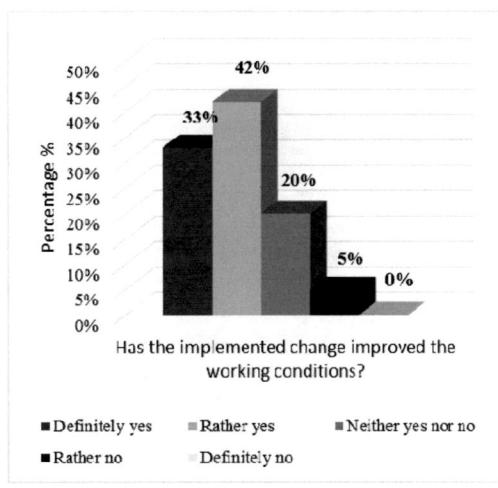

Figure 1. Subjective feelings of workers after the change of safety and health culture (N = 40).

trucks that they did not feel the improvement of working conditions, and the overload of the skeletal muscle system did not decrease.

The data in Figure 2 show that the objective intended during the introduction of changes was achieved. Feeling of improvement in the work comfort among 2/3 of employees is a very good result. It should also be noted that the real noticeable reduction in the overload of musculoskeletal can be expected after several months of the use of electric pallet trucks.

By introducing the electric pallet trucks, executives have to adapt workplaces, as well as prepare the organizational structure. The process of changing the culture of the organization was divided into three stages, which had been described previously. The efficiency of the process is the key to achieve the desired objectives. In the case of a change of hand pallet trucks into the electric ones, some changes in the new working process needed to be introduced in the plant (adapting premises for electric trucks, the introduction of new safety and occupational procedures and the policy) and employees (specialized training with station instruction, the use of personal protective equipment for employees of battery room, update the occupational risk assessment).

Employees of transshipment terminal assess positively the whole process of changes. 75% of employees said that the process of change had improved the safety of the work environment. It is worth noting that most of the staff determined that the changes applied to the transshipment terminal would increase quality and productivity.

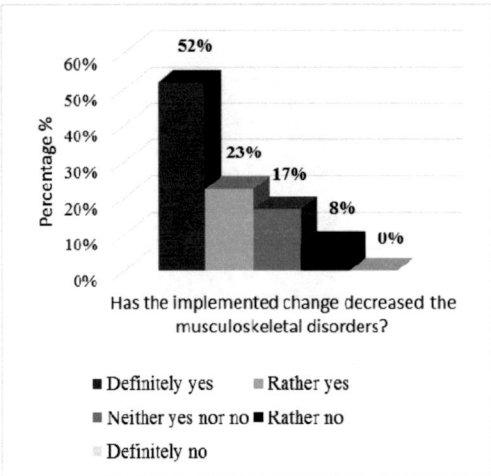

Figure 2. Efficiency evaluation of the implemented change (N = 40).

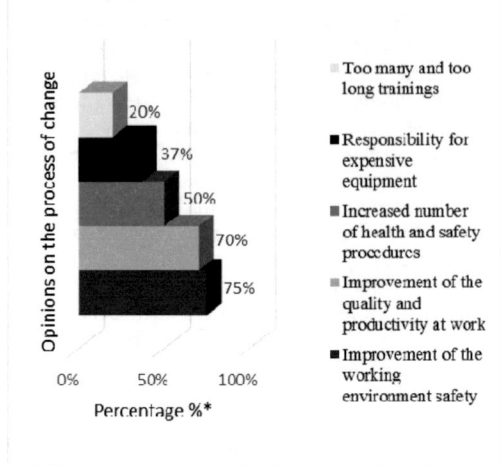

*Percentages do not account for 100% because respondents could give more than one answer.

Figure 3. Evaluation of the changes (N = 40).

One should also refer to the negative reviews on the process of change—Figure 3. Half of the employees referred negatively to the increased number of health and safety procedures. More than one-third is afraid of additional liability. In addition to the dangerous goods they are responsible for electric pallet trucks. A small percentage of terminal workers (20%) responded negatively to the amount and duration of training directly related to the process of change of safety and health culture at work. It is worth pointing out that the trainings carried out by a team of advisers were necessary to start the process of working with electric trucks pallet.

An analysis of quantitative research indicated that the process of change of the organization culture and safety at work is very complex. Although the workers were an impulse to produce the changes, but some negative opinions about the process can be noticed. It should be noted that the employees know only the end result of teamwork advisers for health and safety. Awareness of improving working conditions, as well as a change of hand pallet trucks into electric ones would have accomplished without the help of inspectors. Valuing of the change process by building awareness and purposefulness through training and safety procedures had a negative impact on the assessment of the whole process of change.

Implementation and improvement of the process of safety management and occupational health affects the quality and productivity. This is shown by subjective assessment of the workers who relate to the process of change positively, with particular emphasis on improving safety and quality and productivity of work. Updating

of safety policy should go hand in hand with the update of risk assessment, as potential new hazards do not relate only to the employee, but also are associated with the policy of occupational health and safety.

4 CONCLUSIONS

The process of improving the safety and health at work is very complex. Therefore, it should be integrated with the organizational culture change.

Regardless of the type or size of the organization all the company units must be involved in the process of development. Determination of employees, as well as team advisers for health and safety contributed significantly to the improvement of the working environment by replacing HPT with the electric ones.

An analysis of quantitative research after the introduced process of changes in working environment showed that after the introduction of the proposed changes the employees' satisfaction with their working conditions increased significantly. It is worth noting that the transshipment terminal workers referred to the quality and performance. According to the vast majority of them change in the culture of organization, and at the same time improving the safety and health at work increased quality and productivity.

A positive transformation process of working conditions was made through the involvement of all units operating in the organizational structure of the transshipment terminal. Important was the fact that emerged from the process of change, namely, one could notice synergistic cooperation between employers, employees and inspectors for health and safety of the workplace described.

The initiative of changes proposed by the workers themselves testified of the high probability of efficiency of the process itself. Change in the culture of the organization made that the awareness of the safety and health of workers and employers is growing.

REFERENCES

Cameron K. & Quinn R. 2006. *Organizational culture—the diagnosis and change,* Ed. Economic outbuilding, Kraków 2006. p. 133–135.

Pacana A. 2013. *Implementation, auditing and improvement of safety management systems and occupational health in line with PN-N-18001.* Rzeszów University of Technology Publishing House, Rzeszow 201. p. 53.

Pacana A. 2015. *Audit of management systems bhp.* Rzeszów University of Technology Publishing House, Rzeszow.

Pacana A., Bednárová L., Pacana J., Liberko I., Woźny A. & Malindžak D. 2014. *The influence of some factors oriented film production process for its resistance to puncture.* Chemical Industry, 93 (12), 12/2014, Publishing SIGMA-NOT, Warsaw 2014. p. 2263–2264.

Regulation of the Minister of Labour and Social Policy of 18 March 2009 amending the regulation on occupational health and safety manual handling (Journal of Laws of 2009 No. 56, item 462).

Regulation of the Minister of Labour and Social Policy of 26 September 1997 on general safety and health at work. (Journal of Laws of 2003, No. 169, item 1650 later amended).

Rovnak M., Chovancová J. & Bednárová L. 2013. *Managing environmental risks in production companies.* In: Ecology, economics, education and legislation: conference proceedings, volume II: 13th international multidisciplinary scientific geoconference SGEM 2013:. -Sofia: STEF92 Technology, 2013. -ISBN 978-619-7105-05-6. -ISSN 1314-2704. -pp. 651–658.

Benefits of EMAS easy implementation in SMEs

A. Feranecová, M. Ivaničková, N. Jergová & M. Sabolová

Faculty of Business Economy with Seat in Košice, University of Economics in Bratislava, Košice, Slovak Republic

ABSTRACT: Eco-Management and Audit Scheme, EMAS, is one of the approaches by which organizations can actively reduce the impact of its activities on the environment and also can increase the efficiency of its operations. The result is a positive impact on the environment, increasing of competitiveness and economic effects. The "EMAS easy" is focused on the implementation of EMAS in Small and Medium-sized Enterprises (SMEs). EMAS easy is an easy access and a set of tools that enable simplified, time-consuming and less costly implementation of EMAS. Wider application of this methodology also supports the European Commission, which had already begun in the third round of the program for consultants in various member states. Main objective of this paper is to identify the benefits of implementing EMAS Easy in SMEs by investigation of referred benefits in the literature and comparing them with the facts in practice of particular enterprises.

1 INTRODUCTION

1.1 Basic EMAS characteristics

The Eco-Management and Audit Scheme (EMAS) is a voluntary environmental management tool for organizations that want to enhance and improve their environmental performance (Hillary 2004). Its aim is to promote the improvement of the environmental performance of organizations implementing and maintaining environmental management system, a systematic and periodic evaluation of its effectiveness, publishing verified information on their environmental performance, open dialogue with the public and stakeholders, not least the active involvement of its employees (Bednárová et al. 2013, Bosák & Olexová 2013.). Enterprise with EMS represents Table 1.

A project officially called "Capacity Building for EMAS easy in Northern Europe" was included also in the Slovak Republic.

The main advantages of EMAS easy implementation can be summarized easily. The whole process should take no more than 10 days, may not require the participation of more than 10 people, the documentation should be no more than 10 pages, all in the context of transparent divided into 30 steps. Creators of the EMAS easy method argue that the real obstacles of the environmental management systems implementation are not due to the demanding standards, but access to these standards (Constantini & Crespi 2008).

Therefore a method EMAS easy has been developed to reduce red tape barriers resulting from the necessary knowledge and the costs of consultation and certification (Steger 2000, Biondi et al. 2000,

Table 1. Number of enterprises with EMS

Year	ISO 14001 in the world	ISO 14001 in Europe	EMAS in the world
2004	90 554	39 805	3 930
2005	111 163	47 837	4 137
2006	128 211	55 919	4 800
2007	154 572	85 097	5 671
2008	188 815	78 118	6 119
2009	223 149	89 237	7 404
2010	251 548	203 126	7 707
2011	267 457	175 280	8 090
2012	285 844	100 550	8 174
2013	298 253	121 687	8 376

Ammenberg & Hjelm 2003, Fresner & Engelhardt 2004, Constantini & Crespi 2008).

To overcome barriers to implement EMS in SMEs, there are a number of special requests from which we can choose. They may be classified into the following four categories (Dalhammar 2000).

- Piecemeal approach to rewards on the way to certification (Dalhammar 2000).
- Coaching (Tunnessen 2000, Wells & Galbraith 1999).
- Standardized solutions for the deployment of EMS (Wells & Galbraith 1999).
- Joint EMS and group certification (Ammenberg et al. 1999, Ammenberg & Hjelm 2003).

In EU, for example in Sweden, a joint EMS and group certification gained much attention and today there are many companies where they work with shared EMS. Some of them completed the certification or planning to do in the near future (Ammenberg & Hjelm 2003).

1.2 *EMAS easy benefits identification*

EMAS is often presented as an appropriate tool for larger companies (Palmer & van der Vorst, 1996 Wells & Galbraith, 1999). It is also clear that the EMS has been largely adopted by large companies (Merritt 1998, Gribble & Dingle 1996). The suitability of the EMS and its most common standard ISO 14001 for SMEs is often discussed (Hutchinson & Hutchinson 1995, Palmer & Van der Vorst 1996, Gerstenfeld & Roberts 2000). There were even opinions and attitudes that SMEs were not at all taken into account in establishing standards and therefore unsuitable for these companies (Gleckman & Cruel 1997). Table 2 represents enterprise with EMAS in EU according to size.

The EMS is an increasingly diffused tool among enterprises operating in different sectors, thanks to the drive and impulse coming from the voluntary certification schemes (such as EMAS and ISO 14001) in which they are mainly applied (Iraldo, Testa & Frey, 2010). Unfortunately, the strategies and tools designed especially for large organizations are often uncritically transferred to smaller organizations (Dandridge, 1979, Welsh & White, 1981, Holt et al, 2000). EMAS is no exception (Palmer & Van der Vorst 1996, Gleckman & Cruel 1997).

Since the appropriateness of environmental management systems as specified in ISO 14001 may be disputed, SMEs are worth finding resources and time to implement EMAS Easy. Some of these SMEs may not have the opportunity to decide whether to accept and perform EMAS certification. This may be a requirement for one or more customers. Fig. 1 represents increasing number of the enterprises with EMS.

Many authors (Steger 2000, Biondi et al. 2000, Ammenberg & Hjelm 2003, Fresner & Engelhardt 2004, Constantini & Crespi 2008) identified the following benefits of EMAS Easy implementation:

- Introduction of quality environmental management system (EMS).
- Environmental risk management.
- Reliable assessment and verification EMS, the environmental statement.
- Environmental information accredited independent verifier.
- Guarantee full compliance of the applicable environmental legislation and thus reduce the risk of potential charges for sanctions in connection with its failure to saving resources and energy, reduction of operating costs.
- Reduce the risk of pollution incidents for which the organization is responsible.
- The use of the EMAS logo as a marketing tool.
- Increase business opportunities in markets having regards to the conduct of the enterprise to the environment (e.g. public sector).
- Improved relations with customers, the public and administrative authorities.
- Increase business confidence to investors and other entities.
- Improved working environment and increased employee commitment.

Costs of the EMAS and EMAS easy implementation represent Table 3.

Because of growing interest in EMS around the world, the demand for research is also increasing. The world found it necessary to evaluate its effects. Along with the increasing number of evaluations, reviews, empirically based studies began to be carried out at the end of the 90s. In a review performed by Ruth Hillary in 1999, it was at the heart of SMEs (Hillary 1999, Hillary 2004). It focused not only on the possible advantages, but also the obstacles for the adoption of EMS management in carrying out a detailed survey of 33 studies conducted in the EU. Hillary study showed that SMEs can find real and valuable benefits for the implementation of the EMS. Channels of communication skills, knowledge and attitudes have improved, though small businesses have switched to the system. They were also published experience and positive results in terms of improved environmental protection, ensuring compliance with laws

Table 2. Enterprise with EMAS in EU.

Enterprise according to size	The number of the enterprise with EMAS
Micro	560
Small	594
Medium-sized	899
Sized	543

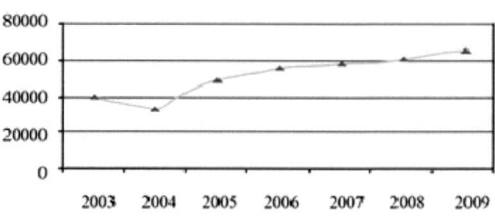

Figure 1. Graphical representation of the increase in the number of enterprises with EMS.

Table 3. Costs of the EMAS easy implementation.

Size	Costs in the first year of EMAS implementation	Annual costs of EMAS	Costs in the first year of EMAS easy implementation	Annual costs of EMAS easy
Micro	21000	10700	11500	2500
Small	33000	21000	20100	3300
Medium	37600	20500		
Sized	40000	33000		

and energy and material efficiency. In addition, the individual businesses strengthen the image, communication, relations. The main disadvantage of introducing EMS was that it required a higher level of cost and time than expected.

In 2000, Ulrich Steger published a report covering the 24 original studies (Steger 2000). In this paper the authors attempt to assess the impact that EMS had on organizations that focus on environmental targets and impacts. Moreover, they search for economic costs and benefits of EMS in the organizations. The survey is based mainly on smaller EMS empirical research. Steger (2000) states that the use of EMS leads to more efficient organization of the flow, a higher degree of legal compliance and comprehensive utilization of resources "win-win" a potential environmental and economic benefits.

Impact on improving specific areas of the environment is often problematic and has certain restrictions. Most companies would achieve its environmental objectives, regardless of EMS.

Two years later was performed Austrian evaluation study (Pecher et al. 2002). This study focused on the environmental effects and financial benefits. No general conclusions can be drawn from this study, because the EMS in some of the evaluations showed a good strong effect, while in others it showed none at all. One of the conclusions of this study was that future evaluation studies should choose their research methodology more cautiously than the previous assessment.

A study submitted Ammenberg & Hjelm (2003) contained a summary of a study on the effects of EMS systems. He found that the results concerning the effects of their EMS systems have been divergent. According to him, it is not possible to determine whether organizations that have implemented EMS have a better environmental impact than organizations without EMS, research was based on a large statistical study on indicators of environmental impact and is based mainly on the perception of environmental managers.

It appears that the generally positive results are presented. These studies also show that the EMS led to a number of environmental management activities, but scientists have been unable to find a causal relationship between these activities and the reduction of environmental impacts.

Two-thirds of controlled trials involving impact on compliance with legislation indicate that the introduction of EMS results in improved compliance. It seems that the EMS has a positive impact on emergency preparedness and the actual number of environmental accidents.

Ammenberg & Hjelm (2003) concluded that the instrument EMS can be used to effectively reduce the environmental impact. However, existing evidence about the positive impact on the environment is not sufficient to conclude that the introduction of EMS generally leads to a reduction in environmental impact. It is also proposed that more attention should be paid to methodological questions when effects are analyzed by EMS. Summing up the findings in the evaluation of the abovementioned studies on the effects on the environment have a wide range of results arising from the application of EMS. It appears that there is an agreement between the reviewers that EMS helps to improve the efficiency of the organizational view of environmental issues.

2 THE RESEARCH PRACTICE

2.1 Phases of research

Using ec.europa.eu and register of the EMAS implementation we found enterprises which have already a system EMAS implemented. Each of these enterprises we contacted through a questionnaire in which we asked about the benefits from the introduction of EMAS. We received back 33% of the questionnaire.

But these were enterprises with EMAS and also EMAS easy implementation. Then we must choose only the enterprise with EMAS implementation. Therefore on the head of the questionnaire was basic information about EMS implementation. From these enterprises we have only 34 enterprises with EMAS easy.

In the second phase of research, we re-tried to address the companies that have not responded for the first time and parables responses from other

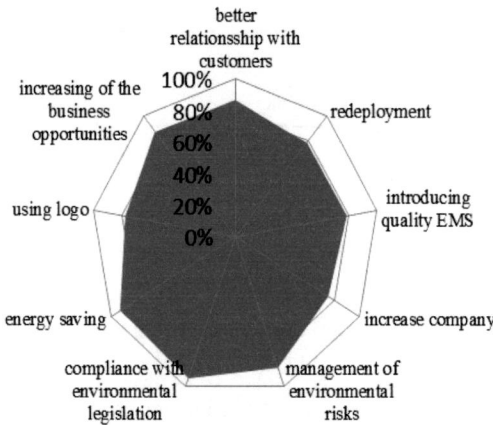

Figure 2. Spider graph of the research results.

companies. Finally we achieved 61% of research sample.

Subsequently, questionnaires were evaluated. There were closed questions in which the company replied to only yes or no. After the evaluation we got clear answers in the benefits of implementing EMAS Easy.

2.2 The survey results

From our research we concluded that more than ½ of enterprises confirmed benefits from implementing EMAS Easy. For displaying results we used the spider chart (Figure 2).

- Management of environmental risks.
- EMS assessment and review compliance with environmental legislation.
- Energy saving.
- Reduce operating costs.
- Better relationships with customers.
- Increase company.
- Redeployment.
- Increasing of the business opportunities.
- Logo and its use for marketing purposes.
- Introducing quality EMS.

Only 25% of companies give recognition to human resources in the enterprise—namely, the improvement in working environment and the increasing of employee commitment. Most of the small and micro enterprises have many problems with financing implementation EMAS easy costs. And many of these enterprises have no people who are interested in this area, so they must teach them. Human resources represented the most important but also most expensive component for implemented EMAS easy.

3 CONCLUSIONS

A common approach has been developed in recent years and it seems that its popularity is growing among SMEs. To overcome barriers to implement EMAS easy in SMEs, there are a number of special requests, from which you can choose.

In this article we investigated the benefits from the introduction of EMAS Easy, which are reported in the literature. Then we compared benefits from literature with real conditions in enterprises. From the results we have concluded that human resources area is often neglected.

ACKNOWLEDGMENT

Article is the outcome of the project: I-15-109-00.

Increasing of the economic and environmental performance with a focusing on integrating management systems.

REFERENCES

Ammenberg, J. & Hjelm, O. 2003. Tracing business and environmental effects of environmental management systems—a study of networking small and medium-sized enterprises using a joint environmental management system. *Business Strategy and the Environment.* 12(3): 163–174.

Ammenberg, J. & Hjelm, O. 2002. The connection between environmental management systems and continual environmental performance improvements. *Corporate Environmental Strategy* 9(2): 183–192.

Bednárová, L., Liberko, I. & Rovňák, M. 2013. Environmental benchmarking in small and medium sized enterprises and there impact on environment. In. *Geo-Conference on ecology, economics, education and legislation SGEM 2013, 16–22 June, 2013*, Albena, Vol. 2., p. 141–146.

Biondi, V., Frey, M. & Iraldo, F. 2000. Environmental management systems and SMEs: motivations, opportunities and barriers related to EMAS and ISO 14001 implementation. *Greener Management International.* 29 (Spring 2000): 55–70.

Biondi, V., Frey, M. & Iraldo, F. 2000. Environmental management systems and SMEs: barriers, opportunities and constraints, *Greener Management International.*

Bosák, M. & Olexová, C. 2013. Recent environmental trends and innovations in the Slovak small and middle enterprises, In. *GeoConference on ecology, economics, education and legislation SGEM 2013, 16-22 June, 2013*, Albena, Vol. 2., 247–254.

Constantini, V. & Crospi, F. 2008. Environmental regulation and the export dynamics of energy technologies. *Ecological economics* 12(1): 447–460.

Dandridge T.C. 1979. Children are not 'little grown ups': Small business needs its own organizational theory. *Journal of Small Business Management* 17(2): 53–57.

European Commission. 2008. Proposal for a Regulation of the European Parliament and of the Council on the voluntary participation by organizations in a Community eco- management and audit scheme (EMAS), Brussels: European Community. (COM 402/2)

Fresner, J. & Engelhardt, G. 2004. Experiences with integrated management systems for two small companies in Austria. *Journal of Cleaner Production* (12): 623–631.

Gerstenfeld, A. & Robert, H. 2000. Size matters: barriers and prospects for environmental management in small and medium-sized enterprises. In: Hillary R. *Small and Medium-Sized Enterprises and the Environment.* Sheffield: Greenleaf Publishing Ltd.

Gleckman H, & Krut R. 1997. Neither international nor standard: The limits of ISO 14001 as an instrument of global environmental management. In Sheldon C. ISO 14001 and Beyond: Environmental Management Systems in the Real World. Sheffield: Greenleaf Publishing Ltd.

Gribble, N. & Dingle, J. 1996. Environmental Management Systems: A Western Australian Perspective. Perth, WA: Murdoch University School of Biological and Environmental Sciences.

Hillary, R. 2004. Environmental management systems and the small enterprise. *Journal of Cleaner Production* 12(6): 561–569.

Hillary, R. 2004. Environmental management systems and the smaller enterprise. *Journal of Cleaner Production* 12(7): 763–777.

Hillary, R. 1999. Evaluation of Study Reports on the Barriers, Opportunities and Drivers for Small and Medium-Sized Enterprises in the Adoption of Environmental Management Systems. London: Department of Trade and Industry, 1999.

Holt, D., Anthony, S. & Viney, H. 2000. Supporting environmental improvements in small and medium-sized enterprises in the U.K. *Greener Management International* 30 (7–8): 29–49.

Hutchinson, A. & Hutchinson, C. 1995. Sustainable regeneration of the U.K.'s small and medium-scale enterprise sector: some implications of SME response to BS7750. *Greener Management International* 9(1): 74–84.

Merritt, Q. J. 1998. EM into SME won't go? Attitudes, awareness and practices in the London Borough of Croydon. *Business Strategy and the Environment* 7(2): 90–100.

Pecher, A., Tschulik, A. & Martinuzzi, A. 2002. EMAS— an instrument for environmental communication (examining EMS-evaluation studies).

Wells, R. P. & Galbraith, D. 1999. Proyecto Guadalajara. *Greener Management International* 28(12): 90–103.

Welsh, J.A. & White, J. F. 1981. A small business is not a little big business. *Harvard Business Review* 59(7–8): 18–32.

Production Management and Engineering Sciences – Majerník, Daneshjo & Bosák (Eds)
© 2016 Taylor & Francis Group, London, ISBN: 978-1-138-02856-2

Flood warning and forecasting system

V. Ferencz
The Ministry of the Environment of the Slovak Republic, Bratislava, Slovak Republic

J. Dugas
Faculty of Business Economy with Seat in Košice, University of Economics in Bratislava, Kosice, Slovak Republic

ABSTRACT: Among its key priorities, the environmental policy of the Government of the Slovak Republic lists the implementation of programs aimed at preventing climate change and mitigating its effects within the framework of the measures adopted under the Climate-Energy Package, to which end, an appropriate organizational, economic, legislative and institutional landscape is being. In flood prevention, the Government of the Slovak Republic has been consistently implementing effective flood prevention measures and programs promoting comprehensive river basin management.

1 FLOOD CONTROL

Flood control plans prepared by the state flood control authorities consist of flood control plans for

a. Flood prevention works, submitted by an administrator of a watercourse of water management importance.
b. Emergency and rescue operations, submitted by municipalities and the Fire and Rescue Service.

The National Council of the Slovak Republic passed Act No. 129 on the integrated Rescue System (IRS) in 2002. This legislation governs the organisation of the IRS, competences and tasks of state authorities and rescue services participating in the IRS, stipulates the rights and obligations of municipal authorities and other legal persons, natural persons licensed to conduct business and other natural persons with respect to the coordination of actions related to the provision of help in the case of an imminent threat to life, health, property or the environment (Act No. 129, 2002).

The purpose of the IRS is to immediately and without delay provide the necessary help to persons whose life, health or property is at threat.

The IRS involves the following:

a. Ministry of the Environment;
b. Ministry of Health;
c. District offices in the seat of region;
d. Rescue services—basic, other, police departments.

Basic rescue services comprise:

a. Fire and Rescue Service;
b. Medical rescue service providers;
c. Civil protection control chemical laboratories;
d. Mountain Rescue Service;
e. Mine Rescue Service.

Other rescue services comprise:

a. Army of the Slovak Republic;
b. Municipal fire services;
c. Company fire brigades;
d. Company fire services;
e. Units performing state supervision or activities under separate regulations;
f. Civil protection units;
g. Municipal police;
h. Railway Police departments;
i. Slovak Red Cross;
j. Other legal and natural persons involved in the protection of life, health and property.

The basic IRS infrastructure comprises coordination centres at district offices in the seat of the region whose primary task is to ensure coordination of activities of IRS stakeholders in the territory falling within the authority of the district office, and to operate emergency lines to receive calls in case of life, health or property hazard.

State crisis management is governed by Act No. 387/2002 Coll. on state management in crisis situations outside wartime and warfare. The Act defines the competences of state authorities with respect to state management in crisis situations outside wartime and warfare, the rights and obligations of legal and natural persons with respect to the preparation for crisis situations outside wartime and warfare and their solutions, and penalties for the breach of obligations laid down by that act.

The Act specifically defines the basic terms such as:

- *crisis situation* outside wartime and warfare (hereinafter only referred to as "crisis situation") is a period of an imminent threat to or breach of the security of the state during which the state authorities are entitled to declare a state of emergency or an extraordinary situation, provided that the conditions laid down in a constitutional act or a separate regulation are met;
- *crisis management* outside wartime and warfare (hereinafter only referred to as "crisis management") is a set of management actions of crisis management authorities designed to analyse and assess security risks and threats, to plan and adopt preventive measures, to organise, implement and control activities carried out during the preparation for crisis situations and for their solution;
- *crisis staff* is an executive body of a crisis management authority whose task is to analyse risks related to the crisis situation related, propose measures to resolve the situation and coordinate operations of bodies and services falling within its authority during the crisis situation;
- *civil emergency planning* means preparation and coordination of measures to ensure the functioning of state authorities, public order and security and civil protection during the crisis situation.

2   BASIC CLASSIFICATION OF FLOODS

Flood means:
- a temporary substantial rise in the water level of a watercourse with an imminent threat of water spilling out of a water channel or with water already spilling out of the channel;
- a situation when the natural runoff of precipitation water to a recipient 1) is temporarily blocked and land is flooded by internal waters;
- a situation when due to an ice drift, formation of ice dams, ice gorges and other obstacles in the watercourse channel there is a threat of the water spilling out of the watercourse channel or the water is already spilling out of the watercourse channel;
- a situation when the land is flooded due to extreme precipitations; or
- a situation when due to a breakdown of or accident on a water structure, there is a threat of the water spilling out of the watercourse channel or the water is already spilling out of the watercourse channel;

Flood hazard occurs when the situation has, in particular, the following characteristics:

- long-lasting heavy atmospheric precipitations accompanied by a rapid runoff to watercourses;
- meteorological warning of extreme precipitations;
- increased runoff from the melting snow and dangerous ice drift;
- fast rise in water level of a watercourse which is likely to reach flood alert levels;
- occurrence of extraordinary circumstances.

A flood situation is a situation when there is a risk of flood hazard or when a flood has already occurred. It is a situation characterised by the attainment of individual flood alert levels on watercourses or water structures. A flood situation also means a situation when the stability or safety of a water structure is at risk or disturbed.

Flood alert levels:

1st flood alert level—a state of alertness identified (but is not declared):

- when the water level or discharge specified in a flood control plan is reached and when the water level of the watercourse follows an upward trend. On regulated, embanked watercourses, it is a situation when water rises out of the watercourse channel and reaches the toe of embankment. On natural unregulated watercourses, it usually is a situation when water level rises and approaches a shore line;
- when an increased runoff from the melting snow is expected based on meteorological forecasts;
- when internal waters occur, if the water level on adjacent watercourses exceeds the level of internal waters.

2nd flood alert level—a state of alertness is declared:

- when the water level or discharge specified in a flood control plan is reached and when the water level of the watercourse follows an upward trend. On non-embanked watercourses, it is a situation when the water level in the channel has reached a shore line and follows an upward trend at the beginning of a snow melt if, according to a flood forecast service, a rapid rise in water levels can be expected, or where an ice drift is expected;
- when internal waters occur, if the maximum level of internal waters as defined in the operational rules of a water structure is preserved by intensive pumping.

3rd flood alert level—a state of emergency is declared:

- when the water level or discharge specified in a flood control plan is reached; in the case of a lower water level, if the state of alertness on an embanked watercourse has been in force for twenty days, or if the water starts to seep through a dam, and/or if other unexpected circumstances occur which may cause damages. On non-embanked watercourses, it is a situation when water has risen out of the watercourse channel and may cause damages.

- during an ice drift, if there is a direct risk of formation of ice gorges or if ice gorges have already started to form;
- when internal waters occur, if, with a pumping station running at its full capacity in constant operation, the water level rises above the maximum level specified in the operational rules of a water structure;
- in the case of flash floods caused by extreme precipitation and expected progress of a flood wave;
- when an area downstream from a water structure has been inundated as a result of a breakdown or accident on water structure installations.

The states of alertness and emergency are declared and recalled, upon proposal by a watercourse administrator or at own initiative, by:

- a municipality mayor, applicable for the territory of the municipality;
- a head of a district environmental office, applicable for the territory of several municipalities and for the territory of the district;
- a head of a district environmental office in the seat of the region, applicable to watercourses that run through two and more territorial districts of the region, unless they have already been declared by heads of district environmental offices;
- a minister of the environment of the Slovak Republic, applicable to cross-border sections of watercourses, unless the minister has delegated this task to another state flood control authority.

The declared states of alertness or emergency may remain in effect even after the water level subsides below the applicable grade line on the water gauge, if the reasons leading to its declaration persist. Such reasons may include, for example, a failure on a water structure (a damaged dam), the need to carry on with flood rescue operations (pumping water out of cellars, mud removal, restoration of essential utility networks, etc.) or flood prevention works.

The description and breakdown of flood alert levels, procedures applicable to the declaration and recall of flood alert levels and a list of major and auxiliary water measuring stations are included in flood control plans for prevention works.

3 CAUSES OF FLOODS

Floods represent a natural phenomenon which cannot be prevented, albeit its implications can be alleviated by certain measures that are typically implemented afterwards. Floods are destructive, being a natural phenomenon that has always existed in the past and will continue to exist in the future.

A flood means a situation when water cannot flow away from terrain or when it suddenly runs off from a reservoir or the flow in riverbed is constricted temporarily.

The floods are broken down based on their causes as follows:

- *snow*—caused by snowmelt or rainfall occurring in the spring,
- *rain*—caused by heavy rainfall in the summer or in the autumn,
- *ice*—caused by clogging of the river channel by ice jam,
- *mixed*—a combination of multiple factors.

Every flood has its specific characteristics, parameters, causes and progress which are unique and unrepeatable. However, there certain other types of floods categorised by their cause:

- *regional floods*—caused by long-lasting precipitation spreading over a wide area or by snow melting within river basins; these floods are sufficiently abundant, affect large territories and last for several hours or days. These are the most typical floods which affect the entire river basins and river systems. They are usually occurring during the first warm days in the spring due to fast snowmelt. When warm weather is accompanied with heavy rainfall, the precipitations in combination with meltingsnow can cause sudden flooding,
- *flash floods*—caused by local short-lasting heavy rainfall, their impact is confined to a relatively small area. They start with a typical storm cloud which can pass above several valleys or travel a longer distance whereby accumulating water vapour caused by evaporation from land surfaces or valleys,
- *outburst floods*—caused by temporary natural or artificial damming of streams.

After a flood alert has been declared, the administrators and owners of watercourses or the administrators and users of water management structures submit to state flood control authorities interim reports on the flooding causes and progress and on measures undertaken as part of flood control and security works.

The flood hazard cannot be eliminated in its entirety; however, it can be minimised by taking preparatory measures for potential flood situation and preventive flood control measures.

In recent years the region of central Europe was hit by several catastrophic floods which were extraordinarily destructive, causing material damage and casualties. The Slovak Republic is situated on the continental watershed divide bordering with the Carpathian range, its population is concentrated in mountain valleys along the four principal rivers (Danube, Váh, Hron and Bodrog) which are characterised, inter alia, by frequent occurrence of floods. The area of some 50 000 km² is populated by more than 5 million people, which implies a high population density (100 people per km²). In fact, the density is even higher due to the fact that only a smaller part of the territory is inhabited. In addition to flood hazard on large river basins, flash floods pose a significant threat as well. Flash floods, which are starting to occur almost regularly every year, have been hitting mostly small catchments in the sandstone flysch mountain range of the northern and eastern Slovakia. Due to their extremely fast response to intensive rain, practically no warnings by the so-far used conventionally methods can be issued.

The underlying priorities of the Government's environmental policy will include implementation of programmes to prevent climate change and mitigate its consequences as part of the measures adopted under the climate energy package by providing necessary organisational, economic, legislative and institutional conditions (Majerník et al. 2013).

Results of preliminary flood risk assessment.

After analysing available data, a total of 559 areas (1,286.445 km) have been identified in the territory of the Slovak Republic, of which:

1. 378 were geographical areas where potentially significant flood hazard exists,
2. 181 were geographical areas where significant flood risk is likely.

Flood hazard in the individual partial river basins: (http://www.minzp.sk, 2015).

• Dunajec a Poprad—31 areas/73.400 km
• Morava—51 areas/125.290 km
• Dunaj—no significant flood hazards have been identified
• Váh—192 areas/460.050 km
• Hron—54 areas/169.650 km
• Ipeľ—9 areas/23.750 km
• Bodrog—129 areas/237.400 km
• Slaná—31 areas/57.705 km
• Hornád—57 areas/122.00 km
• Bodva—5 areas/17.200 km

At present there are 34 automatic weather stations. These stations fully comply with the recommendations of the World Meteorological Organisation. The programme of observation uses the SYNOP messages which are part of the international meteorological data exchange program. Such messages are extremely important for the coverage of meteorological forecasts due to the complex terrestrial meteorological information. This set of stations also provides online information about precipitation intensity and totals.

At present there are also 74 Hydrological Forecasting Stations (HFS) currently monitoring surface waters in accordance with the Flood Protection Act. In the case of flood situation, the warnings, alerts and hydrological situation evaluations are issued for these stations at shorter intervals (12, 6 and 3 hours). During floods, most of the HFSs forecast the time and the level of flood crest.

There are also 59 Hydrological Operative Stations (HOS) in use for the same monitoring purposes as HFS, with the only difference being that they are operated in the so-called "sleeping mode". After the set limits have been exceeded, the stations will issue a warning and switch to the active mode. In case an intense rainfall is expected, the regular query mode for analysing the hydrological situation is activated on these stations. In a standard situation, the stations are queried twice a month to populate the database and check their functioning.

4 POVAPSYS PROJECT

One of the basic components for improving the quality of protection against floods is the "Flood Warning and Forecasting System of the Slovak Republic (POVAPSYS)", representing an integrated flood warning and forecasting system.

The implementation of the POVAPSYS project can facilitate a more significant alleviation of damages caused by floods, in particular casualties, injuries and property damage sustained by the population. POVAPSYS will allow issuing timely and high-quality meteorological and hydrological forecasts, including warnings about extreme flood situations and operative provision of information to state authorities responsible for flood control. Timely identification of the critical flood level allows for preventing or alleviating flood damage.

Objectives of the POVAPSYS project:

• timely and high-quality meteorological and hydrological forecasts;
• warnings about extreme flood situations and operative provision of such information to flood control authorities;
• developing instruments using hydrological forecasts, warnings and alerts for a more significant alleviation of damages caused by floods, in particular casualties, injuries and property damage sustained by the population;
• integrated automated forecasting and warning system allowing to issue hydrological forecasts in approximately 100 forecast profiles, as well as warnings and alerts about flood hazards in at-risk areas.

POVAPSYS—planned activities:

- building a network of terrestrial stations
- building the systems for distance monitoring methods
- building information systems and technology, including the telecommunications system
- building the systems for forecasting models, methods and methodologies.

The presently used 240 hydrological stations of the POVAPSYS system are placed along 11 river basins, as shown in Table 1.

Table 1. Hydrological, meteorological and precipitation monitoring stations broken down by river basin.

Number of stations by type

River basin	HOS	HFS	MET	PREC
Morava	7	2	1	4
Váh	14	19	9	20
Dunaj	6	7	3	0
Nitra	8	5	3	7
Hron	7	8	5	13
Ipeľ	5	2	2	3
Slaná	2	6	0	0
Bodva	1	2	0	1
Hornád	4	9	3	8
Bodrog	4	13	4	16
Poprad	1	1	4	1
Slovakia	59	74	34	73

Legend:
HOS hydrological operative station.
HFS hydrological forecasting station.
MET standard weather stations complying with WMO regulations.
PREC any station monitoring the precipitation intensity and totals.

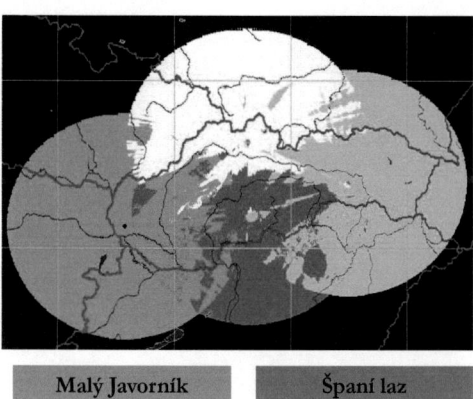

Malý Javorník	Španí laz
Kubínska hoľa	Kojšovská hoľa

Figure 1. The required placement of weather radars. Source: Original design by SHMI.

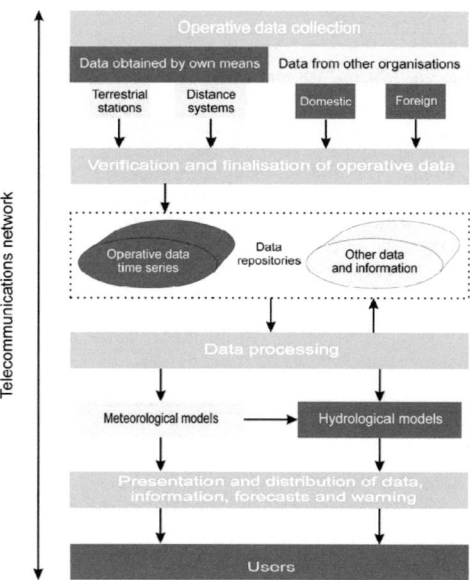

Figure 2. POVAPSYS—basic scheme. Source: Original design.

5 CONCLUSIONS

The activities aimed at launching the POVAPSYS system should be regarded as the first step towards improving the flood prevention in the territory of the Slovak Republic. However, we should bear in mind that this system is not designed to directly improve flood protection. In essence, it should improve the forecasts and provide a sufficient amount of data required for evaluating extreme situations (Hellman & Liu 2013). Even though it includes a warning system, this does not mean that the direct threat to life and property will be eliminated (Kekäle & Helo, 2014).

Over the period of the last ten years, technology gaps in flood prevention have narrowed very slowly. Hand in hand with enhancing the flood warning and forecasting system of the Slovak Republic, the POVAPSYS project will significantly contribute to improving the operative meteorology and hydrology in the Slovak Republic as a whole.

POVAPSYS—expectations:

- a greater amount of higher-quality and up-to-date hydrological and meteorological data in near real time
- faster hydrological and meteorological forecasts and warnings—allowing more time for the preparation of flood prevention measures
- more accurate and reliable hydrometeorological forecasts and warnings
- raising the awareness of the population about flood hazards.

REFERENCES

Act No. 129 on the integrated rescue system (IRS) in 2002.

Act No. 387/2002 Coll. on state management in crisis situations outside wartime and warfare.

Hellman, P. & Liu, Y. 2013. Development of Quality Management Systems: How Have Disruptive Technological Innovations in Quality Management Affected Organizations? QIP Journal: Quality Innovation Prosperity, 17 (1): 104–119.

Kekäle, T. & Helo, P. 2014. The Tipping Points of Technology Development, QIP Journal: Quality Innovation Prosperity, 18 (1): 1–14.

Majerník, M., Mihok, J., Tkáč, M., Bosák, M., Szaryszová, P. & Tarča, A. Environmentálne manažérstvo v integrovanom systéme. 2013. ISBN 978-80-971555-1-3.

Ministry of the Environment of the Slovak Republic. 2015. (http://www.minzp.sk/).

Slovak Hydrometeorological Institute (SHMI). 2015. http://www.shmu.sk/sk/?page=1.

Production Management and Engineering Sciences – Majerník, Daneshjo & Bosák (Eds)
© 2016 Taylor & Francis Group, London, ISBN: 978-1-138-02856-2

Oil tank fire modelling for the purposes of emergency planning

J. Glatz, M. Gorzás & M. Hovanec
Department of Safety and Quality, Technical University of Kosice, Kosice, Slovak Republic

ABSTRACT: Accidents on large capacity oil storage tanks may present several alternatives such as pool fire—heat flux, pool fire—dispersion of toxic combustion products, flash fire and vapor cloud explosion. The presented paper describes a modelling of crude oil storage tank fire and it considers three possible fire scenarios. The worst scenario is the fire of storage and holding tank at the same time, when the largest area burns. At pool fire calculations, heat flux boundaries are considered for surrounding technology and intervening personnel. With respect to radiant heat during fire, it is necessary to determine the boundaries for the intervening team as well as for the location of fixed and mobile monitors. Graphic representation using available maps showing the heat flux range with different intensity provides the intervening team and the commander of the intervention with information necessary for a safe operation.

1 INTRODUCTION

Emergency scenarios must be elaborated in order to successfully handle fires of large capacity tanks. Relevant elaboration requires a modelling of possible consequences, which means heat flux range at large capacity tank fire. By plotting the ranges of individual values of heat fluxes on maps we obtain the data for the needs of efficient and safe placement of intervening forces and means, training of executive and management bodies, for the purposes of verifying the adequacy of planned measures, staff training and so forth.

2 CHARACTERISTIC SCENARIOS OF THE FIRE OF 30 000 M³ OIL TANK

Heat flux density was calculated according to relations specified below, for a tank with a capacity of 30000 m³. The heat flux density for individual scenarios with 30000 m³ tanks may be determined from the surfaces, Figure 1.

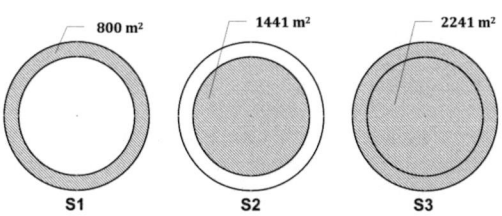

Figure 1. Scenarios of large storage tank fire.

S1—pool fire in emergency tank—surface 800 m²
S2—pool fire in storage tank—surface 1441 m²
S3—pool fire in emergency and storage tank at the same time-surface 2241 m².

The calculations consider the heat flux boundaries for surrounding technology and intervening personnel in the event of fire (boundaries from 44 kW/m² up to 1.8 kW/m²). The size of the storage tank was rounded up in each scenario.

3 POOL FIRE

Pool fire means burning of material vaporizing from the bounded or unbounded liquid layer. The pool fire is conditioned by the existence of burning liquid, initiating source and oxidizing agent. The fuel poured over a large surface has a high heat release rate in a short period of time (Casal 2008).
    The pool fire is evaluated if the following conditions are met:

– the given substance falls into the category of hazardous properties flammable (liquid), very flammable, very flammable liquid or extremely flammable,
– the duration of the fire is at least 15 minutes (Kandráč 2001), (Metodologie 2002).

At the surface velocity of burning oil 0.033 kgm²/s¹ and corresponding to the surface on which the oil may spill in space, the above mentioned conditions are met at the amount released at a continuous leak lasting 10 minutes. At heat flux calculation a model of bounded liquid layer was used, stated in (CPR 14E, 2005).

Table 1. Physicochemical properties of oil.

Boiling point [°C]	>30
Flash point [°C]	<−25
Hazard class for flammable liquids	I
Lower explosion limit/upper explosion limit	[% vol.]
Steam phase 20°C	1.74/8.94
Steam phase 40°C	1.60/8.60
Steam phase 60°C	1.52/8.35
Reid vapor pressure [kPa]	49
Relative vapor density	3
Relative liquid density	0.7–0.9
Density of liquid at 20°C [kg/m³]	700–900
Dissolution in water	slightly soluble

The following relations were used for the calculation of the heat flux range:

For the ratio of flame length L and diameter D it holds that:

$$\frac{L}{D} = 55 \cdot \left(\frac{m''}{\rho_{air} \cdot (g \cdot D)^{1/2}} \right)^{0.67} \cdot u^{*0.21} \qquad (1)$$

where D = pool diameter [m]; m'' = burning or mass loss rate per unit area per unit time [kg/m²/s]; ρ_{air} = density of air [kg/m³]; g = gravitational acceleration [m/s²] and u^* = dimensionless wind velocity [-].

The target received radiation is given by:

$$q'' = E_a \cdot F_{view} \cdot \tau_a \qquad (2)$$

where q'' = heat flux received by receptor [kW/m²]; τ_a = atmospheric transmissivity [-]; E_a = average heat flux [kW/m²] and F_{view} = geometric view factor [-].

Thermal radiation is absorbed and diffused in the atmosphere. This causes reduction of radiation received by the target. The parameter quantifying this fact is the atmospheric transmissivity. Some literary sources ignore transmissivity and state it as equal to 1. At larger heat flux distances (more than 20 m), the absorption may be 20–40%. Transmissivity may be calculated according to the following relation:

$$\tau_a = 2{,}02 \cdot (p_w \cdot x)^{-0{,}09} \qquad (3)$$

where τ_a = atmospheric transmissivity in the range of 0–1 [-]; p_w = partial pressure of water [Pa] and x = distance between the flame surface and the target [m].

The hydrocarbon products burning are accompanied by soot formation. The emitted surface flux E_a for hydrocarbons can be calculated according to (CPR 14E, 2005):

$$E_{act} = E_{max} \cdot (1 - \varsigma) + E_{soot} \cdot \varsigma \qquad (4)$$

where E_{max} = maximum heat flux [kW/m²]; E_{soot} = heat flux of soot [k/m²] (using the value of 20 kW/m² and ζ = the fraction of the surface of the flame covered by soot [-], (for hydrocarbons 0.8.

The maximum emitted thermal flux from the source E_{max} is calculated according to:

$$E_{max} = \frac{F_s \cdot m'' \cdot \Delta H_c}{1 + 4 \cdot L/D} \qquad (5)$$

where F_s = fraction of the combustion energy radiated from the flame surface (in the range of 0,1–0,4) [-] and ΔH_c = the heat of combustion of the flammable material [J/kg].

Geometric optic factor needed to relation (2) was calculated for a cylindrical shape of flame with an elongated base downwind.

Knowing the "q", the surface receptor temperature Ts may be calculated from the equation according to (Guidelines, 2003):

$$q'' = q''_{rr} + q''_{conv} = \sigma \cdot (T_S^4 - T_\infty^4) + h \cdot (T_s - T_a) \qquad (6)$$

where q'' = incident heat flux to the target [kW/m²]; q''_{rr} = heat flux reradiated from the target to the surroundings [kW/m²]; q''_{conv} = heat flux convected from the target to the surroundings [kW/m²]; σ = Stefan-Boltzmann constant ($5{,}67 \cdot 10^{-11}$ kW/m²K⁴); T_s = surface temperature of the target [K]; T_∞ = temperature of the surroundings [K]; T_a = ambient temperature of the air around the target [K] and h = heat transfer coefficient (0.015 kW/m²K can be used as a first estimate).

The following table 2 shows the resulting values of calculation of parameters of a 30000 m³ tank pool fire.

For the probability of death caused by thermal radiation the probit function applies according to the relation (CPR 14E, 2005):

$$Pr = -36{,}38 + 2{,}56 \cdot ln(t \cdot q^{4/3}) \qquad (7)$$

where Pr = probit corresponding to the probability of death [-]; t = exposure time [s] and q = heat radiation [W/m²].

The thermal radiation effect depends on the duration of exposure and the level of thermal radiation. The duration of exposure, in accordance with the procedure stated in (CPR 18E 1999), is 20 seconds. This standardized time was decided on the basis of the generally accepted assumption that the persons leave the exposed area within 20 seconds. At the given duration of exposure, the lethal level for protected and unprotected persons is when thermal radiation reaches 35 kW/m² (probability of death > 98%).

Table 2. Overall results of calculation of parameters of a 30,000m³ tank pool fire.

Equipment						
Variant of pool fire			S3—pool fire in emergency and storage tank at the same time-surface 2241 m²			
Receptor at the terrain level (T) or at the crude oil level (H)			T		H	
Atmospheric stability class F2/D5			F2	D5	F2	D5
Fire characteristics:						
Amount of dangerous good needed for duration of 15 min fire	m_{15} min	[t]	55.7	55.7	55.7	55.7
Duration of the fire (burn the entire amount)	t_{max}	[h]	112	112	112	112
Diameter of equivalent circular pools	D	[m]	48.5	48.5	48.5	48.5
Initial Level (base of fire)	h_i	[m]	15.6	15.6	15.6	15.6
The average length of the flame	H_f	[m]	30.9	25.5	30.9	25.5
The angle of inclination of the flame	Θ	[°]	35.2	49.2	35.2	49.2
Average emitted heat flux from the source	E_a	[kW/m²]	24.0	25.2	24.0	25.2
Impact to equipment:						
8 kW/m² (unprotected equipment)	x_8	[m]	*	*	48.0	50.1
Impact to personnel:						
The distance from the center of the pool to the level of E = 20kW/m², (100% mortality for protected and unprotected people in exp. 60s)	x_{20}	[m]	*	*	25.6	28.7
The distance from the center of the pool to the level of E = 5 kW/m², (Lethality and start a domino effect in exp. 60s)	x_5	[m]	38.3	47.0	58.7	58.0
The distance from the center of the pool to the level of E = 3kW/m² (Irreversible in exp. 60s)	x_3	[m]	62.9	63.2	71.1	67.5
The distance from the center of the pool to the level of E = 1.8 kW/m², (Reversible in exp. 60s)	$x_{1.8}$	[m]	81.4	78.1	85.6	79.0

Table 3. Levels of thermal radiation effects.

Level of effects	q with exposure time of 60 s [kW/m^2]	q with exposure time of 30 s [kW/m^2]
1—negligible or no effect	<1,8	<3
2—reversible effects	1,8–3	3–5
3—start of irreversible effects	3–5	5–9
4—threshold of lethality or domino effect	>5	>9

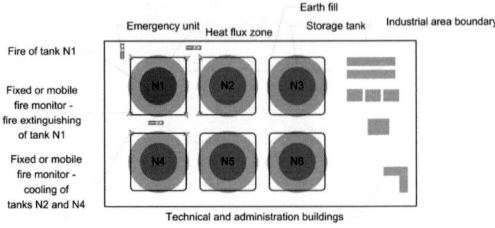

Figure 2. Schematic representation of thermal flux ranges and placement of emergency forces and fire equipment.

Figure 3. 3D representation of a thermal flux range and its transmission to the head-up display of the commander of the intervention.

In contrast, the methodology by the (ARAMIS 2004) project considers the exposure time 30 or 60 seconds and divides the effects into four levels (Table 3).

Applying the probit function and exposure time 60 seconds, the lethal level for protected and unprotected persons is when thermal radiation reaches 15 kW/m^2 (probability of death 98%). As a lower limit of thermal radiation level at a given exposure of 60 seconds, the set value was 5 kW/m^2. The average probability of death for unprotected persons in the zone over 5 kW/m^2 is 50%. The correction factor 0.14 for clothing applies to unprotected persons in the zone 5–15 kW/m^2. The fatal consequences for life are not anticipated for persons in this zone protected by buildings.

4 GRAPHIC REPRESENTATION OF HEAT FLUX RANGES FOR EMERGENCY PLANNING

The map data shall provide necessary information at emergency planning as well as intervention actions during an accident, primarily about the infrastructure of the plant and wider surroundings. The graphic representation of thermal flux values is only one of the parts constituted by map data for emergency planning. Figure 2 shows an example of thermal flux range for the purposes of safe and efficient placement of forces and means during the assumed N1 tank fire.

Today innovative technologies enable the representation of thermal flux ranges on mobile devices as well, which can directly provide relevant information to the intervention forces about whether they are in safe zone or in a zone with a certain heat flux.

Figure 3 shows a 3D representation of a thermal flux range for a better visualization for the intervention forces and overall coordination of the action. The GPS technology allows the commander of the intervention to have control over the whereabouts of each member, and using the head up technology (Fig. 3) he can immediately send information on the current situation or on the instructions for the intervention.

5 CONCLUSIONS

The modelling of the impacts of potential adverse scenarios and using these results for the purposes of emergency planning should become a part of each business which may pose a risk to the surrounding. With the use of modern technologies we are increasingly prepared to overcome these adverse scenarios.

The modelling of the impacts may not only serve for use at emergency planning but also at designing new technologies, or possibly at moving the existing technology to a safe distance on the basis of new knowledge.

ACKNOWLEDGEMENT

Paper is the result of the Project implementation: University Science Park TECHNICOM for Innovation Applications Supported by Knowledge Technology, ITMS: 26220220182, supported by the Research & Development Operational Programme funded by the ERDF.

This contribution is the result of the project implementation APVV-0337-11 "Research into new and newly emerging risks of industrial technologies within integrated safety as a precondition for management of sustainable development".

REFERENCES

ARAMIS. 2004. Deliverable D.2.C "The Risk Severity Index". (Accidental Risk Assessment Methodology for Industries in the Context of SEVESO II Directive)

Casal, J. 2008. Evaluation of the Effects and Consequences of Major Accidents in Industrial Plants. Industrial Safety Series Volume 8. Barcelona: Elsevier Science Limited.

CPR 14E. 2005. Methods for the Calculation of Physical Effects ("Yellow Book"). 1997. Third edition. The Hague: Second revision print.

CPR 18E. 1999. Guidelines for Quantitative Risk Assessment—("Purple Book"). First edition. 1999. Den Haag: Sdu Uitgevers.

Guidelines for Fire Protection in Chemical, Petrochemical, and Hydrocarbon Processing Facilites. 2013. New York: CCPS.

Kandráč, J., Skarba, D. & Úradníček, Š. 2001. Metodická príručka pre zaraďovanie podnikov s podprahovými množstvami vybraných nebezpečných látok a pre predbežný odhad rizika v podnikoch podliehajúcich režimu zákona o závažných haváriách. Bratislava: RISK CONSULT.

Metodologie pro identifikaci a vyhodnocení synergických a kumulativních jevů. 2002. MŽP ČR

Production Management and Engineering Sciences – Majerník, Daneshjo & Bosák (Eds)
© 2016 Taylor & Francis Group, London, ISBN: 978-1-138-02856-2

The application of magnetic materials for a neodymium-based thermal fuse in sprinklers

M. Hovanec, M. Gorzás & J. Glatz
Department of Safety and Quality Production, Technical University of Kosice, Kosice, Slovak Republic

ABSTRACT: The subject matter of this paper is a neodymium-based thermal fuse for sprinklers based on specific changes in the area of magnetism. A neodymium-based thermal fuse may be used for sprinklers with a glass bulb. In the field of technical safety, materials with reversible or irreversible processes may be used to identify the required technical parameters of particular states. Neodymium-based thermal fuse for sprinklers makes use of the change of magnetism of neodymium materials. The magnetic properties, may be defined by a material composition of a neodymium. By means of a material composition of a neodymium we can define the so-called opening temperature in advance which the neodymium reaches through gradual heating with an increasing ambient temperature, and at which force F_1 is bigger than force F_{koerc}, causing the release of neodymium and safety peg, resulting in the flow of the fire extinguishing medium through the sprinkler.

1 INTRODUCTION

Sprinkler is a part of an automatic fire extinguishing system belonging to the group of water fire extinguishing systems which reduce the risk of fire spreading. Sprinkler systems are activated automatically when a certain ambient temperature is reached. Sprinkler systems are used primarily in residential premises, shopping centres, office premises, as well as in various industries.

Currently the following types of sprinklers are used with respect to installation (Reliable 2015):

- *Standing*—mounted upright on the pipe.
- *Hanging*—mounted on the ceiling, facing downward.
- *Horizontal*—mounted on side walls.
- *Dry (hanging)*—installed in places with danger of freezing.
- *ESFR*—special heads for rack storage.

2 THERMAL FUSES IN SPRINKLER HEADS

In most constructions, the thermal fuses in sprinklers currently used are based on the principle of liquid expansion in a glass bulb located between the seal cap of the valve and the head frame, see Figure 1 and Figure 2. The glass bulb is supposed to keep the sprinkler head in closed position.

When the ambient temperature increases, the liquid in glass bulb expands, which results in the

Figure 1. Sprinklers—type F1FR, left standing and right hanging variant (Reliable 2015).

Figure 2. Sprinklers—type F1FR, left vertical wall and right conventional variant (Reliable 2015).

destruction of the glass bulb at a set temperature. The destruction opens the sprinkler head which results in the flow of the fire extinguishing agent.

When the ambient temperature increases, the liquid in glass bulb expands, which results in the destruction of the glass bulb at a set temperature. The destruction opens the sprinkler head which results in the flow of fire extinguishing agent.

Thermal fuses for ESFR sprinklers are of a different type since they use fusible metal plugs. The thermal fuse in this case consists of a two-part metal element that is fused by a heat-sensitive alloy. Once the ambient temperature around the sprinkler head reaches a specified temperature, the alloy releases and the metal elements separate, which causes the seal cap to fall away. Water is then released.

3 MAGNETIC AND THERMAL QUALITIES OF NEODYMIUM MAGNET

3.1 Neodymium magnets

Neodymium magnets currently present the strongest type of magnets with excellent qualities such as remanence and energy density (Neodýmové magnety 2015). They belong to the group of rare-earth magnets (lanthanides). Their main element is iron with the admixture of neodymium (Nd) and boron (B). Other elements added to the final alloy are mainly cobalt (Co), dysprosium (Dy)—these elements improve the magnetic properties (remanence, coercive force) and thermal resistance (maximum working temperature) of magnets.

Operating temperature is max. +60 up to +240°C depending on the material class. The neodymium magnets have excellent resistance to external demagnetizing conditions and in normal conditions they maintain permanent magnetism.

3.2 Thermal influence on magnetic properties of neodymium magnets

At first, it is necessary to explicate the Curie-Weiss law, which says that the magnetic susceptibility χ of a paramagnetic substance depends on its temperature by the relation:

$$\chi = \frac{C}{(T - T_c)} \tag{1}$$

By the value of magnetic susceptibility the magnets can be divided to:

- Diamagnetic ($\chi < 0$)
- Paramagnetic ($0 < \chi < 1$)
- Ferromagnetic ($\chi > 1$).

3.3 Selected magnetic materials and their properties

Neodymium (NdFeB) is a specific magnetic material. Neodymium magnets to be used in

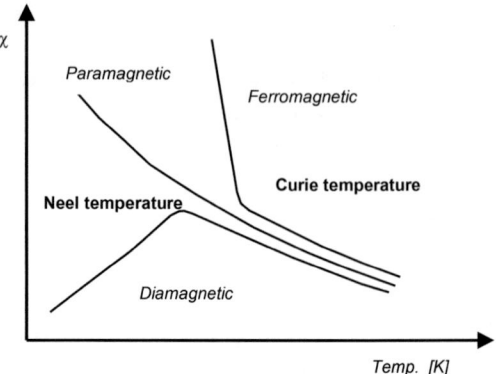

Figure 3. Reduction of magnetization of ferromagnetic materials as a function of temperature.

Table 1. Technical characteristics of selected neodymium magnets (Neodýmové magnety 2015).

Class	Remanence B_r[T]	Coercive force H_c [kA.m^{-1}] Normal [H_{cb}]	Intrinsic [H_{cl}]	Max. energy product [kJ.m^{-3}]	Max. working temperature T_w [°C]
N33	1.13 ÷ 1.17	≥812	≥1353	247 ÷ 271	80 ÷ 240
N35	1.17 ÷ 1.22	≥868	≥955	263 ÷ 287	80 ÷ 200
N40	1.25 ÷ 1.28	≥907	≥955	302 ÷ 326	80 ÷ 180
N45	1.42 ÷ 1.38	≥923	≥955	342 ÷ 366	80 ÷ 150
N48	1.38 ÷ 1.42	≥923	≥955	366 ÷ 390	80 ÷ 120
N50	1.40 ÷ 1.45	≥796	≥796	382 ÷ 406	60 ÷ 100
N52	1.43 ÷ 1.48	≥796	≥876	398 ÷ 422	60
N54	1.45 ÷ 1.51	≥939	≥875	405 ÷ 437	60

Table 2. The most commonly used ferroelectrics (Pietriková & Bánsky 2009).

Name	Formula	T_c [°C]
Barium Titanate	$BaTiO_3$	135
Lead Titanate	$PbTiO_3$	490
Potassium Niobate	$KNbO_3$	435
Lithium Niobate	$LiNbO_3$	1210
Lithium Tantalate	$LiTaO_3$	665
Potassium Dihydrogen Phosphate	KH_2PO_4	−150

unfavourable environment are sintered HAST NdFeB magnets. Compared to classic neodymium magnets, these are corrosion-resistant (there is no need for protection coating). Endurance testing is performed at a temperature above 100°C, relative humidity (>85%, atmosphere is saline solution) and high pressure (up to 4 bar) for the period of

20 days. These conditions correspond to around 10 years lifetime of a magnet, therefore these magnets are designed for heavy use.

Basic parameters of selected neodymium magnets are specified in Table 1.

The principle is that the higher the material class (N35 to N52), the lower the recommended working temperature Tw but higher magnetic product (Prasol 2014).

For utilization in safety area (dependence on temperature) it is possible to further use the Curie temperature Tc and a magnetism change, Table 2.

4 THE CORE OF THE TECHNICAL SOLUTION

The core consists in utilization of the change of magnetism of neodymium magnets, depending on temperature, to which they are exposed. By means of material composition of a neodymium magnet it is possible to define the change of coercive force depending on the working temperature. Increasing the temperature in the vicinity of neodymium magnet leads to its gradual heating, which changes its magnetic properties that hold the safety peg in place.

$$F_{coerc} \geq F_1 \qquad (2)$$

F_{koerc}—coercive force of neodymium *[N]*
F_1—force applied on safety peg in the contact place with the seal cap through the spacer washer *[N]*.

The required value of coercive force of neodymium, which can be influenced by material composition, is primarily given by the size of contact surface of the neodymium and an element of ferromagnetic material. Heating the neodymium leads to gradual reduction of coercive force of neodymium.

Once the so-called opening temperature is reached, which can be influenced by material composition, the safety peg is released and fire extinguishing agent starts to flow as a result of the difference between the coercive force of the neodymium and the force applied on the safety peg.

4.1 The application of the technical solution

Neodymium (9) with a defined required thermal dependency between the working temperature and the coercive force is attracted by coercive force to the element of a ferromagnetic material (8). The system is composed of a neodymium (9), an element of ferromagnetic material (8) and a safety peg (6), serving as a thermal fuse in the sprinkler head which secures a closed position of the sprinkler.

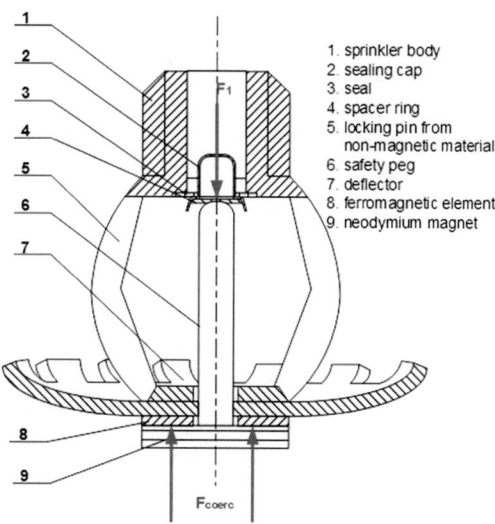

1. sprinkler body
2. sealing cap
3. seal
4. spacer ring
5. locking pin from non-magnetic material
6. safety peg
7. deflector
8. ferromagnetic element
9. neodymium magnet

Figure 4. Schematic view of sprinkler head with neodymium based thermal fuse.

The increasing ambient temperature results in neodymium heating (9) and thereby the change of its magnetic properties. Once the so-called opening temperature is reached, the neodymium (9) and safety peg (6) is released and fire extinguishing agent starts to flow as a result of the difference between the coercive force of the neodymium (9) and the force applied on the safety peg (6).

5 DISCUSSION

The advantages of the application of the technical solution based on neodymium are:

- Defining the temperature range of the safety peg release by means of neodymium composition.
- Magnetism, compared to other phenomena, binds the classic force and and field quantities depending on the temperature. In the given technical equipment it is possible to utilize the change of coercive force depending on temperature.
- High reaction rate at temperature change.
- Simple thermal fuse structure.

The disadvantages of the technical solution:

- The irreversible effect of neodymium magnetism change.

6 CONCLUSION

The aim of this paper was to highlight the utilization of the properties of specific magnets (neodymium magnets) described above in order to identify thermal states in technical safety. The neodymium-based thermal fuse for sprinklers is based on specific changes in the area of magnetism.

In the field of technical safety, materials with reversible or irreversible processes may be used to identify the required technical parameters of particular states.

The model of the technical solution of a thermal fuse was described by means of a schematic description of a thermal fuse as a substitution for the use of a conventional glass bulb. It is characteristic of this neodymium-based fuse for sprinklers that once the so-called opening temperature is reached, the safety peg is released and fire extinguishing agent starts to flow as a result of the difference between the coercive force of the neodymium and the force applied on the safety peg.

ACKNOWLEDGEMENT

Paper is the result of the Project implementation: University Science Park TECHNICOM for Innovation Applications Supported by Knowledge Technology, ITMS: 26220220182, supported by the Research & Development Operational Programme funded by the ERDF.

This contribution is the result of the project implementation APVV-0337-11 "Research into new and newly emerging risks of industrial technologies within integrated safety as a precondition for management of sustainable development".

REFERENCES

Neodýmové magnety. 2015. Information on: http://www.magnety.sk/magneticke-materialy/neodymy/.
Pietriková, A. & Bánsky, J. 2009. Základy inžinierstva materiálov. Košice: Elfa.
Prasol, T. 2014. Magnetické materiály a ich využitie v technickej bezpeènosti. Košice: TUKE.
Reliable—Sprinklery Model F1FR. 2015. Information on: http://www.reliablesprinkler.com/sites/default/files/content/files/pdfs/products/Czech/136CzechRevO.pdf.

Production Management and Engineering Sciences – Majerník, Daneshjo & Bosák (Eds)
© 2016 Taylor & Francis Group, London, ISBN: 978-1-138-02856-2

Certification of overall quality of alternative toner cartridges

P. Hrdlicka & S. Burian

Faculty of Economics and Management, Czech University of Life Sciences Prague, Praha, Czech Republic

ABSTRACT: The paper deals with the design of comprehensive certification of toner cartridges in determining the quality of the final print, utilization rate of toner cartridge, exclusion of copyright infringement and determining the criteria of health risks of using toner cartridges, through one general certificate for the manufacturer itself with the option of online validity verification of the certificate and supplied assortment using personal identification number or other unique identifier, and further authentication of products themselves via QR code or barcode.

Certification is targeted for use particularly by state administration in public procurement of tenders. According to Article 67 of Directive 2014/24/EU (which must be implemented by April 2016 into the national legislation of each Member State), the economic advantage of the offer is set as the sole criterion for awarding the public contract.

1 TARGET & METHODOLOGY

1.1 Target

The aim of this paper is to propose a methodology for granting certification and follow-up system for verifying the authenticity of granted certification in the sphere of production and distribution of toner cartridges for laser printers and copiers. The actual goal of the certification itself is to create an instrument to eliminate the sale of toner cartridges infringing patent right, representing a risk to health or excessively polluting the environment. Certification is focused mainly on purchase system of toner cartridges by the public sector, which is in Central Europe one of the key customers of toner cartridges, which does not meet the legislative requirements.

1.2 Methodology

At first there will be defined the basic types of toner cartridges by their manufacturers and method of production, as well as a contemporary form of certification of manufacturing of toner cartridges will be analyzed. Basic areas of limits of the production of toner cartridges and methods of their violations will be identified. The impact on the environment by each type of toner cartridges will be identified. In the next section, the analysis of method of individual purchases of toner cartridges by the public sector and by the private sector will be carried out and the impacts of the current status will be determined.

The main part of the paper will propose a solution using a certified production.

2 BASIC CLASSIFICATION OF TONER CARTRIDGES

The toner cartridge is a consumable used in laser printers. By the manufacturer, it is possible to distinguish three basic types of toner cartridges.

2.1 OEM toner cartridges

OEM toner cartridges are offered under the brand name identical to the brand name of the printer itself, and they are distributed within the trading network consistent with the trading network, within which the printing device itself is distributed. These toner cartridges are simply marked as original toner cartridges, so if the printer is labelled Epson, for example, the toner cartridge will be labelled with the name Epson.

2.2 Compatible (alternative) toner cartridges

Compatible or alternative toner cartridges are produced by manufacturers that are not connected in any way with the manufacturers of printers, for which these toner cartridges are designed. For example, as for the laser printer Epson, the manufacturer of toner cartridge is "Zhuhai Someway Electronic Science and Technology Co., Ltd.", and the cartridge is then offered either as a compatible toner cartridge without given manufacturer or is offered under the brand name of a particular distributor operating in the local market.

2.3 *Properly renovated toner cartridges*

Renovated toner cartridges are OEM (original) toner cartridges, which are completely renovated. This means that toners are not newly produced, but they are being completely disassembled, cleaned and completed OEM toner cartridges. It is possible to replace some worn parts of toner cartridge. Renovation is carried out both by the OEM manufacturers themselves, and by the manufacturers that are not connected in any way with the OEM manufacturers.

3 BASIC DIFFERENCES BETWEEN TONER CARTRIDGES

3.1 *Methods of patent protection*

When designing toner cartridges, the OEM manufacturers register a number of patents protecting these toner cartridges. Most of the toner cartridges are protected by numerous patents, which prevent their secondary manufacturing. Without the disposition of patents, it is possible to only renovate these toners without violating the patent law, because during renovation there are used original parts protected by patents, and so there is no secondary manufacturing of parts protected by patents (Adams 2006, Kingston & Scally 2006).

3.2 *Health risks*

Toner powder that is contained inside the toner cartridge is transferred during printing on paper, in which it is "baked" using a very high temperature (about 200 degrees Celsius). When handling the print cartridge (particularly when replacing the print cartridge in the printer) and even during the actual printing, there occurs the release of toner powder particles into the air and its inhalation.

Toner powder thus poses a health risk for users of printers and its composition must be in compliance with applicable legislative measures for the content of individual substances that can cause health complications. This compliance is documented by a document named "Material Safety Data Sheet" (MSDS), which must be issued for each type of toner cartridges in the official language of the country in which the toner cartridge is being sold.

"Material Safety Data Sheets" (MSDS) usually accompany the OEM toner cartridges and renovated toner cartridges; the consumers have the option to request these safety data sheets and

identify the determination of health risks associated with the use of toner cartridges. In contrast, the safety data sheets usually do not accompany the compatible toner cartridges, and thus it is not possible to determine their composition and health risks. Because of the maximum pressure to minimize the production costs and the absence of the analysis of composition and risks, it is moreover possible to assume the failure to comply with the required limits (Dikshith 2013).

3.3 *Waste management*

In terms of waste generation, the renovated toner cartridges are the most environmentally friendly. Thanks to the renovation, one OEM toner cartridge is thus reused and, thanks to the renovation of toner cartridge there will be savings of about 300 grams of especially plastic waste. It is possible to renovate OEM toner cartridge approximately 5 times, then the toner cartridge is not any more suitable for further renovation. Companies involved in the renovation of toner cartridges re-buy these toner cartridges, and thus motivate users of toner cartridges for environmentally friendly behaviour.

Manufacturers of the OEM toner cartridges are usually simultaneously the renovators of toner cartridges, and they offer the renovated toner cartridges as another series of their products on selected markets.

It is not possible to renovate compatible toner cartridges. The reasons are technical—OEM toner cartridges are designed so that it is possible to renovate them; compatible toner cartridges do not allow disassembly and reassembly. Another reason is the expectation of patent infringement in the production of compatible toner cartridges from the side of renovators of toner cartridges. A counterfeit would again be created thanks to the renovation of a counterfeit.

Compatible toner cartridges thus represent the highest burden on the environment.

4 CURRENT MARKET SITUATION IN THE SECTOR OF CENTRAL/EASTERN EUROPE

Within the sector of Central and Eastern Europe (especially Poland, Czech Republic, Slovak Republic, and Hungary), the OEM, compatible and refurbished toner cartridges are normally distributed. The compatible toner cartridges are usually distributed regardless of their protection through patents applicable to the given region. Thus, a substantial part of the market supply of toner cartridges is made up of toner cartridges that infringe patents of the OEM manufacturers. In comparison with other groups of toner cartridges, these toner cartridges are characterized by a significantly lower market price. It is possible to say that the price distribution is roughly as follows: the toner cartridge price is the highest price (100%), the properly refurbished toner cartridge by the other side is about ½ of price of original cartridge (50% of the price of OEM toner cartridge), the compatible toner cartridge represents about ½ of price of refurbished toner cartridge (25% of the price of OEM toner cartridge).

4.1 Way to buy toner cartridges by public sector

Public sector typically realizes the purchases of toner cartridges using the electronic auctions (e-auction of small-scale for small-scale public contract). The public contract means buying goods by the governmental entity (public authority), which is the state, a government unit, municipality, as well as the organizations founded by them, but also other entities that manage money or values derived from taxes, fees or other public sources. The basic evaluation criterion of the competition is then typically the lowest bid price with automatic evaluation method. The winner of the competition thus becomes a supplier who offers the lowest price for implementation. Assuming that it is possible to correctly determine the subject of the contract by entering the name of the subject of the contract, this criterion gives the correct result of the competition. In practice, suppliers of alternative toner cartridges of the lowest quality become the winners of the competition due to production costs. The safety data sheets do not accompany such toner cartridges; their health risks are therefore not detected, there occur the infringement of patents, increased burden on the environment, and the violation of the warranty terms of printing devices may occur.

The public sector is thus becoming one of the most important customers of toner cartridges that infringe patents, are not accompanied by the safety data sheets and cause an excessive burden on the environment. During a transparent competition for the purchase, it is impossible to exclude such toner cartridges from the purchase, because the contracting authority of the competition is not equipped with the technical means for verifying the patent purity of toner cartridges, authenticity and correctness of safety data sheets. The contracting authority of the competition is unable to distinguish whether the toner cartridge is renovated, or whether it is a completely newly manufactured cartridge.

According to Article 67 of Directive 2014/24/EU (which must be implemented by April 2016 into the national legislation of each Member State), the economic advantage of the offer is set as the sole criterion for awarding the public contract. The setting of methodology of economic advantageousness is based on the areas of quality, environmental and social area (EUR-LEX. 2013).

4.2 Way to buy toner cartridges by private sector

It results from a questionnaire survey involving manufacturers of toner cartridges in the Czech Republic that major customers of toner cartridges in the private sector (mainly in the automotive industry) are aware of the risks associated with the acquisition of compatible toner cartridges, and eliminate the risks involved directly through inspections of manufacturing processes of the supplier. Medium and small customers or end consumers have no information about the risks, or are aware of the risks, but they lack a simple tool to eliminate risks. They deal with risks either through the purchase of OEM toner cartridges or through the acceptance of compatible cartridges and acceptance of risk.

5 CERTIFICATION OF TONER CARTRIDGES

5.1 Existing certification of toner cartridges

It is possible to certify all types of toner cartridges using the ISO 9001 certification, which guarantees to the buyers in time still the same quality of toner cartridges and related supply services. However, this quality is determined by the manufacturer itself, and it is indifferent in the course of obtaining this certification, whether or not there is a violation of patent law (Myhrberg 2009).

Furthermore, the standard used is ISO 14001 (ISO 14001:2004), which determines the continued fulfilment of the objectives set by the manufacturer in the field of emissions. The standards ISO/IEC19752, ISO/IEC19798 determine in details a very precise methodology for determining the utilization rate of toner cartridges. (Whitelaw 2004)

The industry standard DIN 33870 specifies the exact procedure of renovation of toner cartridge to be comparable in quality with original toner cartridge (Kiehl 2001).

5.2 Proposed measures

The measure is a proposal for a new quality standard intended for toner cartridges from the secondary manufacturing. The standard is designed with maximum emphasis on its clarity and comprehensibility of benefit to the end user of toner cartridge.

1. Environmental area:
   - Exterior part of the toner cartridge must come from previously used OEM toner cartridge, where the term "used" is understood as a toner cartridge that comes from the scrap collection or repurchase of toner cartridges.
   - The manufacturer must have concluded a contract for ecological disposal of waste according to the specific legislation of the given state.
2. Area of the patent harmlessness:
   - No part of the toner cartridge may not infringe patent protection
   - All used parts must be used from the OEM toner cartridges without interfering with the technical implementation of these components, except for the removal of the original identification signs of the OEM manufacturers or their trademarks, and cleaning of these components; or all of the components used must be purchased from the secondary manufacturing from manufacturers certified as "patent pure manufacturer of parts for printing consumables", while the organization granting certification by this certificate states on its internet presentation an updated list of such suppliers with updates always the first calendar day of the month
   - All identifying marks and trademarks of the OEM manufacturer must be removed from the toner cartridge
3. Area of the utilization rate of the toner cartridge:
   - Before filling with a new toner, the cartridge part of the toner must be completely cleaned with compressed air
   - If the toner cartridge contains an optical cylinder, this cylinder must be replaced with a new one
   - If the toner cartridge contains a squeegee of the remaining toner, this must be replaced with a new one
   - If the toner cartridge contains a waste container, this must be completely cleaned with compressed air
   - The toner cartridge must be refilled with the amount of toner powder similar or greater than the amount of toner powder in the OEM toner cartridge
4. Area of social relations:
   - "Material Safety Data Sheet" (MSDS) must be issued for each toner cartridge, demonstrating its health safety
   - Renovation of the toner cartridge must be performed in the country for which the certificate is granted.

The precise guidelines for manufacturers, precisely defining the term "renovation", will be created to this general part.

5.3 *Organization that manages the certification*

Professional union or professional association is an organization granting certification. Only organizations that meet the requirements for certification may be members of a professional union or professional association. Certification will be issued free of charge to members of a professional union or professional association, and it is always for a particular product.

The organization administering the certification shall be entitled to inspect the accounting records of the organization to which the certificate will be granted, namely for the purpose of control of the purchase of components for the production of toner cartridges and a comparison of the amount of purchases with the amount of sales. Allowing access to production within the organization, to which the certificate was granted, is also a part of the grant of certification, without prior notice. The organization administering the certification will be authorized to conduct anonymous test purchases of toner cartridges in order to verify the compliance of certification. The organization's activities are funded by membership fees.

5.4 *Form of certificate*

The certificate itself will consist of two parts—the public part and non-public part:

1. Graphical and textual expression with a registered trademark.
It will be possible to publish this registered trademark on the packaging of certified product. The holder of certification will be allowed to publish the acquisition of the trademark in printed or electronic materials, and show the trademark during tendering and electronic competitions. It will be possible to online verify the authenticity of granting of certification in the internet presentation of organization granting the certification. The URL for verification of trademark will be part of the trademark; it will be expressed both in the form of a QR code as well as a text URL address. Validation granting of the certification will be traceable by the business identification number and by trade name. A part of the statement will be the exact specification of products, for which the certification was granted.

2. Verification of the authenticity of the product itself using QR code and its alternatives.
Each product, to which certification will be granted, will contain a label with a unique identifier of toner cartridge located on the body itself of toner cartridge. The identifier will include a QR code with a URL address that displays information about the toner cartridge. Basic information will identify the manufacturer and product identification. Customer identification is an optional

parameter. At the first reading of the QR code, the toner cartridge will be marked as "authentic", at the second reading of the same QR code, the toner cartridge will still be marked as "authentic", but information about previous reading of information on this toner cartridge will be displayed, including the exact date, time and IP address from which the request was entered. At the third reading of the same QR code, the cartridge will be marked as "counterfeit", and a call to contact the issuer of certification will be displayed. The issuer will also be notified. At the reading of the QR code, which is not registered in the system of issuer, the toner cartridge will be marked as "counterfeit", and the issuer of certification will be notified at the same time. Along with the QR code, the label will also contain information of textual expression of the code and the URL address for verification of the authenticity. The organization issuing the certification implements the proper printing of QR codes, together with the storage of a unique identifier into the internal database. Methods of database security against unauthorized entry and data storage are not described for safety reasons. Distribution of labels to the sellers is realized in the normal way of sending, delivery is verified. Activation of codes will be implemented as late as the moment of confirmation of manufacturer on the sale of toner cartridges with a specific series of QR codes.

6 ECONOMIC ADVANTAGES OF PUBLIC PROCUREMENT

The individual EU Member States must implement from April 2016 into their national legislation the Directive 2014/24/EU, which introduces as the sole criterion of the implementation of public procurement, which is understood as the summary of qualitative, social and environmental aspects. Currently, the status of implementation of purchase of toner cartridges is not practically resolvable for the public sector so that they would not buy products in violation of applicable legislation (infringement of patents, lack of safety data sheets). Purchase of toner cartridges will become even more elusive due to the introduction of economic advantage criterion rather than the lowest bid price. It is possible to say that the toner cartridges with the lowest bid price are at the same time the toner cartridges with definitely the highest environmental impact (it is not possible to recycle or refurbish them, and they become municipal waste after first use); at the same time these toner cartridges reduce employment rate and production in the country, into which they are imported—see earlier study of the author, from which it follows that due to the import of compatible toner cartridges, a loss of at least 140 jobs occurred in the

Czech Republic, and annual production of 1.2 million kg waste occurred which would not have been produced in case of elimination of these toner cartridges. (EUR-LEX, 2013).

It is therefore evident that the lowest bid price need not to be simultaneously the economically advantageous price; evident contradiction of these two parameters is evident here.

7 CONCLUSION

The public sector usually realizes the purchases of toner cartridges using parameter of the lowest bid price. However, the lowest bid price in this commodity may be in direct conflict with economically advantageous price, which will be the sole criterion for public procurement in the European Union member countries, from at least April 2016. The methodology of determining the economic benefit, but also verifying the veracity of the information submitted by individual bidder entities is very difficult even for specialists in the given branch; correct assessment of the economic advantage is extremely difficult for the public sector. Solving the problem lies in the certification of toner cartridges, the proving of which can be a criterion for participation in public procurement. Compliance with certification represents an optimal fulfilment of the criterion of social sphere (a substantial part of the production process takes place in the country, in which public contract is realized; fulfilment of the requirements for the health risks of use is proven), environmental sphere (previously manufactured parts are used, thereby minimizing the generation of waste), while the qualitative aspect is tackled by setting mandatory production processes. Certified product thus meets all the parameters, and it is possible to decide between the individual competing parties solely on the basis of the parameter of the lowest bid price. Certification moreover eliminates the purchase of products infringing patent protection of the OEM toner cartridges manufacturers, from the side of public sector.

In the private sector, the certification provides a simple tool for eliminating purchases of toner cartridges infringing patent protection, presenting increased health risks and increased burden on the environment. The certification also guarantees the utilization rate of toner cartridge at least equal to the OEM toner cartridge.

The system of records and granting certification enables an online verification of the authenticity of the certification granted to specific manufacturer and specific product. Thanks to this, the online verification of the authenticity minimizes the risk of counterfeit of the certification itself.

REFERENCES

Adams, S.R. 2006. *Information sources in patents. 2nd completely new ed.* München: K.G. Saur.

Dikshith, T. 2013. *Hazardous chemicals: safety management and global regulations.* Boca Raton: CRC Press Inc.

EUR-LEX. 2013. *Directive 2014/24/EU of the European Parliament and of the Council of 26 February 2014 on public procurement and repealing Directive 2004/18/EC Text with EEA relevance.* Published 1st July 2013. [online]. <http://eur-lex.europa.eu/legal-content/EN/TXT/?uri=uriserv:OJ.L_.2014.094.01.0065.01.EN G>.

Kiehl, P. 2001. *Klein Einführung in die DIN-Normen. 13 Auflage.* Stuttgart: Teubner.

Kingston, W. & Scally K. 2006. *Patents and the measurement of international competitiveness: new data on the use of patents by universities, small firms, and individual inventors.* Northampton: Edward Elgar.

Myhrberg, E.V. 2009. *A practical field guide for ISO 9001:2008.* Milwaukee: ASQ Quality Press.

Whitelaw, K. 2004. *ISO 14001 environmental systems hand book.* Boston: Elsevier/Butterworth Heinemann.

Production Management and Engineering Sciences – Majerník, Daneshjo & Bosák (Eds)
© 2016 Taylor & Francis Group, London, ISBN: 978-1-138-02856-2

Life-cycle assessment of product through of the SWOT analysis

B. Hricová, E. Lumnitzer, M. Piňosová & A. Goga Bodnárová
Faculty of Mechanical Engineering, Technical University of Košice, Košice, Slovak Republic

ABSTRACT: Any entity entering into the life cycle of the product must also bear responsibility for its influence and be involved in the reduction of the overall negative impact on the environment. Not only in the world but also in Slovakia, have companies attempted to apply analytical methods for assessing the life cycle of products. This method compares the different environmental impacts of the products with respect to the different stages of their life cycle. Because of the complexity and inaccessibility of some information that is necessary for a full LCA, we decided to use SWOT analysis. SWOT analysis was carried out within an organization engaged in the manufacture of aluminum rims and was carried out separately for each phase of product's LC, starting from raw material extraction and processing of materials to the end phase of product's life.

1 INTRODUCTION

Life cycle assessment is divided into the areas of raw material extraction and processing of materials, production, and product distribution to consumer, its use by consumers and final disposal or re-use of raw materials—recycling. Each area of the life cycle of brings serious impacts and threats to the environment, which should be continuously monitored and minimized. Effective means of protecting the environment itself is a pollution prevention measures that each organization should take before production (Badida et al. 2001, Králiková et al. 2008).

2 NEGATIVE IMPACTS IN THE PROCESS OF ALUMINUM RIMS PRODUCTION

As each life cycle of a product has a negative impact on the environment, aluminum rims production has also a number of negative effects.

Table 1 identifies the overall negative impact of aluminum rims' life cycle.

3 SWOT ANALYSIS OF ALUMINUM RIMS'LC

SWOT analysis is needed to understand company's strengths and weaknesses and through this identification we are able to see threats that the company must face. It is also helpful in revealing opportunities that can be well used and eliminate threats far ahead. SWOT analysis separates information from environment analysis into internal (strengths and weaknesses) and external "issues" (opportunities, threats). Once these aspects are identified, SWOT analysis determines which of them can help the company to meet their goals and what obstacles must be overcome or minimized to achieve desired results. SWOT is popular also because it assesses the impact of internal

Table 1. Negative impacts of aluminum rims'LC.

Negative impacts of aluminum rims' LC			
Extraction and processing of raw materials	• Disruption of the natural environment • Creation of tailings • Creation of dumps and their decommissioning	• Pollution of ground water and reduced water quality • Weathering • Loss of agricultural land and disruption of landscape relief	• Emissions and dust • Drainage of rock structures • Reduced use of groundwater's

(Continued)

Table 1. (*Continued*).

	Negative impacts of aluminum rims' LC		
	Air pollution: • Particulate matter emissions discharged from melting chimneys and combustion plants • VOC arising from powder paint line • PS from paint shop—xylene, toluene, butyl acetate • VOC from coolants	*Water and soil pollution*: • Waste water from the paint shop • Oils from quenching • Cooling liquids and oils from the machining process	*Impact on employees*: • Noise and vibration from technological processes • VOC from chemical processes • Emissions and particulates from smelting processes
Packaging and distribution	• Consumption of packaging material • Air pollution by exhaust emissions	• Surface water pollution • Fossil fuels consumption	• Land take by construction of storage space and communications
	• Surface water pollution when cleaning aluminum wheels	• Land take by storing	• Dumps full of impaired rims
Disposal	• The consumption of energy and natural resources	• CO_2 production • Land take	• Waste heat • Emissions from disposal process

and external environment on the foreseeable development of the company through algorithm solutions that are based on the implementation of intuitive approaches. Important part of the process is to determine significance level of each factor by numbers (1–5)—number 1 is for negligible importance and number 5 denotes significant importance. (Swot analysis 2015) General SWOT analysis of the organization, whose production activity is aluminum rims production, is shown in Fig. 1. The following figures and schemes include a SWOT analysis for each stage of aluminum rims' LC with graphical representation of the results obtained (Fig. 2–Fig. 6).

Figure 1. General SWOT analysis of the company.

4 MEASURES TO REDUCE THE NEGATIVE EFFECTS

Based on the analysis of the life cycle of aluminum rims specified in Sec. 2, we have proposed the measures for elimination of the effects observed, which are shown in Table 2.

5 CONCLUSION

When developing a SWOT analysis for the assessment of the life cycle phases it was necessary to pay attention not only to one phase (Bednárová

Figure 2 — General SWOT analysis of the company

Strengths	scale of significance	Weaknesses	scale of significance
• Highly qualified of the manpower	4	• Damage to the relief of the country	5
• The use of modern technologies	5	• Financial cost	2
• Compliance with legislation	4	• Energy and water consumption	3
		• Land take	4

Opportunities	scale of significance	Threats	scale of significance
• Contribution to reducing dependence on supply of aluminum	5	• Erosion and landslide	5
• Increasing of competition in the market	4	• Seismic influence	5
• Creation of new jobs	3	• Contamination of soil and groundwater	5

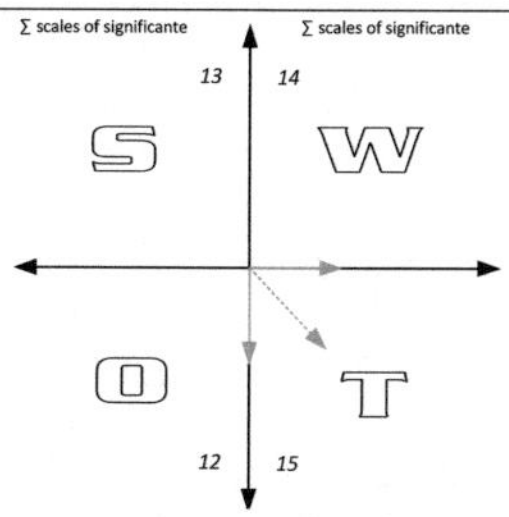

Σ scales of significante   Σ scales of significante

13   14

S   W

12   15

O   T

Figure 2. General SWOT analysis of the company.

Figure 4 — SWOT analysis of packaging and transport

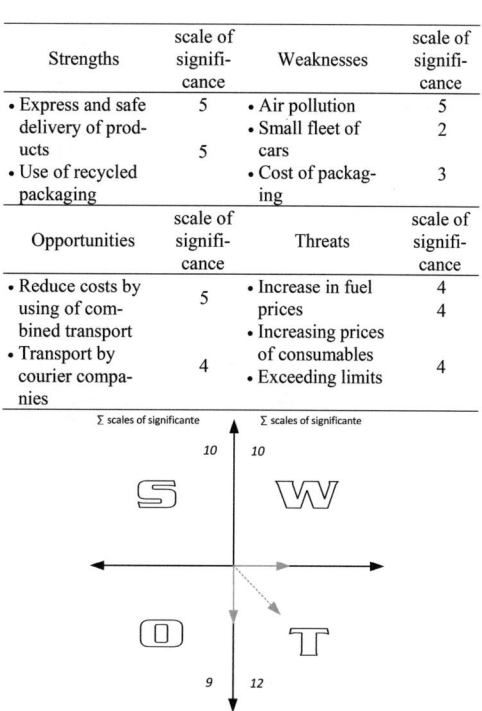

Strengths	scale of significance	Weaknesses	scale of significance
• Express and safe delivery of products	5	• Air pollution	5
• Use of recycled packaging	5	• Small fleet of cars	2
		• Cost of packaging	3

Opportunities	scale of significance	Threats	scale of significance
• Reduce costs by using of combined transport	5	• Increase in fuel prices	4
• Transport by courier companies	4	• Increasing prices of consumables	4
		• Exceeding limits	4

Σ scales of significante   Σ scales of significante

10   10

S   W

9   12

O   T

Figure 4. SWOT analysis of packaging and transport.

Figure 3 — SWOT analysis of production

Strengths	scale of significance	Weaknesses	scale of significance
• Know-how	4	• Production cost	3
• Strategic management of the production process	3	• Limited financial resources	3
• High quality of production	5	• Environmental pollution in the production process	5
• Compliance with legislation	4	• Low preventative measures	4
• Qualified manpower	3	• Strong dependence on sales	3
• Innovation of production processes	4	• Poor awareness about the company	3

Opportunities	scale of significance	Threats	scale of significance
• Increase in fuel prices	5	• Competition in the market	4
• Positive approach to the ecodesign	5	• Increases in energy prices	4
• Production capacity utilization	4	• Rising raw material prices	4
• Job creation	3	• Changes in legislation	3
• Increase in the technological intensity	4		

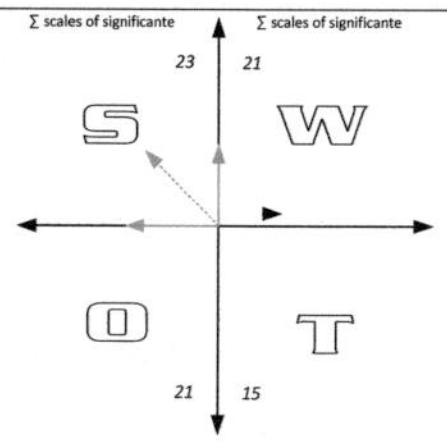

Σ scales of significante   Σ scales of significante

23   21

S   W

21   15

O   T

Figure 3. SWOT analysis of production.

Figure 5 — SWOT analysis of use

Strengths	scale of significance	Weaknesses	scale of significance
• Long life	5	• The initial investment	5
• Reduction of fuel consumption	5	• The cost of nurturing products	3
• Product design	3		
• Guaranteed quality	4		

Opportunities	scale of significance	Threats	scale of significance
• Exchange of replacement parts	4	• Deformation of discs	3
• Assistance Services	5	• The prices increase	4
• Servicing	4		

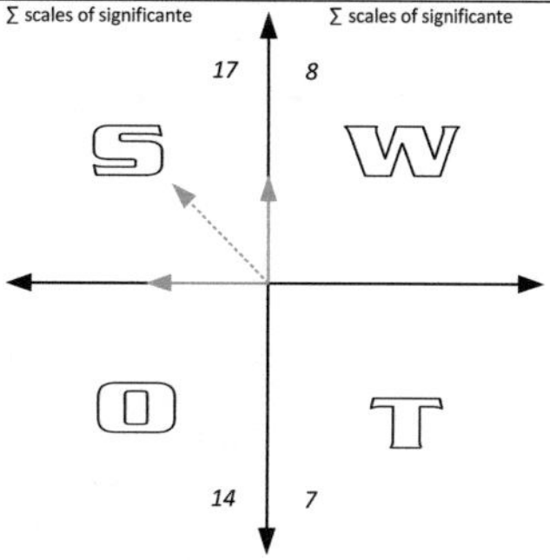

Σ scales of significante   Σ scales of significante

17   8

S   W

14   7

O   T

Figure 5. SWOT analysis of use.

Strengths	scale of signifi-cance	Weaknesses	scale of signifi-cance
• Saving of materi-al	5	• Energy consump-tion	5
• The use of sec-ondary raw mate-rials	5	• Land take	4
• The minimum load on the envi-ronment	5	• Emissions from the recycling process	5
• Energy saving	5	• Air pollution	5
• Saving of natural resources	5		
• Reduction of CO_2 production	4		

Opportunities	scale of signifi-cance	Threats	scale of signifi-cance
• Increase in tech-nological com-plexity	5	• Change of of leg-islation	3
• The introduction of cleaner tech-nologies	4	• Compliance with the emission lim-its	4
• Providing of pre-ventive measures	5		

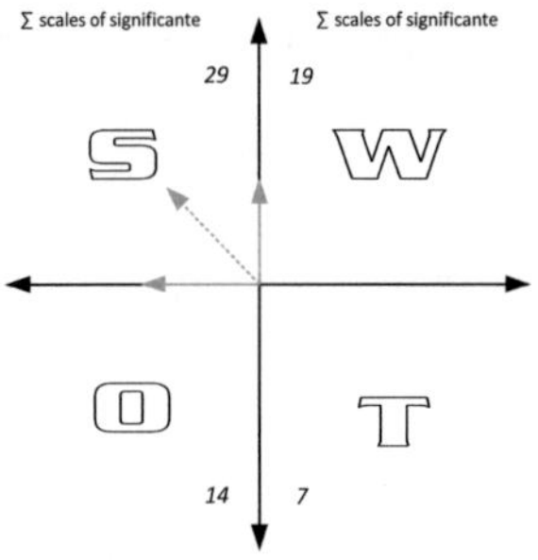

Figure 6. SWOT analysis of recycling.

2008; Bednárová et al. 2012), but globally to entire life cycle of aluminum rims. Taking into account all factors and weighting the importance of the individual life cycle phases, the analysis shows that "threats" influence extraction and processing of raw materials phase and packing and shipping phase. On the other hand, production and use phase showed strong points. Extraction and pro-cessing of raw materials phase is greatly affected by "threats" such as contamination, erosion and landslides. These threats can be reduced by using modern technologies and complying with legisla-tive. Production phase is dominated by strengths. Nevertheless, it is necessary to seize opportunities in the form of modernizing and correct environ-mental approach to the production. Great weight of significance at the packing and shipping phase bear "threats" that has more economic than envi-ronmental nature. Strengths and weaknesses are at the same level. Therefore we should take the opportunity and differentiate them. Use phase has visible strengths which indicate that the mere use of aluminum rims have no negative impact on the environment.

We see recycling as a positive assessment of the product and as the benefits for the production pro-cess by reusing material and conserving natural resources. By recycling aluminum we save up to 95% of the energy needed for its production from baux-ite. In the final evaluation, extraction and process-ing of raw materials phase has the biggest impact on the environment of all life stages of aluminum rims.

From the available and gathered information we evaluated the economic cost ratio of aluminum

Table 2. Measures to be taken to eliminate the negative effects of aluminum rims.

Measures to be taken to eliminate the negative effects

Extraction and processing of raw materials	• Compliance with applicable legislation • Monitored and controlled tailings Preventing the creation of dumps	• Limit the occurrence of leachate • Compliance with pollution limits	• Making financial reserves necessary for elimination of emergency situations • Containment of dust and gas leakage from storage
Production phase	• Ensure the implementation of corrective measuring of the amount and concentration of particulates before putting the source into operation • Develop a set of technical and operational parameters and measures • Develop air pollution-transfer assessment • Install separator of gas and dust pollutants	• Regularly monitor the quality and quantity of discharged WW from WWTP • Ensure trouble-free operation of automatic pH sensors • Measure, record and evaluate water abstractions for different purposes • Strict adherence to security measures	• Drain surface water into drains • Use cleaning agents based on aqueous solutions • Introducing environmentally friendly coolants • Use of PPE • Noise regulations in work environment

(*Continued*)

Table 2. (*Continued*).

Measures to be taken to eliminate the negative effects			
Packaging and distribution	• Increase the use of recycled packaging • Recycle used packaging or dispose them in the eco-friendly way	• Use combined transport • Consumption of fossil fuels	• The use of vehicles with hybrid or electric drive
Use	• Use environmentally friendly detergents	• Eco-friendly dispose of waste water resulting from cleaning the rims	• Observe manufacturer's instructions for extending the useful life of the product
Disposal	• Modernization of recovery technology	• Observance of safety measures	• Using BAT technology in raw material recycling

rims. One of the drawbacks from the economic point of view is the initial investments into production facilities, which represent 19.5 million. €—for the organization with estimated production of 500 thousand pieces of aluminum rims per year.

Many manufacturing companies are coming to market with an established system of product life cycle assessment. A new trend in the form of technological innovation for production solutions is expected to be coming to market in near future, concerning the elimination of pollution in the environment, which is, based on the survey, reduced by only a small extent.

ACKNOWLEDGEMENT

This paper was supported by the Slovak Research and Development Agency under the contract No. APVV-0432-12 and KEGA No. 039TUKE-4/2015.

REFERENCES

Badida, M., Majerník, M., Šebo, D. & Hodolič, J.: Strojárska výroba a životné prostredie, Košice: Strojnícka fakulta TU, 2001, 253s. ISBN 80-7099-695-1. [In Slovak]

Bednárová, L., Chovancová, J. & Sirková, M.: Medzinárodný manažment. 3. doplnené vydanie. Prešov: Fakulta manažmentu, 2012. 160 s. ISBN 978-80-89568-55-0. [In Slovak]

Bednárová, L.: Ekonomická efektívnosť environmentálneho manažérstva, Environmentálne účtovníctvo, Grafotlač Prešov, 2008, ISBN 978-80-8068-733-5. [In Slovak]

Králiková, R., Rovňák, M. & Králik, M. 2008. Visualization of environmental data. 8th International Scientific Conference SGEM 2008: Modern Management of Mine Producing, Geology and Environmental Protection, 16.-20. June 2008, Albena, Bulgaria., pp. 577–584, ISBN 954918181-2.

SWOT Analysis. Manktelow, J. & Carlson, A. [cit. 2015-20-03]. Dostupné na internete: <http://www.mindtools.com/pages/article/newTMC_05.htm>.

Production Management and Engineering Sciences – Majerník, Daneshjo & Bosák (Eds)
© 2016 Taylor & Francis Group, London, ISBN: 978-1-138-02856-2

Alternative biological approaches to improve the environment in deteriorated areas

O. Hronec, P. Adamišin, J. Bejda, E. Huttmanová, J. Vravec & R. Vavrek
Faculty of Management, University of Presov in Presov, Presov, Slovak Republic

ABSTRACT: As a result of long term extraction and processing of magnesite ore, in some regions of the Slovakia soils have been severely deteriorated. Soil reaction in these areas reaches values pH 9, and more. Challenges to redeveloping these sites include finding methods of their revitalization. Though, technical recultivation is under current economic conditions difficult to achieve. In contrast, biological method of recultivation is economically acceptable with potentially significant environmental benefits. The possible way of innovative biological method of recultivation is cultivation of Phragmites australis (Cav.) Trin on deteriorated soils. In addition to positive revitalizing effect, the plant produces large-scale biomass which can be used as renewable energy source. The paper is focused on assessment of possible ways of cultivation of Phragmites australis (Cav.) Trin on alkaline soils and assessment of its possible use as an energy crop. Regression analysis and selected moment characteristics are used to achieve stated objectives.

1 INTRODUCTION

The quality of soil resource as one of the fundamental components of the environment as well as agro-ecological conditions in the Slovak Republic is diversified. As a result of previous orientation of the national economy to build heavy industry based on high energy and raw material intensity, occurred in our country the (excessive) contamination of soil, which has negatively affected the quality of the environment including the soil characteristics. Most of these activities have toxic effects on living organisms, when permissible concentration levels are exceeded. (Singovzska & Balintova 2014)

Long-term mining activities represent a part of anthropogenic activities that have a significant impact on the landscape and its changing structure. Mining has a significant impact on the landscape, creating a whole new country, which is related to change and microclimate, adversely affect the biota, is accompanied by a considerable loss of agricultural and forest land, affects the hydrological system and pollutes the air with dust particles. (Vrábliková & Vráblík 2002)

The consequences of industrial activity resulted mostly in soils acidification, alkalization and metallization. Alkalinisation of soils is largely the result of alkaline, mostly particulate imissions and despite the fact that in Slovakia it is not so widespread phenomena, from the milder impact on soil, e.g. surroundings of cement and lime plants,

it can have highly devastating effects (e.g. in the areas of magnesite processing plants). Endogenous resources of soil alkalinisation are particularly heavily mineralized groundwater, causing soil salinization associated with alkalization. Salinization and alkalinisation of soils greatly reduces agricultural production. Alkalinisation of soils in Slovakia is caused mainly by anthropogenic activities.

2 OLD ENVIRONMENTAL BURDENS IN SLOVAKIA

Slovak republic is a country extremely rich in the natural crystal magnesite, the reverse base is 3,400 million tonnes there. (Csikosova, Culkova & Antosova, 2013)

The Slovak Republic is the fourth largest producer of magnesite in the world. On the territory of Slovakia is produced more than 6.5% of world production of magnesite. Mining and processing of magnesite is also one of the major economic sector of the Slovak national economy. The production of magnesite is according to Cicmanova (2002) mainly localized on sites Jelšava (Magnesite Works in Jelšava) and Lubeník (SLOVMAG Lubeník). Extraction of magnesite and subsequent processing is a very dusty process. Production of magnesite clinkers is conducted by thermal decomposition and clinker process. These companies have affected by its production not only air quality, but mainly the quality of soils, which are due to the extraction and processing of magnesite highly alkalized. A strong alkalinisation of soils caused heavy deterioration of soils in some sites placed in the immission field of above mentioned companies to the extent that the microbial life there has disappeared. In these areas are soil also significantly metalized (with the high doses of heavy metals, in particular Hg, Mn, As, Cd, Pb, Cu, Al, Fe). Heavy metals are a leading group of contaminants, which is involved in changing soil properties and significantly interferes with the processes occurring in the soil environment. (Javoreková 2008)

Jelšava-Lubeník area where the mining and processing of magnesite is concentrated, is the area, where strong alkalinisation of the soil is present. Mainly the dominant anthropogenic activity determined the inclusion of the area to the environmentally burdened areas, resp. according to environmental regionalization of the Slovak republic—among one of the six districts of Slovakia with heavily deteriorated environment. In this area, during the processing of magnesite, magnesium oxide is emitted into the atmosphere. Magnesium oxide causes the alkalinisation of more than 12 000 hectares of agricultural land and more than 6,600 ha of forest land. In this area, the soil pH is around 8–9, which corresponds to a strong alkalinisation of soils. Magnesium imissions cause many undesirable phenomena on soil, vegetation and animals, reflected also in many adverse events such as poorer production and economic results. These events led to the collapse of indigenous plant communities on soils in imission field of above mentioned companies and only a few resistant species insignificant from production, agricultural, forest and aesthetics point of view are present there. Alkalinisation of soils and accumulation of other environmental problems in this area have arisen primarily to reduced crop production. In some areas, strong alkalinisation caused loss of production ability of soil where any plant is hard to be grown.

2.1 Possibilities of bio-remediation of contaminated soils

Adjusting soil pH towards neutralization is quite difficult and long-term process. Adjusting soil pH is economically but also very time consuming. One of the options in the process of adjusting soil properties and revitalization of damaged soils is the use of biological processes.

Natural process of reducing contamination is one of the ways to remove pollutants from the environment. This process is defined as the sum of processes naturally occurring in the natural environment, which without human intervention

lead to limit the amount toxicity, mobility or concentration of contaminants. By Ouyang (2002), phytoremediation presents new technologies and methods of using green plants and their associated rhisospheric microorganisms to accumulation, fixation or complete removal of contaminants occurring in soil, groundwater, surface water or the atmosphere. Phytoremediation uses green plants and their associated microorganisms as well as agronomic techniques to eliminate or transform contaminants from the environment.

Mechanisms of phytoremediation of heavy metals can be simply interpreted as removing metals from the soil via their transport to the root system and the plant body. Subsequently the plants are harvested and the surface is in turn planted by new plants to the time until the concentration of metals in the soil is reduced to acceptable levels. (Cluis 2004)

These processes are based primarily on the use of more environmentally friendly practices and processes than for technical remediation. This is particularly useful especially in those areas whose environment is severely disrupted. In their application can be more widely used less skilled workers. As stressed areas are usually simultaneously less economically developed areas with limited job creation, implementation of procedures there may also contribute to the sustainability of the development area, to increase the quality of life also caused other secondary effects.

By the long-term observation and conducted research (Angelovičová & Fazekašová 2014, Fazekašová 2012, Hronec et al. 2012), we found that in Jelšava-Lubeník burdened area, respectively Jelšava district where soils are significantly contaminated, *Phragmites australis* (Cav.) Trin appeared in recent years. *Phragmites australis* (Cav.) Trin is originally humid plant, but in this area it grows literally in dry sites, where ground water is in the depth of several metres. Striking vitality of *Phragmites australis* (Cav.) Trin was found, as mega population in more sites, where the pH value reached more than 9 (which is on the border of strongly alkalized soil) and in such sites, where it does not occur and according to the published statements its presentation was not recorded in the past. The leaves of *Phragmites australis* reflected the different gradient of PGEs emissions, and may thus be considered as potential biomonitors of atmospheric pollution (Bonnanno & Pavone 2015) but its spread is determined by many factor including reduced salinity and nitrogen runoff (Fussell et al. 2015). Based on the results of former research (Hronec & Hajduk 1996, Hronec et al. 2012) we can state that it is hopeful, dominant, resistant, anti-erosive and technically available kind of plant, which provides alternative solution of bio-remediation and fertilization of contaminated soils.

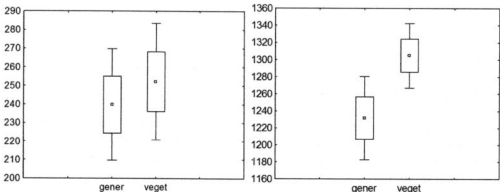

Figure 1. Plant height (in mm) at the beginning and end of the experiment (after 2 years).

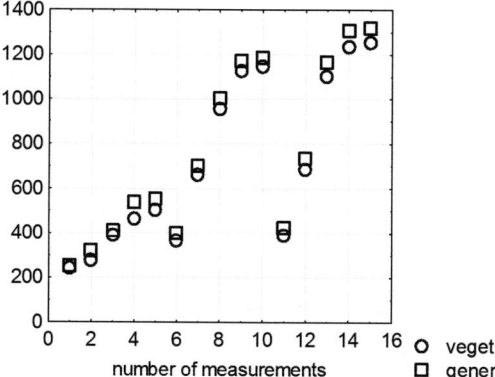

* Note: decrease at 6, respectively 11 measurements of both methods is due to the start of the growing season (i.e. the first measurement in the respective year)

Figure 2. The average plant height for individual measurements in both methods of reproduction.

The immission field of exhalation sites of Slovak magnesite works Jelšava and Slovmag, Lubeník we conducted a further research to survey the species, which should be characterized by increased resistance and could have potential properties such as Phragmites australis (Cav.) Trin. Phragmites australis (Cav.) Trin, however, proved to be most suitable crop for remediation purposes for several reasons. Therefore, we verified the various methods of reproduction of Phragmites australis (Cav.) Trin, so the growing on contaminated soils is the most economically efficient. We verified the vegetative and generative methods of reproduction, and we came to the conclusion that the generative reproduction appears to be more efficient (and also in terms of next generation biomass). The strength (speed) of growth and biomass production was subsequently verified in the process of growth and it has been proved that the method of reproduction does not affect the formation of biomass (the formation of leaves and sprouts).

At the beginning of the measurements, the differences in plant height in both methods of reproduction were not statistically significant. But the observed measurements in the last year (5 last measures) always led to a statistically significant difference.

3 PHRAGMITES AUSTRALIS AS A SOURCE OF BIOMASS

Course of events in the field of energy efficiency in EU suggests that it is the topic with high priority. EU commission is aiming to sustainability, competitiveness and safety in all stages of the energy chain. The European Commission, for the promotion of energy efficiency, proposed binding targets 20/20/20 for every member state. (Directive 2012/27/EU 2012) These are the set of measures, e.g. climate protection, development of renewable and other energy sources (Heller 2006) with the aim to decrease greenhouse gas emission by 20% and increase the share of renewable energy sources to 20%.

The energy use of biomass in Slovakia is currently lags far behind potential opportunities—energy, economic and environmental. According to Vilcek (2013) the highest energy yields can be expected from the biomass of plants grown in arable land (approximately 11.0 $MJ.m^{(-2)}$).

Despite inconsistencies in the data and the lack of a uniform methodology for calculating technically usable potential of biomass (as well as other renewables) represents the biomass—after solar and geothermal energy—the source with the third largest potential. Estimates of the total technical potential of biomass (forest, agricultural and rest) range from 75.6 PJ to 120.3 PJ. (Ministry of Economic SR, 2006) However, the utilization of biomass share of total consumption of primary fuel-energy sources of the Slovak republic is currently only 1%. (Horbaj 2006)

Recent trends in the use of phyto-biomass for energy purposes head towards the use of cereals, but also other alternative and economically efficient sources are sought. E.g. maize as energy crop has many disadvantages—high inputs, fluctuations in harvest, the risk of soil erosion and limited area for cultivation. (Jamriška 2007) From traditional crops are best cereals (triticale, rye), including straw, while the energy efficiency of straw is higher than the combustion of entire plants. Other crops which can be grown for energy purposes are e.g. *Brassica napus, Helianthus annuus* but also the grasses—*Festuca arundinacea, Arrhenatherum elatius, Phragmites communis* etc.

Some alternative studies have been conducted on some biomass samples such as cotton stalks Coates (2000), tea waste Demirbaş (1999), waste paper and wheat straw Demirbaş (1998), olive refuse Yaman et al. (2000) were used to obtain biobriquettes. However, the existing information about one type of biomass sample cannot be applicable to another one.

Pedroli et al. (2013) conclude that increased demand for biomass for bioenergy purposes may lead to a continued conversion of valuable habitats into productive lands and to intensification, which both have negative effects on biodiversity. On the other hand, increased demand for biomass also provides opportunities for biodiversity, both within existing productive lands and in abandoned or degraded lands.

Laboratory research of biomass briquetting possibilities was conducted by Bejda et al. (2002). Research was focused on the impact of selected factors on the quality of the briquettes made from wood chips and plants including Phragmites australis, namely the impact of the size of external pressure forces and temperature in the compaction process, the density of the briquettes, which should be at least 1000 kg.m⁻³.

The oak and spruce sawdust and also Phragmites australis (Cav.) Trin were selected for the experiment. Phragmites australis (Cav.) Trin was dried in laboratory conditions to humidity of approximately 10%. It was then chopped, crushed and sieved. For the experiment only sifted fraction (up to 2 mm) was used. Samples were prepared from 100% oak and spruce sawdust and Phragmites australis (Cav.) Trin as well as multi-component samples consist of 80% wood chips and 20% Phragmites australis (Cav.) Trin.

The experiment was conducted at room temperature (21°C) but experiment was conducted also with the pre-heated sample of Phragmites australis (Cav.) Trin. Since the ignition temperature of wood with bark or without bark as well as the brown or hard coal is approximately 220°C Prokeš (1999), pre-heating temperature of the sample in pressing container was set to 200°C, and after pre-heating, the sample was pressed.

Briquetting was realized in the laboratory compactor with a maximum load of 2000 kN and it was conducted in the steel pressure vessel of cylindrical shape. During briquetting from stated materials, the force necessary to compression was recorded. After removing the briquette from pressing device its weight and volume was measured.

Following from experimental as well as other known data, compacting pressure p and specific weight ρ—briquettes density, were calculated. Subsequently, dependencies between variables were processed, which are shown in figures 3 and 4. In addition to the experimental data points, logarithmic dependence for particular materials was calculated by the method of least squares and their relevant equations and regression coefficients.

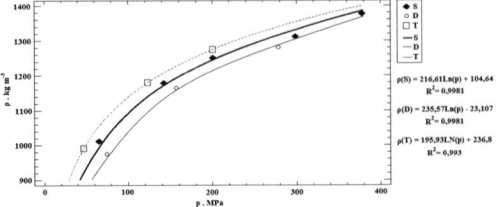

Figure 3. Dependence of density on pressure for briquettes from spruce sawdust (S), oak sawdust (D) and Phragmites australis (Cav.) Trin (T), Bejda et al. (2002).

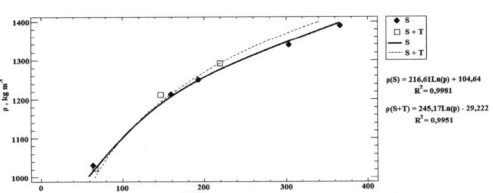

Figure 4. The dependence of density on pressure of spruce sawdust briquettes (S) and multi-component sample (S + T), Bejda et al. (2002).

Fig. 3 shows the dependence between briquettes density and compacting pressure of single-component materials. The results of the research show that the density of Phragmites australis achieved under the same pressures higher values than the oak and spruce; hence achieving the same briquettes density of Phragmites australis is possible at a lower pressure, which is less demanding on the technical characteristics of the pressing device. However, single-component briquettes from Phragmites australis were inconsistent; they were disintegrated at lower pressures, whereas at higher pressure they were breaking, which for the further energy use is unacceptable. For comparison, the experiment was repeated at the spruce and oak sawdust, which are coherent and stable.

Subsequently, multi-component samples consisting of 80% wood sawdust and 20% Phragmites australis. Figure 4 shows the dependence of density from the pressure of spruce and Phragmites australis (S + T), and for comparison also single-component sample of spruce (S). From the picture it is clear that the density of the multi-component material at the same pressure reaches higher values than the density of spruce. Briquettes from a multi-component material were consistent and stable.

Figure 5 illustrates dependences resulting from the experiment conducted at the higher temperature (T = 200°C). It is obvious that the effect of pre-heating of the material significantly affected

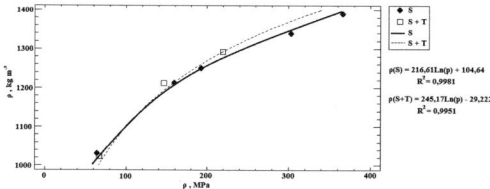

Figure 5. The dependence of the density on the pressure of the Phragmites australis (Cav.) Trin briquettes at temperature T = 20°C and on the pre-heated sample at temperature T = 200°C, Bejda et al. (2002).

the quality of the sample. While to reach density of about 1100 kg.m^{-3} at room temperature (T=20°C) requires a pressure of about 80 MPa, this pressure is significantly reduced when sample is pre-heated before compression to only 31 MPa.

Technical and energy demands of briquetting device, thus significantly decreased. The briquettes made from pre-heated Phragmites australis samples were consistent and stable.

Growing Phragmites australis for biomass purposes on strongly contaminated soil can open other relating problems, regarding heavy metal content in biomass and its transmission to the atmosphere during the combustion process. In all uses of biomass it is fundamental to know both chemical composition and thermal behaviour of the given type of biomass in oxidizing and inert atmospheres. Research conducted by Ghetti et al. (2013) suggest that evaluation of the thermal behaviour of biomass sample should be focused on the quality of the pyrolysis product, also known as tar or bio-oil and should consist of following steps:

- Characterization and chemical analysis of biomass sample,
- Thermal analysis performed in an oxidant and inert atmosphere,
- Thermal analysis of biomass-derived pyrolysis products,
- Evaluation of the experimental data, identifying the main correlations.

These steps should be followed during the next research activities focused on cultivation and reproduction of Phragmites autralis for biomass purposes on strongly alkalized soils.

4 CONCLUSIONS

As a consequence of the mining and processing of the magnesite ore, some areas of Slovakia are characteristic by strongly alkalized soils, when pH increased to 9 and more. The physical and chemical characteristics of soil were changed, the ero-

sion increased and phytocenoses decreased. Large areas are impossible to use for agriculture. Technical recultivation is under current economic conditions difficult, but biological method of recultivation is economically acceptable with potentially significant environmental benefits. Cultivation of Phragmites australis (Cav.) Trin in such damaged soils presents the innovative biological method of revitalisation. In addition to positive revitalizing effect, the plant produces large-scale biomass which can be alternatively used for energy purposes. Experiments showed that multicomponent samples (e.g. spruce sawdust + Phragmites australis (Cav.) Trin as well as preheated sample of Phragmites australis (Cav.) Trin can be used for briquetting and briquettes meet requirements for their density and the stability of shape.

From the analysis presented so far, the multilateral benefits of revitalization of specific areas by biological methods are obvious. Although the main effect is to improve the quality of the environment in affected regions, produces biomass can be further serve as renewable source of energy, which increases the added value of biological revitalization of the area and at the same time it can improve the economic aspects of the processes.

Despite the positive findings of presently conducted research, it is necessary to continue. The next steps should be aimed at safety of using briquettes from Phragmites australis (Cav.) Trin for energy purposes and possible risk of transmission of heavy metals to the atmosphere during the combustion process.

ACKNOWLEDGEMENT

"Paper is the result of the Project implementation: University Science Park TECHNICOM for Innovation Applications Supported by Knowledge Technology, ITMS: 26220220182, supported by the Research & Development Operational Programme funded by the ERDF."

REFERENCES

Angelovičová, L. & Fazekašová, D. 2014. Contamination of the soil and water environment by heavy metals in the former mining area of Rudňany (Slovakia). Soil & Water Res 9(1): 18–24.

Bejda, J., Kádárová, J., Krepelka, F. & Miklúšová, V. Laboratórny výskum briketovateľnosti bioodpadu. In STU Bratislava (ed.), Technika ochrany životného prostredia (TOP) 2002, 22.-23. máj. Častá Papiernička: STU Bratislava.

Bonnanno, G. & Pavone, P. 2015. Document Leaves of Phragmites australis as potential atmospheric biomonitors of Platinum Group Elements. Ecotoxicology and Environmental Safety, 114: 31–37.

Cicmanova, S. 2002. Geoenvironmental assessment of magnesite deposit Jelšava—Dúbrava massif. *Geoenvironmental assessment of magnesite deposit Jelšava—Dúbrava massif* 50: 87–95.

Cluis, C. 2004. Junk-greedy Greens: phytoremediaton as a new option for soil decontamination. *Biotechnology journal* 2004(2): 61–67.

Coates, W. 2000. Using cotton plant residue to produce briquettes. *Biomass and Bioenergy* 18(3): 201–208.

Csikosova, A., Culkova, K. & Antosova, M. 2013. Magnesite industry in the Slovak Republic. *Gospodarka Surowcami Mineralnymi-Mineral Resources Management* 29(3): 21–35.

Demirbaş, A. 1999. Evaluation of biomass material as energy sources—upgrading of tea waste by briquetting process. *Energy sources* 21(3): 215–220.

Demirbaş, A. & Şahin, A.: Evaluation of biomass residues—briquetting waste paper and wheat-straw mixtures. *Fuel processing technology* 55(2): 175–183.

Directive 2012/27/EU on energy efficiency, amending Directives 2009/125/EC and 2010/30/EU and repealing Directives 2004/8/EC and 2006/32/EC.

Fazekašová, D. 2012. Evaluation of soil quality parameters development in terms of sustainable land use In: InTech (ed.), *Sustainable development—authoritative and leading edge content for environmental management.* Rijeka: InTech.

Fussell, S., Dionne, M. & Theodose, T. 2015. Expansion Rates of Phragmites australis Patches in a Partially Restored Maine Salt Marsh. *Wetlands* 2015(35): 557–565.

Ghetti, P., Ricca L. & Angeliny, L.: Thermal analysis of biomass and corresponding pyrolysis products. *Fuel* 75(5): 565–573.

Heller, W. 2006. New green paper on European energy policy. *ATW—International Journal For Nuclear Power* 51(4): 266-+.

Horbaj, P. 2006. Možnosti využívania biomasy v SR. *Acta Montanistica Slovaca* 2006(11): 258–263.

Hronec, O. & Hajduk, J. 1996. Significant resistence of Phragmites australis Cay. Trin. On the soils intoxicated with magnesium immissions. *Ecology* 1996(2): 117–124.

Hronec, O., Vilček, J., Adamišin, P., Andrejovský, P. & Huttmanová, E. 2012. Use of Phragmites australis (Cav.) trin and its reproduction in the revitalization of contaminated soils. In University of Novi Sad (ed.),

MMA 2012—Advanced production technologies: proceedings of the 11th international scientific conference. Novi Sad: University of Novi Sad.

Hronec, O., Vilček, J., Torma, S. & Lisnyak, A. 2012. Environmental aspects of Phragmites australis use at fertilization of contaminated soils. *Ľudyna ta dovkilľa. Problemy neoekolohiji* 2012(3–4): 113–119.

Jamriška, P. 2007. Rastlinná výroba—zdroj obnoviteľnej energie. In SAPV (ed.), *Predpoklady využívania poľnohospodárskej a lesníckej biomasy na energetické a biotechnické využitie.* Nitra: SAPV.

Javoreková, S., Králiková, A., Labuda, R., Labudová, S. & Maková, J. 2008. *Biológia pôdy v agroekosystémoch.* Nitra: Slovenská poľnohospodárska univerzita v Nitre.

Ministry of Economy SR. 2006. Stratégia vyššieho využitia obnoviteľných zdrojov energie v SR.

Ouyang, Y. 2002. Phytoremediation: modeling plant uptake and contaminant transport in the soil-plant-atmosphere continuum. *Journal of Hydrology* 266(1): 66–82.

Pedroli, B., Elbersen, B. & Frederiksen P. 2013. Is energy cropping in Europe compatible with biodiversity?—Opportunities and threats to biodiversity from land-based production of biomass for bioenergy purposes. *Biomass & Bioenergy* 2013(55): 73–86.

Prokeš, O., Ciahotný, K. & Krzack, S. 1999. Odhad produkce tekavých látek při spalování pevných paliv v malých topeništích—uhlí, rudy. *Geologický pruzkum* 47(4): 14.

Singovzska, E. & Balintova, M. 2014. Metal pollution assessment in sediments of the Smolnik creek, Slovakia. *Pollack Periodica* 2014(9): 115–127.

Vilcek, J. 2013. Bioenergetic potential of agricultural soils in Slovakia. *Biomass & Bioenergy.* 2013(56): 53–61.

Vráblíková J. & Vráblík P. 2002. Zkušenosti s revitalizací antropogenně postižené půdy. In SAV (ed.), Zborník z 3. mezinárodní konference "Život v pôdě" 29.1.2002. Bratislava: Ústav krajinnej ekologie SAV.

Yaman, S., Şahan, M. Haykiri-açma, H., Şeşen, K. & Küçükbayrak. S. 2000. Production of fuel briquettes from olive refuse and paper mill waste. *Fuel Processing Technology* 68(1): 23–31.

Production Management and Engineering Sciences – Majerník, Daneshjo & Bosák (Eds)
© 2016 Taylor & Francis Group, London, ISBN: 978-1-138-02856-2

Monitoring of competitiveness indicators of the controlling enterprise

Z. Chodasová & Z. Tekulová
Slovak University of Technology, Bratislava, Slovak Republic

ABSTRACT: Business management is often in a situation where must decide whether to produce or not a particular product, which products or services prefer in a given period. Such decisions require flexibility to respond to market demands, so managers need flexible information system with a high quality information choice. Implemented for example financial controlling as a management tool in the company creates the basis for ensuring quality and effective management of corporate finance. Therefore it is important to focus on creating a database of financial indicators in the business entity, such as the company's liquidity, where monitoring and control helps managers to perform many important economic decisions. The question of monitoring liquidity and capital structure optimization, as an indicator of competitiveness plays an important role in business management, which is also described in this article.

1 INTRODUCTION

Trader in the market economy is under intense competitive pressure and it is forced to continually improve it's internal processes, management systems and to respond to new situations and new features management methods, which also provides corporate controlling. A prerequisite for achieving the optimal liquidity in the company is to balance revenue and expenditure in the short as well as in the long term. Solution for these issues usually brings financial plan and its control. The process of balancing revenue and expenditure has very specific nature and is marked by certain degree of inaccuracy, which means that although the balancing financial plan is a precondition for achieving the interim and permanent liquidity, it is not a guarantee. An exception can occur due to different trends of actual and projected cash flow. Although the company's liquidity planning is not a 100% guarantee to achieve continuous liquidity, it is important, because it significantly increases the likelihood of it's achieving and helps avoid predictable errors (Horvath & Partnes, 2004).

Several factors affect the financial performance of the company. One of them is the structure of the financial resources of business assets. It is necessary in managing the financial structure to take into account the basic rules for the financing based from long experience of financial managers. A significant role plays the typical structure of financial resources in the sector also or possibly the financial structure of the strongest competitor. Equally important are the leading experts on finance and financial structure views, developed in financial theory. Next article provides an overview of opinions, observations and possible factors affecting the management of the financial structure, the practical part focuses on the practical evaluation of the selected indicators of sectoral environment of industrial production in subclassifications of subject production activities Manufacture of motor vehicles.

2 LIQUIDITY MANAGEMENT

Liquidity management has a specific character and it is up to the company how to settle this problem. There are certain general principles of how to manage the liquidity of the company. They can be summarized as follows:

- to avoid delays in need of commitments,
- to optimal use of credit limits,
- to avoid exceeding credit limits,
- to avoid losses from inaction funds,
- to regulate the speed of the flow of funds,
- to ensure the availability of flexible short-term resources,
- to build information systems to support monetary disposition.

Liquidity is conditional to sufficient funds capacity whose condition and development is contingent upon cash flow in the company. This statement captures cash flow which is an important element of corporate financial management not only for the management of continuous liquidity. It helps identify and solve financial problems and reduce financial risks. For company liquidity planning to the future are a very important elements past economic phenomena that can, based on a thorough analysis, provide important information concerning the events and factors that have had in history positive or negative impact on the liquidity of the company. Based on this information the enterprise can more effectively plan its cash flow and secure liquidity in the future, which is related to effective production (Chodasová & Tekulová 2014).

Cash-flow statements must be distinguished from the indicators of cash flow. Cash flow indicator is often used in evaluating the company's liquidity mainly by creditors. It is a measure of the company's ability to form own economic activities the cash surpluses usable for the financing of essential needs. This is an important qualitative information for the assessment of not only the present but also the future solvency and financial stability of the company. Suitable ratios of cash flow for the assessment of liquidity are mainly total no-debts flow and short-term no-debts flow.

The total of no-debts flow tells the ratio between the funding and the ability of foreign capital to offset the liabilities funding from its own activities. The recommended value of this ratio is in the range <0.2; 0.3>, however, depends on the business sector, the size of the company. Greater predictive value expresses the degree of debt reduction in development time while the declining value indicates problems with protection of investments from its own resources and the worsening financial situation of the company (Ručková 2011).

$$Total\ no\text{-}debts\ flow = cash\ flow/other\ sources \qquad (1)$$

Short-term no-debt flow focuses on short-term liabilities and reflects on for how long is the company able to pay its current liabilities by the surplus funds from its own activities.

$$Short\text{-}term\ no\text{-}debts\ flow = \\ cash\ flow/short\text{-}term\ commitments \qquad (2)$$

There are several old saying about the business. One of them says "cash is the king". The other is "a positive cash is happiness". There is continuous research on unsuccessful companies. 60% of the unsuccessful businesses say that all or most of their failures were related to cash flow. As a guide for its positive future, companies should not forget another saying. "Nothing is greater than cash. Profit is nice, cash is necessary"(Kucharčiková, A., 2011).

3 CAPITAL STRUCTURE COMPANY CONTROLLING

Controlling is a system that helps to achieve business goals, prevents surprises and timely light on red when danger occurs requiring the manage-

ment to take appropriate action. It is an element of business management which allows identifying gaps and through its tools and methods to correct the development of business processes towards the prevention of adverse conditions and to ensure the efficient running of the organization. It is a system by which management receives the necessary information about the company activities and prepares analysis for its management. It may therefore be considered as an important support member in the process of corporate decision-making. Capital structure controlling has priority in company controlling system because of the large liquidity problems, poor law enforcement and secondary corporate insolvency (Chodasová, Z., 2005).

The structure of corporate financing sources assets in financial theory also called financial structure. The financial structure means ratio of individual components of equity and loan capital in total capital which financially covers the assets of the company. Static financial structure state is characterized by enterprise liability side of the balance sheet. Process dynamics represents the share of equity and loan capital on financial cover of company gain assets for certain period of time (Chodasová 2012).

From the perspective of some authors financial structure is a broader term. Indeed, some authors define the capital structure term that can be seen as a part of the financial structure. Others say they identify both of these concepts. A number of factors affects the financial structure, namely the acquisition cost and components of capital binding, the risks associated with increasing loan burden, the composition of company assets, the level, fluctuations and perspectives of the cash flow, the manner and intensity of the corporate profits taxation, the need to maintain the chosen level of liquidity and more. For consideration of the optimal capital structure are important following:

- Expenses on acquisition and binding on own equity are higher than the cost of obtaining foreign capital, it is because the owner's risk is higher than the risk of the lender and further because the share of profits are paid to the owner after taxation, while interest on loans are part of the cost of company, thus reducing profits and the tax base for income tax.
- Increasing the share of foreign capital in the total capital of the company—so called financial leverage raises a downward trend in the average cost on acquisition and related capital in turn increases the risk of corporate insolvency and other financial problems that can culminate in the liquidation of the company. The owners recognize the growth risks and as his compensation demand a higher return on their capital, likewise increasing the share of debt lenders demand higher inter-

est arises therefore contra-tendency—increasing the share of foreign capital pushes to increase the cost of obtaining a binding internal and external capital (Hilmar 2008).

The criterion for optimizing the financial structure is the proposal for a share of own and borrowed funds, when the cost of obtaining a binding either equity or debt capital are the lowest. Based on practical experience significant financial managers were compiled over time more funding rules which financial theory called basic funding rules. Basically it is a kind of tool in the management structure of corporate liabilities. Literature provides various breakdowns of basic funding rules. Another text refers to one of the possible breakdowns, namely breakdown of the rule of vertical financial structure and regulation of horizontal property and capital structure. The vertical financial structure based on the vertical analysis of the balance sheet. According to this rule should be equity (own equity) to the business commitment (outside capital) in a 1:1. In the corporate practice may happen despite that the company does not comply with this rule, prosper anyway. The reason is simple. The ratio of equity and debt capital depends primarily on the sectors in which it operates (Ďurišová & Jacková 2007).

The rule of horizontal property and capital structure has two variants. The first is the golden rule of financing a second golden balance rule. According to the golden rule of financing should be individual assets of the company financed from such sources which will be available to businesses throughout the useful life of certain assets. The ratio of long-term capital and long-term sequestration of assets should be greater than or equal to one. Otherwise, part of a long-term committed asset is financed by short-term capital, which can lead to paralysis of the financial health of the company. According to the golden rule of balance sheet should be the long-term assets financed by equity of the company. From a broader perspective the long-term assets should be financed by long-term commitment of financial resources which means that long-term assets can be financed by foreign sources also long-term business. From the widest perspective of the golden rule of balance sheet should non-current assets including long-term sequestration of current assets include long-term capital.

Based on the financial practice was defined other financing rules, such as gold proportional rule, Rule 1:1 or Rule 2:1. According the golden, for example "pari" rule, should be long-term assets financed by equity which is rather conservative approach. According to the golden rule of the ratio, the pace of investment growth did not outstrip growth in revenue, even in the short term (Chodasová & Tekulová 2014). Under Rule 1:1 the sum of treasury bills and debts should not be lower than the amount of short-term foreign capital. Rule 2:1 says that the amount of liquid assets should double the amount of short-term foreign capital.

The task of controlling is to create a comprehensive system of information about costs and performance, economic indicators, which should not only evaluate the development of the company as a whole, but to look at the operation of the partial views that are critical in terms of management. The basic mission of controlling capital structure will be to solve fundamental questions of the cost of capital, the optimal structure of own and external resources from the perspective of financial indicators, as well as funding rules and the use of capital. As mentioned above, the mission of the capital structure controlling is the assessment of the current financial structure, debt levels of the company and its ability to cope with this indebtedness. As the underlying basis for evaluating developments and propose measures we used indicators of financial debt that are designed to measure the financial risk of the company. It can be shown that the increased use of credit financing increase return on equity as far as the return on assets is far higher than the interest rate of the loan. But heavily loaded companies are also sensitive to the drop in business activity and therefore there is a higher probability of not being able to fulfil their contractual and non-contractual obligations, as with businesses that have a lower ratio of debt to equity. Moreover, since shareholders receive only the residual cash flow, return on equity riskiness increases with financial debts.

Financial analysts usually use two approaches to measuring the financial debt. The first approach is to create ratios of balance sheet items, which determines to what extent the company uses loans to finance their activities. The second approach is to use data from the profit and loss account for the calculation of coverage indicators that measure the company's ability to repay loans. Indicators of the balance sheet are mainly Total debt assets and Debt to equity. The indicator of coverage include coverage of interest costs, fixed payments coverage and cash flow coverage (Ďurišová 2012).

There are several reasons why the company prefers equity against liabilities (foreign funding sources), respectively, on the contrary. Some influence on the choice of forms of financing has mainly legal forms of business, size and profitability of the company, the sector in which it operates, but also macroeconomic environment, legal framework and business credit policy. Not less important factor is the quality of management processes, quality of managers, quality of information system, control over the company, quality of assets (mainly long-term) or organizational culture. In countries where is capital market developed businesses can use the initial subscription of shares on the stock market to raise capital for the issue of additional shares or issue debt securities (Tokarčíková 2011).

The optimal capital structure is influenced by the cost of capital which represents the expenditure for the company needed to obtain different forms of capital, respectively, rate of return required by investors investing their money in the company. Till examining the cost of capital we are distinguished:

- The cost of different types of company capital
- The average cost of the total company capital
- Optimal indebtedness.

The optimal capital structure is determined by the minimum cost of the total capital in the company while the total cost can be expressed by the following formula:

$$TCc = Cbc \times (1 - Tr) \times TCe/TC + Occ \times Oc/TC \ (3)$$

where:

TCc—total cost of capital,
Cbc—cost of borrowed capital,
Tr—rate of income tax,
TCe—esternal capital in total,
TC—total capital,
Occ—own capital costs,
OC—own capital.

Optimal financial structure of the company significantly determines financial security (Certain Financial) given by the expected liquidity, financial freedom (financial discretion) expressing such a company's ability to repay the commitments in which the company can exploit investment opportunities in any form of financing, inflation (inflation) and the situation on the financial market (Tekulová & Chodasová 2012).

Among the factors influencing capital structure is mainly the commercial risk associated with the private existence of company, corporate tax position (deductibility of the interest burden from the tax base of income tax), financial flexibility and last but not least, managerial conservatism and aggressiveness.

4 MODIFICATIONS OF THE CAPITAL STRUCTURE CONTROLLING OF COMPANIES IN THE AUTOMOTIVE INDUSTRY

The automotive industry becomes very dynamically developing sector in Slovakia and the driving force behind the economic development of the country. Extensive investment into the sector should result in Slovakia's top position in the car production. However, if as a country we want to maintain a competitive position in the global market, it is necessary to capture the global trends in this area and apply measures that follow from them (Kucharčiková 2011).

The article maps the situation in the automotive industry at global, regional and national level and aims to make recommendations for the management of Slovak companies—suppliers of the automotive industry. Despite a relatively small market, Slovakia is becoming one of the main automobile production centres in Central and Eastern Europe. Automotive industry directly employs more than 57,000 SR inhabitants and indirectly can go up to five times the number of employees. Production of VW, PSA and KIA should soon climb to 800,000 units a year, of which should go 90% for export. This level of dependence on exports is not unprecedented in the world industry.

Thus Slovakia could gain leadership in automotive production. An investment of Volkswagen, PSA, Peugeot, Citroen and KIA Motors in Western Slovakia attracted and still attracts a network component supplier. Answering the question of what is attractive in Slovakia also raise a significant investment in the automotive industry among others, the geographical location in Central Europe with a combined market potential of over 350 million people. Another factor should be the availability of skilled and relatively cheap labour. At the same level of productivity of labour costs in Slovakia are about 30% lower than in the Czech Republic, Hungary and Poland, and 6.5 times lower than in most Western European countries.

Among the Slovakians pros we can be attributed a favourable environment for investment and automobile production. The Slovak government has proceeded very proactive in providing infrastructure solutions and attractive incentives. This approach combined with convincing reforms has helped secure these prestigious projects. The Slovakia's economy has shown clear focus on development activities, industrial production and it can be assumed that this favourable business environment for companies in the automotive industry should continue.

Given the above favourable industry circumstances the question is how the selected firms processes. The Capital Structure Controlling and if selected benefits provided by the government affecting the company's financial management in the automotive industry. On the basis of data from the annual report 2013 of one of the automotive companies in Table 1, following results confirm the established fact that overall average total cost of capital is 27.26%, in the overall perception of the business environment is needed to compare the results to the average value sectors.

In any case the cost of foreign capital represents 1.41%, which is abnormally low percentage of the

Table 1. Results one of the automotive companies.

Indicator	Acronym	Result
The costs of equity	Occ (%)	20,33
Total capital costs	Tcc (%)	27,26
The cost of external capital	Cbc (%)	1,41
Total indebtedness	(%)	69,03

Source: Own calculations, The Annual Report of the Company 2013 (Becker 2013).

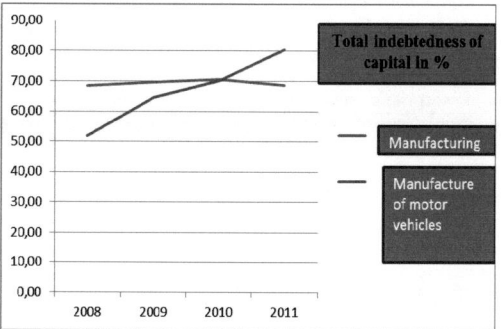

Figure 1. Total indebtedness of capital in %.

load, unrealistic number in the current business environment from the position of the candidate in obtaining external sources of capital, which demonstrates the impact of benefits from the state and not the same business conditions in an environment of financial markets.

The total debt of the company is 69.03%, Figure 1, which is on the border of the recommended values of debt compared to Segment environment of industrial production and vehicle production, development follows the overall debt situation in sectoral environment. The listed situation is caused by high capital intensity of the industry. Optimal capital structure affected by cost of capital which represents the expenditure for the enterprise to obtain different forms of capital, respectively rate of return required by investors investing their money into the company, is determined by the minimum cost of the total capital in the company (SCBs.r.o.2011).

5 CONCLUSIONS

The current economic situation motivates the companies regardless of the type of activity to the most effective financial management in order to ensure liquidity. This is due to a contraction in demand coupled with the instability of material and financial flows. Permanent and stable liquidity is essential for every business. If this condition is not satisfied it is likely that in the near future would the company gets into a trouble associated with the inability to pay obligations, which may lead not only to loss of competitiveness and reputation but also to its existence problems. Insolvency fact brings a loss of suppliers, banks and even customers confidence. Therefore is necessary the continuous provision of liquidity for the company.

The aim of optimal provision of liquidity is to have a volume of funds that the company is able to fulfil its obligations and also not to bind surplus funds which in turn represent a loss of potential opportunities. Business finance is a set of financial relationships routed to the interior and to the surroundings of the company. The purpose of these relationships is:

• effective use of potential resources and their rational investment,
• effective sourcing from the surrounding economic area and their use in achieving the objects of the company,
• rational redistribution of created profit with the needs of enterprise development,
• eliminate business risks and potential fluctuations.

Corporate finance is a system of monetary relations to which the enterprise enters in obtaining of financial resources, allocating them at a bound in individual asset, by productive use of assets and by the distribution of achievements. Reliable statement about the financial situation can be obtained by analyzing its ability to pay its obligations—liquidity. It is obvious that a financially stable company is able to meet their obligations and company in difficulties have problems with it, so the managers must realize that the quality of financial management reduces business risk, so it is important to create, monitor and evaluate the individual variables that helps to eliminate this.

The task of controlling is to create a comprehensive system of information about costs and performance, economic indicators, which should not only evaluate the development of the company as a whole but to look at the operation of the partial views that are critical in terms of management. The basic mission of controlling of the capital structure will be to meet the basic questions about the cost of capital, the optimal structure of own and external resources from the perspective of financial indicators, as well as funding rules of the capital use.

This contribution was wrought with the support of VEGA č. 1/0335/13.

REFERENCES

Becker H. 2013. Správa o konjunktúre Inštitútu pre hospodársku analýzu a komunikáciu 19. februára 2013. Bratislava: p. 8.

Chodasová, Z. 2005 Asset of Modern Management Methods in Development Process, *Vadyba management*, Univerzity Vilnius Nr. 1 (9): 13–20.

Chodasová, Z. 2012. *Podnikový controlling—nástroj manažmentu.* Bratislava: STATIS.

Chodasová, Z. & Tekulová, Z. 2014: Monitoring of indicators of competitiveness to improve the business strategy. In *Kontrolling na malych i srednich predprijatijach: sbornik naučnych trudov 4. meždunarodnogo kongressa po kontrollingu. Praga, Česká republika, 25 aprelja 2014. NP,* Moskva: Obedinenie kontrollerov.

Ďurišová, M. 2012. Process model of managerial functions applied in business performance. *Economics and management spectrum* no. 1(2)/ VI. Žilina Faculty of Operation and Economics of Transport and Communications, University of Žilina: 7–12.

Ďurišová, M. & Jacková, A. 2007. *Podnikové financie.* Žilina: Vydavateľstvo EDIS.

Hilmar J.V. 2008. *Controlling a new management tool,* Praha: Profess.

Horváth & Partnes. 2004. *Nová koncepce controllingu.* Praha: Profess Consuling.

Kucharčiková, A. 2011. *Efektivní výroba.* Brno: Computer Press.

Ručková, P. 2011. Finanční analýza—metody, ukazatele, využití v praxi. Praha: GRADA Publishing.

SCB s.r.o.2011. Stredné hodnoty finančných ukazovateľov ekonomických činností v SR za rok 2011, Bratislava: SCBs.r.o.

Tekulová, Z. & Chodasová, Z. 2012. Controlling the capital structure in a company. In: *Scientific proceedings 2012: Faculty of Mechanical Engineering, STU in Bratislava.* Bratislava:ed. Slovak University of Technology.

Tokarčíková, E. 2011. Influence of social networking for enterprise's activities. *Periodica polytechnica Office. Social and Management Sciences* 19 (1), Hungary: 37–4.

Production Management and Engineering Sciences – Majerník, Daneshjo & Bosák (Eds)
© 2016 Taylor & Francis Group, London, ISBN: 978-1-138-02856-2

Selection of adequate environmental policy tools—case study of production company

J. Chovancová
Department of Environmental Management, Faculty of Management, University of Presov, Presov, Slovak Republic

J. Čorba & M. Tomčíková
Faculty of Business Economics, University of Economics, Bratislava, Kosice, Slovak Republic

ABSTRACT: Voluntary tools of environmental policy present a set of activities primarily used by business entities to reduce their negative impact on the environment. These tools provide systemic approach to environmental issues and their implementation can bring many advantages such as cost reduction, competitiveness, enhancing innovation, creating green image etc. Nowadays companies in all economic sectors and areas can chose from the wide range of voluntary tools which can help to improve the environmental performance of the company. Though, decision making process in selection of adequate environmental policy tool requires consistent analysis of major environmental issues of the company, setting criteria and specific demands of the company and expected benefits related to particular environmental policy tool implementation. Paper presents the possible approach to selection of environmental policy tools, which responds to specific demands of the company. Proposed approach is verified on the case study of the production company.

1 INTRODUCTION

The current world of innovative technologies, growing demands of the market and competitive race, got companies into a position where, without forcing, all available tools to prevent pollution of the individual components of the environment are activated, reduce waste production, eliminate negative aspects and promote production and technologies with positive environmental impact (Annandale et al. 2004). Attitude of companies from various sectors of business to environmental protection does not longer appear as barrier to business development, but conversely. More and more businesses that want to be competitive in the domestic and global market care about reduction of material and energy costs and they are actively involved in the management of its environmental aspects.

At present, the voluntary tools for improving environmental performance and their application, especially in small and medium-sized enterprises, is increasingly discussed topic (Chovancová & Rusko 2010, Ramus 2002, Zorpas 2010). Voluntary tools, as opposed to standardized environmental management systems, represent simpler, less formal and less difficult approach to environmental protection (Ayuso 2006, Chovancová & Hudáková 2009). Their advantage is lower financial and administrative burden (Paulíková & Kopilčáková 2007),

which pre-determinates them for implementation especially in small and medium-sized enterprises.

The presented contribution, based on the case study, provides guidance which companies can follow when selecting an instrument of environmental policy, taking into account the specific needs of the company and the requirements for the environmental management tool. By the analysis and suggestion of an appropriate tool, we were supported by knowledge gained from expert sources, internal information sources of Nexis Fibers, Inc., brainstorming, benchmarking conducted with company representatives and spider chart by which we graphically evaluated the results of analysis.

2 MATERIALS AND METHODS

The diversity of activities of individual companies implies that they have different environmental problems and therefore approach to solving them is different, too. This of course considerably complicates their unification in terms of environmental protection, particularly in industrial or manufacturing enterprises. These are generally focused on very different activities, but they also have many common characteristics.

Manufacturing enterprises are major polluters of the environment (Veber et al. 2006). In addressing their environmental problems, they must overcome a series of obstacles mainly related to lack of funds and expertise. The EU is trying to make all the activities and strategies consistent with the requirements of environmental protection and sustainable development, and therefore also invests significant resources in an effort to help enterprises to apply a range of environmental management tools, such as environmental audits, environmental assessment and products labelling, the environmental performance evaluation etc. These tools were created mostly as voluntary internal initiatives of business and organizations. They are already affecting the production and distribution of products in the European Union and in other countries (Rusko et al. 2007). However, it is understandable that businesses around the world carefully consider not only the financial benefits of this orientation (identification of savings, increase the efficiency of processes and activities, new market opportunities, etc.), but also assess the risks arising from the defensive behavior towards environment (accidents, inability to obtain a bank loan etc.) (Ramus 2002, Masanet-Llodra 2006).

Environmental management tools may have more forms which can be divided into formal and less formal tools. Formal ones are represented by two internationally recognized forms: European EMAS (Eco-Management and Audit Scheme) and the international ISO 14 001 (Majerník et al. 2009). In addition, however, there are more types of less formal environmental management tools used in Europe and worldwide. It is difficult to state that any of these forms would be one and only adequate for all companies. Less formal tools of environmental management can be considered as a first step towards the introduction of a formal EMS (Kollár & Brokeš 2005). It should be noted that EMS is the only means to achieve the goal and not a goal itself.

For improving the environmental performance of the business sector, a number of other active voluntary tools in different countries and sectors have been developed (Pauliková & Kopilčáková 2007). These tools are oriented to a specific business activity (products, services) and are tailored to specific areas and technical conditions, which in many cases increases their effectiveness. They are applied under specific conditions, either by choice at the enterprise level, or in some cases, for example on the basis of agreements at sectoral level, or as agreement with local or regional government. Most of them can be applied without being certified or do not require the existence of environmental management system according to the ISO standard or EMAS. They may therefore be a precursor or functional training for future introduction of EMS. They can be used also for companies with EMS in order to improve the already implemented management system.

Literature offers us a broad overview of voluntary tools of environmental management, their categorization and detailed characterization (Chovancová & Hudáková 2009). In this broad spectrum, however, for the purposes of this paper we focus on only three of them, which we selected as the most suitable for application in our case company—Nexis Fibers, Inc.

2.1 Environmental Management System (EMS)

Environmental Management System (EMS) is the best known and widely used tool, suitable for all types of organizations allowing integrating environmental issues into their activities. ISO 14001, which is a key document in the implementation of the environmental management system defines this as "part of the overall management system that includes organizational structure, planning activities, responsibilities, practices, procedures, processes and resources for the preparation, implementation, achieving, reviewing and maintaining environmental policy" (ISO 2004). It serves as a protective factor against the negative impacts on the environment, and promotes activities aimed at enhancing environmental quality (Kucko 2002, Majerník et al. 2009, Morris 2004). Environmental management systems can be implemented according to ISO 14001, or by the European scheme of environmental management and audit scheme—EMAS.

2.2 Environmental Performance Evaluation (EPE)

The Environmental Performance Evaluation (EPE) is a voluntary tool which can be implemented in companies with certified environmental management system as well as those that they do not have the system certified (Chovancová & Rusko 2010). This tool works with relevant, verifiable and quantifiable information and enables management of the company to monitor and improve the environmental performance of the company. The key document is the standard EN ISO 14031 and its supplementary document ISO 14032, refering the specific examples of the application of environmental indicators in business practice (Majerník et al. 2009).

2.3 Bluesign

Bluesign can be included among the relatively new voluntary tool of environmental management. It is independent standard for the textile industry, which supports manufacturing processes focused on maximum productivity of resources while respecting the environmental protection and occupational health and safety. It provides confidence for all cells involved in the textile chain through suppliers, manufacturers to consumers that the quality of products and services managed by "intelligent control system of input stream" complies with the applicable regulations and limits of this standard. Best Available Technology (BAT) is a term defined by European Community Directive no. 96/61/EC and precisely this Regulation is adopted by the bluesign® standard, which guarantees that textile products are made from the resources and components with the lowest possible content of hazardous substances.

Standard Bluesign stands on five pillars, or principles:

– Resource productivity,
– Consumer safety,
– Air emission,
– Water emission,
– Occupational health & safety. (Bluesign 2010).

Bluesign standard gained international recognition and thanks to its strict requirements and global business and legal measures facilitate global trade flows (Ferencová 2010) and has a lead over environmental labeling of products it textile industry.

Other voluntary tools of environmental management also include environmental benchmarking, environmental management accounting, environmental reporting, Ecodesign, Life Cycle Analysis, Ecolabeling, etc.

3 CASE STUDY RESEARCH

According to Yin (1994) the case study is one of several ways of doing social science research. The findings of case study research, like experiments, are able to be generalised to theoretical propositions and not to population or universes, because with this methodology the aim is to generalise theories (analytic generalisation) and not to enumerate frequencies (statistical generalisation.

The criteria for choosing our case study research' firm were related to three main issues:

– the company does not have certificated environmental management system but is aware of its environmental impacts and wants to improve its environmental performance via suitable tool of environmental policy,
– due to the fact that case study research is a interactive process, it is necessary to detect motivation by the company to participate in it, as research demands several fields visits, which imply many time dedication from the participating company,
– company should consider that environmental disclosure means a real commitment with its stakeholders or interested parties; therefore they are keen to disclose environmental information.

For our case study research, Nexis Fibers Inc. was selected. The headquarter of the company is located in Emmenbrücke, Switzerland. The company is a major global player in the market for industrial fibers. The company designs, develops and delivers a wide range of its product portfolio to customers in Europe and around the world. Key products of the company are airbags, tires and MRG, industrial fibers for ropes and nets including polyamide polymers 6.66 and 6.10. In addition to Switzerland parent company, there are manufacturing plants also in Latvia and Slovakia.

In the next part of this paper we will focus on the issue of voluntary tools of environmental management in the Slovak manufacturing plant of this company.

The production program is focused on fibers intended for the manufacture of tires, rubber-coated products, ropes, nets, fabrics for the manufacture of airbags and technical fabrics. The actual production portfolio program pre-determinates the pro-export orientation, mainly to the markets of Western Europe.

As the core business of company is in automotive industry, it follows the trend of quality management system certified according to technical specification ISO TS 16949, designed for companies working in the automotive industry.

4 RESULTS AND DISCUSSION

Suitable methodology for the selection of voluntary tool of environmental policy can be offered by spider analysis. Spider analysis allows quick and transparent assessment of the status of the monitored voluntary tools of environmental policy through the selected system of parameters. The principle of this method is to construct the spider chart usually on the basis of a number of indicators. Graph is created by concentric circles, where degree of compliance with the established requirements for the voluntary tool of environmental policy is applies.

For the needs of individual organizations wishing to use spider analysis in decision making process of selection of voluntary tool of environmental policy, it is necessary to modify the selection of indicators so they allow easy assessment of the compliance of voluntary tools with specifics requirements of the company.

Spider chart at first glance gives an idea of evaluated voluntary tools. Decision of the enterprise using spider analysis is simple. If the "spike" of circles are beyond average values, it is the tool that best meets the company's requirements and vice versa. When comparing two or more instruments, determining is surface area of the spider chart.

Based on the product portfolio, active management of the company Nexis Fibers inc., and its current approach to environmental issues, we choose these voluntary tools of environmental policy: Environmental management system, Environmental performance evaluation and Bluesign, described above.

These three tools were selected after initial diagnosis of the company, and brainstorming with representatives of the company. From this point of view, these tools give a presumption of effective improvement of environmental performance, and consequently enhance competitiveness of the company and create new business opportunities and new markets. With the help of acquired company's data and knowledge from expert sources we present decisive criteria for the selection of the most appropriate voluntary tools and through benchmarking we assess their suitability for improving environmental performance of Nexis Fibers. Spider cart provide fast spatial comparison of the data obtained.

Each of the presented tools can bring some benefits in the road of continuous improvement of environmental performance of the company. In order to select the most appropriate instrument we established criteria set by company and using benchmarking we evaluate the relevance of voluntary instruments for Nexis Fibers Inc.

The scale for criteria assessment is from 1 to 5, wherein:

1. unsatisfactory level, significant deficiencies,
2. needs to improve,
3. meets requirements
4. exceeds the greater part of the requirements,
5. significantly exceeds all requirements.

It is immediately apparent from the spider chart that assessed voluntary tools are balanced, but closer examination of the individual evaluation criteria indicates some differences. The biggest difference is reflected in the environmental policy where EMS compared to other instruments—EPE and Bluesign meets the most of criteria set for comprehensive environmental policy formulation and implementation. We can conclude that all three voluntary tools meet requirements of our case company and reflect to particular needs of environmental management. Though, based on results of spider analysis, it seems that EMS meets the widest spectrum of these requirements.

Finally, even though our case fulfils the requirements for being considered as a rigorous case study research we are aware of the limitations of such methodology as possible bias in findings due to the researcher perceptions and values, the difficulty in designing case study research and the self-election of the participating company. Furthermore, this methodology enables to generalise findings to theory but not to the whole population.

Table 1. Evaluation criteria of selected voluntary tools.

Evaluation criteria	EMS	EPE	Bluesign
Environmental policy (formulation of environmental impacts of company's activities and products, determination of environmental objectives, implementation of environmental policy of the company)	4	1	1
Environmental aspects (defining the most significant environmental aspects)	4	4	3
Compliance with the environmental legislative	4	4	3
Reducing the use of hazardous substances in the production process (homologation of chemical components with national and international consumer safety and environmental measures, regulations for Occupational Health and Safety, the EU REACH Regulation No.1907/2006, SVHC—Substances of Very High Concern, ETAD—the Ecological and toxicological Association of dyes and organic pigments manufacturers; licenses, permits, limits on permissible quantities)	3	2	4
Environmental protection, occupational health and safety (processes affecting the environment, eliminating the risk of environmental accidents and ecological resilience, emergency plans, working conditions—noise, dust, ergonomic, training, employee care)	3	2	3
Comprehensive, easy and compatible system structure	4	4	3
Control and Corrective Action (periodical control and evaluation of the environmental objectives and tasks, measurement, monitoring of environmental aspects and activities with significant effects on the environment, corrective and preventive action, documentation and records, periodical internal audits)	4	4	4

(Continued)

Table 1. (*Continued*).

Evaluation criteria	EMS	EPE	Bluesign
Continuous improvement of the environmental performance (innovations and technological efficiency, financial, raw material and resource saving)	3	4	3
Employees and stakeholders engagement	4	2	3
Competitiveness and strengthening of Marketing (acquisition of new business partners, business opportunities, increase product competitiveness, public relations, image of the company and products, credibility towards investors, banks, insurance companies, responsibility towards environment, customers, clients, the public)	3	2	3
Customers (responding to the environmental requirements of customers)	3	2	3
Overall Rating (mean)	3,5	2,8	3,0

5 CONCLUSION

Environmental protection is increasingly entering into a relatively fixed and stable organizational structure of the management of enterprises. Rapidly changing and tightening legislation in the field of environmental protection and management, and the consequent threat of economic sanctions and loss of image, increase motivation for undertaking positive changes in relation to the environment.

The experience from the implementation of EMS in countries with developed market economies clearly confirm the growing interest of customers in suppliers with certified environmental management systems, respectively consumers interested in environmentally friendly products, even for possibly higher price. In the near future, at national as well as international level, more and more enterprises voluntarily (such as the need to ensure the prosperity) approach to building formal and less formal tools for improving environmental performance.

Based on the above result, companies can gain many advantages resulting from the implementation of voluntary tools of environmental management—direct (quantifiable and non-quantifiable) as well as indirect (quantifiable, non-quantifiable).

Following the results of data analysis of the corporate practice of Nexis Fibers, Inc. and theoretical framework, we conclude that voluntary instruments EMS, EPE and Bluesign by their content and structure have potential to support the efforts already made in undertaking to continuous improvement of environmental performance. Individual tools were assessed in order to compliance

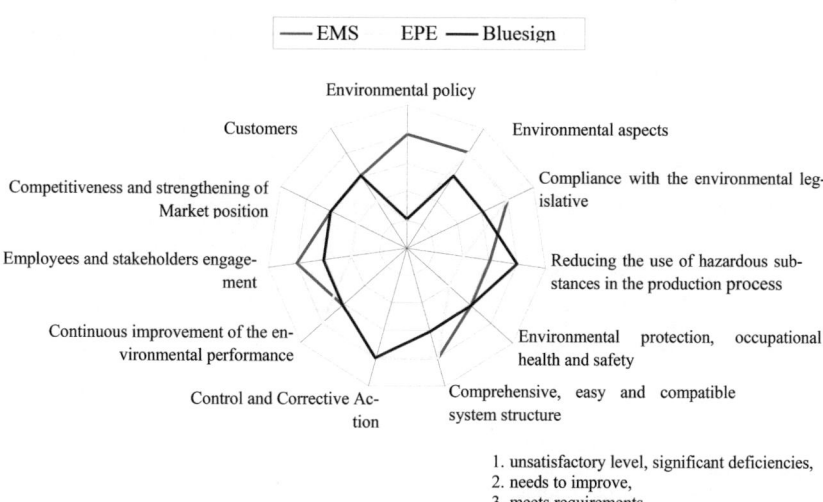

1. unsatisfactory level, significant deficiencies,
2. needs to improve,
3. meets requirements
4. exceeds the greater part of the requirements,
5. significantly exceeds all requirements.

Figure 1. Spider chart—evaluation criteria of selected voluntary tools.

with expectations and needs of Nexis Fibers, Inc. and we conducted benchmarking and spider analysis to determine the most accurate voluntary tool.

All three instruments—EMS, EPE and Bluesign based on selected evaluation criteria showed a balanced eco-efficiency for business practices of Nexis Fibers Inc. We emphasize and draw the attention to EMS and EPE, which can lead the company towards complex improvement of current environmental performance. These tools with their content and wide scope complement each other. Not omit the effectiveness of Bluesign, which offers businesses the opportunity to increase the competitiveness and flexibility to respond to the environmental requirements of customers.

The presented decision-making process may be applicable in a similar way in other companies seeking ways to improve their environmental performance.

So how can we achieve an increase of interest and the consequently also implementation of voluntary tools of environmental management in business practice? This obviously requires a fairly extensive and comprehensive program covering a wide range of factors and actors (Huttmanová 2008). One important component should include adequate information campaign, the aim of which would achieve at least a basic level of environmental awareness of the greatest number of stakeholders and generating interest in the activities in environmental field by the business community.

ACKNOWLEDGEMENTS

The study was supported by KEGA 032PU-4/2014.

REFERENCES

Annandale, D., Morrison-Saunders, A. & Bouma, G. 2004. The impact of voluntary environmental protection Instruments on company environmental performance Business Strategy and the Environment. Business Strategy Environment. vol. 13, pages 1–12. Published online in Wiley InterScience (www.interscience.wiley.com)

Ayuso, S. 2006. Adoption of voluntary environmental tools for sustainable tourism: analysing the experience of Spanish hotels. In. Corporate Social Responsibility and Environmental Management. Volume 13, Issue 4, pages 207–220.

Blusign. 2010. independent standard for the textile industry. St.Gallen: Environment, Health & Safety, Bluesign technologies EHS commitments, Benefit from a unique standard!, Look inside the bluesign® standard, (2010) [17.2.2015]. available online: http://www.bluesign.com.

Chovancová, J. & Hudáková, Z. 2009. Systémy environmentálneho manažérstva a ich uplatnenie v podnikovej praxi na Slovensku /.—In: Manažérstvo životného prostredia 2009 [elektronický zdroj]: zborník príspevkov z 9. konferencie so zahraničnou účasťou; Bratislava 10.-11. december 2009.—Žilina: Strix, 2009.—ISBN 978-80-89281-56-5.—S. 52-57.

Chovancová, J. & Rusko, M. 2010. Environmentálne správanie malých a stredných podnikov na Slovensku [Environmental performance of the small and medium sized enterprises in Slovakia].—In: Manažérstvo životného prostredia: zborník príspevkov z 10. konferencie so zahraničnou účasťou; Bratislava 16.-17. december 2010.—Žilina: Strix, 2010.—ISBN 978-80-89281-67-1.—S. 35-39.

Ferencová, M. 2010. Modern Marketing and Value as One of Its Key Terms. In: Ministrare. Tom I. Warszawa: Katedra Filozofii Wyźsej Finansóv i Zarzadzania w Warszawie, 2010. P. 231–235. ISBN 978-83-61087-88-5.

Huttmanová, E. 2008. Budovanie znalostnej spoločnosti a jej význam pre zvyšovanie národnej a regionálnej konkurencieschopnosti v problémových oblastiach Slovenska. In Identifikácia zmien zložiek životného prostredia problémových oblastí východného Slovenska. Zborník z medzinárodnej vedeckej konferencie. Nitra: Slovenská poľnohospodárska univerzita, 2008. s. 82–87. ISBN 978-80-552-0087-3.

ISO (International Organization for Standardization). 2004. ISO 14001 Environmental Management Systems: Specification with Guidance for Use.

Kollár V. & Brokeš P. 2005. The Barriers for Implementation of Environmental Management System to Small and Medium Enterprises and Possibilities how to Overcome them, Business review, volume IV., Number 7.

Kucko, L. 2002. Minimum o systémoch environmentálneho manažérstva. Bratislava: ASPEK, 2002. 18 s.

Majerník, M., Chovancová, J. & Hodolič, J. 2009. Environmentálne manažérske systémy. Skalica: Stredoeurópska vysoká škola v Skalici, 2009. ISBN 978-80-89391-05-9.

Masanet-Llodra, M.J. 2006. Environmental Management Accounting: A Case Study Research on Innovative Strategy, Journal of Business Ethics, Vol. 68, pp. 393–408.

Morris, A. S. 2004. ISO 14000 Environmental Management Standards: Engineering and Financial Aspects. John Wiley & Sons Ltd. 2004. ISBN: 0-470-85128-7.

Pauliková, A. & Kopilčáková, L. 2007. Hodnotenie faktorov pracovného prostredia priemyselnej prevádzky z hľadiska systémovej dynamiky. In: RUSKO, M. – BALOG, K., [Eds.] 2007. Manažérstvo životného prostredia 2007.— Zborník z konferencie so zahraničnou účasťou konanej 5.-6.1.2007—Jaslovské Bohunice.—Žilina: Strix et VeV, Prvé vydanie. ISBN 978-80-89281-18-3.

Ramus, C.A. 2002. Encouraging innovative environmental actions: What companies and managers must do. Journal of World Business, 37, pp. 151–164.

Rusko, M., Chovancová, J. & Duchoň, M. 2007. Spectrum of voluntary tools used in application of environmental policy in organization's practice.—In: Machines, technologies, materials: international journal.—ISSN 1313-0226.—is. 6-7, p. 28–31.

Veber, J. et al. 2006. Managment kvality, environmentu a bezpečnosti práce. Legislatíva, systémy, metódy, praxe. Managment Press, Praha, 2006. 358 s. ISBN 80-7261-146-1.

Yin, R.K. 1994. Case Study Research. Design and Methods. Applied Social Research Methods Series 5, Sage Publications.

Zorpas, A. 2010. Environmental management systems as sustainable tools in the way of life for the SMEs and VSMEs. Bioresource Technology 101, 1544–1557.

Production Management and Engineering Sciences – Majerník, Daneshjo & Bosák (Eds)
© 2016 Taylor & Francis Group, London, ISBN: 978-1-138-02856-2

Use of management accounting in business management

A. Jacková

Faculty of Management Science and Informatics, University of Zilina, Zilina, Slovak Republic

ABSTRACT: In market economics it's necessary to control the subject and value aspects of processes in a company. We can judge whether a company will thrive amongst its competition, remain on a level of simple reproduction or lead to an unwanted demise. A successful company needs a good accountant, who can provide management with objective information on the companies progress and economic development. Management accounting is created for the purpose of management, which represents its main purpose. It gives managers and leading employees a better understanding of the economic operations in their company and helps them to make the correct assessments from account records.

1   INTRODUCTION

In order to successfully manage a company, executive managers need to evaluate complex information about a companies activities, which are given to them by the financial accounting.

"Company accounting focuses on the ad valorem aspects of economic functions in a company and records its quantitative progress. In accounting, economic functions are measured and expressed in relation to a specific company. As a result of this accounting is mainly a business activity. It was created and developed as a tool for company management. Its information may be the most detailed, but not sufficient. That is the reason we see the term managerial accounting used in practice as it incorporates a range of complex information needed for the efficient work of managers. Managerial accounting is one aspect of accounting, which gathers data about economic occurrences, their analysis and processing into a form useful for management in the economic management" (Tumpach 2008).

However, the success of a company depends not only on the field in which it does business, but also on the tools and methods it utilizes.

2   DEFINING MANAGERIAL ACCOUNTING

Managerial accounting is created for the purpose of management and should first and foremost serve the administration in decision-making regarding functions and future development of the company. "Compared to financial and tax accounting, where users require a unified output and comparability of accounting information, the information used by managers is not subject to outside regulation. This accounting subsystem is not sharply defined regarding its structure nor does it have a globally accepted term. In the Anglo-Saxon region, this subsystem is commonly referred to as Management Accounting or Managerial Accountancy. French speaking countries identify it as accounting for administration (Compabilité de Gestion) and in latest German literature the name accounting of expenses and earnings aimed at decision-making is used (Entscheidungsorientierte Kosten—und Leistungsrechnung)" (Král 2010).

Before World War II and closely after analog goals on a low level were monitored by department company economics and company accounting. However their development under a centrally planned management was limited by legislative changes, which resulted in a separation of accounting subsystems and budget calculations followed by a subsidization to nation-wide economics regulation of industry and scholar relations. Not even current adaptations in organizational accounting for managers abet a unification of the Anglo-Saxon, German and French comprehension. It merely reduces the mission of organizations accounting to a narrow field of information, which it offers for the needs of financial accounting.

Managerial accounting in terms or company management affects strategy development, planning and control functions, decision-making, optimization of assets, presentation of main indicators to company employees and the protection and development of resources. A requirement for efficient functioning of managerial accounting in a company is the existence of defined SMART goals (Specific, Measurable, Assignable, Realistic, Time-related), strategic and operative planning, system of asset and financial management, downward revision, measurable goal achieving and a system of managerial audit. Thus a managerial accountant has to focus on the future of a company using analysis, predictive models and outputs of financial accounting and make recommendations for the upper levels of management so that a company can achieve its goals. To this purpose he devises complex analysis such as SWOT, in which he compares factors like development of costs, future transactions, inflation development, interest rates, social contingencies, political developments, environmental hazards as well as market analysis and an analysis of competitors. Based on SWOT analysis a companies management can then make decisions about strategies, values, company mission and modify company management to embrace these.

3 THE POSITION OF MANAGERIAL ACCOUNTING IN A COMPANIES ACCOUNTING INFORMATIONAL SYSTEM

Accounting informational system is an important part tool in the framework of managing a company as a whole, its individual systems and managing processes. In terms of market economies it's divided into financial and managerial accounting. Managerial accounting is treated as a subsystem of information systems and merges different information that is important for effective conduct of a company. Management accounting involves cost accounting, calculations, budgeting, company accounting, tax accounting and environmental management accounting.

3.1 Cost accounting

A detailed framework of managerial accounting is based on the type of information it provides to managing employees. The main defining feature is the relation between information and the stage of decision-making process it relates to. This relation was typical for the development of management accounting. In the first stage, accounting with a this focus looked mainly at the costs and the relating incomes. In the second stage, costs divided in this fashion could be compared to budgeting or the calculated balance. The result was a foundation for short to middle term decision-making based on the digressions. Based on its content, cost accounting was traditionally conceived as production accounting and responsibility accounting. The main goal of production accounting, in close relation to calculating production is to answer the question: what are the costs, profits, gross margin and other characteristics of company products, work and services. Responsibility accounting in relation to planning systems, budget and company prices

monitors the answers to questions: how do individuals company bodies contribute to company goals, how to manage company bodies to have their functions lead to optimal fulfillment of company goals as a whole. Under entrepreneurial conditions and in relation to significant changes cost accounting saw new developments, also defined as Activity Based Accounting, which main goal was to provide a framework for the management of entrepreneurial process and its individual subsystems.

"The basic endeavor of management accounting in the third stage, which is accounting for decisions, is to broaden the amount of provided information to include information, with the help of which it will be possible to judge different options of future company development, answering the question "What if...?", This enables the use of accounting information not only for basic management processes, in which the basic parameters were already decided, but also to decide on future procedures. The third stage is marks the transitioning point from cost accounting to managerial accounting" (Král 2010).

3.2 Calculations

The basic purpose of calculations is to manage the economic efficiency in terms of products. Effective company management requires a calculating system. A company calculating system in terms of function and period consists of preluseive (planned, operative, conversion) and final calculations. The relation between calculations and management accounting becomes apparent while formulating prices of company products. The prices have to be more preferable to the competition, which is seen as a basic requirement of economic success. While formulating a price, the management has to have access to information about the company costs, because based on them they can adjust the prices to the needs of the market and the customer. This information is gathered thanks to cost accounting and is the basis of management accounting.

3.3 Budgeting

It consists of economic information regarding future economic events in the company over a specific period of time. Budgeting belongs to planning information, which appoint tasks for company departments. Detailed systems of company budgeting are carefully protected by successful companies in a competitive environment. Budgeting is one of the tools in management accounting and constitutes a system of company budgets. That is based on a companies strategic goals transformed into smaller tasks, defined by economic data. Company and department budgeting is closely connected to the company calculating system and cost accounting. Amongst costs, budgeting also entails other economic categories, such as profit, income, expenses, inventory, assets and so on. The connection between budgeting and management accounting becomes apparent when formulating a budget. When appointing norms of overhead expenses based on their actual development in the past it is necessary to gather information from financial accounting. Another relation comes from the control of budgeted costs and earnings of a company and economic centers and comparing these to data from financial accounting. Estimated calculations provide data for budgeting about primary costs and in return budgeting provides information about indirect costs for the formulation of calculations.

3.4 Intercompany accounting

Intercompany accounting is the most important sectional tool of management accounting. A companies organizational structure should take into consideration the economic responsibilities of individual intercompany departments.

Responsibility accounting keeps track of the economic responsibilities of intercompany departments using costs, earnings, profits, assets and so on. Intercompany departments are in practice identified as centers, divisions, branches of enterprise. In order to efficiently manage a company it is not necessary to correctly identify intercompany departments, but to create a balance in responsibilities and privileges in such a way that intercompany departments are motivated to contribute to the prosperity of the company. Intercompany costs also have an important role for responsibility accounting. These can be defined based on costs, profits or fixed payment tariff. All the while these prices have to compatible with the overall valuing of intercompany departments (burden center, benefit center). Intercompany accounting keeps track of intercompany costs and is part of financial accounting. A company can manage its intercompany accounting based on its needs.

3.5 Tax accounting

Even tax accounting is part of management accounting. Every company manager can save his company hundreds of euro by monitoring its tax expenses. But a incorrect tax examination and the resulting sanctions can cost a company thousands. Tax accounting is conducted in such a manner to take into account the analytic record, i. e. the transformation of budgeted trading outcome before taxes to the tax base defined by law, mainly in terms of expenses regarding obtaining, securing and preserving revenue. For this reason we interpret every expense which is objectively spent for

the purpose of obtaining, securing and preserving revenue which is charged to the company accounting as tax expense. Based on these claims we can say that financial accounting mainly build the basis of taxes. Accurately conducted accounting remains the basis for correct tax evaluation and is a jumping-off point for asserting expenses in a demonstrable manner. Tax accounting can be considered as a specific approach to managing financial accounting, in which individual bookkeeping cases or balance calculations are evaluated based on their effect on taxes. Its goal is to continuously and consistently create a framework of files for the year-end tax calculations. An analytic record doesn't always entail an analytic account to a synthetic one. It can be an analytic account or any other form of analytic record, such as registration documents, supporting analytic records, accounting systems and calculations. Tax accounting encompasses entire accounting divisions concerning individual taxes—income tax, value added tax and their forms of payment.

3.6 Environmental management accounting

The foundation of environmental accounting is the environment. Its main concern is defining, valuing, evaluating and showing environmental bonds and financially relevant costs relating to the impact of the companies activities, products and services on the environment. The subject of environmental management accounting is information about substance and energy flows and information about environmental expenses.

Environmental management accounting focuses on the accounting of environmental expenses of a company and includes basic components of sustainable development, namely the environment and economy. Considering that financial accounting doesn't provide accurate information about environmental expenses a need to implement environmental management accounting into the practice of manufacturing companies has incurred. Environmental management accounting monitors and evaluates information from financial and management accounting with data concerning substance and energy flows in relation to each other with the goal of increasing the efficiency of material and energy utilization, decreasing the impact of manufacturing company activities on the environment, reducing the environmental hazards and enhancing the companies economic success.

4 MODERN METHODS OF MANAGEMENT ACCOUNTING

Global trends of building world-class companies is from the standpoint of management accounting focused on the reduction of expenses and finding means and ways to optimize expense types. Using modern methods creates space for the optimization of three basic factors: optimizing quality, costs and time.

Costs are an important economic category, which impacts the functioning of a company, which is why it is necessary for this category to be analyzed, evaluated and optimized. Optimizing expenses is an activity, which is provided by optimal expenses. In this process it is necessary to accept the influence of different factors on the level of expenses, while it is necessary to eliminate the negative influences. The process of optimization monitors complementary effects and so-called compensating effects (the effect of mirrored positive and negative influences from the progression of expenses). Optimal expenses represent the most positive, profitable and lowest costs of expenses with the acceptance of corresponding conditions, such as the level of production capacity, work organization, technical equipment, ability to provide production factors and so on. Reaching a level of optimal expenses represents a difficult process and managers are often faced with the question how to optimize the amount of expenses in a company. This growing trend in the cost department forces companies to put more focus on this problem and that is why managers are looking for various tools and methods to cut the companies expenses and thus increase the ability to compete on the market. Management accounting gives mangers the space to apply various new methods, because this form of accounting is not hindered by the law. The main problem however is the application of modern methods and approaches, as the introduction of these requires a significant time investment, a change in company policies, change of employee spirit and the motivation for constant improvements. Some companies however are too old-fashioned for these new methods. The concepts of building a world class company and the means for gaining and advantage over the competition are specific for every field and can be applied at any stage of development considering the time and, quality and costs. The main step however is the willingness to change and allow new options, new trends in the company, which can lead the company to achieving the efficiencies of world-class companies.

The pyramid of modern methods has individual approaches, i.e. methods on different levels of management, from operative to strategic. Their significance can be applied on every level of management. They are methods, which can be approached from different viewpoints and different company activities, as their benefits aren't only in the department of expense optimization, but more in the techno-logic, technical, personal, economic, social and material side of the manufacturing process. Seeing as how every method minimizes the company expenses, it can be applied in the field of management accounting. Famous and daily applied methods are amongst other Activity Based Costing (ABC), Target Costing (TC), Kaizen Costing (KC), Quality Costing (QC), Environmental Management Accounting (EMA) and Cost Controlling (CC).

Activity Based Costing (ABC)—calculation of expenses based on partial activities. The ABC method monitors the need for expenses in specific products in relation to individual company activities. The goal of this calculation is to describe activities in detail, sort them into sub-activities, processes, assigning resources to individual assets and allocating funds to these resources. The method allocates indirect costs to individual activities. The ABC method gives information on product costs, services, clients, regions, distributive channels, and so on (Stanek 2003).

Target Costing (TC)—method of managing target costs is a modern method focused on the market, which uses cost accounting as a means of defining the target costs. The basis of this idea is a marketing research aimed at the specification of a customers needs regarding the product. The target market cost is established based on the actual market demand. This method is mainly used on newly introduced products with a high expected production volume. The TC method represents an active managerial tool, as it estimates a harmonization of all assets on all company departments involved in the production of the new product. Target costs are reflected in the reduction of the targeted retail price by the profit margin. The targeted retail price is established with the goal of securing long-term profitability of the product.

Kaizen Costing (KC)—continuous improvement is a system based on identifying and removing waste in the manufacturing process. Constant improvement is realized by innovative steps in the form of smallest details, while the fundamental idea is to identify the place where a given process should improve. The impact of this method is reflected in the minimization of company production costs.

Quality Costing (QC)—is a managerial and regulatory activity focused on the monitoring and evaluating of quality costs. These entail product quality from in terms of fulfilling customer expectations. Main goal of this approach is the optimization of quality costs in order to fully satisfy customer needs, reduce failed products and refunds.

Environmental Management Accounting (EMA)—system of environmental accounting aimed at the gathering, recording, monitoring and evaluating environmental expenses of a company, which were a result of manufacturing activity. The goal of EMA is to optimize environmental costs via reduction of the negative on the environment and reduction of energy and resource consumption.

Cost Controlling (CC)—is a system focused on the monitoring and evaluating of company costs, revealing imperfections in their calculation, budgeting, analysis and evaluation. This method allows for the reduction of costs and exploration of cost-saving options on all levels of management in all company activities.

These mentioned methods can, after evaluating their costs and benefits applied, be applied in manufacturing companies. Objectively, the use of these methods brings companies and improvement in their financial indicators as they are focused on the reduction of costs. One negative factor about the implementation of these methods is the low awareness between managers of these methods, low willingness to change, insufficient knowledge in terms of applying these methods and financial costs of its implementation. Despite these barriers we can claim that these methods have a bigger effect on the company from a long-term perspective. In terms of the fundamental company goal to achieve profit, we can state that the application of these methods is a necessity for the company. Benefits of these methods have a broad impact, as every one of the aforementioned has a specific application and use in manufacturing companies. Integration of modern trends is noticeable primarily in cost management and its optimization, as the main contribution of each of these methods is the decrease in different cost types.

An essential part of management accounting is the question of applying new modern methods, that help managers make the right decisions. There are many approaches that in a significant way affect decision-making. But they are based on subjective evaluation and opinions. Objective tools for managers are these aforementioned trends based on information in forms of indicators, functions and parameters. Economists and field experts recommend the application of these methods, as they see their true potential.

5 MANAGEMENT ACCOUNTING AS A TOOL OF COMPANY MANAGEMENT

Management tools are based on the evaluation of information and data relevant to the decision-making, planning, organizing, leading and control, i.e. management. Fundamentally management tools consist of information gathering and evaluation. The degree to which information is evaluated depends on the needs of specific levels of management.

The basis of management tools is standardized information systems. Accounting is one of them. Data collection in accounting is formalized based on the legislatives in accounting, but other guidelines as well. These can sometimes even be internal company guidelines.

Accounting is used in company management, but it does not provide a transparent information for bigger companies with a structured portfolio and complex process map. Big and middle-sized companies use accounting in operative management. Even from this perspective it is mostly considered a specific form of management and not operative financial management. Tools used in management accounting are an extension of accounting. They provide management with information about aggregated variables, financial indicators and in some specific calculations managers have access to accounting information but assigned to individual products and clients.

Management accounting is an important tool of tactical decision-making and provides a method of measuring the progress in plans, leading up to strategic goals. Similarly to how accounting enters into management accounting and completes and interprets the information for the decision-maker, management accounting enters other management methods. One example of these methods is Balanced Scorecard, which adds the client relations to accounting. Balanced Scorecard is a tactical decision-making tool, but also monitors plan progress, be it tactical or strategic. In order to strategically manage a company, it is necessary to enrich the levels of analysis of internal relationships in a company from financial accounting, management accounting and Balanced Scorecard with an analysis of external fields with an impact on the company.

6 CONCLUSION

The end of the previous century was the time when financial theory came together with new trends in the entrepreneurial field. Ongoing globalization processes are changing and strengthening the competition on an international scale. Different forms of integration on a microeconomic scale are taking place. The need to monitor long-term prosperity is coming to the forefront.

With the Slovak Republic entering the European Union our companies can feel the pressure of globalization ever so greatly. They are perceiving globalization as a microeconomic phenomenon, which they are adjusting their mission and strategy to. The spotlight is on the newest means of communication and information as well as more successful and environmentally friendly technologies.

Companies are becoming aware of the possibilities of a free market, glowing competition, shifting focus on human resources, their knowledge, wisdom and skills. Generally speaking there are two ways of dealing with these changes. They can either choose to passively adjust, or to actively take part in the globalization process. The first approach is typical for smaller companies without the means of expanding abroad. The second approach is characteristic for bigger companies with access to greater financial resources. They don't perceive globalization as a threat, but as an opportunity.

Achieving a profit and maximizing it as a main purpose in entrepreneurship is no longer considered as the highest goal of a company. Current globalization processes accompanied by a strengthening competition are forcing company management to effectively identify which actions create the highest value for the company. Functions leading to profit maximization are certainly not one of them. Development activities should therefore support growth of performance from a financial as well as non-financial perspective, while only the financial indicators reflecting the economic growth are relevant.

Management represents individual activities with the purpose of defining and reaching goals. The success of management is based on the use of every available resource leading to that goal. This knowledge concerns the information gained from management accounting amongst others.

While perfecting management it is necessary to take a look at the managerial approach, realized by the manager on the company level or the level of an internal department. It seems easy, but it isn't. First it is necessary to realize, that management is not a monotonous activity, but a constantly shifting process, affecting other shifting processes, with their own logic and goal. If management fails to realize this, it leads to an internal conflict caused by the collision from wasted energy. That is why it is necessary to realize, that management is a focused effort towards a goal with the least amount of energy being wasted. It consists of several partial activities defined and partial managerial activities.

The most important and most significant partial activity is decision-making. A decision is made whenever it is necessary to choose one of several options. In order for this to happen, information is required. Processing information transcends throughout all managerial activities. One might say that information is the nervous system of management. Information is encountered on every level of management and they need to be constantly worked with. Without information processing it is impossible to manage.

Problems, which need to be solved, are encountered daily in the process of management. We perceive them as a dissatisfaction with the current state, the need to solve problems and reach a removing the conditions that lead to the problem in the first place. Problems and their solutions penetrate into all partial activities, most of all into decision-making. Some companies prove that their success is based on superior information system, which main part is a management oriented accounting. Some companies are becoming aware that their only chance to survive against their competition, is to efficiently manage factors contributing to the profitability of their assets and economic standing. That is the reason they have decided to implement such a system into their company.

ACKNOWLEDGEMENT

This article is one of the outputs of the VEGA 1/0421/13 project.

REFERENCES

Blažek, L. & Landa, M. 2006. Ekonomika řízení podniku. Brno: Masarykova univerzita.

Čechová, A. 2006. Manažerské účetnictví. Brno: Computer Press.

Drucker, P. 2000. Výzvy managementu pro 21. Století. Praha: Management Press.

Fotr, J. et al. 2006. Manažerské rozhodování. Praha: Ekopress.

Grasseová, M. 2010. Analýza podniku v rukou manažera. Brno: Computer Press.

Jacková, A. & Chodasová, Z. 2011. Analýza nákladov podniku a metóda ABC. Ekonomicko.manažérske spectrum, V(2): 103–108.

Král, B. 2010. Manažerské účetnictví. Praha: Management Press.

Sedlák, M. 2009. Manažment. Bratislava: Iura Edition.

Slávik, Š. 2005. Strategický manažment, Bratislava: Sprint.

Stanek, V. 2003. Zvyšování výkonnosti procesním řízením nákladů. Praha: Grada Publishing.

Tumpach, M. 2008. Manažérske a nákladové účtovníctvo. Bratislava: Ekonómia.

Production Management and Engineering Sciences – Majerník, Daneshjo & Bosák (Eds)
© 2016 Taylor & Francis Group, London, ISBN: 978-1-138-02856-2

Stakeholder management as part of integrated management system

J. Jaďuďová & J. Zelený
Faculty of Natural Sciences, Matej Bel University, Banská Bystrica, Slovak Republic

J. Hroncová Vicianová
Faculty of Economics, Matej Bel University, Banská Bystrica, Slovak Republic

I. Marková
Faculty of Natural Sciences, Matej Bel University, Banská Bystrica, Slovak Republic

ABSTRACT: Objective of the paper is to analyse the application of the stakeholder management as part of integrated management. Subject of the analysis are medium and large the construction companies in Slovakia, selected from company database supplied by the Information and Statistics Institute—INFO-SAT. As of 31.12.2014 our criteria were met by 196 medium and 14 large companies, comprising the major set. The survey was conducted on a selective sample of 98 companies during January–February 2015. Subject topic of the survey was selected integrated management systems: quality management system ISO 9001, environmental management system ISO 14001 and workplace safety and health protection management system OHSAS 18001. We evaluated the results using Spearman and Kendal coefficient, Kruskal-Wallis test and the Chi quadrant STATISTICA 5.5 programs, module: Nonparametrics/Distrib and SPSS 2.0. We came to the conclusion that the companies apply non-systemic stakeholder management.

1 INTRODUCTION

Quality and reliable construction with a minimum negative environmental impact and securing employee health protection and safety should currently belong among basic standards of every construction company. The tool for effective management of these areas is implementation of integrated management system (hereinafter IMS). Integrated management system can be conceptually viewed as a set of mutually interlinked systems, sharing same resources (personnel, information, material, infrastructure and finance), with the aim of satisfying stakeholder needs. Considering the above, the integrated management system should contain currently formalised systems, according to international ISO norms, which focus on quality, environment, safety and health protection (Krsmanovic et al. 2014). Integration of these systems is only possible based on similar structure of norms and presence of several identical elements and tools. Despite the aforementioned compatibility of individual management systems in the norms ISO 9001, ISO 14001 and OHSAS 18001, we have identified differences: focus of systems on different subjects, different objectives, and requirements of different stakeholders.

Marková (2011), Korimová & Hroncová Vicianová (2014) assume that company operating in the market cannot produce quality, unless it manages its losses and takes care of environmental protection, safety of its employees in the process management system over and above legislative requirements. Implementation of IMS in a construction company is a means for the company to adopt preventative behaviour and to avoid financial losses due to the lack of quality, alternatively stemming from penalties for legislative breaches in the area of environmental protection and workplace safety and health protection.

2 RESEARCH OBJECTIVE, MATERIAL AND METHODS

Subject topic of our survey was selected integrated management systems: quality management system ISO 9001 (hereinafter the QMS), environmental management system ISO 14001 (hereinafter the EMS) and Occupational Health and Safety Management Systems OHSAS 18001 (hereinafter the OHSAS). Scientific objective was to analyse the level of application of the stakeholder management as part of integrated management in selected medium and large companies in Slovakia, included according to their main activities in the construction segment. The survey involved companies, which had implemented at least two of the examined integrated management systems, since the management integrated system is defined as a merger or combination of minimum two or more management systems into a single system (Chovancová et al. 2010).

The sector selection in evaluating implementation of various norms and at the same time, identification of relationship with various stakeholders is closely linked to the nature of the sector and its energy and material demands. Despite the impact of economic cooling-off in the private, as well as public investment segments, the construction industry is still considered one of keys sectors of the Slovak economy over the past several years. It is a major consumer of various types of energies, mineral resources, materials and products, the sector produces an enormous amount of building waste, demolition materials and emissions. Its status in overall economy is determined as the share in gross domestic product (GDP) and overall employment in national economy, produced by construction industry (Fig. 1).

In 2013 its share of GDP was around 6.9 %, construction industry created 7.6 % of overall added value and employed approximately 7.3 % of the total number of workers employed in Slovak economy (ÚEOS—Komercia 2014). These facts led us to conduct research on a sample of medium and large companies operating in this segment, having a share of 20.1 % and where we anticipate that they use systems, which are the subject of our examination (QMS, EMS and OHSAS). Small companies were excluded from this survey, due to low level of reporting of the use of various norms and systems on their websites. According to INFOSTAT as of

31.12.2014 the selected segment comprised 196 medium and 14 large companies, which represented the basic set. Our research sample consisted of 91 medium and 7 large companies. The survey was conducted in January—February 2015. We verified the representativeness of the selection set using Chi-squared test in program SPSS 2.0. Based on p-value = 0.850 we can confirm that the selection set is representative in accordance to the company's size.

As part of the primary research we draw information from annual reports, policies and other strategic documents available on the survey companies' websites. Next we used also secondary information sources, such as available results from research into the topic, conducted at home and abroad. Using the Spearman and Kendall coefficient we examined the dependency between the number of integrated management systems and identified relationship with selected stakeholder groups (owners, employees, customers, suppliers and the general public). Applying the Kruskal-Wallis test, Spearman and Kendall coefficient we examined whether there is a dependency between IMS implementation and the company size. As part of improvement of the studied data we used two statistical programs STATISTICA 5.5, module: Nonparametrics/Distrib and SPSS 2.0.

3 FINDINGS AND DISCUSSION

The most implemented system among the examined companies was QMS, followed by EMS and OHSAS (Fig. 2). Kováčová (2012) also came to the same conclusion, that in comparison with other management systems the QMS is the most frequently certified system worldwide, as well as in Slovakia. In 2010 a total of 1,109,905 certificates was issued worldwide, representing a year-on-year

Construction sector as a GDP and overall employment percentage contribution

in 2010 - 2013 in

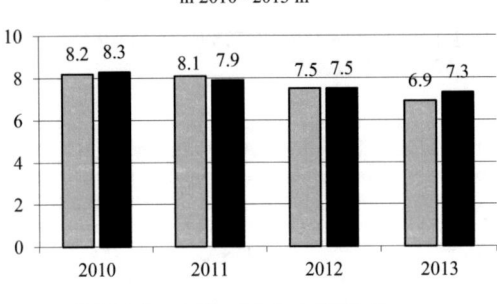

□ share of construction industry in GDP in %

■ share of construction industry in employment in %

Figure 1. Share of construction industry in GDP and employment in Slovakia (Source: ÚEOS—Komercia 2014).

Number of integrated management systems in the examined sample expressed in absolute values

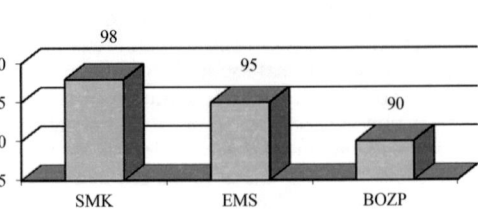

□ systems of integrated management

Figure 2. Number of enterprises using integrated management systems studies.

growth by 4% when compared with 2009. At the end of 2009 there were 3,475 certificates issued in Slovakia, which is an increase by 73% when compared with year 2004. Similar finding was published by Čierna & Danková (2012), which quote that 84% of medium and large companies operating in Slovakia have implemented quality management system under the norm ISO 9001.

In terms of company size, large companies had implemented all examined integrated management systems. We examined the dependency between IMS implementation and the company size also by statistical testing using Kruskal-Wallis test (p-value = 0.3314), Spearman (p-value = 0.334) and Kendall coefficient (p-value = 0.331). Based on the results we can conclude that in the examined sample the number of implemented IMS systems has no impact on the company size. We are inclined towards the opinion presented by Gašparík (2000), that QMS implementation, according to ISO 9001 requirements is the most effective means to improve the quality of construction companies and building sites.

The company management bears responsibility towards owners, investors, creditors, employees, suppliers and customers, arising from legislation, alternatively regulated contractually (Varcholová & Dubovická 2011). In this respect, we analysed selected groups of stakeholders, which represent primary and secondary groups (Hroncová Vicianová 2014, Zelený et al. 2010). We selected them, based on Chovancova & Tejo (2013), who assert, that key entities affecting management systems' integration are the customers and suppliers (determining the quality production requirements), the company and the general public (determining environmental requirements), employees (determining safety, protection and social requirements) and owners (determining resources and their allocation).

The most identified stakeholder by companies are the customers of (54.08 % of companies) and employees (43.88 % of companies) (Fig. 3). Currently it is important for companies to act in a manner, when together with taking into account the company interests, they also responded to interests of customers themselves, who represent an important partner for them. In line with their requirements, the customers are mostly linked to the QMS area, safety of products/services, supplied by the organisation (Džupina & Bosmanová 2013, Marková et al. 2014). Accepting this assertion, we carried out testing, determining the dependence between the number of implemented integrated management systems and the defined relationship towards customers. Testing, using the Spearman coefficient (p-value = 0.504751) did not reveal such dependency. Despite this result we

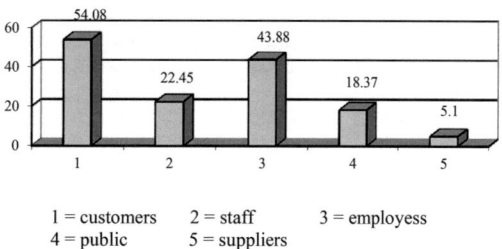

Number of companies, who have defined stakeholder in %

1 = customers 2 = staff 3 = employess
4 = public 5 = suppliers

Figure 3. Number of companies according to the defined stakeholder quoted in %.

agree with the conclusion by Karkalíkova (2010), that the customer is the key element of the IMS implementation, as the basis of input and output processes, supplier of suggestions and the end product consumer, since also the companies involved in our examination had their customer relationships defined most frequently.

Employees were the next preferred group for the examined companies. In any IMS system they represent a key stakeholder, since they have to be involved in the process in order for the system to function and the organisation must have a defined relationship towards them (Schiffel et al 2014, Jaďuďová 2013, Hronec et al. 2014). Subsequently, we examined the dependency between the number of IMS systems and the defined relationship towards employees (Table 1). This dependency was not confirmed either (p-value = 0.454392). The examined sample did not reveal any dependency between the number of implemented systems and the suppliers either (p-value = 0.68). It is often the case that large companies subcontract significant portion of their operations to small and medium companies. As a result, one of the primary business activities is the enhancement of relationships within the supply chain (Hroncová Vicianová 2014).

The only dependency (Tab. 2) revealed, was between the number of implemented IMS and the general public (p-value = 0.013521, r = −0.248740), i.e., a correlation between the increasing number of implemented IMS systems and decreasing identification of relationship towards the general public in the company's strategic documents. In creating IMS policy, the surveyed organisations inclined towards the first principle method as described by Zelený et al. (2010), since these are based only on own needs and requirements and restrictions stipulated by relevant legislation and norms. Despite the strong to medium strong negative dependence, we

Table 1. Dependencies between the number of implemented IMS and the various stakeholders..

		Owners	Staff	Customers
IMS	p-value	0.419611	0.454392	0.504751
	Spearman coefficient	0.082448	−0.076441	−0.068177
	p-value	0.229060	0.264786	0.319936
	Kendall coefficient	0.082448	−0.076441	−0.068177

Source: Own processing according to Statistica 5.5.

Table 2. Dependencies between the number of implemented IMS and the various stakeholders.

		Suppliers	Public
IMS	p-value	0.687758	0.013521
	Spearman coefficient	−0.041108	−0.248740
	p-value	0.548704	0.000285
	Kendall coefficient	−0.041108	−0.248740

Source: Own processing according to Statistica 5.5.

do not consider this result as positive and we are of the opinion that even the largest companies in the sector should identify their relationship towards the general public and within the intentions of the integrated management. They should advance closer to the third model by Zelený et al. (2010), when the company also includes the stakeholders in the process of policy drafting. The general public, as a secondary stakeholder represents a key partner for every company, since positive perception by the general public improves the company competitiveness (Hronec et al. 2013, Tomaškinová et al. 2014).

The least identified group are the owners (5.1 % of companies). The test results of Spearman coefficient (p-value = 0.42) prove, that there is no dependency between the number of implemented integrated management systems and identified relationship towards the owners. In IMS the owners (as one of the primary stakeholders) affect the organisation's direction, and therefore, even despite the result produced, we incline towards the opinion that companies should define their relationship towards this group in their strategic documents (Hroncová Vicianová 2014, Tomaškinová et al. 2010).

4 CONCLUSIONS

Quality, and environmental protection, workplace safety and health protection are worldwide key factors leading to success of construction companies and towards positive perception of these companies by customers, employees, suppliers and the general public, where the construction takes place (Gašparík 2000). In view of the given statement, we have analysed the application of the stakeholder management by selected groups under the umbrella of integrated management as applied by medium and large construction companies. Subject topic of our survey were selected integrated management systems: quality management system ISO 9001, environmental management system ISO 14001 and workplace safety and health protection management system OHSAS 18001.

Results of the analysis performed, did not confirm dependency between the number of implemented IMS systems and the defined relationship towards customers, employees, suppliers and the owner. Dependency was shown only between the number of implemented IMS systems and the general public, although we can treat this dependency as weak to medium strong negative dependency. We do not consider our findings as positive and we would recommend to the companies to define their relationships towards stakeholders, who represent their key partners and to involve them in the process of goals definition for integrated management (Zelený et al. 2010, Ubrežiová et al. 2013).

Among the examined integrated management systems the most commonly implemented was the quality management system. Construction plays an important role in the country's economy, but at the same time it is significantly contributing to negative effects with respect to the environment, being a by-product of all construction phases. In order to mitigate these environmental impacts companies implement the environmental management system (Hroncová & Ladomerský 2003, Drimal & Balog 2012). Construction is a high-risk activity and therefore one of the most important duties of construction companies is to implement their workplace safety and health protection principles. As the results have revealed, out of 98 examined medium and large companies 98 had implemented QMS system, 95 EMS and 90 OHSAS system. The integrated management system is a rather advanced field of operations in construction companies, however a link to stakeholder management is absent.

ACKNOWLEDGEMENTS

This work was supported by the Cultural and Educational Grant Agency of the Ministry of Education, Science, Research and Sport of the Slovak Republic project no. KEGA 035UMB-4/2015 "Environmental management in sphere of production".

REFERENCES

Čierna, H. & Danková, A. 2012. Quality planning in Slovak manufacturing organisations. Acta Facultatis Technicae XVII(3): 131–138.

Chovancová, J. & Tejo, J. 2013. Implementation of quality, environment and safety management systems with emphasis on their integration. In Rusko, M. (ed.), Environment management system 2013; Proc. intern. confer.,Bratislava, 18–19 April 2013. Žilina: Strix, p. 11–15.

Chovancová, J., Majerník, M. & Juríková, J. 2010. Integrated management systems. Communications 12(1): 7–74.

Drimal, M. & Balog, K. 2012. Health Risk Assessment in Environmental Impact Assessment Framework. Životné prostredie 46(2): 89–92.

Džupina, M. & Bosmanová, Z. 2013. Relationship of Slovak consumers towards corporate responsibility. SELYE e-studies 4(4): 12–24.

Gašparík, J. 2000. Quality management in construction. Bratislava: Jaga group v.o.s.

Hroncová, E. & Ladomerský, J. 2003. Environmental management micro-regions and municipalities. In Integrated safety. Trnava: Royal Union, s.r.o. p. 247–249.

Hroncová-Vicianová, J. 2014. Application of the corporate responsibility concept in selected segments of the Slovak national economy. Banská Bystrica: Publishing house UMB in Banska Bystrica—Belianum, Faculty of Economy.

Hronec, Š. & Štrangfeldová, J. 2013. Socio-economic effects of education in the context of economic return. The New educational review 32(2): 172–183.

Hronec, Š., Mikušová Meričková, B. & Hroncová Vicianová, J. 2014. Social non-economic effects of education on the level of crime. The New educational review 38(4): 43–56.

Jaďuďová, J. 2013. Perception of Corporate Social Responsibility in selected sector of national economy of the Slovak Republic. In Mokryš, M. & Badura, Š. & Lieskovský, A. (eds), EIIC 2013; Procc. intern. confer., Slovak Republic, 2–6 September 2013. Žilina: Edis, p. 339–342.

Karkalíková, M. 2010. Product quality in services. Bratislava: Publishing house Ekonóm.

Korimová, G. & Hroncová Vicianová, J. 2014. Tools of state support in selected areas of Corporate Social Responsibility in Slovakia. In Kubátová, K. (ed.), Theoretical and practical aspects of public finance 2014; Proc. intern. Confer., Prague, April 2014. Prague: Wolters Kluwer, p. 126–136.

Kováčová, N. 2012. Integration possibilities of selected management systems. In Merkúr 2012; Proc. intern. confer., Bratislava, 6–7 December 2012. Bratislava: Ekonóm, p. 446–460.

Krsmanovic, M., Horvat, A. & Zivkovic, N. 2014. Analysis of the experiences in the implementation of an integrated management system. In Markovic, A. & Rakocevic, S.B. (eds), SYMORG: XIV International symposium new business models and sustainable competitiveness; Proc. intern. symp., Zlatibor, 6–10 June 2014. Serbia: Smederevo, p. 1505–1511.

Marková, V. 2011. Selected aspects of Corporate Responsibility. In Pokorná, D. & Sojková, J. (eds), Corporate responsibility of companies—transfer of scientific findings into practice, Olomouc, 17 February 2011. Olomouc: Moravian University Olomouc, p. 28.

Marková, V. et al. 2014. The concept of corporate social responsibility in selected economic sectors. Radom: Instytut naukowo-wydawniczy "Spatium".

Schiffel, L., Šmida, Ľ. & Sakál, P. 2014. Application of sustainable HR management in industrial company. Information transfer 29: 86–93.

Tomaškinová, J., Tomaškin, J. & Dávidíková, Z. 2010. The increase of economic gains through implementation of environmental management system in the company Metsä Tissue inc. Žilina. Acta regionalia et environmentalica 7(2): 46–50.

Tomaškinová, J., Tomaškin, J. & Rákaiová, M. 2014. "FIT" model as a key tool of stakeholder education in the context of education for sustainable development in protected areas in Slovakia. In INTED 2014: 8th International technology, education and development conference; Proc. intern. confer., Valencia, 10–12 March 2014. p. 3719–3728.

Ubrežiová, I., Stankovič, L. & Mihalčová, B. et al. 2013. Perception of Corporate Social Responsibility in companies of Eastern Slovakia region in 2009 and 2012. Acta Universitatis Agriculturae et Silviculturae Medelianae Brunensis LXI(7): 2903–2910.

ÚEOS—Komercia. 2014. Slovak building almanac. Bratislava: Ministry of Transport, Construction and Regional Development.

Varcholová, T. & Dubovická, L. 2011. Evaluation approach to the impact of corporate responsibility programmes on financial performance. Economic outlooks 40(3): 377–386.

Zelený, J., Fabian, G. & Jaďuďová, J. 2010. Environmental policy and management of organisationsPart 6. Environmental policy, management and stakeholder management. Banská Bystrica: Matej Bel University, Faculty of Natural Sciences.

Production Management and Engineering Sciences – Majerník, Daneshjo & Bosák (Eds)
© 2016 Taylor & Francis Group, London, ISBN: 978-1-138-02856-2

Process map creation with regard to compliance with specific customer requirements

L. Kamenicky & J. Sinay
Faculty of Mechanical Engineering, Technical University of Kosice, Kosice, Slovak Republic

ABSTRACT: Current production and business in the automotive industry is subject to meeting the requirements of management systems and specific customer requirements and needs. The aim of any organization, regardless of their activity, is to meet these requirements, which are an integral part of customer needs and expectations and are specified in the purchase agreements or contracts by means of appropriately divided and set processes. Process map is used for visual display of processes, grouped into three basic groups (management, customer oriented and supporting). It represents an individual matter for each organization, i.e., it cannot be transferred from one organization to another. Organizations differ in opinions on division and subsequent creation of process maps of processes within the organization. This article describes the creation, logic and division of processes in the organization as well as a tool for assessing the efficiency and effectiveness of individual processes.

1 CUSTOMER REQUIREMENTS

Several supplier organizations for automotive production have already introduced integrated management system (ISO/TS 16949, ISO 14001, OHSAS 18001) at a very high level. One of the drawbacks with respect to changing specific requirements of individual customers (CSR—Customer Specific Requirements) is that the system is dependent on the administration, where the current trend is to eliminate the human factor, automatic check of deadlines, highlighting and prioritizing and etc. Another challenge is the weak relationship between the departments, where in case of a change in problem solution (internal problem, claims ...), the change was lengthy and difficult to document. Consequently, for the customer, the process is too rigid and lacks transparency, and in many cases there are no updates of the necessary documents and accessing information becomes complicated. There are no serious deviations or large number of deviations with respect to certification or control audits in accordance with ISO/TS 16949. However, a problem arises with respect to audits oriented on or linked to a specific product, for example, according to VDA 6.3 (Hoyle, D. et al. 2009).

Another problem is that the current trend of top management of the organization is constantly pushing for cost optimization, where one of the most frequently used solutions how to reduce costs is the synergic effect of individual functions and subsequent change in organizational structure.

Therefore, it is important to structure processes in such way so as to prevent the loss responsibilities for individual processes and disturbance in updating management documents in case of any change in the structure or a change in ownership of processes and transfer of competencies.

2 PROCESS APPROACH—A MODEL OF QUALITY MANAGEMENT SYSTEM

ISO/TS 16949, chapter 0.2 Process approach says: "This International Standard promotes the adoption of process approach when developing, implementing and improving the effectiveness of the quality management system to enhance customer satisfaction by getting to satisfy their needs."

The creation of process maps within an organization together with their identification and their interactions as well as their management to fulfill the intended purpose can be seen as a process approach.

Model of quality management system shown in Fig. 1, based on the process approach, illustrates the links between the processes specified in chapters of ISO/TS 16949 4-8 (quality management system, management responsibility, resource management, product realization and measurement, analysis and improvement). It is obvious from the model that the parties concerned, and in particular the customer, play a significant role in defining requirements as inputs. Monitoring of customer satisfaction requires evaluation of information on how the cus-

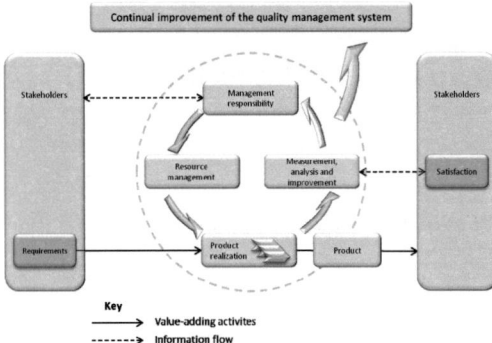

Figure 1. Model of a process-based quality management system.

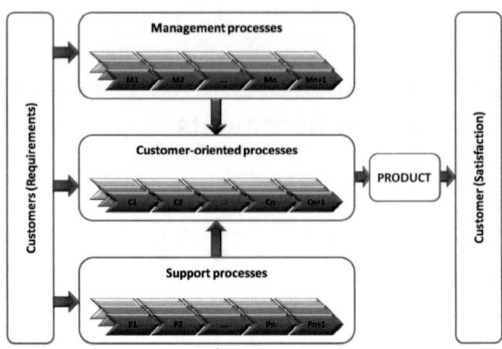

Figure 2. Process map.

tomer perceives the organization and whether the organization complied with their requirements.

The model of the quality management system contains an application of Deming's four-step methodology sequence of continuous improvement PDCA (Plan—Do—Check—Act) which lists the following stages:

- Plan; performance and determination of the objectives and processes necessary to deliver results in accordance with customer requirements and the organization's policy.
- Do; defines the implementation of processes.
- Check; uses monitoring and measurement of processes and products, their comparison with the policy, objectives and requirements for product and communication of the results.
- Act; implements activities on the continuous improvement of process performance.

3 CREATION OF PROCESS MAPS

The general requirements for quality management system (section 4.1) states: "The organization shall determine the sequence and mutual interaction of processes." In response to this particular requirement, a method for graphic visualization of processes was developed, the so called process map. Process map may be defined as a display of processes and their interaction in visual form. Fig. 2 depicts process map as a tool by which the organization describes its processes in an established hierarchy: management, head (customer focused) and supporting processes.

3.1 Customer

Creation of a process map must be based on the model of quality management system where the

position of the customer on each side declares its primary function.

Position of the customer is important since the beginning of the creation of process map. What is important, is the input from their part, i.e. formulating requirements for the implementation of the product based on their requirements (CSR). It means accepting the offer from their side, and asking for a quotation also expresses their confidence in the supplier, where the suppliers must comply with CSR and must contain specified objectives for quality. New suppliers must be audited—by potential analysis of established suppliers by process audits typically focused on a specific product (group of products). Development suppliers must undergo audit of development. The audit is carried out according to specific customer requirements (e.g. VW group according to VDA 6.3).

On the other hand, customer feedback is important, whether in form of evaluation of quality of supply (PPM), the evaluation of delivery performance (OTD), special status and claims. Evaluation of process audits using scorecards carried out by the customer based on its CSR, and various "soft facts" evaluations (the level and method of communication, problem solving, IT support, meeting deadlines etc.) for each department, which is in conjunction with the client (project management, quality, logistics, development, ...).

3.2 Product implementation

The implementation of the product shown in Fig. 3, resulting in the product itself, is in practice called customer-oriented process. These include all of the processes which affect the product itself at all stages, i.e., since the receipt and processing of demand through project management, industrialization, manufacturing of the product itself and

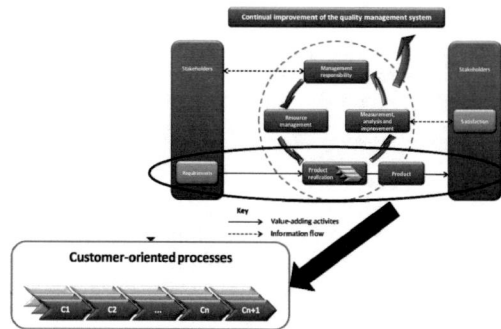

Figure 3.   Product implementation.

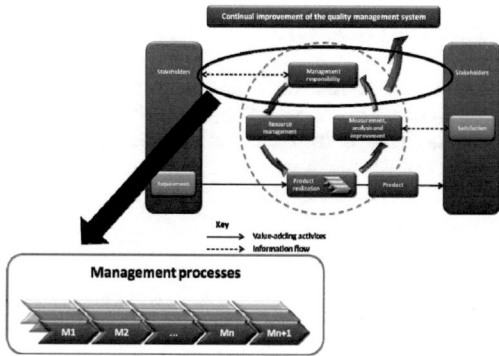

Figure 4.   Management processes.

the very important customer feedback. It may be a self-assessment of supplier by supplier themselves or an audit by the customer—at least release of series production. Customers QS-9000 (US automotive industry) require the approval process for production parts (Production Part Approval Process—PPAP) or customers according to VDA 6.1 require quality assurance of supplies according to VDA 2 (PPF—the release of the production process and the product).

3.3   *Management responsibility*

Fig. 4 shows management responsibility, which is on top of the process map marked as managerial processes. These processes significantly affect product implementation as well as support processes. These processes may include management, strategy and planning, documentation management, evaluation of QMS and SEM, auditing, continuous improvement, corrective and preventive actions.

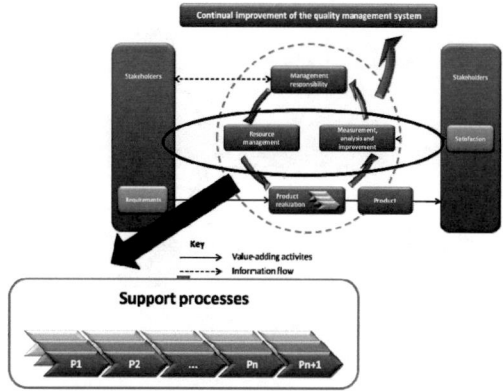

Figure 5.   Supporting processes.

3.4   *Supporting processes*

Changes of organizational structure and transfer of competences between departments most often affect the seemingly "less essential" departments or processes that by their nature can fit into different areas, which means that the process as such would not suffer by their transfer from department to department, however, there would be a mismatch in the documentation and updating. These processes shown in Fig. 5 are called supporting processes. Supporting processes are divided into as small logical subsystems as possible. If there is a change of ownership, organizational regulations are updated, where the responsibilities are transferred to another owner and the owner is updated in the manual. Other documentation, correctly created at the beginning, does not need updating.

4   TURTLE DIAGRAM

A tool typical for auditing process—turtle diagram shown in Figure 7, is used in describing various processes. Turtle diagram or "turtle model" allows you to identify, examine and, reduce the risk associated with the process as far as possible. This is understood as risk analysis in relation to output. The results of the process risk analysis are supporting process that must develop and maintain customer oriented processes.

Organizations are usually trying to use flowchart diagrams to describe the mutual relationships. During the creation, the following must be considered when describing:

– how the process works (the individual steps),
– what functions, departments, staff, process owner belong to the process or are responsible for it,

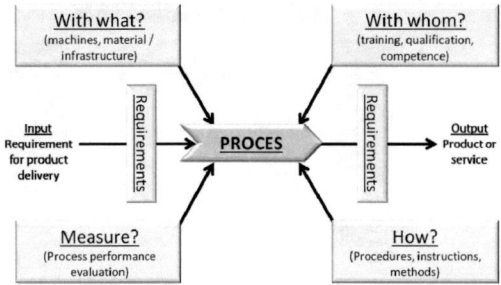

Figure 6. Turtle diagram.

– what resources are available,
– how effectively is the process carried out (indicators for evaluating the effectiveness of the process)

should be sufficient to understand the given process and further the fact that one must undergo trainings before performing given activity.

5 EVALUATION OF THE EFFECTIVENESS AND EFFICIENCY OF INDIVIDUAL PROCESS

When creating a process map, it is also important to consider whether, and how the process can be evaluated, or how its effectiveness and efficiency can be evaluated. Organizations use for evaluation BSC (Balance Scorecard) tool, which includes all processes and their evaluation in individual, predefined intervals e.g. months. BSC itself forms a basis for annual evaluation system as such, that is a report on the activities of the organization is prepared based on BSC. It uses simple logic of traffic lights used for easy visualization, therefore even an uninterested person can recognize what shape the organization itself is in at a glance. In prac-

tice, each process has its evaluation, where, if the objective is not met, the process owner must define corrective actions to eliminate the occurrence and recurrence of the deviation.

6 CONCLUSIONS

When creating a process map of an organization, it is very important to divide the processes not it the way that is currently set up in the organization, i.e. according to divided competences, but so that the processes are described as simply as possible, and no loss of responsibility for the processes occurs in case of any change, both in the structure or competences.

ACKNOWLEDGMENT

This contribution is the result of the project implementation: VEGA No.1/0150/5—Development of new method of implementation and verification of integrated machinery safety systems equipment systems and industry technology.

REFERENCES

EN ISO 9001: 2008. 2008 Quality management system. Requirements.
Hoyle, D. 2005. Automotive Quality System Handbook: Incorporating ISO/TS 16949: 2002. Second edition 2005. ISBN 0-7506-6663-3.
Markulik. Š. & Nagyová, A. 2009, Quality management sytem, TU Košice, Faculty of mechanical engineering, 79p.
Pardy, W. & Andrews, T. 2010, Integrated management systems: Leading Strategies and Solutions, ISBN 978-0-86687-196-0.
STN ISO/TS 16949:2009. 2009 Quality management system. Particular requirements for the application of ISO 9001: 2000 for automotive production and relevant service part organizations.

Comparison of devices based on psychoacoustic parameters of noise

D. Knežo

Department of Manufacturing Technologies, Faculty of Manufacturing Technologies with a Seat in Prešov,
Technical University in Košice, Slovak Republic

ABSTRACT: Accompanying phenomenon of devices operation is noise, which is sense by people who are located near to subject devices. Psychoacoustic parameters are one of possibilities to characterized inflection to people. Presented article is focused on proposed method for evaluation devices based on multicriterial comparison of psychoacoustic parameters of noise. Main part of article is aimed to study devices, which work in closed loops and consist of several phases. Method is based on assumption, that noise has cumulative character in influence to people, and therefore is monitored not only values of psychoacoustic parameters and also duration of exposition. Described method allows to compare devices in individual working phases and also to compare whole working cycles. Result of comparison is also recommendation and description of device, which by its operation is at least negatively affect people and environment.

1 INTRODUCTION

Noise is part of parameter by which is characterized function of devices around people. Noise is undesirable sound, which negatively affect people from physical and psychical aspects. Using psychoacoustic parameters can be defining negative effects on human psychic (Howard & Angus 2009, Lumnitzer, Badida, & Polačeková 2012). Comparison of devices monitoring influence of noise is necessary to use method, which set values of psychoacoustic parameters, what is added to multiciterial deciding.

Ižaríková & Džoganová (2014) describe method of multicriterial comparison of devices using which was realized comparison of several types of tires based on psychoacoustic parameters of noise caused by moving vehicle tires. Lumnitzer & Behún (2013) deals with possibilities of the optimization of the psychoacoustic sound emitted by automobiles.

Article presents multicriterial method for comparison of devices with the same functionality, which work in closed loops, where one cycle consists of several phases.

2 FORMULATION OF THE PROBLEM

Consider n devices $Z_1, Z_2, ..., Z_n$ with the same functionality, where accompaniment is noise. Functionality of devices Z_k, $k = 1, 2, ..., n$ consist of closed loops, while length of one cycle Z_k is T_k. One cycle of each device consist of l subsequence phases $F_k^{(1)}, F_k^{(2)}, ..., F_k^{(l)}$, while phase $F_k^{(i)}$ occurs in time from $\tau_k^{(i-1)}$ to $\tau_k^{(i)}$, $i = 1, 2, ..., l$.

If individual lengths of phases are marked $t_k^{(1)}, t_k^{(2)}, ..., t_k^{(l)}$, where $t_k^{(i)} = \tau_k^{(i)} - \tau_k^{(i-1)}$, is evident, that for each $k = 1, 2, ..., n$ is valid

$$\sum_{i=1}^{l} t_k^{(i)} = T_k. \tag{1}$$

Each device $Z_1, Z_2, ..., Z_n$ produce noise, which is characterized by psychoacoustic parameters $P^{(1)}$, $P^{(2)}, ..., P^{(m)}$.

Comparison of devices $Z_1, Z_2, ..., Z_n$ mean determination of device sequence

$$Z_{k_1} \preceq Z_{k_2} \preceq \cdots \preceq Z_{k_n} \tag{2}$$

from the sight of negative impact of produced noise to environment. Enrollment $Z_i \preceq Z_j$ is negative influence of produce noise by device Z_i to environment and is at the utmost as device Z_j.

Comparing of devices is based on assumptions as follows:

Assumption 1 The higher the value of parameter $P^{(j)}$ is, the negative influence of noise is higher.
Assumption 2 The higher of value of exposition time the negative impacts on environment have wider range.

3 WEIGHTING COEFFICIENTS OF PSYCHOACOUSTIC PARAMETERS

In general psychoacoustic parameters have variable impact on human being. Take into account differences are define weighting coefficients. The higher weighting coefficient is the selected parameter has more significance impact on negative affection to human. Sense of noise has subjective character and therefore to set weighting coefficient is too complicated. In practice to determinate weighting coefficients are used several methods as follows some examples:

- method of classification criteria into classes (Křupka et al. 2012),
- method of sequence (Fotr and Dědina 1997, Fotr et al. 2000, Fotr and Hořický 1988, Křupka et al. 2012),
- pointing method Metfessel allocation (Fotr and Dědina 1997, Křupka et al. 2012),
- method of evaluating scale (Fotr and Hořický 1988, Křupka et al. 2012),
- method of comparison significance of criteria by preference sequence (Křupka et al. 2012),
- method of pair comparing (Křupka et al. 2012 Černý and Glückaufová 1982, Fotr and Dědina 1997, Fotr et al. 2000, Fotr and Hořický 1988),
- method of quantitative paired comparison criteria (Saatys method) (Křupka et al. 2012, Fotr and Dědina 1997, Fotr et al. 2000),
- analytic hierarchy method (Křupka 2004, Ramík 2000, Saaty 1980).

All mentioned methods are included into multi criteria deciding in general and are based on subjective evaluating of criteria. In the case of define weighting coefficients for psychoacoustic parameters of noise, would be suitable to set individual values by statistical research.

Presented article expect, that are known weighting coefficients of psychoacoustic parameters. Mark of symbol $w^{(j)}$ represent normative weighting coefficient of parameter $P^{(j)}$ and is valid:

$$w^{(j)} > 0, \quad j = 1, 2, \ldots, m, \tag{3}$$

and

$$\sum_{j=1}^{m} w^{(j)} = 1. \tag{4}$$

4   COMPARISON OF DEVICES

Assumption for time parameter $P^{(j)}$ during one working cycle Z_k is valid:

$$P_k^{(j)}(t) = \begin{cases} f_k^{(1j)}(t), & t \in \left\langle \tau_k^{(0)}, \tau_k^{(1)}, \right\rangle \\ f_k^{(2j)}(t), & t \in \left\langle \tau_k^{(1)}, \tau_k^{(2)}, \right\rangle \\ \vdots \\ f_k^{(lj)}(t), & t \in \left\langle \tau_k^{(l-1)}, \tau_k^{(l)} \right\rangle, \end{cases} \tag{5}$$

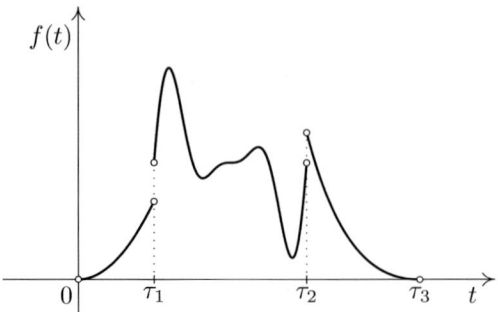

Figure 1:   Graphical dependence of function $f(t)$ for $l = 3$.

where $f_k^{(ij)}(t) \geq 0$ (see Fig. 1).

Devices are compared by numerical evaluation. Mark $h_k^{(ij)}$ as evaluation of device Z_k in phase $F_k^{(i)}$ for parameter $P^{(j)}$, $h_k^{(i)}$ is overall ranking of device Z_k in phase $F_k^{(i)}$ and h_k as overall ranking of device Z_k in one working cycle.

4.1   Theoretical model

Taking into account Assumption 1 and Assumption 2 coefficient $h_k^{(ij)}$ can be calculated using formula

$$h_k^{(ij)} = \int_{\tau_{i-1}}^{\tau_i} f_k^{(ij)}(t)\,dt. \tag{6}$$

For $h_k^{(i)}$ than follows

$$h_k^{(i)} = \sum_{j=1}^{m} w^{(j)} \cdot \int_{\tau_{i-1}}^{\tau_i} f_k^{(ij)}(t)\,dt, \tag{7}$$

respectively

$$h_k^{(i)} = \sum_{j=1}^{m} w^{(j)} \cdot h_k^{(ij)}, \tag{8}$$

where $w^{(j)}$ is normative weight for parameter $P^{(j)}$. For h_k is obtained formula

$$h_k = \sum_{i=1}^{l} \sum_{j=1}^{m} w^{(j)} \cdot \int_{\tau_{i-1}}^{\tau_i} f_k^{(ij)}(t)\,dt, \tag{9}$$

respectively

$$h_k = \sum_{i=1}^{l} h_k^{(i)}. \tag{10}$$

Is obvious that

$$h_k^{(ij)} \in (0, \infty), \tag{11}$$

$h_k^{(i)} \in (0, \infty)$ (12)

and

$h_k \in (0, \infty).$ (13)

Can be proof that

1. if $h_{k_1}^{(ij)} \le h_{k_2}^{(ij)}$ in that manner is for parameter $P^{(j)}$ in i-th phase functionality $Z_{k_1} \preceq Z_{k_2}$,
2. if is $h_{k_1}^{(i)} \le h_{k_2}^{(i)}$ is in i-th working phase $Z_{k_1} \preceq Z_{k_2}$,
3. if is $h_{k_1} \le h_{k_2}$ is during working process cycle $Z_{k_1} \preceq Z_{k_2}$.

4.2 Application of the model

Practical comparing of devices usually do not contains function describing time course of noise parameters during one working cycle (5), but we can obtain discrete values of individual psychoacoustic parameters and therefore compare devices respectively calculated values for assessment devices using equation (6), (8) and (10) is impossible.

Suppose that we want to calculate value

$$\int_a^b f(t)\,dt,$$ (14)

where function $f(t)$ is not known, but are obtained values of $y_i = f(t_i)$ in points $t_i = a + i \cdot h, i = 1, \ldots, s$, where $h = (b - a)/s$. In that case is valid

$$\int_a^b f(t)\,dt \approx \sum_{i=1}^{s} \frac{b-a}{s} \cdot y_i = (b-a) \cdot \bar{y},$$ (15)

where $\bar{y}$ is arithmetical average of value $y_1, y_2, \ldots, y_s$.

Assumption that each parameter is for individual devices and phases characterized by one value, which is approximate of mean value for function describe dependence of chosen parameter in time (Fig. 2). Value of psychoacoustic parameter $P^{(j)}$ of device Z_k in phase $F_k^{(i)}$ is marked $p_k^{(ij)}$. Using (15) from (6) for evaluation device from the sight of one parameter in one phase is obtained

$$h_k^{(ij)} = t_k^{(i)} \cdot p_k^{(ij)}.$$ (16)

Analogically for complex evaluation of device Z_k in phase $F_k^{(i)}$ from equation (8) is obtained

$$h_k^{(i)} = \sum_{j=1}^{m} w^{(j)} \cdot h_k^{(ij)} = \sum_{j=1}^{m} w^{(j)} \cdot t_k^{(i)} \cdot p_k^{(ij)},$$ (17)

and for evaluation for one working cycle from (10) is get

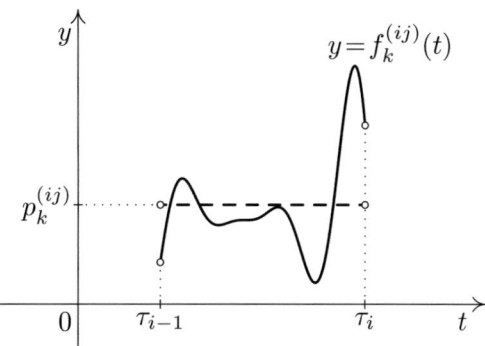

Figure 2.   Mean value of function $f_k^{(ij)}(t)$.

$$h_k = \sum_{i=1}^{l} h_k^{(i)} = \sum_{i=1}^{l} \sum_{j=1}^{m} w^{(j)} \cdot t_k^{(i)} \cdot p_k^{(ij)}.$$ (18)

4.3 Comparison for the same length cycles and phases

Assumption, that corresponding phases are for individual devices with the same length, i.e.

$$t_1^{(i)} = t_2^{(i)} = \cdots = t_n^{(i)} = t^{(i)}, \qquad i = 1, 2, \ldots, l.$$ (19)

Thus also working cycles for individual devices

$$T_1 = T_2 = \cdots = T_n = \sum_{i=1}^{l} t^{(i)} = T$$ (20)

are the same. In that case is possible to set normative evaluation for each device.

Evaluation of $H_k^{(ij)}$ for device Z_k from the sight $P^{(j)}$ in phase $F_k^{(i)}$ is described by equation

$$H_k^{(ij)} = \frac{p_k^{(ij)}}{p_{\max}^{(ij)}},$$ (21)

where

$$p_{\max}^{(ij)} = \max\{p_1^{(ij)}, p_2^{(ij)}, \ldots, p_n^{(ij)}\}.$$ (22)

Complex evaluation for $H_k^{(i)}$ of device Z_k in phase $F_k^{(i)}$ is prescribed by mathematical formula

$$H_k^{(i)} = \sum_{j=1}^{m} w^{(j)} \cdot \frac{p_k^{(ij)}}{p_{\max}^{(ij)}} = \sum_{j=1}^{m} w^{(j)} \cdot H_k^{(ij)},$$ (23)

where $w^{(j)}$ is normative weight of parameter $P^{(j)}$.

Evaluation of device Z_k based on parameter H_k for one working cycle T is defined

$$H_k = \sum_{i=1}^{l} \frac{t^{(i)}}{T} \cdot H_k^{(i)} = \sum_{i=1}^{l} w_p^{(i)} \cdot H_k^{(i)},$$ (24)

131

where

$$w_p^{(i)} = \frac{t^{(i)}}{T}, \quad i = 1, 2, \ldots, l, \tag{25}$$

are normative weighting coefficients of individual phases in working cycle, which determinate relative lengths of mentioned phases. It is obvious, that values $H_k^{(i)}, H_k^{(i)}$ and H_k are normative, what means there are from the range $(0,1\rangle$.

5 CONCLUSIONS

Presented method allows compare of devices based on calculated values from the mathematical equations (16), (17) and (18) to achieve absolutely assessment. Result is that add or remove the device from the function group of ordered device is sufficient to calculate parameters only of added or removed device.

Compare of devices based on values obtained from equations (21), (23) and (24) gives relative information about devices. Adding or removing device from the group is necessary to additionally calculate all evaluation for all devices and create new ranking of devices. Mutual rank is without changes.

Method was applied in comparing wash machines (Moravec et al. 2015), which are typical representative of devices working in close loops and consist of several phases. Working cycle consist of three phases: wash, rinse and spin. Comparing was based on psychoacoustic parameters: roughness, sharpness, loudness, tonality and fluctuation.

REFERENCES

Černý, M. & Glückaufová, D. 1982. *Vícekriteriální vyhodnocov ání v praxi (Multicriterial evaluation in practice)*. Praha, Czechoslovakia: Státní nakladatelství technické literatúry.

Fotr, J. & Dědina, J. 1997. *Manažerské rozhodování (Managerial decision)*. Praha, Czech Republic: Ekopress Praha.

Fotr, J., Dědina, J. & Hruzová, H. 2000. *Manažerské rozhodování (Managerial decision)*. Praha, Czech Republic: Ekopress Praha.

Fotr, J. & Hořický, K. 1988. *Rozhodování. Řešení rozhodovacích problému řízení (Deciding. Solution of decision problems in managing)*. Praha, Czechoslovakia: Institut řízení.

Howard, D. & Angus, J. 2009. *Acoustics and psychoacoustics, fourth edition*. Oxford, UK: Focal Press, Oxford.

Ižaríková, G. & Džoganová, Z. 2014. Analýza psychoakustických parametrov metódou váženého súčtu poradí (Analyzing of psychoacoustic parameters using mean sum order method). *Transfer inovácií 30/2014*, 86–89.

Křupka, J. 2004. Porovnání metod multikriteriálního rozhodování (Comparing methods based on multicirterial deciding). *Sborník z konference Public Administration and Informatics within Public Administration*, 191–195.

Křupka, J., Kašparová, M. & Máchová, R. 2012. *Rozhodovací procesy (Deciding processes)*. Pardubice, Czech Republic: Univerzita Pardubice, Fakulta ekonomicka—správní, Ústav systémového inženírstvi a informatiky.

Lumnitzer, E., Badida, M. & Polačeková, J. 2012. *Akustika. Základy psychoakustiky (Acoustics. Base of psychoacoustics)*. Košice, Slovak Republic: Technická univerzita v Košiciach, Strojnícka fakulta.

Lumnitzer, E. & Behún, M. 2013. Optimization of the psychoacoustic effects of the automobile sound. *17th International Conference on Intelligent Engineering Systems, proceedings, June 19–21, 2013, Costa Rica*, 173–176.

Moravec, M., Liptai, P., Badida, M. & Knežo, D. 2015. Approaches for measurements and evaluations psychoacoustic properties of the products. *in press*.

Ramík, J. 2000. *Analytický hierarchický proces (AHP) a jeho využití v malém a středním podnikání (Analytical hierarchy process (AHP) and its using in small and middle business)*. Opava, Czech Republic: Slezská univerzita v Opavě, Obchodně podnikatelská fakulta.

Saaty, T.L. 1980. *The Analytic Hierarchy Process*. McGraw—Hill International Book Company.

Production Management and Engineering Sciences – Majerník, Daneshjo & Bosák (Eds)
© 2016 Taylor & Francis Group, London, ISBN: 978-1-138-02856-2

Research of environmental labelling in the Moravia-Silesia region (Czech Republic)

J. Kodymová, R. Kučerová & A. Király
VŠB—Technical University Ostrava, Ostrava, Czech Republic

ABSTRACT: The paper presents the development of environmental labelling in the frame of National Programme of environmental labelling of products and services in the Czech Republic. It also includes findings from the research on environmental labelling type I, which has been carried out by students of VSB—Technical University of Ostrava in the Moravian-Silesian Region since 2009. This research was done in randomly selected retail shops from the five major retail chains (CR Schwarz, Ahold, REWE Group and Tesco Stores CZ). We monitored selected fast moving consumer goods (specifically the detergents, toilet paper and soap), which have environmental labelling type I (ecolabel). In all stores we observed the decrease in the products officially labelled in accordance with the National Programme of environmental labelling of products and services and EU Ecolabel, which corresponds to the countrywide trend in terms of number of registrations of these products.

1 INTRODUCTION

Environmental labelling is a type of marketing, which enables producers or service providers to differentiate (or environmentally define) their products from the other products. First ecolabelling certification program was launched in the Federal Republic of Germany—ecolabel Blue Angel—in 1978. According to Bjørner et al. (2004) a lot of sceptics thought that the price was the only regulator of the market, and such labelling would have no significant effect on the sale of the product. However, the rapid expansion and success of environmentally friendly products led to the establishment of other programs, e.g. in Canada, USA, Japan, the Scandinavian countries, New Zealand and other countries. (Bjørner et al. 2004)

According to Bjørner et al. (2004) the main purpose of the environmental labelling is the verifiable environmental information about the product, which should help the companies to obtain more market share. Regarding the above mentioned facts, two main entities enter the process. On the one hand it is a customer, who prefers certain product, and on the other hand, it is the company that can respond to the customer's demand or induce the demand for a particular environmental product.

If we assume that the utility value of the product is essential for the consumer, and eco-labelled products and products without the eco-labels have the same functional properties and differ only in the way they had been made, then the customer who buys the ecolabelled product demonstrates that is interested in welfare friendly production and consequently the whole society. (Brouhle & Khanna 2012). Bjørner et al. (2004) observed the impact of the Scandinavian ecolabel "Nordic Swan" on household. The study found out that households are willing to pay about 13–18% higher price for goods with the ecolabel. They monitored the following fast moving consumer goods: toilet paper, paper towels, laundry detergent (Bjørner et al., 2004).

A number of scientific works has analysed the market from the viewpoint of the producer. Dosi & Moretto (2001) used a dynamic model of investment decisions, which describes environmental labelling in terms of the enterprise. They found out that the ecolabel helps stimulate the markets in terms of environmental innovation, and that market may limit the supply of products, which are not manufactured in accordance with environmental protection. Their results suggest that ecolabelling improves consumer's confidence in the company, which may lead to increase in demand for other products of the company, even if the other products are less environmentally friendly. De & Nabar (1991) focus on the imperfection of the third-party certification in the competitive market, while Mason (2006) studies disparate information within the market, under which firms decide, whether they adopt ecolabel or not. Roe & Sheldon (2007) created a model of vertical product differentiation and analyzed the role played by the quality of communication in decision-making on environmental labelling. Roe and Sheldon (2007) created a model of vertical product differentiation and analyzed the role played by the quality of communication in making decision on environmental labelling. Baksi & Bose (2007) studied producers who would choose to label their products as well as ways that they have chosen for their labelling (self-declared claims against designating third parties—e.g. in accordance with ISO 14021 (ISO 14021).

In some European countries coexistence of multiple types of environmental labelling is common. Most European countries introduced their own national environmental labels in 80–90 s of 20th century (e.g. Germany—Blue—Angel, Norway, Sweden, Denmark and Iceland—Nordic swan, etc.). In 1992 all EU countries began to implement the EU Ecolabel (the European flower) into their programs. (Ecolabel EU, online)

The Czech Republic introduced a system of eco-labelling, initiated by the Minister of the Environment and the Minister of Trade and Industry, in 1993. In 1994 ministries declared the National program for labelling products with trademark Environmentally Friendly Product (now the National Programme of environmental labelling of products and services) (ME 2006). In 2004, when the Czech Republic became a member of the European Union, the goods started to be labelled with the EU Ecolabel (also known as the Flower/The Flower). (Regulation (EC) No. 66/2010)

Both of these programs that are currently used in accordance with the technical standard ISO 14020 (ISO 14020). In the context of environmental labelling we can distinguish three types of environmental labelling: environmental labelling Type I (ISO 14024), environmental labelling Type II (ISO 14021), and environmental Type III (ISO 14025). (ISO 14024; ISO 14021; ISO 14025)

Ecolabel—environmental labelling Type I has major advantage over other types of environmental labelling. It gives the customer a very clear and simple report on its overall environmental performance. In the case of the National Environmental Labelling Programme of the Czech Republic (Environmentally Friendly Product and Services—EFPaS) and the EU the following types of marks are supported (Fig. 1).

Within the next chart we can see the number of licenses of environmental labelling supported by the National Programme for labelling environmentally friendly products and services—EFPaS and the EU Ecolabel. (Fig. 2)

Since 2008 the ratio between the EU Ecolabel and a EFPaS has changed. In 2011 the EU Ecolabel hold 22% of the total valid licenses (since 2007, its share of total marking has increased by almost 14%). Since 2012, there has been a significant decrease in the total number of licenses.

Figure 1. List Type I environmental labeling program supported by the National Eco-labeling in the Czech Republic (Environmentally Friendly Product and Services—EFPaS) and the EU (CENIA, online).

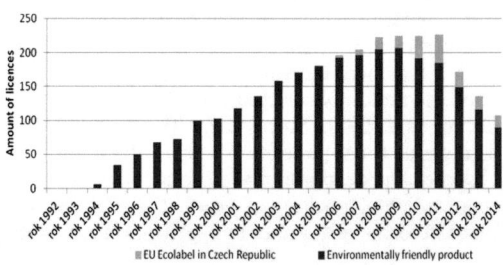

Figure 2. Development of environmental labelling products EFPaS and the EU Ecolabel (CENIA, online).

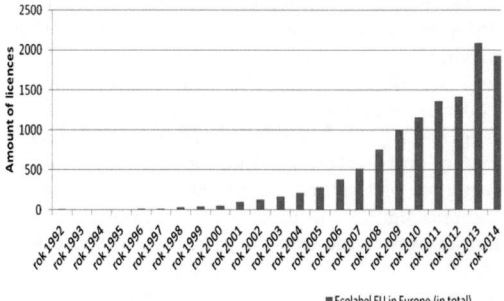

Figure 3. Development of ecolabelling in the European Eco-labelling Programme—EU Ecolabel (Ecolabel EU, online).

It is caused by the change of strategies of the computer producer whose 57 licences expired in 2011. The decision of the computer producer has led to a significant decrease in the total number of licenses (the number of licenses accounted for more than ¼ of the total number of licenses). However, the website of the national certification authority (CENIA) shows that the producer has just changed type of environmental labelling and he has been using the environmental labelling Type III (CENIA, online)

The following Fig. 3 shows that development trends of EU Ecolabel in Europe are quite different.

The figure 3 shows that the quantity of licenses with the EU Ecolabel in Europe increased in 2012. However, there was a significant rise (the annual change reaching 30%) in 2013. Suddenly, there is a break point in 2014. The absolute number of licenses decreased. There was a long-term growth of EU Ecolabel products and services and the volume of their purchases was increasing, but the amount on the market was still relatively low.

This means that the development of Czech and European environmentally friendly labelling is different. In case of the Czech Republic there is a steady decline since 2012, while the EU Ecolabel recorded the first decline in 2014.

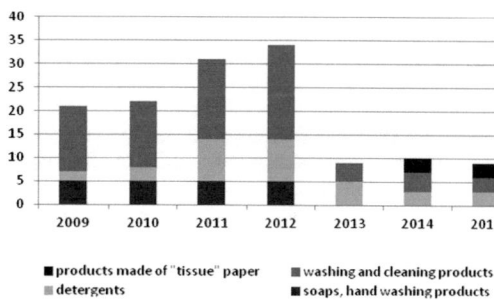

Figure 4. Development of environmental labelling of Type I (ecolabel) in 2009–2015. (CENIA, online).

2 THE BACKGROUND AND METHODOLOGY OF MONITORING OF ENVIRONMENTALLY FRIENDLY PRODUCTS

Monitoring of selected products began in 2009, and has been carried out annually by students of VSB—Technical University of Ostrava. The main objective of the research was to determine the availability of Fast Moving Consumer Goods (FMCG), which was represented by hand washing detergents, cleaning products and toilet paper. We monitored FMCG with environmental labelling in hypermarkets and supermarkets in the Moravian-Silesian region.

According to the organization CENIA [9] there was about 8% of FMCG of the total number of products with the valid environmental labelling (Fig. 4). There were registered 21 licences in the above-mentioned product groups in 2009 according to the National Programme of environmental labelling of products and services (EFPaS) and the EU Ecolabel. In 2010–2012 there was a gradual increase in licenses of all products (FMCG consist of 12% of the total number of licenses of EFPaS and the EU Ecolabel in 2012). In 2013, however, the number of licenses decreased to 9 (FMCG consist of only 8% of the total number licences of EFPaS and the EU Ecolabel in Czech Republic). In 2014 and 2015, the number of licenses of the selected FMCG settled down on 9 licenses. The following graph presents the evolution of the number of licenses of EFPaS and the EU Ecolabel in the period.

In the Fig. 4 you can see visible reduction in the number of licenses, which occurred one year later (in 2012), against the total number of licenses in the Czech Republic (Figure 1). The decline of the number of licenses was caused by the expiration of detergent product licenses (in 2012 about 14 to 20 licenses were registered and in 2013 there we only 4 ones). In some cases, the licensees were not willing

to renew a license or they disagree with the new product criteria or they were not willing to bear the higher costs associated with the new criteria (CENIA, online).

Due to the significant decrease in environmental labelling license in 2013, we have expanded the monitoring to products marked with a label from the global environmental library of environmental labels ECOLABEL INDEX (ECOLABEL INDEX, online). In the database there are 459 of different ecolabels, therefore students photographed all labels found in the shops and then they were compared photographed labels with the labels from the database ECOLABEL INDEX. Based on this comparison we decided to monitor (except FMCG with EFPaS label and EU Ecolabel) also mostly used label AISE Charter for Sustainable Cleaning.

We have chosen hypermarkets and supermarkets (retail chains) according to the analysis of Incoma plc. The Incoma (INCOMA, 2009) refers that the Czechs are losing interest in small and traditional stores, and they prefer hypermarkets and supermarkets. In 1997, 62% of families did shopping in the traditional and small shops, while in 2008 64% of families spent the majority of their expenditure on food and drug goods (FMCG) in hypermarkets and supermarkets (INCOMA 2015). In 2008, the most popular retail chains were mainly Schwarz group companies consisting of chain Kaufland and Lidl, followed by other companies in the following order: Tesco, Ahold Czech Republic, Makro Cash & Carry ČR, and Globus (INCOMA 2009). In some cases the owners of the companies changed during the research, but the composition of top-ten companies with the largest turnover in FMCG remained the same. In 2014 46% of FMCG were sold in the 5 largest supermarket chains in the Czech Republic (Kaufland, Penny Market, Tesco, Albert and Lidl). These companies controlled more than 2/3 of the total markets in terms of customer references (INCOMA 2015).

Regarding these facts, we have chosen the most preferred and widely used supermarkets and hypermarkets (retail chains) in the Czech Republic for our monitoring. On average, there were always 16 stores, which were monitored per 1 year.

The monitored shops of retail chains were selected randomly within the Moravian-Silesian Region, which is one of the 14 regions of the Czech Republic. The Moravian-Silesian region is situated in the north-east of the country and borders on Slovakia and Poland. The region has the third highest population of the Czech regions and highest population density within the Czech Republic after Prague.

Several diploma theses led by Jana Kodymová, which were related to this research, were defended in 2011 and 2012. This paper summarizes the information on customer and companies surveys (Konstantinidu 2011), and expand monitoring on randomly selected stores in several cities in the Czech Republic (Prague, Pilsen, Olomouc, Brno—a total of 15 shops were followed) (Srokova 2012).

3 RESULTS AND DISCUSSION

There is a very low correlation between the amount of the offered product and number of products with environmental labelling (the coleration index is just 0,357). The products with Czech EFPaS label and EU Ecolabel were found only in supermarkets and hypermarkets and we did not find any product with Czech EFPaS label and EU Ecolabel in discount stores during the time of our research. The results (average value for each company) can be seen in the following Fig. 5. We pick up just chains, where we found product with EFPaS label and EU Ecolabel (for better clarity).

In 2009–2012 we observed that license products (Environmentally friendly products and EU eco-labelling products) were linked to mainly two products (detergents LENA and REAL, which used EFPaS label). These results are inconsistent with the overall decrease amount of licenses in the Czech Republic (Fig. 4). In 2009–2012 there is significant correlation between amount of all EFPaS licenses of FMCG and amount of products with EFPaS label in hypermarkets and supermarkets (see Fig. 6). The correlation index is 0,99.

In 2011 the results were compared with the results of the research of other major cities of the Czech Republic (Prague, Pilsen, Olomouc, and Brno). All results were comparable and did not differ from each other (Srokova, 2012). In 2013, the licenses of these two products expired and detergent Frosch has become dominant in the supermarkets. But in some stores two products, LENA and REAL, were still being sold. At that time they

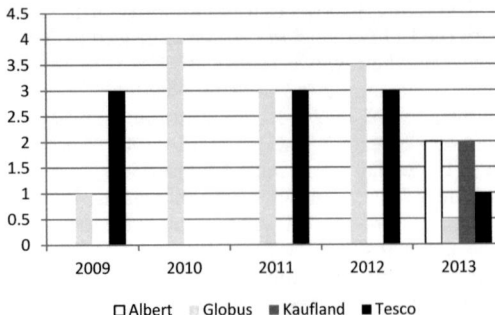

Figure 5. Presence of labelled product (Environmental friendly product and EU ecolabelling product).

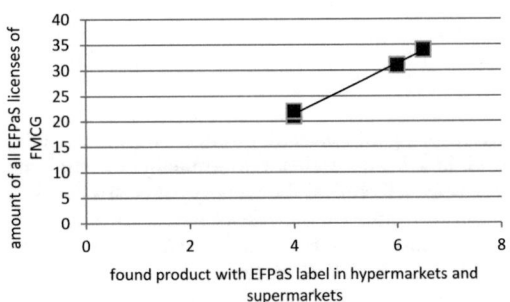

Figure 6. Linear regression of amount of all EFPaS licenses of FMCG and amount of found product with EFPaS label in hypermarkets and supermarkets.

were distributed equally to almost all observed stores.

Since 2013 significant reduction in the number of EFPaS and EU Ecolabel licenses has been observed in the selected product categories (see Fig. 4). Therefore we have decided to include research on the other types of environmental labelling.

We found several other national environmental label among the selected product category (Der Blaue Engel from Germany (Der Blaue Engel, online)), Umweltzeichen from Austria (Das Österreichische Umweltzeichen, online) and Nordic Swan form Scandinavia (Nordic Ecolabelling, online)). These brands have a similar certification system like the EFPaS and the EU Ecolabel. However, these are brands that are not primarily targeted at the Czech customer.

We also found that a significant number of foreign producers have registered trademarks AISE Charter for Sustainable Cleaning (see Fig. 7). (Responsible Care, online)

This label is granted by the Association for Soaps, Detergents and Maintenance Products (AISE), and it is similar to self-declared environmental claims according to ISO 14021 (ISO

Figure 7. The brand AISE Charter for Sustainable Cleaning (Responsible care, online).

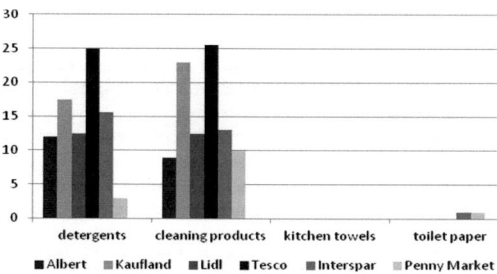

Figure 8. Proportion of individual producers in selected product categories.

14021). The certification is based on compliance with the criteria of the lifecycle product assessment, but it is not confirmed by an independent authority. The proportion of the main categories of these products in various shops in 2013–2014 is shown in Fig. 8.

We conclude that the products with environmental labelling, which appear in the Czech hyper- and supermarkets, are not primarily targeted at Czech customers. The stores provide environmentally labelled products which are registered under national environmental label programs of Germany, Austria and Scandinavia. There is another important fact. The companies producing detergents are willing to invest in environmental labelling of their products. However, producers are currently developing the labelling mainly within their cartels of the producers (e.g. Responsible Care (Responsible Care, online) for the chemical industry.

A significant analysis of customers' preferences of environmental labelling was presented by Hungarian company Flash Eurobarometer 256—The Gallup Organization (2009). This study was focused on the attitudes of Europeans related to sustainable consumption and production. This study reported that 22% of Czech customers are interested in eco-label products, but only 12% of the respondents correctly recognized EU Ecolabel. The study shows that 30% of Czech respondents have never seen the EU Ecolabel and another 48%

heard about EU Ecolabel, but they do not prefer labelled products.

These results correspond to the conclusion of the research by Konstantinidu (2011). She found that 77% of respondents had never heard about any type of environmental labelling and 13% of respondents thought that environmental labelling equalled to BIO label. The BIO label refers only to food products and is subjected to completely different certification process (Council Regulation (EC) No. 834/2007).

This approach of costumers correlates with the approach of producers. Konstantinidu (2011) contacted all the main selected retail chains and asked them if their top products were labelled. The majority of the companies (except for the chain Globus) did not pay any special attention to environmentally friendly products. Globus is the only one, which sells labelled environmentally friendly products. Nevertheless, there are products that are labelled as environmental friendly only by the producer and have no environmental labelling or validation given by an independent authority.

We can see the similar negative development in the other product with the exception of "Furniture" category. The category "Furniture" has been systematically supported by the state since 2010, when the Czech government adopted the "Rules for the implementation of environmental requirements in public procurement and purchases of state and local authorities". These rules oblige authorities to apply environmental requirements in selected product groups to all public purchasing (ME, online). Unfortunately, these rules are not consistently applied in practice (among other things because there is no enforceability of these rules) and we do not have any available information about their further development.

There are limitations of the research. First, the research is not systematic and we observed only randomly selected stores of the most favourite retail chains in the Czech Republic. The other limitation is that our research focused only on retail chains, but environmentally conscious customers can buy EFPaS and EU Ecolabel products in specialized stores.

4 CONCLUSION

In 2012 decrease in the number of licenses with EFPaS and EU Ecolabel was recorded in the Czech Republic. The number of products in product categories has decreased since then as well as the certified companies and licenses issued. Consequently, the amount of the licensed products in the monitored shops has also decreased. The results of our monitoring suggest that ecolabelling is not perceived as too strong issue Moravian-Silesian region in the Czech Republic (60–70% of customers either do not know it or do not identify with it). Our results suggest that there is no significant demand for the ecolabelled products in supermarkets and hypermarkets in Moravian-Silesian Region (Czech Republic), so the multination producers do not consider EFPaS (mainly) significant potential and use it only rarely. For this reason multination retail chain are not motivated to devote special attention to EFPaS and EU Ecolebel products in the Czech Republic.

The main limitation of this research is that it only affects the possibilities consumers purchasing in retail chains. In the future, it would be appropriate to carry out a survey in a more systematic way and expand our research to other retailers.

REFERENCES

Baksi, S., & Bose, P. 2007. Credence goods, efficient labelling policies, and regulátory enforcement. *Environmental & Resources Economics* 37(2): 411–430.

Bjørner, T. B., Hansen, L. G. & Russell, C. S. 2004. Environmental labeling and consumers' choice—an empirical analysis of the effect of the Nordic Swan, *Journal of Environmental Economics and Management* 47(3): 411–431.

Brouhle, K. & Khanna, M. 2012. Determinants of participation versus consumption in the Nordic Swan ecolabeled market. *Ecological Economics* 73: 142–151.

Cenia: *Environmentally friendly products.* Available at: http://www1.cenia.cz/www/ekoznaceni/ekologicky-setrne-vyrobky (in Czech).

Council Regulation (EC) No. 834/2007 of 28 June 2007 on organic production and labelling of organic products, 2007. *Official Journal L 189*: 1–23.

Das Österreichische Umweltzeichen. Available at: https://www.umweltzeichen.at.

De, S., Nesbar P. 1991. Economic implications of imperfekt quality certification. *Economics Letters* 37(4): 333–337.

Der Blaue Engel. Available at: https://www.blauer-engel.de/

Dosi, C. & Moretto, M. 2001. Is ecolabeling a reliable environmnetal policy measure? *Environmental & Resource Economics* 18 (1): 113–127.

Ecolabel EU, available at: http://ec.europa.eu/environment/ecolabel/.

Ecolabel Index—global directory of ecolabels. Available at: http://www.ecolabelindex.com/.

Flash Eurobarometer 256—The Gallup Organization. 2009. *Europeans'attitutdes toward the issue of sustainable consumption and production.* Analytical report. Available at: http://ec.europa.eu/public_opinion/flash/fl_256_en.pdf.

Incoma GfK. 2009. *Shopping monitor 2009.* Prag: Incoma GfK (in Czech).

Incoma GfK. 2015. *Shopping monitor 2015.* Prag: Incoma GfK (in Czech).

ISO 14020—Environmental labels and declarations—General principles.

ISO 14021—Environmental labels and declarations—Self-declared environmental claims (Type II environmental labelling).

ISO 14024—Environmental labels and declarations—type of environmental labelling—Principles and procedures.

ISO 14025—Environmental labels and declarations—III. type of environmental declarations—Principles and procedures.

Konstantinidu, P. 2011. *Environmental friendly products in the Czech Republic,* diploma thesis. (in Czech).

Mason, C. F. 2006. An economic model of eco-labeling. *Environmental Modeling and Assessmnet,* 11(2): 131–143.

ME. 2006. 17. communication of Ministry of environment. The rules of the Ministry of the Environment on the implementation of the National Programme for Labelling Environmentally Friendly Products and Services Available at: http://www.cenia.cz/share/web/opzp_pravidla_npesv_20080821.zip (in Czech).

ME. Rules applying environmental requirements in public procurement and purchases of state and local autority. Available at: http://www.cenia.cz/web/www/web-pub2.nsf/$pid/CENMSG0 KHR90/$FILE/Pravidla.pdf (in Czech).

Nordic Ecolabelling. Available at: http://www.nordic-ecolabel.org/.

Regulation (EC) No. 66/2010 of the European Parliament and of the Council of 25 November 2009 on the EU Ecolabel, Official journal L 27, 30.1.2010, p. 1–19.

Responsible Care. Available at: http://www.icca-chem.org/en/Home/Responsible-care/.

Roe, B. & Sheldon, I. 2007 Credence good labeling: the efficiency and distributional implications of several policy approaches. *American Journal of Agricultural Economics* 89(4): 1020–1033.

Schumacher, I. 2010. Ecolabeling, consumers' preferences and taxation. *Ecological Economics,* 69 (11): 2202–2212.

Sroková, D. 2012. *Monitoring of environmental friendly products in the Czech Republic,* diploma thesis (in Czech).

The A.I.S.E. Charter For Sustainable Cleaning. Available at: http://www.sustainable-cleaning.com.

Production Management and Engineering Sciences – Majerník, Daneshjo & Bosák (Eds)
© 2016 Taylor & Francis Group, London, ISBN: 978-1-138-02856-2

Comprehensive emergency management for airport operator documentation

J. Kraus, P. Vittek & V. Plos
Department of Air Transport, CTU in Prague, Prague, Czech Republic

ABSTRACT: This article deals with unification of the required documentation of an airport operator, which relates to processes and activities taking place at an airport in response to realization of extraordinary or unusual event. The purpose is to introduce a system into these often overlapping documents to improve the quality of airport operations. This applies to airport guide, airport security program, airport emergency plan, airport contingency plan, safety management manual and quality manual. Each of these documents is to certain extent concerned with reactions to the above mentioned events. However, it is important to have all phases of the comprehensive emergency management cycle affected so that the transitions between phases or procedures in the concerned documents were flawless. In this article the airport operator documentation is thoroughly analyzed; each phase of the comprehensive emergency management is assigned to appropriate document which should ensure the closure of the cycle.

1 INTRODUCTION

Extreme importance is currently put into ensuring safety in aviation. When there is some incident or accident, this is widely disseminated thanks to the media attention and it does negatively influence otherwise good air transport reputation. Nevertheless, air transport is the safest means of transport according to statistics compared to the distance traveled. Despite the overall high level of safety, there is one flight segment identified as clearly having the highest accident rate according to statistics available. This is the final approach and landing segment (Fig. 1), i.e. the part of flight on an airport or in its vicinity.

In the first chapter of this article there is the fundamental issue defined that we are trying to solve. It is a relationship between legislation, quality and safety and the potential solution.

The next two chapters focus on definition of environment in which the defined issue is located. They are aviation safety documents specified by legislation that need to be established by airport operator. Specifically this involves airport security program, safety management manual and airport emergency plan.

The next chapter describes the comprehensive emergency management used to solve the defined issue.

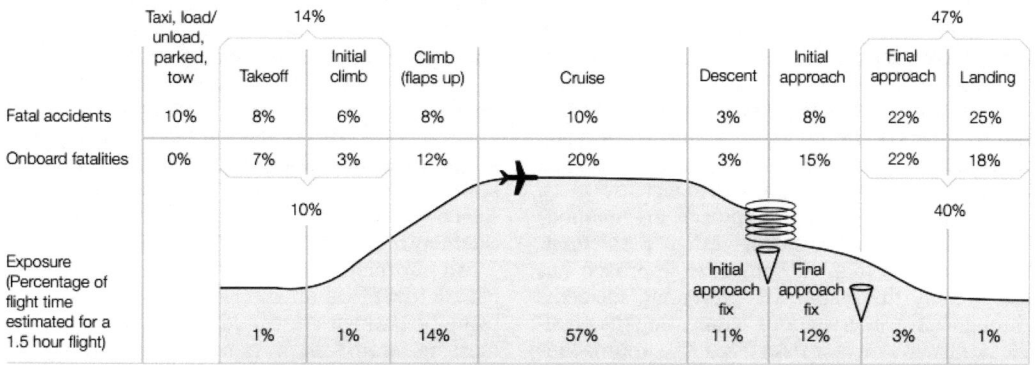

	Taxi, load/unload, parked, tow	Takeoff	Initial climb	Climb (flaps up)	Cruise	Descent	Initial approach	Final approach	Landing
			14%					47%	
Fatal accidents	10%	8%	6%	8%	10%	3%	8%	22%	25%
Onboard fatalities	0%	7%	3%	12%	20%	3%	15%	22%	18%
			10%					40%	
Exposure (Percentage of flight time estimated for a 1.5 hour flight)		1%	1%	14%	57%	11%	12%	3%	1%

Note: Percentages may not sum precisely due to numerical rounding.

Figure 1. Percentages of fatal accidents and onboard fatalities (Boeing, 2015).

In the last chapter there is the comprehensive emergency management applied to the defined issue of airport emergency planning. There are also positive effects outlined that the application would have on the safety and quality of air transport.

2 DEFINING THE FUNDAMENTAL ISSUE

In aviation, safety and quality have clear rules of relationship in which they interact together. These rules are as follows:

- Standards must be followed.
- Standards are here to ensure quality.
- In aviation quality is safety.

From this rules it is evident that everything in aviation is related to aviation safety. It can be therefore deduced that all the documents dealing with emergency planning are here to ensure, or rather increase aviation safety. Unfortunately, the influence of legislation is limited and solves only selected parts of airport emergency planning issues. Hence as a solution there must be established a system capable of gripping and controlling whole aviation safety or any integrated part of it. Proposed solution of utilizing comprehensive emergency management is such a system.

3 AVIATION SAFETY

Safety in civil aviation is achieving stable results during the last years provable by the fatal accident probability which today equals in aviation approximately to $1*10-7$. There has been a long journey to achieve this number where the increasing of aviation safety has been a long process from the very beginning until today. In the history of approach to safety we can identify three main directions—reactive, predictive and proactive. The first approach is based on response to potential hazard when there was a realization of an incident or accident and within its investigation there were some corrective actions proposed and then implemented. This is the principle of aircraft accident investigation which is in use today and without this approach it would not be possible to increase safety.

The second and third approach are methods that use to greater or smaller extent predictions. On the top of that, the proactive approach has additionally the feature of preventing incidents through active interventions, eliminating the possible occurrence of new risks. Proactive approach is thus the most effective but also the most expensive and time-consuming. The process of increasing safety is consequently based on a combination of all three methods.

Currently the proactive approach is being improved by introducing new tool—safety indicators. Safety indicators are performance indicators established for a specific area of application. Contribution of safety indicators to increased safety lies with the fact, that properly configured system of safety indicators provides guidance to safety management for the implementation of precisely targeted safety measures on a specific problem area.

Safety indicators are one component of safety management system, which is used to effectively control the safety of organization at an acceptable level. The principle of safety management system is a closed cycle of hazard identification, risk assessment, risk management. An essential part of a functional SMS is a presence of safety library which contains all the information concerning hazard identification, risk management and in the event of a safety incident it contains the results of investigation etc. The safety library may use its own data resources, but also data from the "outside". Here comes across another fact and that is the sharing of safety data between users and operators in the commercial air traffic. Safety information sharing, free of personal data, is the precondition of safety knowledge and safety awareness dissemination.

In addition to the investigation reports and own safety management data sources it is suitable to use an information system gathering reports from employees. Reporting systems are of two types—mandatory and voluntary. The supposed principle of these systems is to gather events reports that lead or could have led to the occurrence of events inducing impact on safety. The reporting system should not be perceived as a means of repression, but it should be seen rather as having positive effect on propagation of safety awareness. Staff willingness to provide information by the means of reporting system depends on the "safety culture" in the organization, i.e. on the level of safety awareness, perceived potential for contribution of each worker to increase the level of safety and so on. There is also a close connection to the concept of "just culture" which refers to fair approach of management to the events investigation. It is not only crucial to determine the fault and responsibility of the operational staff, but it is often more important to reveal the true cause of the event in such a way that it we can prevent it or at least considerably reduce the possibility of recurrence.

All information on safety occurrences, reports, hazard identification and risk management documentation which are stored in the safety library, must be stored in a standardized format, i.e. safety data must be unified adhering to the given specifications for records, having accurate content and description so that it is easy to navigate user through the records.

Basic document for the effective SMS establishment is ICAO Doc. 9859: Safety Management Manual.

4 AIRPORT OPERATOR DOCUMENTATION

4.1 *Security program*

The objective of the airport security program is similar to the National Security Program, i.e. it is supposed to prevent acts that may induce adverse consequences to the security of civil aviation, notably the security of passengers, flight personnel and the general public by establishing the necessary security measures, determining responsibility for their assurance and laying down operational procedures for their implementation. (Czech Republic, 2006)

It is an airport operator's hierarchically most important document. According to the legislation there are precisely specified requirements for the minimum that must be included in security program, so that its purpose can be achieved. The document has several parts such as description of buildings, areas and airport security devices, their equipment and location, security measures, procedures and security checks, emergency planning. And right here is the important link to other documents mentioned, since emergency planning required by security program is related to the domain of aviation security, but the airport emergency plan involves both safety and security domains. There is clearly some duplicity. An interesting point is that there are only physical attacks mentioned, but cyber defense is currently also very important and its importance grows proportionally to the size of the airport. (Czech Republic, 2012; Pitas, et al 2014)

4.2 *Safety Management Manual*

ICAO Doc 9859 Safety Management Manual (SMM) is the basic manual for a commercial air transport operator company for establishment and effective utilization of safety management system. This manual follows the Accident Prevention Manual (Doc 9422).

SMM is divided into several parts. The Manual's introduction contains basic familiarization with aviation safety issues including terms definition so that terms like hazard and risk are properly understood by safety personnel. Then there is a summary of the regulatory requirements including requirements stemming from the Annexes to the Chicago convention on aviation safety (parts of Annex 1; 6; 8; 11; 13; 14 vol. I, 19).

An important chapter in SMM in relation to emergency response planning is a chapter dedicated to safety management system. Its content is descrip-

tion of the procedure for hazard identification, risk assessment and risk management. Emphasis is placed on attention to proper communication of safety towards operational personnel and efforts to continuously improve the level of safety in the organization. One of the safety management system's items is the Emergency Response Planning (ERP). Here SMM states that it is necessary that the operator who compiles ERP must communicate with all relevant departments and assemble own ERP with respect to the other department's ERP to ensure the interdependence of activities during emergency. In the appendix to this chapter there are described more in detail various activities, there is laid down an implementation plan and also there is mentioned the utilization of safety indicators as an element of measuring organization's safety performance.

4.3 *Airport emergency plan*

Airport emergency plan is a document preparing airport and non-airport units for the reaction to emergency occurrence. Its purpose is to provide coordination of all units intervening so that the reaction takes place quickly and effectively. Therefore, it is necessary to create procedures for response of every rescue unit for each type of emergency event related to aviation safety and aviation security. Airport emergency plan and other documentation of airport operator compatibility is very important, because airport emergency plan should follow the guidelines laid down by airport operators safety management manual and should also include procedures for emergency planning of security events defined by security program. (ICAO, 2007)

Unfortunately, the importance of the proper content of the document is not visible from the legislation. For instance Decree No. 108/1997 Coll. states in section 4.3 only that it is necessary to have an airport emergency plan as mandatory content of airport operation manual: "Description of the airport emergency plan for emergencies at the airport, including:

- plans for action in emergencies at the airport and its surroundings, including aviation accidents, malfunctions and aircraft emergencies during flight, fire of airport equipment, unlawful acts, environmental accidents and all non-standard situations, having the character of an emergency under specific legislation (Act No. 239/2000 Sb.),
- description of the procedures of training and testing of facilities and equipment which will be used in emergency situations (Act No. 240/2000 Sb.), including the frequency of such tests.
- description of the test procedures training in emergency situations, including the frequency of such exercises,

- list of the legal and physical entities at the airport and outside, carrying on activities related to the operation of the airport, their telephone and fax numbers, email addresses, addresses of the Society for Aeronautical Telecommunications and Information Services (SITA), or their allocated radio frequencies within an Integrated Rescue System,
- establishment of the airport command staff for the organization of partial exercise of emergencies,
- nominating the person accountable for conducting the intervention on site emergency,
- list of organizations operating at the airport, contact persons and phone numbers on which are continuously available." (Czech republic, 1997)

European Union regulation Part-ADR (Easa, 2014) does not clarify emergency planning adding more accuracy. Uniform structure across Europe is important thing indeed, but the differences compared to the original national documents and of individual airports will lead to every airport establishing an emergency plan not being exactly what the airport would need. This regulation also does not cover all areas that should be in the emergency plan.

5 COMPREHENSIVE EMERGENCY MANAGEMENT

Comprehensive Emergency Management (CEM) poses a modern approach to the issue of making emergency plans, and various other documents/regulations addressing response to emergency/crisis situations. This approach is stemming from the modern concept where CEM is a closed loop between four constantly recurring activities. These activities include:

- Mitigation
- Preparedness
- Reaction
- Recovery.

Each of these activities is linked to the previous and together they form a closed loop of activities related to the elimination of the impact of emergencies on airport operations. The entire cycle is shown in Figure 2.

5.1 *Mitigation*

This process can be characterized as a search for potential risks and efforts to maximize their mitigation or complete elimination of risks resulting from an emergency occurrence. Mitigation process can be understood as brainstorming aimed at finding relevant risks which need to be addressed with regard to the particular airport's environment. Mitigation is very important for other processes of CEM. It consists of the following sub-activities:

- Identification and elimination of hazards.
- Risk Management.
- Mitigation.
- Distraction of the risk involved.

5.2 *Preparedness*

Preparedness phase consists of preparation for the realization of an emergency event and training of all procedures that need to be performed when the emergency occurs. The phase therefore lies with the preparation of written procedures, i.e. task cards, responsibilities and rights assignment for the intervention, but also in their own training for emergency situations in real conditions. During procedures preparation for specific interventions, the communication with other non-airport rescue units is necessary because perfect cooperation between airport and non-airport units is significant for the smooth progress of action.

5.3 *Reaction*

Reaction phase refers to the ability of rescue teams to intervene in case of realization an emergency and mitigate its effects. This phase is practical application of the two phases above during the realization of an emergency and in return we get feedback of the highest quality on the procedures functioning, their effectiveness etc. However, reaction phase does not represent only activities on site of the emergency, but it also deals with all the supporting processes, i.e. logistic support of the units at place of action, the supply of necessary equipment and transportation of survivors, wounded or dead persons. Then there are covered also activities associated with airport management, communication with media, survivors etc.

Figure 2.   Comprehensive emergency management cycle.

Phase could be divided into three areas:

– Support for on-site intervention.
– Informing the public.
– Helping the needy.

5.4 *Recovery*

The last of the four phases is the recovery phase after emergencies. This phase is dedicated to recover the operation of an airport to the state before emergency where the airport is capable of normal operation again. This phase is partly overlapping in terms of the meaning of some actions from the previous phase. On the example of an airplane crash at an airport, the phase of recovery can be imagined as an activity of removing the airplane wreckage from the airport surface so that it would not hinder normal airport operations. Furthermore, it will be about a recovery of the operating areas (damaged signs, signals lights, etc.) to the normal serviceable state.

For a successful recovery phase, it is necessary to secure:

– Well trained staff.
– Support of various parties.
– Technical support.

6   APPLICATION OF COMPREHENSIVE EMERGENCY MANAGEMENT INTO AN AIRPORT OPERATOR DOCUMENTATION

In the chapters above was described environment for airport emergency planning, in which airport operator staff must currently work.

When comparing the comprehensive emergency management with individual documents of airport operator, mutual connections and relationships between documents and CEM phases of the cycle could be identified. Mitigation as a part of the comprehensive emergency management dealing with impact limitation has some analogy in the safety management manual. That deals with increasing level of safety as a whole in given aviation organization which also applies to the emergency planning.

Preparedness as a part of preparation for dealing with emergencies is covered directly by airport emergency plan, since its creation (in any form) is a preparation step itself. Here is conducted a basic decomposition of various types of events into individual activities and a specific communication mode is set up for the event of emergency.

Part of comprehensive emergency management dealing with the very reaction is only an application of the airport emergency plan into the real situation. This depends on the quality of the plan and consequently on the quality of the preparedness

phase to ensure dealing with emergencies in the shortest possible time and with the least losses of life and damages to properties.

The very last part of the cycle—recovery is for most airports (for all small, most of the medium ones and some of the large ones) very vague and does not have its counterpart in the airport operator documentation. The recovery itself is after the emergency occurrence systematically not addressed and it is carried out according to the ad hoc decision of relevant persons only. Due to that it is necessary to have part of the recovery processed in detail and because of its important relation to emergency event investigation, the recovery process should be part of an airport emergency plan.

Interface between individual cycle's parts must be also thoroughly analyzed. Interface between mitigation and preparedness phases (safety management manual and airport emergency plan) and between preparedness and reaction (airport emergency plan) are easily covered and identifiable. However, interface between response and recovery within the comprehensive emergency management is not that precisely given unlike between the above mentioned preparation and response phases. Because of that it is always necessary to thoroughly analyze that interface and ensure a smooth and flawless transition between the end of reaction to emergency and process of recovery to normal airport operation.

This means mainly to ensure communication with the non-airport units which are not part of the emergency response units, i.e. police investigation and air accidents investigation authority. Another point is the transfer of command and control to respective personnel/unit (in most cases

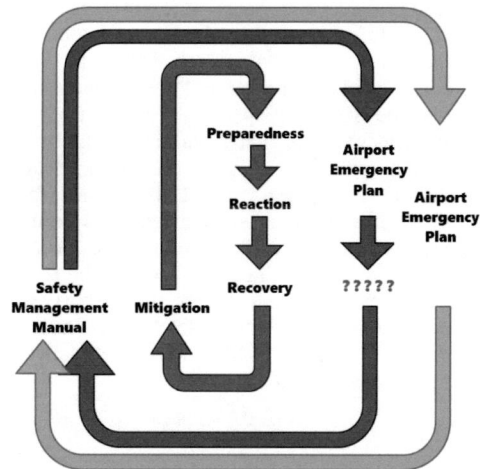

Figure 3.   Comprehensive emergency management cycle and airport operator documentation.

to airport's crisis task force) after the emergency and also ensuring proper transfer of information.

When creating a response plan it is equally important to consider its relationship with the recovery process what makes the procedures optimization possible to ensure the shortest time requirement from the occurrence of an emergency to the return to normal airport operation.

Proposal of comprehensive emergency management application into an airport operator documentation and the extension of an airport emergency plan are shown in Figure 3.

7 SUMMARY

In this article there was discussed the utilization of comprehensive emergency management for airport operator documentation so that the CEM cycle could be completed. After CEM application is clear that the documentation currently does not deal with the process of recovery, which is necessary for increasing aviation safety. This status can also be identified in the legislation; the airport is not mandated to have any plan for the recovery process.

Taking into account the current state, unification of emergency plans across Europe via EU regulations seems to be partly advantageous, but it also allows dangerous situations in establishment of an airport emergency plan, because it could be created without having look on the specific needs for that particular airport. The regulations also do not mention the need for the implementation of the recovery process in the airport emergency plan what is a proposal of our research.

The benefit of such implementation would be increase in the level of safety due to the establishment of different stages of the comprehensive emergency management which forms a closed cycle for safety improvement, as it would: 1) improve the identification and analysis of the types of emergency events that should be described in an airport emergency plan, 2) set up an effective collaboration between airport and non-airport response units in the phase of reaction and during the recovery process which would allow speeding up the whole process. The result would be reduction in losses of life and damage to property and also reduction in financial losses from the airport unavailability, which would increase the quality of air transport as a whole.

ACKNOWLEDGMENT

This paper was supported by the Ministry of the Interior of the Czech Republic, grant No. VG20132015130.

REFERENCES

Boeing. Statistical Summary of Commercial Jet Airplane Accidents. Worldwide Operations 1959–2013. [cit. 31-03-2015].2013 [online]. Available from: <http://www.boeing.com/news/techissues/pdf/statsum.pdf>

Clerk Maxwell, J. A Treatise on Electricity and Magnetism, 3rd ed., vol. 2. Oxford: Clarendon, 1892, pp. 68–73.

Czech Republic. Act nr. 410/2006 Sb. Available from: <http://www.zakonyprolidi.cz/cs/2006-410#prilohy>

Czech Republic. Konsolidované znění vyhlášky Ministerstva dopravy a spojů č. 108/1997 Sb., kterou se provádí zákon č. 49/1997 Sb., o civilním letectví a o změně a doplnění zákona č. 45/1991 Sb., o živnostenském podnikání (živnostenský zákon), ve znění pozdějších předpisů. In: 108/1997 Sb. 1997.

Czech Republic. L 14: Letiště. 2009, 641/2009-220-SP/4.

Czech Republic. L 17: Ochrana mezinárodního civilního letectví před protiprávními činy. 2013, 465/2013-220-AVS/2.

Czech Republic. National Security Program. Ministry of transport of the Czech Republic, 2012.

Draft AMC and GM to Part-ADR. EASA. Köln, 2014. Available at: <http://www.easa.europa.eu/system/files/dfu/agency-measures-docs-agency-decisions-2014-201X-XXX-R-AMC-GM-ADR-(DRAFT).pdf>

Hrůza, P., Soušek, R. & Szabo, S.: Cyber-attacks and attack protection. In Proceedings of the 18th Word Multi-Conference on Systemics, Cybernetics and Informatics. Orlando, Florida: International Institute of Informatics and Systemics, 2014, pp. 170–174. ISBN 978-1-941763-04-9.

ICAO. Doc. 8973—volume V: Security manual for Safeguarding Civil Aviation Against Acts of Unlawful Interference. In: ICAO, AC 150/5200-31C. Canada, 2007,

ICAO. Doc. 9859: Safety management manual. In: ICAO. Canada, 2013.

Pitas, J., Němec, V. & Soušek, R.: Mutual Influence of Management Processes of Stakeholders and Risk Management in Cyber Security Environment. In The 18th World Multi-Conference on Systemics, Cybernetics and Informatics. Orlando, Florida: International Institute of Informatics and Systemics, 2014, vol. II, pp. 94–97. ISBN 978-1-941763-05-6.

USA. SLG-101: Guide for All-Hazard Emergency Operations Planning. FEMA. USA, 1996.

USA. Advisory Circular: Airport emergency plan. In: AC 150/5200-31C. USA, 2010.

Production Management and Engineering Sciences – Majerník, Daneshjo & Bosák (Eds)
© 2016 Taylor & Francis Group, London, ISBN: 978-1-138-02856-2

Contractual insurance of industry as a dynamic development policy factor

V. Kyseľová

Faculty of Economics, Technical University of Košice, Košice, Slovak Republic

ABSTRACT: The aim of the Europe 2020 framework is to ensure economic growth. Important role in this process plays the industry whose economic importance is growing and is one of the keys to prosperity, economic recovery and competitiveness of the European Union. A new approach to an integrated European industrial policy based on the principle of a social market economy leads to higher productivity, growth and creation of jobs, to the strengthening of competitiveness and the transition to a circular economy. Industry brings with it new risks that threaten its stability and economic activity. One from possible means of eliminating the negative impact of risk factors in the industry and p is the use of insurance, which represents the alternative of financial compensation of the potential consequences of random and unpredictable events that are the subject of the insurance contract and related damages and losses.

1 GLOBALIZATION AND INDUSTRIAL POLICY

Globalization as a today's phenomenon is a complex dynamic processes intervening in all areas of society and is considered a key factor in the further development of the world economy. Globalization is a structured process with lots of dimensions and levels, composed of its actors, their mutual interactions and institutional relations. The impact of globalization leads to the development of international economic relations, the interconnectedness and interdependence of national economies. It is interesting to point out that many European countries have achieved high levels of globalization (see Table 1).

Globalization as a complex interdisciplinary social phenomenon with its positives and negatives brings many changes. It not only creates new job opportunities, but brings with it new risks whose effects may have a significant impact on all countries of the world and every individual.

Growing economic interconnectedness of individual countries, the emergence of transnational corporations, which through foreign capital and know-how offer a wider range of services and a greater ability to meet the needs of potential customers around the world, the development of technologies, particularly information technologies, which simplify and speed up communication, are positive aspects of globalization. The high concentration of economically active entities brings healthy competitiveness and within the fight for the customer also a wider range, higher quality of services and reducing prices.

Table 1. 2015 KOF index of globalization—most globalized countries.

Nr.	Country	Index
1.	Ireland	91.30
2.	Netherlands	91.24
3.	Belgium	91.00
4.	Austria	90.24
5.	Singapore	87.49
6.	Sweden	86.59
7.	Denmark	86.30
8.	Portugal	86.29
9.	Switzerland	86.04
10.	Finland	85.64
11.	Hungary	85.49
12.	Canada	85.03
13.	Czech Republic	84.10
14.	Spain	83.71
15.	Luxembourg	83.56
16.	Cyprus	83.54
17.	Slovak Republic	83.52
18.	Norway	83.30
19.	United Kingdom	82.96
20.	France	82.65

Globalization brings along also negative aspects such as power asymmetries influence on economic, cultural and political life, regional conflicts, tensions between companies, loss of national identity, standardization and other consumer public.

The consequences of negative aspects of globalization are for example:

- strong interconnection and interdependence of national markets may cause that economic crisis in one country indirectly hits other countries as well;
- emergence of large retail chains will ruin small businesses;
- multinational companies may not be subordinated to any authority, in case of strict legal measures to protect the domestic market or the environment that are inconsistent with the company's vision, they may move their operations somewhere, where the rules are more lenient for them;
- massive transfer of capital, labor and employment opportunities in some regions causes problems with unemployment, which leads to aggravation of social inequality in the society (Suša 2010).

Industrial policy in the era of globalization is oriented to the needs of the global market and promoting competitiveness. Increasing competitiveness is linked with the promotion of entrepreneurship. One of the main conditions for maintaining competitiveness and economic growth of companies is the ability to transfer innovation and technology in the production of products or the provision of services. The essence of this philosophy is to increase production efficiency by modernizing machinery and equipment and introducing innovative and advanced manufacturing technology, leading to product innovation, in the context of the reduction of energy consumption and negative environmental impacts of industrial enterprises. The added value within the competition is the support of industry automation, requiring IT support second to none. An important part of ensuring a high cooperation and interoperability of open single market is to increase the level of services through e-commerce implementation.

(Ecommerce Europe 2015)

In order to strengthen the competitiveness of European industry the following key measures were taken:

- careful monitoring of new legislative measures;
- revision of existing legislation, in order to break down excessive administrative burden in industries;
- financial support for SMEs, facilitating access to credit and financial support in the pursuit of cross-border activities;
- reinforcement of European standardization in order to meet the industry needs;
- improving transport, energy and communications infrastructure and services;
- a new strategy for the management of raw materials and their sustainability;
- stimulating innovative activities in the field of advanced industrial technology, construction, bio-fuels and road and rail transport;

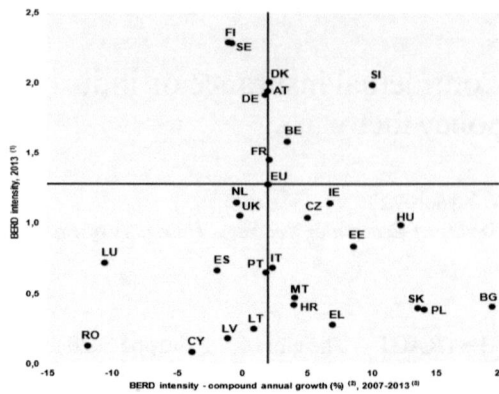

Figure 1.   BERD intensity (business enterprise expenditure on R&D as % of GDP).

- promotion of innovation and solving the challenges of energy intensive industries;
- strengthening of space policy in collaboration with the European Space Agency and Member States;
- transparent provision of relevant information on the competitiveness of Member States in the form of annual reports.

The aim of this policy should be healthier business environment. Entrepreneurs in pursuit of cross-border activities will not have to overcome a number of legislative and administrative barriers, accelerating the implementation of industrial innovation, leading to timely deployment and commercialization of technologies (European Commission 2014).

Despite high competition governmental authorities estimate that in 2025 the proportion of European manufacturing output and trade will constitute fifth of world production and trade and production will constitute 15% of value added. A significant impact on these shares will have adequate investment in innovation, research and development and strengthening of the internal market. Trends show that most Members States experience strong growth in business R&D intensity (Figure 1).

2   INDUSTRY AS A KEY FACTOR OF ECONOMIC GROWTH

Industry is a key factor in achieving a competitive Europe. It extends beyond the industry framework and is closely linked to raw materials, energy, business and consumer services, as well as to tourism.

The European Commission in the adopted document "The renewal of European industry" as of 22.1.2014 in Brussels called on Member

States of the European Union to acknowledge the crucial importance of the industry in creating job positions and economic growth and the implementation of accepted theses on industrial competitiveness into all areas of national policy. Modernization of industry is to be implemented through investments in innovation, resource efficiency, new technologies, skills and transparent funding. This policy aims to create a more favorable environment for business. Although the current investment situation in Europe is still recovering from the crisis, investment in equipment, metal products and machinery has remained relatively resilient until today.

A strong and stable industrial base is an essential element of economic resilience. The industry accounts for more than 80% of European exports and private research and innovation activities. Almost every fourth employee in the private sector is working in the industry and that often at highly skilled positions, while each additional job in manufacturing creates 0.5 to 2 jobs in other sectors. The share of industry on GDP of the European Union in 2020 is expected to be at 20%.

(European Commission 2014)

The European Union has more than 500 million consumers, 220 million workers and 20 million entrepreneurs. The executive notes that one out of four jobs in the private sector in the European Union is in the manufacturing sector and at least one out of four jobs is in services, which are related to the industry and depend on the industry, whether it is positioned as a supplier or a client. Up to 80% of R & D activities in the private sector are carried out in the industry that is a driver of innovation and search for solutions of challenges. Up to two-thirds of industrial employers are small and medium-sized enterprises.

(EurActiv.sk 2012) (European Commission 2015)

In 2000, at the Lisbon summit, representatives of the Member States of the European Union adopted the Lisbon Strategy, whose main objective is the development of economic competitiveness, which can be achieved by substantial reforms and adequate development policies. To build a prosperous European Union it was necessary to adopt a set of policies, which allows the European Union as a whole to maintain competitiveness, social justice, to protect its citizens and promote their freedoms in order to ensure employment.

An important part of the Lisbon Convention is to make Europe an attractive environment for investment and work, which assumes:

– a healthy business environment—the business environment of the European Union is dependent on the functioning of the common market,

where it is necessary to promote measures in the field of the state aid for small and medium enterprises and to ensure increase in job creation;
– simplification and rationalization of tax laws and the creation of a new generation of customs and tax programs in cross-border business in order to ensure better and easier mutual functioning of the national systems and speeding up the process in the fight against tax frauds;
– competitiveness in key sectors of the economy;
– adequate transport infrastructure—the creation of trans-European networks for transport, energy and telecommunications;
– creation and implementation of new technology solutions with a focus on clean coal technologies and technologies based on renewable energy, in order to respond to energy needs;
– revision of existing legislation on the free movement of goods, persons, services and capital within the single market in order to promote continuity and support growth.

The European Union aims to build a smart, sustainable and inclusive economy. Mutual correlation of these three complementary priorities is essential to achieve higher employment rates, productivity and social cohesion. To support research, technology and innovation, in order to ensure economic growth is in focus of the Europe 2020 framework strategy, which represents a vision of the European market economy in the 21st century.

The concept of the Europe 2020 strategy is based on three priorities:

– Smart growth—developing an economy based on knowledge and innovation;
– Sustainable growth—promoting a more ecological and competitive economy that is resource efficient;
– Inclusive growth—fostering a high-employment economy delivering social and territorial cohesion.

(European Commission 2010) (European Commission 2015)

The European Union will achieve its objectives only if they will be jointly realized by its institutions, national and regional governments, local authorities and the citizens themselves.

3 CONTRACTUAL INSURANCE OF INDUSTRY AS A KEY ELEMENT OF PROSPERITY

A strong and stable industrial base is a fundamental element of economic resilience. Industry as the sector that goes beyond the manufacturing sector has an impact on the market of raw materials and energy, on business and consumer services as well as on tourism.

Industry accounts for four fifths of European exports and four fifths of private sector investment into research and development are coming from industry, and so it plays an important role in promoting sustainable growth, creating valuable job positions and in realization of challenges of the society. Europe is in certain strategic sectors such as automotive, aerospace, engineering, space, chemical and pharmaceutical industry a global leader, which with properly selected industrial policy instruments enables to achieve even better results.

Part of the industrial policy is the promotion and improvement of the business environment. SMEs (Small & Medium Enterprises) in the European Union represent 98% of companies, provide 67% of jobs (see Figure 2.) and are the main driver of economic growth, innovation, employment and social integration. Increasing the competitiveness of the Member States' economies cannot be ensured without increasing the competitiveness of businesses (European Commission 2013).

Within an active and responsible approach to business it is important to ensure oneself against unforeseen events, whose adverse financial consequences could jeopardize the business activity.

A specific instrument of financial elimination of the negative consequences of random events is insurance.

Insurance can be understood as protection against insurance risks, where the policyholder transfers his risks, whose potential loss consequences are from his individual situation unacceptable, to the insurer, who by a sufficiently large group of risks of a similar nature is generally able not only to manage the taken risks using insurance premiums, but they become also the subject of a profitable commercial activity.

Insurance is a specific type of monetary service where the insurer provides for a premium insurance cover for the risk taken, so in the case of

insurance event occurrence, the insurer provides the policyholder with insurance claim (Majtánová et al. 2005).

Commercial insurance companies have a comprehensive range of insurance products for small and medium businesses that offer effective solutions of insurance cover within basic and catastrophic hazards in the case of service interruption or disruption of entire business.

The industry insurance is among other oriented on conceiving insurance protection for things tangible, intangible, insurance of machinery and equipment, fleet car insurance, transportation insurance, construction and installation insurance, electronics insurance, ships and aircraft insurance, business interruption insurance, management insurance, liability insurance caused by operational activities, liability insurance for damage caused by a defective product, professional liability insurance and other (see Figure 4).

Part of the activities of insurance brokers is a detailed analysis of insurance risk, audit of exist-

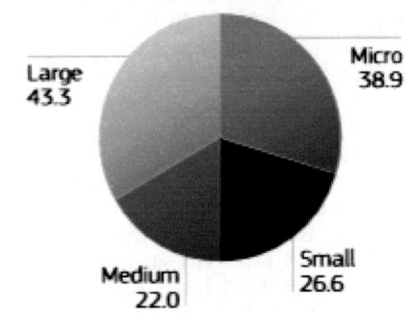

Figure 3. Persons employed in different enterprises according to size (in millions).

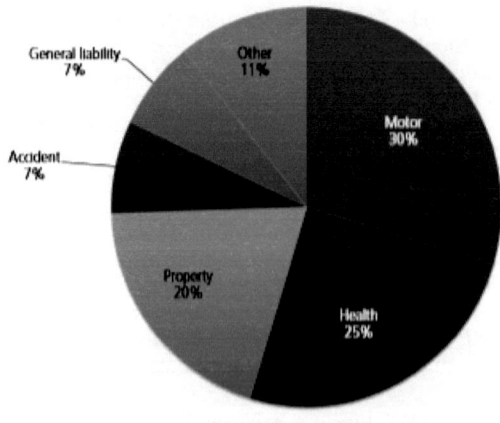

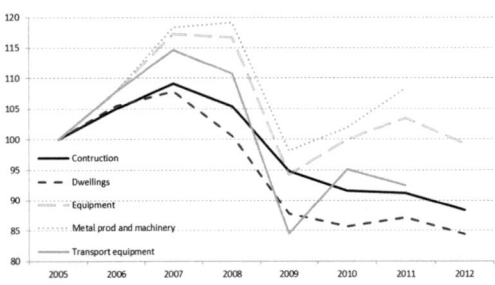

Figure 2. Evolution of investment components in the EU (2005 = 100).

Figure 4. Breakdown of total non-life premiums in 2012 (in percentage).

ing insurance contracts, proposal of insurance protection solution, management of insurance contracts and supervision over the settlement of claims.

Mutual cooperation between the broker and the entrepreneur on the design of insurance, including procedures and risk-management principles application, as well as methodological guidance of the client by a qualified specialist and broker, are important attributes of effective business activity through securing insurance and elimination of potential danger of financial impact.

4 CONCLUSION

The new industrial policy of the European Union should create favorable conditions to maintain and develop a strong, competitive and diversified industrial base and promote the transition of manufacturing sectors to more efficient use of energy and resources.

The objective of a horizontal approach to industrial policy is to promote the competitiveness of fundamental sectors of Europe's industry, sectors of manufacturing industry and services and take advantage of opportunities presented by globalization and the green economy.

The main initiative of the European Union in the field of internationalization of industrial policy in the era of globalization is the improvement and creating a favorable business environment, especially for small and medium-sized enterprises, for example by reducing transaction costs of business activities within the European Union, by promoting clusters and by improving access to credit.

Within the framework of this policy is at the same time taken into account the promotion of social responsibility of entities doing business, which is a key factor in attracting long-term confidence of employees and consumers.

A part of this policy is to maintain a strong industrial and knowledge base and provide leadership within the global sustainable development.

REFERENCES

Bertelsmann Stiftung 2014. Globalization report 2014: Who benefits most from globalization?. Online. <http://www.ged-project.de/wp-content/uploads/2014/03/Globalization_report_2014_final.pdf>.

Ecommerce Europe 2015. A European initiative to improve the functioning of the single market for companies and consumers. Online. <http://www.ecommerce-europe.eu/news/2015/a-european-initiative-to-improve-the-functioning-of-the-single-market-for-companies-and-consumers>.

EurActiv.sk 2012. Priemyselná politika vo veku globalizácie. Online. <http://www.euractiv.sk/podnikanie-v-eu/zoznam_liniek/priemyselna-politika-vo-veku-globalizacie-000289>.

European Commission 2010. Európa 2020: A strategy for smart, sustainable and inclusive growth Online. <http://eur-lex.europa.eu/LexUriServ/LexUriServ.do?uri=COM:2010:2020:FIN:SK:PDF>.

European Commission 2013. The European Union explained: Enterprise: A new industrial revolution. Online. <http://europa.eu/pol/pdf/flipbook/en/enterprise_en.pdf>. ISBN 978-92-79-42048-1.

European Commission 2014. Globalisation. Online. <http://ec.europa.eu/economy_finance/international/globalisation/index_sk.htm>.

European Commission 2014. Press release: Commission calls for immediate action for a European Industrial Renaissance. Online. <http://europa.eu/rapid/press-release_IP-14-42_sk.htm>.

European Commission 2015. Európa 2020. Online. <http://ec.europa.eu/europe2020/index_sk.htm>.

European Commission 2015. European semester thematic fiche: Research and innovation. Online. <http://ec.europa.eu/europe2020/pdf/themes/2015/research_and_innovation.pdf>.

European Commission 2015. Evolution of investment components in the EU (2005 = 100). Online. <http://europa.eu/rapid/press-release_MEMO-13-816_en.htm>.

European Commission 2015. Pracovný dokument útvarov komisie: Správa o krajine, Slovensko 2015. Online. <http://ec.europa.eu/europe2020/pdf/csr2015/cr2015_slovakia_sk.pdf>.

Europe Insurance 2014. European Insurance in Figures: Statistics N°48. Online. <http://www.insuranceeurope.eu/uploads/Modules/Publications/european-insurance-in-figures-2.pdf>.

KOF, ETH Zürich 2015. KOF Index of Globalization. Online. <http://globalization.kof.ethz.ch/>.

Kováč, J. & Mihok, J. 2011. *Priemyselné inžinierstvo*. Košice: SjF TU v Košiciach. ISBN 978-80-553-0806-7.

Majtánová, A., Krátka, Z., Littvová, Z. & Vachálková, I. 2005. *Poisťovníctvo*. Bratislava: Ekonomická univerzita v Bratislave, Národohospodárska fakulta. ISBN 80-225-1940-5.

Suša, O. 2010. Globalizace v sociálních souvislostech současnosti: Diagnóza a analýza. *Filosofia*. Praha. ISBN: 978-80-7007-320-9.

Production Management and Engineering Sciences – Majerník, Daneshjo & Bosák (Eds)
© 2016 Taylor & Francis Group, London, ISBN: 978-1-138-02856-2

Quality triad and indicators of profitability of quality

A. Linczényi
Faculty of Materials Science and Technology in Trnava, Slovak University of Technology in Bratislava, Trnava, Slovak Republic

ABSTRACT: Quality management has to consider a lot of new aspects regarding the latest trends in the field. Authors of proposed paper will point at the interconnection of three major aspects: technical, communication and economical aspect. Authors proposed from this aspects Triad of quality. The main aim of quality management is essentially increasing the effectiveness indicators in organization rather than satisfying customers by certain product properties or focusing on increasing the market share. Authors proposed indicators of profitability quality-Returns on Quality ROQ. Authors proposed triad of quality as a new way of perceiving the economical aspects of quality. The paper will comprise a proposal of system of indicators based on communication aspect and economical aspect of quality.

1 INTRODUCTION

It is interesting that neither the Slovak literature on quality nor professional community do mention that majority of the projects aimed at building the quality management systems and various quality programmes have failed. This obviously does not mean that the acquisition of a certificate or another support in the field of quality is considered a failure. The aim should be to achieve the results better than those achieved before the application of the abovementioned projects, rather than a pure acquisition of a certificate. That means that the costs for the projects must be lower than the yields achieved due to the implementation of such project. As documented in sources, numerous enterprises thus either abandoned those programmes or dramatically reduced them; there were even some cases of bankruptcy due to the application of such programmes. Neither do the Slovak sources mention how many enterprises, having won a quality certificate or award, went bankrupt. The bankruptcy under the conditions of economic crisis definitely cannot be ascribed to the application of quality projects. On the other hand, if quality projects are effective, the enterprise should be protected to certain degree from the impact of the economic crisis.

There is a question whether quality does or does not have a direct influence on economic results of an enterprise /2/. Authors of the current paper are convinced that such direct influence does exist, yet certain principles (e.g. those accented by Juran, contact with customers in particular) have been neglected in the field of quality management. Just look at the contents of ISO standards. Customers are mentioned, yet the methods of contact, assessment of post-production stages and particularly their feedback on research and development are missing in the standards.

The first version of ISO standards defined the role of marketing in quality management (Juran's idea that marketing represents both inlet and outlet of quality management), while this was omitted in the later versions. Similarly, the subject of economy quality was stated just marginally in ISO standards. The target should not be enhancing the technical properties of a product, or eliminating the active approach of employees towards quality by directive standards, but rather increasing the profitability of organisation. Companies seem to forget that a product must meet customer demands, yet the main aim is increasing the profitability of organisation; otherwise all the effort to increase quality is fallacious. Increased attention should be therefore paid to communication with customers while emphasising the economic returns on quality (Linczényi, 2006).

2 QUALITY TRIAD

Doubts regarding whether quality has or does not have a direct impact on the economic results of a company leads to the fact that management in many enterprises, instead of seriously dealing with quality, declares the increase of quality and acquisition of certificate just formally, thus considering the task accomplished.

Deming, one of the major gurus of quality, claimed (correspondingly to the concept of quality management in his time), that there is not a direct connection between financial results and quality, as financial returns on quality are invisible and unrecognisable. Such (mis)concept of management has been revealed by the US General Accounting Office claiming that just a minority of enterprises-finalists of the Malcolm Baldridge award proved some savings or better economic results achieved due to quality programmes /1/. Authors of this contribution take the liberty to claim that the situation in Europe is even worse in this context (Synek & Co., 1996).

Quality is a very complex phenomenon influenced by numerous factors. When elaborating any quality programmes, three major aspects have to be taken into account:

- technical aspect; a product must be designed and manufactured with the properties assuring that the customer satisfaction will be met,
- communication aspect; customers must be convinced about the advantage of an offered product's purchase; thus the acquisition of new customers and retention of current ones are the matter of communication aspect, yet it is the communication aspect which is not sufficiently regarded in quality programmes,
- economic aspect; the aim of the quality programmes should be neither increasing the technical level of individual properties of the manufactured products, nor increasing the level of satisfying the customer needs, but achieving the advanced technical level of the manufactured products and satisfying the customer needs, i.e. achieving better economic results and profitability of the enterprise (Linczényi & Nováková 2011).

The authors of this paper will attempt to express the relationships between these aspects by Quality Triad (this triad is contribution of authors to new understanding of quality economics). Since technical aspect is primarily a matter of constructers, developers and managers of production processes, this paper focuses on the communication, economic aspects in particular. As for the communication aspect, it is worth to emphasise that the achieved economic results of quality depend on effective communication with customers, and the economic aspect must therefore express the main aim of quality programmes and quality increase. Neglecting these facts necessarily leads to the failure of programmes or projects of quality increase.

Sides of the triangle in the triad of quality express the activities that must be implemented in order to assure the success of quality programmes, while the basis expresses the technical aspect of quality (a product must be designed and manufactured with certain properties), and the legs express the communication aspect of quality (only an effective communication with customers and monitoring their demands and satisfaction with the supplied products can help retain current customers and attract new ones). Communication is generally carried out by the department of marketing, while it is the communication itself which is a pre-requisite of effective quality programmes.

Angles of the triangle express the results of activities. The results of activities in the field of design are products with certain properties, while the result in the field of communication with customers is the manufacturer's market share, and subsequently the increase of market share is the supposition of good economic results achieved via quality programmes. The top of the triangle expresses the economic result of the previous activities and can be thus indicated as Return on Quality (ROQ).

The sequence of the quality increase process can be expressed by four basic steps:

- Step 1: carrying out the research targeted to determining the customer requirements and assessing the organisation's ability to meet those requirements; elaborating the list of requirements and harmonising the customer requirements with the organisation processes.
- Step 2: carrying out the communication with customers in order to convince them about the organisation's ability to meet customer expectations,
- Step 3: assuring the impact of the manufactured product's quality on customer satisfaction,
- Step 4: measuring the market share and the impact of quality on the achieved profit. Within

QUALITY TRIAD

Figure 1. Triad of quality.

this step, it is necessary to determine the quality programme related costs, Net Present Value (NPV) due to the increased market share and to compare the profit improvement with the costs associated with the implementation of quality programmes (Nováková 2006).

Triad of quality provides a new insight into quality economics. Most companies currently apply the approaches based on PAF model in quality economics. The model which appeared in 1946 does not regard the changes having taken place in quality management and the concept of quality itself. PAF Model is exclusively focused on technical aspect, as its original objective was to seek an optimum level of low-quality production. This also defined the structure of so called quality costs used in the model, which is focused on low-quality rather than quality, though losses due to low-quality production and product quality are in fact caused by wasting material, energy and workforce involved in particular production process, thus having nothing in common with quality. In these conditions if the ratio of wasters decreases, quality is better and vice versa. This does not mean that the losses of the wasters should not be decreased and monitored, yet by modern definition of quality these costs are not quality costs. Similarly appraisal costs are in fact a component of production process costs, while prevention costs are a component of the costs for training the staff /3/.

Regarding the abovementioned triad of quality, we are presenting a brand new structure of quality costs focused on costs for quality assurance. The structure of quality costs comprises the following groups:

- costs for research, development and preparation of production,
- costs for retaining current customers (defensive strategy) /1/,
- costs for acquiring new customers (offensive strategy) /1/.

As for group 1, it actually expresses the slogan saying that 80% of quality is created in pre-production phases. If this is true, then costs in this field are quality costs. Costs to retain current customers actually represent the total of all benefits an organisation provides to loyal customers including the costs for post-production phases. Those costs may be considered the ones for defensive strategy of a company. Costs for acquiring new customers represent the costs particularly for advertising, as well as the costs for market research, identifying the customer requirements etc. Those may be considered the costs for offensive strategy of a company.

3 INDICATORS OF PROFITABILITY OF QUALITY

The effectiveness of such approach requires building a system for monitoring and assessing the quality costs, comprising the following steps:

- defining the cost issues that will be included into particular groups of quality costs,
- determining responsibility for issuing the initial documents for individual cost issues,
- establishing a system for collection and summarisation of quality costs,
- assessing the impact of quality costs on the company profit (Nováková & Ovsenák, 2013).

There are several options to assess quality. ROQ indicator is one of them. Authors of this paper propose that this indicator takes the following form:

$$ROQ = \frac{P}{QC} \tag{1}$$

where $QC = CRD + CD + CO$.

P is profit from the production of particular product, costs for research and development (CRD), costs for defensive strategy (CD) and costs for offensive strategy (CO).

Denominator in the formula says that profit is not created barely by quality cost. The ratio does not directly express the effectiveness of quality system. However, if to examine the ratio in a sequence of time, we can indirectly deduce the effectiveness of quality management system from whether the variations of the ratio exhibit positive or negative development. When creating this indicator, the influence of time factor should be taken into account. If a product is manufactured for more than 1 year, costs for research and development are single-shot; ROQ formula should then involve only the ratio of the costs attributable to 1 year of the product manufacturing.

The abovementioned indicator applies to total production. If a company manufactures several products and needs to express ROQ indicator for the one product, this indicator will include the profit and the quality costs of this product:

$$ROQ_i = \frac{P_i}{QC_i} \tag{2}$$

The quality triad allows also to express the return on innovation activities with the indicator ROIA.

$$ROIA = \frac{P}{CRD} \tag{3}$$

If we follow the evolution of the profitability of quality this development could we express by means of the proportional indicator.

$$\frac{RQ_{x+1}}{RQ_x} = \frac{\dfrac{P_{x+1}}{QC_{x+1}}}{\dfrac{P_x}{QC_x}} \qquad (4)$$

This indicator should be greater than 1. If it does not, the produce products cease to meet customers requirements and manufacturer should change the properties of product (Rust et al. 1996).

In the next interpretation of indicators of quality returns we will come back to already mentioned indicators of Net Present Value (NPV) and to indicator of Economic Value Added (EVA). But both cases are based on the precondition that the quality is basically influenced by the costs on research and development. Moreover, these costs are not single, but they have the character of operative investments, therefore they may be included to these considerations as investments. If we use the first method, then the present net value equals to:

$$NPV = \sum_{1}^{n} \frac{CF_t}{(1+k)^t} - (CQ + C_i) \qquad (5)$$

where CF is a cumulated cash flow value,
CQ are quality costs,
Cri are investment costs caused by the research of certain product,
k is the company discount tariff,
t means year 1 up to n years.

If the indicator of economic value added is used, then

$$EVA = NOPAT - (WACC.C1t + WACC.C2t) \qquad (6)$$

where EVA is an economic value added,
NOPAT is an operative profit after taxation,
WACC is a weighted average costs on foreign and own capital
C2t is own and foreign capital invested for a long term for research and development
C1t is own and foreign capital invested for a long term for other areas.

If the indicator value is positive, the development is favourable if the indicator value is negative, the development is unfavourable, non profitable. Both indicators (NPV,EVA) may be use to express quality returns and they are modern indicators similar to others indicators of profitability. If the

first method is used, the influence on the present net value may be expressed as follows:

$$ROQ = \frac{NPV}{QC} \qquad (7)$$

And when the economic value added is used:

$$ROQ = \frac{EVA}{QC} \qquad (8)$$

If we want to express the return on quality including losses from low-quality production, such extension may lead to the expression of a "completion ROQ indicator" which would represent the sequence of following items:

– costs on research and development (CRD),
– losses from the low-quality production (CL),
– costs on defensive market strategy (CD),
– costs on offensive market strategy (CO).

Completion quality returns would be then equal to the ratio of profit (or present net value, or economic value added:

$$KROQ = \frac{P}{CRD + CL + CD + CO} \qquad (9)$$

4 CONCLUSIONS

Besides technical aspect, quality management should involve also communication and economic aspects in order to be successful. It is essential regarding the effectiveness of quality management, the profit achieved and profitability of a company. Triad of quality discussed in the paper provides a new aspect of quality costs, focusing on quality itself rather than on low-quality as in case of PAF model. The quality costs used in PAF model should to be monitored because it reduces the company profit, but having nothing common with the modern understanding of quality. In contrast, the approach used in article and the designed structure of quality costs allows to track the relationship between quality and the company profit and directly expressed the profitability of the quality. The information in the article are based on previous analyses of foreign and domestic literature. Application of indicator ROQ in practice of industrial companies has not yet been comprehensively evaluated and is currently the subject of grant tasks. All information contained in the article represent a new trend in the field of evaluation of profitability of quality. In practice and in literature this approach is unknown and not used.

REFERENCES

Linczényi, A. 2006. Indicators of Benefits from Quality. *Proceedings Quality for Life*, Ostrava.

Linczényi, A. & Nováková, R. 2011. Returns on Quality—ROQ model. *55. Congress EOQ*. Budapest, Hungary.

Nováková, R. 2006: Indicators of Profitability in Quality Management and their Influence on Competitiveness of the Company. *Proceedings Quality for Life*, Ostrava.

Nováková, R. & Ovsenák, V. 2013. Awards for design quality of evidence = (Design awards = proof of quality). *Marketing Identity Design, which sells: Proceedings of the International Scientific Conference*. Trnava: FMK UCM.

Rust, R., Zahorik, A. & Keiningham,T. 1996. *Return on Quality*. Irwin Professional Publishing, Chicago.

Synek, M. & Co. 1996: *Managerial Economy*. Grada Publishing. Prague.

Production Management and Engineering Sciences – Majerník, Daneshjo & Bosák (Eds)

The proposal of innovation support in Small and Medium-sized Enterprises

S. Lorincová & M. Potkány
Technical University in Zvolen, Zvolen, Slovak Republic

ABSTRACT: Small and Medium-sized Enterprises (SMEs) are significant part of businesses in Slovakia. They participate in the creation of gross domestic product, export, import or other parameters of prosperity evaluation. Competitive fight is forcing businesses to bring new ideas, products or services to the market. Therefore, it is essential that SMEs should deal with innovations, which are considered as an essential part of business activities that contribute the development of competitiveness. SMEs can provide innovation activities at their own, in case that they have own funds. SMEs can use financial support from programs designed to support innovation. Another way to support the innovation in SMEs is the use of aid from supporting institutions. In Slovakia there are many institutions which support the sector of SMEs. The aim of the paper is to propose a single institution that would comprehensively support the sector of SMEs and would cooperate with other players.

1 INTRODUCTION

If the firms do not innovate, their products would become unattractive and they would close down their business. It is not the goal of any entrepreneur. On the other hand, if a company is interested in realizing their business as long as possible, it is necessary to carry out various types of innovation. Many SMEs provide these activities alone. But there are various methods that facilitate the innovation process, in the case that the company does not have funds. It is possible to support the sector of SMEs through various European programs or through various support institutions. In Slovakia, there is a lot of institutions that deal with the issue of supporting of SMEs such as Slovak Investment and Trade Development Agency, Slovak Research and Development Agency and many others. These institutions deal with the sector of SMEs, they are providing the support through the use of various programs aimed at research, development, innovation activities, or other activities. The Slovak Republic understands that research and development creates a presumption of knowledge-based economy, which provides competitiveness and is a source of new knowledge. However, it is necessary that this issue is not only discussed, but it is necessary to take steps to enable this form of support to get in a practice, because those institutions are fragmented and their help is too crowded, fragmented, since each institution supports SMEs by different way.

2 THE PROPOSAL OF INNOVATION SUPPORT IN SMALL AND MEDIUM-SIZED ENTERPRISES

In Slovakia, there are a number of institutions that deal with the issue of innovation support in SMEs. A single institution would be sufficed for closer link to the activities of all institutions which support SMEs. We propose a creation of The Institute for Support of Small and Medium-sized Enterprises, which would have its branch in each region, it would support SMEs through innovation centers with nationwide coverage and coordination, because according to Slovák (2008) the headless construction of supporting institutions is not the key to innovation and the knowledge economy.

This institution would provide financial and non-financial assistance at the highest level of quality, which would create conditions for innovation. We propose that institution would cooperate with foreign institutions of similar purpose and would assist in involving the Slovak enterprises in international projects. The advantages of this cooperation we see in support of other activities carried out by the institution. Institution would seek out the foreign partner or sponsors for cooperation with Slovak companies. This would result in an influx of foreign experience, innovative approaches, ideas and thoughts in different areas. Obtaining of valuable information at a higher level of quality, would help in the removal of barriers that discourage SMEs in innovation activities. At the same time, the successful cooperation with foreign partners could be the inspiration for Slovak businesses to take initiatives themselves, to take actions to search for foreign partners, which would result in improving the business environment.

In the financial area, we recommend the collaboration with financial institutions, which would reflect in support of SMEs financing. It would dispose by trained personnel who would provide assistance in obtaining funds from European Union programs. The institution would provide assistance in obtaining these funds, because inexperience and unpreparedness is the current problem of Slovakia. Most of the problems with the implementation of projects is based on non-binding conditions, obligations, time schedules and technical provisions contained in the agreement. The beneficiaries have underestimated their study and therefore here, we see the greatest potential of the institution, in the form of assistance to entrepreneurs. According to Novota (2006) strict control requires projects to be processed without a single error. Therefore we propose the cooperation with employees who are skilled in the field of funding, with the aim to eliminate the percentage of failed projects.

The institution would stimulate cooperation of SMEs with universities and regional government. It would be an intermediary between SMEs, the scientific sector and the regional government, which is confirmed by the experience from Finland, where more than 70 telecommunications and radio-communications projects have been developed for the last 15 years through cooperation between businesses, universities and research centers.

Another proposed activity of institution would be an advising the general public in financial and non-financial areas. Furthermore, it would participate in the creation of new legislation and its adjustment to ensure transparency and avoid abusing, unfair treatment or tunneling.

The institution would spread the philosophy of the SMEs support through business incubators, conferences, seminars and other activities. The Finnish innovation strategy is aimed at developing innovation. Similarly Slovak strategy could consist of innovation support with business incubators and university sector. Our recommendation is confirmed by the experience of the UK, where 32 incubators were linked to the university environment already in 2008.

The proposed institution would be responsible for business incubators in Slovakia and institution would coordinate their activities. International experience shows that systematic support of inventions, discoveries and knowledge acquired by research—development activities in the economic and social practice is essential to ensure sustainable development of the knowledge-based society.

The part of building a knowledge society is to focus on the development of innovative potential. Sustainable economic growth must be based on the latest knowledge. The innovative potential of universities plays a key role in this process. Therefore, our next proposal is to encourage closer co-operation of three players. We propose the creation of cooperation between universities, the business sector and regional government, which in the current period is realized in all regions of Slovakia mainly through the regional innovation strategies. We expect an equal participation of all three players.

It is essential to promote and strengthen the cooperation with the three players, as in Slovakia, the cooperation between universities and the business sector is not developed at the same level, as at the University of Cambridge in the UK, which dealt with this issue already in the 60s of the 20th century. At that time, about 25 companies cooperated with universities. The number of participating companies and universities are currently being increased more than 10 times.

Universities would be included to this system because they are considered to be a driving force of the society. They have human capital in the form of students. Universities educate people with their own opinions. They must not forget to modernize their curricula to get new methods of education into the practice. Then the students could have access to innovations in various fields. The development of students would be supported through innovative practices. Their talent is the most important asset of student because it creates space for new innovations.

The cooperation between the three players, we see in implementing of research projects through which it is possible to improve the knowledge of students in universities. Better conditions would be created for the development of new innovative products, services or technologies. The application of research results in practice, it would mean a competitive advantage for SMEs and the benefits would be passed to the region in which the company operates.

With an aim to sale of innovative products by universities, they would carry out the research projects as same as the marketing research in various areas that would be targeted to the needs and requirements of customers. Students would focus their attention to the needs and demand of customer or by their own initiative they would come up with proposals that would be verified through the research. Since every business is trying to look for opportunities and ways how to improve its products, services or processes, there is space for collaboration with universities in the creation of new products and services. After verification of customer needs, SMEs would help in the implementation of students ideas in practice. SMEs would ensure the realization of the students' ideas by producing innovative products and services to the market.

Horníková & Kampová (2014) give the examples of successful cooperation between universities and SMEs. It can be find in Dublin City University, which has developed technology to capture images and sound via a mobile phone. Now is used for football matches and Irish mobile operators have expressed interest in licenses.

Another opportunity we see in ordering the university services by the business sector. Enterprises would participate in the financial support of universities by project funds. On the other hand the universities themselves would apply for various grants. Inspiration we can find in a Finnish technology and innovation agency that support innovation in enterprises by cooperation with the universities in funding innovation, science and research.

Finland could be an example of the innovation and research, which is at a very high level of quality. Finland is the only country among all countries of the European Union which spends the largest expenditure on science. The funds allocated for this area has increased, on the other hand, even in times of crisis, the volume has not decreased, which helped to start up the information and communication technologies, through the most famous manufacturers of ICT, which is Nokia.

In addition to cooperation between universities and the business sector in research projects, we see the space for cooperation of three actors in the education. Students have the potential, which can be used for innovation. We propose to restore the model, which works in Slovakia in the form when secondary schools participated in teaching students in cooperation with businesses in which students could work after graduation.

The initiatives in training its own employees may be based on the businesses as there are growing demands for quality education of its employees. Our proposal is supported by the fact that this model already works in Ireland. There is another experience when the foreign company was interested in openning a branch in Slovakia, but they met with the fact that although we have a lot of applicants but their qualification is insufficient. It is necessary to increase the level of funding for universities as well as in Finland, where the largest volumes of funds are allocated for the field of education as well as science and research. There is scope for regional governments to ensure sufficient funding for this area.

It is important that universities start to offer something interesting, specific, what do not provide the others. In this context, we propose that the university would prepare the study programs within the framework of lifelong learning, which would accept the requirements and needs of practice, as it is in the Ireland, where universities offer different types of courses in the form of short and long courses, distance learning, evening school, distance learning, e-learning, and other forms of education.

The representatives of the business sector and regional government would participate in the teaching process, where students could directly improve their knowledge under the guidance of professionals. Students would implement projects in enterprises. Cooperation between universities, SMEs and regional government would help all players. University as well as SMEs and regional government would benefit from this cooperation, because teachers as well as representatives from business, regional government would be in contact with the practice, which would enrich their practical skills.

We see the role of regional government in providing of financial support, in the legislative enclosure of the issue, in the competences to develop cooperation of three actors, in the support of innovation in SMEs, in elaboration and implementation of innovative strategies into the practice, in providing further benefits for enterprises that participate in such projects, for example by providing tax relief, various fees, etc. For us it is important that regional government would take action in cooperation with the universities and SMEs in closer mutual cooperation.

It would bring a positive effect accruing to all players by applying our proposed system. Developing of closer cooperation between universities and SMEs would help mutual relations and the commercialization of knowledge itself, moreover, it would create natural competition and benefits would accrue to the trader who participates in the cooperation as well as to regional government, which would create space for development of economy and improve the situation of the whole region.

Furthermore, we believe that it would be necessary for regional governments to cooperate with universities, SMEs and business incubators through university incubator which is considered as a tool for innovation support.

The proposed cooperation of universities, SMEs, regional government and business or university incubators is confirmed by Luger (2008), who stresses the relationships between governance, universities and the private sector at different levels. Skokan (2010) has a similar view to this issue.

159

He states that regional cooperation is essential for creation of institutions that would produce and distribute knowledge for preparing a skilled workforce. Business incubators and universities support technology transfer and diffusion of innovations. The demand for knowledge comes from SMEs and organizations that create innovative products and processes. These include suppliers, cooperation partners and other players.

According to Kao (2009) the access to new ideas, information, opportunities would be arrange by the creation of a cooperation of these players. It would eliminate the barriers to the further development of innovation in the region. The proof of this cooperation can be found in Ireland Trinity College in Dublin Innovation Center. The specific aim was to help teachers to become entrepreneurs. International companies established in the university were managed by academic staff. They attract the international corporations which could set up laboratories and use college human resources and technological equipment. The government set up public laboratories and research centers in the area of university to ensure cooperation with enterprises.

There is another example of real operation of innovation support centers abroad. It could be an inspiration for Slovak Republic. Kao (2009) summarized the four basic models that are available:

– Model 1: The Focused Factory—This model of innovation center is characterized by narrow specialization in a particular area, whether it is a biomedical or technological solutions. The focused-factory innovation model combines a clear strategic intent with a concentration of infrastructure and high-octane talent in an effort to discover and deploy new solutions to big challenges. Countries such as Singapore and Denmark, for instance, focus their innovation investments on a handful of industries or research fields. Singapore created Biopolis, a biomedical research center is recognized in medical research. Biopolis has become a globally recognized center for stem cell research. Biopolis attracts top scientists around the world, as well as the best graduate students. There are provided the best conditions for their research. Organizations that set up shop at Biopolis participate in relationships with government agencies, venture capital firms, global pharmaceutical companies, academic research labs, and other institutions. There is a lot of synergies and many benefits for example from sharing resources. In innovation centers there work linking between research centers and business perfectly. GlaxoSmithKline, for example, founded its Centre for Research in Cognitive

and Neurodegenerative Disorders at Biopolis. Centralization of technology and best scientific capacity is a huge benefit for companies. Singapore is just one of several countries hosting focused factories. Companies that want to leverage their existing patents and intellectual property, say, in wireless technology, precision manufacturing, or clean technology, might also look to Finland and Denmark. Emerging economies such as Chile and Vietnam have focused factories as well.

– Model 2: Brute Force—This model is the other model which may inspire Slovak Republic. The brute force model is an innovation version of the law of large numbers. By applying massive amounts of low-cost labor and capital to a portfolio of innovation opportunities, countries (most obviously China and India, but also Brazil) hope that a huge quantity of ideas from a substantial number of talented people will eventually yield valuable discoveries. China, currently the world's center of outsourced manufacturing, will be the next hub of brute force innovation. The Chinese Politburo has set itself the concrete goal of turning China into an innovation-driven country by 2020. To that end, China has chosen 10 of its leading universities to receive extra funding in order to achieve world-class status. The goal is to churn out well-educated specialists in every area of science and technology. The Chinese automobile industry offers a glimpse of the brute force model in action. Thanks to an outpouring of educated innovators from Chinese universities, there are now an estimated 50 car companies in China, producing a Precambrian explosion of new business models and automobile designs. Many of these companies will fail, but some may prove to be world-beaters. Warren Buffett's recent $230 million investment in BYD Company, a Chinese maker of batteries for electric cars, signals his awareness of the potential for Chinese R&D. China offers innovation advantages to other kinds of companies, as well as Microsoft. The company has found that the center allows it to tap expert and junior Chinese talent at a comparatively low price. Microsoft supports the work of top Chinese academics and encourages researchers to publish their work and participate in academic conferences. It also funds projects selected by the National Research Fund of China. In return for all this, Microsoft can gain access to a trove of IP and build invaluable collaborative relationships.

– Model 3: Hollyworld—is all about providing opportunities to build a "global creative class." It leverages "the increasing returns law of cool community". As more and more smart entrepreneurs gather in one place, the more attractive that place becomes to other like-minded people. In the 1990s, Silicon Valley pioneered this law to excellent effect. Today, urban centers as diverse as Bangalore, Helsinki and Toronto have adopted a Hollyworld model. Entire countries are also moving in this direction. India, for example, is shifting its role as the world's back office to that of innovation epicenter. The country is doing this by partnering the best graduates of its Indian Institutes of Technology with Indians who have trained at such universities as Stanford, MIT, and Cambridge and are now thriving in Western economies. Indian entrepreneurs who have already made their mark in Silicon Valley are now cementing commercial ties to their homeland in globalized technology enterprises. As this occurs, India's resident creative class is becoming more influential, cosmopolitan, and skilled. Another country that is adopting this model successfully is Singapore, where Hollyworld and the focused factory meet. Singapore is willing to generously fund life-sciences graduate students, regardless of nationality, providing that they maintain a suitable grade point average and return to Singapore for the equivalent of national service. In this way, the country will enlarge its overall population of the creative class.

- Model 4: Large-Scale Ecosystems—Several countries have adopted this model such as Finland, where the innovation system was designed, in part, as a response to the cataclysmic economic change in 1991, occasioned by the collapse of the Soviet Union. In one stroke, a significant percentage of Finland's foreign trade vanished, plunging the country into recession. Faced with an economic near-death experience, the country decided to focus on education, science, and technology and to improve its innovation capability. Today, Finland enjoys a well-run innovation system benefiting from strong governmental stewardship. It can be an inspiration for the Slovak Republic.

3 CONCLUSIONS

It is evident that the Slovak business incubators are not as known as the mentioned centers abroad. They have a long way to become well-known. The Slovak business incubators can inspire by foreign experience and then become famous, such as incubators in Singapore, Finland or Denmark.

Slovakia can inspire by model 2 "Brute Force" through education support in different areas that we identified in the previous proposals of closer cooperation between universities, SMEs and regional government. Our proposal for closer cooperation of three actors is supported by literature, where Slovák (2008) states that systematic cooperation does not exist in Slovakia yet and the investments supports from government is inadequate in Slovakia. The author adds that one of the traditionally strong European scientific—research sector is energy. This could be very interesting for Slovakia for example in geothermal energy. The future industry will be water and waste water treatment. According to Slovák (2008) water and energy will be the most important themes in the time of ten years and we can expect changes in the logistics flows in the coming years. Therefore we incline to Kao (2009) stating that it is necessary to consider the capabilities and resources of country and create a strategy to provide businesses with innovation services and how to create an attractive environment for specialists. SMEs should consider benefits offered by business incubators and capabilities of individual models. A world of global innovation is here for everyone but only well prepared businesses will be successful.

ACKNOWLEDGEMENTS

This paper is the partial result of the Ministry of Education of Slovak Republic grant project VEGA No. 1/0268/13.

REFERENCES

Horníková, Z. & Kampová, S. 2014. Cooperation between universities and business? Of course! *Trend* 24(26).

Kao. J. 2009. Tapping the World's Innovation Hot Spots. *Harvard business review* 89(3): 109–115.

Luger, M.I. 2008. *On the government's role in regional economic development.* New York: Routledge.

Novota, M. 2006. Effective tool of regional development. *Eurobiznis* 6(10): 24–26.

Skokan, K. 2010. Innovative paradox and regional innovation strategies. *Journal of Competitiveness* 2(2): 30–46.

Slovák. K. 2008. Niilo Tapani Saarinen: Innovation and bureaucracy do not mix. *Trend* 18(17): 56–58.

Production Management and Engineering Sciences – Majerník, Daneshjo & Bosák (Eds)
© 2016 Taylor & Francis Group, London, ISBN: 978-1-138-02856-2

Level of protection of critical infrastructure in the Slovak Republic

T. Loveček, A. Veľas & M. Ďurovec
Faculty of Security Engineering, University of Žilina, Žilina, Slovak Republic

ABSTRACT: The article deals with possible ways to specify the European regulations and the Act No. 45/2011 Coll. on critical infrastructure, due to ambiguity of the determination method for the required protection level of critical infrastructure elements of energy network. Existing legislation in the European Union approaches the physical and object protection of it declaratively and does not specify proposals for its solution. We define three basic general approaches to the protection of constructional objects. Further, in more detail, we describe the possibility to apply already existing relevant standards of physical and object security used in the energy sector. The article includes a categorization and specification of the protective measures divided into four main groups. Currently there are no parameters set by technical regulations of particular measures which create conditions for the use of new approaches, evaluation methods and proposal of protection elements of critical infrastructure.

1 INTRODUCTION

According to the §2, letter a) of the Act No. 45/2011 Coll. on critical infrastructure (hereinafter the "Act"), the element of critical infrastructure is mainly engineering structure, public service and information system in the sector of critical infrastructure, disruption or destruction of which would have serious adverse consequences on economic and social function of the state and thus on the quality of life of citizens, considering the protection of life, health, safety, property as well as environment.

In the competence of the Ministry of Economy of SR, it concerns the sources of the power system, facilities of the electricity distribution systems, selected elements of the gas distribution network, storage tanks of oil and oil products, oil refineries, pipelines etc.

According to the §9 of the Act, the owner is obliged to protect the element against the disruption or destruction. For that purpose, when modernizing the element, the owner is required to apply technology that ensures its protection and according to §10 of the Act implement a security plan.

The security plan of the element of CI is, in accordance with the §10 of the Act, a document which includes a description of various threats of disruption or destruction of the element, vulnerabilities of the element and security measures for its protection. Minimal procedure for the elaboration of the security plan is presented in the Annex 2 of the Act.

According to (Act No. 45/2011), security measures for the protection of the element of CI are divided into:

a. Permanent security measures which include investments and procedures to ensure the protection of the element, namely:
- Mechanical barriers
- Technical security devices
- Security features of information systems
- Organisational measures with emphasis on the process of notification and warning as well as on crisis management
- Training of personnel responsible for protection of the element
- Inspection measures to abide the continual security measures
b. Special security measures which are applied according to the intensity of the threat of disruption or destruction of the element.

The law does not specify to what extend and what level (e.g. security class) should be the security measures implemented.

Existing legislation in the European Union approaches the physical and object protection of the critical infrastructure elements of energy network declaratively and does not specify concrete proposals for its solution. Although the document (Green paper 2008) lists possible ways (tools) to improve prevention, protection, readiness and response within the critical infrastructure protection in the conditions of the EU, it does not specify them further. In the documents "The concept of critical infrastructure in the Slovak Republic and the method of its protection and defence" and "National program for protection and defence of the critical infrastructure in the Slovak Republic" are again generally characterised tools which can be used to reduce the threat to critical infrastructure. These could be technical means to deter, detect, verify, signalise and eliminate an intruder (mechanical and electronic) and activities of security services (e.g. intervention of the security forces and the armed forces).

2 GENERAL APPROACH TO THE PROTECTION OF THE STATE STRATEGIC OBJECTS

Protection of the strategic facilities/elements or objects of the state is, in Slovakia, solved individually in a variety of law regulations, however, with a different philosophy of approach to their protection. It concerns for instance strategic objects of the defence infrastructure, mainly objects of special importance and other important objects (Act No. 319/2002), nuclear facilities (Decree No. 51/2006), buildings and premises for storage and handling with classified information (Decree No. 336/2004) or objects of financial institutions (Presidium 2006).

Under these general statutes, three basic approaches to define the extent and the level of protection of strategic objects of the state are used (Boc et al. 2013):

a. Directive approach, where the subject must accept exactly specified protection system, regardless of the specifics of the operation and the environment it is located in (e.g. protection of classified information in SR until 2004).
b. Variant approach, where the subject can choose from limited number of variations, combining different technical security means and organisational or regime measures (e.g. protection financial institutions or protection of classified information in SR since 2004).
c. Variable approach, where the subject has to take measures within the protection system which take into account breakthrough resistance of the mechanical barriers, reaction times of the intervention force and probability of the detection of alarm systems (e.g. protection of nuclear facilities).

Nowadays, the most effective approach is considered to be the variable approach, based on the assumption that it is necessary to use so many technical security means that the intruder would be detected and detained by the intervention unit even before reaching its target. However, in the EU, there has not yet been established methodology, standard or a software tool that would define in detail and enable the application of variable quantitative approach to protection. Certain starting solutions are offered by software tools which were created in the USA, Russia or South Korea primarily for the need to protect nuclear material and equipment.

In general, we can talk about qualitative and quantitative approach. The qualitative approach to design or evaluation of the protection system of the object is based on the expert estimation of the evaluators where it is not possible to precisely demonstrate the efficacy, reliability or efficiency of these systems and it is necessary to rely on the professional competence of the creators of technical standards, generally binding regulations or software applications (e.f. RISKWATCH—Campus Security, RISKWATCH—Nuclear Power, RIS-KWATCH—Phys. & Homeland Security, RISKWATCH—NERC, from company Risk Watch International develops, USA). (Boc et al. 2013).

Quantitative approaches are based on mathematical and statistical methods which allow us, by the use of measurable input and output parameters, to exactly demonstrate the efficacy, reliability or efficiency of the security system. The software tools using these methods include for example SA-VI/ASSESS (Sandia National Laboratories, USA), Sprut (ISTA, Russia), Vega-2 (Eleron, Russia), Analizator SFZ (FRTK MFTI, Russia), SAPE (Korea Institute of Nuclear Non-proliferation and Control, South Korea) or SatANO (University of Žilina, Slovakia). (Boc et al. 2013).

3 STANDARDS OF THE PROTECTION OF CRITICAL INFRASTRUCTURE ELEMENTS IN THE ENERGY SECTOR

In 2008–2010 the project "APENCOT: Analysis of the Protection of Energy Networks' Crucial Objects against Terrorism and Proposal of Security Standards" was implemented. This project was a part of the program "The Prevention, Preparedness and Consequence Management of Terrorism and other Security-related Risks (CIPS)". This programme was announced in 2007 by the European Commission—Directorate—General Justice, Freedom and Security under the Council Decision No. 2007/124/EC (Council 2007). The main researcher in this project was the company F.S.C. BEZPEČNOSTNÍ PORADENSTVÍ, a.s., Ostrava. The project was aimed to protect critical objects of generation, transmission and distribution energy networks. Within this project, security standards of physical protection for the energy network were created.

The above mentioned security standards have an objective to set appropriate levels of individual protection measures of energy network assets in order to prevent the threats of terrorism and further criminal activities threatening the assets of the energy network. Their purpose is to prevent forced entry into protected buildings and facilities and related risk of destruction and damage of technological equipment and other assets. These security standards apply directive approach, i.e. the subject has to adopt exactly defined security measures. For practical application of these standards "methodology for assessment of protection of energy facilities against crime and terrorism" was created. Under this methodology, the assessment of physical protection is done by the analysis of the level of the current state and its comparison with the minimal standard of physical protection of an object in corresponding category (APENCOT 2010).

Each object of the energy network is significant in regard to ensuring the safety of systems of generation, transmission and distribution of electricity, whereby the significance of these objects reaches different levels which must correspond to different level and extend of security measures. Within the mentioned methodology, 4 categories of objects were defined, while the category I includes the objects and technological units of crucial importance, whereas the category II includes objects with major importance and the category III includes the objects with important meaning. Classification of the objects into given categories is specified in the following tables 1–2.

In 2009–2011 the company F.S.C. BEZPEČNOSTNÍ PORADENSTVÍ, a.s. implemented another project CIPnES: Critical Infrastructure Protection in Energy Sector which was also a part of CIPS. This project is closely related to the previous project and resulted into security standards for physical protection of gas and oil industry. (CIPnES 2011).

These projects were followed by the project of grand scheme by the Ministry of the Interior of the Czech Republic "Objectification of Threats and Risks to Facilities for Generation and Transmission of Electricity" (2011–2013). The main outcomes of this project were:

a. Methodology for unified identification of facilities for generation, transmission and distribution of electricity by national and European critical infrastructure and assurance of the physical protection of these facilities.
b. Software tool for individual determination of criticality of the facility for the generation, transmission and distribution of electricity.
c. Scientific monograph "Critical Infrastructure of Electricity Sector: Identification, Assessment and Protection".

Single outputs of the project are publicly available at the website of VSB—Technical University of Ostrava, Faculty of Security Engineering: https://elektrina.fbi.vsb.cz/page/o-projektu/.

In Slovakia, the elements of CI are ratified by the Resolution of the Government of SR and are assigned into the sectors of critical infrastructure. The list of elements of critical infrastructure in competence of the Ministry is classified information in accordance with the law (Act No. 45/2011) on the protection of classified information, as amended.

In 2013, the Ministry of Economy of SR established a working group to prepare guidelines for owners of the elements of critical infrastructure in the sector of energy and industry. The purpose of this is to establish standards to assure the protection of the elements of CI and define the minimal necessary level and extent of the measures of physical and object security (hereinafter "FOS") for each category of objects and premises of energy network and industrial sector (Ministry 2014).

Within the processing of this regulation the outputs of the project CIPnES: Critical Infrastructure Protection in Energy Sector were used. Methodological guidance of the MoE SR on security measures for the protection of the critical infrastructure elements in the sectors of energy and industry is currently in the final stage of processing.

According to the (Ministry 2014) four zones of protection were defined for each type of CI element, the same as in the project APENCOT:

a. Extra secured zone I
b. Secured zone II
c. Protected zone III
d. Controlled zone IV.

Corresponding minimal standards of physical and object security were defined for all four zones and for each subsector of energy and industry sector separately (subsectors of electricity industry, gas industry, oil and oil products, mining, metallurgical, chemical and pharmaceutical industry).

For each security zone I to IV, the extent and the level of security is defined in four basic groups of security measures:

a. mechanical barriers,
b. technical security devices,
c. physical protection,
d. regime and organisational measures.

Mechanical barriers are characterized by mechanical protection means, which complicate or prevent an intruder (unauthorised person) to enter the protected area or manipulate with protected objects. The application of mechanical barriers lies in their mechanical solidity, resistance of used materials and a link to the other form of protection (e.g. alarm systems). It is possible to use mechanical barriers in three basic protection zone perimeters:

Table 1. Categories of transmission network assets and corresponding security zones (APENCOT 2010).

Cat. of assets	Assets of transmission network	Security zones

I.	Technical dispatching of the transmission network owner
	Commercial dispatching of the transmission network owner
	Premises accomodating facilities for processing and transmission
	    of transmission network data (IT department, servers, computer
	    rooms, etc.)
		Extra secured zones

II.	High voltage substation
	Main switchgears of high and low voltage
	Fuse board cabinet of high-frequency converter (HF room)
	Cable end-bog room and LAN data switchgears
	UPS backup power supply, including diesel generators
	Ventilation engine room
	Office of management, owner of the transmission network
	PABX and telecommunication room
	Service room for control units for intrusion and hold-up alarm system
	Other areas assigned for the owner of the transmission network
	    (e.g. cash register, workstation of the physical protection, issuing
	    and administration of ID cards according to the Act No. 412/2005
	    Coll. on protection of classified information and security
	    competence, etc.)
		Secured zone

III.	Buildings of joint operations Central objects and objects of secondary techniques Diesel generator station and auxiliary substations Garages and selected repair shops Inspection towers and selected storage space Corridors of technological infrastructures of the building sublevels of the owner of the transmission network	Protected zone

IV.	Interior areas defined by the fencing of the facility of the owner of the
	    transmission network
	Interior areas defined by the outer shell of the building of the owner of
	    the transmission network
	Offices of the administrative subdivisions of the owner of the
	    transmission network
		Controlled zone

	Poles, constructions and cabling of the transmission network Outdoor parking area for vehicles (parking) Outdoor air conditioning units	Outer zone

a. Perimeter protection provides security of the surroundings of the protected object. Its circuit (perimeter) can be defined by nature (rivers) or artificially (fence, wall, etc.). Basically, we talk about mechanical barriers designed for the outdoor use, main factor of which is spatial separation of the protected object form the external environment. Mainly, it concerns various types of fencing, driveways and entrances into the protected area—gates, wicket gates, turnstiles, security sluices, bars, spike barriers, etc.

b. Building envelope protection prevents violation of the shell of the object and all of its opening fillings. They mainly protect the construction openings of the building (doors, windows, etc.) to prevent the penetration of an intruder. It is necessary to consider the walls, floors, ceilings and roofs of buildings, which are also the object of an attack.

c. Object protection ensures the protection of objects in the protected place and the protection of items saved in storage facilities in various places of interest within the object. These are mainly facilities, purpose of which is to protect valuable objects, documents, financial cash and other important records. In addition to the mechanical breakthrough resistance, they are required to be fire resistant. This group can include security and commercial depositories.

According to (Act No. 45/2011), the technical security device is understood as an access control system, electronic (electric) security system, CCTV system, fire detection and alarm system, equipment to detect substances and objects, bugging device detector and a device for physical destruction of the information media.

Table 2. Categories of assets of power generation and distribution network and corresponding security zones (APENCOT 2010).

Cat. of assets	Power generation and distribution network assets	Security zones
I.	Dispatching department Corporate server rooms, transit and control PABX Replacement and backup sources of electrical energy of class I asset Selected central offices (surveillance centres, cash registers with limit over 200 000Kč, dept. of important documents, etc.)	Extra secured zone
II.	Heating and Power plants Hydro-electric power plants with installed capacity exceeding 10 MW Transformer stations of distribution network VHV/HV VHV Switching stations Types of office buildings such as company residence and customer service Regional and local server rooms and PABX Replacement and backup sources of electrical energy of category II asset Departments handling protected data, archives and register offices, tests, measurements, cash etc. Storage space for the category II assets	Secured zone
III.	Transformer stations of distribution network HV/LV HV Switching stations Hydro-electric power plants with installed capacity of less than 10 MW Wind power plants Office buildings of the operation management and multipurpose type Other IT equipment and other PABX Garages, workshops, warehouses for the category III assets	Protected zone
IV.	Distribution substations HV/LV Training centres, recreation centres, lodging houses, welfare facilities etc. Storage space for the category IV assets Shelters, local boiler-rooms and parking areas	Controlled zone
	Outdoor parking area for vehicles (parking), poles, constructions and cabling of the transmission network	Outer zone

Access Control System is a system containing all structural and organizational measures related to the equipment necessary for the access control (EN 50133-1).

Intruder Alarm System is an alarm system detecting and indicating the presence, entry or attempt of an intruder to enter the protected area. Hold-up Alarm System is an alarm system providing the user with means for intentional generation of emergency alarm situation. Intrusion and Hold-up Alarm System is a combined intrusion alarm system and hold-up alarm system. Intrusion and hold-up system may consist of several subsystems. (EN 50131-1)

Fire Detection and Fire Alarm System is a group of components including a central unite which is capable of detecting fire indications and send signal for appropriate action. (EN 54-1)

CCTV: Closed Circuit Television System is a system consisting of camera equipment, monitoring and accessory equipment for transmission and management purposes, which may be necessary for the supervision over the protected space. (EN 50132-1)

Physical protection can be divided into self-protection (e.g. neighbourhood watch) and protection provided by security services which are divided into public and private.

Regime and organizational measures ensure correct and efficient function of the aforementioned technical and physical protection measures.

In the upcoming methodological guidelines, unlike in the standards of the projects APENCOT or CIPnES, no specific requirements for the security level of mechanical barriers and technical security devices are defined, i.e. required security classes are not defined according to a relevant technical standard.

4 CONCLUSIONS

Following the adoption of the guidelines of MoE SR, the owners of CI elements will have more accurate vision of which security measures from the field of physical and object security they should enforce in their premises. However, by the adoption of these measures, only those risks are minimized or reduced which are caused by a deliberately acting person whose objective is to damage or destroy the CI element. According to the law, all the risks and threats having impact on functionality, integrity and continuity of the CI element must be assessed in the security plan. It implies that merely the implementation of the security measures referred to above may not be sufficient. It is necessary to take into account the security measures of other areas such as:

a. Personal protection system (e.g. occupational safety and health protection and protection of government and state officials)
b. Protection of information systems
c. System of technical (technological) security (e.g. prevention of major industrial accidents)
d. Fire protection system
e. Environment protection system.

The above mentioned systems/fields of security are partially solved in number of generally binding legal regulations and technical standards, and therefore we can assume that the legislators no longer mentioned them in the law (priority was given to the protection against terrorist attacks from the outside).

However, there is a wide variety of possible risks and threats which cause disruption or limit the activity of CI element. Therefore, the owner should preferentially address the security measures which do not primarily deal with the probability of the occurrence of a negative event but with those, which reduce the impact of these events, i.e. security measures of the Business continuity management (e.g. impact analysis, incident management plans, business continuity plan, disaster recovery plan, service level arrangements, etc.).

ACKNOWLEDGEMENTS

This article is part of the project Critical Infrastructure Protection Against Chemicals Attack. The project is co-financed by the EU program Prevention, Preparedness and Consequence Management of Terrorism and Other Security-Related Risks Programme of the European Union.

REFERENCES

Act No. 45/2011 Coll. on critical infrastructure
Act No. 319/2002 Coll. on defence of the Slovak Republic, as amended
APENCOT—Security Standards for Physical Protection, F.S.C. Bezpečnostní poradenství, a.s., 2010, 24s.
Boc, K., Hofreiter L., Jangl Š., Loveček, T., Mach, V., Seidl, M., Selinger, P. & Veľas, A. 2013., Object Protection of the Critical Transport Infrastructure, Žilina: University of Žilina, 2013. 237 s. ISBN 978-80-554-0803-3.
Council Decision 2007/124/EC, Euratom, Council decision of 12 February 2007, establishing for the period 2007 to 2013, as part of General Programme on Security and Safeguarding Liberties, the Specific Programme Prevention, Preparedness and Consequence Management of Terrorism and other Security related risks, Available at: http://eur-lex.europa.eu/LexUriServ/LexUriServ.do?uri=OJ:L:2007:058:0001:0006:EN:PDF.
CIPnES—Methodology for Assessing the Physical Protection of the Assets of the Electrification Network, F.S.C. Bezpečnostní poradenství, a.s., 2011, 20s.
Decree of the Nuclear Regulatory Authority of the Slovak Republic No. 51/2006 Coll. which assigns details on requirements for physical protection.
Decree of the National Security Authority of the Slovak Republic No. 336/2004 Coll. on physical security and building security.
EN 50133-1 Alarm systems. Access control systems for use in security applications. Part 1: System requirements.
EN 50131-1 Alarm systems. Intrusion systems. Part 1: System requirements.
EN 50132-1 Alarm systems. CCTV surveillance systems for use in security applications. Part 1:System requirements.
EN 54-1 Fire detection and fire alarm systems. Part 1:Introduction.
Green paper on a European programme for critical infrastructure protection, Accessed on 8 June 2008, Available at: http://eur-lex.europa.eu/LexUriServ/site/en/com/2005/com2005_0576en01.pdf. Cited 10.6.2013.
Ministry of Economy SR 2014. Methodological Guidance of the MoE SR No. 29014/2014-1000-53190 on security measures for the protection of the critical infrastructure elements in the sectors of energy and industry.
Presidium of the Police Force, the Office of Judicial and Criminal Police 2006. Methodological Guidelines to ensure uniform approach of the Police corps units when considering risk analysis related to the security of premises of banks and branches of foreign banks, Bratislava 2006.

Production Management and Engineering Sciences – Majerník, Daneshjo & Bosák (Eds)
© *2016 Taylor & Francis Group, London, ISBN: 978-1-138-02856-2*

Generalized inventory models of EOQ type and their implementation in mathematica

L. Lukáš & J. Hofman
Faculty of Economics, University of West Bohemia, Pilsen, Czech Republic

ABSTRACT: The paper deals with generalized inventory control models which are linked to the well-known EOQ model. We present models assuming another than constant rate of demand within an inventory cycle. The demand rate takes polynomial form which is specified by particular interpolation conditions. These conditions give a chance to include various intermediate data to describe inventory level during the cycle in detail. Optimal replenishment lot size and optimal total cost per unit of time yielded from these models with different interpolating conditions are compared with similar quantities provided by classical EOQ model. Second, we discuss another model which allows both stock dependent demand rate and holding cost, which also allows deteriorating items under inventory to accept. This model uses maximization of an objective function on opposite to EOQ type models. We use Mathematic not only for numerical calculations but also for symbolic derivation of analytic formulae specific for models discussed.

1 INTRODUCTION

There are two main overall reasons for carrying inventory which are well-known—to reduce costs and to improve customer service. The main aim of any inventory system is that when and how much to order.

The EOQ (Economic Order Quantity) model was developed to find the optimal order quantity for one commodity. Theoretical platform for building standard inventory models, the EOQ and POQ (Production Order Quantity) ones in particular, belongs to topics covered by standard management science textbook, e.g. Taylor (2004). More elaborated and sophisticated models one can find in special monograph, e.g. Axsaeter (2006), which among other topics concerns also with inventory cost money both in sense of tied up capital and running and administrating inventory itself. To run an inventory efficiently it is important to balance the service to customers against these costs. This is the purpose of inventory management, which focuses on the decision on when to replenish the inventory and how much or how many units to replenish it with. Such topics are discussed in Beullens (2014), Chung & Huang (2014), Glock et al. (2014), and Tripathi et al. (2014), as well. However, we will basically follow theoretical frameworks presented both in Lukáš (2012) and Yang (2014), respectively, for our inventory models formulation.

2 GENERALIZED EOQ TYPE MODELS

The generalized EOQ type models release restrictive assumptions of constant demand rate, and constant unit holding cost c_1, both being required by EOQ model exclusively thus giving the well-known formulae

$$q_{opt} = arg\ min\ C(q) = \sqrt{(2c_3 Q/c_1 T)},$$
$$C(q_{opt}) = \sqrt{(2c_1 c_3 Q/T)}, \tag{1}$$

where Q is an accumulated demand, T gives the total inventory control period, e.g. one year, and c_3 is fixed cost per order.

For any generalized EOQ type model, the total inventory cost per cycle $N(t)$ has the following form

$$N(t) = c_3 + \int_0^t c_1(\tau) z(\tau) d\tau. \tag{2}$$

Though, a question arises which functional form to select for $c_1(\tau)$, and $z(\tau)$, respectively. We have proposed the simplest possible form—algebraic polynomials. Denoting $p_n(\tau)$ a polynomial of n-th degree, we will assume $c_1(\tau) = p_i(\tau)$, $i \geq 1$, and $z(\tau) = p_j(\tau)$, $j \geq 2$, while linear function $p_1(\tau)$ suits to EOQ model.

The purpose of such generalization is at least two-fold one. First, variable unit holding cost $c_1(\tau)$ allows to handle inventory cases of storing deteriorating items, perishable goods, or any other products with their holding costs sensitive to elapsing time during an inventory cycle. Second, a higher degree of demand rate $d(\tau)$, starting even with the first degree which yields corresponding inventory level $z(\tau)$ to be quadratic polynomial, thus enables us both to cope more flexibly with time dependent demand during inventory cycle, and also to use prospectively additional interim information of inventory level being detected during inventory cycle course. Last but not least, we would like to express our opinion that using such versatile inventory models will be advantageous for building adaptive inventory control systems.

In general, the total inventory cost per unit time $C(q)$ serving as prospective objective function to be minimized is formulated using expression (2) as follows

$$C(q) = N(t)/t, \quad q = tQ/T. \tag{3}$$

Thus, we may write a problem for determination of optimal replenishment lot size q_{opt} in well-known form

$$q_{opt} = arg\ min\ C(q), \quad q \in]0, +\infty[. \tag{4}$$

At present, we have elaborated full family of generalized EOQ type inventory models with $i = 0$, 1, and $j = 1, 2, 3$, and we propose GEOQ(i, j) to be a suitable denotation of them particularly. In general, GEOQ(i, j) models are equipped with various additional interpolation conditions applied upon inventory level function $z(\tau)$, and/or $c_1(\tau)$, too. These conditions concern either a function value (called Lagrangian condition, denoted by L), or a derivative value (called Hermitean condition, denoted by H), respectively.

Symbolic derivation of analytic formula for q_{opt} for particular GEOQ(i, j) model follows the out-lined steps which are all performed by Mathematica:

1. express $c_1(\tau)$ being represented by $p_i(\tau)$ in pure analytic form using its interpolation data given in symbols,
2. express $z(\tau)$ being represented by $p_j(\tau)$ in pure analytic form using its interpolation data given in symbols, too,
3. calculate $N(t)$ using (2), which is possible in full analytic form forasmuch integrand having been assumed as polynomial exclusively,
4. use the substitution $t = qT/Q$ to recast $N(t)$ into $N(q)$,
5. express objective function, i.e. the total inventory cost per unit time $C(q)$, using $C(q) = N(q)/t$ and the substitution mentioned above,

Table 1. Additional interpolation conditions of the model GEOQ(1, 3).

Case	a)	b)	c)	d)
Type	L	LH	LL	HH,H

6. calculate $dC(q)/dq$ in pure symbolic terms,
7. solve the equation $dC(q)/dq = 0$ which expresses necessary condition of optimality to get q_{opt} formula desired,
8. express formula for minimal value of total inventory cost per unit time $C_{opt} = C(q_{opt})$ by substituting q_{opt} into $C(q)$.

The proposed procedure is quite straightforward one, however it is rather tedious technically and very error-prone one when doing by hand. Hence, we use symbolic calculation power of Mathematica to perform such challenging task.

For illustration, assuming $\tau \in [0, t]$ we define $c_1(\tau)$, and $z(\tau)$ of model GEOQ(1, 3) in following form, where z'(.) denotes a derivative

$c_1(\tau) = p_1(\tau) = \gamma h + (\gamma d - \gamma h)\tau/t$, $c_1(0) = \gamma h$, $c_1(t) = \gamma d$,

$z(\tau) = a_0 + a_1\tau + a_2\tau^2 + a_3\tau^3$, $z(0) = q$, $z(t) = 0$, and with two additional conditions being applied

a. $z(s_k) = w_k$, $s_k = \theta_k t$, $w_k = \omega_k q$, θ_k, $\omega_k \in]0,1[$, $k = 1, 2$, $\theta_1 < \theta_2$, $\omega_1 > \omega_2$,
b. $z'(0) = -\varphi_h q/t$, $z(s_2) = w_2$, $s_2 = \theta_2 t$, $w_2 = \omega_2 q$, θ_2, $\omega_2 \in]0,1[$, $\varphi h \in [0,3]$,
c. $z'(t) = -\varphi_{dd} q/t$, $z(s_1) = w_1$, $s_1 = \theta_1 t$, $w_1 = \omega_1 q$, θ_1, $\omega_1 \in]0,1[$, $\varphi d \in [0,3]$,
d. $z'(0) = -\varphi_h q/t$, $z'(t) = -\varphi_d q/t$, φ_h, $\varphi_d \in [0,3]$.

For illustration, the simplest case of model GEOQ(0, 2) with one additional interpolation condition of type L: $z(t/2) = \omega q$, $\omega \in [0.25, 0.75]$ yields following expressions for optimal quantities

$q_{opt} = \sqrt{(6c_3 Q/(c_1(1+4\omega)T))}$,
$C_{opt} = C(q_{opt}) = c_1(1+4\omega)\sqrt{(2c_3 Q/(3c_1(1+4\omega)T))}$,

$$(5)$$

which corresponds with formulae given in Lukáš (2012), p. 64, precisely.

3 MODEL WITH BOTH STOCK-DEPENDENT DEMAND RATE AND HOLDING COST RATE

Now, we will follow Yang (2014) in formulation of another model. It posses specific and interesting feature that its objective function is formulated in form of total profit per unit time which is to be maximized for retailer, on contrary to EOQ type models which all need total cost per unit time to be minimized for inventory holder. The model assumes infinite replenishment rate and zero lead-time, too, but allows shortages in general. Since the model accents role of retailer by specific objective function we need to consider both purchasing cost per product unit p and the selling price s, as well, assuming $s \geq p$.

Shortages are represented by stock-out period of cycle when $z(\tau) \leq 0$. Let t_1 denote a time in which inventory level drops to zero, i.e. $z(t_1) = 0$, assuming $t_1 \leq t$. Inventory level at the end of cycle $\tau = t$ will be denoted $z(t) = r$. Fraction of demand during the stock-out period which will be backordered is assumed to be $\delta \in [0,1]$, while $(1 - \delta)$ will be lost. Further, c_2, and c_4 will denote shortage cost per-product unit per unit time, and cost of lost sales perproduct unit, respectively.

Demand rate $d(\tau)$ is defined as a power function of inventory level of inventory level

$$d(\tau) = \lambda \cdot (z(\tau)^\beta, z(\tau) > 0, = \lambda, z(\tau) \leq 0, \qquad (6)$$

where $\lambda > 0$ gives the demand rate level during stock-out period and serves as a scalar multiplier during a stock-in period, too, and $\beta \in [0,1]$ determines the power dependence of demand rate upon inventory level, which is also called inventory level elasticity of demand rate.

Holding cost $c_1(\tau)$ of currently stored amount of product is defined in similar analytic way within a stock-in period as follows

$$c_1(\tau) = h \cdot (z(\tau))^\gamma, z(\tau) > 0, \qquad (7)$$

where $h > 0$ gives a scaling constant, and γ gives the power dependence of holding cost upon inventory level, respectively, with $\gamma \geq 1$ being assumed. This constant is also called holding cost elasticity. Note that expression (7) looks as more compact expression than the integrand in (2) does, however a similarity therewith is simple to show

$$c_1(\tau) = h \cdot (z(\tau))^\gamma = h \cdot (z(\tau))^{\gamma-1} \cdot z(\tau), \qquad (7a)$$

when having separated $z(\tau)$, we get $h \cdot (z(\tau))^{\gamma-1}$ directly as to express the unit holding cost in standard form.

Since both demand rate $d(\tau)$ and holding cost $c_1(\tau)$ depend upon inventory level $z(\tau)$, an important role in model is played by relations between derivatives dd/dz and dc_1/dz at different time instants τ, in general. Moreover, Yang (2014) explains also their economic meaning as follows

1. at the beginning of replenishment cycle, i.e. at $\tau = 0$, one assumes to satisfy

$$(s - p)dd(\tau)/dz(\tau)|_{\tau=0} \leq dc_1(\tau)/dz(\tau)|_{\tau=0}, \qquad (8a)$$

$$(s - p)dd(\tau)/dz(\tau)|_{\tau=t} \geq dc_1(\tau)/dz(\tau)|_{\tau=t}, \qquad (8b)$$

2. at the end of replenishment cycle, i.e. at $\tau = t$, if $z(t) = r \geq 0$ holds, to satisfy an opposite

where expression (8a) maintains that the increase of sales profit due to the unit increase of inventory level doesn't exceed the increase of holding cost at the same argument, thus preventing retailer to order an infinite quantity of product at $\tau = 0$, while expression (8b) maintains the retailer to sell all stock available, i.e. to get $z(t) = r = 0$, as the sales profit exceeds that increase of holding cost at $\tau = t$.

In general, two cases of this model can be distinguished depending upon assumption regarding a shortage—either with shortage, i.e. $r < 0$, or without shortage, i.e. $r \geq 0$, respectively.

We shall follow the last case as it stands in correspondence with the GEOQ(i, j) models discussed in previous paragraph.

Reminding general inventory balance condition written in infinitesimal form

$$dz(\tau)/dt = s(\tau) - d(\tau), \quad z(0) = z_0, \qquad (9)$$

with initial inventory level $z(0)$ given, $s(\tau)$, and $d(\tau)$, representing supply process, and demand process, respectively, we may write an initial value problem for EOQ model as follows

$$dz(\tau)/d\tau = -q/t, \quad z(0) = q. \qquad (9a)$$

Moreover, due to EOQ model assumptions the following hold: $d(\tau) = q/t = Q/T = $ const, $s(\tau) = q \sum_{i=0}^{\vartheta} \delta(\tau - it)$, where $\delta(.)$ denotes the Dirac function, and ϑ represents the greatest integer obeying relation $\vartheta t < \eta$. Finally, solution of equation (9a) is $z(\tau) = q(1 - \tau/t)$, which also gives $z(t) = 0$, where t denotes the replenishment cycle period.

Now, assuming supply process of discussed model to be the same as for the EOQ model, we may write the differential equation describing the inventory level $z(\tau)$ of the model in following form

$$dz(\tau)/dt = -\lambda \cdot (z(\tau))^\beta, \quad z(0) = q, \qquad (10)$$

which yields solution

$$z(\tau) = (q^{(1-\beta)} - (1 - \beta)\lambda t)^{1/(1-\beta)}, \quad \tau \in [0,t]. \qquad (11)$$

Hereby assuming $q > r$ and setting $z(t) = r$, solution (11) provides an expression for replenishmen cycle period

$$t = (q^{(1-\beta)} - r^{(1-\beta)})/((1 - \beta)\lambda) > 0. \qquad (12)$$

171

Now, we write the objective function, i.e. the total profit per unit time $P(q,r)$, as follows

$$P(q,r) = (s(q-r) - p(q-r) - N(t))/t, \qquad (13)$$

where the first term expresses sale revenue and the second one the purchase cost, thus being combined together they give sale profit per cycle, and finally, the third term represents the total inventory cost per cycle based upon (7). As usual, c_3 is the ordering cost per cycle, t is given by expression (12), and the quantity $N(t)$ is calculated by

$$N(t) = c_3 + \int_0^t c_1(\tau) d\tau = c_3 + h \int_0^t (z(\tau))^\gamma d\tau$$
$$= c_3 + h \int_0^t (z(\tau))^{\gamma-1} z(\tau) d\tau. \qquad (14)$$

As we focus ourselves to numerical procedure for solving maximization problem

$$(q_{opt}, r_{opt}) = \arg\max P(q,r), q, r \in]0, +\infty[, q > r, \quad (15)$$

we refer to Yang (2014) for more technical details, thus providing an analytic expression of $P(q, r)$ as follows

$$P(q,r) = (1 - \beta)\lambda((s - p)(q - r) - c_3 \\ - h(q^{(\gamma+1-\beta)} - r^{(\gamma+1-\beta)})) / \\ (\lambda(\gamma + 1 - \beta))/(q^{(1-\beta)} - r^{(1-\beta)}). \qquad (16)$$

The second case of the model allowing shortage during cycle t is presented in Yang (2014), as well. The objective function proposed therein is the total profit per unit time again, being denoted $R(q, r)$ now. It looks like expression (13), however including two terms more as for shortage cost and lost sales cost, respectively. Its analytic form is following

$$R(q,r) = \lambda((s - p)(q - r) - c_3 - hq^{(\gamma+1-\beta)} / \\ (\lambda(\gamma + 1 - \beta) - c_2 r^2/(2\delta\lambda) + c_4(1 - \delta) r/\delta) / \\ (q^{(1-\beta)} / (1 - \beta) - r / \delta), \qquad (17)$$

where c_2, c_4, and δ in sequel are parameters expressing the unit shortage cost per unit time, lost sales cost per unit time, and fraction of the demand during the stock-out period that will be backordered, $\delta \in [0,1]$, respectively.

4 IMPLEMENTATION OF MODELS IN MATHEMATICA

First, we concern briefly with implementation of GEOQ(i, j) inventory models in Mathematica. It is realized in two stages:

1. Derivation of selected model dependent formulae for q_{opt}, and C_{opt} in symbolic forms.
2. Having such formulae, we substitute particular numerical values of corresponding parameters therein, calculate desired quantities, and issue plots.

In general, we have already mentioned procedure steps for symbolic derivation of such formulae. First thing to do, there is solution of interpolation problem using Mathematica function Solve and application of replacement rules. Further, symbolic integration, i.e. step 3), is performed via function Integrate, and symbolic derivation and solution of equation $dC/dq = 0$, i.e. steps 6) and 7), are realized via functions D and Solve, respectively. Here, a lack of paper space prevents us to give more details thereon. However, our Mathematica code is available upon request on the first author.

For illustration, we select model GEOQ (1,3) case b), i.e. $c_1(\tau)$ is linear function, and $z(\tau)$ is cubic polynomial with two additional interpolation conditions of type (H,L), in particular.

The Figure 1 demonstrates that inventory level plots represent an important checking instrument in practice. At once, we inspect two of constructed functions $\zeta(\tau)$ depicted with full and dashed lines respectively, to be not acceptable as feasible inventory levels for the GEOQ(1,3) model since both violate the necessary condition therefor, i.e. to be non-increasing function on whole replenishment cycle period t.

In Figure 2, the optimal results q_{opt}, $C(q_{opt})$ calculated by the model GEOQ(0,3) case a) are compared with corresponding EOQ results, i.e. $q_{opt} = C(q_{opt}) = 40$, being issued by formulae (1), setting $Q = 800$, $T = 1$, and c_1, $c_3 = 1$. The sequence of values q_{opt} is decreasing, on opposite to $C(q_{opt})$ values, when sequence of parameters (θ_1, θ_2) is

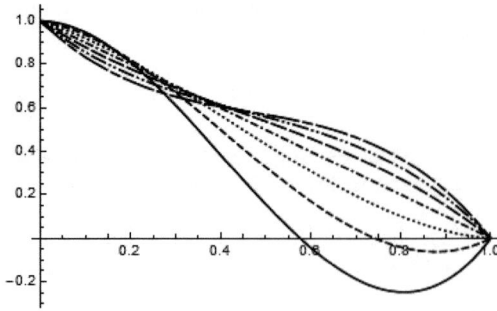

Figure 1. Plot of GEOQ(1,3) normalized inventory levels $\zeta(\tau)$, defined by ratios $\zeta(\tau) = z(\tau)/q$, on normalized replenishment cycle [0,1], using data: case b): $\varphi_h = (0, 1/3, 2/3, 1, 4/3, 5/3, 2)$, $\theta_z = (0.35, 0.40, 0.45, 0.50, 0.55, 0.60, 0.65)$, and $\omega_z = 0.5$, constant in all cases.

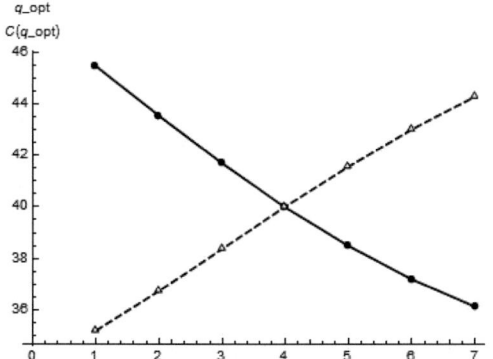

q_opt
$C(q\_opt)$

Figure 2. Plot of qopt (full line), and $C(q_{opt})$ (dashed line), for GEOQ(1,3) → GEOQ(0,3), i.e. $c_i(\tau) = c_i$, constant, and case a) with $(\theta_1, \theta_2) = \{(0.15, 0.55), (0.2, 0.6), (0.25, 0.65), (0.3, 0.7), (0.35, 0.75), (0.4, 0.8), (0.45, 0.85)\}$, $\omega_1 = 0.7$, $\omega_2 = 0.3$, and $\gamma_h = 1$, $\gamma_d = 1$.

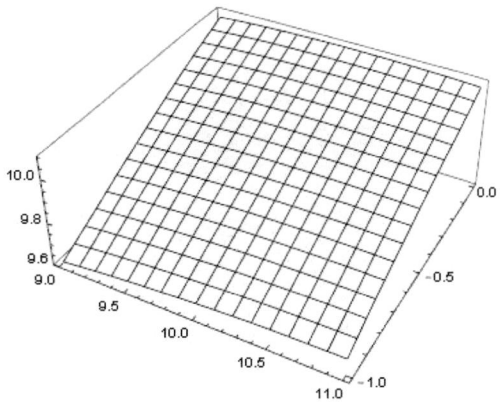

Figure 3. Plot $R(q, r)$ at vicinity of the optimum (q_{opt}, r_{opt}) = (10.0348, 0) on the region $(q, r) \in [9,11] \times [-1,0]$.

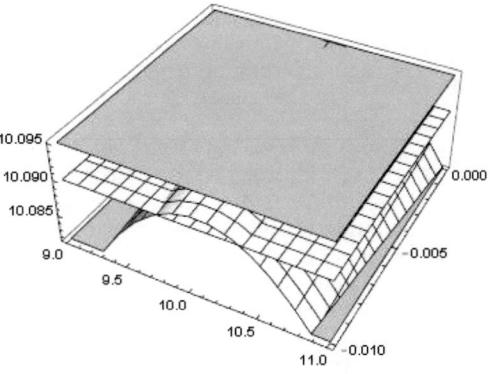

Figure 4. Plot $R(q, r)$ at vicinity of the optimum $(q_{opt}, r_{opt}) = (10.0348, 0)$ on the region $(q, r) \in [9,11] \times [-0.010,0]$, together with horizontal plane at the attitude $R_{opt} = 10.095$.

increasing while keeping parameters ω_1, ω_2 constant all the time.

Now, we present implementation of the second model which assumes both stock-dependent demand rate and holding cost rate. Its two variants being different with respect a shortage are represented particularly by objective functions $P(q, r)$ (16), i.e. no shortage, and $R(q, r)$ (17), i.e. with shortage, respectively.

Yang (2014) presents iterative procedures for numerical solution of these models being considered as constrained optimization problems, in general. These procedures are based upon solution of FOC (First Order Conditions) for $P(q, r)$, and $R(q, r)$, respectively.

We have selected another approach based upon utilization of Mathematica function FindMaximum directly to objective functions $P(q, r)$, and $R(q, r)$ with corresponding constraints applied.

In sequel, we present our comparative results of cases given in Yang (2014). The input data are following: $\lambda = 1$, $h = 0.5$, $s = 62$, $p = 50$, $\delta = 0.8$, $\beta = 0.1$, $\gamma = 1$, $c_2 = 2$, $c_3 = 10$, $c_4 = 1$.

First, we solve the model with shortage, i.e. the problem

$$\max R(q,r), \quad q \geq 0, r < 0, \quad (18)$$

and we get solution: $R_{opt} = 10.095$, $q_{opt} = 10.0348$, and $r_{opt} = -1.19282 \times 10^{-6}$, which stands in perfect coincidence with those one given in Yang (2014).

Moreover, we make 3-D plots of the objective function $R(q,r)$ to see it graphically. The Figure 3 shows clearly that there are very different slopes, i.e. the first derivatives, in horizontal and lateral direction, respectively, which make difficult to search

optimum by simple graphical inspection if the vicinity is 'natural', i.e. it does not zoom any direction. On the contrary, the Figure 4 depicts the same objective function $R(q,r)$ on a region with the r axis being zoomed 100-times, thus making detection of the optimum easier. We plot also a horizontal plane at the attitude $R_{opt} = R(q_{opt}, r_{opt}) = 10.095$ for better inspection precisely, so we may find a spot on the edge thereon to localize the optimum quite easy.

Second, we solve also the model without shortage, i.e. the problem

$$max\ P(q,r), q \geq 0, r > 0, \quad (19)$$

and we get solution: $P_{opt} = 10.1239$, $q_{opt} = 9.95129$, and $r_{opt} = 0.201697$, standing again in perfect coincidence with corresponding one given in Yang (2014).

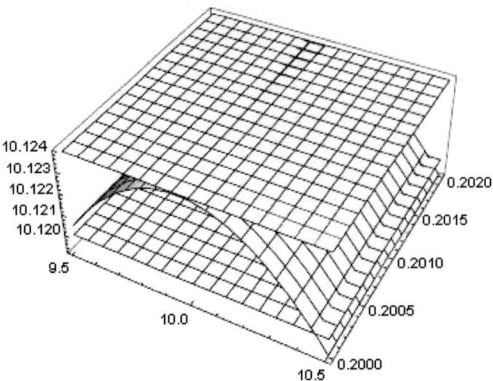

Figure 5. Plot $P(q, r)$ at vicinity of the optimum $(q_{\mathrm{opt}}, r_{\mathrm{opt}}) = (9.95129, 0.201697)$ on the region $(q, r) \in [9.5, 10.5] \times [0.200, 0.202]$, together with horizontal plane at the attitude $P_{\mathrm{opt}} = 10.1239$.

The Figure 5 shows the objective function $P(q,r)$ on a region with the r axis being zoomed 100-times, again. Plotting horizontal plane at the attitude $P_{\mathrm{opt}} = 10.1239$ makes detection of maximum easier, too.

Even though short, we may quote our just positive experience with using Mathematica function Find Maximum for finding solution of maximization problems which we tried directly instead of programming cumbersome iterative procedure proposed in Yang (2014). Sure, there will be necessary to solve even other examples to check full versatility and efficiency of that Mathematica function to be applied for that type of inventory model and/or prospectively for other closely related models, as well.

5   CONCLUSIONS

– Framework of generalized EOQ type models, called GEOQ(i, j), has been developed which enable both time dependent unit holding costs and demand rate. Analytic description of such models is based upon additional interpolation conditions expressing intermediate information during replenishment cycle.

– Alternative inventory control model assuming both stock-dependent demand rate and holding cost rate is presented, too. This model provides an important additional feature to include or to neglect stock shortage during replenishment cycle, thus leading to two different variants.

– In general, inventory control models providing dependency of holding cost either on cycle time or stock available are necessary management instruments when storing commodities with deteriorating items.

– We present implementation of both GEOQ(i, j) models and alternative one with both its variants in Mathematica, Wolfram Research, Inc., uniquely. Albeit short, we believe that it also demonstrates versatility and symbolic power of this large software package within a field of advanced inventory control models development. Furthermore, we use Mathematica for both numerical calculations of all examples presented, and issuing of plots, as well.

– Near future research will be focused on thorough numerical experiments with discussed models and their practical use. Further, we would like to concentrate ourselves also upon a role of inventory, its financial issues in particular, within a broad framework of valuation of firms. There is generally known that an amount of firm asset allocated in inventory might cause awkward and unexpected effects on financial position of company, especially when storing deteriorating items, in particular.

ACKNOWLEDGEMENT

The research project was supported by the grant no. 15-20405S of the Grant Agency, Prague, Czech Republic.

REFERENCES

Axsaeter, S. 2006. *Inventory Control*. 2nd ed., New York: Springer.

Beullens, P. 2014. Revisiting foundations in lot sizing—Connections between Harris, Crowther, Monahan, and Clark. *Int. J. Production Economics* 155: 68–81.

Chung, K-J. & Huang, Ch-K. 2009. An ordering policy with allowable shortage and permissible delay in payments. *Applied Math. Modelling* 33: 2518–2525.

Glock, Ch.H., Grosse, E.H. & Ries, J.M. 2014. The lot sizing problem: A tertiary study. *Int. J. Production Economics* 155: 39–51.

Lukáš, L. 2012. *Probabilistic models in management—inventory theory and statistical description of demand* (in Czech Probabilistic Models in Management—Theory of inventory and statistical description of demand). Prague: Academia.

Snyder, L.V. 2014. A tight approximation for an EOQ model with supply disruptions. *Int. J. Production Economics*155: 91–108.

Taylor, B.W. 2004. *Introduction to Management Science*. 8-th ed., New Jersey, Upper Saddle River: Prentice Hall, Pearson Education.

Tripathi, R.P., Singh, D. & Mishra, T. 2014. EOQ Model for Deteriorating Items with exponential time dependent Demand Rate under inflation when Supplier Credit Linked to Order Quantity. *Int. J. Supply and Operations Management* 1(1): 20–37.

Yang, Chih-Te. 2014. An inventory model with both stock-dependent demand rate and stock-dependent holding cost rate. *Int. J. Production Economics* 155: 214–221.

Thermovision application for modulating of local anesthesia

M. Majerník & J. Šimonová
University Hospital Louis Pasteur, Košice, Slovak Republic

J. Živčák
Technical University in Košice, Košice, Slovak Republic

T. Grendel
East Slovak Institute of Cardiovascular Disease, Košice, Slovak Republic

ABSTRACT: Local anaesthesia of brachial plexus (nerves of upper extremity) is commonly used to anesthetize upper extremity. Local anaesthesia causes sympathetic block with subsequent vasodilatation and increased blood flow leading to local increase in skin temperature. This warming can be detected by thermo vision (thermo camera FLIR 320) and appears in 5–10 minutes after regional anaesthesia. Sensory block usually appears in 30–45 minutes after regional anaesthesia administration, i.e., thermo vision can answer the question of successful local anaesthesia 30 minutes before onset of sensory block in every patient with regional anaesthesia.

1 INTRODUCTION

Regional anaesthesia of brachial plexus (nerves of upper extremity) is commonly used to anesthetize upper extremity. It is used for management of postoperative pain and for treatment of some chronic neuropathic pain states as well. Galvin et al. (2006) has confirmed that local anaesthesia causes sympathetic block with subsequent vasodilatation and increased blood flow leading to local increase in skin temperature.

This increase in temperature could be a useful method to evaluate block effectiveness with higher specificity and sensitivity than changes in sensation to pinprick and cold stimuli. Minville et al. (2009) showed that there is complete sensitive nerve blockade of upper limb within 5 to 10 minutes after administration of regional nerve block if there is local skin temperature increase of at least 1°C. If the temperature rises less than 0.5°C, nerve blockade is insufficient. We are applying these principles in our study.

Measurement and evaluation of temperature changes using thermocamera is early and appropriate indicator for successful local anaesthesia. Therefore we recommend implementation of this technique into the clinical practise as an effective tool for intensification and optimization of local anaesthetic techniques.

Table 1. Demographic data of 15 patients.

Demographic data	Patients (n = 15)
Age (years)	42 ± 14
Sex (female/male)	5/10
Weight (kg)	80 ± 11
Site of operation	Hand: 4 Wrist: 6 Forehand: 4 Elbow: 1
Success of regional anaesthesia	15 patients (100%)

2 METHODS

This study was conducted with approval of the ethics committee of University Hospital of Louis Pasteur in Košice, Slovakia.

Procedures were carried out at the preparation rooms with a constant temperature (22 ± 6°C) and humidity (61 ± 7%) measured with thermometer Casio. Premedication was administrated 10 minutes prior to the procedure.

Patients were in a supine position with the hand laying on their stomach. Procedures were carried out under standard monitoring of vital signs—ECG, NIBP, Pulse Oxymetry (Fig. 1). First image with thermo camera (FLIR 320) was made 20 minutes after premedication (Fig. 2). Brachial

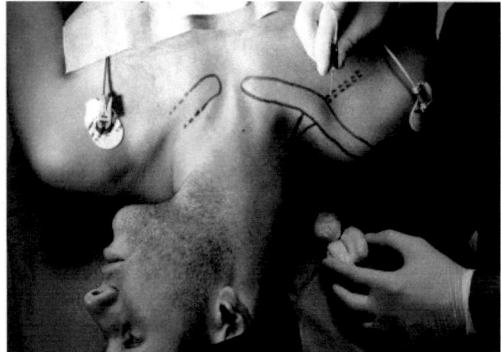

Figure 1. How the procedure is carried out.

Figure 2. Images were taken with thermocamera FLIR 320.

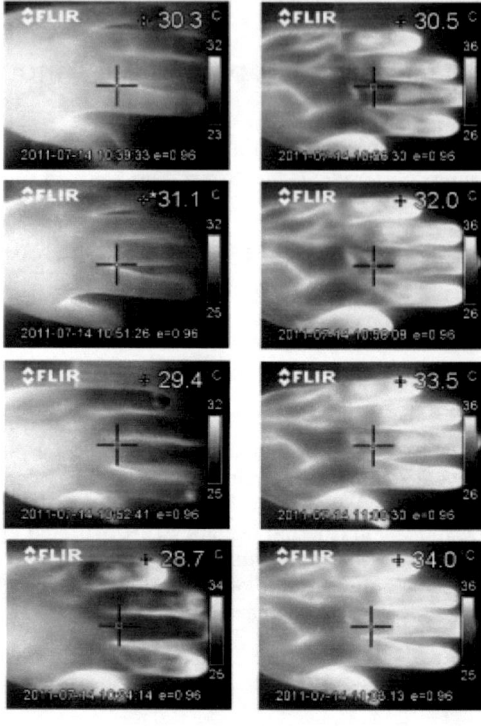

Figure 3. Thermo-images of the hand. Notice the change in temperature after regional anaesthesia.

plexus nerves were identified with nerve stimulator Stimuplex HNS 12 B Braun. Stimulating needle B Braun was inserted vertically to the base after topical anaesthesia with local anaesthetic. Twitching of fingers during nerve stimulation with current of 0.3 mA was considered to be a successful response to stimulation. Local anaesthetic was injected.

Second figure (Fig. 3) was made immediately after local anaesthetic injection and subsequently each minute afterwards until 5th minute, than every second minute until 13th minute post injection. Thermo-images were processed with free software taken from Flir internet home page. We evaluated thermograms by the software FLIR Reporter 8.5. Acquired temperatures of fingers and dorsum of the hand (minimal, maximal, average) were manipulated by software SPSS Statistic 17.0, Minitab 15, MS Excel. We calculated when the skin temperature increase stopped.

Thermovision is a method of visibility of objects by detecting the objects' infrared radiation and creating an image based on this information. All objects produce and emit infrared energy as a function of their temperature. The infrared energy emitted by an object is known as heat signature of this object. In general, the hotter an object is, the more radiation it emits. Thermovision camera (thermal imager) is a heat sensor that is capable of detecting differences in objects' temperature. This machine collects the infrared radiation from objects in the scene and creates an electronic image based on information about the temperature differences.

In our study measurement of temperature differences after regional anesthesia was taking place at University hospital, Department of Anesthesiology and thermovision camera FLIR 320 was brought by Department of Biomedical Engineering and Measurement, Technical University. Our thermovision camera was newly calibrated (Technical University) and fixed in a stabile position 60 cm vertically above the hand (hand in prone position). Thermovision camera FLIR 320 is long wave, handheld, Focal Plane Array camera with size $25 \times 15 \times 7$ cm and weight 0,9 kg. It has operating temperature: $-15°C$ to $50°C$ and measurement

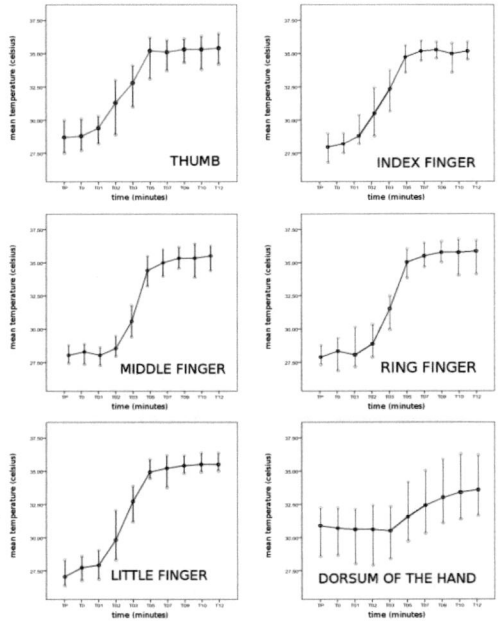

Figure 4. Rise in the skin temperature of various fingers and dorsum of the hand.

Figure 5. Thermovision camera FLIR 320 used in our study.

range: −20°C to 500°C. FLIR 320 has a thermal resolution of <0.07°C, an accuracy of ± 2%, and a picture resolution of 320 × 240 pixels. Because the emissive factor of the skin is 0.98, the measured temperature values can be evaluated as skin temperature values. Infrared imaging was performed with the forearm in the prone position (picture of dorsal hand) after regional anesthesia.

No interventions were performed during infrared imaging, and the patients were instructed not to move during the measurement period. We defined the specific areas of interest for every forearm and measured temperatures of fingers and dorsum of the hand (minimal, maximal, average temperature). Later we compared these temperatures and detected local increase of skin temperature of every hand after regional anesthesia. All of the patients included in our study were undergoing elective hand surgery, but there was no pathology placed on the hand, which could affect skin temperature of the hand.

3 RESULTS

We confirmed the results of Minville et al. (2009) meaning that if skin temperature increased in the upper extremity (warming) by at least 1°C within 5–10 minutes after regional anaesthesia (nerve blockade) administration, there was complete sensory blockade in the upper extremity 30 minutes after administration. We found out that skin temperature of fingers increased in time and stopped in 5th minute. Skin temperature of dorsum of the hand decreased in the first 3 minutes, but than increased until 13th minute (Figure 3, 4).

We also identified time to warming by at least 1°C. 73.3% of patients reached 1°C warming in 5 minutes and 100% of patients reached 1°C warming in 10 minutes. This time was shorter than time in Minville et al (2009)'s study. Average time necessary to reach warming by at least 1°C in our study was 4.7 ± 2.3 min (20% shorter than in Minville et al. (2009) study).

The diversity of applications infraclavicular block expressed in a separate injection was already published in the comparative study (Minville et al. 2009), in other site of puncture needle—1 cm below the clavicle and 1 cm medial to the processus coracoideus, then in a different inclination of the needle when the needle is pointing at 45° angle to the skin in the axilla. It is further distinguished individual nerve stimulation. Nerve was blocked musculocutaneus 10 ml LA as the first and then it was stimulated. Secondly, the needle is pulled out and redirected media after the event to elicit responses median nerve stimulation or radial nerve or ulnar nerve. It was blocked 30 ml LA. The total amount of the LA, one patient was 40 ml, which is 100% more than the amount of LA used in our clinic (I. KAIM). Last difference lay in the LA used when the use of the comparative study, instead of the mixture of bupivacaine (0.5%), and trimecaine (1%), lidocaine (1.5%).

4 CONCLUSIONS

In our study all the patients with "well-done" regional anaesthesia of the upper extremity reached 1°C warming in 5–10 minutes (caused by sympathetic block). This 1°C warming predicts success of sensoric block, which usually appears 30–45 minutes after regional anaesthesia administration. It means that, thermo vision can answer the question of successful regional anaesthesia 30 minutes before onset of sensory block in every patient with local anaesthesia.

Measurement and evaluation of temperature changes using thermocamera is early and appropriate indicator for successful local anaesthesia. Therefore we recommend implementation of these techniques into the clinical practise as an effective tool for intensification and optimization of local anaesthetic techniques.

REFERENCES

Galvin, E.M., Niehof, S., Medina, H.J., Zijlstra, F.J., van Bommel, J., Klein, J. & Verbrugge, S.J. 2006. Thermographic temperature measurement compared with pinprick and cold sensation in predicting the effectiveness of regional blocks. Anesth Analg. 102: 598–604.

Galvin, E.M., Niehof, S., Verbrugge, S.J., Maissan, I., Jahn, A., Klein, J. & van Bommel, J. 2006. Peripheral flow index is a reliable and early indicator of regional block success. Anesth Analg. 103: 239–243.

Hopf, H.B., Weissbach, B. & Peters, J. 1990. High thoracic segmental epidural anesthesia diminishes sympathetic outflow to the legs, despite restriction of sensory blockade to the upper thorax. Anesthesiology; 73: 882–889.

Majerník, M. 2011. Využitie termovíznej techniky pri diagnostike a terapii pacientov na intenzívnom lôžku, p. 24–30.

Minville, V., Gendre, A., Hirsch, J., Silva, S., Bourdet, B., Barbero, C., Fourcade, O., Samii, K. & Bouaziz, H. 2009. The efficacy of skin temperature for block assessment after infraclavicular brachial plexus block. Anesth Analg. 108: 1034–1036.

Production Management and Engineering Sciences – Majerník, Daneshjo & Bosák (Eds)
© *2016 Taylor & Francis Group, London, ISBN: 978-1-138-02856-2*

Process innovations and quality measurement in automotive manufacturing

M. Majerník, L. Štofová, M. Bosák & P. Szaryszová
*Faculty of Business Economy with Seat in Košice, The University of Economics in Bratislava,
Košice, Slovak Republic*

ABSTRACT: According to ISO/TS 16949, production quality is focused on achieving business goals by means of continual improvement processes of automotive manufacturing and car parts. The authors present a more complex case study and evaluation methodology of standard implementation effectiveness in connection with optimizing the measurement system. The methodology for evaluating the quality of business processes on the basis of metrological system has been developed on the basis of defining the advantages and specific business factors in the particular company. However, it may be useful in automotive manufacturing companies in general, for determining the development strategies in the production quality measurement and management. The paper provides a reference book for evaluating the metrological system of the company and creation of conditions for continual improvement of a comprehensively approached production quality in the supply chain of automotive manufacturing.

1 INTRODUCTION

In recent years, the quality management system significantly expanded throughout the business world, be it in the sense of the original concept of Total Quality Management (TQM) or the model ISO 9001 (for the purpose of standardization and certification). This has led to a series of studies and prediction of quality management development in general and also for specific operating conditions (Tomčíková & Živčák 2012).

Currently there are several strategies or approaches to the quality management which can be implemented by the company in order to maintain and improve the quality of its processes, products, services and overall business performance. Among the most important are (Majerník et al. 2015):

- Total Quality Management (TQM),
- Malcolm Baldrige,
- EFQM Excellence Model,
- The international standard ISO 9001:2015.

The international standard ISO 9001 is derived from the TQM concept and is primarily concerned with the Quality Management System (QMS) for all the partial processes of the company, such as design, development, sales, production, assembly and servicing of products and services. The literature introduces various opinions on the effect of implementation of quality management system by ISO 9001 on the business environment. Previous studies generally confirmed that the quality management systems have significantly influenced the performance of companies. On the other hand, according to the International Organization for Standardization, the management systems may also increase the customer satisfaction, minimize the costs and handle the related business risks, which improves competitiveness in the market (Rovňák et al. 2013, Agarski et al. 2012).

Quality management systems provide an opportunity to thoroughly evaluate the goods and services of suppliers as well. The technical Specification ISO/TS 16949 is a QMS standard which defines specific requirements for the use of the international standard ISO 9001:2015 for automotive manufacturing and relevant service organizations (Majerník & Štofová 2014).

The ISO 9001:2015 specifies the requirements for QMS which are to be applied for both, the internal business use (certification) and the contractual purposes with suppliers. Both standards are based on the process approach with reference to the process systems (interconnected business activities) which enable the transformation of business inputs into outputs in car manufacturing and related assembly components (Sila 2007).

ISO/TS 16949 recognizes the singularity of processes of each automotive supplier and provides important tools to help better meet the customers' requirements which are specific to each automotive business. At the same time it eliminates the redundancy, costs and administrative burden caused by the influence of competitive businesses. The main motive for companies, obtaining the certificate for a functional system built according to these standards, is to achieve a significant increase in sales through increasing the product quality.

The measurement system analysis methods and more exact methods of analysis are used at the examination of effects of ISO/TS 16949 on the quality enhancement.

2 ADVANTAGES AND OBSTACLES OF ISO/TS 16949 IMPLEMENTATION

Apart from the benefits resulting from the quality management system by ISO 9001, ISO/TS 16949, there are other advantages, such as:

- maintaining the high level of manufacturing process and thereby stable and high quality of goods and services provided to customers,
- possibility to optimize the costs—reducing the operating costs, reducing the costs associated with poor-quality products, saving in materials, energy and other resources,
- through effectively set processes increasing the sales, profit, market share and thereby increasing the customer satisfaction,
- thanks to providing high-quality production it is the only opportunity to gain supply contracts from the car manufacturers,
- improvement of management system, enhancement of organizational structure, of organization of work,
- improvement of the system and of the performance throughout the organization,
- creating the system which flexibly responds to changes in market requirements, individual customers, legislative requirements and internal changes (e.g. when implementing new technologies, organizational changes, etc).
- ISO/TS 16949, compared to ISO 9001, is much more specific and its application does not require special interpretation. The costly implementation, however, presents a disadvantage especially for small suppliers of car parts.

The advantages of ISO/TS 16949 may be demonstrable at the evaluation, comparison at the audit and at implementation of management system through the measurement of indicators. In order to achieve the set goals, it is possible to adjust the calculations for evaluating the benefits (indicators).

3 MEASUREMENT SYSTEM AS A KEY ELEMENT MANAGEMENT SYSTEM ACCORDING TO ISO/TS 16949

There has been a number of different methods of measuring indicators aimed to assess the accuracy and quality of products, whereas these are primarily focused on the overall measurement deviation, commonly known as measurement uncertainty.

In many production processes the parts, components and pieces are measured to ensure compliance with certain specifications. However, these measurements may be misleading in the event that the measurement system analysis itself is insufficient (Szabo et al. 2013).

Measurement errors may be caused by measurement devices, readings of meters or estimations, or by environment. In such context, variability plays the key role in quality improvement, since only a scale with acceptable repeatability and reproducibility may present the adequacy of measurement processes of a given product. The measurement system analysis is an essential part of the overall quality assurance programs, which should be used by all companies (Bednárová et al. 2013).

Evaluation of measurement system analysis is an important aspect of quality management and processes improvement. In recent years, particularly in connection with ISO 9001, ISO/TS 16949 and a Six Sigma method, for example quality control by personnel began to focus more on measuring systems, balanced with respect to repeatability and reproducibility (Majerník et al. 2014).

Repeatability is defined as the "variation in measurements obtained with one gauge being used for several times by one appraiser while measuring a characteristic on one part". Reproducibility is defined as the "variation in the average of the measurements made by different appraisers using the same gauge when measuring a characteristic on one part".

Montgomery and Runger (1993) argue that the system measurement analysis should play an active role within precautions for quality improvement monitored by the company, and it should be used for identification of the source of variations within the processes measurement, including quantification of these variants. In order to reduce the uncertainty in terms of the ISO/IEC 17025, the assessment is formulated to evaluate the type I (α) errors, where the risk of error rate of products is identified by producers, and type II (β) errors, where the risk of error rate is identified by consumers. In practice, the role of the measurement system analysis is not always to get the exact characteristics (dimensions) of a component, but to provide measurements which deviate from the actual value only to a certain extent.

In addition, there is also some uncertainty about the measurement inaccuracy since it is very hard to obtain exact measurements of these errors. Type II errors require special attention since they may directly affect the subsequent production processes and lead to customer complaints or an increase in marginal costs.

AIAG stated that Statistical Process Control (SPC) is also commonly used to monitor the ability of the measurement process, and that SPC is not only a useful tool, but also an essential part of quality assurance because without good evaluation, measurement and reliability it is not possible to consider the measured data as reliable.

Previous studies on this subject focused primarily on how companies used measurement systems analyses, or on how they compared various measurement methods in the evaluation. For example statistical methods were used, how to apply the evaluation methods, or whether there were any interactions between researchers and products. The results of this analysis could serve as a basis for evaluation and continual improvement of measurement systems, which enables a more efficient product monitoring and achievement of better quality specifications and thereby greater customer satisfaction.

3.1 Measurement system analysis

Measurement System Analysis (MSA) is generally a systematic procedure which identifies variation components in accuracy and the correctness of assessment of measuring instrument used in the measurement system. MSA objectives are:

- determining the range of found differences caused by gauges,
- identifying the sources of variability in the testing system,
- evaluation of the ability of the tested gauges.

MSA is an important feature of the Six Sigma method and system implementation according to ISO/TS 16949, which is used to evaluate the reliability of important input and output data in the production process, for understanding the changes caused by workers, instruments, materials, procedures or environment, and by means of data analysis it also serves as a basis for improvement of business activities (Majerník et al. 2014).

MSA primarily concerns the statistical processing and graphs for implementation of simple experimental design and statistical analysis of measurement system errors, it evaluates possible changes in measuring instruments and in the work of operators on site. However, the reliability of the data recorded in the measurement process is varia-

ble. An ideal measuring system should statistically produce "zero error rate" of a measured product, which is impossible in practice.

Measurement error for different instruments is approached with regard to accuracy and precision. Accuracy refers to the difference between the measured value and the actual value of the sample, whereas the measurement accuracy is the change resulting from the same device by repeated measurements of the same sample. However, it is likely that one or two errors appear in each measurement system. In addition, accuracy is divided into two parts: the repeatability and reproducibility. It has already been pointed out that repeatability is a change caused by measuring instrument, while reproducibility results from the measurement system. Measurement system analysis proposes three variables to confirm the repeatability and reproducibility:

- range,
- average range,
- ANOVA (Analysis Of Variance).

With increasing demands on more and more complex electrical wiring in cars, there is a requirement for higher quality contacts and stronger anchoring of a contact fixed on a wire, for their subsequent mounting into wiring harness connectors. As an effective quality assurance tool, evaluations and checking of production process eligibility are performed using process capability indices.

Table 1 shows field-proven acceptable general ratios for measurement errors.

For example McNeese and Klein (1991) indicated that the change of analysis and sampling in measuring systems are features which could be

Table 1. Acceptable general ratios for error.

MSA	Evaluation	Recommendation
<10%	Generally regarded as acceptable measurement system.	Complexity of processes for organizing or classifying parts is recommended.
10%–30%	Considered acceptable in certain applications only.	They take into account the cost of measuring instruments and their maintenance; cooperation with the customer.
>30%	Unacceptable.	It is needed to support the selection of the measuring system. Some suitable measuring strategies may be used for troubleshooting, e.g. reduction of final change in measurement.

Table 2. Evaluation of the process eligibility.

Equipment Variation EV:	0.04943	EV = Rbar * K1
Appraiser Variation AV:	0.04585	AV = Sqrt((XbarDiff * K2)^2 – (EV^2)/10 * 3))
Repeatability and Reproducibility R&R:	0.06742	R&R = sqrt(EV^2 + AV^2)
Part Variation PV:	0.15660	PV = Rp * K3
Total Variation TV:	0.17050	TV = sqrt(R&R^2 + PV^2)
Equipment Variation EV[%]:	28.99	EV[%] = 100 * EV/TV
Appraiser Variation AV[%]:	26.89	AV[%] = 100 * AV/TV
Repeatability and Reproducibility R&R[%]:	39.54	R&R[%] = 100 * R&R/TV
Part Variation PV[%]:	91.85	PV[%] = 100 * PV/TV
Number of distinct categories	3.28	ndc = 1.41 * PV/R&R

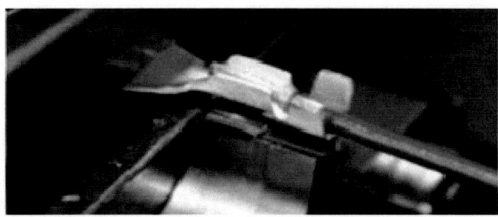

Figure 1. Crimp contact and a wire in the course of fixing.

better used to improve the measuring abilities, and that the product and measurement variations are two factors that form the overall change. In this case, the changes in the variant are to be emphasized in the measuring process.

4 MEASUREMENT SYSTEM ANALYSIS IN A PARTICULAR AUTOMOTIVE PRODUCTION PLANT

The object of research were wiring harnesses of an automotive industry company according to SK NACE 29320. The purpose of verifying the eligibility of a process is a wire crimp contact (type N 907 647 01) with a cross section 0.35 mm, fixed on a crimping machine (RN 071) in the manufacturing process, and the course of actions connected with its output to a subsequent process.

Release pertains to the manufacturing process which has an effect on the final product quality. Output from the process must be carried out in the shortest possible time following the application of the manufacturing process to the serial procedure.

We selected 10 samples representing the actual or expected range of process variability and we selected the operators 1, 2, 3 and tagged samples 1–10 so that to operators these numbers were not clear and they were in random order. Before measurement, each operator performs a calibration of a measuring instrument. This procedure was repeated 3 times with altered order of samples. The measured values are processed in Table 3.

Measuring instrument is not eligible (NDC value < 5). Measurement system is not sufficiently sensitive (does not sufficiently distinguish individual categories). Measurement system needs to be improved and a problem has to be identified and fixed with subsequent preventive measures.

Using the statistical methods, the results may be estimated from the standard deviation calculated using real data, while others are estimated from the standard deviation calculated using the considered distribution of probability, based on experience or other information, Table 2.

MSA is performed by testing (measuring) a chosen parameter by an operator or group of operators. The effect of repeatability (one operator repeats the measurement of a monitored product parameter) and reproducibility (a group of operators measure the same parameter) on the total variance is monitored.

The variance of the monitored product parameter may be caused by the product itself (ovality, deformation ...) or by measurement system. Measurement system consists of operator, gauge and measurement method.

The result of the analysis is the determination of accuracy (to what extent the measurement system contributes to the resulting variance) and suitability of use of such measurement system for the monitored parameter. If the effect of repeatability prevails, it is necessary to re-evaluate the suitability of the gauge and the compliance with the standard procedure at the measurement.

The protocol on Graph 1 shows the calculated ranges and standard deviations for each variability component. The repeatability or variability of product was set EV = 0,04943 and the reproducibility or variability of operator AV = 0,04585. The variability of the measurement system was calculated R&R = 0,06742. Once the variability of individual factors was determined, it was compared with the tolerance value T. The total proportional variability of the measurement system was % R&R = 39,54%. The stated value is higher than the required standard % R&R > 30%. Based on that,

Table 3. Perform crimping device.

Pcs	Measure 1	Measure 2	Measure 3
Operator 1			
1	0.66	0.60	0.67
2	1	1	0.95
3	0.85	0.8	0.81
4	0.85	0.95	0.90
5	0.55	0.45	0.43
6	1	1	1
7	0.95	0.87	0.91
8	1	1	1
9	0.90	0.91	0.91
10	0.96	0.84	0.90
Operator 2			
1	0.55	0.55	0.50
2	0.95	0.80	0.99
3	0.76	0.70	0.75
4	0.80	0.88	0.90
5	0.45	0.43	0.40
6	0.99	0.95	0.80
7	0.70	0.78	0.70
8	0.90	0.80	0.85
9	0.80	0.88	0.85
10	0.95	0.95	0.85
Operator 3			
1	0.69	0.77	0.80
2	0.95	0.94	0.86
3	0.55	0.55	0.60
4	0.70	0.65	0.70
5	0.40	0.45	0.40
6	0.95	0.80	0.75
7	0.70	0.88	0.75
8	0.95	0.90	0.90
9	0.88	0.90	0.84
10	0.95	0.90	0.88

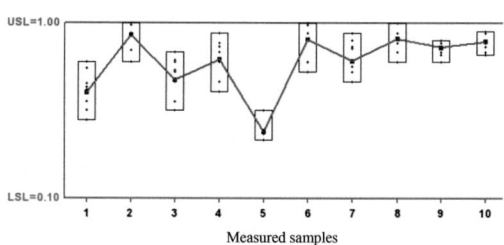

Legend: LSL (Lower Specification Limit) USL (Upper Specification Limit) Tolerance = (USL − LSL)—zone of satisfactory values for a customer.

Figure 2. Measurements of individual parts.

we can say that in this case the measurement system is ineligible.

In case the measurement system is ineligible (error rate over 30%), we standardize it by multiple measurements and averaging the results. Despite the time and costs it provides reliable measurement

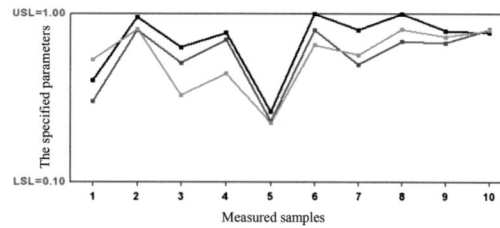

Figure 3. Interaction operators * parts.

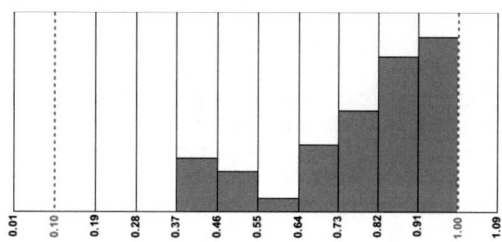

Figure 4. Graph 1 histogram of measurement system analysis.

results, the measurement process of the system is improved.

The aim of our research was eligibility verification of the process of fixing the wire crimp contact before entering the serial production, which is a part of wiring harness designed as an intermediate material for the automotive industry. A crimping machine is used for fixing, which fixes the crimp contact on a wire. Problems could arise, if the wires made with crimp contact, which are further used for assembly of intermediate material—the wiring harness, did not meet the required values and tolerance and for that reason the process was not eligible and stable for a following process in serial production. This creates the need for verification of eligibility of partial manufacturing process.

5 CONCLUSIONS

The paper presents a systematic procedure of innovation within company organizational measurement processes based on implementation of internationally recognized standard ISO/TS 16949—Quality Management System—Special requirements for the use of the standard ISO 9001:2015 in companies producing cars and spare parts and a relevant measurements system as a key feature of product quality management. The systematic procedure is based on the evaluation of the gauge using the Measurement System Analysis and a percentage determination—the combined value of repeatability and reproducibility.

This method is useful especially:

- for products of automotive industry, where a control plan is referred to it,
- where a gauge is installed before eligibility of machinery and process is performed,
- before installation of measuring instrument.

Within the context of this paper a range of measurement system analysis is more effectively considered, in order to create a model of prediction for evaluation of capacity of measurement system during the measurement. Moreover, if the analyses performed within this paper were not constantly improved, they would not meet the requirements. According to STN EN ISO 9001:2008 (2015), the quality is defined as the degree to which a set of inherent characteristics fulfils requirements (of customers). The demands on metrological support of main companies' activities increase with a natural pressure to increase the quality of products. The system of metrological support is an important part of quality management systems within the network of suppliers of automotive industry.

The eligibility of the measurement process may evaluated e.g. by analysis of control processes which is based on requirements of a VDA 5 standard (or DIN EN V 13005) and takes into consideration the measurement uncertainties. Its disadvantage is the restriction of use to measuring geometrical quantities. Evaluation of measurement process eligibility based on the analysis of system eligibility has not been standardized yet. MSA is an additional document to the standard (technical specification) STN ISO/TS 16949:2009. As far as the measurement process is carried out in an eligible system, it is assumed that the process itself is eligible too.

ACKNOWLEDGEMENTS

This contribution is a result of the Project for young teachers, researchers and PhD students, no. I-15-110-00, 2015: Methodology for implementation of Integrated Management for small and medium-sized enterprises in SR and EU levels.

REFERENCES

Agarski, B., Kljajin, M., Budak, I., Tadic, B., Vukelic, D., Bosák, M. & Hodolič. J. 2012. Application of multicriteria assessment in evaluation of motor vehicles' environmental performances, *Technical gazette*. ISSN 1330–3651, 19 (2), pp. 221–226.

Bednárová, .L, Pacana, A. & Chovancová, J. 2013. Managing environmental risks in production companies *In: Ecology, economics, education and legislation*, Conference proceedings, volume II: 13th international multidisciplinary scientific geoconference SGEM 2013:.—Sofia: STEF92 Technology, 2013. ISBN 978-619-7105-05-6. ISSN 1314-2704, pp. 651–658.

ISO/TS 16949:2009. Quality management systems–Particular requirements for the application of ISO 9001:2008 for automotive production and relevant service part organizations.

Lakhal, L., Pasin, F. & Limam, M. 2006. *Quality management practices and their impact on performance*, International Journal of Quality & Reliability Management, 23 (6), 625 p.

Majerník, M., Bosák, M., Szaryszová, P., Tarča, A. & Štofová. L. 2014. Model for assessing the environmental performance of product systems within sustainable development. VEGA 1/0292/13. In *Geoconference on ecology, economics, education and legislation SGEM 2014*: 14th international multidisciplinary scientific geoconference, 17–26 june 2014, Albena, Bulgaria. volume III.—Sofia: STEF92 Technology Ltd., 2014., pp. 285–292. ISBN 978-619-7105-19-3.

Majerník, M., Szaryszová, P., Bosák, M., Štofová, L. & Kabdi, K.. 2015. Integrated management of environmental—safety and technical risks of plant producing automobiles and automobile components. In *Communications: scientific letters of the University of Žilina*.—Žilina: University of Žilina, 2015.—ISSN 1335-4205.—Vol. 17, no. 1 (2015), pp. 28–33.

Majerník, M. & Štofová, L. 2014. Quality management in the integrated system as a tool for business excellence and sustainability. In *Interdisciplinarity in Theory and Practice*: Journal for Presentation of Interdisciplinary Approaches in Various Fields.—Arad: Editura Adoram, 2014.—ISSN 2344-2409.—No. 1 (2014), pp. 59–63. Online: http://itpb.eu/index.php/ct-menu-item-3/15-economics/83-quality-management-in-the-integrated-system-as-a-tool-for-business-excellence-and-sustainability.

Mateides, A. 2006. *Manažérstvo kvality*. Bratislava: EPOS, 2006. 751 p. ISBN 80-8057-656-4.

McNeese, W.H. & Klein R.A. 1991. *Measurement System Sampling and Process Capability*, Quality Engineering. 1991; 4(1), pp. 21–39.

Montgomery, D.C. & Runger, G.C. 1993. Gauge Capability Analysis and Designed Experiments Part II: Experimental Design Models and Variance Component Estimation. In *Quality Engineering*. 1993; 6(2), pp. 289–305.

Rovňák, M., Chovancová, J., Bednárová, L., Adamišin, P. & Huttmanová, E. 2013. Managing environmental risks in production companies. In *13th international multidisciplinary scientific geoconference SGEM 2013*: Ecology, economics, education and legislation: conference proceedings, volume I. Sofia: STEF92 Technology, 2013. ISBN 978-619-7105-04-9. ISSN 1314-2704. pp. 651–658.

Sila, I. 2007. Examining the effects of contextual factors on TQM and performance through the lens of organizational theories: An empirical study. In *Journal of Operations Management*. 25(1), pp. 83–109. doi:10.1016/j.jom.2006.02.003.

Szabo, S., Ferencz, V. & Pucihar, A. 2013. Trust, Innovation and Prosperity. QIP Journal: Quality Innovation Prosperity, 17 (2): 1–8.

Tomčíková, M. & Živčák, P. 2012. *The practical contribution of information systems in times of globalization*. In: Management 2012: research management and business in the light o fpractical needs/vedec. red. Róbert Štefko, vedec. red. Miroslav Frankovský, vedec. red. Ján Vravec (Eds.). Prešov: Bookman, 2012. pp. 253–256. ISBN 978-80-89568-38-3.

Production Management and Engineering Sciences – Majerník, Daneshjo & Bosák (Eds)
© 2016 Taylor & Francis Group, London, ISBN: 978-1-138-02856-2

New methodology for crisis management RM/RA CRAMM and its legal frame

M. Mamojka
Academy of the Police Force in Bratislava, Bratislava, Slovak Republic

J. Müllerová
University of Zilina, Zilina, Slovak Republic

ABSTRACT: The paper deals discusses the application of RM/RA CRAMM methodology to use in system of crisis management and its legal limits. This new methodology comes out from the CRAMM method used by many large institutions for risk assessment of IT risks. After its original modification leading to new methodology, it is suitable to cover floods risk assessment. The RM/RA CRAMM is for first time used in practise in Slovakia where its modification was developed to cover wide range of risk management issues very common in Slovakia such the floods. Legal and legislative aspects of its practical use are enhanced in the paper as well as whole process of its use. The methodology will be used by crisis management government employees within the preventive phase of crisis management.

1 CRISIS MANAGEMENT

1.1 Risk management prevention

The protection of society can be divided into two areas: the internal security and external security. External security falls under the Ministry of Defence, under the Ministry of Internal Affairs.

Aim of the internal security it the protection against criminal activities and to protect people from the effects of emergency situations like natural disasters, industrial accidents and fires.

Dealing with incidents and crisis phenomena is a major social problem. Authors see the large field of knowledge and activities to be implemented in the risk management science and practice, particularly in the field of prevention of emergencies. The trend of crisis management is to focus on prevention, in order to prevent incidents and huge losses of life and property. Using minimum resources and to reach maximal effect, avoiding unnecessary loss of human life and material values is impossible without effective prevention actions. Crisis phenomena is therefore necessary to solve before they arise, before they their destructive effects appears on the affected system and the surrounding environment. Compliance with this principle considerably eases and streamlines the rescue work and increases their effectiveness.

Generally, it is estimated that €1 invested in effective prevention will save € 100 in any possible damage. E.g. example: a few months before massive floods in the Czech Republic, 2002. Members of City Council Prague criticized the Mayor for unnecessary purchase flood barriers worth several 100 thousand crowns. At the end, this barrier prevented the spillage of the Vltava River in the historical center and thus protected billions in value.

The greatest negative example is related to Catastrophic Asian tsunami in 2004, which killed 320,000 people. Only after this catastrophe happened some of the affected governments started to invest into warning signalization system against Tsunami.

The investment in the preventive action represents marginal costs comparing to the potential lost and damages, in spite of fact that human life values is impossible to measure in monetary units.

Prevention of incidents can be defined as the implementation of measures which prevent incidents resp. minimize their impact. It can be described as a continuous process to an emergency preparedness. Takes place both at the preparation stage in order to prevent Emergencies during the intervention itself, when we try to prevent the spread of Emergency situations. After repairing the damage on lessons learned from the shortcomings in the action, it is necessary to innovate and implement effective preventive measures.

1.2 Current limits of emergency management theory implementation

Looking at the experience of crisis management in the Slovak Republic is an obvious gap between theory and practice.

The fact remains that the majority of government employees has only limited theoretical knowledge, and space and receive private organizations included in the budget. The outputs of their studies are mostly descriptive quantitative- and too general, which considered sufficient.

On the other hand, developing technologies of simulation, training devices, geographical information systems, communication technology is very promising. There are also many applications that can accelerate decision-making in crisis management. Good manager should be able to use the modern technologic means to improve the readiness of the officers for crisis and emergency events.

Authors find it useful to develop a theoretical field transformed into practical actions. However, there are some barriers. At the first place there are professional qualifications of personnel, their knowledge skills and attitudes as well. Secondly, and the most common mentioned limiting factors for the development of crisis management are limited budgets, time and lack of quality technologies.

1.3 Legal limits of crisis management

Very significant limiting factor in the practice of crisis management is the legal framework for crisis management, which must be respected. Field of activities, rights and limitations of rights are clearly enshrined in the applicable legal standards.

Constitution at the first place it is especially Constitutional Law 227/2002 and other seven laws related generally and dealing with issues of defense and security, National crisis management and Law 179/2011 of economic mobilization such a Law 570/2005 of military emergency or very important and exact Law 387/2002 of governance in crisis situations outside the time of war and war state including later amendments to certain laws.

The Constitutional Law 227/2002 describes the functioning of the Security Council of the Slovak Republic in peacetime and National crisis Management in crisis situations except of the time of war and state of war. There are also seven Regulations currently valid.

These legal standards create the legal frame—space for crisis management head-officers, public officers, citizens and people living in Slovak republic in time of Emergency situations. The legal limitation of personal freedoms are described e.g. When announcing the state of emergency in the affected area are some limited rights of citizens, officials and citizens should be able to explain what is allowed and what is not allowed.

These legal limits have to be respected throughout the Crisis management of country and by each citizen as well also in the Peace time.

2 METHODOLOGY RM/RA CRAMM THEORY

2.1 Aim and resources

When looking for a suitable methodology for the main criterion we established—a broader applicability in risk management. European Agency for Networks and Information Security—ENISA has released a list of methods and methodologies for risk assessment. RM means Risk Management, Risk Assessment RA (Enisa 2014). One of the methods was CRAMM. The mere abbreviation methodology means Risk Management/Risk Assessment CCTA Risk Analysis and Management Method.

The initial focus of CRAMM, which was established in 1987 is Information Technology security (IT) within the UK government agency for development of information and communication technologies.

Another important theoretic and practice source is a Standard ISO 31010 Risk management, in addition to the definitions of key terms offers a number of techniques and methods that can be used in assessing risks. Among them are well-known methods such as fault tree analysis or multi-criteria decision making. The object of research output has become CRAMM respectively. Its appropriate change and modification treatment and subsequent application to different types of risk environment was set as a main objective.

2.2 Target group of use

Above mentioned facts of legal and qualification limits there are also reflected in one of the long-term objectives Academy of Police force where-in our new methodology RM/RA CRAMM was created and first-time used.

Methodology RM/RA CRAMM is intended primarily for government entities in the field of crisis management, Police, Military forces, Fire forces, Integrated rescue system,

Using this methodology, it is possible to assess and evaluate risks in the selected area within the scope of the subject of public administration. It may be a risk of technological operations, transport hubs, water-ways, black landfills, maladjusted social communities and other risk objects within a given territory. (An example would be securing the site against occasional crime in times of crisis (violent crimes, property crimes—theft, robbery, looting).

2.3 From CRAMM to RM/RA CRAMM

Marguis modified CRAMM with aim that it could be used by everyone without purchasing an expensive license. Thus losing the advantage linked against the process and use the extensive database. On the other hand, there are savings of a lot of resources. Marguis has introduced a simple 10-point implementation procedure CRAMM while maintaining the original focus on information technology IT (Marguis 2008).

2.4 Modification, creation of new method

Marguis's process has become the basement of our modification. Changing the focus from IT to the general risk management—RM is pointless original database focused only on IT systems. Modification RM/RA CRAMM is designed to risks from floods, the fires, other natural disasters, industrial accidents, leaks NL, mass traffic accidents and social riots, or other risks to the monitored area. The focus of this methodology is also useful for the active employment policy as well as many educational institutions.

2.5 Risk assessment

The clarification of the risk assessment process, is the core of our methodology. The diagram shows the process of risk management (Risk management). It is based on the risk assessment process (Risk Assessment), which precedes the definition of scope and determine the relationship between elements of the system. An integral part of any management is continuous communication, monitoring and evaluation of changes. Communication within the team and externally throughout the process under way is continuous and crucial for each team cooperation. In case of new evidence appearance it can be always implemented into the process of risk assessment, then select and adopt appropriate measures, which reflect the actual facts.

In the first stage of risk assessment i.e. at the stage of identifying risks apply universal support techniques brainstorming or using the Delphi method, which are able to take into account the individual views of several stakeholders.

The second phase is risk analysis. Under the risk analysis we understand the consequences of expression, the likelihood that there will be an extraordinary event and the expression level of risk. The starting point is the standard IEC/FDIS 31010 Risk management—Risk assessment techniques providing 24 techniques and methods that are very useful for risk analysis (ISO 2011). The analysis also apply other procedures and methods.

The third step is the evaluation of the risk assessment. Outputs of the risk evaluation are quantitative and qualitative. The preferred attitude for assessing is the quantitative and semi-quantitative output as result. Semi-quantitative output is numerically expressed in words, ie evaluation. each

number (interval) corresponds to a predetermined narrative description of the scenario 0—negligible risk, 10—very high risk.

It need to be considered it necessary to emphasize that the methodology RM/RA CRAMM covers the period of preventive measures in preparation for potential emergencies. The outputs are dedicated to the implementation of effective preventive measures against threats most at risk. If the estimated incident occurs before the adoption of the recommended measures will need to take emergency measures, crisis nature (evacuation, sheltering, rescue units intervention, intervention by the ingredients etc.).

3 PRACTISE OF RM/RA CRAMM

The implementation consists of several fundamental phases including ten steps first time introduced by project realized by authors at Academy of Police Force (Müllerová & Mamojka 2014).

3.1 Preparation phase

STEP 1. Exploration of the area/field, establishment of a work team, finding sources of information. Creating a basic working document—risk assessment table.

3.2 Risk identification and risk analysis phase

STEP 2. Identification of the elements (entities) e.g. population, tangible, intangible assets, monuments, other values in the area under consideration, using the data from the census, municipalities, autonomous regions, market prices according to realtors, other databases to market prices of assets to be found on web. Only qualified expert estimations are to be applied.

STEP 3. Preparation of the table to which the. For each asset (property, machinery, equipment, software, territory) state the name of its owner—a man who best knows its use and value.

STEP 4. Interview with the "owner" of the assets was to provide an answer to the question, how will change the value of assets at the incident. What is the cost of replacement assets to cover losses incurred. Evaluate the following attributes:

a. Element value (assets): What is the cost of compensation for the element under consideration? Scale:
  - Very low (0–1), [10 €]
  - Low (2–3) [100 €]
  - Medium-high (4–5) [1,000 €]
  - High (6–7) [10,000 €]
  - Very high (8–9) [100,000–1,000,000€]
  - Countless (10) [∞].

b. Integrity—the impact of disablement or destruction of an element to the scheme in Categories:
- Low (1–3),
- Intermediate (4–7),
- High (8–9)
- Very high (10).

c. Irreplaceableness stage—after which time you can leave the element shutdown, respectively. How quickly it is possible to replace this subject. Is the object of strategic importance? Does it belong to critical infrastructure? Irreplaceableness grades:
- Low (1–3) easily replaceable [weeks]
- Middle (4–6) [days]
- High (7–8) [hours]
- Very high (9) [min]
- Critical (can not fail) (10).

3.3 Risk evaluation phase

STEP 5. Use a support team of employees, professionals and other employees determine the likelihood that there is a threat referred to the previous phase. Use a structured questionnaire. Consider disasters (terrorism or natural) human factor, procedural errors. Create a column for each threat. Use the category negligible (0), low (1–4), moderate (5–7), high (8–9) and a very high probability (10). Update the evaluating table for each element.

STEP 6. Risk calculation as (1):

$$Ri = \left(\frac{a \cdot Value + b \cdot Integrity + c \cdot Substitution}{3} \right) \times pi \quad (1)$$

a, b, c parameters represent the weights of attributes of risk analysis. $a + b + c = 3$. Assign the marks as low (1–33), medium (34 to 67) or high (68–100) for each element (object). This procedure results in to the list of the most endangered areas! The value, integrity and substitution are variable in the range from 0 to 10, than probability for each element shall be calculated. It one of these three variables is found or considered as more important than others, the value from 0 to 3 can be used as a parameter of the weight-importance. According to the simple relationship applies that the sum of the parameters.

3.4 Risk response

STEP 7. Start with selecting countermeasures in assets with the highest risk. Make a list of possible countermeasures. To create the list, use the following techniques: brainstorming, interview, Best Practices.

STEP 8. Consider countermeasures and ways to alleviate the threat. Focus on the threat of a higher level. Do not prefer fast, easy and inexpensive measures for lower level threats. Give priority to counter-measures which:

a. protect against multiple threats
b. the protection of high risk assets
c. Can be applied where there are no established countermeasures
d. They are less costly
e. They are more effective in the prevention or mitigation of threats
f. I can foresee threats
g. They can be made quickly, easily and inexpensively (even for low risk).

STEP 9. Apply a Multi-Criteria Decision MDCA—AHP Online Help Make Rational it. This step is necessary when too many solutions are to be considered. After creating a list of measures again we apply the methods of multi-criteria decision and choose the most effective measures which subsequently recommended to implement. The effective application of a Multi-Criteria Decision recommend it online tool Make Rational. (Thus avoiding repeated manual calculation of nuts immediately receive graphical output according to our criteria and weighting that can be updated at any time).

STEP 10. Get resources and implement effective measures for top-level threats. Take advantage of your evaluation to justify the need to adopt the proposed measures.

The results of all the risk assessment steps are to written in the Risk Assessment (RA) table. It easies the use of methodology in public administration institutions connected to crisis management.

4 CONCLUSIONS

Methodology RM/RA CRAMM has the ambition to become a comprehensive tool for risk assessment and management (RM/RA). Its suitability for use in the risk assessment in crisis management was verified within the research inter-institutional task in Academy of Police Force. Modified method RM/RA CRAMM does not have the original database or appropriate software, but builds on the platform of Marguis adjustments that is adapted and rebuilt for the requirements of the Risk Management (Müllerová & Mamojka 2014).

Former CRAMM's focus on the risks of information and communication technologies was extended thank to modifications of algorithm into the wide range of risks.

Application of methodology in Risk management is limited by certain legislation. There are seven law and seven regulations which must be respected. An appropriate mix of methods under the methodology has a decisive influence on its effectiveness was allowed thanks to existence of IEC ISO/FDIS 31010 Risk management—Risk assessment techniques. The present basic version RM/RA CRAMM uses e.g. Interview, MCDA, brainstorming, and Best Practices. In specific cases, they can be used other methods from the portfolio within the 10 steps described in the paper.

Linking applications of methods of risk assessment methodologies and RM/RA CRAMM will be one of the main benefits of its future development.

REFERENCES

City Council Prague, 2013. *City Council Prague Minutes 33/1 from 12.12.2013.*

ISO IEC/FDIS 31010 Risk management—Risk assessment techniques, 2011.

Marguis, H. 2008. *10 Steps to Do It Yourself CRAMM,* DITY Weekly Newsletter, USA Vol. 4/50.

Müllerová, J. & Mamojka, M. 2014. *VYSK 205/2014-RM/RA CRAMM method for emergency situation prevention* Bratislava: Academy of the Police force.

RAC—Risk Analysis Consultants: CRAMM.

Production Management and Engineering Sciences – Majerník, Daneshjo & Bosák (Eds)
© 2016 Taylor & Francis Group, London, ISBN: 978-1-138-02856-2

Environmental assessment of technologies applying powder fire extinguishers

I. Marková, J. Zelený, M. Drímal & J. Jaďuďová
University of Matej Bel, Banská Bystrica, Slovak Republic

ABSTRACT: A large number of high-technologic equipment especially in sectors "highly susceptible" to fires is equipped with various types of fire extinguishing equipment using various substances. The paper presents a brief overview of the most commonly used methods and approaches of environmental technology assessment and comparison between the performance and characteristics of selected extinguishing powders through a standard procedure. This comparison provides basic information required for the environmental assessment of technology-equipped fighting equipment on the basis of extinguishing powders, so that the evaluation can be carried out by any of those methods and assessment approaches. For the purposes of the experiment, there were used samples of fire extinguishing powders, which are currently used for filling of fire extinguishers and fire extinguishing equipment. Methods were applied by standard: STN EN 615:2001. Based on the results by the optimal choice of fire extinguishing powders are new powders as ABC and BAVEX.

1 INTRODUCTION

A large number of machinery, mainly in sectors that are "highly susceptible" to fire, are equipped with various types of fire extinguishing devices that use different fire extinguishing agents. Nowadays, powder fire extinguishers are used for rapid and initial fire fighting. Powder fire extinguishers are a group of chemical substances with a mixed composition. They are part of portable fire extinguishing devices, fixed fire fighting equipment and mobile fire extinguishers. Currently, they are the most widely used form of fire extinguishers. According to the type and method of manufacturing, powder fire extinguishers have various physical, chemical and toxicological properties. As the fire fighting equipment and fire extinguishing devices are an integral part of technologies, they must be also included in the processes of suitability assessment, particularly in the assessment of environmental suitability (Zelený et al. 2014, Andráš et. al. 2013).

The aim of this paper is to present a brief overview of the most commonly used methods and approaches of environmental technology assessment. Based on this overview, we want to compare the production and performance characteristics of selected powder fire extinguishers through a standard procedure. This comparison should provide basic information required for the environmental assessment of technologies equipped with a fire fighting technology based on powder fire extinguishers so that the assessment could be performed by any of the above methods and assessment approaches.

1.1 Environmentally sound technology

When talking about Environmentally Sound Technologies (EST), we must realize that we need not strictly speak of a technology as such. EST can be defined as overall systems that include the know-how, products, service and accessories as well as organizational and management practices, etc. The criteria according to which the technology is assessed may change as a result of new information, changes in the state of knowledge, views of the institutions involved and a number of alternatives may arise as a result of technological progress. Generally, an environmentally sound technology is a technology that is not only acceptable from the environmental point of view, but it is also economically viable and socially acceptable (Figure 1).

Environmental technology assessment is a specific assessment procedure. Currently, the EST assessment uses various methods such as EnTA, DICE, ESTP, etc.

1.2 Environmental Technology Assessment (EnTA)

EnTA is a method used as an assisting means in the decision-making process aimed at assessing the likely impact of using the proposed or existing

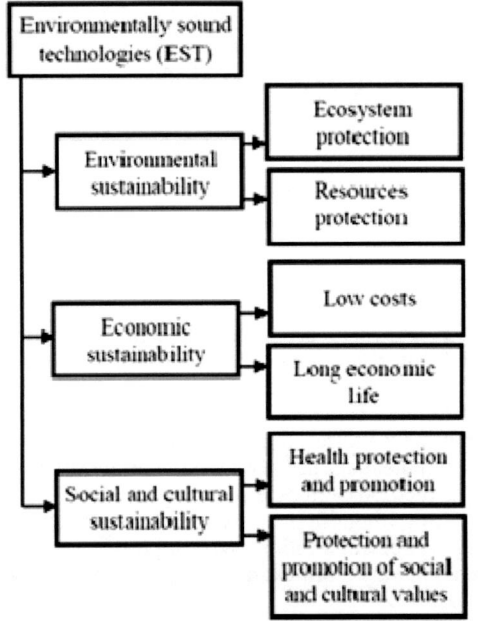

Figure 1. Sustainability criteria characterizing the EST.

technology on the environment. The assessment carried out by this method takes into account the financing costs of the technology, financial advantages and disadvantages of the application of the proposed technology in the relevant company/site and the environmental, social and political impacts of its future operating (Zelený et al. 2014).

The EnTA method is mainly a qualitative tool that minimizes the need for detailed technical data. It is designed to allow a comprehensive dialogue leading to a consensus. It is intended for prevention, i.e. it should be used mainly ex ante, i.e. to avoid an environmental problem, and not ex post, i.e. when an environmental problem has began to manifest itself. It is a multidisciplinary method enabling to consider technical, economic and environmental conditions as a complex set and simplifies the mutual relationship between the technology and its environment and the consequences of this relationship. It examines the environmental impacts of the whole technological system, including the use of resources and waste production throughout the whole life cycler.

1.3 DICE method

This method enables to characterize and assess the extent of environmental impacts associated with the technology, resp. with various alternative technologies, via four steps: Describe—description of technological proposals, alternatives, their requirements, operating conditions, Identify—

identification of environmental aspects of the technology on the environment as well as the use of natural resources, waste management, human resources, infrastructure, Characterise— defining possible environmental impacts of the aspects, Evaluate—assessing the overall impact of these impacts under local conditions, characterizing the proposed technology in terms of its environmental suitability and economic availability production throughout the whole life cycler.

1.4 Environmentally sound technology performance assessment

This method is intended to assess the performance of an environmentally sound technology through a verification protocol. The assessment process consists of three stages, namely preliminary examination of the technology, detailed assessment of the technology and data verification. ESTPA facilitates the assessment of environmentally sound technologies in internationally recognized technical protocols containing an appropriate verification technique and statistical analysis (Zelený et al. 2014, Andráš et. al. 2013).

1.5 MET matrix

Risk analysis of technological processes has an important role in assessing the suitability of a technology. It is a systematic application of the methods of hazard identification and assessment, conditions or causes of an accident, with appropriate assessment of possible consequences. The so called MET matrix is often used for this purpose. It consists of: M—identification of the occurrence and qualitative and quantitative analysis of the flows of all elements of material nature occurring in the assessed technology, E—identification of the occurrence and qualitative and quantitative analysis of all flows of energy occurring in the assessed technology, and T—identification and analysis of toxic properties of elements occurring in the previous flows, but in the context of the whole life cycle of the assessed technology.

2 POWDER FIRE EXTINGUISHERS

Powder fire extinguishers are a group of chemical substances with a mixed composition that are used for the purpose of immediate fire fighting. They are categorized into BC, ABC and M powders. Powder fire extinguishers are part of portable fire extinguishing devices, fixed fire fighting equipment and mobile fire extinguishers and they are used for rapid and initial fire fighting. Opinions on the use of powder fire extinguishers have been gradually developing with the development of powders as extinguishing agents. These powders consist of solid particles in the form of powder or dry chemicals. Based on the conclusions of experiments, the size of powder particles was set to 0.1 mm (Orlíková 1995, Kalousek 1999, Balog 2005, Marková, 2008).

Dynamics of the effect depends on the technical design of the fire fighting equipment. The amount of kinetic energy of individual powder particles must be sufficient to affect the entire centre of fire (i.e. the particles penetrate into the centre of fire with great vigour) and the number of particles per unit volume must exceed the critical amount for thermal reactions (Ewing et. al. 1989, Schroll 2002, Marková 2011).

The development of powder fire extinguishers is directly connected with the development of powder fire extinguishing devices by which the powder is transported into the centre of fire (Schroll 2002).

The principle of extinguishing is the assumption: Burning is the reaction in which short-term radicals, active molecules and atoms are made. These cause branching the chain and subsequently burning.

The powder wall that takes energy of the atoms, radicals and active molecules is made when the particles of powder are brought to flame. Their active energy will be so small that it is impossible to influence another process of the chain reaction. If the chain reaction is interrupted, the flame goes out. Therefore it is said about the "wall effect" eventually about the negative catalysis in heterogeneous phase (Saito et. al. 1996).

All powder fire extinguishing devices operate on the same principle: the powder is extruded from a vessel by the pressure of the carrier (extrusion) gas, such as argon, carbon dioxide, CO_2, helium and nitrogen (Kalousek 1999, Balog 2005, Marková 2008, Orlíková & Štroch 2002).

3 EXPERIMENTAL PART

3.1 Samples and methods

For the purposes of the experiment, there were used samples of fire extinguishing powders, which are currently used for filling of fire extinguishers and fire extinguishing equipment. It was a Furex 40 Furex 75, 70 and Neutrex pills called BC powder ABC powder, BAVEX and MAP. Brief characteristic of samples is enclosed in the Table 1.

Experimental methods were applied by standard: STN EN 615: 2001 with a specific description of the various experiments, such as the determination of density, water repellency, moisture content test, resistance to jolting, abrasivity powder test at low temperatures and resistance to caking and clumping and their application in fire fighting equipment is stable. For the description and identi-

Table 1. The chemical compounds of samples.

Extinguishing powder samples	Chemical compounds
Bavex	Ammonium phosphate (monobasic), Ammonium dihydrogen phosphate, Ammonium sulfate, (2: 1), Diammonium sulfate; Sulfuric acid, Diammonium salt silica, mica, Muscovite magnesium aluminum silicate
Furex	Ammonium dihydrogen phosphate, Ammonium sulphate
Neutrex 40	A mixture of inorganic salts (etylenediamine tetraceptique soda), silicic acid, corrosion inhibitor Lubricant, processing aids
Neutrex 70	Mixture of inorganic salt, silicic acid, corrosion inhibitor, lubricant, processing aid

fication of microscopic structure of samples it was used light transmission scope Olympus BX 50.

3.2 Results and discussion

When assessing technologies as environmentally sound technologies EST, we took into account the manufacturing characteristics of the technologies and the performance characteristics of powder fire extinguishers to mean the criteria of the DICE method. The fire extinguishing efficiency depends on the development of the powder cloud, which is directly related to transportability. Regarding transportability, the shape of crystals (Figure 2) significantly influences this characteristic and is therefore very important. The size of crystals is determined by sieve analysis.

The size of crystals determines the powder density and size of the specific surface. The highest percentage of dust particles was in the range <60 μm, 125 μm. This interval corresponds with the size of particles of 0.1 mm, when powder fire extinguishers have optimum fire extinguishing efficiency and optimum flow properties (Figure 2).

In the production it is necessary to find a compromise solution to the detriment of the fire extinguishing effectiveness and in favour of the fluidity and shelf life. On the other hand, dry powder is only useful if clumping does not occur during its storage (Figure 3).

The basic substance of these powders are ammonium phosphates namely diammonium phosphate $(NH_4)_2HPO_4$ and ammonium phosphate $NH_4H_2HPO_4$.

To be possible to use ammonium phosphates in the field of fire technology it is important to produce them in the form of small crystals (with the diameter about 0,1 mm), they have to be enough liquid and resistant against humidity. Their

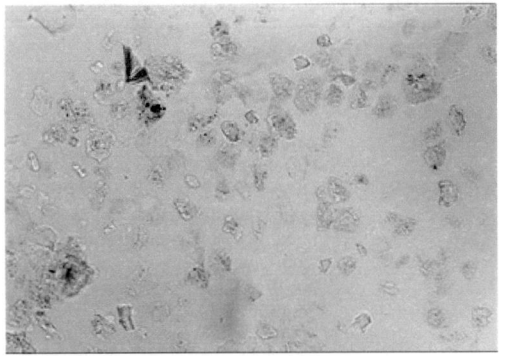

a)

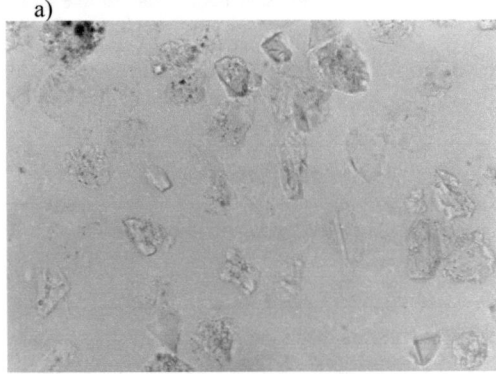

b)

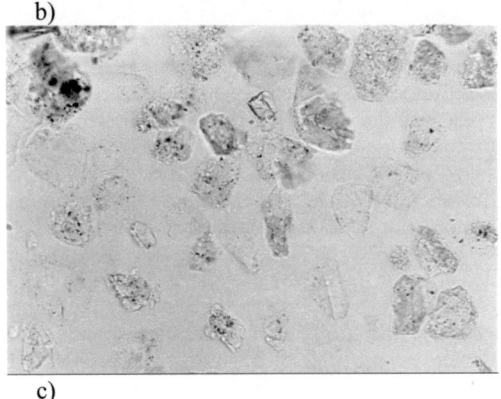

c)

Figure 2. Microscopic analysis samples of the powder extinguisher FUREX 40. Legend: a) at 100x magnification, b) at 200x magnification, c) at 400x magnification.

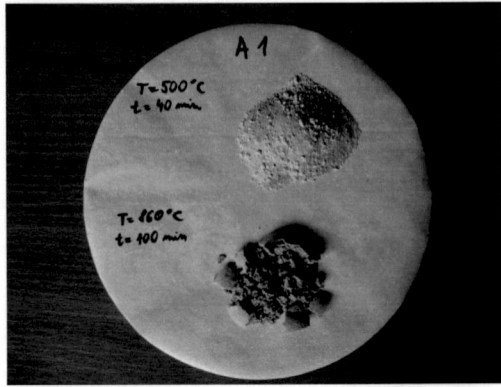

a)

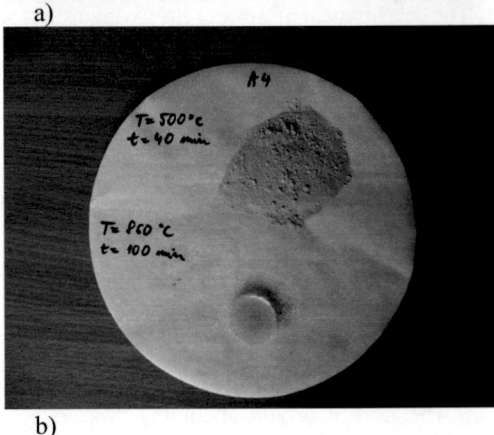

b)

Figure 3. Samples of FUREX 40 (A1) and ABC powder (A4) after being removed from the furnace at a temperature of 500 °C and 860 °C.

$$HPO_3 + heat \rightarrow polyphosphates\ make\ glaze \quad (3)$$

Shelf life is then determined by the treatment and water-repellent properties of the powder (Figure 4).

The powder fire extinguishers with the prescribed moisture content do not have corrosive properties in fire extinguishing devices, piping of these devices and in the extinguishing of solid objects.

BC powders in admixture with water, generally at an elevated temperature, corrode unprotected iron and attack coatings sensitive to alkalis. At fire temperature, ammonia may separate from the ABC powder and cause corrosion of ferrous metals. It is commonly referred to that at high temperatures of the burning material these gases are extruded from the centre of fire and are therefore ineffective. Corrosive effects of powder fire extinguishers were confirmed by our long-term experiment. The results are significant in the presence of water. As part of fire fighting, technical and mobile devices used

properties mustn't be changed even after long-term storage. The extinguishing effect of powders is based on "wall effect" but there comes also another effect that makes glaze on burning surface areas. ABCD powders on base phosphates are decayed; see Equations 1, 2 and 3 below:

$$(NH_4)_2HPO_4 + heat \rightarrow H_3PO_4 + 2NH_3 \quad (1)$$

$$H_3PO_4 + heat \rightarrow HPO_3 + 2NH_3 \quad (2)$$

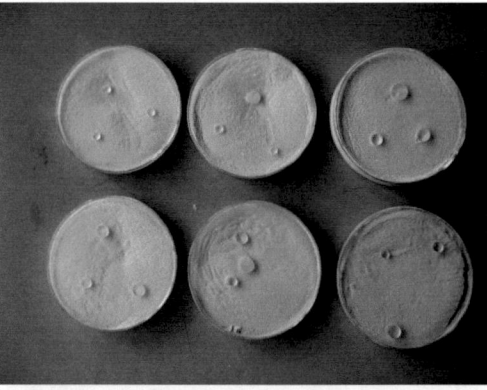

a)

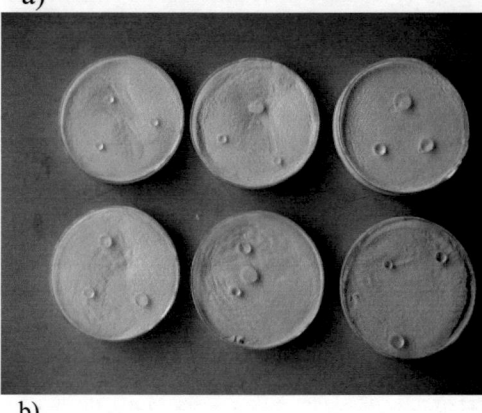

b)

Figure 4. The FUREX 40, FUREX 75, NEUTREX 70 in the top row and ABC powder, BAVEX, MAP in the lower row before being inserted into the desiccator, b) after being removed from the desiccator.

Describe

- technological procedure
- alternatives – system of extrusion gases
- unlimited requirements of the technology

Identify

Aspects related to:
- only human resources,
- infrastructure,
- supportive EPS technology in case of automation

Characterise

Resulting impacts on:
- the health and safety are minimal,
- local environment regenerates easily from the effect of the powders,
- company protection system

Evaluate

The technology fulfils the EST requirements in terms of environmental suitability and economic availability

Figure 5. Application of the structure of the DICE method.

for fire fighting, powder extinguishers have significantly positive effects thanks to their manufacturing and performance characteristics. In view of the technological assessment of the above chemicals, they can be used in all types of fires, their shelf life is in the range from −50 °C to 60 °C in fire extinguishing devices, they are non-toxic to living organisms and thanks to their good flowability they can be transported by hoses and pipes (Ewing et. al. 1989, Orlíková 1995, Balog 2005, Marková 2008).

At the same time it is necessary to accept that during their application and in the absence of a cooling effect the environment becomes covered with dust. The fire extinguishing effect of powders can not be characterized as a single effect, but as the interaction of several phenomena where the anticatalytic (inhibitory) and insulating (wall) effect are of crucial importance. This fact must be taken into account in the EST assessment. The effect of the powders depends on the tactics of the fire fight-

ing intervention, weather conditions, powder cloud density (number of particles per volume unit of the cloud) and pressure conditions under which the powder reaches the center of fire. A fixed fire extinguishing device firmly attached to the building or technological objects with a permanent risk of fire is a fire extinguishing device containing particularly a stable source of the fire extinguishing medium, distribution pipes, drain fittings, triggering mechanism and a signaling device. According to the requirements or the level of protection it is possible to apply powder fire extinguishing devices which reliably extinguish the fire in the protected segment.

Aerosol fire suppression systems with thermal decomposition are considered the most modern group of fire fighting equipment. They are based on the formation of fire suppressing aerosol at the site of fire due to thermal decomposition of the extinguishing agent. The resulting aerosol enters into reactions with flame reactants, i.e. with active particles—radicals and it terminates the combustion process in a chemical way. Basically, during fire fighting there is a reaction which changes the aerosol into a producing mixture with fine particles of 0.001 to 0.1 mm. Aerosol fire suppression with powder is accompanied by a number of side effects, e.g. energy consumption for chemical reactions that take place in the flame. During the application of the above fire extinguishing system the aerosol is formed at the site of fire from chemical compounds contained in the so-called generator of fire suppression aerosol. The extinguishing system is initiated by the initiating temperature in the process of burning or fire (Schroll 2002).

Powder extinguishers as chemical substances are the filling of fire extinguishing equipment. It is a system of chemical substances for the purpose of rescue, but it is important to stress that they are also a potential source of risk and we assess their environmental parameters for the purpose of monitoring the environmental impact of using them for the purpose of risk assessment. In the application of the used EST assessment methods such as EnTA, DICE, ESTP, the technologies of powder fire extinguishing meet the above requirements. Our research of the technology of powder fire extinguishing is presented by the DICE method based on selected parameters of powder extinguishers via the 4 steps specified in Figure 5.

4 CONCLUSION

The Based on the presented methods and approaches of environmental technology assessment EST and the comparison of production and performance characteristics of selected powder fire extinguishers by a standardized procedure STN EN 615: 2001 [15] we obtained basic information needed for the environmental assessment of technologies equipped with fire extinguishing systems based on powder extinguishers. Application was carried out by the DICE method.

For the purpose of the experiment we used samples of the powder extinguishers Furex 40, Furex 75, Neutrex 70 and BC powder, ABC powder, BAVEX, which are the filling in the currently used fire extinguishing devices and fire fighting equipment. Based on the results described, the optimal choice of powder fire extinguishers are newer powder extinguishers such as ABC powder, BAVEX that meet the requirements specified in the standard, such as density determination, water repellency, moisture content test, bouncing resistance, powder abrasivity, test at low temperatures and resistance to sintering and clumping and their application in the fire extinguishing device is stable.

ACKNOWLEDGEMENTS

This work was supported by the Cultural and Educational Grant Agency of the Ministry of Education, Science, Research and Sport of the Slovak Republic Project No. KEGA 035UMB-4/2015 "Environmental management in sphere of production".

REFERENCES

Andráš, P., Zelený, J. & Ladomerský, J. 2013. Where are heading demands of society in the environmental sector, there has also tend research and training efforts department of environmental management. [on-line] *ACTA UNIVERSITATIS MATTHIAE BELII Section Environmental management*: 15(2): 6–10. [cit.2015–02–12] Available on the internet: http: //sparc.fpv.umb.sk/kat/ken/akta/attachments/article/254/ACTA%202013%2015–2.pdf.

Balog, K. 2005. *Safe disposal of extinguishing agents.* Bratislava: STU Bratislava—Faculty of Materials Science and Technology in Trnava, Ministry of Environment of the Slovak republic, Fire Equipment Manufacturers' Association of the Slovak Republic, 150 pp.

Ewing, T., Faith, F.R., Hughes, J.T. & Carhart, H.W. 1989. Flame extinguishment properties of dry chemicals: Extinction concentrations for small diffusion pan fires. *FIRE TECHNOLOGY (1989)*, pp: 134–149 Maj.

Kalousek, J. 1999. Fundamentals of physical chemistry of fire, explosion and extinguish. Ostrava: SPBI, 1st edition, 115 pp.

Marková, I. 2008. *Fire extinguishers.* Monography. Zvolen: BratiaSabovci (eds.), 1st edition. 45–110.

Marková, I. 2011. Extinguishing efficiency of gaseous extinguishing agents set by cup burner test. In *SPEKTRUM*, X (2): 37–40.

Orlíková, K. 1995. *Extinguishers.* Ostrava: SPBI, 1st edition, 65 pp.

Orlíková, K. & Štroch, P. 2002. *Extinguishers – classic and modern.* Ostrava: SPBI, 1st edition, 84 pp.

Saito, N., Ogawa, Y., Saso, Y., Liao, C. &, Sakei, R.1996. Flame-extinguishing concentrations and peak concentrations of N_2, Ar, CO_2 and their mixtures for hydrocarbon fuels. *Fire Safety J.* 1996, 27: 185–200.

Schroll R.. 2002. Fire Behavior. Portable Fire Extinguishers. In: *Industrial Fire Protection Handbook.* 2 sd edition. CRC Press LLC.

STN EN 615: 2001 *Fire protection. Fire extinguishing media. Specifications for powders* (other than class D powders). Test methods: Annex A: Test method for determining the volume weight. Annex D: Water repellency test.

Zelený, J. et al., 2014. *Treatment of risk.* University textbook. BanskáBystrica: Bellianum, 1st edition, 375 pp.

Practical experience with the application of 3MU method in the automotive production

Š. Markulik

Faculty of Mechanical Engineering, Technical University of Košice, Košice, Slovak Republic

T. Páricsiová

Magneti Marelli Slovakia, Kechnec, Slovak Republic

ABSTRACT: Today the pressure of competitive struggle allows customers in supply chains push the product price as low as possible. Expressed mathematically Price = Cost + Profit. This means that the price is no longer a sum of costs and profit. The price is fixed and determined by the customer and the organization that wants to succeed in the competition does have other choice than accept the price to maximize its profits by minimizing its cost. Expressed mathematically, Price – Cost = Profit. Therefore, many organizations apply various methods to minimize their costs. The cost reduction can therefore be understood in the broader sense as minimizing waste and increasing efficiency. One of these instruments, which points out the waste in production is a Japanese tool 3MU. The aim of this paper is to highlight the importance of analyzing internal processes of the organization to identify waste and then eliminate it.

1 3MU—METHOD

The strategic aim of any organization should be to achieve continual improvement of processes to enhance overall performance. Before the improvement itself, it is necessary to carry out a thorough analysis of individual processes (or the process itself). The aim of improvement is not only generating improvement proposals, but it is important to track the results (impacts) of the proposals.

An important aspect of identifying waste is mapping all activities in the process and assess their contribution to value creation. In the event that these activities do not add value and do not directly support other activities, it can be stated that this is an activity that does not constitute value and this is the source of waste. Such analysis of processes uses also audiovisual recording for better and deeper analysis. Any measure taken to eliminate waste should be turned into money to be able to evaluate the effectiveness of the measure.

Under Toyota theory, waste is recognized as all factors, which prevent a company from achieving a perfectly efficient production system. 3MU is method incorporates three Japanese words. They are: MUDA—waste, MURA—unevenness, MURI—overburdened (Figure 1).

MUDA—means the status of activities in which materials, equipment and manpower do not contribute to the value of the product. Eliminating waste can result in fluency of workforce, while also creating working conditions under which employees are able to give the required long-term work performance. Expert sources define a different number of types of loss, however, there is broad agreement on seven basic types of loss—waste, identified across various organizations. It is the following types of waste (Yamazumi Workload Charts, 2015).

MURA—diversion or unevenness, it also expresses the loss incurred by missing or incomplete concordance of capacities within the management of production. It arises where human labor lacks fluency, or the material is delivered irregularly, which for example disturbs the balance of constant material flow in production, continuous

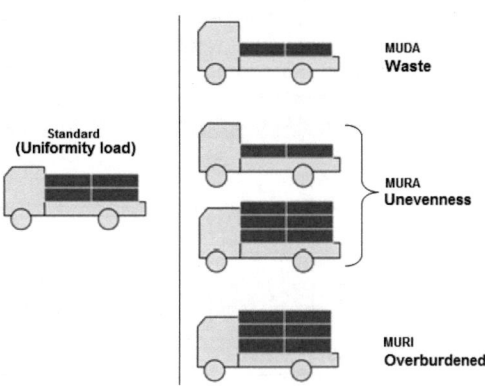

Figure 1. Principle of 3MU method.

production supply and so on. Specific manifestations of MURA are:

- losses created by queues,
- losses due to unequally balanced capacities.

Other tools such as Andon, Heijunka, Kanban, etc. can be used for effective application of 3MU.

MURI—is unreasonable, unnecessary burdening of the body of an employee or performances that often result from MURA. Overburdening describes the losses incurred by high demand within the labor process. Meanwhile it is necessary to distinguish between overburdening by means of manipulation and overburdening in the manufacturing process. Losses due to manipulation occur by physical and psychological overload of employees, which is manifested by fatigue, stress, increased errors and dissatisfaction with work. These losses can be eliminated from ergonomic point of view as well. Reducing the burden on a person in work process and adjusting workplace layout by applying knowledge of ergonomics can effectively eliminate losses due to improper manipulation.

2 3MU METHOD PLACE OF APPLICATION

Application of 3MU tool was implemented in the production organization of mechanical engineering nature. It was an organization engaged in the production of propulsion components for three specific car manufacturers.

3MU Tool was applied to the assembly line. The line was used to assemble propulsion mechanisms for cars (hereinafter the product). The assembly line is divided into five successive processes. The total duration from input of material to the assembly process up to the output of final product from the assembly line was repeatedly measured to be 6: 30 min. Assembly line processes, including operations which constitute them are shown in Figure 2.

Due to objective analysis, the individual processes and their activities were repeatedly measured in terms of time duration and recorded on camcorder with respect to the analysis of operators' movement. Individual activities were analyzed

by Yamazumi method. By this means the activities were analyzed in terms of significance and evaluated as an operation:

- with added value,
- necessary,
- with no added value.

The duration of individual's process on the assembly line was as next Table 1.

Process yield of the assembly line was determined to value of 30 assembled products per hour. One operator worked in the process 1 and 2 before introducing changes. By means of observing the site and subsequent analysis of repeated video recordings it has been found out that their workload is severely unbalanced compared with each other.

In Process 1, waste has been identified—waiting of the operator. This process is largely automated what caused the operator wait until the automatic operation was completed. The average measured duration of the process was 1:07 min.

Waiting time amounted to 49 seconds (up to 73%! of the total duration of the process is represented by the waiting time of the operator). During this time the employee did not perform any other work.

The total time duration of this process is limited by the production capacity of automated equipment. Analysis arrived at the conclusion that the time of operator is not effectively used.

In Process 2, heavy workload on the operator has been identified. In Figure 1 it can be seen that Process 1 is formed by two operations, while Process 2 is formed by ten. The analysis of the process has found that there was no waiting time of the operator, he was permanently burdened. The measured duration of time showed 1:43 min. This confirms the argument that the unevenness (Mura) and overburdening (Muri) are the source of waste (Muda).

Given that Processes 1 and 2 are directly related to successive workstations, the aim was to evenly burden the two operators in order to minimize the waiting time of the operator in Process 1 and reduce the level of workload of the operator in Process 2.

Table 1. Duration of processes.

	Duration min
Process 1	1:07
Process 2	1:43
Process 3	0:56
Process 4	1:39
Process 5	1:05
The total duration	**6:30**

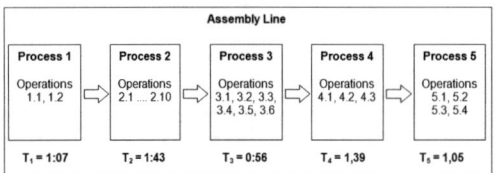

Figure 2. Scheme of assembly line.

In Process 3, the analysis of on-site observation and repeated video recording showed waiting time, but it was caused by the follow-up process. The average duration of the process was measured in value 0: 56 min. Waiting time in this case amounted to 15 seconds (i.e. 27% of the total duration of the process). The operator of the process is the operator of the following Process 4.

In analysis of Process 4, no waiting time was identified. The average duration of the process was measured in value 1:27 min. Analysis of the video recording, however, identified unnecessary movements of the operator (Figure 3) that consumed 35 seconds of the total time of the process. In this case, unnecessary operator movements were caused by improper layout of the workplace and accounted for 40% of the total time of Process 4!

Analysis of Process 5 did not identify any downtime of the operator, or any other types of waste. The average duration of the process was measured at 0: 55 min.

Conclusions of the analysis of processes of the assembly line were as follows:

• three out of five processes on the assembly line included waste,
• the first process involved waste of operator's time waiting for the end of automated operations,
• in the second process heavy burdening of the operator (up to 10 operations in the process) has been identified,
• possibilities to reduce wasted time and looking for evenly distributed burdening of operators of process 1 and 2,
• the third process involved waste due to waiting for the end of the previous process,
• the fourth process involved waste due to unnecessary movements of the operator due improper layout of the workplace,
• no waste has been identified in the fifth process.

3 PROPOSAL OF MEASURES

The analysis of the individual processes shows that no waste due to unnecessary waiting and movements has been identified in Process 3 and 5, as it was in processes 1, 2 and 4. Therefore, the proposed measures focused primarily on shortening waiting times, even the workload of workers as well as the elimination of unnecessary movements of the operators.

Processes 1 and 2 are subsequent processes, whose working environment is represented by one common space. Proposal of measures was based on this knowledge and was therefore designed to find a solution that on one hand minimizes waiting of the operator in Process 1 while reducing workload of the operator in Process 2. Getting familiar with these processes during the on-site observation was an incentive to change the workload of operators of both processes in such way that the three operations of Process 2 operator have been moved into work of Process 1 operator. After mutual evaluation and measuring the times the following measures have been introduced (Table 2).

Proposal of measures influenced Process 4 as well. Floor plan layout of the workplace has been changed—relocation of the assembled components in the workspace in order to eliminate unnecessary movements of the operator (Figure 4).

Table 2. Comparison of Process 1 and Process 2.

	Operations	Duration	Waiting time
Before changes			
Process 1	2	1:07	0:49
Process 2	10	1:43	0:00
After changes			
Process 1	5	1:07	0:17
Process 2	7	1:24	0:00

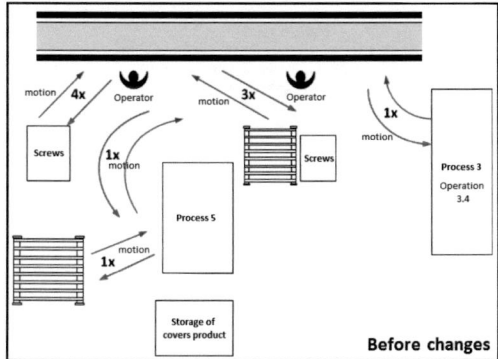

Figure 3. Analysis of operator movement and assembly line layout.

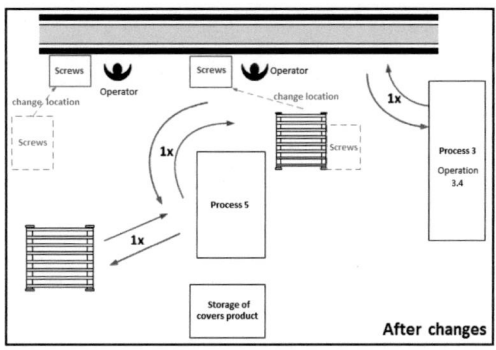

Figure 4. Operator movement after changes.

Table 3. Process 4 (before and after changes).

	Operations	Duration	Unnecessary movement
Before changes			
Process 4	3	1:39	0:49
After changes			
Process 4	3	1:04	0:17

After repeated analysis of introduced changes the average process time was measured in value 1: 04 min. This shortened the duration of the process by 13 seconds. This is not only with regard to saving time but also the reduction of the ergonomic burden of operator by elimination of unnecessary movements. After the introduction of changes in the workplace the following changes occurred in the next Table 3.

4 CONCLUSION

When analyzing the processes, waste in processes must be seen from a broader perspective. Proposed measures may sometimes cut across numerous processes that can bring a comprehensively positive effect, as was the case in the above mentioned two processes (Process 1 and 2). Also, it should be borne in mind that not every waste of time of operator in the process can be removed, because it can be caused by, for example its level of automation. However, it is possible to look for the possibility of using operator's labor pool for other actions as far as the nature of work and the specificity of the process allow. With regard to preservation of the anonymity of the organization and not disclosing sensitive economic data, the following assessment can be made.

After calculation of the proposed savings, the assembly line will save more than twenty thousand Euro each year. This article points out the fact that even the seemingly small imperfections in the process may be the source of great waste in long-term perspective. It is therefore necessary to identify the weaknesses in the process, analyze and introduce measures that will bring the effect in the form of savings. Organizations are offered various tools, application of which can improve processes and thereby increase overall production efficiency often with negligible investment, which on the contrary, can bring significant savings.

ACKNOWLEDGEMENTS

This contribution is the result of the project implementation: VEGA 1/0150/15 Development of implementation and verification methods for integrated safety systems of machines, machine systems and industrial technologies.

REFERENCES

Bauer, M. 2012. *KAIZEN—The way to lean and flexible enterprise.* (Albatros media, Brno).
Imai, M. 2011. *KAIZEN—Method to introduce efficient and flexible production company* (Computer Press, Brno).
Paricsiová, T. 2014. *Elimination of waste in the manufacturing process using the tool 3MU.* (Master Thesis).
Yamazumi Workload Charts. 2015. http: //www.acsco. com/yamazumi.htm.

Production Management and Engineering Sciences – Majerník, Daneshjo & Bosák (Eds)
© 2016 Taylor & Francis Group, London, ISBN: 978-1-138-02856-2

An essay of firefighting vehicles' reliability

M. Monoši
Faculty of Security Engineering, University of Žilina, Žilina, Slovak Republic

L. Jánošík
Faculty of Safety Engineering, VŠB—Technical University of Ostrava, Ostrava-Výškovice, Czech Republic

ABSTRACT: This paper summarizes the results of the study and evaluation of primary data on the operation of firefighting equipment with a focus on exit vehicle type water tenders. A total of 75 vehicles were evaluated within the project. Vehicles on the chassis TATRA, Renault, MAN and Mercedes-Benz have been selected for the monitoring. Vehicle operation was evaluated for the period 2010–2013. Monitored fire equipment is operated in professional units of the Fire Rescue Service (FRS) of the Czech Republic in Moravian-Silesian, South Moravian and Zlín Regions. Primary analysis focused on the utilization of exit and other activities was performed at first. There were mileage and hours of machine work, i.e. operation of vehicle engines on the place, evaluated at the beginning. Availability of the monitored equipment was evaluated in terms of reliability. Selected characteristics of reliability and maintainability in terms of professional fire units' activity were determined.

1   INTRODUCTION

The paper reassumes on the previous author's publications focusing on evaluation of the operation and maintenance of firefighting equipment on the chassis of Mercedes-Benz Atego (Jánošík et al. 2010), TATRA 815 in Jánošík & Melichar (2010) and more recently Renault Midlum in Jánošík & Pecina (2013), which were deployed at Fire Rescue Units in Moravian-Silesian, Hradec Králové and Zlín Regions. Results of detailed analysis that emerged from the evaluation of the operation and servicing of selected fire equipment are summarized in this paper. Primary data have been exported from electronic records at Fire Rescue Service. Operating diaries kept in paper form were used for vehicles register previously. Electronic Information System IKIS II fulfills this function since 2010. Data on the vehicles' operation for the subsequent evaluation were obtained thereof. Information about the operation was processed only for the 2010–2013 period primarily due to the availability, credibility and completeness of the input data. The system was put into full operation from January 2010. The incompleteness and, in some cases, even lack of credibility of paper records within observing vehicles TATRA and Renault Midlum were the reason for this decision, as stated in Jánošík & Melichar (2010) and Jánošík & Pecina (2013).

2   CHARACTERISTICS OF FIRE AREAS

The Moravian-Silesian Region, which has an area of 5.4 thousand km² and a population of about 1.2 million, is in terms of organization of fire brigades divided into 6 Area Departments (AD), with a total of 22 fire stations. Seven fire fighting vehicles on MB Econic chassis are placed at six fire stations in the Area Department of Ostrava. Eleven fire fighting vehicles on MB Atego chassis are located at three area departments outside of Ostrava.

South Moravian Region (area of 7.2 thousand km², population about 1.2 million people) is in terms of fire brigades' organization divided into 6 Area Departments. There are 24 fire stations in the region, 8 of which are incorporated under Area Department Brno. Fire equipment at MAN chassis is at number 24 vehicles deployed at 20 fire stations. In this 9 vehicles are situated in Area Department Brno. Fire equipment at TATRA 815 chassis is at number 9 vehicles located at seven fire stations.

Zlín Region, which has an area of 3.9 thousand km² and a population of about 0.6 million people, is in terms of organization of fire brigades divided into 4 Area Departments. Individual Area Departments represent a total of 13 fire stations. Thirteen busiest fire engines of the CAS type (Fire Fighting Vehicle—Water Tender) on the chassis Renault Midlum were selected to monitor the operation and failure rate. Renault Midlum vehicles are located at 8 different stations. Vehicles on TATRA 815 chassis are deployed at 8 fire stations.

3   MONITORED FIRE EQUIPMENT

Among the monitored vehicles dominates fire equipment on chassis MAN TGM 13.240 4 × 4 BL that is located in the South Moravian Region. These vehicles are based on mixed chassis, of middle weight category, with the engine output of 176 kW, pump capacity 1500 l/min at 10 MPa water pressure at the output. Volume of the water tank is 2200 liters. Fire superstructures are made by companies THT Polička Ltd, Polička, Czech Republic, SPS Ltd, Slatíňany, Czech Republic and SZCZESNIAK Pojazdy Specjalne Ltd., Bielsko-Biala, Poland.

Mercedes-Benz vehicles are represented by two types of chassis in Moravian-Silesian Region. The first model is Econic 1833 LL 4 × 2. This is a typical urban chassis, heavy weight category, engine output of 240 kW, pump capacity 2000 l/min at 10 MPa water pressure at the output, volume of the water tank 2700 liters. Model Atego 1528 4 × 2 F represents the second type of chassis. This is a mixed chassis, in middle weight category, engine power 205 kW, pump capacity 2400 l/min at 8 MPa water pressure at the output, volume of the water tank is 2500 liters. Fire superstructure is made by THT Polička Ltd, Polička, Czech Republic, for all vehicles in the region.

Renault vehicles in the Zlín Region are represented by models of Renault Midlum 270.14 P 4 × 4. The vehicle chassis is mixed, middle weight category, engine output of 200 kW, pump capacity 2400 l/min at 8 MPa water pressure at the output, volume of the water tank is 2500 liters. Fire bodies are made by companies THT Polička Ltd, Polička, Czech Republic and Wawrzaszek ISS Ltd, Bielsko-Biala, Poland.

Vehicles on the chassis TATRA 815 were chosen into monitoring from the South Moravia and Zlín Regions. There are two types of chassis. The first one is the classic TATRA 815-2 4 × 4 known as TerrN°1, cross-country gears, middle weight category, engine output of 325 kW, pump capacity 2400 l/min at 8 MPa water pressure at the output, volume of the water tank 4000 liters. The second type is a special model 815-7 TATRA 6 × 6 variant used in the water tender, off-road chassis, heavy weight category, engine output of 325 kW, pump capacity 3000 l/min at 10 MPa water pressure at the output, volume of the water tank 9000 liters.

Fire superstructures are made by THT Polička Ltd, Polička and SPS Ltd, Slatíňany, both from the Czech Republic, and Wawrzaszek ISS Ltd, Bielsko-Biala, Poland.

Detailed tactical-technical characteristics of the equipment mentioned above can be seen at the Web sites of fire superstructure manufacturer THT Polička Ltd, Polička or in (Monoši et al. 2013).

4   STATISTICS OF INCIDENTS

To illustrate the utilization of fire brigades in the mentioned regions, the number of exits during years 2010 to 2013 is shown in Table 1 (Vonásek et al. 2013). These figures do not include other journeys, i.e. non-exit firefighters' activity. It only serves to illustrate the image of the region from the standpoint of the exploitation of fire brigades.

5   EVALUATION OF OPERATIONAL WORKLOAD

The results of the evaluation of the operational workload monitored vehicles are summarized in Tables 2–4.

Table 1.   Number of fire unit's interventions.

Year	2010	2011	2012	2013
South Moravian Region	13,622	12,180	13,879	14,032
Zlín Region	7151	5847	6381	6308
Moravian-Silesian Region	14,835	14,008	15,005	21,011

Table 2.   Overview of exit activity during 2010–2013.

Average annual operating characteristics on a vehicle	Number of vehicles	Average age [km]	Number of interventions	Other exits
M-B Atego	11	8	1146	567
M-B Econic	7	2	531	181
MAN TGM	24	13	583	272
TATRA 815 (FRS SMR)	9	9	125	63
TARTA 815 (FRS ZLR)	11	13	40	42
Renault Midlum	13	7	152	76

Table 3. Overview of the service work on firefighting vehicles during 2010–2013.

Average servicing the vehicle in the monitored period	Repair after failure		Preventive maintenance	Repair after damage
	The chassis base	Fire super-structure		
M-B Atego	6	8	23	46
M-B Econic	15	6	12	39
MAN TGM	11	9	74	277
TATRA 815 (FRS SMR)	5	2	22	0
TARTA 815 (FRS ZLR)	5	2	2	1
Renault Midlum	10	8	11	2

Table 4. Summary of vehicles' workload during 2010–2013.

Average annual operating characteristics on a vehicle	Mileage [km]	Converted mileage [hour]	Machine work at the site [hour]	Converted machine work at the site [hour]
M-B Atego	4864	97	99	4943
M-B Econic	8704	174	68	3391
MAN TGM	11,227	225	241	12,065
TATRA 815 (FRS SMR)	3096	62	77	3866
TARTA 815 (FRS ZLR)	1209	24	34	1723
Renault Midlum	5564	111	77	3866

6 RELIABILITY

Reliability can be generally characterized as a property of the object consisting in the ability to perform required functions under defined conditions in a defined time period in Stodola (2002). The so-called test plan method (Famfulík et al. 2010) was chosen for the failure flow analysis in Daněk & Široký (1999) to evaluate this indicator. By this method, mean times to failure can be determined for a small group of products. The test plan censored by time-to-failure, so called t—plan, was used for failure rate evaluation. The duration of the test is the limit and the number of detected failures is a random variable. The assumption of the test is that the products are repaired after a failure. The accumulated working time of the vehicle T_{AKU} is a time variable representing the test procedure. T_{AKU} is the total time during which all products were in operation. The accumulated working time for the chosen t—plan is calculated according to the following equation:

$$T_{AKU} = \sum_{i=1}^{r}(\tau_0 - \theta_i) + (n - r) \cdot \tau_0 \quad (1)$$

where τ_0 = test time, from the beginning to the r_0-th failure; n = number of tested products; r = number of fault units; θ_i = time needed to repair the i-th product during the test interval.

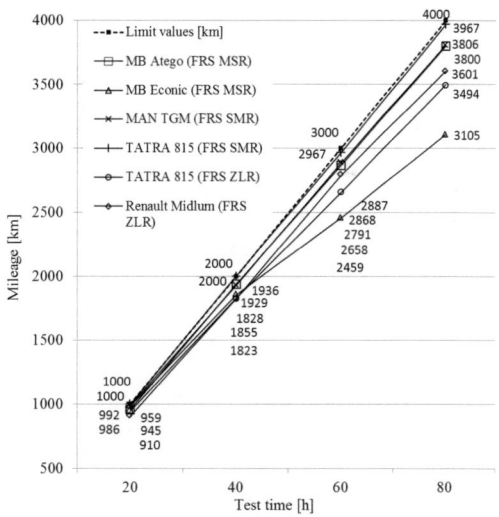

Figure 1. Summarized results of accumulated working time.

The assessment of the failures' severity preceded the calculations. The criteria were used by the method FMEA. Insignificant failures were excluded of the group of failures. The calculation of the accumulated working time according to the equation (1) was conducted for four values of the test time: 20,

40, 60 and 80 hours. These intervals, after calculating by the average speed of 50 km/h, represent the mileages of 1000, 2000, 3000 and 4000 km. In standard practice, indicators related only to the 1000 km mileage interval are used. The decision to use three other longer intervals was based on operating vehicle loads stated in vehicle characteristics. The reason was to register at least a six-month deployment of vehicles in exits due to the different sizes of kilometric annual mileage between area departments. The results of calculations of the average accumulated working time are shown in Figure 1.

7 MEAN TIME BETWEEN FAILURES

Mean Time Between Failures (MTBF) is the most frequently used indicator in practice for assessing the reliability of repairable systems. It is the mean operating time between two consecutive failures. The indicator is determined as the sample mean of the measured operating times according to in Stodola (2002):

$$T_s^* = \frac{1}{n}\sum_{i=1}^{n} t_i \tag{2}$$

where t_i = the i-th vehicle operation time during the reporting period; n = the number of vehicles included in the test.

In the calculation, all monitored vehicles are included, i.e. both vehicles with failures, and those without failures during the reporting period.

Results of calculations for all observed vehicles are given in Figure 2.

8 AVAILABILITY

Availability is the ability of an object to perform the required functions under given conditions at a given time interval while ensuring the desired ambient conditions in Stodola (2002). This state of the object can be characterized by a number of complex indicators of reliability. Availability coefficient K_p defined by equation (3) was used in our calculations:

$$K_P = \frac{\sum_{j=1}^{n} t_{pj}}{\sum_{j=1}^{n} t_{pj} + \sum_{i=1}^{n} t_{oi}} \tag{3}$$

where Σt_{pj} = the sum of times of failure-free operation; Σt_{oi} = the sum of service time during the period under review.

We can rewrite the equation (3) with using former applied symbols into the equation (4):

$$K_P = \frac{T_{AKU}}{T_{AKU} + \sum_{i=1}^{n} \theta_i} \tag{4}$$

Calculation results for defined time intervals of tests are shown in Figure 3.

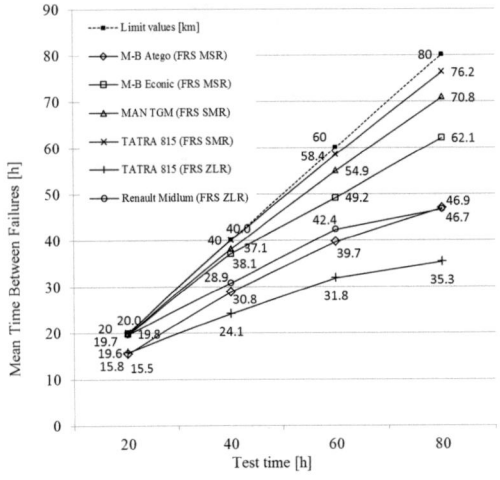

Figure 2. Summarized results of the mean time between failures.

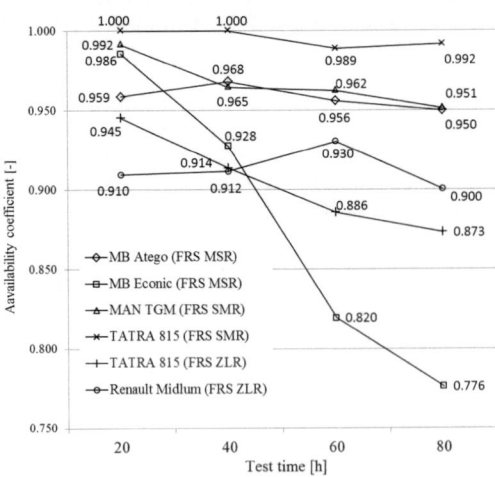

Figure 3. Summarized results of the availability coefficient.

9 MAINTAINABILITY

Maintainability is a vehicle characteristics consisting of ability for the prevention and detection of failures by the specified maintenance in Stodola (2002). In essence, it is a complex of features expressing the easiness, simplicity and non-sophistication of maintenance work. These attributes are determined by vehicle design, quality staff, diagnostic and automated equipment, labor organizations and many others. These factors cannot be clearly and precisely evaluated and therefore partial indicators of maintainability are commonly used. In our case the mean time of maintenance was determined by calculations.

The mean maintenance time is the time interval during which the maintenance intervention is performed on the technical system, including technical and logistical delays. It is the most often used indicator for assessing the reliability of systems under repair in practice. The mean maintenance time is calculated as the mean value of the operating time between two successive failures. In practice, the mean maintenance time is determined as the sample average of measured operating times according to the equation (5) as specified in Stodola (2002):

$$t_{su} = \frac{1}{n}\sum_{i=1}^{n}t_{iu} \qquad (5)$$

where t_{iu} = maintenance time of the i-th vehicle; n = the total number of vehicles in operation in the monitored period.

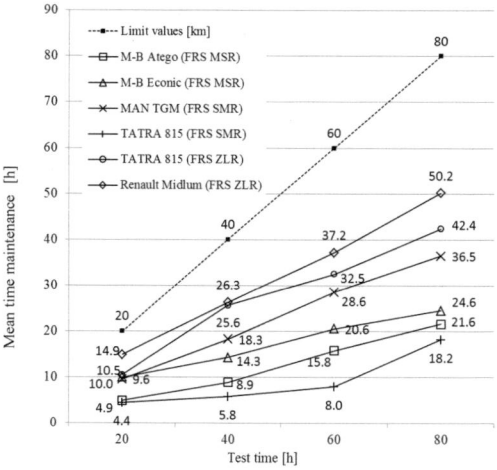

Figure 4. Summary results of mean time maintenance calculations.

All monitored vehicles were included into the calculation again, i.e. both vehicles with some failure, and those without any failure during monitored period.

The traditional problem of incomplete and inaccurate records of preventive maintenance and operational control emerged again when evaluating. This fact was confirmed in consultations with engineers. All scheduled preventive maintenance and operational checkups were performed, but not all were recorded. It depends on particular man. Therefore, that scheduled and regularly performed activities (monthly, semi-annual and annual maintenance of vehicles) have been included into the calculations. The calculation results are summarized in Figure 4.

10 CONCLUSION

Some direct conclusions can be drawn of the tangible results of research, in which the most important is the fact that for vehicles' reliability evaluation is generally useful to characterize the operation mileages by longer interval than the commonly used 1000 km mileage is. It has been declared on other characteristics that at longer intervals the results vary, both in the negative, and positive direction. The imaginary winner among monitored vehicles in a given period were firefighting vehicles of first and second exit, based on the chassis TATRA 815-2 TerrN°1, deployed in the South Moravian Region.

This result should be treated with some limitation, however, in terms with the recommended range of the paper. Firstly, monitored vehicles in the particular group were divided into several age categories. Further, repairs after the failure were sorted by the place of origin (see Table 4) and even there were differences in fire superstructures, unrelated to the type of the chassis. However, another more detailed study should be performed for describing this distinctiveness.

REFERENCES

Daněk, A. & Široký, J. 1999. *Theory of Replacement of Conveying Vehicles* (in Czech). Ostrava: VŠB—Technical University of Ostrava.

Famfulík, J. et al. 2010. *Reliability Tests: Selected Stochastic Methods* (in Czech). Ostrava: VŠB—Technical University of Ostrava.

Jánošík, L. et al. 2010. Operational Reliability of Vehicles Mercedes-Benz Atego. *TRANSACTIONS of the VŠB—Technical University of Ostrava. Safety Engineering Series* (in Czech). Ostrava: VŠB—Technical University of Ostrava.

Jánošík, L. & Melichar, D. 2010. Operational Reliability of Vehicles TATRA. In Michail Šenovský (ed.), *Fire Protection 2010. XIX. Annual International Conference, Ostrava, 8–9 October 2010* (in Czech). Ostrava: VŠB—Technical university of Ostrava.

Jánošík, L. & Pecina, L. 2013. Operating and Maintaining Analysis of Fire Appliances on Renault Midlum Chassis In Ladislav Šimák (ed.), *International Scientific Conference RKS 2013; 18. Annual International Conference, Žilina, 5-6 June 2013* (in Czech). Žilina: University of Žilina.

Monoši, M. et al. 2013. *Firefighting Equipment. University Textbook* (in Slovak). Žilina: EDIS, University of Žilina.

Stodola, J. 2002. *Operational Reliability and Diagnostics.* (in Slovak), Brno: University of Defence.

Vonásek, V. et al. 2013. *Statistical Yearbooks 2013. Czech Republic* (in Czech). Praha: Ministry of Interior—General Directorate of Fire and Rescue Service of the Czech Republic.

Production Management and Engineering Sciences – Majerník, Daneshjo & Bosák (Eds)
© 2016 Taylor & Francis Group, London, ISBN: 978-1-138-02856-2

Fire safety characteristics of oak wood and medium-density fiberboard

J. Müllerová & J. Vácval
University of Zilina, Zilina, Slovakia

ABSTRACT: The aim of this paper is to assess the extent to which the type of wood and its way of manufacturing affects the resulting fire properties of material. There is mainly monitored and compared the behavior of solid wood and fiberboard under conditions of impact of heat flux simulated on a cone calorimeter. The paper studies the behavior of wood materials with a similar density and the same mechanical and performance characteristics in conditions of thermal load. Oak wood here represents the original natural material with high density. The medium-density fiberboard represents a technological material which has similar properties as wood with high density, but is made from wood with low density. Part of this paper is also a description of the properties and possibilities of use of a both materials. In description of medium-density fiberboard is also mentioned the manufacturing process.

1 INTRODUCTION

1.1 *Material description*

Wood is a natural material with good properties which have already found its application in almost every sector of human activity and the gradual development of the technologies, nowadays, topical issue of economic and ecological use of natural resources which is also related to the use of forests and wood. The result of the development of technology and also the need for efficient use of resources are in the wood industry materials, which are manufactured from wood particles, forming essentially waste from the processing of solid wood. These materials are found due to its favorable properties and low price just wide range of applications such as solid wood (Müllerova & Pus-kajler 2014). One such material is a medium density fiberboard. This is the material, the production of which use knowledge of processing and pulp and paper production process is more demanding than the original materials were produced from wood waste in the last century.

The current MDF is made from high quality white chips and precision technology, resulting in a board material of high density and good properties. Just to be mentioned, the density MDF compensated for example, oak, which is among the most solid and dense woods. Some physical and mechanical properties of these materials are also similar (Bomba et al. 2012). In favor of MDF worth the price which is lower than that of solid oak wood of the same dimensions, and preferably also the MDF production possibilities dimensions according to specific requirements. Oak wood again its strength,

resistance to external influences and of course, aesthetics is a material that can be difficult to replace. On the one hand, therefore, stands the oak wood, representing a natural material, on the other hand, there is MDF, agglomerated material made of wood technological process. This study follows the properties of the two materials, focusing on their behavior under conditions of thermal stress, which evaluates the test results on a cone calorimeter.

1.2 *Oak characteristics*

Generally favorable properties are the reason why oak found wide application in various industries. Made into durable solid elements (bearing parts of structures). Oak wood is just as beech wood an important raw material for mechanical processing. Oak lumber and blanks are processed for the needs of the furniture industry, mainly for the production of seating and table furniture. Laminated wood as a large-scale material may be intended for the needs of the production of furniture, precious interior products (e.g. Stairs, railings and other building - joinery, windows and doors). It is important for the production of classic and mosaic floors.

Top quality oak wood assortments are used to make decorative veneers. Due to its strength and durability are used for shipbuilding, water works, bridge construction, manufacturing barrels for quality wines and spirits, also different vats and columns. It is a suitable material for carving and lathe work. It is used also for sports equipment in products for vehicles.

Thin and poor selections oak find their use in the pulp and paper industry, the manufacture of large-area materials etc. Oak belongs in our conditions among one of the best utility plant which provides moderate and hard wood. It is a strong, flexible wood that is very well fissile. General properties of wood provide a wide range of processing and surface treatment. One of the biggest advantages of oak wood is its durability expressed by tensile modulus and pressure modulus value (Table 1).

Flexural modulus is also very high. In terms of durability in contact with the ground, where there is a risk of damage to all kinds of rot, is classified as a durable oak trees, exposed and impregnated wood will last 40 to 120 years. Under the roof has a shelf life of 100–200 years. Underwater, it is 300–800 years and always dry oak wood lasts up to 600 to 1000 years. Among some of the few disadvantages are that relatively poorly impregnated and seas and in drying the structure often form cracks. Due to the high content of tannins in contact with the iron black. Additional physical properties are described in the Table above (Table 1).

2 MEDIUM DENSITY FIBERBOARD

2.1 Use and properties

Medium density fiberboard is a fiberboard of medium density. It is a sheet material made from lignocellulosic fibers under heat and/or pressure. The binding of the individual fibers of the board subsequently formed either felted fabrics and their own tack or adding glue. Density of the medium density fiberboard ranges of 600 to 800 kg·m^{-3} and substantial impact on the physical and mechanical properties. Production of MDF is a part on the findings from the pulp and paper. Feedstock is a wood without bark, which is processed first on the interstage blank—chip. This was followed in the process of production than it is divided into the individual fibers, followed by compression forming the board with a fine homogeneous texture (Ye et al. 2007).

Table 1. Oak physic properties.

Property	Parallel to fiber [MPa]	Vertically to fiber [MPa]
Tensile strength	90	8.1
Compressive strength	65	9.6
Shear strength	11	–
Flexural strength	110	–
Tensile modulus	14000	1000
Pressure modulus	11778	2046
Shear modulus	1320	9100
Flexural modulus	13000	-

In terms of fiber production the length of wood particles is an important parameter, further their meshability, lignin and hemicellulose. More often, the production of fibers and then pressing MDF used softwoods. There are several reasons, hardwoods are primarily heterogenous structure. While softwood is composed primarily of vein, hardwoods further comprise a vessel, libriform fibers and a greater proportion of parenchymatous (stock) of cells that in some hardwoods can join up to 15%, while that of conifers is approximately 1%. The number of elements in wood space unit is on broadleaf trees, about 3–5 times higher than in conifer species, but because of their different size and structure it is possible to produce fibers of uniform size which is required for the production of fiber boards.

Therefore, a production company for fiber coniferous preferred raw material (Kollmann et al. 1975). Specifically softwood spruce has favorable ratio of cell length to thickness of the cell walls than pine. Pine further includes more accompanying substances and resin. For higher resin content, higher hardness and very hard bulge is not suitable for production of fibers and larch. In the case of a favorable price and availability of raw materials may also be encountered with the production of fibers from hard deciduous trees (eg. Beech and oak), which are usually added to a mixture of fibers from conifers. Plants with higher density are also generally thicker cell walls and therefore have a higher yield of fiber, but fiber is made of lower quality than plants with more uniform composition of wood and, moreover, its production is energy-intensive (Ockajova 2007).

2.2 Properties of MDF

The most important feature of the MDF is already mentioned homogeneity across all the boards, which allows clear chamfered relief to areas and profiling hips boards. Ye accents especially great possibility finish and edges is one of the main advantages of this material over other agglomerated materials (DTD, OSB, etc.). Some physical and mechanical properties of MDF are shown in Table 2 (Ye et al. 2007).

Comparing the performance on the basis of tests on a cone calorimeter described by Babrauskas (Babrauskas 1997). Prior to the evaluation of the properties of both materials and comparing them to the results of tests on a cone calorimeter is shown a short summary of the foremost common and different characteristics:

– Density—the two plates have approximately the same density. In both cases, the density is within limits and depends on the particular piece of wood or material. For these measurements are approximate density values determined on the basis of sample sizes and weights. Sample oak had calculated a density of about 780 kg . m^{-3} density MDF was 750 kg . m^{-3}. Mechanical properties—the values given in the tables along with experience in use indicate that oak has very good mechanical properties. MDF board achieves similar although lower values. The MDF is a problem of resistance to moisture, which can be solved using suitable adhesives in the production process.

– Structure and species of trees—oak plank structure depends on its composition, which affect the conditions under which tree species grows. MDF is homogeneous throughout its cross section. Although the densities of the two boards are almost identical, oak plank is made from high-density wood, while the MDF is made of soft wood and its density is obtained in the production process.

– Reaction to fire—two boards are classified in reaction to fire class D s2 d0. Given that there are in both cases, the wood material is not in the same rank of the ordinary.

Those characteristics of transfer key elements that will be used in the evaluation and comparison of the features panels on a cone calorimeter track.

Table 2. Fire-technical characteristics of some wood species and their particles.

Oak wood	Flash-point	[°C]	360 – 370
	Ignition temperature	[°C]	400 – 410
	Heat of combustion	[MJ.kg^{-1}]	19.8
	Fire reaction Class	[-]	D s2 d0
Spruce wood	Flash-point	[°C]	350 – 360
	Ignition temperature	[°C]	390 – 400
	Heat of combustion	[MJ.kg^{-1}]	19.9
Oak wood sawdust	Flash-point	[°C]	229
	Ignition temperature	[°C]	342
Spruce wood sawdust	Flash-point	[°C]	214
	Ignition temperature	[°C]	306
Wood dust from fibre board	Temperature	[°C]	310
	Ignition temperature of dust	[°C]	410
		[g.m^{-3}]	60
	Lower explosion limit	[MPa]	0.96
	Maximum explosion pressure		
MDF board	Fire reaction Class	[-]	D s2 d0

3 CONE CALORIMETER TESTS

3.1 Tests

Every board is to carry out the tests made one a representative sample of 100×100 mm. The thickness of the two boards was 12 mm. These treated samples were then loaded in a cone calorimeter with a heat flux of 40 kW · m^{-2}, corresponding to a temperature of around 790°C at a standard distance from the sample surface of the conical emitter. Controlled burning intensity was mainly defined by the heat release rate further, significant times of testing and weight loss rate. The results of the measurements are summarized in Table 3. The most important variable measured by a Cone Calorimeter is Heat release rate (Babrauskas & Peacock 1992). HRR together with the ignition time and other characteristics are very valuable value in order to describe real fire characteristics of flammable material (Müllerova & Vacval 2014).

At the results of the test times can be seen that the initiation of MDF occurs earlier. The difference is about 30 seconds, which is half the time of initiation of the boards of oak wood. This trend is also retained in the transition to phase flameless combustion and end of the test, the overall combustion cycle MDF faster. For the major parameter of heat release rate reaches MDF significantly higher than the mean value of the intensity of burning. Due to the longer duration of the test oak boards but the total heat released during combustion oak boards above. When monitoring the percentage of weight

Table 3. Cone calorimeter test results on Oak and MDF board.

Material		Oak	MDF
parameter	unit	value	value
Ignition time	s	99	56
Non-flame burning	s	812	738
End of test	s	965	891

parameter	unit	value	time	value	time
Heat release rate	kW.m^{-2}	230.4	122	318.94	90

parameter	unit	value	value
Total heat released	MJ.m^{-2}	101.41	97.19
Average HRR 180 s	kW.m^{-2}	150.26	184.42
Average HRR 300 s	kW.m^{-2}	132.45	150.67
average mass loss rate	g.s^{-1}m^{-2}	10.66	9.89
mass loss	g	76.05	67.44
proportional mass loss	%	80.62	75.61

loss can be seen that more material was combusted during the test blocks; the difference is 5%.

Based on the observation of a graph can be formulated combustion cycle valid for both boards. In the introduction to the material reacts to the action of heat flux gradual release of flammable pyrolysis products and the rate of heat release increases only very slowly. The moment a sufficient concentration of flammable substances from the surface occurs initiation and rapid kinetic combustion, which is documented in the chart steep upward curve of heat release rate. At the same time during combustion beginning to surface plates generate carbon layer. This layer is gradually increasing, and extends deeper into the material, thereby preventing the rapid release of gaseous pyrolysis products. Originally rapid kinetic combustion thus ceases and burning intensity decreases rapidly. At the time lightening speed reach the second stage of combustion, which is characterized essentially stationary combustion. In this phase occurs first, followed by a gradual decline gradually increasing intensity. This phase ends about 500 second tests, which creates distortions created carbon layer.

The carbon layer that gradually during the second phase penetrated deeper into the material is now due to cracks violated and increasingly prevents release of gaseous pyrolysis products from lower layers. Consequently, the HRR rapidly increased and reached a second peak test. After this peak the intensity has decreased, the gaseous products are evolving constantly in small amounts until there is a transition phase in flameless combustion. The intensity still slowly decreases until the end of the trial.

4 RESULTS AND DISCUSSIONS

It can be seen that in general the flow of combustion of both materials and the same differences can be found in the allocation to the flow of the several stages referred above (Figure 1).

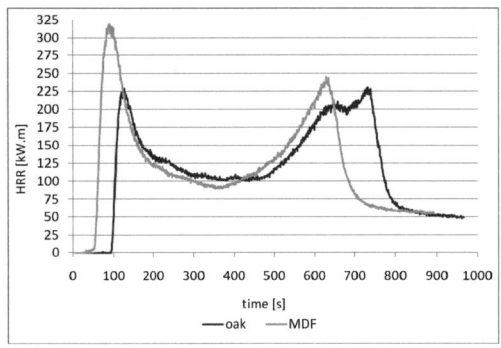

Figure 1. Heat release rate of the oak board and MDF.

MDF initiation occurs earlier about half a minute. This may be due to two factors. The first is that the action of heat disrupts bonds between MDF heat labile particles. Consequently, the individual fibers are more prone to ignite as if it is structured in a stronger mutual relation. The second is formal-dehyde present in the board in the form of adhesive, which facilitates the initiation of the release. The second difference at this stage is the maximum intensity of burning, where oak plank reaches nearly 30% lower level of intensity.

The reasons are the same, altering the structure and release formaldehyde, which help to intense burning MDF. On the other hand, compact structure oak boards prevents sufficiently rapid release of gaseous pyrolysis products and consequently the maximum intensity level is lower.

Creating a carbon layer and the decay continuing the large release of gaseous products is the same for both materials, but with oak boards, the impact of cross-links in the material stronger. It mentioned resistance oak boards is the reason why after burning stationary phase, in which the only difference is the fact that MDF has a larger difference between the maximum and minimum intensity, the intensity of the onset of the second peak test more gradual. The resulting carbon layer in the case of oak board does not interfere in such widely than is possible with MDF. Similarly, due to the fact that the values of the intensities of peaks burning in oak boards comparable, we can say pretty much the same, while the MDF board has a second peak compared to the first significantly lower. Noticeable shift over time to achieve the second peak intensities and their almost identical further ensure that the total heat generated during combustion of the two samples of materials released is almost the same (see Table 3). The transition to the phase of flameless combustion is already in the two boards nearly identical, the difference is only a time lag, which gradually arose during testing.

Both tested materials has proven their flammability after certain time. Unlike the heat treated wood described by Müllerova (Müllerova 2013). Those common untreated materials started to burn after some time by visible flame. The comparison of the intensities of traces of burning oak and MDF board is difficult to see the differences between natural wood and technologically processed material made from wood particles. The origin of these differences comes out primarily from the structure of boards, which are manufactured by different particles and bonds between the individual particles. Although the density of the two boards is about the same, differences that can be observed in the characteristics of the mechanical properties are retained even if the burden of these panels heat flux, even increasing. This implies already fire characteristics of oak and pine woods. Better coherence oak boards with the result that, although it is released from the same (even higher) amount of heat as MDF, its course is smoother and more balanced. We can say that the oak plank more resistant to heat while maintaining their properties. The MDF was observed in terms the influence of heat, there was a large disturbance of the integrity of the board, allowing intense combustion and rapid loss of its original characteristics. Its share of that has a formaldehyde-releasing from the plate during the application of heat. This fact confirms the advent especially the second peak intensity, which is the influence of emerging cracks in the carbon layer in MDF significant.

It can be stated, therefore, and also appreciated that although both materials in many respects have very similar characteristics under conditions of heat a natural material wins. It should however be noted that there is no trend of replacement MDF by oak or vice versa. With oak e.g. on the carrier structures and the application of the MDF as backing layer of laminate flooring in both cases an appropriate action, which takes into account the properties of both materials and the use of which the most preferred for the production of quality products. It should be also stated that in this test no smoke analysis where done and no toxicity was measured. This important issue is well described in (Müllerova 2014).

REFERENCES

Babrauskas, V. & Peacock, R.D. 1992. *Heat release rate: The single most important variable in fire hazard.* Fire Safety Journal. 18, pp. 255–272.

Bomba, J., Böhm, M. & Reisner, J., 2012. Materiály na bázi dřeva/Wood based materials, *Česká zemědělská univerzita v Praze*: 31–81.

Kollmann, F., Kuenzi, E. & Stamm, A. 1975. *Principles of wood science and technology. Wood based materials.* Springer Berlin.

Müllerová, J. 2013 *Fire safety of heat treated wood.* Research Journal of Recent Sciences, Vol. 2, no. 12, pp. 80–82.

Müllerová, J. 2014. Health and safety hazards of biomass storage. In *Int. Multi. Sci. GeoConference SGEM*, STEF Bulgaria, pp. 261–266.

Müllerová, J. & Puskajler, J. 2014. *Review of health and safety risks of wood chips use.* In: Advanced Materials Research. ISSN 1022–6680. Vol. (2014), pp. 426–431.

Müllerová, J. & Vácval, J.: Sledovanie správania materiálu na základe výsledkov skúšok na kónickom kalorimetri./Material behaviour monitoring based on cone calorimeter test results In: *Advances in Fire and Safety Engineering,* MTF Trnava, Alumni Press 2014.

Očkajová, A. 2007. Materiály a technológie 1: Vlastnosti dreva/Materials and Technologies 1: Wood properties, *Univerzita Mateja Bela v Banskej Bystrici, Fakulta prírodných vied, Katedra techniky a technológií*: 17–21.

Ye, X., Julson, J., Kuo, M., Womac, A. & Myers, D. 2007. *Properties of medium density fiberboard made from renewable biomass.* 98(5): 1077–1084.

Production Management and Engineering Sciences – Majerník, Daneshjo & Bosák (Eds)
© 2016 Taylor & Francis Group, London, ISBN: 978-1-138-02856-2

Analysis of the causes of nonconforming product in supplier-customer chain

A. Nagyova

Faculty of Mechanical Engineering, Technical University of Kosice, Kosice, Slovak Republic

M. Palko

1ENERGY, Ltd., Kosice, Slovak Republic

ABSTRACT: There are many examples of management of nonconforming products in practice. Generally, if the product does not meet the requirements of the technical standards, internal regulations, legislation or the customer, or the end user, it represents non-conformity. The basic objective of the management of nonconforming product is to capture such a product before it reaches the customer or before it is otherwise inappropriate used. In the case such situation occurs, it is the responsibility of the manufacturer or the responsible person to carry out adequate measures to eliminate the identified non-conformity while implementing appropriate preventive measures to prevent occurrence of the same or similar mistakes. This article illustratively highlights the analysis of occurrence of non-conformity in functional management system, where insufficient surface finishing of the packaging material caused depreciation of the product itself. Metrological verification confirmed the cause of non-conformity at the supplier, which was an incentive for complaint procedure.

1 CONTROL OF NONCONFORMING PRODUCT ACCORDING TO EN ISO 9001

1.1 *Identification of nonconforming product*

The technical inspection of the product or conformity verification of a shipment from a supplier may reveal that the product does not meet the pre-specified requirements and expectations. If nonconformity is identified, it is necessary to implement a number of additional steps, which identify the cause of this nonconformity.

1.2 *Labelling and separation of nonconforming product*

Labelling is one of the basic requirements of a quality management system. The organization has to clearly identify all materials, tools, instruments, information, documentation etc. Identification and traceability removes anonymity form all major elements of the process. Labelling nonconforming products as elements that should in no case reach the next process cycle or customer is an absolute necessity.

1.3 *Collecting data on nonconforming product and defining the causes of nonconformity*

Data collection is the main tool for further control of nonconforming product and identification of the cause of nonconformity. If possible, data collection is carried out directly on site where the nonconformity occurred. It consists of various measurements, analyses, examinations, tests, etc., aimed at identifying the cause of nonconformity.

1.4 *Determination of corrective and preventive actions and their implementation*

Upon finding the causes of nonconformity supported by relevant evidence, the quality management system requires the determination of corrective and preventive actions. It is a logical step whose aim is not only to solve a particular nonconformity or complaint but to prevent its recurrence in general, and thus ensure continuous quality improvement.

1.5 *Documentation*

Documentation of the control of nonconforming product is one of further activities required by EN ISO 9001: 2008. The documentation includes the aims and procedures and systematizes activities. Its use contributes to (Hrubec et al. 2009, Bosák & Olexová 2013):

- achieving conformity,
- staff training,
- repeatability and flow of work,

- provision of objective evidence,
- evaluation of the effectiveness and sustained suitability of nonconforming product control.

1.6 *Assessment of damage*

If nonconformity is identified during the production process, or on the premises of the organization, the financial costs are much lower than if it is identified by the customer. Organizations should therefore pay more attention to output control and thus minimize the risk of releasing and shipping nonconforming product to the customer.

2 CASE STUDY—PROCEDURE FOR THE IDENTIFICATION OF NONCONFORMING PRODUCT USING THE 5W2H METHOD

2.1 *Principles of 5W2H method*

This study describes the particular steps related to the identification of nonconformity in drinks packaging in a company selling energy drinks. The organization operating in Slovakia is a company established in August 2012 aimed at manufacturing and distributing its own brand energy drink 1Energy on the European market. The product is packed in 330 ml aluminium cans of the company's own design. The can consists of two parts: the lid and the body of the can. Both parts are manufactured by separate processes on high-speed automated lines. The production takes place in the manufacturing plants of a Dutch supplier. The cans are then transported to a subcontracting organization operating in Poland, which is responsible for filling.

Each consignment (product), received at the warehouse of the Slovak company is subject to the entry inspection. In this case, it is a simple visual inspection of all pallets that are delivered to the warehouse and random inspection of several unpacked pallets and particular cans. The purpose of this inspection is to reveal any nonconformity. The most commonly found nonconformity is mechanical damage caused during transport (damaged packaging, mechanically damaged/ visibly destroyed cans). This type of nonconformity is considered as part of the product transporting process. The worker who is in charge of the inspection and collection of the goods not only performs the entry inspection, but also to creates the necessary photo documentation.

2.2 *The use of 5W2H method and proposal for immediate action*

Not all nonconformities are or may be revealed by a simple entry inspection of the supplied products.

Some arise over time in specific conditions or during certain activities. The Slovak company also revealed nonconformity some time after receiving the consignment. About three weeks from the receipt of the consignment (of 62,400 1Energy beverage cans), damp or wet packaging (cartons) were observed in the warehouse. This phenomenon was noticed by a worker responsible for the warehouse and immediately reported to employees in charge, who started to deal with the issue.

The Slovak company uses a variety of tools and methods that help to clearly identify nonconformities. In this case, it opted for the 5W2H method.

The applied procedure consisted of answering seven basic questions. Based on the responses, it was possible to identify the cause of nonconformity as follows:

What is the problem? The problem is wet cardboard and moisture discovered in individual packs, or among cans. Why is it a problem? It poses a problem because the deteriorated product is no longer tradable. Items with no visible damage are not tradable either, until the cause of the moisture is revealed. Where do we encounter the problem? The problem was revealed in the company's own warehouse. Who is impacted? The problem was discovered by a warehouse worker. When did we first encounter the problem? The problem was discovered on 25th August 2013. How did we know there was a problem? The problem was dis-covered visually when handling the goods. How often do we encounter this problem? Initially about 34 wet cardboard packs were detected (each containing 24 beverage cans).

3 POSSIBLE CAUSES OF NONCONFORMITY IN CAN PACKAGING AND PRODUCTS

Although the release of the product to the customer was suspended immediately, it was necessary to determine the cause of nonconformity. Some of the aspects influencing the occurrence of nonconforming product include the human factor, systemic or random errors, poor prevention, and low frequency of inspections between operations. It was therefore necessary to perform a thorough analysis and identify the cause of nonconformity.

Option 1: Nonconformities in the stage of aluminium sheet manufacturing.

Nonconformities could arise from the mere fact that can is a metal product. Rolled sheets of special aluminium alloy must meet strict requirements. The final thickness of the can is only 0.13 mm, so any flaws in the semi-finished product may result in nonconformities. Nonconformities of aluminium sheets may include:

- chemical properties of the material (contained impurities, low content of alloying elements),
- material structure,
- material surface quality (roughness),
- strength properties of the material,
- mechanical properties of the material,
- their shapes and sizes.

Option 2: Nonconformities in the stage of can lid production.

The can lid is not a difficult part to produce, so it is necessary to consider only three possible nonconformities:

- dimensional nonconformity in cutting blanks of the sheet metal,
- dimensional nonconformity in bending the edges of the lid,
- insufficient lifetime of the sealing element—silicone.

Option 3: Nonconformities in the stage of can body production.

The process that gives the can its typical shape is called deep-drawing. The basic prerequisite for the process is the quality of the material to be drawn, while the important variables in the process of deep-drawing, which may affect the conformity of the final product are:

- the ratio of the diameter of the blank and the punch,
- sheet thickness,
- clearance between the punch and the die,
- radii of curvature of the punch and the die,
- holding force,
- friction and lubrication,
- speed of the punch.

In the manufacture of the can body, nonconformities may arise also while it is being imprinted. Possible nonconformities of the imprint include:

- inaccuracy of colours in comparison to the pattern,
- inaccuracy in design—blurred edges, blurred letters in the information for consumers,
- insufficient paint drying period to increase production speed.

The decorative printing is followed by protective coating. The protective coating is applied from the outside and from the inside of the can. The inside of the can is coated to prevent any contact of the beverage with metal. It also serves as a protection against corrosion. External coating also protects the can against corrosion. Primary corrosion (internal, after the contact of beverage with metal) or secondary corrosion (external, caused by the ambient humidity) in 0.13 mm thick cans results in leaks of the beverage. Nonconformity at this stage could be caused by insufficient internal and external protective coating.

Option 4: Nonconformities in the stage of can filling.

Empty cans are automatically removed from pallets and move along the conveyor belts of the line. The first problem related to the protective coating of the cans may occur there. Older or cheaper filling lines are equipped with stainless steel guard rails along the conveyor belts, which gradually rub the protective coating off the cans. Similarly, the conveyor belt itself can be made of cheaper materials and when the cans stop moving, its constant movement rubs off the protective coating at the lower rim of the can. To prevent this phenomenon, modern lines are equipped with the rails and belts made of materials with a low coefficient of friction, usually silicone. Nonconformity in the stage of filling may also be caused by an unwanted object in the can—a hair, a speck of dirt etc. This can be prevented by a sensitive sensor that rejects such cans, and they automatically fall off the line. Another nonconformity which is removed off the line by operation inspection is inaccurate filling. Cans that are out of tolerance are rejected automatically. The most important phase in the filling process is closing of the can lid. The lid is joined with the body of the can in their predetermined shape, by a contact force of the press and silicone seal of the lid. All this is performed data production rate of 2,000 pieces per minute. Nonconformity that may arise at this stage is insufficient tightness caused by inappropriate setting of the press for the type of cans, which may result in leakage of beverage from cans.

Option 5: Nonconformity in the stage of packaging, transport and storage

The filled cans are traditionally packed by 24 pieces in cardboard trays. These are then wrapped in foil. Such packs are then palletized according to the type of cans. 330 ml cans are usually packed 10×10. When packing, it must be borne in mind that the cans take a long way until they reach the end customer. Sometimes it may last a month in the shipping container in saltwater environments.

Possible nonconformities in the packaging include:

- unfitting cardboard trays causing movement of cans (Figure 1).
- thin packaging film—also poor stability,
- uneven distribution on the pallet—poorly distributed weight, deformation of cans.

There are no special requirements concerning shipment and storage, as long as the packing process is performed well. It is important to avoid freezing temperatures, in which the volume of the beverage changes; it expands, which mechanically

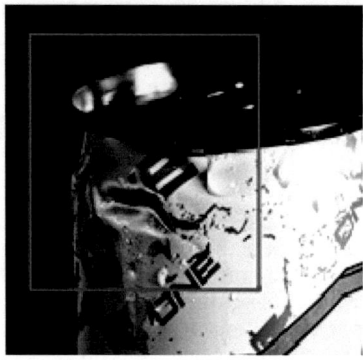

Figure 1.   Mechanically damaged can.

Figure 2.   Secondary corrosion.

destroys the can. On the other hand, it is recommended to avoid direct sunlight and high temperatures.

3.1   *Determining the cause of nonconformity*

Moisture, which caused damage to the cans and packaging comes directly from the drink itself, since the storage conditions comply with the requirements of the supplier. It is caused by leaks from beverage cans. Considering the technical features of the can and a method of its manufacture, there are only three possible causes of beverage leakage, namely:

• mechanical damage
• leakage from poorly sealed can lid,
• leakage right through the can body.

The cans that were mechanically damaged during transport are rejected at the entry inspection when receiving the shipment. If the nonconformities were caused by insufficient sealing of the cap, it would be possible to identify the cause by vertical strips of dried beverage residues, starting below the edge of the rim. After a thorough analysis, both possibilities were rejected. However, the third type of leak was identified after inspecting a large number of randomly selected cans. To clarify the causes, it should be added that the leak of the beverage through the body of the can may be caused only by two reasons, namely:

• primary corrosion,
• secondary corrosion.

Primary corrosion occurs when the inner surface of the aluminium can is not sufficiently protected by the protective coating applied during the manufacturing process, as described in the previous chapter. The beverage, which contains citric acid, reacts with the thin material of the exposed can and consequently causes a leak. The secondary

corrosion occurs when the can is not well protected from the outside—the thickness of the outer coating is insufficient. Corrosion then arises from external moisture. The secondary corrosion was observed on a large number of randomly selected samples (Figure 2).

This type of corrosion was observed on all samples in almost identical areas, indicating a systemic error. Moisture which caused the secondary corrosion came from some randomly mechanically damaged cans. These had been removed during the initial inspection. The period lasting three weeks before the problem was noticed indicates the fast rate of the reaction. Had the whole batch not been immediately resorted, the chain reaction could have deteriorated the entire consignment.

4   VERIFICATION OF NONCONFORMITY BY METROLOGY

Initial analyses revealing secondary corrosion on the surface of the cans lead to a presumption that the protective coating on the outside is not sufficient and there are various causes of incorrect parameters (nonconformities). It was necessary to confirm or reject this assumption on the basis of measurements of the thickness of protective coating.

The object of measurement consisted of 12 samples of empty unused can bodies that were randomly selected from the inventory of the Dutch supplier of cans. The prescribed thickness of the coating is the internal know-how of the can supplier and it is not featured in any particular standard or other technical documentation. Therefore, besides the samples of 1Energy cans, also 12 sample cans of competing brand Control, which are produced by the same Dutch factory, were used. Samples of both types of cans were individually labelled by numbers 1to 12 to ensure repeatability of the measurements. The measurements were carried out in precisely identified sectors of the cans. Specifically, in the three different sectors in a horizontal direction (circumferentially) and three different sectors in the vertical direction, designated A1–C3 see Fig. 3. There were 9 measurements on each individual sample in total.

There are not many manufacturers of sensitive measuring devices, which are capable of measuring the thickness of the protective paint of a can. For verification was used certified equipment from Sencon, which is capable of measuring the weight of the paint coat on surface from the inside and outside of the can. The output of the measurement can be automatically connected to a PC and specific software. It was obvious after the measurement of the first samples that the assumption of nonconforming protective coating would be confirmed. The Control brand reference sample cans contained about 3 g of protective coating per m², while in the sample cans of 1Energy, the amount of protective coating was lower, slightly below 1 g/m², or even absent. Low values or zero was measured in those areas where the secondary corrosion was present in the stored filled cans.

4.1 Complaint procedure and design of corrective and preventive actions

The results of measurements of the protective coating confirm the nonconformity of a measurable and quantifiable technological parameter. The first immediate action after this finding was separating the damaged cartons from the undamaged ones. The results of the analysis showed that the primary cause of nonconformity was the insufficient surface treatment of the can body, or the outer protective coating. This process is a part of canim printing. Decorating is an automatic process of imprinting cans and applying the protective coating at a speed of 2000 cans per minute. The Dutch factory uses a Stolle Concord decorator. The decorating works on the principle of offset printing. Each colour has a precise numerical designation and precise ratio for mixing various hues, which is important e.g. when printing logotypes and accurate designs, where it is necessary to use the same colours consistently.

Besides colours, also protective coating is applied in this manner. Such technology is fast and precise, if the machine is in proper working order. The decorator contains many rotating parts, precision of which is related to the condition of the bearings. The bearings, as elements transferring the rotary motion, may affect the precision of the machine and its vibrations. Upon receiving the complaint related to the nonconformity of protective coating, the Dutch producer conducted a comprehensive inspection of its equipment and processes. The inspection revealed that the bearings used in operation were worn, i.e. past their prescribed lifetime. These bearings are located on the shaft of a rubber cylinder which applies protective coating on offset printing transfer elements. The layers of paint are very thin and so even minimum wobbling of the cylinder causes individual deviations (lack of coating). This is also the reason why the individual low values were always located in the same can sectors.

The technical condition of the supplier's equipment directly affects the quality of the product. The discovery of bearings that are no longer able to fulfill their function was unacceptable. The first corrective action was to exchange all shaft bearings, which exceeded their lifetime. Another corrective action that prevents shipment of nonconforming product is a more thorough final inspection. Each batch of cans produced on a given day has to undergo rigorous metrological processes.

The above corrective actions were made by the supplier. The Slovak customer and contractor had no direct impact on the technical support and could not directly influence the production processes. Nor is it the intention of the organization. However, the company had to insist on the required quality of the product delivered. Therefore, the preventive action involves requiring a quality certificate, which has to be submitted for each new consignment.

5 CONCLUSION

Management of nonconforming product is an important element of the quality management system. The resulting non-conformity may result from lack of organization approach but also from failure of individuals. In the search for the causes, an organization should rely on generally accepted practices and methods that can eliminate such situations. However, effective and user-friendly methods of management of nonconforming product are

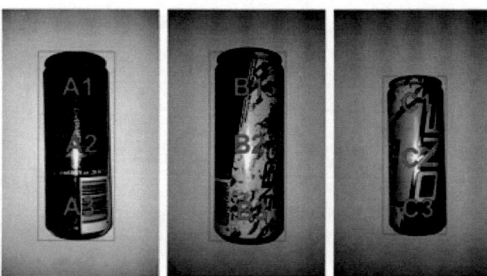

Figure 3.   Selected measurement sectors.

relatively rare. In this case, the search for evidence of causes of non-compliance was on the part of the customer (Slovak distributor of energy drinks), even though the non-conformity was caused by the manufacturer. Such system of "evidence" is not optimal in terms of mutual cooperation and maintaining proper relationships. It is necessary to ensure not only correct setting of the system of management of nonconforming product within the organization but also ensure a way to assess the resulting claims without affecting mutual trust and without affecting future cooperation.

ACKNOWLEDGMENT

This paper is the result of the project implementation: VEGA 1/0150/15 Development of implementation and verification methods for integrated safety systems of machines, machine systems and industrial technologies.

REFERENCES

Bosák, M. & Olexová, C. 2013. Recent environmental trends and innovations in the Slovak small and middle enterprises, In 13th International Multidisciplinary Scientific Geoconference SGEM 2013: Ecology, Economics, Education and Legislation, Volume II, Sofia: STEF92, Technology Ltd, 2013, 247–254 p.

EN ISO 9001: 2008.2008 Quality management system. Requirements.

Hrubec, J. 2001. Quality Control. Nitra. SPU.

Hrubec, J. et al. 2009. Integrated Management System. Nitra. SPU. 543 p.

Majerník, M. Kollár, V. & Juríková J. 2010. Accreditation and certification in environment. SVŠ in Skalica. 122 p. ISBN 978–80–89391–07–3.

Markulik, Š. & Nagyová, A. 2009. Quality management system. TU Košice, Strojnícka fakulta, 79 p.

Mateides, A. et al. 2006. Quality management. Bratislava: EPOS, 2006. 75 p.

Nenadál, J., Noskievičová, D. Petříková, R., Plura, J. & Tošenovský, J. 1998. Modern Quality Control. Praha: Management Press, 280 p.

Nenadál, J. 2008. Quality management systems II. Ostrava. Department of Quality Management. 2008, 163 p.

Plura, J. 2001. Planning and continual improvement of Quality. Praha 2001: Computer Press. 244 p.

Sutoova, A, & Grzinčič, M. 2013. Creation of Defects Catalogue for Nonconforming Product Identification in the Foundry Organization In: Quality Innovation Prosperity. Vol. 17, no. 2, p. 52,58. Access: http: // www.qipjournal.eu/index.php/QIP.

Teplická, K. & Kadárová, J. 2013. Effectiveness achievement of maintenance process by the controlling approach, In: Annals of Faculty Engineering Hunedoara—International Journal of Engineering. Vol. 11, no. 1 (2013), p. 233–236.-ISSN 1584-2665.

Vadász, P. 2007. Testing and Quality Control. TU Košice, Faculty of Metallurgy, 118 p.

Production Management and Engineering Sciences – Majerník, Daneshjo & Bosák (Eds)
© 2016 Taylor & Francis Group, London, ISBN: 978-1-138-02856-2

Utilisation of quality circles in the assembly process of final product

J. Namešanská & Š. Markulik

Faculty of Mechanical Engineering, Technical University of Košice, Košice, Slovak Republic

ABSTRACT: Quality circles are an undemanding method, based on the PDCA principle, and they enable organizations to achieve cost saving, improvement of workplace relationships, teamwork, brainstorming, innovation, and the like. This case study examines the final cost of the assembly process product and measures, based on the basis of teamwork in a quality circle, leading to its reduction. Specifically, it involves the implementation of three measures, which have brought the organization cost savings of up to 88% on the assembly line. Quality circles of the producer of audio-visual equipment were successful and will be used to improve other processes and reduce costs as well.

1 INTRODUCTION

Quality is very important and necessary nowadays, therefore it should not be neglected. Customers demand quality products that meet their requirements, to meet their needs. Improving quality does not only affect customers, it has an effect on suppliers as well, or, in other words, product providers, through factors such as: price of product, higher efficiency, lower production costs. A satisfied customer helps organizations stabilize their position on the market, thereby ensuring job security.

Quality itself cannot be managed. However, the processes that lead towards the desired quality can be managed. People and equipment that are involved in the creation of quality can be managed in the processes. Quality management is a process which begins with customer requirements on a product. It continues with product implementation and ends with customer care.

In order for all the processes in an organization to be of good quality, it is necessary to apply a whole range of appropriate tools and methods. Range of quality assurance cannot only be seen as external quality for the end-user. Quite the contrary. Internal process quality is a prerequisite to the final one.

To ensure quality is a major challenge for an organization. If an organization wishes to achieve above average results, quality at all levels, ranging from the delivery of materials to the delivery of the final product to the customer, must be assured.

2 QUALITY CIRCLES AND THEIR REBIRTH

Quality circles are a method of quality control originating in Japan. The author of the concept is Kaoru Ishikawa in cooperation with the Union of Japanese Scientists and Engineers (JUSE) (Management Mania 2013). This simple method is based on the principle of brainstorming. It reached its biggest success in the 60's and 70's last century. Despite the results it brought, its utilisation gradually disappeared from the management of organizations. However, in the modern history of the management of organizations, quality circles are again finding their place. This is mainly because of their financial and time simplicity and the results they bring. Fundamentally, the idea is to create small groups of 5–11 members who, within the scope of their organization units, long-term and voluntarily focus on quality improvement.

Only employees with very good results and high discipline are recruited to the quality circles. Membership to circle is therefore a matter of prestige. Circles in every way should be supported by the management and accepted suggestions for improvements put into practice immediately. The principle is that, if the management does not accept some proposals, reasons should be presented to the members of the circle.

Utilization of quality circles in practice: the circles act as an incentive, have an impact on work performance, and also affects the self-realisation and personal development of its members. They significantly increase the participation of its members in the running of an organization and increases the sense of belonging to an organization (Keblerová & Vykydal 2011).

They are derived from Deming's PDCA cycle. These groups meet at least once a week for 1–2 hours during business hours under the leadership of a team leader, who can be elected by the members. Furthermore, the members identify, analyse and solve work-related problems. The process of selection, solving and solution is shown

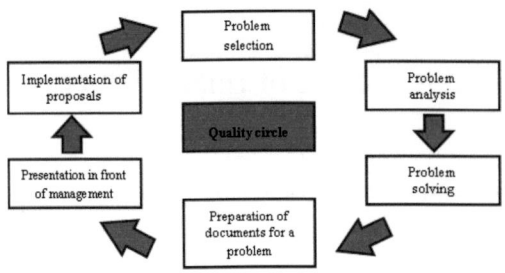

Figure 1. Quality circles and their procedures.

in (Figure 1). The groups self-select problems they want to solve, but which are not related to salaries and working conditions. Among the basic areas, to which the circles dedicate their time, are the improvement of health and safety at work, design and product quality improvement, and the improvement of the working environment.

Quality circles are usually not financially appraised. The funding received through the improvements are used to improve the working environment and working conditions (Hutchins 1985), (Hutchins 2008).

Experience shows that the number of quality circle members usually starts to decline within three years. Larger quality circles have a better chance of survival. A significant decline in membership indicates a failure of a circle in a given organization (Tang 1996).

3 CASE STUDY–DESCRIPTION OF PROBLEM PROCESS

The project was performed in an organization engaged in the production of audio-visual equipment. Among all the production processes in the organization, the most critical assembly process appeared to be the assembly of the final product. Therefore, using quality circles, attention was specifically given to this phase.

The team consisted of 6 people, one of whom was the head of the circle—the team leader. This member was a member of the production department. There were two members from the technical support department on the team. The quality control department, the process engineering department and the maintenance department were each represented by one member. The final product assembly was determined as problematic, based on monthly results for the cost of poor quality. The production line, where this process took place, consists of a line as well as a cell system. The input consists of components from previous processes as well as components from external suppliers.

Thirteen production workers, one material worker and one foreman worked on the production line. Taking into account ergonomic parameters of each workstation, each one had a work desk and space to store material. Apart from this, there was an exact working procedure, which described the manufacturing process and all of its important specificity, on each workstation. Apart from employees, there were also manufacturing machines, robots, adjusting ancillary appliances, test cubicles and a bending machine on the production line.

The process was monitored by an automatic system able to track and record the entire manufacturing course of a particular product.

3.1 Problem analysis

Start of the analysis was the first step when solving problems. Data was collected for a period of one month (January 2014), and the cost of poor quality was observed. The total cost (Figure 2) was €1345.99. The largest share of the cost in Euros was the component—"top plastic foil." It is a component that gives a more attractive look to the final product. This is why the price of one component is high.

After the analysis of available data, it was found that the largest proportion among all errors on the top plastic foil were bubbles (Figure 3).

35 out of the total of 70 errors were due to a bubble (Table 1).

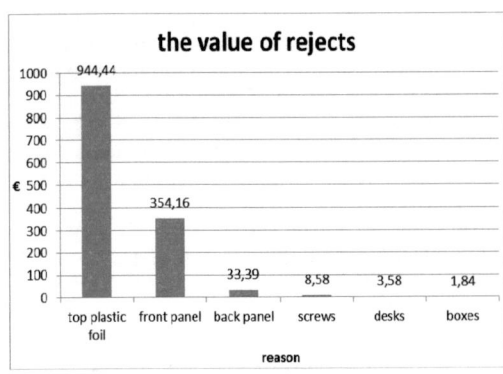

Figure 2. The value of rejects in January 2014.

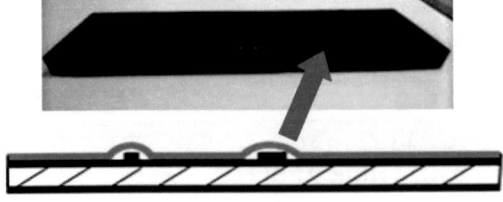

Figure 3. Bubbles under the top plastic foil.

Table 1.   Evaluation of errors on the top plastic foil for January 2014.

Assessed items	Number of items (pc)
Bubbles under the plastic foil	35
Incorrect foil position	22
Faulty foil from production	7
Scratches	4
Damage due to bending by machinery	2

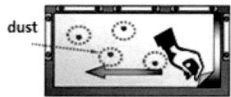

Figure 4.   Removal of impurities using a blow gun.

Figure 5.   Deposit of material from the supplier.

Table 2.   Evaluation of errors on the top plastic foils for February 2014.

Assessed items	Number of items (pc)
Incorrect foil position	22
Damage due to bending by machinery	15
Scratches	4
Faulty foil from production	2
Bubbles under the plastic foil	1

3.2   Solution of cause no. 1 on the top plastic foil

Based on brainstorming, Ishikawa diagram was drawn. Potential causes were established on the non-compliance with the rules (5S) on the line. Furthermore, there was the possibility of rejects caused by poor packaging. The working method was also checked.

The Ishikawa diagram implied that bubbles under the plastic foil may have been caused by the production process. Nevertheless, they may also have come from the supplier. A detailed analysis revealed that the bubbles formed as a result of small impurities that got between the upper section and the top foil, which was glued onto the part.

Therefore, it was necessary to revise the production process. It was found that there was potential for product contamination in the workplace. The production line was located by the main passage way with many people passing every day. People's movement caused dust particles to stir and later fell onto the components. Since the relocation of the production line to another part of production was not possible due to capacity and economic reasons, other solutions, which would be able to eliminate such impurities, were sought.

The first measure was to introduce a blow gun to the production line (Figure 4), which was used to blow away all potential impurities from the upper part of the plastic foil.

During the inspection of the manufacturing process, the working team noticed that impurities could reach the top parts from supplier boxes, tops of which were not closed.

Material concerned (Figure 5) is imported and stored at the organization for some time. It is at this time when, by not closing the tops of boxes, small dust particles are deposited on the top parts of components. Therefore, it was suggested for the material in the boxes to be covered by an extra layer. Naturally, it was necessary to discuss the whole matter with the supplier. This was due to the addition of extra material, which would increase the cost and would be reflected in the price of the given component. Finally, a common solution was found—the second measure. As the organization

had many unused covers, the covers were lent to the supplier for this component.

After the implementation of these two measures, the following period was evaluated—February 2014 (Table 2).

The result was surprising—the problem of bubbles was minimized. Only one reject, which was caused by impurity between two parts, was recorded in February. Further analysis showed that another reason for the high number of rejects on the top plastic foil was the wrong positioning of the foil glued to the component.

3.3   Solution of cause no. 2 on the top plastic foil

In analyzing cause no. 2—incorrect position of the foil, an Ishikawa diagram was drawn. Regarding the wrongly glued component, it was thought that the biggest problem may have been the ancillary appliance and its setting. Apart from this, the working procedure was again reviewed. Not only the working procedure, but also the facility itself and its settings, were verified by direct observation in the workplace.

As this was the same workplace where the proposed measures had already been implemented, the team had an opportunity to check the working procedure again after some time had passed. The work procedure consisted of a removal of the imported component and blowing air over it. Apart from this, the operator used a fixation agent. First, the top foil was inserted onto a device. The exact position was defined by an ancillary slat.

After a short observation period, it was sometimes found that when inserting the upper component, the operator did not precisely lay the part, regarding the shape and size of the foil. In this case, the part shifted, consequently leading to the incorrect positioning of the given component. However, this error could not visually be assessed by the operator, as it had shifted by 2.2 mm. The semi-product was evaluated as a reject only in the final product assembly stage. A gap occurred there when all the components had been fitted. For the customer, this was unacceptable according to the specifications for this model.

Therefore, it was suggested to add clamps in the front (Figure 6), which would fix the part during its fitting into the specified positions. Naturally, the stops had to be made of a material that would not quickly wear out. Teflon was recommended as the most appropriate material.

The team jointly evaluated the proposal after the introduction of countermeasures. The result was the wrong position not occurring even once over the following month, March 2014 (Table 3).

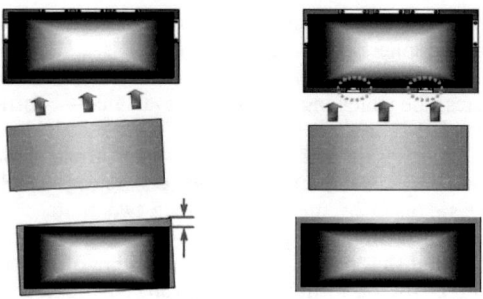

Figure 6. Gluing the top plastic foil into the correct position.

Table 3. Evaluation of errors on the top plastic foils for March 2014.

Assessed items	Number of items (pc)
Scratches	10
Damage due to bending by machinery	7
Faulty foil from production	2
Bubbles under the plastic foil	1
Incorrect foil position	0

4 CONCLUSIONS

Overall, the reduction of the number of rejects was managed by more than 88% over the project period. At the same time, the initial cost of adjustments on the line was € 437 for the implemented ideas. The main cost to the investment was the blowing gun at a cost of € 299. The remaining measures were at a very low cost. The project was undemanding in terms of time and expenses, and due to its implementation, the 5S situation on the line has improved. The whole project lasted three months. However, quality circles found their use in this organization even after this project. Attention continued to be given to problems with product storage, and the like.

ACKNOWLEDGEMENTS

The research reported in this paper was conducted within projects APVV-0337-11, Research into new and newly emerging risks related to industrial technologies within integrated safety and security as a precondition for sustainable development management.

REFERENCES

Hutchins, D.C. 1985. *The Quality Circles Handbook.* New York: Pitman Press.
Hutchins, D.C. 2008. *Hoshin Kanri: the strategic approach to continuous improvement.* Burlington: Gower.
Kelblerová, M. & Vykydal, D. 2011. Application of QFD and SPC Methods in the Processes of Design and Products Manufacturing. *Proceedings of 12th International Carpathian Control Conference ICCC' 2011.* New York: Institute of Electrical and Electronics Engineers. 181–186.
Management mania. 2015. https: //managementmania. com/sk/kruzky-kvality.
Tang, T. 1996. The case of active and inactive quality circles. *Journal of Social Psychology.* Abington: Taylor & Francis Group. 57–67.

Production Management and Engineering Sciences – Majerník, Daneshjo & Bosák (Eds)
© 2016 Taylor & Francis Group, London, ISBN: 978-1-138-02856-2

Recycling of liquid and solid waste into fuel pellets and briquettes

A. Nikulin & S. Kovshov
Mining Faculty, National Mineral Resources University, Saint Petersburg, Russian Federation

E. Mráčková
Faculty of Wood and Sciences and Technologies, Technical University in Zvolen, Zvolen, Slovak Republic

ABSTRACT: The article deals with new approaches for using organic wastes, such as recycling them directly in the place of storage, using the least energy-intensive technology solutions based on new concepts of structure formation and kinetics of the briquette molding. The rational formulae of cells preparation (briquettes, pellets) based on organic waste are demonstrated. The paper presents the experimental samples of fuel briquettes and pellets. The results of the qualitative analysis of the finished product, fuel pellets based on peat, are shown as well. The practical value of the results of research is represented by recommendations for the treatment of solid and liquid organic waste for different types of industries. The most important result is the ability to transition the small power plants (boilers) to multicomponent alternative fossil fuels (briquettes, pellets). Wood processing and waste processing enterprises represent the main consumers of the proposed technological solutions.

1 INTRODUCTION

1.1 The energy

Economical and industrial developments of Russia are almost impossibly without growing of consumption of energy resources. At the same time all kinds of energy must be available in all far places of country in spite of their remoteness of energy infrastructure. Construction of atomic power stations which planned in government program "Development of Russian's power energy until 2030 year" will guarantee energy safety of the country. However this program is not protect small power energy and local businesses (for example farms, saw-mill, etc.) (Sedliačik et al. 2010).

It is possible to assure them by renewable energy sources and secondary using materials (wastes). Introduction in fuel and energy complex new kinds of type of fuel it is a way to solve energy, economic and social problem.

Nowadays in Russian Federation is producing less than 1% in total energy production different kinds of energy by treatment biomass. For example, in Sweden is ensured up to 20% of energy necessary by processing biomass resources. Another European country, like Republic of Slovakia, in direct with European Union Directive 2008/28/EU "On the promotion of energy from renewable sources" owe until 2020 increase part of energy from renewable energy resources to 14% in total power consumption (Tereňová and Krajčovičová 2014).

According to official position of Government of Slovakia biomass can be effective alternative of energy which get from traditional energy sources and may make considerable deposit in decrease of natural gas consumption for producing heat energy. Republic of Slovakia import about 90% necessary energy resources (Ružinská et al. 2014).

One of the most effective ways to decrease consumption traditional energy resources is processing solid combustible waste, including biomass from farms to produce heat and electric energy.

2 MATERIAL AND METHODS

Biomass fuel is wood and waste of its processing as bark, spill, sawdust, wood dust, branches of tree, nonstandard wood, wood pellets and briquettes (Salo and Keränen 1995). From farm is waste sunflower husk, straw, waste of a sugarcane (bagasse), peel of rice, millet, peanuts, etc. and from special plantations of the is "power" wood, bush, etc. and from the liquid and gaseous fuel received from biomass fuel one way or another (Nikolaisen et al. 1998).

In certain cases carry peat to biomass fuel. This fuel holds intermediate position between renewable and non-renewable power sources. The annual gain of peat, for example, in Russia exceeds its con-

Table 1. Qualitative characteristics of different types of biomass fuel.

Biomass fuel	Moisture [%]	Ash content [%]	Sulphur content [%]
Wood fuel			
Spill, sawdust, wood dust, branches of tree	8.60	0.4–0,6	0–0.3
Wood pellets and briquettes	9.10	0.4–0.8	0–0.3
Branches of tree	35–55	1.5	0.02–0.05
"Power" wood of plantation	25–50	1.5	0.005–0.03
Bark	21–65	2.6	0–0.1
Farm waste (fuel)			
Straw	10.20	4.10	0.05–0.2
Hemp	15–75	1.6–6.3	0.03–0.07
Peel of rice, millet, peanuts, etc.	12	10	0.2
Fuel waste			
Water purification slime	53–77	35–50	0.2–5
Manure, dung, etc.	4.92	15–42	0.3–1.1
Waste of hide processing (belkozin)	54	2.5	2.6
Livestock fuel			
Bone meal	7	30	0,7
Animal fat	0.1–0.6	0.1	0.02
Other			
Vegetable fat	0–1,2	0–0,1	0
Tall oil	0.1–0.3	0.2–0.4	0.2–0.3
Peat	38–58	2.9	0.1–0.5
Peat (pellets and briquettes)	10.15	2.8	0.25

Table 2. Qualitative characteristics of different types of biomass fuel.

Biomass fuel	Chlorinity [%]	Heat of combustion [MJ/kg]	Unit weight [kg/m^3]
Wood fuel			
Spill, sawdust, wood dust, branches of tree	0–0,05	16–18	200–350
Wood pellets and briquettes	0–0.05	19–21	550–700
Branches of tree	0.02–0.05	19–21	200–350
"Power" wood of plantation	0.01–0.1	18–20	200–350
Bark	0–0.02	20–25	300–550
Farm waste (fuel)			
Straw	0.05–1.5	18–20	low
Hemp	0.04–0.1	19	low
Peel of rice, millet, peanuts, etc.	0.2	20	high
Fuel waste			
Water purification slime	0.05–1.5	15–24	low
Manure, dung, etc.	0.6–2.4	19–21	
Waste of hide processing (belkozin)	1.2	19	
Livestock fuel			
Bone meal	0,5	23	high
Animal fat	0	39	
Other			
Vegetable fat	0	39	
Tall oil	0	40	
Peat	0–0.1	19–27	300
Peat (pellets and briquettes)		20–27	

sumption that allows considering it as a renewable source though time restoration of peat on a place of production exceeds 200 years (Ushakov 2011).

Biomass fuel is divided into groups on various types. Division depending on initial material for 26 types of biomass fuel are presented in Table 1, 2.

The list of types increases and detailed as more and more species of plants and waste becomes possible to use as fuel, because of development of technologies of burning (Shuvalov et al. 2011).

Prospects using biomass fuel are caused by opportunity to briquet it and using as fuel for obtaining heat and electric energy (Shuvalov et al. 2008).

For increase of caloric content of fuel briquette is offered to input into structure of biomass briquette carboniferous waste. It is important to control temperatures of burning and existence of the polluting substances in combustion gases. Therefore in the conditions of availability of different types of waste there is possible a variation structure in dependence on appointment and types of the boiler equipment.

Using of multicomponent fuel briquettes will allow (Izmalkov 2004): to attract renewable energy resources in energy; to gain ecological effect due to decrease in emissions of CO_2, SO_2, NO_x, elimination of dumps and recycling; to reduce the cost of briquette fuel due to using of low-liquid waste and to increase power safety of productions.

3 RESULTS AND DISCUSSION

Research in the field of mixture industrial waste (coal slime, oil processing waste), solid and liquid biomass waste needs to be begun with determination of optimum parameters of fractional structure and moister of initial components (Shevchenko 2011).

Preparation on fractional structure of wood waste is offered to be carried out on hammer disintegrator. A rational ratio of quality of crushing of wood waste with high rates of productivity depends on a choice of the size of a cell of a final lattice. During research were used a cell with size: 1, 3, 5 mm.

Briquetting of the crushed wood waste was estimated on plasticity of mix. As a result established that the maximum fineness of particles of wood shouldn't exceed 3 mm. The fractional analysis showed that at the size of a cell of 3 mm the following maintenance of particles turns out: 3 mm-29,5%, 1 mm-17,6%, 0,5 mm-17,3%, 0,3 mm-29,5%. Briquetting of mix was carried out on an extruder press (Figures 1a, 1b).

Briquetting technology is in Figures 2a, b, c, d, the components prepared on moisture and fractional structure mix up in drum mixers where in strictly certain quantity water is pumped.

Further mix the batcher gets to an extruder press where it mixes in addition up and bricketed. Ready briquettes give all the best on pallets and go on a warehouse for drying at ambient temperature not less than 25°C. After a set of final weight, briquettes are packed and go to the consumer. Briquettes have a form of the punched cylinder (length—300 mm, diameter—70 mm) (Kovshov et al. 2012).

Figure 1b. Technology of extruder press.

Figure 2a. Preparation of technological process of briquetting of multicomponent briquettes.

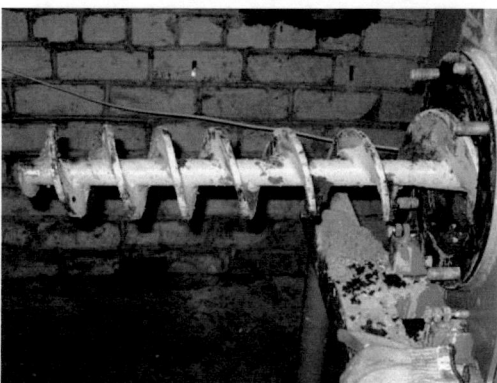

Figure 1a. Extruder press.

Figure 2b. Manual technological process of briquetting of multicomponent briquettes.

Figure 2c.   Briquet.

Figure 2d.   Final products of technological process of briquetting of multicomponent briquettes.

Table 3.   Characteristics of multicomponent briquettes..

No.	Indicator	Working order	Dry	Dry ashless
		Condition of the fuel		
1.	Mass fraction of the general moisture in working order (%)	10.35	–	–
2.	Test moisture analytical (%)	7.31	–	–
3.	Ash content (%)	15	17	–
4.	The lowest warmth of combustion (kcal/kg)	3623	3957	4928
	The highest warmth of combustion (kcal/kg)	3875	4250	5380
5.	Mass fraction of the general sulfur (%)	0.2	0.3	0.4
6.	Volatile-matter content (%)	41.7	49.8	76.7
7.	Type of coke	bound together	–	–

Table 4.   Ratio of components of mix.

Test sample	Ratio coal/sawdust
1	100/0
2	75/25
3	50/50
4	25/75
5	0/100

On it technologies are received prototypes of briquettes. The rational percentage ratio of components, masses is defined by wood waste 35–40%; manure (softener) 10%; coal trifle 15%; oil slimes 5% and the rest water.

The received briquettes were sent to analytical laboratory (JSC Lenenergo) for definition of qualitative characteristics of fuel. Results are presented in Table 2.

According to calorific characteristics briquettes are between firewood (2200–2700 kcal/kg) and coal (4800–5500 kcal/kg) that allows using them in coppers of small productivity and private heating systems. The content of sulfur (0,3%) in briquettes is 2–3 times less, than in coal that leads to reduction of emissions of oxides of sulfur that is a necessary condition when using in a housing estate.

For research of qualitative characteristics of briquettes on the basis of wood sawdust (d < 3 mm) and a coal trifle (d < 1 mm) were prepared 5

structures of prototypes with a ratio of the components specified in Table 3.

On the basis of the received results it was succeeded to define relation of emissions of CO_2 and calorific ability on percentage of coal and sawdust as a part of charcoal briquettes (Figure 3).

We structures the sample No. 2 with a ratio in structure coal/sawdust—1/3 as at its burning emissions of CO_2 decrease by 26% in comparison by coal briquettes is optimum, and calorific ability decreases more than by 20% and makes 4500 kcal that surpasses calorific ability of sawdust more than on 2000 kcal and wood briquettes on 500–800 kcal. Such fuel cuts energy carrier consumption when obtaining thermal and electric energy for 40% without considerable modification of the boiler equipment. Application of multicomponent

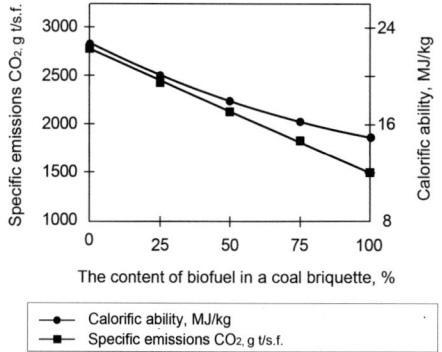

Figure 3. Relation of parameters of briquettes on their qualitative structure.

briquettes in household furnaces and private heating systems won't lead to their overheat and will provide ecological effect for achievement of the demanded concentration of CO_2 in inhabited sector (Pokorný et al. 2014).

4 CONCLUSION

Optimum fractional structure of wood waste for its briquetting together with a coal trifle: 3 mm-29,5%, 1 mm-17,6%, 0,5 mm-17,3%, 0,3 mm-29,5%.

The structure of household briquettes with the following ratio of components of mix is received, mass %: wood waste—35–40, manure (belkozin)—10, a coal trifle—15, oil slimes—5, the rest water.

The received charcoal briquettes possess the following qualitative characteristics: a mass fraction of the general moisture in working order—11%, an ash content—15%, warmth of combustion the lowest—15,8 MJ/kg, an exit of volatiles—41,7%, a mass fraction of the general sulfur—0,2%.

An optimum ratio of coal to wood waste as a part of charcoal briquettes—1/3 that allows to reduce when burning briquettes emissions of CO_2 by 26% in comparison with coal briquette.

ACKNOWLEDGEMENTS

The paper was supported by grant of the President of Russian Federation (2015–2016).

REFERENCES

Izmalkov, A.V. 2004. *Environmental Risk and Safety in Technological*. Moscow: House Science World.
Kovshov, S.V., Kovshov, V.P., Nikulin A.N. & Epifancev K.V. 2012. Technological Solutions to Improve the Efficiency of Secondary Energy Resources Utilization. *Energy economics, technology, ecology* 2012(6): 22–27.
Nikolaisen, L., Nielsen, S., Larsen, M.G., Nielsen, V., Zielke, U., Kristensen, J.K. & Holm-Christensen, B. 1998 Straw for Energy Production. *Technology—Environment—Economy*. Trøjborg Bogtryk: BioPress.
Pokorný, J., Kučera, P. & Vlček, V. 2014. Specific knowledge in assessment of local fire for design of building structures In: *Advanced Materials Research* Trans Tech Publication Switzerland AMR.1001.362: 362–367.
Ružinská, E., Mitterová, I. & Zachar, M. 2014. Evaluation of thermal degradation of wood with environmentally problematic application of coatings. In: *Advanced Materials Research* Trans Tech Publication Switzerland AMR.1001.300: 300–305.
Salo, K. & Keränen, H. 1995 Biomass IGCC. Seminar on *Power Production from Biomass Espoo*, Finland: VTT. 1–17.
Sedliačik, J., Bekhta, P. & Potapova, O. 2010. Technology of low-temperature production of plywood bonded with modified phenol-formaldehyde resin. *Wood Research* 55 (4): 123–130.
Shevchenko, D.Y. 2011. Determination of screw volumetric combustion burner parameters for transporting pellet fuel into the gasifier gasification car. *VISNIK Naukovi Journal* 14 (168): 31–38.
Shuvalov, Yu.V., Nikulin, A.N., Veselov, A.P. & Bulbashev, A.P. 2007 Preparation of fuel briquettes. Patent 2337131 RF. Reference number: 2007130872/04. Moscow: Ministry of patent.
Shuvalov, Yu.V., Tarasov, Yu.D. & Nikulin, A.N. 2011. Justification of Rational Technology of Obtaining a Fuel and Energy Resources Based on Solid Combustible Carbonaceous Wastes. *Mining Information-Analytical Bulletin* 2011 (8): 243–247.
Tereňová, Ľ. & Krajčovičová, J. 2014. How structure of materials influence the fire resistance of the wooden panel house construction. In PTEÚ (ed.), *Advances in Fire, Safety and Security Research*, Bratislava: Fire Research Institute of the Ministry of Interior SR.
Ushakov, A.G. 2011. Preparation of Solid Waste Fuels. Problems and Ways to Implement. *Alternative Energy and Ecology*. 2011 (7): 106–114.

Production Management and Engineering Sciences – Majerník, Daneshjo & Bosák (Eds)
© 2016 Taylor & Francis Group, London, ISBN: 978-1-138-02856-2

The importance of marketing for effective management of quality indicators

R. Nováková & V. Ovsenák
University of Ss. Cyrill and Methodius, Trnava, Slovak Republic

ABSTRACT: Marketing and communication have always been and will be a part of quality management. Experts dealing with this area show the importance and impact of marketing tools for effective management of quality indicators. However, the revised ISO standards do not include the element of marketing. The marketing element is represented only by customer orientation. Content of the contribution points to the need to go back to the fundamental basis of quality management, as well as the trends preferred in the present. An important assumption to produce the products of high quality level is the understanding of customer's expectations and the moment when it is necessary to use all marketing tools. Relevant information acquired by marketing research can be used in the design and development of products. Most qualities are added to products currently in the phase of their design and development. An important indicator is added value expected by customers.

1 INTRODUCTION

In quality management, if the organizations want to survive in the market, they produce goods and services having the characteristics required by customers. In order to achieve this objective, a set of activities has to be carried out. The exact structure can be found in the Juran's quality spiral.

If you take a closer look at the individual activities lying on the spiral of quality, we can say that each of them has a connection to the marketing activities of the organization. Quality in marketing is more focused on the detection of customer's requirements; brief description of the product's characteristics, advertising and print media about the product, the instructions for use and the information regarding feedback (Linczényi et al. 2004).

In literature we may find several approaches and definitions of marketing. To illustrate it we choose the definition by the American Marketing Association. The definition says: "Marketing is the process of planning and practical implementation of the concept of development, pricing, promotion and distribution of ideas, goods and services, which leads to the implementation of a mutual exchange the satisfaction of individuals' and organizations' needs."

Modern marketing is based on the knowledge of the market, emphasizes the importance of focusing on the products, the relationships and networks. The reasons are the rapidly evolving information technologies and globalization of markets (Gregson 2008).

Modern approaches to business marketers often use the following categories:

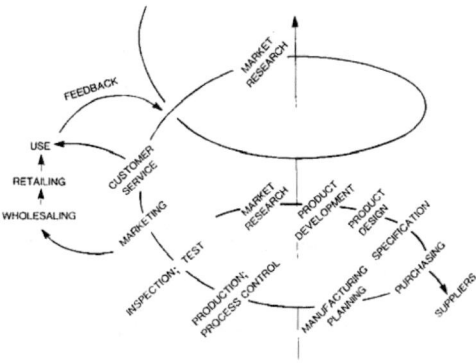

Figure 1. Juran quality spiral.

1. *Transaction Marketing*—the aim is to obtain a customer, orientation on product's qualities, short-term time horizon, less emphasis on customer service, quality is a matter of production, it is focused on the growth of market share and even more on the orientation towards the competition.
2. *Relational marketing*—relationship marketing—the aim is to keep the customer, orientation on the benefits of the product, long-term time horizon, more emphasis on customer service and high responsibility towards the customer, quality is a matter for all areas of marketing, focus on customers' loyalty (Hrubec et al. 2008).

As marketing guru Philip Kotler states in Marketing Memorandum for a Comprehensive Quality Management, he identifies the following principles of marketing approach to total quality management:

- Quality must be perceived by a customer.
- The effort to achieve quality must be reflected in all business functions.
- All employees of an organization must be engaged in quality.
- Quality requires top partners.
- Quality can be permanently increased.
- Improving the quality sometimes requires radical changes.
- Quality does not always mean higher costs.
- Quality is necessary, but sometimes inadequate.

Although marketing is no longer in the model of quality as an independent element, it must be said that the ISO 9000 standards contain certain requirements related to the scope and responsibilities of marketing. In this context, we present the following:

- *Section 7.2.1.* Determination of requirements for the product—the organization must determine requirements specified by the customer, including the requirements for delivery and post-delivery activities, requirements not specified by customer but necessary for specified or intended use, if known, statutory and regulatory requirements related to the product.
- *Section 7.2.2.* is a review of the requirements related to the product- focuses on the commitment to deliver the product to the customer and checks whether the contractual requirements have been met and whether the organization is able to meet them, etc.
- *Section 7.2.3.* Communication with the customer—the organization must implement effective measures allowing communication with customers regarding product information, handling contracts, orders and changes, including complaints and feedback from the customer.
- *Section 7.3.2. and 7.3.3.* Inputs and outputs of design and development—requirements for a product must be precisely defined and they must provide appropriate information for purchase, production and provision of services, which specify product's characteristics important for its safe and proper use.
- *Part 8.2 and 8.2.1.* Monitoring and measurement of customer's satisfaction—the organization must measure and monitor the information related to customer's perception, whether the requirements and expectations have been met and the methods for acquisitions of such information must be chosen.

The fact that the revised ISO standards do not include marketing as an independent area that should be dealt with often causes that not all organizations can correctly identify marketing processes and fail to set quality indicators properly so that they can monitor the effectiveness of these processes.

2 INDICATORS AS MEANS FOR MONITORING THE IMPROVEMENTS IN QUALITY MANAGEMENT

Before we start dealing with detailed indicators focused on an efficient quality management, we should define the indicator itself. The indicator is primarily a foreign word for measurable indicator of an outcome.

There are more explanations of the word's content, for closer identification we use the following definitions:

a. According to the National Strategic Reference Framework indicators are measurable indicators and are an essential means for monitoring the progress and impact.
b. According to the Ministry of Finance indicators are tools for monitoring and evaluation of the objective.
c. These indicators are used to assess the economy, efficiency and effectiveness of the use of financial resources for the area in question (Linczényi et al. 2004).

Based on the qualitative point of view, we can say that the indicators consist of quality attributes, i.e. specific or desired characteristics and conformity criteria for particular outcomes of the process, or system-wide quality management, which the process owner or system owner guarantees to the customer provided he reaches them in mutually negotiated conditions.

We can distinguish:

a. quantitative and qualitative indicators of quality
b. measurable and non-measurable
c. subjective and objective
d. economic and non-economic.

Following the above, we can generate quality indicators for different areas such as:

1. *Field of the quality of life*—the indicators would be divided into indicators focused on personal dimension (subjective and psychological) and spatial (objective) and we could profile the following: health, financial security, education, life chances and opportunities, etc.
2. *Field of education quality*—these are the most commonly divided into quantitative and qualitative, we can include indicators such as

attendance, curriculum implementation, implementation of educational standards, the number of admitted students, the quality of knowledge, quality relationships based on motivation, innovation in teaching, etc.

3. *The area of health service quality*—here we can focus on indicators that are measurable and non-measurable, e.g. number of hospitalized, number of rehabilitations, the level and quality of prevention, quality of health, diagnostics, etc.
4. *The area of quality in tourism*—here we could include indicators such as the number of foreign tourists, foreign exchange share in GDP, number of nights spent in hotel facilities, tourists' satisfaction with the services provided, the number of businesses related to tourism network assessing the quality of service, etc.

In this way we could continue with other areas within our national economy. The aim of this paper is not to focus solely on indicators of quality in marketing. Although we can proceed to generate quality indicators in marketing from various viewpoints, we have chosen a method based on marketing mix. Marketing mix is known both to broad professional and general public and creates a platform for basic division of indicators focused on effective monitoring of activities in marketing activities. Marketing mix is understood as—*product—price—place* and *promotion* (Oblak & Glavonjić 2014).

3 THE CLASSIFICATION OF QUALITY INDICATORS BY MARKETING TOOLS

3.1 *Quality indicators in relation to the product*

At first, we will characterize the indicators of responsibility for products. They are more focused on performance and focus on the following areas:

1. *health and safety of customers*—these indicators are assessed in relation to the stages of the product life cycle and are designed to improve the impact of products on the health and safety of consumers.
2. *labelling of products and services*—these indicators are based on the type of product information, procedures and percentage of product's significance related to the requirements to spread information.
3. *marketing communication*—includes programs for adherence to laws, standards and voluntary commitments, in this context the generation of indicators develops. They are mostly focused on communication tools.
4. *customer's privacy*—from the title it is obvious we speak about indicators related to violation of customer's privacy and loss of customer's

personal data, e.g. the number of legitimate complaints.
5. *compliance with legislation*—such indicators would include non-compliance with laws and regulations aimed at the acquisition and use of products, e.g. monetary value of significant fines.

In general, we would like to state that marketing communication should be in accordance with the ethical and cultural standards, protection of privacy. If it is not the case they become risky to organization of any type in terms of loss of customers and other stakeholders, damage of reputation, increased financial costs and legal consequences.

The concept of responsibility in marketing communications is besides the context of national and international law expressed in the voluntary and self-regulation volumes such as ICC—International Code of Advertising Practice or the OECD Guidelines for national companies (GRI—2000 to 2006).

3.2 *Quality indicators in relation to price*

The most common method used by customers when purchasing is comparing. We compare individual quality characteristics and parameters. However, the basis of comparison characteristic for all customers, manufacturers and products is the price.

Table 1. Contradictions in the perception of price seen from producer's and customer's positions.

Manufacturer	Customer
The price is increased by the use of high quality materials	He seeks to reduce the price since he does not have sufficient financial funds—low basic salaries
The price is increased by investment into new technologies	He seeks to reduce the price by the use of additional service
The price is increased by tax payments	He seeks to reduce the price by using various marketing resources such as loyalty programs, discounts, etc.
The price is increased by staff salaries	He seeks to reduce the price due to the potential repairs—warranty period
The price is increased by e.g. costs for marketing communication, etc.	He seeks to reduce the price with bulk discounts, etc.

The price is usually expressed precisely and thereby ensures a transparent comparability. However, in practice there is a contradiction in the requirements of customers and manufacturers in price. While the producer is interested in the highest price possible, customer wants the exact opposite (Nováková & Habiňáková 2012).

Unsuitably chosen price can bury even the highest quality product. For proper pricing strategies we can use the following pricing strategies:

a. premium pricing
b. penetration pricing
c. price skriming (these are mostly products having the excellence nature)
d. economy pricing (with the economic, resp. discount price).

In practice there are many case studies showing the relationship between quality and price. Their common message is that not always does a cheap, and in many cases poor quality product bring the desired effect in the form of savings. On the other hand, experts in the field of quality say that if the product is sold, i.e. it finds its customer, it can be regarded as good, as for some people the price is the major indicator of quality.

3.3 Quality indicators in relation to the distribution channels and sale site

If we focus only on the sale site, we could point to surveys that show that 70% of customers make their decision directly on-the spot. Therefore, the quality indicators should be closely linked to place where the sale takes place. In the marketing mix, we understand also the place as the process of

distributing the product from the place of origin to the sale site for the customer. Due to this reason this area most commonly uses the quality indicators focused on logistics.

These indicators are focused on the following objectives:

- *Objectives of economic nature* (liquidity, profitability, optimization, etc.)
- *Objectives of performance nature*—they are focused on provision of the proper quality level of services (sufficient amount of high quality material, semi-finished products, purchased parts, finished products starting with the entry into the organization up to distribution to the customer in the right amount, type, quality, time, price and proper place).

If we want to structure the indicators of logistics, we could follow the model created by the German author Shutle. He organized the logistics indicators into the following blocks:

a. structural and framework and indicators
b. productivity indicators
c. quality indicators
d. performance indicators.

3.4 Quality indicators in relation to promotion— communication with customers

The general goal of marketing communications is to provoke customer's attitude to the product resulting in a purchase. To be able to keep the quality focused on communication effect we have to keep certain sequence of development and implementation steps of marketing communication support.

Table 2. The system of logistics indicators with focus on quality.

Purchase	Transport	Storage	Planning	Distribution
Average time spent when receiving goods	Level of service	Number of errors	Intensity of stocks	Average delivery time
Number of defective deliveries	Meeting deadlines	Level of shortage	Number of deferred missing deliveries	Supplying availability
Number of return shipments	Accident rate	Observance of deadlines	Stationary stocks	Number of incorrect deliveries
Number of delaying deliveries	The incidence of damage	Level of storage and service	Average stocks	Delivery precision
Average time for re-ordering		Structure of supplies	Committed capital	Rate of extension
		Store losses for a certain period	Structure of stocks according to their age	Number of deferred deliveries
		Proportion of unusable stocks	Proportion of complaints	

The sequence of steps is as follows:

– Identification of target beneficiaries
– Setting the goals of communications
– Create a message for the recipient
– Division of the total budget for marketing support
– Decision about communication mix
– Choosing the communication channel
– Monitoring the results of communication
– Management and coordination of the whole process of marketing communication.

However, we must take into account that the marketing communication differs in its criteria in consumer markets and industrial markets (B2B, B2C).

The effectiveness of marketing communication is most often assessed on the basis of:

a. communication effect
b. sales effect.

Methods that measure the efficiency of marketing communication include:

1. *Test of recognition*—based on the identification of correct and incorrect identifications
2. *Test of remembering*—assessing the ability to remember an advertising campaign or advertising message spontaneously or provably
3. *Measuring the feedback*—we identify how marketing communication has affected the level of awareness of the advertised message (Matúšová 2013).

In practice the best known quality indicators within the communication mix tools are e.g. GRP index—the ratio of ads impact on the total number of people in the audience, per-click parameters, the number of visits to web pages, return on investment in advertising and PR, the cost for acquiring new customers and retaining the existing customers, etc. Digital Marketing Agencies Association (ADMA) has developed a database of indicators on the basis of which it is possible to measure individual means of communication mix. This refers to economic but also non-economic indicators.

4 CONCLUSIONS

The aim of this paper is to show the need for a wider focus on quality management marketing activities in the organization. As already mentioned above, instructions contained in the revised ISO standards are general and therefore many organizations do not realize that the quality parameters are adjusted practically in the early stages of quality spirals and they are an important gateway for the future direction in assuring and improving the quality within organization. If low-quality inputs enter the process, it produces low-quality outcomes and the same is true for the process of providing marketing activities. It is obvious that due to the extent of the contribution, we could not point out all the areas and quality indicators, which are possibly connected to the process of assuring and improving marketing activities. However, we have tried to point out various points of view and a wider range of information we can acquire by generating a specific portfolio of indicators and which can serve as an important prerequisite assuring customers' satisfaction. At the same time it becomes a means for monitoring the improvements and progress in the organization.

REFERENCES

Gregson, A. 2008. *Pricing Strategies for Small Business.* Self Counsel Press.
Hrubec, J., Virčíková, E., Bajla, J., Dufinec, I., Girmanová, L., Grmanová, E., Hekelová, E., Janošcová, R., Jašková, D., Julény, A., Kneppo, P., Kureková, E., Lakatoš, P., Majer, I., Majerník, M., Nováková, R., Pačaiová, H., Palfy, P., Petrík, J., Sinay, J., Šalgovičová, J., Tkáč, M., Zgodavová, K., Žabka, J. & Žarnovský, J. 2008. *Integrovaný manažérsky systém, vysokoškolská učebnica (Integrated Management System, university textbook).* Nitra: SPU Nitra.
Linczényi, A., Paulová, I., Nováková, R., Šalgovičová, J. Kučerová, M. & Mĺkva, M. 2004. *The project of distance education in quality management, output from grant Open Society Foundation IDEP č. G/276/02/61300.* Bratislava: STU.
Matúšová, J. 2013. *Budovanie a komunikácia značky: značka v PR a Reklame (Building a brand communication: brand in PR and commercials).* Trnava: Univerzita sv. Cyrila a Metoda v Trnave, Fakulta masmediálnej komunikácie.
Nováková, R. & Habiňáková, E. 2012. Process-oriented marketing activities in woodworking industry. *Wood and Furniture Industry in Times of Change—New Trends and challenges: Proceedings is the outcome of the International Conference WoodEMA 2012*: 66–71. Trnava: Faculty of Mass Media Communication UCM in Trnava.
Oblak, L. & Glavonjić, B. 2014. Model for the evaluation of radio advertisement for the sale of timber products. *Drvna industrija* 65 (4): 303–308.

Production Management and Engineering Sciences – Majerník, Daneshjo & Bosák (Eds)
© 2016 Taylor & Francis Group, London, ISBN: 978-1-138-02856-2

The process of personnel marketing in management production

J. Novotný
Faculty of Economics and Administration, Masaryk University, Brno, Czech Republic

ABSTRACT: Business management uses sales marketing to implement a company's sales functions. A set of tools is used for this, which structure the connection of their production function through links to the sales market. These tools are product policy, pricing policy, communication policy and distribution policy. These sales-policy tools can also be used as a starting point for a definition of personnel marketing-mix tools. The aggregate of personnel activities can be divided into the product variables area of operations, the price variables area, the communication variables area and the place variables area of operations.

1 INTRODUCTION

In a market economy, the effective management of a business's production process requires its inter-connectedness with the market. With the business's sales function, the management uses the tools of sales marketing when selling its products. However, in the acquisition of production factors management does not have a similar comprehensive set of tools linking the supply market with production.

This mainly causes difficulties in the supply of personnel on the external labour market. A specific characteristic of the work production factor is that its agent is a person—an employee—who carries out work through his work performance. In order to carry out the work according to the needs of his employer, he has to be prepared for it—qualified—as well as motivated for the work process. All of this, naturally, assumes that the employee will be procured on the external labour market for employment in the business.

2 MARKETING IN THE THEORY OF BUSINESS ECONOMICS

A marketing approach (marketing paradigm) is characterized by Raffée (1984) as *"a practical-normative, content-specific expression of a decision-making and system approach, with a vision of a business management market"*. In this sense marketing is defined as a "concept of business management, where in order to achieve the business objectives the business activities are aimed at the current and future demands of the market" (Bedlingmaier 1993).

In the specialist literature, according to Wöhe (1996), there are three interpretations of marketing, differing in the scope of their conception:

- Marketing is designed for the optimal arrangement in the area of sales.
- Marketing is a market-oriented business theory.
- Marketing is a separate scientific discipline.

In contemporary theories of business management, sales-oriented marketing, which is the narrowest interpretation of marketing, has a universally respected presence. Here the concept of marketing is as a sales function of the business and sales, or rather the sales function of the business, is interpreted in a marketing sense.

3 THE PROCESS OF PROBLEM-SOLVING IN SALES MARKETING

If we observe marketing as a business function, it involves the recognition of various specific issues and challenges which may surface when solving marketing problems. According to Thommen (1996), the following stages can be distinguished in connection with the general process of problem-solving:

1. *Identifying the situation:* In this stage it is necessary to acquire important information about the current and future development of the business. Of particular importance are:
   - business objectives
   - the general conditions of the environment and the relationship between the business and its environment
   - the needs of actual and potential customers (the business takes into account appropriate markets), which have to be established by market research.
2. *Establishing marketing objectives:* Deducing marketing objectives from internal-business (assumptions about values, business objectives, disposable production potential) and exter-

nal-business realities (environments). Typical marketing objectives are related to turnover, market share and geographically defined markets, products and customers. It is generally the case that marketing objectives are derived from business objectives. However, due to the great importance of the marketing objectives for a business, there are no clear boundaries between the business's goals and the marketing goals.

3. *Determining marketing tools:* The marketing goals, which are specified by the previous stage, are the starting point for further development at this stage. From these are derived sub-objectives for the individual sub-areas of marketing, and consequently means and measures are derived which lead to the achievement of these objectives. In business management theory these sub-areas (Hopfenbeck 1992) are described as sub-sales policies or marketing tools. These are:

- production policy
- pricing policy
- promotional policy
- distribution policy.

The term "sub-sales policy" has its origin in the overall sales policy of the business, formed by sales objectives, measures and means with a higher degree of generality, but also objectives, measures and means from sub-areas, which combine to create a company's sales policy. The term marketing tools reflects the fact that marketing theory uses the term marketing tools to describe identical sales concepts—product, price, promotion and distribution. In its own way, marketing research has a superior, or rather integrating status in relation to marketing tools. It provides information for the specific creation of marketing instruments.

4. *Determining the marketing mix:* The term marketing mix is used in marketing literature for the optimization of sales-policy tools. As is stated by Wöhe (1996), its purpose is to achieve efficiency by using these tools. This is dependent on whether the tools are:

- purposefully selected
- carefully balanced
- used at the right level.

This attempt at an optimal combination does not only relate to the differentiated use of individual marketing tools, i.e. sub-sales policy as such, but also to the differentiated use of marketing sub-tools, i.e. individual measures in sub-sales policy. The structure of the individual components of the relevant marketing tools is close to the downstream overview according to Wöhe (1996):

Product policy is seen as the core of marketing. It has to form an offer that satisfies the needs of the buyer, and the objective is the positive differentiation from their competitors' offers. The following measures (marketing sub-tools) are used:

- product innovation
- product quality
- choice
- customer services.

Pricing policy is a marketing tool used in price competition with competitors. This is based on the following marketing sub-tools:

- prices
- deductions
- payment terms.

Communication policy is a sub-area of sales policy that normally has to present homogenous goods as a special type of product. The following measures are used:

- advertising
- sales support
- public relations.

Distribution policy is the final marketing tool of the four for sub-sales policy. It has to ensure that products are sold at the right place at the right time. In order to achieve a competitive advantage it makes use of these distribution policy measures:

- physical distribution (marketing logistics)
- distribution acquisition (sales bodies)
- distribution channels.

5. *Marketing implementation:* The stage of implementing a sales policy, synonymous with marketing in business-management theory, involves the implementation of the marketing objectives and measures formed in the previous steps, which have an analytical-project character, via specific actions. This could be the implementation of a business's new sales policy involving all the marketing tools, or it could be an advertising campaign or the establishment of a sales network.

6. *Marketing results:* This concluding stage involves an assessment of the results achieved and comparing them with the initial objectives. The information on fulfilling the marketing objectives provides concrete results for use in the process of marketing problem-solving.

As stated above, the process of marketing problem-solving deals with the monitoring and description of the business's sales function. In keeping with the decision-making paradigm, the theory of business management examines the relevant area of economic activities in the business (business function) as a sequence of decisions taken by a person in order to achieve the business's objectives.

4 PERSONNEL MARKETING

Manpower is the most important production factor. As the agent of this manpower, a person is the most important production factor. In the process of working, a person, as the agent of this production factor, invents constructs, produces and sells. He also manages and implements the production and sales processes.

The business-management production factor of work can be understood in the theory of business management as the use of the physical and psychological characteristics of a person to realize a business's goals. In this interpretative conception, a person's work performance depends both on the physical and psychological abilities of the workforce and on the efforts to fully utilize these abilities.

The physical and psychological abilities of the workforce are conditional upon fitness and talent. They are affected by age, the development of aptitudes, specialist training and the acquisition of practical work experience. The capability of the workforce to perform certain activities results from their fitness, talent, age, specialist training and practical work experience.

The result of work—the productive contribution of an employee towards fulfilling the business's objectives—depends on whether the business is able to use the specific capabilities of the workforce and whether the workforce or employee is willing to fully contribute his abilities. In business management theory (Wöhe 1996) the four decisive factors affecting productive effort are normally:

- personnel selection
- working conditions
- remuneration for labour
- voluntary corporate social benefits.

However, as Wöhe (1996) warns, performance efficiency can be significantly reduced if the employees are not used according to their capabilities (in the right positions), or do not have the appropriate working conditions (organization of work, physical working conditions, employee relationships), or if they are not fairly rewarded (equivalence of pay and performance), or if there is no personal or professional development support from the business (training provision).

In relation to decision-making about the work production factor, where decision-making is the principle of selecting recognized problems applied in the theory of business management orientated towards decision-making, then four major problem areas can be inferred:

- the best possible selection and placement of workers
- the creation of optimal conditions for work performance
- providing fair remuneration for work, i.e. an equivalent level of difficulty of work and level of performance from the employee
- providing voluntary corporate benefits which positively influence the business climate.

As part of personnel problem-solving, these areas can be expressed in a more structured manner and it is possible to further subdivide and expand upon them. However, this subdivision is more static and orientated inwardly in the business. A clearer definition of the relationship towards the job market is lacking.

Personnel marketing, which in recent years has formed a kind of marketing paradigm of the business production factor of work, overcomes this problem. As can be inferred from the rather sparse research sources, e.g. Bleis (1992) and Batz (1996), where Bleis (1992) still speaks of personnel marketing as a controversial concept, personnel marketing is seen as learning about the optimal implementation of the business's personnel function. There is, therefore, a noticeable similarity to the marketing concept of the sales function (mentioned in the introduction to this chapter), and in order to emphasise this, personnel marketing will be discussed in this article in the same manner as the theory of an optimal arrangement of the personnel area.

The basis of the marketing arrangement of personnel is the definition of the tools of personnel marketing. Here, as with sales marketing, the starting point is the product. It is due to this that the position, or employment in the business, is offered on the job market. Other tools include pricing, promotion and place, which substitutes for distribution in the otherwise identical tools of sales marketing.

All four of these tools are interconnected in the marketing mix. However, at the same time, each of these tools is also divided internally and other sub-tools are formed—with their own internal variables.

The product variables in personnel marketing are: corporate vision, career design and the business's value system. The rewards system, the cafeteria system, working-time models and concepts of incentives are known as price variables, i.e. sub-tools of the marketing tool. Advertising, personnel relations, public relations and sponsoring are variables of the promotional or communicative factor. The variables of the factor of place are target groups, personnel leasing, franchising and outplacement (Meffert 1996).

In the personnel marketing mix a business can use a greater or smaller number of factors affecting the labour market. A summary of these variables comes from (Meffert 1996), and the differentiation can be described using the following questions:

- How can the supply of job performance be characterized from the perspective of a business's personnel policy? (product variable)
- What are the terms and conditions offered by the labour market for job performance? (pricing variables)
- What information channels need to be used in terms of personnel policy to communicate with the production supply? (communication variables)
- In terms of personnel policy, what working conditions should be in place for the supply of job performance? (place variables).

Personnel marketing, as a theory for the optimal arrangement of the personnel area, takes a wider view of the business-theory production factor of work. Unlike standard definitions of the production factor of work in the theory of business management, which are based on the readiness and willingness to perform at work, personnel management gives a four-factor breakdown. The factors of product, price, promotion (communication) and place, with further divisions for sub-tools—variables—allowing businesses to create specific personnel marketing mixes that actively respond to the situation in the labour market.

5 THE PROCESS OF PROBLEM-SOLVING IN PERSONNEL MARKETING

If we start from Batz's (1996) concept of personnel marketing, where the subject under research is personnel marketing as a business function, then here, in parallel with the explanation of sales marketing, we can use the plan for the process of problem-solving to deduce the problems and challenges of this business function. Here there is an emphasis on this parallel, because until now personnel marketing had not been considered to be part of business-management theory, and the process mentioned for problem-solving is based on the decision-making paradigm that is characterised by the modern theory of business management.

1. *Identifying the initial situation* in terms of personnel marketing requires the gathering of vital information on the current and future develop-

ment of the business. Of particular importance are:
- corporate vision
- corporate objectives
- corporate culture
- general conditions, the relationship between the business and its environment
- the attractiveness of the company as an employer for current as well as potential employees, or for target groups in the labour market.

2. *Establishing the goals of personnel marketing:* The general objective of personnel marketing is to acquire the right people at the right time to fulfil the business's goals. Achieving this is aided by the specific operable marketing objectives, which are derived from the business's internal goals, and which respond to the conditions surrounding the business. According to Staele (1996), the objectives of personnel marketing can be divided into:
- material objectives, e.g. the optimization of the creation of values which ensure the business has effective and attractive labour conditions for its employees. Here one desirable objective could be the transformation of the employees into co-entrepreneurs in a quantitative and qualitative sense.
- personnel objectives, e.g. the evaluation and development of employees is a measure of the implementation of personnel work. The desired objective here could be the measurable improvement of the corporate climate in a quantitative and qualitative sense.
- economic objectives, e.g. productivity and profitability are measures of personnel work performance. A desired objective would be a rise in productivity and profitability by x percent.

3. *Specifying the tools of personnel marketing:* In this progressive step, the marketing objectives, which were specified in the previous stage, serve to define the sub-objectives for the individual tools of personnel marketing. These are then used to extrapolate the measures and means needed to achieve these objectives. The personnel marketing tool forms groups of variables—a component of the personnel marketing mix—and it indirectly defines the scope of its operation. We can subsequently talk about areas of personnel marketing:
- product area, comprising the operational space of product variables
- pricing area, defined as the space for the operation of pricing variables
- promotional area, formed by the operational space of promotion or communication variables

- place or placement area, comprising the operational space of placement variables.

Therefore, defining the areas or tools of personnel marketing differs from sales marketing in only one case, but even here there is an obvious similarity in content.

Defining the sub-objectives and extrapolating the measures and means within this stage, therefore, means concretization in comparison with the objectives of personnel marketing determined in the previous step with respect to the areas mentioned. This concretization is based on an analysis of personnel marketing, which provides the necessary information about employees, customers, the business environment and the competition.

4. *Establishing the personnel marketing mix*: As with sales marketing, defining the marketing mix also means the optimization of personnel-marketing tools. Here too it seeks to utilize the measures selected for the effective establishment of the specified objectives. Also, in the case of the personnel-marketing mix its effectiveness depends on the purposefulness in the choice of the individual tools, their balance and quantitative suitability. Through the observed objective, the determined structure of factors therefore comprises an appropriate choice, from a quantitative and qualitative perspective, of variables—the components of the personnel marketing mix. The division of these components, according to Meffert (1996), is related to the following overview:

Product variables are components of the marketing mix in personnel marketing, which are designed to help shape the attractiveness of the company as an employer, offering interesting options in relation to participation in performance creation. This covers:

- corporate vision
- system of management
- career design
- system evaluation (assessment).

Pricing variables are factors of the marketing mix which create interest in acquiring employment and in connecting fully with the company through the creation of lucrative extra-professional conditions. These are:

- remuneration system
- cafeteria system
- models of working hours
- incentive concepts.

Promotional variables of the personnel marketing mix contribute to a company's attractiveness for job seekers and for employees by stimulating interest in the area of communication. These are:

- advertisements
- personnel relations
- public relations
- sponsoring.

Place variables are factors of the personnel marketing mix which contribute to the attractiveness of working for a company through specific forms of employee placement, or specific ways of connecting them to the business's activities. These are:

- placement in a target group
- personnel leasing
- franchising
- outplacement.

5. *Realization of personnel marketing:* Part of the process of problem-solving in personnel marketing is the implementation of measures formulated in the previous stages. By optimally combining the factors of the marketing mix, there is a greater chance of achieving the business's marketing, performance and financial goals. However, presentation of the personnel-marketing-mix variables and their characteristics as part of the appropriate groups only accentuated one area of their competence. The offer of interesting options had an active effect on existing and potential employees. However, there is another effect within the practical application of these incentives: candidates showed an interest in these options to the extent that they tried to acquire the required knowledge and skills, and their application in performance, which will allow them to utilize these options.

6. *The results of personnel marketing:* Personnel marketing contain measures which concern people, who are the most important production factor, and that is why it is necessary to pay sufficient attention to the evaluation of the results. As was shown by Batz (1996), in order to monitor its measures, personnel marketing has developed the concept of personnel controlling. According to the author, personnel controlling has various functions—regulatory, evaluating, integrating, as well as information and servicing. It is orientated in three directions: towards employees, customers and the business's environment.

6 CONCLUSION

As has been described above, the process of problem-solving in personnel marketing uses a decision-making paradigm to observe and describe a business's personnel activities using the principles of marketing. The progressive stages of the decision-making progress are supplemented by business activities which aim to provide attractive employment opportunities to the particular type of people it needs. They actively influence them and provide them with attractive options. Job applicants then try to acquire the knowledge and skills that are required. The principle of supply and demand is in evidence in the labour market and the quality of the worker and the appreciation of it play very important roles. This reinforces the market concept of the production factor of work in a business's practices, while respecting the subjectivity of its agent—the worker—and the social dimension.

REFERENCES

Batz, M. 1996. *Erfolgreiches Personalmarketing.* Sauer-Verlag, Heidelberg 1996.

Bedlingmaier, L. 1993. In W. Hopfenebeck, *Allgemeine Betribswirtschafts- und Managementslehre.* 7. Auflage, Verlag Moderne Industrie, Landsberg/Lech 1993.

Bertram, C.H. 1996. *Qualität in der Personalabteilung.* Reiner Hampp Verlag, München und Mering 1996.

Bleis, T. 1992. *Personalmarketing: Erstellung und Bevertung eines kontroversen Konzeptes.* Hampp, München und Mering 1992.

Hopfenebeck, W. 1993. *Allgemeine Betribswirtschafts- und Managementslehre.* 7. Auflage, Verlag Moderne Industrie, Landsberg/Lech 1993.

Hummel, R.T. & Wagner, D. (Hrsg.). 1996. *Differentielles Personalmarketing.* Schaeffer-Poeschel, Verlag, Stuttgart 1996.

Meffert, M. 1996. In Batz, M. *Erfolgreiches Personalmarketing.* Sauer-Verlag, Heidelberg 1996.

Raffée, H. 1984. Gegenstand, Methoden und Konzepte der Betribswirtschaftslehre. *Vahlen Kompedium der Betriebswirtschaftlehre,* Bd.1 München 1984.

Staele, K. 1996. In M. Batz, *Erfolgreiches Personalmarketing.* Sauer-Verlag, Heidelberg 1996.

Thommen, J. 1996. *Betriebswirtschaftslehre*, Bandl. 4. Auflage, Versus Zürich 1996.

Wöhe, G. 1996. *Einführung in die algemeine Betriebswirtschaftslehre.* 19. Auflage, Verlag Valen, München 1996.

Žák, M. et al. 2002. *Velká ekonomická encyklopedie.* 2nd expanded edition, Linde, Prague, 2002.

Production Management and Engineering Sciences – Majerník, Daneshjo & Bosák (Eds)
© 2016 Taylor & Francis Group, London, ISBN: 978-1-138-02856-2

CMMS as an effective solution for company maintenance costs reduction

P. Poór & M. Šimon
University of West Bohemia, Plzeň, Czech Republic

M. Karková
The Institute of Technology and Business in České Budějovice, České Budějovice, Czech Republic

ABSTRACT: Each manufacturing plant wants to operate their production systems and devices in a reliable manner. What organizations do not want, are manufacturing systems or processes collapsing, leading to production of defective or malfunctioning products. Failures will be reflected, for example, in losses of quality and productivity. One of the most important aspects of well-organized production is machinery maintenance. This article deals first with maintenance as a phenomenon in a manufacturing company, the second part of the article illustrates a practical implementation of a CMMS system.

1 MACHINERY MAINTENANCE IN MANUFACTURING PLANTS

Basic activities of company management include care about basic resources. Each manufacturing plant wants to operate and manage its production systems and equipment in a reliable manner. The fact is that in today's competitive environment when producers want to maintain their position on the market; they have to exploit every possible advantage. One means through which companies seek to gain a competitive advantage is lean manufacturing. Many firms develop an initiative to achieve the best results in the field of maintenance and reliability. Unfortunately, only a few companies realize the importance of synergy, i.e. the strength of the combination of lean manufacturing and maintenance.

The best approach seems to focus on lean manufacturing, maintenance and improvement of reliability at the same time. Simply put, through increasing and improving the reliability of equipment and processes we stabilize the production process. Applying lean manufacturing tools for maintenance will strengthen their mutual synergy (Muharemovic).

When the machine does what it is supposed to and when it is supposed to, productivity and business profitability is maximized. One of the most important aspects of well-organized production is machinery maintenance. The term maintenance here means the "combination of all technical, administrative and managerial actions during item life cycle in order to maintain or restore such a state in which it can perform a required function" (STN EN13306).

Maintenance is no longer limited only by immediate response to special events, overcoming problems of using multiple organs and excessive overtime. If processes and maintenance procedures are appropriately chosen and subsequently adopted actions done well, we can minimize the loss and production becomes stable again. Production capacity will be full and consistently high quality products will become the norm. This condition is called, "excellent care".

But maintenance of machinery can often be quite costly. It will be appreciated that the actual cost savings and profitability can be achieved only by combining the reliability, security and availability in cost-saving process management facilities. Activities and responsibilities of machinery and equipment maintenance in the company are divided into two categories, which are defined as a function of the maintenance transformed into daily operation with regard to solving everyday problems associated with the maintenance of physical facilities (company, machinery, buildings, services) in proper operation.

2 STAGES OF MAINTENANCE DEVELOPMENT

Basic strategies of maintenance are distinguished:

- Proactive maintenance—represents a strategy to prevent the emergence of disorders utilizing corrective action, where the action is directed to the source of the fault. Also includes preventative methods for predictive maintenance (Piotrowski et al. 2006).

• Subsequent maintenance—includes actions that correspond with fault condition of the device (Guizzi 2009). Executing in cases where it is not possible to appropriately select another form of maintenance (Bagadia 2006).

The maintenance activities were not at one time considered as a "necessary evil". Only high costs of maintenance in the range of 15 to 40% of production costs compelled managers to increase the attention to maintenance issues (Galar 2014). Development of maintenance can be observed after World War II (Diaz et al. 2014).

Reactive maintenance can be considered as the first targeted strategy of maintenance management. Maintenance itself was carried out according to the rule "run the device until it goes wrong" and was mainly directed to the damaged parts of the machine. Preventive maintenance is executed at predetermined intervals or in accordance with prescribed criteria and intended to reduce the likelihood of fault or limitation of functionality elements (Figure 1).

The modern era of the information society and computers affect all aspects of everyday life and even in societies. Maintenance and repair processes are currently a complex phenomenon (Kyokai 1996). In many production sites well performed maintenance of machinery and equipment may encounter many problems (Faccio at al. 2013. By implementation of maintenance management with support of IT (CMMS) companies (nearly a half of monitored companies are using CMMS systems—see graph below) are able to make effective use of resources (human resources, inventory, etc.) and ensure an average 8% reduction of maintenance costs in the first year of CMMS deployment (Vlachos 2013).

Most of these systems focus on the basic functions of maintenance, which is primarily used to reduce the number of defects of the production facility. Upgrade of these systems have become systems known as EAM (Enterprise Asset Management), where the function of CMMS is expanded to inventory management of spare parts, spare parts sales and brokering service or tool for predictive maintenance (Mather, 2013). The highest level modules are then maintenance control within advanced ERP systems (Schulz 2013).

3 COMPUTERIZED MAINTENANCE MANAGEMENT SYSTEM (CMMS)

Computer Maintenance Management System (CMMS) is a type of software for management functions such as support management and monitoring operations and maintenance. Computerized maintenance management systems automate most of the functions of logistics provided by maintenance (Bloch 2005; Abdel-Raouf 2014).

Based on the evaluation results of our research, which took place in the form of an extensive survey (about 70 SMEs in Czech and Slovak Republic), it is evident that the incorporation of a system for the computer-aided maintenance sector finds its justification not only for easier monitoring and data archiving. Among the greatest benefits from the use of specialized systems, according to respondents, is the effect on the level of inventory of spare parts. Further, the positive effect of increasing reliability of final products and reducing total cost of production are important (Subben 2013).

Using a computerized maintenance management system in a company means a lot of advantages (Figure 2) (Zhang 2003). One of the greatest benefits of a computer maintenance management system is to remove manual paper work and monitoring activities, which lead to greater productivity. It should be noted that the functionality of CMMS is the ability to collect and store information in an easily researchable form. A computerized maintenance management system does not make maintenance decisions, rather it provides the best information for operations and maintenance managers to adequately service and efficiently run the equipment.

According to a survey in A. Kearney and Industry Week (Giuzzi 2009) 558 companies that currently use computerized maintenance management systems exhibited an average of:

• 28.3% increase in the productivity of maintenance,
• 20.1% reduction in equipment downtime,
• 19.4% savings in the cost of materials,
• 17.8% decrease in inventory maintenance and repair,
• the average payback period was 14.5 months

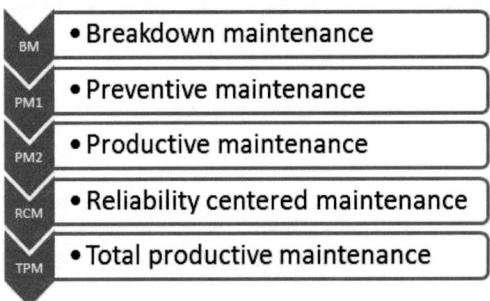

BM	• Breakdown maintenance
PM1	• Preventive maintenance
PM2	• Productive maintenance
RCM	• Reliability centered maintenance
TPM	• Total productive maintenance

Figure 1. Machinery maintenance evolution.

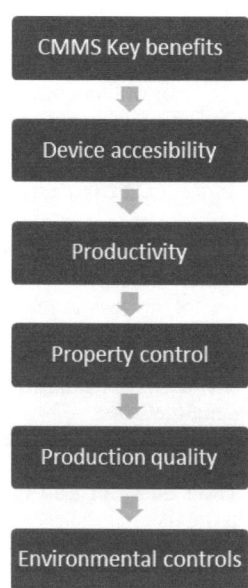

Figure 2.   CMMS key benefits.

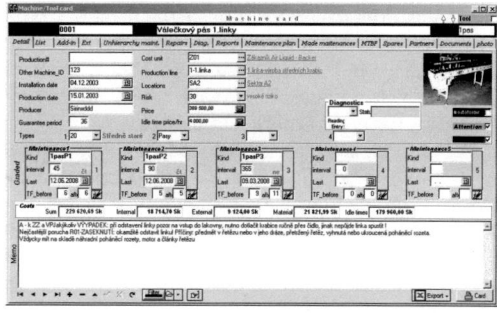

Figure 3.   Machine card.

4   CMMS SELECTION

Planning is essential for the maintenance period. Scheduled events can be manually moved to other data. The plan can be printed in different forms and each action can be printed according to work order. Stepped maintenance means that the higher level in itself always involves a degree lower. Each type of maintenance can be assigned to different workers and their capacity to participate in the tasks (and possibly add even more workers) and what stock items (spare parts) need to be prepared. Repair means action without repetition (Harper 2014).

In a good CMMS, you can define the individual cost centers, enterprise and then watch them on overall maintenance costs. Each machine card machine can be assigned to one cost center. Definition of cost centers is hierarchical, i.e. starting from the highest point of the hierarchy (the company), and it is subordinated to define cost centers (Figure 3).

Machines can be placed in a hierarchical order. By placing the machine it is easy to find in which hall it is located (Cohen 2014).

It is also important to see how the machine follows intervals of preventive maintenances, what the failures were and what was checked, who conducted the previous maintenance, etc. This simply is accompanied by a maintenance status for ISO audit (Lai 2014). Writing in maintenances performed is very simple; usually there is a transfer of work order in the form of maintenance done in one click (Ismail et al. 2013).

5   CMMS IMPLEMENTATION—CASE STUDY

The main objective of this study is, what can a (successful) CMMS implementation bring to a company? Implementation of Profylax CMMS system was used as a model example. The basic component of the overall database for machine maintenance is a machine card. It contains basic information about specified maintenance on the machine. The card is very accurately processed using tabs; each tab contains additional precise information on a given machine. Besides the presence of predefined information we can also use a text note, which is not limited by the length of the text. On this card you can define the level of maintenance. The basic input for the machine card is maintenance manual for the program administrator, where different types of maintenance are specified in detail, and allocated to different time cycles. The Maintenance tab is the main basis for the history and plan of maintenance.

Bookmarks of machines cards are divided into sections: detail, list, banks, distance walked, oil maintenance, repairs, diagnostics, reporting, schedule of maintenance, performed maintenance, parts, partying teenagers, documents, photo.

Summary table shows a list of machines, amount of days to scheduled repair date, repairs and day of the week when the correction is to be realized. All maintenance is described in detail in the text note. For example, after 30 days maintenance is scheduled due to overloads, scheduled maintenance of transport rollers and, deleting the swivel dock, lubrication. After 90 days is planned checking of the oil level, lubrication of machines, control of holder brush, checking and replacing of the filter. After 185 days, cleaning by brush is performed.

Reports tab is used to record all messages from all users for a given machine. The table shows the date and time of reporting, name, day of entry into the program, evidence number of machines, a straightforward description of the message, type of message, maintenance intensity. In this tab, you can write your own messages, postpone, or take over reports. Different message types are represented by different colors and letters.

A maintenance plan is prepared for the review of maintenance that relate to a particular machine. In the next figure (Figure 4), is shown type of service, number of days until the next maintenance, scheduled maintenance date and day of the week. It is possible to recalculate the maintenance plan. After maintenance it is possible to write detailed reports.

Based on successful implementation of CMMS into the system, these benefits are achieved. This summary is a result of our research and study. Benefits are divided into groups:

- immediate,
- long-term,
- benefits for groups of special users,
- Immediate benefits,
- Improving routine maintenance work.
  - simple creation, printing and change work orders
  - easy registration of completed maintenance
  - possibility of immediate action to content planned maintenance
  - enhance communication between people,
  - simple changes in maintenance when new machines or requirements
  - reminders of upcoming events
  - easy routine reporting to higher level management
  - Improved maintenance scheduling
  - plan viewable at any time to print clearly in many forms
  - the plan can include both preventive maintenance, but also planned repairs
  - schedule downtime is available immediately for aligning the maintenance schedule with the production plan
  - the plan can be flexibly changed immediately and seek the optimal variant
  - you can get immediate analysis plan, for example of capacity utilization, which are essential for management

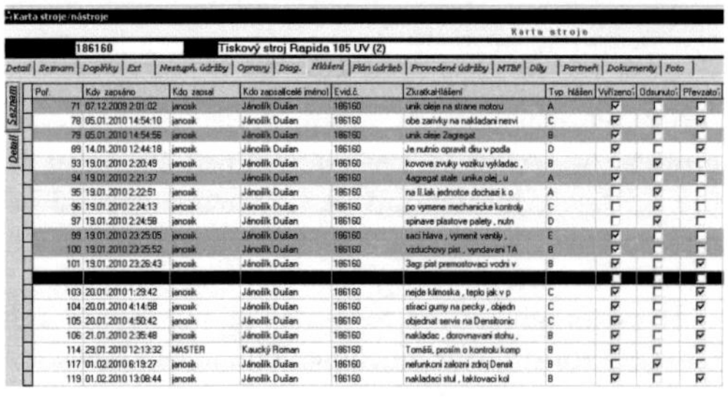

Figure 4.   Report.

Long-term benefits

- Analysis of maintenance for a longer period
  - possibility to control maintenance costs of individual machines, so you can find extremes and gradually build a realistic calculation of production costs
  - possibility to control labor costs and external suppliers, documentation for any outsourcing
  - the possibility of a permanent "non-violent" control capacity utilization
  - simple proof of preventive maintenance (e.g. Audit for ISO)
  - immediate detection of adverse fluctuations in maintenance in extending the period of preventive maintenance
  - when we need to reduce maintenance costs it may be better to find a place where the cost reduction does not lead to an excessive increase in the risk
  - the possibility of testing the effects of business plans for maintenance
- Knowledge about the organization of maintenance
  - reduce the risk of misunderstandings between key people
  - easier administration, faster incorporation of new people
  - ideal for control by "open book management" and to tune the boundaries between the lowest levels of TPM and professional preventive maintenance
- Steadily improving system
  - when it comes to improvements to maintenance procedure, it goes in the description and it is automatically added to work orders
  - easy to update all the consequences when changes are made—the inclusion of a new machine and its maintenance in the already existing plan, extending or shortening maintenance period

The benefits of using CMMS according to the position of those who use it:

- Benefits for routine maintenance worker:
  - clear and bright work orders
  - relevance of design work
  - certainty—simple support in any dispute, what should and should not be done
  - write simple notes, which must not be lost
  - organization of work should reduce wasted time and should reduce interference and unpredictability (Tretten et al. 2014)
- Benefits for maintenance foremen
  - quick planning and replanning
  - plan can never forget something once it is registered
  - faster and more extensive for reporting to higher levels of management

244

- clear evidence management and maintenance staff work through work orders
- reminders of coming events
- organization of knowledge in one place
• Benefits for Production Managers
  - convenient negotiation for production planning and maintenance scheduling
  - transparent and comprehensive reporting service plans
  - immediate availability of current maintenance plan for network
  - much better maintenance work flow
• Benefits for senior management
  - possible analyses of trends in corporate maintenance for a longer period
  - control the amount of maintenance costs to individual machines, people and capacity
  - monitoring the effectiveness of maintenance costs (McGovern et al., 2014)
  - to anticipate consequences of changes in production
  - simple proof of preventive maintenance (e.g. ISO audit)
  - maintenance of easy connection to newly built modern management systems
  - significant risk reduction in maintenance

6 CONCLUSION

Computerized Maintenance Management System (CMMS) Software provides an essential tool for managers aggressively working to decrease downtime, inventory, material and contractor costs.

A 1% improvement in productivity has 10 times the positive impact than a 1% reduction in costs. This is most evident in a production facility that runs 24 hours per day. A conservative estimated gain of 24 hours of production per year pours thousands of dollars to the bottom line. Savings estimate: Downtime cost of $2,500/Hour × 24 Hours = $61,992 annual profit increase (Mapcon 2006).

Without CMMS Software, managers might not know what is in their stockroom, cannot find what they need and, frequently end up buying parts they already have. Of course, CMMS Software allows you to lookup parts, check stock and order only as needed. Savings estimate: Reducing an $800,000 inventory just 10% with a 10% interest rate results in an annual savings of over $80,000. Using a realistic 10% annual reduction in labor and material costs on a budget of $680,000 you return another $68,000 to the bottom line.

Productivity increases of 10% to 20% using modern CMMS Software are standard. Maintenance managers know that saving just 60 minutes (12.5%) per day for each worker in a six-person crew at a $30/hr labor rate for 250 days/year can preserve $45,000 annually for the organization.

In fact, independent surveys indicate the following: industrial average savings result from the initiation of a functional predictive maintenance program: savings of 8% to 12% (Piotrowski 2001). Depending on the type of device to access reactive maintenance and material conditions can achieve savings of 30% to 40%. Average savings in the industry after the start of term predictive maintenance program using:

• Return on investment: up to 10 times,
• reduced maintenance costs: 25% to 30%,
• reduce the number of errors: 70% to 75%,
• reduce downtime: 35% to 45%,
• increase production 20% to 25%.
• provides increased service life and availability of components.
• provides preventive remedies.
• reflected in a reduction in equipment downtime or process.
• reduces the cost of parts and labor.
• integrates technical diagnostics in maintenance management
• provides improved product quality.
• improves the safety of workers and the environment.
• increases employee morale.
• increases energy savings.
• it is reflected in the estimated 8% to 12% cost savings that may result from a predictive maintenance program.

Thanks to CMMS a company gets a complete overview of the functioning of maintenance. Maintenance costs are reduced and the apparatus is always ready at the right time. The efficiency of utilization of the apparatus is improved and the same is true for the use of human and material resources for maintenance.

To sum up, for a successfully operating project of CMMS implementation (Bjorling 2013) several things need to be resolved:

• Human factors characterized by suitable education, salary, suitable personality, commitment.
• Technology—appropriate methods, devices, networking features, price, reports, graphical interface (HMI).
• Processes—integration into the maintenance process, automation of transmission and data collection.
• News—comprehensive evaluation of state, identification of causes, recommended actions, network availability.

A very important success factor is automation of data linking of preventive maintenance, autonomous (TPM) and predictive maintenance using industrial handhelds diagnostic tools and new generation devices, which are intended for operators and maintenance staff.

Another important step for success is operators involved in the process of collecting inspection, operational and diagnostic data (Operator Driven Maintenance).

AKNOWLEDGEMENT

In conclusion, we would like to express thanks for the support of the projects SGS-2012–063 titled "Integrated design of manufacturing system as metaproduct with a multidisciplinary approach and with using elements of virtual reality" and project NEXLIZ—CZ.1.07/2.3.00/30.0038, which is co-financed by the European Social Fund and the state budget of the Czech Republic.

REFERENCES

Abdel-Raouf, A.M.A.L. 2014. Hanafi Maeda. Change Theory: Towards a Better Understanding of Software Maintenance. WSEAS. *Transactions on Computers*, 2014, 13.

Bagadia, K. 2006. *Computerized maintenance management systems made easy: how to evaluate, select, and manage CMMS*. McGraw-Hill Professional, 2006.

Bjorling, S.E., Galar, D., Baglee, D., Singh, S. & Kumar, U. 2013. *Maintenance Knowledge Management with Fusion of CMMS and CM*. 2013.

Bloch, H.P. & Geitner, F.K. 2005. *Machinery component maintenance and repair Volume 3 of Practical machinery management for process plants*. Elsevier, 2005.

Cohen, T. 2014. "The Basics of CMMS." *Biomedical Instrumentation & Technology*. 48.2 (2014): 117–121.

Díaz, V.G.P. & Márquez, A.C. 2014. *Learning from Maintenance Management Models*. After—sales Service of Engineering Industrial Assets. Springer International Publishing, 2014. 55–72.

Faccio, M., Persona, A., Sgarbossa, F. & Zanin. G. 2014. Industrial maintenance policy development: A quantitative framework. *International Journal of Production Economics* 147 (2014): 85–93.

Galar, D. 2014. Context-driven Maintenance: an eMaintenance approach. *Management Systems in Production Engineering* (2014).

Guizzi, G., Gallo M. & Zoppoli, P. Condition Based Maintenance: simulation and optimization. *8th WSEAS International Conference on System Science and Simulation in Engineering*. 2009.

Harper, J. 2014. CMMS work order Management: Closing the gaps around job completion, transactional and financial compliance. *Asset Management & Maintenance Journal* 27.2 (2014): 22.

Ismail, Z. & Narimah K. 2013. *Implementation of information and communication technology (ICT) for building maintenance*. (2013).

Kyokai, N.P.M. 1996. *TPM for every operator*. Japan Institute of Plant Maintenance. Productivity Press, 1996.

Lai, J. 2014. Maintenance Performance: Examination of the Computer-Aided Maintenance Data of a Large Commercial Building. *Journal of Performance of Constructed Facilities* (2014).

Mapcon Technologies, Inc, 2006.

Mather, D. 2013. *CMMS: a timesaving implementation process*. CRC Press, 2013.

McGovern, T. & Sullivan, B. 2013. Journey To CMMS Excellence Manuscript *Proceedings of the Water Environment Federation* 2013.12 (2013): 4404–4410.

Mobley, R.K., Higgins, L.R. & Wikoff, D.J. 2008. *Maintenance Engineering Handbook*. McGraw-Hill Professional, 2008. 1200s.

Muharemovic A., Bisanovic S. & Hajro M. 2014. *Maintenance Scheduling of Thermal Power Units in Electricity Market*, WSEAS TRANSACTIONS on POWER SYSTEMS.

Piotrowski, J. 2001. *Pro-Active Maintenance for Pumps*, Archives, February 2001.

Schulz, D. & Gitzel, R. 2013. Seamless maintenance—Integration of FDI Device Management & CMMS. *Emerging Technologies & Factory Automation (ETFA), 2013 IEEE 18th Conference on (pp. 1–7)*. IEEE.

STN EN13306: Maintenance terminology.

Subhan, A. *2013*. Computerized Maintenance Management Systems. *Journal of Clinical Engineering* 38.3 (2013): 94–95.

Tretten, P. & Ramin K. 2014. Enhancing the usability of maintenance data management systems. *Journal of Quality in Maintenance Engineering* (2014): 290–303.

Vlachis, A. 2013. Rank-Based Ant Colony Algorithm For A Thermal Generator Maintenance Scheduling Problem. *WSEAS Transactions on Circuits & Systems*, 2013, 12.9.

Zhang, Z., Li, Z. & Huo, Z. 2003. CMMS and its application in power systems. *IEEE International Conference on Systems, Man and Cybernetics*, 2003. s. 4607–4612 vol. 5.

Production Management and Engineering Sciences – Majerník, Daneshjo & Bosák (Eds)
© 2016 Taylor & Francis Group, London, ISBN: 978-1-138-02856-2

Implementation of standardized management systems with focus on their integration

J. Chovancová, M. Rovňák, J. Bogľarský & Ľ. Bogľarská
Faculty of Management, University of Prešov, Prešov, Slovak Republic

ABSTRACT: In the last years there has been a significant increase in the number of companies implementing the management systems such as Quality Management System (QMS), Environmental Management System (EMS) and/or the Occupational Health & Safety Management System, or others. Contrary to many other types of standard, management system standards cover multiple aspects, levels and functions of an organization and, therefore, their implementation can have a substantial impact on how an organization operates and manages its business processes. In addition, more and more organizations are applying not only one, but a range of management system standards to satisfy their own needs as well as those of external stakeholders. Authors of the paper present an approach to management system integration, as a key aspect of more effective and sustainable management of a company which secures a better orientation within the legislative requirements, simpler administration and financial savings.

1 INTRODUCTION

In their nature, unsound quality, environmental problems and a risk of injuries have a common cause—a certain degree of irregularity and randomness. These causes can be reduced by effective management systems. A system organizes and simplifies, provides with orderliness and a certain structure. Integration of systems is used in order to create synergy and increase efficiency of management.

Integrated systems represent a very effective way of creating a management system which would take into account not only the quality of products and services (Tarí & Molina-Azorín 2010), but also the approach to the environment and health and safety at work. Besides that, the systematic approach secures a better orientation within the requirements given by legislature and their fulfilment, makes administration simpler and saves financial sources.

Many larger and smaller organizations with foreign capital or with the business primarily in foreign markets have now implemented the Integrated Management System (IMS) in line with international requirements which assists in managing their dominant market position at home and abroad. According to Majernik et al. (2013), thanks to the IMS they in the immediate extent manage their losses from the poor quality of products, negative impacts on the environment, or threat of safety and health at work. At present, Dufinec (2002) assumes that the organization operating in the market can not produce high quality, if it is not managing its losses and doesn't consider environmental protection and safety of its employees beyond the legislative requirements of our country.

In today's difficult economic situation ensuring the success of the organization implementing the various management systems becomes more and more obvious. Among the best known and most frequently applied internationally recognized standards according to Veber et al. (2006), Virčiková (2007), Zelený (2006) etc. can be included: Quality Management System (QMS), Environmental Management System (EMS/EMAS), Occupational Health and Safety Management System (OHSAS).

The short characterization of these management systems and approaches to their implementation is presented in the next part of the paper.

2 CHARACTERISTICS OF THE MOST COMMONLY USED MANAGEMENT SYSTEMS

2.1 Quality management system

In the world, there are altogether several thousands of standards for management systems of quality which are divided into categories and grow in number every year. These standards are divided according to their functions. Among the best known are:

Set of ISO standards ISO 9000—a management system of quality. Standards 9000 consist of several norms that are interconnected and complement each other. ISO standards belong to the most famous standards of quality. It is a set of standards and suggestions together with a manual for auditing of an implemented system. Standards ISO 9000 are an acknowledged standard enabling easier and transparent organization management.

At first the implementation of a QMS was particularly relevant in high demanding activity sectors, like the automotive and aeronautical industries, but it has rapidly extended to every activity sector, becoming a common requisite of any company worldwide and a factor of competitiveness and survival.

ISO 19011—is a manual for auditing management systems, specifically for management and administration of environmental audits and audits of quality.

ISO/TS 16949 (Quality Management System developed by IATF—International Automotive Task Force, working closely with International Organization for Standardization—ISO) specifies the requests for using ISO 9001 in automotive companies and companies producing spare parts.

HACCP standards—Hazard Analysis and Critical Control Points—were developed for food-processing industry.

ISO 22000—a standard developed by the International Organization for Standardization dealing with food safety. This is a general derivative of ISO 9000.

2.2 Environmental management system

Environmental Management System in accordance with ISO 14001 is a part of a complex management system that includes structure, planning activities, responsibilities, practices, procedures, processes and resources for the preparation, implementation, review and maintenance of environmental policy. According to Morris (2004), Majerník & Chovancová (2010) etc. EMS is a part of the company's management system oriented to implementation of activities aiming to environmental protection. This means implementation (integration) of elements of environmental protection in decision-making processes and management of a company.

EMS is a management system related to environmental protection and is based on a series of ISO standards 14000. Another approach to EMS implementation is European scheme EMAS (Eco-Management and Audit Scheme).

In fact, although no reliable references on this matter have been found, it is quite plausible to think that the great majority of ISO 14001—registered companies are also certified in accordance with the ISO 9001 standard. Thus, quality management philosophy and methods have been imported into ISO 14001 from ISO 9001. As a result, it is not surprising that measurement and

evaluation are enshrined as important hallmarks of an effective EMS.

2.3 Occupational health and safety management system

Requirements of quality management system mostly in the area of working environment influence not only the employees of organization, but also public in close neighbourhood. Therefore emphasis is put on meeting requirements of safety and health protection of employees in accordance with new legislative requirements related to industrial accidents and protecting citizens living in the risk area.

The structure of OHSAS Directive is compatible with the essential elements of the legislation of most European countries and above mentioned standards ISO 14001 and ISO 9001 (Rusko et al. 2006).

Similarly to the EMS the basis of the system is searching of threats and evaluation of resulting risks to employees, then underpinning all possible risks and minimizing their impact (Matias & Coelho 2004). Today, regarding our alignment with EU law, this principle is anchored in the legislation of the Slovak republic.

Control system of work security according to Virčíková (2007) is a vast system of organizational structures, procedures, processes and resources, which includes compliance with all legislation requirements.

The ISO 45001—Occupational health and safety is currently being developed by a committee of occupational health and safety experts, and will follow other generic management system approaches such as ISO 14001 and ISO 9001. It will take into account other International Standards in this area such as OHSAS 18001, the International Labour Organization's ILO-OSH Guidelines, various national standards and the ILO's international labour standards and conventions. Expected publication is set on October 2016.

3 IDENTIFICATION OF COMMON AND DIFFERENT PARTS IN ALL SYSTEMS AND THEIR STANDARDS

In some cases, the similarities between QMS and EMS systems can facilitate the integration of the two related management systems (Schwentd & Funck 2002). The people that work in environmental management and at the same time are members of quality teams, assure that quality management goes hand-in-hand with environmental management. The actions that are carried out to achieve quality are, in many situations, the same actions necessary, for example, to achieve effective environmental management. In line with this, ISO 14001 has become compatible with the ISO 9001. Consequently, the integration of environmental issues (including environmental protection and pollution prevention in the management of organizations through the implementation of an environmental management system) allow acquiring a deep insight of the most important environmental aspects associated with its activity, and identifying the processes that need to be improved through the implementation of effective environmental measures (Fresner & Engelhardt, 2004).

Moreover, human resources are the most valuable resource of any company or country, but not always the most valued. Thus, the greatest asset of any organization, any region or any country, are people and their know-how (Santos et al., 2012). Therefore, among others, another system to be implemented in the organizations is the OHSMS. Thus, according to Rebelo & Santos (2012), several fields are showing increasing interest in safety culture as a means of reducing accidents in the workplace. The literature shows that safety culture is a multidimensional concept. Hence, nowadays, companies that search greater profitability and better organization implement the quality systems, aiming at a reduction of defective products and lost time, searching for the loyalty of customers and searching for excellence. From the standpoint of the risk prevention literature, it has been argued that the use of advanced quality management systems help reduce accident rates because quality management methods are based on the principle of prevention rather than corrective actions. Hence, the concept of an OHSMS has become common over the past 20 years Robson et al. (2007).

Standards for QMS, EMS and OHSAS have several elements in common. On the other hand, they differ in some of the basic aspects and requirements. It is possible to determine these requirements and the structure of particular standards by means of analysis of individual management systems. Then, on the basis of the analysis, it is possible to define not only the common elements but also the differences of particular systems. The three systems, the QMS, the EMS and the OHSAS, have almost 80 per cent of the parts in common (Maňko et al. 2005).

Among similarities between the management system standards which enhance the integration may be included:

– The commitment of top management,
– Documentation and records,
– Policy,
– Planning and determining the objectives,
– Processes of the staff training,
– The communication process,

– Audits,
– Managing of discrepancies,
– Corrective and preventive action,
– Management review.

But according to the ISO—IMS publication (2008), a common objective of management system standards is to assist organizations to manage the risks associated with providing products and services to customers and other stakeholders. On other hand the management system of the organizations is frequently split into a number of parts or subsystems, which must be managed separately with relative independence. These parts or subsystems of an organization's management system reflect the different needs and expectations of the stakeholders.

Even though the management systems have many elements in common, there are also parts by which they differ. Among these are:

– specialization of the systems in different subjects and goals,
– need for different partners,
– different degree of transferring of increased costs to customer,
– different degree of importance assigned by a particular system,
– and different emphasis to continuous improvement.

Within this framework, in order to have quality and excellence in products (or services), as well as in the management of the companies that manufacture and provide them, it has become imperative for the companies define and implement quality, environment and safety management system, according to ISO 9001, ISO 14001 and OHSAS 18001 standards, respectively. At the same time the companies must also improve and optimize, continually, these management systems to allow them to true added value for the companies and their stakeholders (Matias & Coelho 2004).

The following questions may arise: How can these three management systems be integrated? Can they be integrated? According to Santos et al. (2012) this is a problem that the most developed companies started to experience some time ago, and it has been discussed by various authors, that we highlight: McDonald et al. (2003), Arifin et al. (2009), Karapetrovic & Casadesús (2009) and Bernardo et al. (2009) among others, who provide a summary of the degrees of integration according to some authors, and Labodová (2004), who reported on the implementation of integrated management systems using a risk analyses based approach.

Many authors deal with the integration of management systems in the basis of common features and differences. But it is necessary to state, that integration into organizational management system is equally important. This kind of integration and issues related to integration into the company core business is discussed in the next part of the paper.

4 HORIZONTAL AND VERTICAL INTEGRATION OF MANAGEMENT SYSTEMS

The word "integration" is generally understood as a connection of parts and thus creation of a whole unit, in other words: uniting, joining, association. Integration of systems is understood as a unification or interconnection of several applications. In the process of implementation of integrated management systems, it is important to be aware of the necessity of horizontal as much as of vertical integration.

Horizontal integration takes place between particular management systems. In this type of integration, it is possible to use common elements of particular standards listed in the previous part of the paper. Unification of the common elements brings cumulative advantages of integration—from this point of view, integration simply enables multiple use of an activity that has already been carried out. It is not necessary to project, implement, test, etc. something new for every new demand, because it is sufficient to use already existing solutions.

Vertical integration proceeds between management systems and the system of an organization itself. An integrated management system is implemented into its functioning management. Without a vertical integration and the connection of IMS to the existing management systems, the introduction of IMS would remain only a formal act and would not bring expected benefits.

Also in this form of integration we can make use of already operating system elements and, on their basis, build an integrated management system.

A graphical illustration of the approaches is in the Figure 1.

To illustrate the integration of management systems, we used a diamond-shaped diagram. Its parts are compact and tightly interconnected in horizontal and also vertical direction. This demonstrates the necessity of a connection between individual management systems creating thus a solid and undivided complex.

IMS is built and implanted within an organization voluntarily and is supposed to lead to the improvement of its overall profile. Therefore among the basic factors decisively influencing such

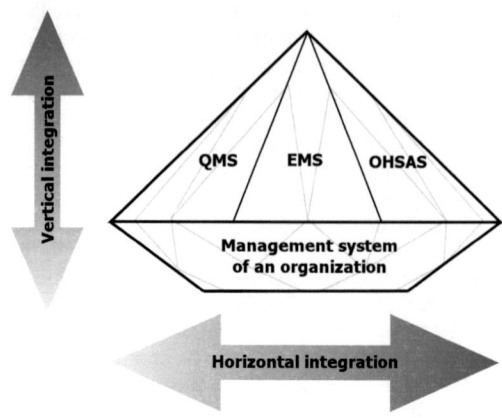

Figure 1. Integration of management systems.

activities are mainly involvement of the management of the organization and the accessibility to essential resources. In the first case, the top management must make a serious decision which may influence the image of the organization. The intention to implement the IMS, sustain and improve it must be organizationally and financially supported so that it would remain functional and fulfil all the predetermined goals.

Such a decision does not differ from any other planned intentions of an organization. In general, the decision must fulfil:

- goals—whether all the systems are to be implemented at once or successively; or whether the final goal is to certificate the systems;
- resources—human, financial, material and other resources that are required by the IMS;
- deadlines and dates—for instance a deadline for the processing of documentation, employee training, certification itself;
- responsibility—choosing a person who would be responsible for the overall administration of the system;
- schedule—is needed for the implementation of a system in accordance with the conditions of the particular organization.

Regarding this, an organization should take into account the fact that the implementation of a system requires sustaining appropriate financial and human resources. Building of the IMS and its integration into an organization is an economically demanding and time-consuming process. It involves considerable knowledge and practical experience. Therefore, the top positions in quality management, environmental or occupational and health and safety management of the organization should be assigned to managers from its top management who are capable of taking up such an important task.

5 BENEFITS OF MANAGEMENT SYSTEMS INTEGRATION AND OBSTACLES OF INTEGRATION

It can be very advantageous to implement and integrate management systems according to ISO 9001, ISO 14001 and OHSAS 18001 into a one functional management system, which will become a useful tool for management and ensure the prevention of all risks related to the activities of an organization (Beranek 2007). This brings many advantages but also many barriers. Among the benefits of integration can be included:

- Avoidance of duplication and conflicts: The requirements of the various systems shall be arranged in particular activities of the organization and then common solutions are adopted.
- Save of time: compared to the implementation of individual systems, IMS is not more time-consuming.
- The benefits of synergies: Solutions that have already been proposed, may also be used for other systems.
- A comprehensive overview of the organization: the activities are analyzed and improved in the integration. If contradictory requirements are revealed, they are clarified and acceptable solutions are designed for all interested parties.
- Optimizing cost management systems: necessary structures are commonly used, for example common documentation in one manual for the integrated management system.

Though the integration of management systems can bring a lot of advantages to the company, the barriers to the integration should be considered too. The main barriers can be divided as follows:

- External reasons:
  • Lack of standards for integrated management systems,
  • Different understanding of existing standards,
  • Lack of tools (or their non-use) in auditing/ management systems evaluation.

- Internal reasons:
  • A formal approach to systems implementation,
  • Promotion of interests of different groups (e. g. quality at the expense of environment or safety),
  • Effort to satisfy certification/consulting companies (also leads to a formal approach to the implementation of IMS).

6 CONCLUSION

An integrated management system represents a universal and effective management tool to achieve stated objectives of manufacturing but also non-manufacturing organizations not only in the question of quality of provided products, the safety of environment and safety at work. IMS is an effective way to maximize the market value and growth of any organization. It has to support its main objectives, which are: minimization of costs, maximization of profits and gaining competitive advantage.

In practice, IMS has to increase efficiency and improve the quality of business, production, information, technological and other processes and thus reduce costs of implementation of the processes. At the same time, the efficiency and quality must be achieved in the shortest time and at the lowest cost.

Vertical and horizontal integration of management systems enables the organization to use synergistic effect which is caused by mutual interconnection and unification of elements from systems, it reduces their administrative demands and saves financial resources. The goal of the paper was to point out the importance of an integration that would take place not only in a vertical direction, that is within the already existing management system of an organization and QMS and/or EMS and/or OHSAS, but also in a horizontal direction—that is the integration of newly implemented systems.

Even though such an approach of an organization towards integration is a time-consuming, staff-demanding and organizationally challenging process, a successful implementation of IMS is connected with many strategic, operating and also economic benefits not only for the organization itself, but also for other concerned parties.

ACKNOWLEDGEMENTS

The study was supported by KEGA 032PU-4/2014.

REFERENCES

Arifin, K., Aiyub, K., Awang, A., Jahi, J.M. & Iten, R. 2009. Implementation of Integrated Management System in Malaysia: The Level of Organization's Understanding and Awareness European Journal of Scientific Research, 2009. ISSN 1450-216X Vol. 31 No. 2: 188–195.

Bernardo, M., Casadesus, M., Karapetrovic, S., & Heras, I. 2009. How integrated are environmental, quality and other standardized management systems? An empirical study. Journal of Cleaner Production 17, 2009, pp. 742–750.

Beranek, Z. 2007. Experiences with integrated systems. In: Planeta 2/2007, Czech ministry of environment, Praha. ISSN: 1801-6898.

Dufinec, I. 2002. Virtual process of quality management of production in integrated management system of company, In: International conference proceedings QUALITY 2002, DT Ostrava, VŠB-TU Ostrava, 2002, ISBN 80-02-01494.

Fresner, J. & Engelhardt, G. 2004. Experiences with integrated management systems for two small companies in Austria. Journal of Cleaner Production, Vol. 12, 2004., pp. 623–631.

ISO (2008): The integrated use of management system standards.

ISO 9001:2008 Quality management systems— Requirements.

ISO 14001:2004 Environmental management systems— Requirements with guidance for use.

ISO 19011:2011 Guidelines for auditing management systems.

ISO 22000:2005 Food safety management systems— Requirements for any organization in the food chain.

ISO 45001 Occupational health and safety—overview.

Karapetrovic, S. & Casadesús, M. 2009. Implementing environmental with other standardized management systems: Scope, sequence, time and integration. Journal of Cleaner Production 17, 2009, 533–540.

Labodová, A. 2004. Implementing integrated management systems using a risk analysis based approach. Journal of Cleaner Production; 2004, 12: 571–80.

Majerník, M. & Chovancová, J. 2007. Environmental management—development and trends. In: Environmental engineering and management: Conference proceedings of 4th international conference EIaM 2007: 22.-24.10.2007, Herľany. Košice: SjF TU, 2007. s. 22–29. ISBN 978-80-8073-894-5.

Majerník, M., Mihok, J., Tkáč, M., Bosák, M., Szaryszová, P. & Tarča, A. 2013. Environmental management in an integrated system. Vol. 1. [Košice: Faculty of Business Economics EU], 2013. 302 p. ISBN 978-80-971555-1-3.

Maňko, M., Liberková, L. & Hricová, B. 2005. The basic access into the integration management systems. In: Engineering sciences 1. oral presentation: 5th interntional conference of PhD students, University of Miskolc, Hungary 14–20 August 2005. Miskolc: University of Miskolc, 2005. s. 129–132. ISBN 963-661-673-6.

Matias, J.C.O. & Coelho, D.A. 2004. The integration of the standards systems of quality management, environmental management and occupational health and safety management in International Journal of Production Research. Vol. 40, no.15 Taylor & Francis 2004.

McDonald, M., Mors. T.A. & Phillips, A. 2003. Management system integration: can it be done? Quality Progress; pp. 67–74.

Morris, A.S. 2004. ISO 14000 Environmental Management Standards: Engineering and Financial Aspects. John Wiley & Sons Ltd. 2004. ISBN: 0-470-85128-7.

Rebelo M.F. & Santos, G. 2012. Integration of the Occupational Health and Safety Management System with the Quality Management System and Environmental Management System—from the Theory to the Action. International Symposium on Occupational Safety and Hygiene SHO—2012. Minho University 9, 10 March 2012. ISBN 978-972-99504-9-0.

Robson, L.S., Clarke, J.A., Cullen, K., Bielecky, A., Severin, C., Bigelow, P.L., Irvin, E., Culyer, A. & Mahood, Q. 2007. The effectiveness of occupational health and safety management system interventions: A systematic review. Safety Science 45, 2007, pp. 329–353.

Rusko, M., Balog, K. & Tureková, I. 2006. Selected chapters of environmental and safety management.— Bratislava: VeV et Strix, Edícia EV-4, First slovak edition, 2006, ISBN 80-969257-5-X.

Santos, G., Rebelo, M., Barros, S. & Pereira, M. 2012. Certification and Integration of Environment with Quality and Safety—A Path to Sustained Success, Intech.

Schwendt, S. & Funck, D. 2002. Integrierte Managementsysteme Konzepte, Werkzeuge, Erfahrungen, Physica, Heidlberg 2002, ISBN 3-7908-1442-3.

Smith, D. 2002. IMS. Implementing and Operating, British Standard Institution, London 2002, ISBN 0 580 33328.

Tarí, J.J. & Molina-Azorín, J.F. 2010. Integration of quality management and environmental management systems Similarities and the role of the EFQM model. The TQM Journal Vol. 22 No. 6, 2010, pp. 687–701.

Veber, J. et al. 2006. Managment kvality, environmentu a bezpečnosti práce. Legislatíva, systémy, metódy, praxe. Managment Press, Praha, 2006. 358 p. ISBN 80-7261-146-1.

Virčíková, E. 2007. Integrated management systems. Faculty of metallurgy, Technical university of Košice. ELFA, s.r.o. Košice. 106s. 2007, ISBN 978-80-8073-761-0.

Zelený. J. 2006. Integrated management system. Technical university in Zvolen, 2006. ISBN 80-228-1576-4.

Production Management and Engineering Sciences – Majerník, Daneshjo & Bosák (Eds)
© 2016 Taylor & Francis Group, London, ISBN: 978-1-138-02856-2

Creation of centres of mining tourism

P. Rybár, M. Molokáč & L. Hvizdák
*Faculty of Mining, Ecology, Management and Geotechnology, Technical University in Košice,
Košice, Slovak Republic*

S. Khouri
*Dean's Office of FBERG, Faculty of Mining, Ecology, Management and Geotechnology,
Technical University in Košice, Košice, Slovak Republic*

ABSTRACT: To create a centre for mining tourism with its inner zones suitable for sociable, family and sports events or possibly for adrenaline activities would be rational solution, since the abandoned mining works would be revived naturally, which might slowly lead to an economic growth of the microregion, in which the centre would be opened. Except for the economic effect, there would be a very positive impact on the maintenance of mining cultural traditions and customs, at the same time it would be possible to offer work for the former miners. Another effect of such a centre of mining tourism is controlled protection of the environment in this region. In the article we propose a model solution based on the real situation in the region Lower Spiš (Slovakia). The tourists will be fascinated by unique technology, nature around the entrance tunnel, already existing infrastructure on the surface, and a mining museum.

1   INTRODUCTION

Paradoxically, together with the stagnation of mining activity in developed countries, mining tourism has won more favors of the public there.

At the end of the 20th century mining has lived up to a rather negative reputation mainly because of huge campaigns of ecological organizations supported by media and by a part of the manipulated public, which has actually never been in real touch with mining activity itself. Despite the fact that the mining technologies are definitely on a different higher level today in comparison with the time of "brownfields", intentions of geological and mining companies to dig for raw materials and the priorities of the public are still in conflict. Environmental organizations have always (successfully) fought against new mining work, asking for various restrictions by legal way regarding geological research of layers, revival or "kick-off" of mining activities and they are strongly insisting on the protection of our Mother Nature "bruised" by the former mining activity (Conesa et al. 2009).

European Union has been caught in a trap, as it supported these protective attempts in a long-term way, mainly for political reasons and today, it changed its policy with an effort to encourage taking advantage of own raw material sources. However, the doctrine of the former development may be changed only slowly and gradually. Despite being closely connected to this article, it is not the main subject of the submitted article.

Going back to the main topic, we kindly call your attention to the ideas of technical intelligence, claiming that the protection of sociodiversity or technodiversity, connected with long-term mining history will only help to preserve all the social and other tracks in the development of the mankind influenced by miners themselves.

In the end, if there is an attempt to preserve biological diversity for the protection of our natural wealth, developed in millions of years than let´s support this philosophy together (Rybár et al. 2013; Klaučo et al. 2012).

2   MINING HERITAGE

Purposeful minerals acquisition for their use in various fields of life has always been a part of our existence, firstly in Neolithic times—in Stone Age, where there were groups of people collecting and later manipulating with such raw materials. This was the time when first barter was done as for example with obsidian or hornstone. In the bronze era there was a trade with copper, tintin and with bronze even among several continents. Similarly, technologies of metal processing were developed and spread. This was a part of mining as well. This was the time when a very special social category of people with certain knowledge—miners—were given certain privileges.

In the Middle Age, miners were given privileges by the king together with the privileges for the mining towns. Miners were free citizens, inferior only to the king. This was the time of mining craft associations, miners acquired a privilege to benefit from the digging and they were the people standing next to the cradle of all sciences, as for example geology, mineralogy, chemistry, geodesy, or cartography, mining was the initial part of establishment of safety at work and it even provoked the establishment and later development of technologies, as for example metallurgy, use of explosives, industry, water resources management, unwatering of underground places or their illumination and mainly one of the most important factors of mining is that it was responsible for the birth and for the further development of secondary and higher education. The list of the main influence of mining activity on the development of our society is still not complete.

Liquidation and recultivation of mining work, together with the standard technological recultivation procedures after the end of mining work in the 20th century, ended up with the liquidation of unique mining work, mining equipment and abandoned former mining villages (Hronček 2011). At that time, and even today, all the solutions were considered to be correct, since they finished the hardwork of miners and unsuitable living conditions in the mining villages.

This was a "victory" with certain failure since we lost unique technologies and works created by miners and we simply sent to the "grave" hundreds or maybe a thousand years with their specific customs of the mining communities and we stopped the developing structure of mining.

Our mining heritage includes in its large part also the heritage of the miner as a human being in the local community, as miners invested their property and even risked their lives because of the very harsh mining conditions at that time (Ballesteros 2009; Vargas-Sanchez at al. 2009).

3 MINING TOURISM

In historically mining countries, as for example countries of the central Europe, mining tourism is based especially on mountain tourism. Surprisingly, also this part of population, who helped with the end of the mining in the past, now shows an interest in its revival. It sounds weird, but together with the decline of mining activity, we can see a large growth of the acceptance of the heritage of historical mining and interest in social, economic and technological mining background (Pretes 2002).

The best example, here, could be Wieliczka in Poland, which is a good illustration of revival and the development of mining tourism, where the number of visitors within a year, in this former salt mine, is now over 1 million (Conlin et al. 2010). This is a great example of sophisticated and elaborated plan with a long-term intention to run a business, by which they gradually carried out its technical and economical intentions of the mine management, obviously supported by the local municipality and by the state government. To do all the necessary work in the underground in order to maintain the security of the visitors and to bring such services that will make this place really attractive for them was a main goal of everyone, obviously within a certain time horizon. It was clear to them from the very beginning what their target group is like and therefore what services will be required. Expectations that the tourists will come, will leave and then they will come back again and again, are fine and optimistic, however in such case we cannot expect more than a few hundred of visitors a year. A few hundred thousand or even a million of tourists are just a myth in such case.

High visiting rate in Bochnia and in Kamienky Opatowskie, which are mines opened for the tourists not far from Wieliczka proves that there is nothing like competition of destinations nearby with a similar intention. Even here, the numbers of visitors per year must be counted in hundred thousand.

It can be said that mining tourism is rather prosperous business in many countries—it is not only already mentioned Poland, but also Germany, Austria, Czech Republic, Spain, United Kingdom, France and others.

So far, the potential of mines, here in Slovakia has not been valued enough and is not appreciated equally. Abroad, we can see that mines are real targets of mining tourism. And, it must be emphasized that it is so not only after the end of the production of the mine, but also during its production and even during the working time and period of running. It is very common to see group excursions of tourists on specially-designed means of transport, by which they can get to the mine, regardless of obstacles—noise, dust, or other risk involving phenomena (Rybár 2015).

It cannot be forgotten that the mine itself changes into a very interesting sight straight after the beginning of the raw materials digging specifically from geological or morfological point of view even without any other research, or antropogenetic changes, conservation, recultivation, revitalisation or building. It can be claimed that it becomes an interesting subject of study not only for mining tourism but also for geotourism and its competitiveness can be evaluated from various points of view (Pavolová et al. 2014). Tourists and people collecting minerals will be allowed to collect some minerals in specific areas of heaps; possibly they can be allowed to practice adrenaline sports and so on.

Maintenance of mining works and technical sights brings problems with the protection of the environment. It is a difficult technical and economical issue to maintain the underground mining works after the end of the mining activity (Drebenstedt et al. 2011).

In Slovakia, our state authorities "want to make us think" that an ordinary citizen without special courses about safety at work, simply cannot be allowed to go down to the mine. Official organs are not aware of the fact that the tourists do not come there-to the mine, to work however to relax and to develop intellectually, or possibly to fulfil its adrenaline needs. It is very similar with the visitors of other places for example caves. A tourist of the mine is reasonably supervised by a professional guide down in the mine, safety is number one in the ladder of importance, but at the same time interesting information for the intellectual development is provided, the tourists will learn about the protection of the underground places or about the equipment which must be kept in its original state in accordance with the mining law, which is superior for the mining tourism (Drebenstedt et al. 2011).

At the end of the 20th century, there was an extreme decline of mining in Slovakia within a couple of years. In reality, it looked like an effort to bring all the mining activities under the control of state authorities. In certain places, selling of mining technological equipment only in a form of scrap metal took place, it was not stopped, just the opposite, it was carried out by the management of the declining mine itself. Consequently, in accordance with a safe liquidation policy of the mine, the premises were flooded. The final phase of the mine liquidation was actually "carried out" through the destruction of these places by our inadaptable people. Finally, these historical objects whether of building or technological character are destroyed and lost forever.

The lack of legal regulations and proper norms for the protection of mining sites after the end of the mining activity at that time is actually the main reason for the impulsive mine liquidation. There were certain solutions as for example bankruptcy proceedings with the opportunity to use the mining sites and already existing infrastructure for other purposes, as for example for the development of mining tourism. Various companies and organizations tried hard to save the isolated technical sights, as for example mining engines, or constructions of funiculars, entrance parts of the tunnels and so on, and this should be offered as the root of mining activity.

This is the beginning of the fight for the revival of mountain tourism in Slovakia as the revival of mining technical sights will attract a lot of tourists here (Rybár et al. 2015b). Various projects were created by lobistic groups, which were competitive for each other. And so Banská Štiavnica and its surrounding, where mining tourism had already existed before, was actually the only place with the state support. In Banska Hodruša nearby, the management of the private mine in activity used its own funds for its gradual building of—in our opinion—successful base of mining tourism.

In 2005, Gelnica was a place, with the first project proposal of its nationwide Slovak character of mining tourism, called Slovak Mining Route. It was during the regional meeting of mining companies, Faculty of BERG and other nationwide Slovak mining organizations. This was the result of the international project INTERREG III C—ENMR (European Net of Mining Regions), solved by Faculty of BERG TU Košice.

For the first time, priorities for the maintenance of historical mining objects were defined with the focus on the:

- protection of history—with a warning saying that together with the liquidation of mines and mining objects we lose precious technical sights,
- maintenance and revival of technical sights,
- change of the projects trying to stop mining activity, proposals of little steps that can save the dying mining activity, determination of the mines that should be left for other purposes (for example mining tourism) if needed and controlled liquidation of mining work,
- making the most of mining sights in Slovakia for the purposes of mining tourism.

4 CENTRES OF MINING TOURISM

The collective comes up with other suggestion how to develop mining tourism in Slovakia, which can be divided into regions of Malé Karpaty, historical central mining towns (with the centre in Banska Štiavnica) and historical Upper-Hungarian (eastern Slovak) mining towns. In the last region—eastern Slovak historical mining towns, we should distinguish various parts. A suitable division can be seen in the project of Upper-Hungarian Mining Route (1). It is Lower Spiš with two microregions of Gelnica and its surrounding, Smolník and its surrounding and Spišská Nová Ves and its surrounding. The second suitable part is Gemer with two microregions of Rožňava and Dobšiná.

The centre of mining tourism is a specialized project, in the middle of which there is historical mining work, or a number of mining works. It includes the features of geotourism, as for example geosites, aquazones with rich wellness offer, or spa treatment, connection with the national parks nearby, natural or artificial features for mountain climbing, mining museums and open-air museums, monuments and memorials connected with famous people and families dedicated to mining, mining schools, but also terrains for offroad cycling and motocycling and other adrenaline sports. Obviously, such centers must offer suitable accommodation and restaurant services, places for games for children and so on. The centre of mining tourism will be in a huge demand, if it focuses on the activities welcomed by the target group. The project itself must have a sophisticated and a suitable business strategy. In the centre, for example, it is impossible to combine the activities for the families with little children and adrenaline sports.

Gradual opening of individual zones within the centre of mining tourism must go hand in hand with the economical proposal of the project. It is crucial whether we build it on a "green meadow" using for example abandoned surface mining objects with only minimal financial demands as for example little changes needed for adrenaline sports-off road cycling, motorcycling, driving terrain motor-cars and military vehicles. In this way, we may have a higher visiting rate with an increase of cash flow after the reconstruction of the terrain for example by having a circle, where people practice adrenaline sports in a safe zone in a form of competition far enough from the audience who is watching them and cheering and yelling. A good strategy may bring a lot of spectators after some time (which means that the visiting rate of the centre will be a couple of ten thousands of visitors a year.).

The zone with the mine itself may be the most demanding part from the financial point of view regarding the centre. In case of strict mining maintenance it may but it may not change into an unprofitable or loss-making part of the centre. In such case we can expect visitors mainly from the groups of (aging) fans of historical mining. Organized presentations for schools will be definitely interesting from the financial point of view. However, this way, we cannot expect such financial gains that would be enough for the survival or possibly further development of the centre. Even here, in this zone, it can be financially effective if we manage to create attractions as for example a 3D presentations of short films in special "film stalls". Presentation on the wall could be very appealing, or possibly "fog effects", or use of holograms for the presentation of individual activities of miners in the underground... Interactive animation and games for children or young people in underground surfaces might help to promote the centre and at the same time to make the visiting rate of this region slowly but constantly higher and higher. "A school of animation" or something similar might be another way how to catch the attention of the visitors, which might be the latest

"smash hit" for the youth. However, history of the region and historical mining must be the common attributes in all cases.

Opening of other zones in the centre must be gradual and continual, of course in such order, in which expenses are lower than the gains of the centre, in order not to get into debts, which would consequently lead to the end of the whole project.

5   CONCLUSION

From the prospective of historical mining, Slovakia is one of the richest countries, at least in central Europe. Conservatism of state administration on one hand and inability to create effective centers of mining tourism on the other hand have brought no fruitful time so far and in comparison with the neighbouring countries we do not keep up with the other states.

Thus, here, at the Faculty of BERG we decided to start with the creation of specialized centers of mining tourism. In the first part, we devote our time to the regions of Spiš and Gemer. We expect that after the formation of the first centre, all—the society itself, mining organizations and mining authorities will join us in our iniciative for the maintenance of such a typical feature of Slovak history—mining.

At the moment we work together with the graduands to finish the next type of a centre of mining tourism, where we map historical mines, technological relicts and remains after the activities of famous families, dealing mainly with mining and development of microregion belonging to a chosen annular area, where we map everything in regular circular structures around the chosen middle point of the centre. The project suggests a unique and interesting product that could be appealing for the tourists mainly from Slovakia, Hungary and Germany.

In 2016 we would like to introduce to the public the third type of the centre of mining tourism, which is already prepared and is based mainly on an up-to-date presentation of vertical and horizontal underground work in this region of historical mining in Spiš region created in various historical periods, together with geological creation of this area. A walk in this terrain with a tablet computer will offer up-to-date presentation and information about the past in confrontation with the present time, which will be depicted for the visitor "in real surrounding", where a walking visitor can see the underground together with the situation on the surface and historical mining work with geological buildings.

We suppose that these projects will "kick off" a huge potential of mining tourism in Slovakia.

REFERENCES

Ballesteros, E.R & Ramirez, M.H. 2007. Identity and community-Reflections on the development of mining heritage tourism in Southern Spain. *Tourism management*, vol. 27, No. 3, pp. 677–687.

Conlin, M.V. & Jolliffe L. 2010. *Mining Heritage and Tourism: A Global Synthesis*, Taylor & Francis e-Library, pp. 158–170.

Conesa, H.M., Schulin, R. & Nowack, B. 2009. Mining landscape: A cultural tourist opportunity or an environmental problem? The study case of the Cartagena—La Unión Mining District (SE Spain). *Ecological Economics*, Vol. 64, No. 4, pp. 690–700.

Drebenstedt, C. & Domaracká, L. 2011. Mountain tourism in Post Mining Region - Case study Lusatian Lignite Region. *Acta Geoturistica*, vol. 2, No. 2, pp. 8–16.

Hronček, P. 2011. Mining of brown coal in the southern Slovak brown coal basin in the inter-war period, *Historicky Caspis*, Volume 59, Issue 1, Pages 57–79.

Klaučo, M., Weis, K., Stankov, U., Arsenovic, D. & Marković, V. 2012. Ecological significance of land-cover based on interpretation of human-tourism impact. A case from two different protected areas (Slovakia and Serbia), *Carpathian Journal of Earth and Environmental Sciences*, Volume 7, No. 3, pp. 231–246.

Pavolová, H., Bakalár, T. & Štrba, Ľ. 2014. Model for the assessment of competitiveness of geotourist destinations in Slovakia. *Acta Geoturistica*, vol. 5, No. 2, pp. 31–36.

Pretes, M. 2002. Touring mines and mining tourists, *Annals of Tourism Research*, Vol. 29, No. 2, pp. 439–456.

Rybár, P. 2015. *Mining heritage*—Polish students work assignment. Unpublished manuscript. Krakow.

Rybár, P., Molokáč, M. & Hvizdák, L. 2015b. Návrh na vybudovanie centra banského turizmu, Manuscript.

Rybár, P., Molokáč, M. & Kovács, K.Z. 2013. Banská cesta hornouhorskými banskými mestami. Barangolás a Felsó-magyarországi Bányavárosokútjan. Upper Hungarian Mining Route. Milagrossa Kft, Miskolc, 224 p.

Vargas-Sanchez, A., Plaza-Mejia A. & Porras-Bueno N. 2009. Understanding Residents' Attitudes toward the Development of Industrial Tourism in a Former Mining Community. *Journal of Travel Research*, vol. 47, No. 3, pp. 373–387.

Production Management and Engineering Sciences – Majerník, Daneshjo & Bosák (Eds)
© *2016 Taylor & Francis Group, London, ISBN: 978-1-138-02856-2*

Transformation of HCS model 3E in IMS in context of sustainable CSR

P. Sakál, G. Hrdinová, H. Fidlerová & M. Šujaková
Faculty of Materials Science and Technology in Trnava, Slovak University of Technology in Bratislava, Bratislava, Slovak Republic

ABSTRACT: The article answers the basic questions about the sustainability of life on the planet Earth: "Why is it important to change the strategy of unlimited growth for the strategy for sustainability?" This article summarizes results and is also continuation of successfully finished projects VEGA 1/9099/02: "Environmental oriented management, marketing and logistics of SBU", project LPP-0384-09 "Concept of HCS model 3E vs. Concept of CSR" and KEGA 037STU-4/2012: "Implementation of the subject" "Sustainable CSR into the study programme industrial management in the second degree at MTF STU Trnava.". Our ambition is to provide the professional community to discuss 'The proposal of the methodology for the concept of sustainable CSR strategies development for SMEs in the context of HCS model 3E" by Gabriela Hrdinová in her dissertation and *"The concept HCS model 3E vs. the concept of Corporate Social Responsibility (CSR)".*

1 INTRODUCTION

Sustainable Corporate Social Responsibility is currently for enterprises and people around the world a great opportunity for sustainable change from their strategies of unlimited growth towards "win-win" strategy.

We call for change of thinking paradigm based on our experiences from Slovak university of Technology and praxis of industrial enterprises in Slovakia.

Therefore as noted in the final report of finished project KEGA and submitted but unauthorized research project VEGA in 2014: "*We will continue in our research activities in the area of corporate social responsibility, sustainable corporate social responsibility and sustainable development.*"

2 HCS MODEL VERSUS CSR

HCS model 3E of local ergonomics program, specific for countries of the Central and Eastern Europe, was created within the USA—Slovak cooperation on the basis of results from 4-years four cooperation in the project APVV no. 019/2001: "*Transforming Industry in Slovakia through Participatory Ergonomic*". Authors K. Hatiar, T. Cook and P. Sakál (Hatiar et al. 2015) focused themselves on man as object and subject of all efforts and the concept where working environment should contribute to building the quality of working life for everybody with sustainable quality of environment and adequate economic conditions for the overall quality of human life. This objective interconnection of three E (HCS model 3E) is not understood or admitted by many stakeholders and lobbyists in Slovak industrial companies for mostly their own economic reasons (Hatiar et al. 2015).

Both mentioned models, HCS 3E and CSR, have the same three pillars: *social, economic and environmental*. The concept of sustainable corporate social responsibility (SCSR) (Hrdinová 2013) is based on the concept HCS model 3E, which can be defined as a micro solution of macro—problem (in terms of known theorem *"think globally—act locally"*).

According Hatiar, Cook, Sakál (Hatiar et al. 2015) difference is that the model HCS 3E is focused more on the material, respectively on the production aspects of the sustainability and an understanding of human being as an object and at the same time as the subject of all efforts, the current CSR is aimed more on the prestigious, philanthropic and marketing site of it in relation with stakeholders.

3 THE INDUSTRIAL (TRADITIONAL) MODEL OF ECONOMY VERSUS THE SUSTAINABLE (NEW) MODEL OF ECONOMY

According to Hrdinová (2013) an industrial model of economy based on the large industrial factories after 150–200 years had no more the possibilities of development.

The new business model is based on the change in the production and distribution of the added value—concept of Created Shared Value (CSV), transition from "strategy win—lost" to "strategy win—win", respectively to "non-zero sum game".

This new business model is the methodological basis of our concept proposal of SCSR system methodology for SMEs in the context of HCS model 3E.

4 CREATED SHARED VALUE (CSV) VERSUS CORPORATE SOCIAL RESPONSIBILITY (CSR)

Professor Zbyněk Pitra in the Czech Management Association (www.cma.cz) discussed the idea of American professors Porter, M. and Kramer, M. *how to save advantages of the capitalist system by modifying the concept of CSR to the concept of CSV (Creating Shared Value).*

According to Pitra and Řezáč (Pitra 2011; Řezáč 2012), *the concept of raising social benefit* "(CSV—Creating Shared Value) differs significantly from the now widely frequented *"concept of social responsibility of organizations"* (CSR—Corporate Social Responsibility) as follows, Table 1.

The concept Creating Shared Value (CSV) is the basis for the proposal of concept Sustainable Creating Shared Value (SCSV) (Hrdinová 2013).

Table 1. Comparison of CSR and CSV.

Concept CSR	Concept CSV
Added value: financial support, socially beneficial activities	Added value: the business and societal benefits of the main activities
Expression: philanthropic activities, care for the environment, corporate citizenship	Expression: Cooperation in the development of shared values (business and social)
Implementation: As a voluntary autonomic response to the demands of temperature, independent of the effort to increase business performance	Implementation: An integral part of the implementation a business strategy, in direct connection with the performance of financial objectives
Agenda: is confined by external standards and requires separate accounts within the department to set up	Agenda: is a part of specific (individual in any organization) actions contributing to the implementation of business plans by standard departments

5 PROPOSAL OF METHODOLOGY FOR THE STRATEGY OF SUSTAINABLE CSR FOR SMES IN THE CONTEXT OF HCS MODEL 3E

The definition of the *HCS model 3E* concept means that (Hrdinová, 2013):

$$\text{HCS model 3E} \approx (\text{SWQ} \wedge \text{SPQ} \wedge \text{SLQ}), \qquad (1)$$

where:
$\approx$—symbol of equivalence,
$\wedge$ —(and), symbol of conjunction,
SWQ—sustainable work quality,
SPQ—sustainable production quality (of goods and services),
SLQ—sustainable life quality.

From the document *"National Quality Programme of the Slovak Republic"* (NQP for the years 2013–2016) considers that:

$$\text{NQP} \approx (\text{SWQ} \wedge \text{SPQ} \wedge \text{SLQ}) \qquad (2)$$

According to the established methodology of the concept of sustainable CSR strategy for SMEs in the context of the concept HCS model 3E, Figure 1 (Hrdinová, 2013):

$$\text{SD/PSD} \supset \text{SCSR} \supset \text{IMS} \qquad (3)$$

i.e., the IMS is the subsystem ($\supset$) of SCSR and SCSR is the subsystem ($\supset$) of SD/PSD.

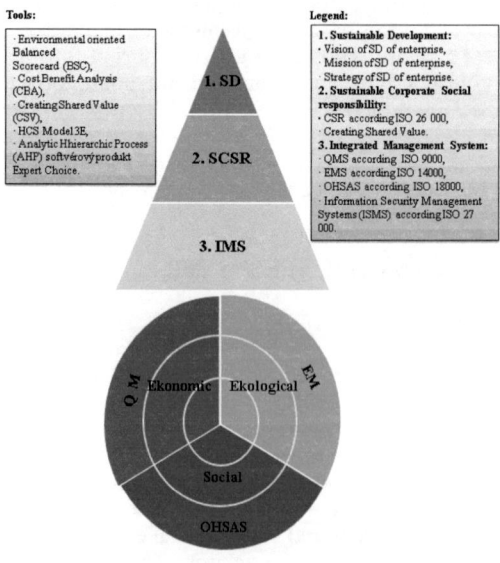

Figure 1. Proposal of the methodology for the concept of sustainable CSR strategies development for SMEs in the context of HCS model 3E concept.

Therefore if SD is according to ISO 26000,

$$SD \approx (Soc, p \wedge Env, p \wedge Econ, p), \qquad (4)$$

where:
    ⊃—symbol of subsystem,
    Soc, p—the social pillar,
    Env, p—the environmental pillar,
    Econ, p—the economic pillar,

then $SCSR \approx (Soc, p \wedge Env, p \wedge Econ, p).$ (5)

The hierarchy of subsystems SD, CSR and IMS is showed Figure 1 (3). Ground plan represents the necessity the (equivalence) SCSR and its three pillars—the relations (4) and (5). This means that all subsystems SD, SCSR and IMS should contain all three pillars: social pillar, the environmental pillar, the economic pillar (it is in our opinion form methodological/system/holistic perspective very significant).

On the left side of Fig. 1 are presented exact tools and concepts for the effective functioning of SCSR (recommended and used in our research). Peter Ponický (2012) indicated in the title of his presentation *"Quality is not result of using the quality tools"*. Our words are: if the fundamental systemic issues are not resolved then in solving partial problems we will always struggle with unresolved systemic issues. This of course means that the desired goal is beyond reach and our effort is already pre-inefficient and doomed to failure. Cited author Ponický asks a question in his article: *"Who and what constitutes the quality?"* and answers: *"People and their relationship!"* We also add "Change of thinking paradigm!*"*

From standards of STN ISO follows that:

$$IMS \approx (QMS \wedge EMS \wedge OSHAS \wedge ...), \qquad (6)$$

where:

    QMS—Quality Management System (STN ISO 9000),
    EMS—Environmental Management System (STN ISO 14000),
    OSHAS—Occupational Safety and Health Management System (STN ISO 18000).

Almost all historically incurred definitions of CSR contain three components/aspects/dimensions: economic, social and environmental.

The definition of CSR in the standard STN ISO 26000 does not contain an economic component. Holistic/systemic approach is in definition of CSR and SD is broken. If a SD is a system, then CSR as his subsystem has to be composed of the same elements. It has to have all three dimensions: economic, social and environmental, which are interconnected!

In the before mentioned the concept (Fig. 1), the sustainable CSR depends on application of concepts and tools as:

$$SCSR = f(HCS \text{ model } 3E; CSV; CBA; \\ BSC; AHP, IMS, ...) \qquad (7)$$

where:
    f—is a symbol of function.
    From modified Porter value chain shows, that:

$$SCSR \approx (SLog, in. \wedge SPro. \wedge SLog, out. \wedge SM \& \\ S \wedge SCS \wedge SSS \wedge SDTg. \wedge SMHR \wedge \\ SBI \wedge SMP),$$

$$(8)$$

where:

SLog, in.—Sustainable Input Logistics,
SPro.—Sustainable Production,
SLog, out.—Sustainable Output Logistics,
SM&S—Sustainable Marketing and Sales,
SCS—Sustainable Customer Services,
SSS—Sustainable Supply Services,
SDTg.—Sustainable Development of Technology,
SMHR—Sustainable Management of Human Resources,
SBI—Sustainable Business Infrastructure,
SMP—Sustainable Margin/Profit.

Then, according (Kuldová 2012) the social impacts of the modified Porter's value chain in formal logical terms are as follows:

$$SLog,in. \approx (emission \wedge load \wedge reduction \ of \\ waste \wedge reduction \ of \ technological \\ downtimes \wedge support \ supplier's \\ activities \wedge ... \wedge) \qquad (9)$$

$$SPro. \approx (emissions \wedge waste \wedge biodiversity \wedge \\ environmental \ impacts \wedge energy \wedge \\ water \wedge OSH \wedge hazardous \ materials \wedge \\ non\text{-}waste/clean \ technology \wedge ... \wedge)$$

$$(10)$$

$$SLog,out. \approx (impacts \ of \ waste \ handling \\ in \ packaging \wedge impacts \ of \\ transport \wedge ... \wedge) \qquad (11)$$

$$SM\&S \approx (marketing \ a \ non\text{-}misleading \\ advertising \wedge pricing \ practices \wedge ... \wedge)$$

$$(12)$$

$$SCS \approx (liquidation \ of \ obsolete \ products \wedge \\ treatment \ of \ motor \ oils \wedge treatment \\ of \ printing \ inks \wedge handling \ with \\ personal \ data \ of \ customers \wedge ... \wedge) \quad (13)$$

$$SSS \approx (public \ procurement \wedge control \ of \ supply \\ chains \wedge rejection \ of \ child \ labor \wedge rejection \\ of \ bribery \wedge fair \ prices \ for \ farmers \wedge ... \wedge)$$

$$(14)$$

SDTg. ≈ (a partnership with universities ∧
          ethical research practices ∧
          product safety ∧ preserving raw
          material ∧ recycling ∧ ... ∧)   (15)

SMHR ≈ (education and training of staff ∧
          diversity management ∧ health care
          and other benefits ∧ out placement ∧
          ... ∧)

                                                  (16)
SBI ≈ (financial reporting ∧ government
          practices ∧ transparency ∧ lobbying
          ∧ ... ∧)   (17)

SMP ≈ CSV   (18)

This means that sustainable margin/sustainable profits can only be achieved by changing the paradigm of creation and distribution of wealth/profit—using the concept of CSV.

Furthermore, according to (Kuldová 2012) its own processing is included sustainable technologies in the modified Porter's value chain in formal logic as:

SLog,in. ≈ (STg transport ∧ STg material handling
          ∧ STg storage and warehousing
          ∧ STg communication system
          ∧ STg testing ∧ STg information system
          ∧ ... ∧)

                                                  (19)
SPro. ≈ (STg basic manufacturing process
          ∧ STg material ∧ STg machine tools
          ∧ STg material handling ∧ STg packaging
          ∧ STg maintenance ∧ STg testing
          ∧ STg building design and operation
          ∧ STg information system ∧ ... ∧)

                                                  (20)
SLog,out. ≈ (STg transport ∧ STg material
          handling ∧ STg packaging
          ∧ STg communication system
          ∧ STg information system ∧ ... ∧)

                                                  (21)
SM&S ≈ (STg advertisig and promotion
          resources ∧ STg audio and video
          ∧ STg communication system
          ∧ STg information system ∧ ... ∧)   (22)

SCS ≈ (STg diagnostic and testing
          ∧ STg communication system
          ∧ STg information system ∧ ... ∧)   (23)

SE ≈ (STg communication systems
          ∧ STg information systems
          ∧ STg transport systems ∧ ... ∧)   (24)

SDTg ≈ (STg products ∧ STg product design
          using PC ∧ STg experimental factory
          ∧ STg pre software development
          ∧ STg information system ∧ ... ∧)

                                                  (25)
SMHR ≈ (STg training ∧ STg motivation
          research ∧ STg information
          system ∧ ... ∧)   (26)

SBI ≈ (STg information systems ∧ STg planning
          and budgeting ∧ STg administrative
          works ∧ ... ∧)

                                                  (27)
where:
    STg—sustainable technology.

Based on the system analysis of relevant documents and concepts can be argued that:

REPORT ⊃ NPQ ⊃ SD ⊃ SCSR   (28)

ie, that SCSR is a subsystem of SD, SD is a subsystem of NPQ and NPQ is subsystem of REPORT.

REPORT ≈ (Soc, p ∧ Env, p ∧ Ec, p)   (29)

In our opinion then are at least formally (de jure) created the basic systematic requirements for the implementation of sustainable strategy of CSR in the context with the HCS model 3E in enterprises (SMEs) in the EU and in Slovakia. Also at least theoretically, are *formed European basic systematic requirements* for the *Proposal of the methodology for the concept of sustainable CSR strategies development for SMEs in the context of HCS model 3E concept* and can be state that it moved from position "utopia" to position of "reality".

5.1 *Proposal for the implementation of SD concept to the business strategy*

In the management and in strategy of the enterprise according to the principles of sustainable development (Hrdinová 2013) is required implementation in the following seven steps:

1. Implementation of stakeholder analysis.
2. Determining the policies and objectives of the SD.
3. Designing and executing an implementation plan.
4. Creating a supportive corporate culture.
5. Establishing limits and performance standards.
6. Message Processing.
7. Improving internal monitoring processes.

A comprehensive sustainable development strategy of an enterprise is then a strategy of SD of enterprise, with the following structure (Hrdinová 2013):

1. Introduction.
2. Analysis of the overall situation.
3. Vision for SD of enterprise based on a new business model, which focuses on improving social benefit of business assets.
4. SD Mission of enterprise based on SD vision.
5. Defining of strategic objectives for SD of company.
6. Identification of the stakeholder's strategic objective.
7. Analysis of the external business environment.
8. Analysis of business environment of sector
9. Analysis of the internal business environment.
10. SWOT analysis of the company.
11. Creating the vision, mission and strategic objectives for the SD strategy of enterprise.
12. Creating the vision, mission and strategic objectives for business strategies of strategic business units – SBU.
13. Creating the vision, mission and strategic goals of sustainable functional business strategies.
14. Sustainable marketing strategy of enterprise.
15. Sustainable manufacturing strategy of enterprise.
16. Sustainable innovation strategy of enterprise.
17. Sustainable strategy for ICT of enterprise.
18. Sustainable human resources strategy of enterprise.
19. A sustainable financial strategy of enterprise.
20. A sustainable environmental business strategy.
21. The organization of the company.
22. Risk management in the enterprise.
23. Change management in the enterprise.
24. Joint ventures.
25. Implementing of the strategy SD.
26. Control a fulfillment of SD strategy.

In short, it is possible to describe the process of creating a comprehensive strategy for SD in enterprise, formally represented in 4 phases of strategic management as a process, according to (Hrdinová, 2013) as follows:

$$PC\,S\,SD \approx (ANALYSIS \wedge CS\,SD \\ \wedge IMPLEMENTATION \\ \wedge CONTROL) \qquad (30)$$

where:

PC S UR—the process of creating a comprehensive strategy for SD in enterprise,
ANALYSIS—analysis of the external business environment, business environment of sector, internal business environment, SWOT analysis,
CS SD—creation strategy for SD (vision, mission and corporate, business, functional strategies),

IMPLEMENTATION—implementation of the strategy SD,
CONTROL—check a fulfillment of strategy for SD.

It is possible to present in process of the development a comprehensive SD strategy in enterprise according to (Hrdinová, 2013) as follows:

$$CS\,SD \approx A1 \to A2 \to A3 \qquad (31)$$

where:

$\to$ symbol of implication ("if"—"then")
A1—operator A1—justify the social meaning of existence of company (specification of long-term social and business mission of the company—the acceptance of change in a thinking paradigm towards SD and sustainable CSR by critical stakeholders)),
A2—operator A2—generate insights into validation in specific socio-economic conditions (creation of vision SD and sustainable CSR about the future—in the medium term—business success),
A3—operator A3—bringing this vision to life by selecting of specific strategic business objectives and implementation scenarios for their success in achieving (formulation of a sustainable business strategy and build implementation plans pursued its goals).

If a Business Entity (BE) needs a prospering Local Community (LC), then the Local Community (LC) also needs a Successful Entity (SE).

Thus, the following applied:

$$BE \approx LC \qquad (32)$$

5.2 *Proposal for the implementation of the concept of sustainable CSR strategy to the strategic management of the company*

Implementation of the concept of sustainable CSR strategy to the system of strategic management of the company (as a subsystem of SD strategy in enterprise) is a long, difficult, time-consuming, effort, new information and knowledge and change of thinking requiring complex task. The main purpose of implementing the concept of sustainable CSR strategy to the strategic management of the company is the successful integration of the various areas of CSR (economic, social and environmental) to the vision, mission, values and strategic objective and values of the enterprise to its culture and operational decisions at all levels of management, so the responsible approach ought to promote a long-term success on the market.

Hrdinová (Hrdinová 2013) suggests using the procedure of implementation of the concept of sustainable CSR strategy into practice of strategic management for (not only industrial) enterprises, according to (Steinerová et al. 2008, Steinerová 2015), which has been modified, implemented and verified in (Hrdinová 2009):

1. Management commitment.
2. Identification of key stakeholders.
3. Determination of values and principles.
4. Analysis of the current situation.
5. Action plan.
6. Implementation.
7. Monitoring.
8. Reporting.
9. Measures for improving.

5.3 The proposal of inclusion of the concept of sustainable CSR strategy in IMS in business practice

Sustainable CSR, according to (Hrdinová 2013) is actually a strategy that an enterprise should create and follow it. Besides three main parts of which it is consists SD, i.e. "Triple-bottom-line", has the following three pillars: economic, environmental and social. There are also other parts of the IMS enterprise that also in coherence with CSR itself consists the overall company strategy. The concept of sustainable CSR strategy should cover all processes within the company. It is an integral part of the strategy for SD in industrial and other companies/ organizations. In article (Hrdinová et al. 2011) the authors propose the inclusion of the concept SD into the strategy SD of enterprise. In this authors presented a so-called. "GRC approach", (Broady et al. 2008) which is the abbreviation of three English words: G—governance, R—risk management and C—compliance, and especially to the uninitiated Greek acronym. G—means Governance—leadership, the government, power. R—means Risk. C—means Compliance with the many laws and guidelines affecting business (and residents).

$$GRC \approx (G \wedge R \wedge C) \tag{33}$$

where:

G—Governance (leadership, the government, power),
R—Risk,
C—Compliance (Control).

From the above entry indicates that sustainable CSR strategy can not only exist as a single strategy in the company. Always represents one of the components of strategic business management. It is an integral part of such a comprehensive strategy for SD in enterprise together with other management systems—IMS. Together, they create in symbiosis and can not fulfill their function without others.

6 CONCLUSIONS

Aim of this article is to offer a solution for application of EU recommendations outlined in Report on corporate social responsibility: accountable, transparent and responsible business behavior and sustainable growth 2012/2098(INI)), p. 11: *"The European Commission proposes a new strategy for a new definition of CSR as 'the responsibility of enterprises for their impacts on society'. In line with this definition, businesses need to have in place a process to integrate social, environmental, ethical, human rights and consumer concerns into their business operations. The aim of that process must be to create shared value for their owners/shareholders and for their other stakeholders and society at large and to identify, guard against and mitigate the possible adverse effects of corporate activities".*

Gabriela Hrdinová in (Hrdinová 2013) proposed the concept and we (at STU MTF with seat in Trnava) are finalized this concept to a successful practical application through cooperation with Slovak industrial enterprises by addressing the bachelor's, master's and doctoral theses.

ACKNOWLEDGEMENT

The paper builds on the results of APVV project No. LPP-0384-09: "Concept HCS model 3E vs. Concept CSR" and KEGA project No. 037STU-4/2012: "Implementation of the subject of "Corporate Social Responsibility Entrepreneurship" into the study programme of Industrial Management in the second degree of study at STU MTF Trnava". The paper is a part of VEGA project No. 1/0448/13: "Transformation of ergonomics program into the company management structure through integration and utilization and QMS, EMS, HSMS".

REFERENCES

Broady, D.V. & Roland, H.A. 2008. SAP GRC for Dummies. Indiana: Wiley Publishing, 2008. 342 p.

Fidlerová, H., Hrdinová, G., Sakál, P. & Šmida Ľ. 2014. Udržateľný strategický manažment vs. udržateľné spoločensky zodpovedné podnikanie vs. integrovaný manažérsky systém strategických podnikateľských jednotiek—Charakteristika predloženého projektu. *Výkonnosť podniku.* Výskumný ústav ekonomiky a manažmentu < http: //www.vusem.sk/public/userfiles/files/VP1_2014.pdf > [Cited: 11.3.2015].

Hatiar, K. Cook, T. & Sakál, P. 2015. *HCS model 3E účastníckej ergonómie. "HCS 3E" Model of participatory ergonomics.* Trnava: Internetový časopis MtF STU. <http: //www.mtf.stuba.sk/docs/internetovy_casopis/2006/3/hatiar.pdf > [Cited: 11.3.2015].

Hrdinová, G., Drieniková, K., Naňo, T. & Sakál, P. 2011 Sustainable CSR—the Integral Part of Sustainable Development Strategy of Industrial Business. *Transactions of the Universities of Košice, Look Days 2011*: International Scientific Conference, 13.–14. October 2011, Aula Maxima, Technical University of Košice, Slovakia. pp. 1–16

Hrdinová, G. 2009. *Analýza možností využitia a návrh systému spoločenskej zodpovednosti firiem (CSR) vo firme CHIRANA PROGRESS, s.r.o. PIEŠŤANY.* [Diploma thesis]—STU MTF in Trnava, – Supervisor: Prof. Ing. Peter Sakál, CSc., Trnava: MtF STU, 87 p.

Hrdinová, G. 2013. *Koncept HCS modelu 3E vs. Koncept Corporate Social Responsibility (CSR).* [Dissertation thesis]—STU MTF in Trnava;—Supervisor: Prof. Ing. Peter Sakál, CSc.—Trnava: MtF STU, 228 p.

Kuldová, L. 2012. *Nový pohled na společenskou odpovědnost firem. Strategická CSR.* Plzeň: NAVA, nakladatelská a vydavatelská agentura, 173 p.

National Quality Programme of the Slovak Republic. Národný program kvality SR na obdobie rokov 2013–2016. <http: //www.unms.sk/?narodny-program-kvality_SR_-na-obdobie-rokov-2013–2016> [Cited: 11.3.2015].

Pitra, Z. 2011. *Jak zachránit přednosti kapitalistického systému?* Praha: Česká manažerska organizace, 2011. 23 p.

Ponický, P. 2012. Kvalita nie je výsledkom použitia nástrojov kvality. *Zborník 9. Národnej konferencie o kvalite "Manažérstvo kvality vo verejnej správe".* Bratislava: MH SR, Spoločnosť pre kvalitu, 2012. 367 p.

Porter, M.E. 1993. *Konkurenční výhoda: jak vytvořit a udržet si nadprůměrný výkon.* VICTORIA PUBLISHING, a.s., 626 p.

Report on corporate social responsibility: accountable, transparent and responsible business behavior and sustainable growth 2012/2098(INI)). http: //www.europarl.europa.eu/sides/getDoc.do?pubRef = -//EP//TEXT+REPORT+A7–2013–0017+0+DOC+XML+V0//EN [Cited: 25.3.2015].

Řezáč, J. 2012. Nový model podnikáni a změna paradigmatu managementu. Praha, 2012.

Steinerová, M. & Makovski, D. 2008. Koncept CSR v praxi – průvodce odpovědným podnikáním. Košice, 2008. 33 p. [Cited: 11.3.2015]. <http: //www.blf.cz/doc/brozura_CSR_web_CZ.pdf >.

Production Management and Engineering Sciences – Majerník, Daneshjo & Bosák (Eds)
© 2016 Taylor & Francis Group, London, ISBN: 978-1-138-02856-2

Options of using the integrated management system

S. Štofko, V. Šoltés & Z. Štofková
University of Žilina, Žilina, Slovak Republic

ABSTRACT: The goal of every entrepreneur is to make a profit. In order to achieve maximal profit, companies have to create and distribute high-quality products and services to their customers. While doing so, companies have to comply with legal requirements; ensure occupational health and safety for their employees; they have not to endanger the environment with their activities. An important factor while making a profit is monitoring of competition and competitive environment; appropriate advertising, well set marketing plan as well as many other business processes. Integrated management system represents an appropriate mean of coping with all of the aforementioned activities as well as efficient management of an organization.

1 INTRODUCTION

Companies and organizations are increasingly forced to deal with the quality and stability of their processes as well as management and continuous improvement of their processes. This enables them to sufficiently fulfill needs of their regular customers, retain them, but also attract new customers. However, customers in general are increasingly more demanding; therefore companies have to upgrade their products to make them more attractive to their customers. Success of company's operations should not be at the expense of overlooking company's level of security, occupational health and safety, environmental protection and at last but not the least also the quality of products and services. The integrated management system is the key element to make the company act responsibly and take care of prevention.

Using aforementioned principles, a company is able to avoid unnecessary financial losses caused by a violation of, or failure to comply with, legal requirements. Establishing and applying the integrated management system provides easier position in terms of today's competitive pressures. With regard to the product development and its continuous improvement, technical and technological development also occurs. These modern technologies, however, have in many cases negative impact on the environment. Therefore, companies, while meeting needs of their customers and manufacturing their products, are being forced by society to emphasize environmental care and protection through legislation.

2 INTEGRATED MANAGEMENT SYSTEM

The basis for the creation of the Integrated Management System is formed by penetrating through the implementation of international standards relating to quality management, environmental management and occupational health and safety management. The Quality Management System, which other management systems based on process approach are bound on, is the main prerequisite of integration.

The key segments influencing the Integrated Management System include:

- customers and suppliers, who determine product quality requirements,
- society and the public, who determine environmental requirements,
- employees, who determine occupational health and safety requirements,
- shareholders, who determine the sources and information (Gašparík 2008).

The trend of multiple companies in the past was to minimize expenses and achieve the greatest possible profit at the expense of safety and failure to comply with legal requirements. High penalties and a risk of losing their reputation meant an overall high risk for companies, who therefore currently invest more resources to provide employee protection and simultaneously make their security policies in accordance with the legislation.

Whether a company sets up the Integrated Management System or supports the three systems as individual units is upon company's own

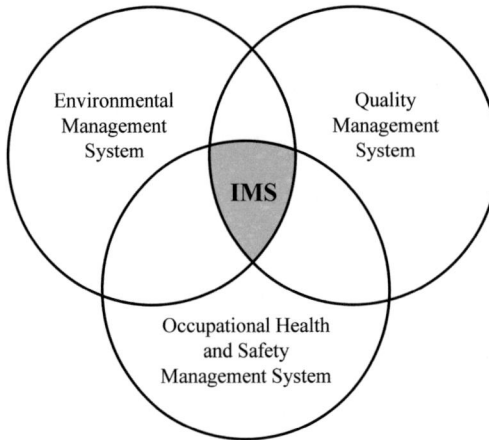

Figure 1. Scheme of the three systems in relation to the integrated management system.

decision. The logical advantage of the Integrated Management System, however, is that it reduces administrative demands saving financial resources. A systematic approach also helps with the implementation of legal requirements to all areas of a company (National 2013). Figure 1 above shows relationship between the three individual management systems and the integrated management system.

Except of reducing expenses, application of the integrated management system also has important role in reducing risks, increasing profit, formalization of the information system, optimizing procedures, improving communication but also eliminating conflict responsibilities and increasing business objectives (Veber 2006). Using this system, a company minimizes its negative impact on the environment, reduces the extent of threatening the surroundings and occurring of injuries, improves company's image through certification of the systems, is able to meet demands of its customers, streamlines its operating expenses and organizational structure, complies with legal requirements and foremost gains advantage against its competition (Majerník 2013b).

The integrated management system thus incorporates three systems, which are governed by international standards. The Quality Management System is governed by standard ISO 9001, the Environmental Management System by standard ISO 14001, the Occupational Health and Safety Management System by standard OHSAS 18001. The universality of these standards is proven by the fact that they may be used in all types of manufacturing or non-manufacturing organizations, as well as in public administration organizations (Vítek 2007).

The aforementioned three key systems may be, if necessary, expanded by supportive management systems, which include:

• Information Security Management System (ISO/IEC 27001),
• Corporate Social Responsibility Management System (SA8000),
• Food Safety Management (ISO 22000) (Hrubec 2009).

2.1 Process of implementing integrated management system

The starting point for building the integrated management system is a resolution of company's top management. However, before the whole process of building the system begins, it is vital to perform an audit which would assess compliance of current systems with international standards. If so, it is possible to proceed to the process of implementing the integrated management system, of which there are two methods (Mateides 2006).

The first method is a gradual implementation of the system with the first step being building the Quality Management System, which is then succeeded by implementing the Environmental Management System and subsequently the Occupational Health and Safety Management System. The second method includes assessment of requirements for each individual system in a form of internal audits and preventive measures integrate these systems and subsequently implement the integrated management system as a whole (Priesol 2007). System implementation is a gradual process which consists of many steps, which include:

• establishing company's policy, objectives and implementation modality,
• preparation of documentation, directives and orders,
• specification of resources (personal, financial, material) and responsibilities,
• ensuring an effective communication, staff training and familiarization with the system,
• gradual implementation, system checking and addressing weaknesses,
• system certification,
• system innovation, continuous improvement and support by the top management (Majerník 2013a).

System documentation must comply with international ISO standards and should be processed with regards on company's size, complexity of company's processes and employees' ability to handle individual processes. The foundation is technological documentation which includes procedures of protecting information in use. In terms of system documentation, archiving, updating as well as liquidation must also be ensured (Spejch-alová 2011).

2.2 *Use of the integrated management system*

A competitive environment, an economic crisis as well as a political situation significantly affects the business climate. In order to retain customers and achieve the maximum profit, companies are being pushed to implement innovative processes into their production. Therefore companies try to modify their organizational structures, reduce their staff numbers and streamline their processes (Štofková 2007). The implementation of the Integrated Management System is an ideal instrument allowing a company to maintain a competitive advantage and thus retain its customers. Therefore many small or large companies have in recent years acceded to the implementation of the integrated management system into their activities.

Well-known companies that have implemented this system include, for instance, Railway Company Slovakia, a.s., Škoda Slovakia, a.s. or the Transport Company Košice, a.s. These companies are the evidence that well-configured integrated management system is an important contributor in company's development. However, this system is not a cure for all problems a company experiences (Štofková, 2012). One example is a certain construction company that has implemented the integrated management system, but has a number of debts and related economic problems nevertheless.

3 SURVEY OF UTILIZATION OF THE COMPANY'S INTEGRATED MANAGEMENT SYSTEM

For the purposes of this paper, a survey was carried out among ten medium and large companies in Žilina Region. Manufacturing companies, wholesales as well as service companies were among those consulted. Figure 2 shows the number of companies with established systems.

In terms of the use of different systems among the addressed companies it may be stated that eight companies had all of the aforementioned systems implemented, one company did not have the Occupational Health and Safety Management System implemented and one company had only the Quality Management System implemented. The following figure 3 expresses the view of the companies with at least two types of systems implemented, addressing a question whether it is more appropriate to institute the system integrated or individually.

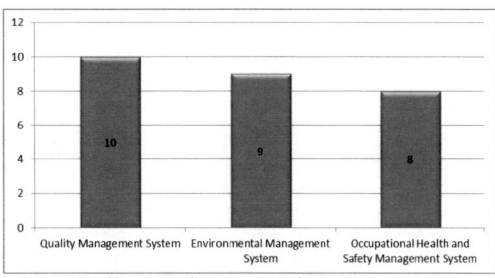

Figure 2. Implemented system in companies.

Large manufacturing companies prefer and recommend instituting the system gradually. The main reason is reported as the ability to adapt to the demands of their customers, business partners and investors. Five large companies of the surveyed companies recommended to implement the integrated management system gradually, four smaller companies recommended instituting the system integrated.

All of the companies reported that the main reason for implementing the system into operation was the promotion of customers' trust. In fact, to fulfil his needs, a customer usually prefers a certified company rather than an uncertified company. Half of the companies note that it is appropriate to introduce the system due to the competitive advantage. Four companies identified cost reduction as the reason for the implementation of the integrated management system and identically four companies identified the possibility of increasing company's security.

The process of acquainting employees with the system is similar in all of the surveyed companies. Detail information is posted on the notice boards and on the intranet in form of directives and Director's decisions. In addition, employees may ask managers questions about the system and are regularly retrained at each change of the system, directive or legislation.

One of the options is to create a consulting body of employees for the system, through which change proposals could get to be negotiated by everyone from employees up to top management, which could lead to streamlining the whole system.

The Quality Management System is the essential element of the integrated management system. In terms of its contribution to the companies, they were able to choose from seven areas they consider the most important. These were the following seven areas:

1. Ensuring corporate competitiveness by improving the quality of production,
2. Optimization of expenses while keeping the required degree of goods quality,

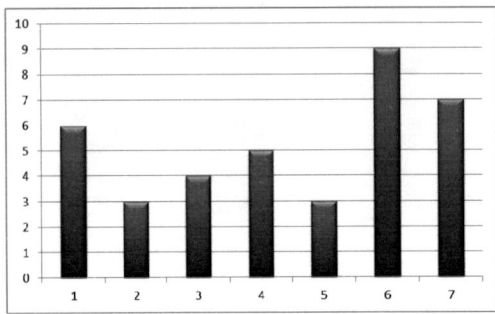

Figure 3.   Contribution of the quality management system to the companies (seven areas).

3. Reducing the number of defective products,
4. Increasing the marketability,
5. Reduction of claims,
6. Effective process management.

The Figure 3 expresses what companies considered as the greatest benefit of the system.

Being a means of effective process management in an enterprise is regarded by companies as the greatest contribution of the Quality Management System. Another contribution is the strengthening of responsibility and self-confidence of workers, followed by ensuring corporate competitiveness by improving the quality of production. The benefit of increased marketability of the products, reducing the number of defective products, reduction of claims and optimization of expenses while keeping the required degree of goods quality were all placed in the latter places of the chart.

There is an interesting finding in an automotive industry company in terms of the system's contribution. Since it is a relatively new company in Slovakia with new technologies and modern management approach, the first five sorts of contributions were recorded before the introduction of the Quality Management System. Only the last two types of contributions were recorded after the system's introduction.

The main reason for implementing this system into practice is to increase the competitiveness of a company, which was reported by 9 companies. Four companies indicated that they implemented the system due to the cost reduction, one company because of the risk of penalties for non-quality products and the observed deficiencies. None of the companies indicated the elimination of security risks as the purpose of the system implementation.

The implementation of Environmental Management System, however, does not have a genuine reason among companies. Logically, the biggest advantage is the protection of human health and the environment. Improvement of the company's image and demonstration of adequate care and compliance with the legislation were placed as lesser reasons. The least significant reason to implement this system into company's operations is a competitive advantage.

The main contribution of the Environmental Management System is, in particular, ability of the system to flexibly respond to any change in the legislation. Another advantage lies in reducing the number of environmental incidents which the company would be held responsible for, as well as minimization of fees and penalties for environmental pollution. The smallest contribution for the surveyed companies is a reduction of the expenses associated with company's liquidation.

The main contribution of implementing the Occupational Health and Safety Management System is mainly associated with the company's ability to manage risks in occupational health and safety area, which improves employees' performance. Smaller number of occupational diseases, minimization of the expenses associated with workplace accidents, improvement of the relations between the company's management and employees, trade unions, public and state institutions all belong to benefits of the implemented system. The smallest number of surveyed companies noted the ability of the system to decrease probability of getting penalized due to violation of legislation associated with occupational health and safety.

Two thirds of the surveyed companies claimed that in the field of Occupational Health and Safety Management System the state applies repression rather than prevention. According to three companies, the level of legal awareness in the field of occupational health and safety is being raised under the influence of the state; only one company claimed that the state is helpful in providing effi-

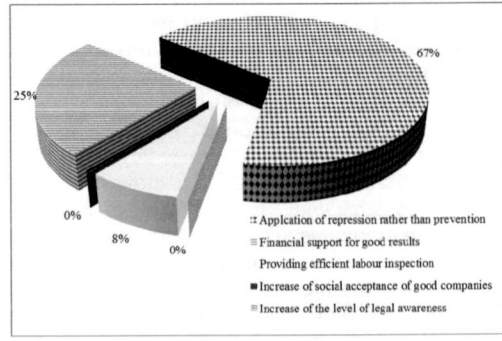

Figure 4.   Influence of the state on the Occupational Health and Safety Management System.

cient labour inspections and state supervisions at the enforcement of occupational health and safety. Figure 4 expresses aforementioned findings.

A very negative fact, however, is that none of the surveyed companies felt any support from the state. The companies would welcome increased interest of the state and would be interested in getting a certain form of reward from the state for having positive results. Naturally, the most intriguing is the financial form of reward.

All of the companies that had at least two of the systems implemented in their activities experienced a minimum of 50% increase of quality of their products compared to before the systems had been implemented. However, many different factors affect the overall contribution of the system. One of these factors is a time period during which the system is in place. The longer the system has been implemented, the better overview of company's processes the management should have and therefore the better results they should achieve. Another factor, however, is the correct setup of the system. No matter how perfect the system may be, it will never work if not set up properly.

4   CONCLUSION

The Integrated Management System is one of the most appropriate instruments for company's efficient functioning. Applying this system enables a company to grow, increase its competitiveness and stabilize itself. Through integration of individual system processes, the system is able to save financial resources and invest them, for instance, to the product quality, thereby meeting customers' needs. Organizational structure of a company is restructured under the influence of the system and a company is able to efficiently perform its activities while achieving objectives.

The integrated management system incorporates the Quality Management System, the Environmental Management System and the Occupational Health and Safety Management System. These individual subsystems are regulated by international standards. After meeting requirements of the standards the final audit may be carried out and then the competent authority may issue a certificate to the company. Approaches to implementation of the Integrated Management System may vary. A company, after carrying out an entry audit, makes a decision whether to implement the system gradually or at once as a whole. Larger companies advise to implement the system gradually, which enables the company to react operatively to the requirements of customers, investors as well as shareholders.

Most of the large companies use the Integrated Management System nowadays. However, not always is the system by itself sufficient enough for stabilization and growth of the company. Only a capable team of managers can set up the system so that it delivers benefits and not a burden. A systematic approach allows the company to increase the quality of their products while paying attention to environmental protection as well as protecting their employees.

According to the carried out survey it is obvious that there is quite a significant interest of implementing a certified integrated management system. At least two of the subsystems constituting the integrated management system were implemented in nine of the ten surveyed companies.

All of the surveyed companies had the Quality Management System implemented as it represents a means of efficient process management. Another advantage is improving the product quality, increasing competitiveness of a company, reducing the number of defective products and claims, thereby increasing satisfaction among the customers as well as increasing the profit.

The main motivation among the surveyed companies for implementation of the Environmental Management System was the protection of the environment as well as human health. Reducing the number of industrial accidents as well as fees and penalties for environmental pollution were other important elements in implementing the system.

The main contribution of implementing the Occupational Health and Safety Management System is the employee protection. The system enables the employees to fully concentrate on given tasks and improve their efficiency. In addition, the system reduces the number of workplace accidents, occurrence of mental diseases and improves overall relationship between top management and employees. A problem regarding this system is a weak support from the state, which, instead of prevention, rather applies repression. The state frequently levies heavy fines for deficiencies, however, uses no form of reward for exemplary fulfillment of the obligations.

From the overall standpoint, the reviewed companies praised the Integrated Management System and argued that the system allowed them to optimize their activities, which has improved their results. An important factor influencing their satisfaction with this system, however, is the length of time period during which the system had been implemented in the company, as well as already mentioned system setup and approach of the top management and employees.

ACKNOWLEDGMENT

This paper was undertaken, as parts of the research project VEGA 1/0787/14 and VEGA 1/0175/14.

REFERENCES

Gašparík, J. 2008. *Integrovaný manažérsky systém.* Tribun EU: Brno, 2008. 158 s.

Hrubec, J. & Virčíková, E. 2009. *Integrovaný manažérsky system.* Nitra. SPU, 2009, 543 s.

Majerník, M., Szaryszová, P., Hakulinová, A. & Markovič, J. 2013a. *Akreditácia, certifikácia, auditovanie.* Bratislava: Úrad pre normalizáciu, metrol ógiu a skúšobníctvo SR, 2013, 204 s.

Majerník, M., Tkáč, J., Bosák, M., Szaryszová, P. & Tarča, A. 2013b. *Environmentálne manažérstvo v integrovanom systéme.* Bratislava/Košice: EU 2013, 302 s.

Mateides, A. 2006. *Manažérstvo kvality.Hist ória, koncepty, met ódy.* Bratislava: Epos, 2006, 751 s.

National Quality Program of SR. 2013 [online], [quoted 17.2.2015]. Available at: http://www.unms. sk/?narodny-program-kvality-sr.

Priesol, J. 2008. *Stratégia implementácie integrovaného manažérskeho systému v organizáciách.* [on-line]. [quoted 26-03-2015]. Available at: http://www.jozef-priesol.sk/domain/ integrovany system/files/prispevok-online-i.pdf.

Spejchalová, D. 2011. *Management kvality.* Vysoká škola ekonomie a managementu, 2011. 211s.

Štofková, J. & Borkowski, S. 2007. *Praktyka zarządzania jakością wyrob ów i usług.* Sosnowiec: Humanitas, 2007, 122 s.

Štofková, K. & Štofková, J. 2012. New trends in management—a way to increase the competitiveness. *Advances in business-related scientific research conference,* Venice. Koper: Edukator, 2012.

Veber, J., Plášková, A. & Hulová, M. 2006. *Management kvality, environmentu a bezpečnosti práce. Legislativa, metody, systémy, praxe.* Management press, s. r. o., Praha, 2006.

Vítek, M. & Vítková, M. 2007. *Management: systémový přístup.* Gaudeamus: Univerzita Hdadec Králové, 2007. 289 s.

Production Management and Engineering Sciences – Majerník, Daneshjo & Bosák (Eds)
© 2016 Taylor & Francis Group, London, ISBN: 978-1-138-02856-2

Data analysis in quality management of the network enterprise

J. Štofková, I. Stríček & K. Štofková
University of Žilina, Žilina, Slovak Republic

ABSTRACT: The focus on quality is a basic supposition to gain a competitive advantage for the company not only by reducing the costs resulting from inefficient production and production errors, but also by gaining more customers resulting from products and services based on customer requirements and from material or substantial anticipated benefit. Before many companies achieve measurable rate of Return on Investments (ROI) of their current analytical capabilities and there are broken strengths which have influence on the resulting ROI and the surrounding pressure factors lowering its resulting value. Each entity engaged in an analysis have to deal now with the arrival of Big Data, not excluding the acceleration of technological innovation, increasing demand for more sophisticated predictive applications and with not an easy task of integration of analytical tools and methods to the majority of processes in the company.

1 INTRODUCTION

For the operation of good analytical practice it is necessary to have the right vision, a steady hand, precise planning and be prepared for flexible process management during implementation. Big amount of unstructured data, nowadays known as "Big Data", managed and analyzed by strong and sophisticated technologies belong to the one of the biggest consequences of the technological revolution of recent years. Constantly increasing volume of data is supported by large amounts of data collected by companies, social networks and the development of Internet applications and by Internet communicating sensors from different areas. These data have the potential to make an enormous value for companies that use the power of "big data" as a decision support tool.

At the conference of the world postal operators in 2014 there was a discussion about the exact definition of "big data" used in logistics or by postal operators, since there is no unified definition until now. In comparison to conventional data "big data" can be characterized by their capacity (which makes them difficult to process and to store), by their rate of production and variety of data sources, whether this data is structured or unstructured. The most commonly used is called "3Vs" definition of "big data" (velocity, variety, volume).

In addition to the basic definition two additional factors should be taken into account namely "value" value of the data for a specific business "Veracity" truthfulness that means quality, reliability and accuracy.

The area is ideal logistics sector for the use of benefits resulting from technological and methodological nature of "big data". Strong indications that the management of data processing is crucial in this business sphere are already hidden in the Greek meaning "practical arithmetic's" 1. Nowadays logisticians process massive flows of goods by which mean big amount of data is produced. Millions of consignments are transported daily and monitored from the place of departure to the place of delivery and record their weight, size, content and location worldwide. But are these data fully utilized?

In this data there is a huge unused potential for improving operational efficiency, customer satisfaction and creating new useful business models. As an example we can show the benefits of an integrated data network from various logistics operators that could remove the current fragmentation of the market and create stronger cooperation and also new types of services (Benninga 2014).

2 TRENDS IN DATA USE

The main users of analyzes are generally called C-suits (Chiefs—Chiefs), managers of department. The use of decision analysis as a tool started to increase since 2009 and is constantly progressing due to increasing implementation of analytical tools in companies. According to a survey carried out in the US and in the UK one third of companies claim that they strongly promote the use of analytical tools in their entire operations.

Less than 10% of companies said they do not promote the use of analysis because of lack of the data suitable for evaluation, lack of technological basis or lack of knowledge of their management.

Two-thirds of companies have for management and data processing in company authorized person of the data "Data chief manager" for the period of more than 18 months. Even companies that do not have anchored this type of job in their organizational structure, they plan to introduce it to more than 70% introduced in the near future.

The first survey in 2009 involved also the question why companies use analytical tools. Only 12% of surveyed companies used the analysis for the predictive monitoring of the development process, according to a new survey, this number has increased to more than 40%. This growth is caused mainly by increasing sophistication in data analysis options that are currently governed by the slogan: "rather to predict tomorrow, as to explain yesterday". Despite this wave of enthusiasm, the demand for predictive analysis is higher than its offer. While about a third of companies apply primarily predictive techniques automatically about twice as many respondents said that their customers require from these techniques in order to predict trends and not just for reporting of current events in the organization (Frost & Savarino 1986).

The situation of growing demand for analysis creates a hole in the market and there is a growing demand for capable analysts able to evaluate data in ever shorter intervals and to resist daily pressures despite limited resources. For example, one of the respondents said "...we had to analyze trends faster than in the past. Customers are more educated and they seek for new technologies that are efficient and flexible. We have analysts who can make fast and accurate decisions".

Another respondent said the impact of social networks on the marketing mix: "Three or four years ago, we did not talk about how to incorporate data from social media into our data structures, but now we do. The requirements are changing almost on a weekly or monthly basis (Hillard 2010).

Each entity should therefore focus on the collection and analysis in the following areas:

– Analysis of social media and digital interactions,
– Voice analysis from call centers
– Monitoring of customer experience through the web analysis,
– View of processes "from the sky" (geo-locational data)
– Understanding the patterns of physical movements from geo-location data,
– Understanding attitudes and behaviors (customers, employees, etc.).

According to the area of business activities, the usefulness of the data analysis is in the following proportions (Fig. 1).

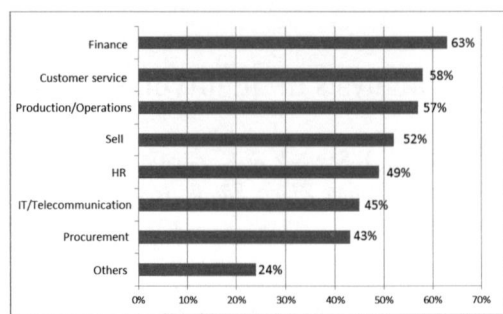

Figure 1.   Use of data analysis in the UK and USA by business area.

In Fig. 1, we can see that the operations in customer service and operations, which include activities of Postal companies, are moving to leading positions. After including financial operations, either through payment orders, delivery or service of postal bank, the operators can be classified to the sectors with the highest need for analysis and data processing (World Bank 2007).

3   BIG DATA AND POSTAL SERVICES

The transition to the so-called data oriented postal company is a continuous process, that is, according to several representatives of the biggest postal companies subject to the following key steps:

– taking into account the needs of customers—data strategies should have balanced structure between increasing efficiency and finding paths to the collection and generation of data increasing customer satisfaction and loyalty. For example, some postal operators began to cooperate with their corporate clients to determine which categories of "big data" should be collected and shared in order to increase their efficiency and productivity.
– raise awareness of the "big data"—DHL and La Poste recommended to postal companies to begin the use of "big data" by creating a senior management and co-conscious opportunities and challenges arising from them.
– Put innovation on the first place—"Big Data" strategy of DHL is the spiritual foundation for customized solutions and innovation division. They correctly identified the "big data" on the radar of logistic trend a few years ago and nowadays on their basis a wide range of applications is offered. With their help the process of new product development is managed from initial analysis and studies up to pilot tests and their discharge.
– Development of new skills—La Poste trained their own marketing experts who are learning to obtain the necessary data from massive

corporate data-bases. Postal companies are looking for recruiting of multidisciplinary experts with comprehensive knowledge—not just data analysis. These can be, for example, data scientists with a background in process, or the marketing area. There is also an increasing demand for lawyers specializing in the area of data privacy, the development of a transparent policy in this area. Finally, it is also necessary to build organizational links "bridges" between IT and other departments to ensure analytical integration across the enterprise.

- creating partnerships and new data services—DHL is building a large scale sharing and processing "big data" with research institutions and customers who actively participate in the research and development of prototypes. We also cooperate with other providers "big data" as telecom operators to promote DHL predictive analysis.
- Process Optimization: operations and sales—cost reduction resulting from more precise directions and estimated demand for shipments (dynamic routing). For example, DHL has developed a tool to analyze the correlation between external factors such as weather, epidemic of influenza, Google Trends and distance data. The result was a model allowing to predict the volume of shipments and better use of personnel and transport vehicles.

La Poste equipped new electric vehicles with sensors to monitor their technical condition.

The postal sector, as well as other industries now is starting to use the analysis of the customer lifecycle. Australian Post uses predictive analysis to create a daily customer sales, and product profitability prediction (DHL 2013).

Italian post office used architecture of "big data" to support financial services in two ways: evaluation of credit risk, real-time creation of tailor offers and discounts for customers using internet banking.

- The variety of products—examples can be DHL MyWays interconnecting individual shippers with occasional deliverer. Potential messengers can book and pick up the shipment delivered by the addressee. By using application service will connect the demand for flexible delivery, with those who offer transport during their time of day.

In England launches program SOPOS combining postal service operators, social networking and data services. This is about sending samples of goods through social media such as Twitter or Facebook. Customer sends a recommendation to the product (possibly in the form of Gifts) your friend (without requiring knowledge of the address), the friend decides whether the sample/gift receipt and provides your contact information. Post subsequently deliver the consignment and become potential customers.

- Promoting economic growth—postal "big data" can greatly promote the business environment. Using global geo-marketing data from Deutsche Post Direct linking socio-demographic and household data, along with statistics of consumption, many German companies are able better to focus on the best placement of their products in foreign markets.

La Poste Datapost currently presents a platform for accessing postal and other data to companies and government agencies to support the creation of new applications that improve their use of the public itself and operator.

- Support for public services—to this goal, many sources refers to the Article of Michael Ravnitzky (2010) from the Postal Regulatory Commission. He introduced a range of potential sensors that are placed on postal vehicles. For example, these could detect chemical and biological substances or radio-active material. Further ones could assess road conditions and inform the potholes and other damage and assist in prioritizing repairs (International Postal Big Data 2014).

4 DATA USAGE NATIONAL POSTAL COMPANY

In recent years Slovak Post has decided to go towards the innovation in data management and data policies by using increased emphasis on the implementation of information technologies in all areas of operation. As mentioned in the previous subsections, innovation is mainly supported by growing demand from consumers increasingly relying on or requiring "smart" applications in everyday life, for saving time and money. Technological development is also necessary to keep up the ongoing informatization of public administration in the programs ESO and description as well as the rapidly developing internet, new electronic services from the competition.

The trend is associated not only with the necessity of adapting the organization and processes, new services, but the main accompanying phenomenon is still bulking accumulation amount of information obtained from different systems. Nowadays the world is considered to be a source of competitive advantage of unused potential in practice, known as "Big Data" (Stríček 2013).

Among some of the major shortcomings of the current model data management include:

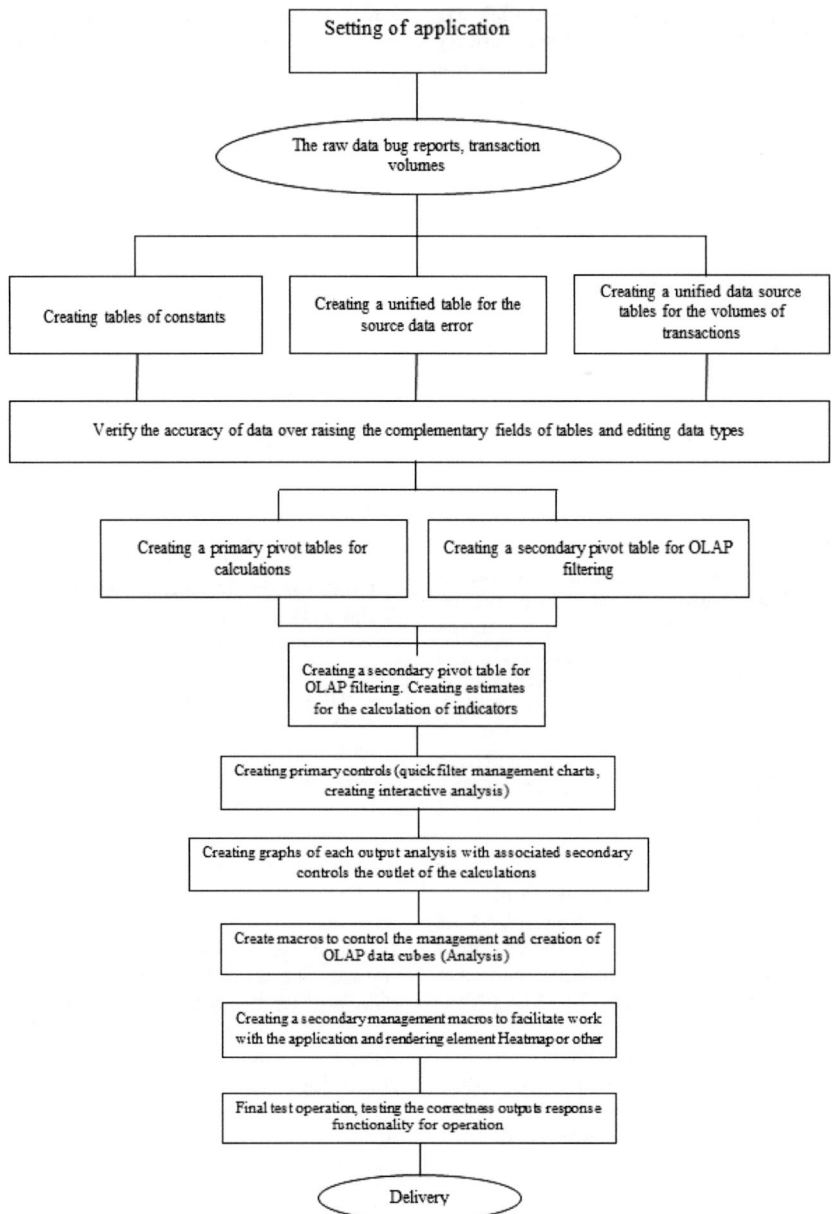

Figure 2. Block diagram of the process design dashboard application.

- Long time lag between the different reports,
- Limited number and variability of these reports,
- Unavailability or ignorance of traffic managers and some statistics.
- Unavailability of data for lower management for creating individual statistics, according to the currently solved situation.

If there is no need for an overall redevelopment of the information system necessary to ensure the full cooperation of all data systems and the availability of all data, AP's local data processing can be solved by a simplified analysis of data. After ensuring the availability of basic data in the system in the form of output data structures in an Excel table, analytical Dashboard application can be created for each process directly in Excel.

Overall evaluation and output is essential for any activity for the processing and analysis of data. It should take into account various factors on which are determined all the elements contained in the resulting analytical application. Therefore it is needed before starting creation of the application to set the basic requirements which should be kept in the design of individual components.

Application requirements:

– Cheap solution in terms of costs for introduction into trial operation and maintenance
– Simplicity and clarity of output graphs for the end user (only minimal instruction are needed and explanation of variables before use)
– Simple and transparent operation without the need for explanatory, intuitive search solutions, control
– Low maintenance (such as recording data correction formulas or graphs, change or add new parameters)
– Fast response (to minimize system requirements and optimization calculations and program cycles to speed up performance)
– Safety against tampering (hide and block access to parts of the application out of its administrator privileges, create a clean user access).

At the beginning, we can determine the first item. These applications are the best form directly as a separate program units operating on the basis of drawing data source SQL database. However, the applications are difficult to maintain, and depends on the size of and the IT department. Therefore, the Slovak Post primary resource procurement before using its own capacities.

Creating applications in Excel program has the advantage of rapid development, easy learning in a familiar environment and familiar functionality and manageability between employees of the Slovak Post. In case of putting into test operation does not require any additional financial burden on software. In the future, of course, the extending of application is expected starting from new features and its incorporation to a full version of a individual program, or at least links to external SQL databases designed in the OLAP structure, where the Excel application did not provide the primary data source OLAP calculations that are performed on the source side server.

Structure Dashboard application on the principle of OLAP system may have the following structure for tracking error of services or products (Fig. 2).

Primarily the entire application would be situated in one of the following Excel workbooks sheets with following names (for compatibility with formulas and macros in English). Name of the document consists of the type of dashboard, its version, and last modified.

Mentioned structure allows building interactive application for simple analytical data evaluation of error rate, according to arbitrarily composite indicators, according to the need of target management.

5 SUMMARY

The implementation of a unified information system with flexible compatibility of individual modules based on reciprocal exchange and availability of datasets is the goal of every organization. The information system should not collect only data, but also allow easy access to them. The ideal would be the direct creation of analytical tools, but threatened by the emergence of new deferred costs when changing data structures or types.

The proposed model of data processing can serve as a basis for guidelines how to build new modules of IS or the total rebuilding to the platform enabling freer and wider access to the data contained. The data would not be lost in the net of "black hole" of different partial systems IS. Direct access to the data reduces the elaborateness and time of their acquisition from different sources, as well as labor intensive and time spent by adapting them to a form suitable for analysis.

In addition to this fact the model of the proposed designed system should be centrally managed by IT department of the Slovak Post, which should be able to carry out ongoing interventions and minor adjustments in the module as well. By simplifying the system and allowing the modification directly by internal IT department, it will save funds of ancillary operations using public procurement orders up to the hundreds of thousands of euros. But more importantly will be flexible to change the desired output data processing and practically immediate application without the development of additional studies to procurement delays and the difficult process of accompanying public procurement.

Designed Dashboard application and its structure can serve as a model for the formation of partial or related analytical applications of this type designed for evaluation data from other areas of activity of the Slovak Post. Alternatively, the structure of the conversion base will serve as a model in the development of analytical tools based on other platforms (SQL,. NET, etc.).

ACKNOWLEDGEMENT

The paper is published within the project VEGA 1/0895/13 a VEGA 1//0733/15.

REFERENCES

Benninga, S. 2014. *Financial Modeling*. Fourth edition, Massachusetts Institute of Technology.

Campbell, J.Y. & Mckinley, A.C. 1996. *The Econometrics of Financial Markets*. Princeton University Press.

DHL 2013. Big data in logistics a DHL perspective on how to move beyond the hype December 2013. <http://www.delivering-tomorrow.com/wp-content/uploads/2014/02/CSI_Studie_BIG_DATA_FINAL-ONLINE.pdf>.

Frost, P.A. & Savarino, J.E. 1986. An Empirical Bayes Approach to Portfolio Selection. *Journal of Financial and Quantitative Analysis*.

Goldman S. 1999. The Intuition Behind Black-Litterman Model Portfolios. *Journal of Finance*.

Hillard, R. 2010. Information-Driven Business, How to Manage Data and Information for Maximum Advantage. Library of Congress Cataloging-in-Publication Data, 2010.

International Postal Big Data 2014. Discussion Forum Recap May 12, 2014. <https://www.uspsoig.gov/sites/default/files/document-library-files/2014/rarc-ib-14-002.pdf>.

Jorion, P. 1986. Bayes-Stein Estimation for Portfolio Analysis. *Journal of Financial and Quantitative Analysis*.

Stríček, I. 2013. *Costs of quality in company and their measurement*. Žilina: Transcom 2013, Section 2.

World Bank 2007. *Education Quality and Economic Growth*. Washington, D.C: The World Bank.

Production Management and Engineering Sciences – Majerník, Daneshjo & Bosák (Eds)
© 2016 Taylor & Francis Group, London, ISBN: 978-1-138-02856-2

Geotourism and geoparks—a sustainable form of environmental protection

Ľ. Štrba, B. Kršák & M. Molokáč
Faculty of Mining, Ecology, Institute of Earth Resources, Technical University of Košice, Košice, Slovak Republic

J. Adamkovič
Department of Management, Faculty of Business Economy with Seat in Košice, University of Economics in Bratislava, Košice, Slovak Republic

ABSTRACT: Natural features and characteristics, e.g. geological structure or landform, are important parts of world heritage. This paper is focused on overview of the concept of geotourism and geoparks as one of relatively new forms how to attract general public, arouse interest of people in geological heritage and inform about its importance and role(s) within natural processes. Uniqueness of geotourism is supported, respecting general principles of this tourism form as presented in this paper, by sustainable utilization of natural heritage, increasing awareness of nature through new and attractive forms of knowledge interpretation to general public. Such activities are the most intensively presented in the area of geoparks, which on one hand inform about sustainable utilization and need of natural resources and on the other hand they encourage to respect natural environment and to preserve original character of the landscape.

1 INTRODUCTION

Landscape, its shape, form and geological structure are the main elements, on which all aspects of natural environment are based. These elements affect such natural geographical features as topography, biodiversity, water distribution or soil type. Moreover, geological structure and landform(s) are significant parts of world heritage. Their protection and conservation ensure that future generations will have possibilities to get knowledge and understanding of geological history of the Earth and the environment in which they live, and to admire natural beauties.

This article is devoted to geotourism and geoparks, not from the tourism point of view but from the perspective of sustainable form of environmental protection. Knowledge, understanding and protection or (geo)conservation of abiotic part of the environment play key role in the process of preservation of the whole system.

2 GEOSITES AND GEOCONSERVATION

The history of systematic nature protection has started in the second half of the 19th century. This was realized in form of local and/or national activities to protect selected areas of specific character

and importance. For the first time, scientific approach was applied by The Royal Society for conservation of forests in British India (Barton 2002).

The first attempt to define and establish protected area was performed in the United States of America, where President Abraham Lincoln signed the Yosemite Grant Act, establishing Yosemite Valley and Mariposa Grove. However, the first real national park in the world is Yellowstone National Park, established in 1872. The concept of the Yellowstone National Park has been subsequently adapted by national parks in other countries (Rybár et al. 2010):

– Royal National Park (Australia)—1879,
– Banff National Park (Canada)—1887,
– Tongariro National Park (New Zealand)—1887,
– Abisko National Park (Sweden), the first national park in Europe—1909.

Important international documents on conservation of nature and natural resources are the UNESCO Convention Concerning the Protection of the World Cultural and Natural Heritage (1972) and the Berne Convention on the Conservation of European Wildlife and Natural Habitats (1979). Propagation and support of biodiversity protection and many other activities did often not consider physical aspect of the environment.

Earth's geological structure (including minerals, rocks, and fossils) and landforms (geomorphology—including remnants and active landforms) are taken as granted. Moreover, these elements have been considered for some kind of background in the process of biodiversity or cultural heritage protection by the vast majority. This situation, thanks to initiatives of several regional, national and international organizations focused on the identification and protection of geological sites, has started to change in 1990s.

Under the patronage of UNESCO, the First International Symposium on the Conservation of our Geological Heritage was held in Digne (France) in June 1991. Here, an International Declaration of the Rights of the Memory of the Earth was unanimously adopted. This can be considered as a crucial step in geological heritage protection. However, one the first recorded attempt to protect a geosite was in the United Kingdom (locality Cheessewring) in 1860 (Hose 2011). But what does it mean 'geological heritage' or 'geosite'?

The term 'geosite' has started to appear in publications (e.g. Cowie & Wimbledon 1994, Cowie 1992) in the beginning of 1990s. Nowadays, when the concept of geosites is well established, these locations has become a subject of interest of many authors (e.g. Gavrila & Anghel 2013; Joyce 2008; Matrínez-Torez et al. 2011; Reynard 2004; Reynard et al. 2007; Wimbledon 1996; Wimbledon et al. 2000; Wimbledon & Smith-Meyer 2012). Several authors have presented their own definitions of geosite, including terms like uniqueness, geological heritage, importance/significance and value. One of the most understandable and accurate definitions were given by Reynard (2004): "Geosites are portions of the geosphere that present a particular importance for the comprehension of Earth history. They are spatially delimited and from a scientific point of view clearly distinguishable from their surroundings. More precisely, geosites are defined as geological or geomorphological objects that have acquired a scientific (e.g. sedimentological stratotype, relict moraine representative of a glacier extension), cultural/historical (e.g. religious or mystical value), aesthetic (e.g. some mountainous or coastal landscapes) and/or social/economic (e.g. aesthetic landscapes as tourist destinations) value due to human perception or exploitation."

Protection of these places is important because: (1) it is our duty to preserve this heritage for future generations—a principle of sustainability, (2) such places offer study and research possibilities which are inevitable for development of science and industry, (3) they help in education of geoscientists, (4) they allow to monitor environmental changes, (5) such places are important leisure and popular tourist sites, (6) geo(morpho)logical elements of the area are one of the most important prerequisites for biodiversity.

While geosites are relatively easy to define, their protection and preservation—'geoconservation'—requires a broader scope to understand in detail. According to Sharples (2002), geoconservation or conservation of geodiversity aims on preservation of geological, geomorphological and soil features and processes and on and the maintenance of natural rates and magnitudes of change in those features and processes. More recent definition given by ProGEO (2011) defines geoconservation as "protection and management of geological sites, areas and specimens for scientific research, education and training, where appropriate, popularization of the Earth's history for a wider public and promotion of good conservation practice".

It can be assumed that primary interest of geoconservation is protection of natural geodiversity, not only to protect direct scientific or aesthetic values perceived by humans, but also to preserve natural ecological (including biological) processes that are subject of interest of the majority of environmental protection programs (Sharples 2002).

3 GEOTOURISM

Geotourism, as a relatively new tourism form, becomes more popular and can be considered for new global phenomenon (Dowling 2008). It can be defined as special tourism form in natural environment with special focus on geology.

The cradle of geotourism is United Kingdom, where, based on (1) increasing number of closed mines and quarries and their not the most appropriate use and revitalization, and (2) destruction of several geo(morpho)logical sites, geologists recognized the need to protect and sustainable use of geologically significant sites. Initially, the importance of geotourism has been associated with promotion and preservation of geosites, mainly in abandoned mines and quarries, by maintaining access to these locations through development of sustainable tourism products and services including leaflets, guided tours led by specialists or construction of new objects (e.g. visitor centers) (Hose 2011).

First definition of modern geotourism was given by Hose (1995). He defined geotourism as "the provision of interpretive and service facilities to enable tourists to acquire knowledge and understanding of the geology and geomorphology of a site (including its contribution to the development of the Earth sciences) beyond the level of mere aesthetic appreciation".

This definition has been updated and/or modified by several authors (Dowling & Newsome 2010, Hose 1996, Joyce 2006, Sadry 2009). At present, generally accepted definition of geotourism is as follows: "Geotourism is a form of natural area tourism that specifically focuses on landscape and geology. It promotes tourism to geosites and the conservation of geo-diversity and an understanding of Earth sciences through appreciation and learning. This is achieved through independent visits to geological features, use of geo-trails and viewpoints, guided tours, geo-activities and patronage of geosite visitor centers." (Dowling & Newsome 2010). According to Rybár et al. (2010), geotourism is based on learning about geological objects and processes with special emphasis on their aesthetic and historical value, and on recognition of technical monuments connected to mining activities (abandoned mines, mining museums, trade routes related to transportation of mining origin goods) as well as technical and cultural monuments connected to historical mining activities. Taking into account broader context of geotourism understanding, several geotourism forms have been defined in last years: underground geotourism (Garofano & Govoni 2012), rural geotourism (Farsani et al. 2013), urban geotourism (Ferreira et al. 2012, Rodrigues et al. 2011), and health & wellness geotourism (Farsani et al. 2012).

Alternative insight on geotourism and its understanding brings National Geographic. On its official web-pages, geotourism is defined as "tourism that sustains or enhances the geographical character of a place—its environment, culture, aesthetics, heritage, and the well-being of its residents" (National Geographic 2014a).

4 GEOPARKS

One of the newest are relatively young form how to attract general public, arouse an interest in geological heritage and inform about its importance, is the concept of geoparks. Moreover, geoparks significantly participate in conservation of geological heritage and its preservation for future generations.

Geopark is territorially defined area with significant geological sites. This geological heritage is used to increase the awareness of important features, which society interacts with. Many geoparks not only promote geohazards, like volcanoes and volcanic activity, earthquakes and tsunamis, but also help to prepare disaster mitigation strategies among local communities. Geoparks provide information on (geo)climate, geographical and other

changes within the Earth's history as well as on sustainable use of renewable resources and effective application of green (environmental-friendly) tourism (UNESCO, 2014).

The concept of geoparks is strongly supported by the UNESCO, which adopted the geoparks program in 1998. In this year, Global Geoparks Network (GGN) was established under the auspices of UNESCO. Subsequently, European Geopark Network (EGN) was established to protect geodiversity, promote geological heritage to general public, and support sustainable development of economic activities in the area of geoparks, mainly to support geological tourism development. Recently, GGN has 111 members from different countries and continents (GGN 2015).

Besides mentioned 111 geoparks, members of the GGN, there are other geoparks in the world, which do not meet criteria to become a GGN member, are in the assessment process or did not send the application form yet. A good example of such geopark is Geopark Banská Štiavnica in Slovakia. This geopark is located in undoubtedly unique area, not only from geological but also from cultural-historical point of view, as this region is closely connected to historical mining activities with several world significant inventions, e.g. mining related water system ('tajchy'). Despite indisputable potential of the area, activities within the geopark do not meet the criteria given by EGN and GGN to become a member.

5 GEOPARKS AND GEOTOURISM AS A ENVIRONMENTAL PROTECTION TOOL

As mentioned above, many environment protection measures had been done long before the definition of the geotourism and/or geoparks. Specific acts devoted to the protection of the environment were adopted at national levels in many countries. These documents predominantly limit actions that can threaten damage or devastate conditions and forms of life, natural heritage, landforms or ecological stability.

In addition to these acts, several international documents (agreements, conventions, etc.) aimed on more effective protection and preservation of world heritage on Earth has been adopted. They protect different area types, which are not law-protected, but comprise important base for science progress and presentation of nature protection abroad. These include e.g.:

– Convention Concerning the Protection of the World Cultural and Natural Heritage (UNESO),

- Man and Biosphere Programme (UNESCO),
- The Convention on Wetlands (Ramsar Convention),
- European Diploma for Protected Areas,
- Natura 2000,
- cross-border protected areas

It can be assumed that there are many documents at national and/or international levels devoted to environmental protection. Logically, a question arises: Do we need to make any other effort to protect the environment? The answer is hidden in the effectiveness of by now adopted acts and other legal norms. The majority of these documents can be interpreted as some kind of restriction, e.g. limitation of the freedom of movement in order to preserve original character of the area. But why do we protect the environment? The most frequent answer would probably be that we protect it from negative human impacts. Here, the legislation is often ineffective. One of the options, how to achieve required effect within the protection, is the support and development of geotourism and geopark concept.

When discussing environmental protection through geotourism and geoparks, we can base on principles on which the geotourism is defined. As assumed by Dowling (2011), geotourism is based on five principles: Geotourism is (1) geologically based (based on geological heritage of the Earth), (2) sustainable (e.g. generates economic profit, supports local community and geoconservation), (3) geologically informative (knowledge gained through geo-interpretation), (4) locally beneficial, and brings (5) tourist satisfaction. From this list, the second and the third principle are the key contribution to the environmental protection. Ideally, geotourism leads to positive interaction between people and geo-environment through effective propagation of the need for geological heritage protection and preservation. Similarly, thirteen geotourism principles defined by the National Geographic include such, which are, at least partially, aimed on environmental protection: integrity of place, protection and enhancement of destination appeal, land use, conservation of resources, interactive interpretation, planning, and evaluation (National Geographic 2014b).

On the first sight two different approaches of geotourism understanding are easy to compare. There are several similarities allowing connecting both approaches (Fig. 1).

Following one or second mentioned group of geotourism principles, the same result will be achieved—successful and (geo)tourist attractive geotourism destination—a place where, among other things, an effective and sustainable environmental protection is ensured.

Geotourism, primarily in the geopark areas, is an active form of tourism. Each tourist has a chance to acquire new information about visited

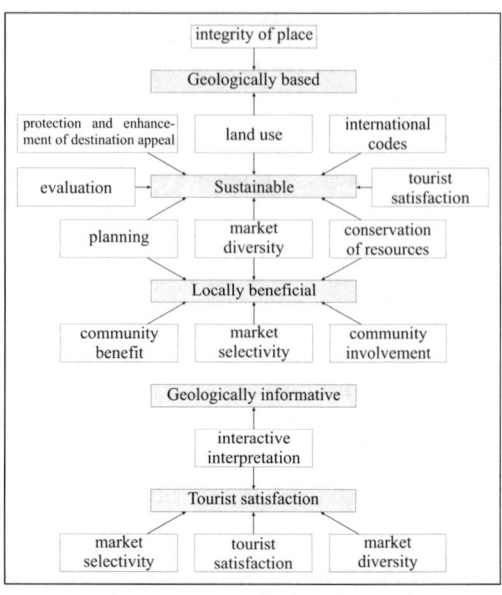

Figure 1. Comparison of geotourism principles after Dowling (2011) (grey rectangles) and after National Geographic (2014b) (white rectangles).

sites. At these sites, not only the importance from the geological point of view but also from broader, context including biotic part of environment is presented. Active education through appropriate interpretation of the knowledge and information, as often applied in geoparks, may positively affect attitude towards the environment, its individual elements and their mutual relations and interactions. As mentioned by Carter (2001), appropriate knowledge interpretation helps others to appreciate importance of something what is considered as unique. Despite the fact, that there are several definitions of the term 'interpretation', each of them is based on the same idea—to share passion for something (e.g. place, relationship, event) what is unique with visitors. Interpretation should evoke absolutely new understanding of visited or seen place (Carter 2001; HDC-Americas 2014).

Change of tourists' attitude towards visited places and real understanding of such places' importance within the ecosystem may ultimately be an effective tool for sustainable environmental protection. Therefore, it is necessary to support activities leading to geotourism development and geopark establishments.

6 CONCLUSION

Environmental protection as a subject of interest of many researchers and organizations consist of different aspects that should be taken into account. Therefore, protection of the environment requires appropriate attention and multidisciplinary approach to be effectively applied. It is necessary to act in several subsequent steps including process of decision making, assessment of selected methods, ecological value of the area and attractiveness (Domaracká et al. 2014; Štrba et al. 2015).

The concept of geotourism and geoparks plays significant role in this effort because, besides "classical" form of protection, predominantly aimed on the protection of abiotic part of the environment; it actively contributes to positive change of general public mind and attitude towards the importance environmental protection. So, adequate attention should be paid to this concept, not only because geotourism is an active form of tourism, but also because it is a sustainable and effective form of environmental protection.

REFERENCES

Barton, G. 2002. *Empire Forestry and the Origins of Environmentalism*, Cambridge University Press, 2002, 48 p.

Carter, J. 2001. *A sense of place (An interpretative planning handbook)*. Inverness: Tourism and Environment Initiative.

Cowie, J.W. 1992. *Report of Task Force Meeting, Paris, France, February 1992, UNESCO World Heritage Convention, Working Group on Geological (inc. Fossil) Sites*. Trondheim: IUGS Secretariat.

Cowie, J.W. & Wimbledon, W.A.P. 1994. The World Heritage List and its relevance to geology. In D. O'Halloran, C. Green, M. Harley, M. Stanley & J. Knill (eds.), In *Geological and Landscape Conservation*. London: Geological Society of London.

Domaracká, L., Cehlár, M. & Muchová, M. 2014. Methods of evaluation process. *SGEM 2014: 14th International Multidisciplinary Scientific Geoconferences: Ecology, Economics, Education and Legislation: Volume 3: 17–26 June, 2014, Albena, Bulgaria*. Sofia: STEF92 Technology Ltd.

Dowling, R.K. 2008. The emergence of geotourism and geoparks. *Journal of Tourism* 9 (2): 227–236.

Dowling, R.K. 2011. Geotourism's global growth. *Geoheritage* 3 (1): 1–13.

Dowling, R.K. & Newsome, D. 2006. *Geotourism*, Oxford: Elsevier Butterworth-Heinemann.

Farsani, N.T., Coelho, C.O.A. & Costa C.M.M. 2013. Rural Geotourism: A New Tourism Product. *Acta Geoturistica* 4 (2): 1–10.

Ferreira, M., Sá Caetano, P. & Patuleia, M. 2012: Below and above the surface: urban geotourism on the Lisabon underground public transportation. *Revista Turismo e Desenvolvimento* 17/18: 345–352.

Garofano, M. & Govoni, D. 2012. Underground Geotourism: a Historic and Economic Overview of Show Caves and Show Mines in Italy. *Geoheritage* 4: 79–92.

Gavrila, I.G. & Anghel, T. 2013. Geomorphosites Inventory in the Măcin Mountains (South-Eastern Romania). *Geojournal of Tourism and Geosites* 11(1): 42–53.

GGN 2015: *Members list—Global Network of National Geoparks*. Available online at: http://www.globalgeopark.org/aboutGGN/list/

HDC-Americas 2014. *Heritage Interpretation and Interpretive Writing*. Available online at: http://hdc-americas.com/hdc-americas-interpretation-writing.aspx

Hose, T.A. 1995. Selling the story of Britain's stone. *Environmental Interpretation* 10 (2): 16–17.

Hose, T.A. 1996. Geotourism, or can tourists become casual rock hounds? In M.R. Bennett (ed.), In *Geology on your doorstep*. London: The Geological Society.

Hose, T.A. 2011. 3G's for Modern Geotourism. *Geoheritage* 4 (1): 7–24.

Joyce, B. 2006. *Geomorphological sites and the new geotourism in Australia*. available online at: http://web.earthsci.unimelb.edu.au/Joyce/heritage/geotourosm-Reviewebj.htm

Joyce, E.B. 2008. *Geosites of Australia: preparing an inventory and framework for the Global Inventory of the IUGS*, Manuscript, 2008, published online, available at: http://earthsci.unimelb.edu.au/Joyce/heritage/GeositesAustraliaJoycems2008.htm

Leafe, R. 1997. Conserving our coastal heritage—a conflict resolved. In J. Hooker (ed.), *Coastal Defence and Earth Science Conservation. Proceedings of the Conference on Coastal Defence and Earth Science Conservation, Portsmouth, 1997*. London: Geological Society.

Martínez-Torez, L.M., Alonso, J. & Valle, J.M. 2011. The Upper Aptian—Lower Albian Amber Deposit of the Peñacerrada II Geosite (Basque—Cantabrian Basin, Northern Spain): Geological Context and Protection. *Geoheritage* 3: 55–61.

National Geographic 2014a. *About Geotourism*. Available online at: http://travel.nationalgeographic.com/travel/sustainable/pdf/about-geotourism.pdf

National Geographic 2014b. *The Geotourism Charter*. available online at: http://travel.nationalgeographic.com/travel/sustainable/pdf/geotourism_charter_template.pdf

Newsome, D. & Dowling R.K. (eds.) 2010. *Geotourism: the tourism of geology and landscapes*. Oxford: Goodfellow Publishers.

Reynard, E. 2004. Geosite. In A.S. Goudie (ed.), *Encyclopedia of Geomorphology*. London: Routledge.

Reynard, E., Fontana, G., Kozlik, L. & Scapozza C. 2007. A method for assessing "scientific" and "additional values" of geomorphosites. *Geographica Helvetica* 62: 148–158.

Rodrigues, M.L., Machado, C.R. & Freire, E. 2011. Geotourism routes in urban areas: a preliminary approach to the Lisbon geoheritage survey. *Geojournal of Tourism and Geosites* 8 (2): 281–294.

Rybár, P., Baláž, B. & Štrba, Ľ. 2010. *Geoturizmus—Identifikácia objektov geoturizmu*. Košice: Edičné stredisko FBERG TU Košice.

Sadry, B.N. 2009. *Fundamentals of Geotourism: with a special emphasis on Iran*. Tehran: Samt Organization Publishers.

Sharples, C. 2002. *Concepts and Principles of Geoconservation*, Hobart: Tasmanian Parks & Wildlife Service.

Štrba, Ľ., Rybár, P., Baláž, B., Molokáč, M., Hvizdák, L., Kršák, B., Lukáč, M., Muchová, L., Tometzová, D. & Ferenčíková, J. 2015. Geosite assessments: comparison of methods and results. *Current Issues in Tourism* 18 (5): 496–510.

UNESCO, 2014. *What is a Global Geopark?* Available online at: http://www.unesco.org/new/en/natural-sciences/environment/earth-sciences/global-geoparks/some-questions-about-geoparks/what-is-a-global-geopark/.

Wimbledon W.A.P. & Smith-Meyer, S. 2012. *Geoheritage in Europe and its conservation*, Uppsala: ProGEO.

Wimbledon, W.A.P. 1996. Geosites—a new conservation initiative. *Episodes* 19 (3): 87–88.

Wimbledon, W.A.P., Ishchenko, A.A., Gerasimenko, N.P., Karis, L.O., Suominen, V., Ohansson, C.E. & Freden, C. 2000. Geosites—an IUGS initiative: science supported by conservation. In D. Barettino, W.A.P. Wimbledon & E. Gallego (eds.), *Geological Heritage: Its conservation and management*. Madrid: ITGE.

Production Management and Engineering Sciences – Majerník, Daneshjo & Bosák (Eds)
© 2016 Taylor & Francis Group, London, ISBN: 978-1-138-02856-2

Multivariate analysis of pollution sources in Moravian-Silesian region

B. Sýkorová, M. Kucbel, J. Drozdová & K. Raclavský

ENET—Energy Units for Utilization of Non-Traditional Energy Sources, Ostrava, Czech Republic

ABSTRACT: Methods of multivariate statistical analysis were used for identification of influence of industrial pollution sources on the air quality in Moravian-Silesian Region (MSR), the Czech Republic. Atmospheric concentrations of Cd, Pb, Co, Cr, Ni, Mn, Fe, Zn, Cu, Sb, V, Sr and As were monitored at 40 localities in MSR during the summer 2014. Cluster analysis divided the localities into 3 clusters. The cluster I is characterized by highest concentrations of all analyzed elements. Cluster I corresponds to contamination produced by element association Co—Cr—Ni—Mn—Fe. The 12 localities in cluster II are situated geographically in the area between Ostrava, Bohumin, Karvina, Orlova, Cesky Tesin and Havirov together with localities from cluster I. The cluster III contains localities with the lowest load. Most of localities are situated in region north-west of Ostrava, region of the Jesenik Mts, with exception of the locality Ostravice, Celadná.

1 INTRODUCTION

In urban air, there is a considerable amount of particulate matter with an aerodynamic diameter less than 10 microns. The particles come primarily from anthropogenic sources (coal, transport, industrial activity), but also from natural sources (resuspension of soil particles) (Loredo et al. 2003). Toxic elements that can cause a number of environmental and health risks are bound to them (Aryal et al. 2013).

Trace elements represent less than 10% in Σ PM_{10}, nitrates, sulphates, and ammonium ions show the highest concentration; they are followed by organic compounds and elemental carbon (Gianini et al. 2012). Fe represents up to 87% of total amount of the 24 monitored trace elements (Muránszky et al. 2011).

Heavy metals from industrial sources are greatly influenced by the type of production; moreover, they are often typical for the area and are not generally useful as a marker. Elements released to the atmosphere during anthropogenic processes are listed in Table 1. To identify the proportion of traffic, some authors use the concentration levels of Cu/Sb, which are released due to brake wear. Sternbeck et al. (2002) indicates a ratio value of 0.22, Pant et al. (2013) shows a higher ratio value of 4.6 ± 2.3. The particles with the Fe—Zn—Cr—Ni are likely formed during re-melting/refining of steel scrap that already contains these metals as alloy components (Choël et al. 2007). To identify the source of contamination, the values of the correlation coefficient, which can identify a common source of the elements, are also used. For metals released from metallurgical processes, Mugica et al. 2002 states a statistically significant correlation dependence in PM_{10} particles between Pb—Cr, Cu—Cr and Mn—Cr (Mugica et al. 2002).

Table 1. Overview of elements and their resources literature data.

Fossil fuel combustion	Fossil fuel combustion: Power engineering	Fossil fuel combustion: Traffic	Combustion from traffic and wear	Industry, metallurgy
Fe, Zn, Pb, Cu[a,b]	Mn, Cr, Cu, Co, As[c]	Cd, Ni[c]	Zn, Cu, Sb, Pb, Mn, Cd, Ni[d]	Clinker-ing: Cd,Pb[e]
	Ba, Mg, Ca, Cu, Fe[f]	Cu, Mn, Sr, Br, Ba, Sb, V[g]	Pb, Cu, Zn, Sb, Ni, Cd[h,i]	Melting processes: Mn, Fe, Zn[e]
	V, Ni, Mn, Co[j]	V, Ni, Cd[a,b]	Cu, Fe, Mo, Sb, Ba[f] Pb, Cu, Sb, Zn[j]	

*[a]Valavanidis et al. 2006, [b]Herngren et al. 2006, [c]Loredo et al. 2003, [d]Shao et al, 2013, [e]Oravisjarvi et al. 2003, [f]Gianini et al. 2012, [g]Pant & Harrison 2013, [h]Johansson et al. 2009, [i]Viana et al. 2009, [j]De Miguel et al. 1999.

In Budapest Muránszký et al. 2011 proved, that the highest concentrations of PM_{10} are shown by Fe > Zn > Pb > Cu > Mn (Muránszky et al. 2011).

The aim of this work is to identify the elemental association's characteristic of pollution sources in the Moravian-Silesian Region with the support of the results obtained by the microanalysis of SEM/EDAX particles, and to determine the enrichment index, which characterizes the importance of individual elements in terms of anthropogenic contamination.

2 MATERIALS AND METHODS

Enrichment Factors (EFs) can be utilized to distinguish between the metals originating from a natural procedure and those from human activities, and to assess the degree of the anthropogenic influence were calculated according to:

$$EF = \frac{C_x\ Sample}{C_x\ Background} \qquad (1)$$

where C_x Sample is the concentration of heavy metal on the sampling location and C_x Background is the concentration of heavy metal measured in the background localities (the lowest measured concentrations of heavy metal in the file).

3 STATISTICAL ANALYSIS

The Exploratory Data Analysis (EDA), the statistical analysis for descriptive statistics, the correlation analysis and the multivariate statistical analysis (Cluster Analysis (CA) and the Principal Component Analysis (PCA)) were performed using OriginPro 9.1 software. By performing EDA it was found that data are not normally distributed. The Box-Cox transformation was performed for ensuring assumption of normality.

The Principal Component Analysis (PCA) and the Cluster Analysis (CA) are the most common multivariate statistical methods widely applied in environmental studies. PCA is generally used to reduce the data and to extract a small number of latent factors (Principal Components, PCs) for analyzing the relationships among the observed variables. Cluster Analysis (CA) was performed to further classify components of different sources based on the similarities of their chemical properties (Chen et al. 2014). CA was performed using Ward's method. The Euclidean distance was employed for measuring the distances between clusters of similar element concentrations. The results are shown in dendrogram representing procedures in the hierarchical clustering solution.

3.1 Sampling sites and analytical procedure

Samples of heavy metals were collected at 40 localities of the Moravian-Silesian Region (Fig. 3). Sampling of heavy metals for each locality always lasted for 24 hours, in the period from June to July 2014. Samples of heavy metals were taken using the high-volume sampler Digittel MD 05 to nitrocellulose filters in accordance with CSN (Czech National standard) EN 12341. After the decomposition in a microwave system in a mixture of acids (HF, HCl, HNO_3, and H_2O_2) metals were analysed by the ICP method in laboratories the Czech Geological Survey (CGS), Prague.

4 RESULTS AND DISCUSSION

SHLU CA results are shown as a dendrogram in Figure 1, wherein the set of 40 localities is divided into three clusters. Table 2 shows that in cluster I, localities with the highest concentrations of the elements are incorporated. In cluster II, contents of elements are lower, and the lowest ones are in cluster III. The lowest concentration differences between clusters are shown by: V, Sr, Co and Ni.

Placement of localities (Figure 2) of cluster I and II defines the area with anthropogenic load, in cluster III, localities without significant anthropogenic load located in the Jeseniky and Beskydy area are concentrated. To identify the elemental associations for individual sources of pollution, the correlation analysis was performed using Spearman's correlation coefficient (Table 3). The values of r = 0.50 were considered statistically significant, at a significance level $\alpha = 0.05$. The clustering analysis that divides the elements into four clusters was also used (Fig. 3). As has a special position.

From Figure 3 it is clear that cluster II, which is probably related to the blast furnace metallur-

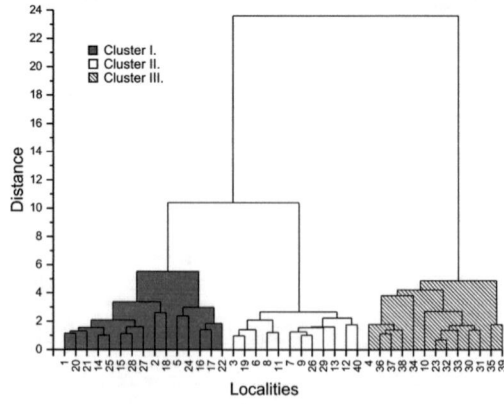

Figure 1. Dendrogram obtained by CA for 40 localities.

Table 2. Average concentrations for individual clusters [ng/m³].

	Cluster I	Cluster II	Cluster III
Cd	0.4	0.3	0.1
Co	0.1	0.1	0.1
Cr	4.4	2.8	2.2
Cu	**10.5**	4.7	3.7
Ni	1.6	1.1	0.9
Pb	**16.9**	8.4	3.5
Sb	4.0	1.9	1.2
V	0.7	0.5	0.6
Mn	25.7	9.1	5.8
Fe	**1027.2**	313.7	188.7
Zn	**94.0**	37.3	12.7
As	1.4	0.6	0.8
Sr	1.4	0.9	1.1

Table 3. Statistically significant correlation dependencies for individual clusters.

	Cluster I	Cluster II	Cluster III
Cr			Co
Cu	Co, Cr		Co
Ni	Cr, Cu	Co	Co
Pb	Cd	Cd	Cd
Sb		Cu	Co, Cr, Cu
V	Sb	Co, Ni	Co, Ni
Mn	Cr, Ni	Co, Ni, V	Cd, Zn
Fe	Cr, Ni, Mn, Zn	Co, V	Co, Mn, Zn

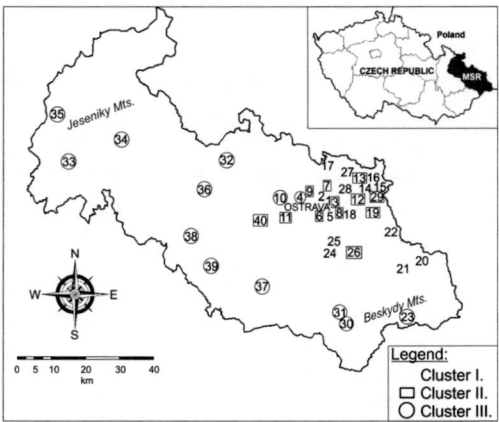

Figure 2. Localities divided according to the results of clustering analysis.. * 1. Ostrava-Radvanice, 2. Ostrava-Marianske Hory, 3. Ostrava-Radvanice OZO, 4. Ostrava-Poruba, 5. Ostrava-Radvanice church, 6. Ostrava-Dubina, 7. Ostrava-Hermanice, 8. Senov, 9. Ostrava-Hostalkovice, 10. Klimkovice-Sanatoria, 11. Klimkovice, 12. Karvina Doly, 13. Karvina Stare Mesto, 14. Karvina Frystat, 15. Karvina-Raj, 16. Karvina, 17. Bohumin-Skrecon, 18. Havirov, 19. Havirov-Sumbark, 20. Trinec, 21. Trinec-Oldrichovice, 22. Ceský Tesin, 23. Horni Lomna, 24. Frydek-Mistek A, 25. Frydek-Mistek B, 26. Nosovice, 27. Orlova, 28. Petrvald, 29. Stonava, 30. Ostravice, 31. Celadna, 32. Opava, 33. Rymarov, 34. Bruntal, 35. Karlova Studanka, 36. Hradec nad Moravici, 37. Novy Jicin, 38. Vitkov, 39. Odry, 40. Bilovec.

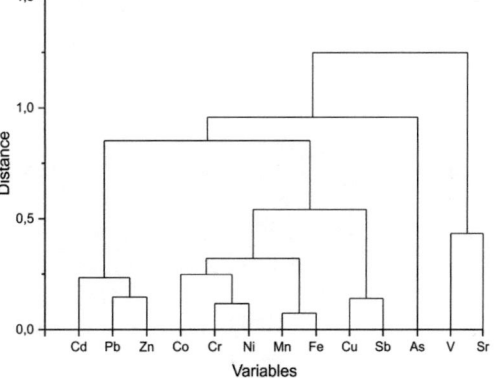

Figure 3. Dendogram obtained by CA for 13 elements.

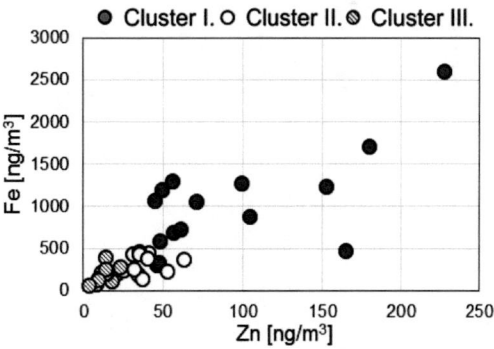

Figure 4. Dependence between Zn and Fe for the individual clusters.

gical processes, is the most significant one (Co—Cr—Ni—Mn—Fe). This element association is connected to a pair of elements Cu and Sb, which can be linked to transport, but the microanalysis results, confirmed the presence of microparticles having a Cu content from fume from the foundries. The triplet of Cd—Pb—Zn is also tied to the metallurgy, but probably to other than blast furnace processes, which is clear from the enrichment index (Table 4).

After performing PCA, two significant principal components were obtained. The first Principal Component (PC1) and the second Principal Component (PC2) accounted for 61.6% and 11.5% of

287

Table 4. Enrichment index values..

Number of locality/ Number of cluster	Cd	Co	Cr	Cu	Ni	Pb	Sb	V	Mn	Fe	Zn	As
1 / I.	5.11	13.89	3.36	8.28	2.81	8.10	2.77	1.89	13.76	19.64	15.02	4.33
2 / I.	6.63	19.39	4.36	17.24	3.35	2.61	12.46	2.49	13.93	16.22	12.01	10.32
3 / II.	4.80	10.70	2.06	4.04	2.11	7.40	2.10	3.21	4.92	6.49	8.26	3.17
4 / III.	1.64	12.77	1.77	7.60	1.52	3.49	4.70	1.60	2.10	2.73	2.87	3.67
5 / I.	11.02	18.47	5.16	13.44	4.13	14.69	2.62	1.87	39.61	25.93	48.24	4.66
6 / II.	2.41	9.08	3.79	4.62	3.41	8.50	3.16	1.57	5.03	3.42	14.05	2.19
7 / II.	5.59	8.65	1.89	6.32	1.75	7.69	4.14	1.17	2.87	3.14	5.77	3.15
8 / II.	6.22	9.49	1.95	5.10	1.97	8.35	2.71	1.81	3.77	3.70	6.71	9.28
9 / II.	4.85	13.39	2.29	7.95	2.33	12.19	5.39	2.23	4.71	6.84	10.89	2.60
10 / III.	2.66	4.39	1.52	2.67	1.63	8.33	2.03	1.07	2.28	1.72	4.91	1.00
11 / II.	4.17	6.27	2.21	3.94	1.83	8.07	2.40	1.10	2.64	2.98	9.16	5.70
12 / II.	9.95	3.91	1.45	2.65	1.26	10.00	2.26	1.01	3.04	2.11	9.90	2.35
13 / II.	7.52	5.85	2.14	11.35	1.47	9.39	7.15	1.09	2.82	5.57	16.85	2.78
14 / I.	9.17	8.32	2.72	10.99	2.26	14.24	4.84	1.38	6.72	13.29	27.87	5.10
15 / I.	10.91	14.92	3.75	8.68	2.81	8.17	6.82	3.32	5.74	10.37	15.18	4.49
16 / I.	11.73	14.65	3.00	7.39	2.89	26.20	9.29	2.48	10.85	18.75	40.89	4.00
17 / I.	14.99	16.09	3.40	10.12	3.59	22.68	10.87	3.69	9.66	19.28	26.46	11.15
18 / I.	14.82	16.21	3.33	46.11	3.08	16.57	19.57	3.99	5.03	8.83	12.91	4.98
19 / II.	3.70	9.05	1.49	4.45	1.86	5.75	3.60	1.98	3.46	3.79	8.46	2.58
20 / I.	6.72	9.61	2.12	5.52	1.97	15.85	2.55	2.26	8.55	16.10	19.09	2.98
21 / I.	7.86	14.19	2.61	7.13	2.71	16.12	3.92	3.44	8.97	18.21	13.18	7.42
22 / I.	19.31	17.17	5.82	14.71	3.59	62.27	8.86	3.13	16.04	39.53	60.71	8.68
23 / III.	2.00	6.68	1.40	3.23	1.67	3.91	1.44	1.58	2.24	3.16	4.02	3.14
24 / I.	6.83	10.44	2.29	8.33	3.29	37.71	6.10	2.08	5.44	7.17	44.18	24.24
25 / I.	5.34	21.12	2.48	13.75	3.22	14.85	6.28	1.93	7.74	10.97	16.35	6.08
26 / II.	6.14	9.41	1.69	4.99	1.86	14.77	3.96	1.32	3.79	6.96	9.34	3.97
27 / I.	15.59	12.55	1.97	5.64	2.07	17.77	9.79	2.44	4.27	4.54	12.38	8.77
28 / I.	4.87	19.07	2.60	9.13	2.00	8.35	7.91	2.61	4.36	5.11	12.67	7.26
29 / II.	7.69	11.10	1.84	7.34	2.14	6.91	3.86	1.85	4.37	6.47	9.39	1.26
30 / III.	5.53	6.75	1.65	1.81	1.43	3.15	1.18	1.92	2.96	3.28	2.97	2.31
31 / III.	3.04	10.77	1.43	2.51	1.72	5.27	1.56	2.26	3.07	4.26	6.32	1.49
32 / III.	1.82	6.81	2.08	4.68	1.60	4.30	1.84	1.57	2.25	3.27	3.09	2.54
33 / III.	2.34	6.91	1.85	4.09	1.96	1.73	1.66	2.60	4.05	3.05	3.78	2.73
34 / III.	1.75	21.32	2.84	12.26	2.54	2.12	6.00	5.42	2.38	5.99	3.80	4.67
35 / III.	1.59	3.18	1.35	1.74	1.66	1.55	1.00	2.15	1.32	1.23	2.26	2.83
36 / III.	2.78	3.61	1.46	3.63	1.49	3.70	1.59	1.61	2.29	2.21	2.59	6.51
37 / III.	2.87	4.37	1.47	4.40	1.65	3.87	2.37	1.10	3.98	3.88	3.74	11.64
38 / III.	2.82	10.04	1.65	5.22	1.89	8.44	2.30	2.23	2.16	1.89	2.51	8.81
39 / III.	1.00	1.00	1.00	1.00	1.00	1.00	1.04	1.00	1.00	1.00	1.00	4.20
40 / II.	18.88	12.96	2.15	3.68	1.75	14.55	3.07	2.50	4.20	5.70	10.79	2.91

EF <2.2–5 moderate enrichment   EF 15–40 very high

EF 5–15 significant enrichment   EF >40 extremely high

288

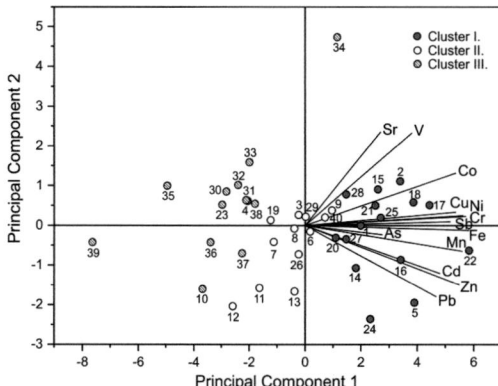

Figure 5. Biplot of localities and element concentrations for PC1 vs. PC2.

the total variance, respectively, and the two components could explain 73.1% of the total variance. A similar grouping of the localities and elements was obtained from PCA. A biplot of the localities and element concentrations, where PC1 vs. PC2 are represented, is shown in Figure 5.

In biplot, three clusters of localities can be easily identified like in the dendrogram (Fig. 1). The first cluster consists of localities with the highest concentrations of monitored elements, the second cluster groups the localities with average concentrations, and the third cluster groups the locations with the lowest concentrations of elements. PC1 is defined by high concentrations of all monitored elements. PC2 is in its positive part primarily loaded by Sr, V and the elemental associations Co—Cr—Ni and Cu—Sb. The negative part of PC2 is loaded by the elemental association Cd—Pb—Zn. From the position of Mn,

Fe, and to a lesser extent As, it is apparent that those elements affect both groups of element associations.

To calculate the enrichment index, the lowest concentration in the set was selected as the background concentration. The largest number of enrichments (11 localities) identified as "very high" with EF 15–40 is shown by Zn, at 4 localities (Ostrava—Radvanice, Karvina, Frydek-Mistek and Cesky Tesin), "extremely high" enrichments >40 were found. With cobalt, 10 localities showed "very high" enrichment coefficient. At 8 localities, lead and iron show "very high" enrichment index, an extremely high enrichment in Pb (62x) was found in Cesky Tesin. From the enrichment index values it is evident that the localities incorporated in cluster II show a "significant" enrichment index for Zn (5-15). The correlation dependencies shown in Table 3 and Figure 4 show that the Zn in cluster II does not show the correlation dependence with

Fe, and its source or form of occurrence must be different than in the case of cluster I (metallurgy—blast furnace operations) and cluster III (traffic) (Shao et al. 2013), or different forms of occurrence. High index enrichment for Cu and Sb was found for the locality of Havirov, the predominant source of pollution in this case is traffic.

In a study of the distribution of Fe in the vicinity of blast furnaces/steelworks in UK it was found that Fe particles constitute up to 12% of PM_{10}. The concentration of Fe in PM_{10} ranged from 0.096 to 2.931 µg/m³. Near the blast furnaces/steelworks, the concentrations of Fe, Mn and Co in PM_{10} were 2.5 to 5 times higher than in more distant surroundings (Mohiuddin et al. 2014). Similar concentrations of Fe were found for four locations in the Moravian-Silesian Region immediately affected by the metallurgical industry. The highest proportion of Fe in PM_{10} was found for the locality Trinec (12.7%) and Trinec-Oldrichovice (9.8%), which are influenced by Třinecké železárny-Moravia Steel, a. s. and for the locality Ostrava—Radvanice (5–7%). In cluster I, the average amount of heavy metals in PM_{10} is 6.24%, 5.4% out of which is represented by Fe. In cluster II, the total amount of metals in PM_{10} is only 2.3%, 1.9% out of which is represented by Fe. In cluster III, the average metal content in PM_{10} is 1.9%, 1.6% out of which is represented by Fe.

At selected locations (Ostrava—Radvanice), the chemical composition of particles was monitored by SEM/EDAX. In the summer period, Fe-oxide particles with Fe content >40% including trace amounts of Cr, Ni, Mn, and alumosilicates with varying the percentage of cations: K, Ca, Mg and Fe (0.5–10%) corresponding to the high-temperature operations. The presence of sulphates was also detected, in which Zn and Fe occurred as the cation, but also other metals: Cr, Ni, Mn, or Zn formed sulphates with Ca. Lead occurred most often in the form of PbO, chlorides or Pb—Ca sulphates. The distribution of elements in dendogram (Fig. 4) is probably associated with the form of the occurrence of elements. The association of Fe—Co—Cr—Ni—Mn is related to the predominant form of occurrence in the oxides of Fe, the association Pb—Zn—Cd is probably associated with a higher proportion of particles in the form of sulphates or chlorides.

The comparison of metal concentrations in PM_{10} in urban environments in Budapest and in the Moravian-Silesian Region is shown in Table 5. The concentration range of elements in the analyzed set is shown in Figure 6. The table shows that of all the monitored metals, only Zn values are up to three times higher for localities included in cluster I, for cluster II, the values are comparable, and for cluster III, they are up to three times lower.

Table 5. The concentrations of elements during the year—basic statistical parameters and concentrations during the seasons after the total decomposition of PM_{10} in aqua regia.

Element	Min.	Max.	Avg.	Med.	Winter	Spring	Summer
Sb	1.5	34	8.2	6.7	7.6	11	7.2
Pb	1.8	136	30	20	36	34	26
V	0.71	12	3	2.4	3	3.7	2.8
Cr	0.29	24	6.9	6.2	5.7	7.6	7.7
Mn	5.6	76	27	25	26	29	29
Fe	396	9495	2115	1493	2241	2883	1460
Co	0.03	1.2	0.32	0.25	0.24	0.46	0.31
Ni	0.02	9.3	2.7	2.3	2.3	3.7	2.6
Cu	12	98	37	33	35	40	39
Zn	8.6	215	53	42	74	43	36
Cd	0.12	6.4	1.1	0.62	1.1	1.7	1.1

	Average—Moravian-Silesian Region Cluster		
Element	I	II	III
Sb	3.98	1.9	1.19
Pb	16.93	8.4	3.53
V	0.72	0.5	0.56
Cr	4.37	2.8	2.21
Mn	25.73	9.1	5.83

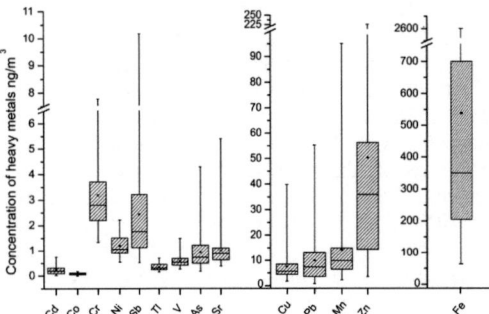

Figure 6. Box plots for 40 sampling points in Moravian-Silesian Region.

5 CONCLUSION

The localities were grouped into clusters according to the total pollution load. In one area, there are areas neighboring areas with high air pollution load as well as lower pollution load, which are composed of the same elements. In locations that are grouped in a cluster I, oxides of Fe containing Co, Cr, Ni and Mn are present in substantial quantities. Zn is not bound to Fe oxides. Besides oxides, PM_{10} particles contain sulphates of Zn and Fe. Pb occurs as $PbCl_2$ or Pb-Ca-SO_4. Localities grouped in cluster II contain a significantly

smaller proportion of Fe-oxide particles (up to 5%), Zn occurs mainly as a sulphate. The enrichment index shows that the most important element in terms of anthropogenic load is Zn, Co, Fe and Pb. A comparison of PM_{10} pollution load in the summer in Budapest and within the Moravian-Silesian Region indicates that the contents of most elements are comparable, there is a predominant trend towards lower levels in the Moravian-Silesian Region, except Zn, which is three times higher in localities grouped in cluster I, and comparable in cluster II.

ACKNOWLEDGEMENT

This paper was supported by research projects of the Ministry of Education, Youth and Sport of the Czech Republic: The National Programme for Sustainability LO1404—TUCENET and SP2015/64—The interdisciplinary study of fuel behaviour, project New creative teams in priorities of scientific research, reg. no. CZ.1.07/2.3.00/30.0055.

REFERENCES

Aryal, R., Kim, A., Lee, B.K., Kamruzzaman, M. & Beecham, S. 2013. Characteristics of Atmospheric Particulate Matter and Metals in Industrial Sites in Korea. *Environment and Pollution* 2 (4): 10–21.

Chen, H., Lu, X., Li, L.Y., Gao, T. & Chang, Y. 2014. Metal contamination in campus dust of Xi'an, China: A study based on multivariate statistics and spatial distribution. *Science of The Total Environment* 484 27–35.

Choël, M., Deboudt, K., Flament, P., Aimoz, L. & Mériaux, X. 2007. Single-particle analysis of atmospheric aerosols at Cape Gris-Nez, English Channel: Influence of steel works on iron apportionment. *Atmospheric Environment* 41: 2820–2830.

De Miguel, E., Llamas, J.F., Chac ón, E. & Mazadiego, L.F. 1999. Sources and pathways of trace elements in urban environments: a multi-elemental qualitative approach. *The Science of the Total Environment* 235: 355–357.

Gianini, M.F.D., Gehrig, R., Fischer, A., Ulrich, A., Wichser, A. & Hueglin, C. 2012. Chemical composition of PM10 in Switzerland: An analysis for 2008/2009 and changes since 1998/1999. *Atmospheric Environment* 54: 97–106.

Herngren, L., Goonetilleke, A. & Ayoko, G. A. 2006. Analysis of heavy metals in road deposited sediments, Analytica Chimica Acta 571 (2): 270–278.

Johansson, C., Norman, M. & Burma, L. 2009. Road traffic emission factors for heavy metals. *Atmospheric Environment* 43 (31): 4681–4688.

Loredo, J., Ordonez, A., Charlesworth, S. & De-Miguel, E. 2003. Influence of industry on the geochemical urban environment of Mieres (Spain) and associated heath risk. *Environmental Geochemistry and Health* 25: 307–323.

Mohiuddin, K., Strezov, V., Nelson, P.F. & Stelcer, E. 2014. Characterisation of trace metals in atmospheric particles in the vicinity of iron and steelmaking industries in Australia. *Atmospheric Environment* 83: 72–79.

Mugica, V., Maubert, M. Torres, M., Muñoz, J. & Rico, E. 2002. Temporal and spatial variations of metal content in TSP and PM$_{10}$ in Mexico City during 1996–1998. *Journal of Aerosol Science* 33 (1): 91–102.

Muránszky, G., Óvári, H., Virág, I., Csiba, P., Dobai, R. & Záray, G. 2011. Chemical characterization of PM10 fractions of urban aerosol. *Microchemical Journal* 98: 1–10.

Oravisjarvi, K., Timonen, K.L., Wiikinkoski, T., Ruuskanen, A.R., Heinanen, K. & Ruuskanen, J. 2003. Source Contributions to PM$_{2.5}$ Particles in the Urban Air of a Town Situated Close to a Steel Works. *Atmospheric Environment* 37 (8): 1013–1022.

Pant, P. & Harrison, R. 2013. Estimation of contribution of road traffic emissions to particulate matter concetrations from field measurements. A review. *Atmospheric Environment* 77: 78–97.

Shao, L., Xiao, H. & Wu, D. 2013. Speciation of heavy metals in airborne particles, road dusts, and soils along expressways in China. Chinese *Journal of Geochemistry* 32: 420–429.

Sternbeck, J., Sjódin, A. & Andrasson, K. 2002. Metal emissions from road traffic and the influence of resuspension results from two tunnel studies. *Atmospheric Environment* 36: 4735–4744.

Valavanidis, A., Fiotakis, K., Vlahogianni, T., Bakeas, E.B., Triantafillaki, S., Paraskevopoulou, P. & Dassenakis, M. 2006. Characterization of atmospheric particulates, particle-bound transition metals and polycyclic aromatic hydrocarbons of urban air in the centre of Athens (Greece). *Chemosphere* 65: 760–768.

Viana, M., Kuhlbusch, T.A.J., Querol, X., Alastuey, A., Harrison, R.M., Hopke, P.K., Winiwarter, W., Vallius, M., Szidat, S., Prévôt, A.S.H., Hueglin, C. H., Bloemen, P., Wåhlin, R., Vecchi, A.I., Miranda, A., Kasper-Giebl, W., Maenhaut & Hitzenberger R. 2008. Source Apportionment of Particulate Matter in Europe: A Review of Methods and Results. *Journal of Aerosol Science* 39: 827–849.

Production Management and Engineering Sciences – Majerník, Daneshjo & Bosák (Eds)
© 2016 Taylor & Francis Group, London, ISBN: 978-1-138-02856-2

Proposal of strategy of building energy management in the public sector

A. Tokarčík & M. Rovňák
Faculty of Management, University of Prešov, Prešov, Slovakia

ABSTRACT: Energy management is understood as influencing subjects in the region to achieve set objectives for the management of energy. The created model of management should be an integral part of public administration in the affected territory. The introduction of energy management at the level management structures is expected to reduce energy costs in buildings and properties that are paid from public funds. The proposed energy management based on an analysis of the current state must define the needs of public entities to reduce energy costs and set tasks, including organizational structure and professional potential competent for the successful completion of the objectives set. In terms of created model, it can then be said, that energy management is a set of tools and activities that serve for the active and innovative development of energy systems in a given area.

1 INTRODUCTION

Energy management system is necessary to be viewed from analysis of raw data view look, as part of management control in local government. From the analysis of local governments in the Prešov region, we can conclude, that currently have management units created energy management only partially. Management is partially documented, implemented, maintained and improved where necessary, in accordance with the activities of the relevant local government, reflecting in the various documents used by the government to manage. The basic driver is resolving of emergency states. Based on our examination of the area of Prešov region, we can conclude, that government leadership has defined subject and borders of major energy sources, that influence it.

2 THE DEFINITION OF THE PROBLEM AND PROPOSAL OF PROBLEM SOLVING STRATEGY

An essential element in the process of energy management is raising the level of perception of optimal energy needs and building competencies to the appropriate management level. Creating of energy management team in the specific conditions of local government with a representative of senior management forms the basis of long-term strategy of maintaining and improving energy policies in the specific government. For maintaining

continuity of energy management activities it is necessary to ensure sufficient resources to maintain and improve energy management. You have to keep in mind, that resources are understood in the context of human resources. Human resources have special skills and knowledge of appropriate technologies in managing and monitoring major sources of energy in order to increase energy efficiency. Correctly implemented energy management must know its reference level, which is based on long-term measurements and their analysis of energy management, which are regularly analyzed and evaluated.

3 THE ROLE AND IMPORTANCE OF SENIOR MANAGEMENT IN BUILDING ENERGY MANAGEMENT

If the senior management of local government is viewed from the perspective of process control in the specified government, then it is necessary to appoint a representative of management (management representatives) with the appropriate skills. (Vlčej 2013). An important task of the appointed representative is to inform senior management about energy management, including the factors affecting the operation of the overall system of energy management. An important task of management representative is to determine the criteria and methods that are necessary to ensure effective operation and control of energy management in the management of devices in control by the local government. For the smooth operation and security of energy management activities, it is needed in terms of the relevant government to draw up plans to keep in line with the energy management of existing operating units. Energy planning must be based on the review of the activities carried out at facilities under the management of the local government. When planning, it is necessary to define the relationship between the use of significant energy resources, their consumption, energy efficiency and under the constant supervision of energy management (Harčarik 2011).

4 ACCESS MONITORING AND TARGETING AND POSSIBILITIES OF ITS IMPLEMENTATION IN ENERGY MANAGEMENT

From a management perspective of energy management, the optimal energy management method seems to be the Monitoring & Targeting (M & T) method. It is based on systematic monitoring of actual energy consumption, analyzing the results and subsequent implementation of corrective actions. In general, the greatest benefits of monitoring and Monitoring & Targeting method are considered (AB Solartrip, s.r.o. 2015):

- cost savings,
- increase the quality of services provided,
- improve the quality of accounting and budgeting,
- improve of preventive maintenance,
- improve the services related to the management of energy,
- coordination of energy management policy,
- reducing emissions and waste,
- increase of level of environmental protection.

The base of Monitoring & Targeting is the system of controlled information flow, which links existing human resources, systems and technologies. Analytical content of M & T can be characterized as continuous energy audit that provides dynamic data about the degree of efficiency in energy consumption and detect the causes of deviations from the expected level of consumption. Implementation component represents a set of management practices that can ensure continuous removal of negative deviations in consumption and the implementation of austerity measures that lead to the continued improvement of energy efficiency in the environment of the specified government. Verification of the savings is a characteristic that distinguishes this method from energy audit (AB Solartrip, s.r.o. 2015).

In terms of overall management processes, that take place in the management of major energy sources in the local government, must be a method

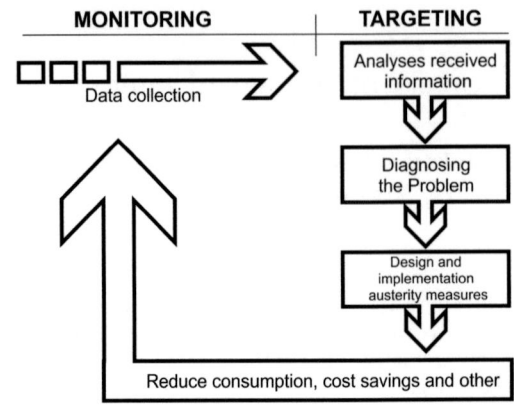

Figure 1. Cycle of Monitoring & Targeting method (AB Solartrip, s.r.o. 2015).

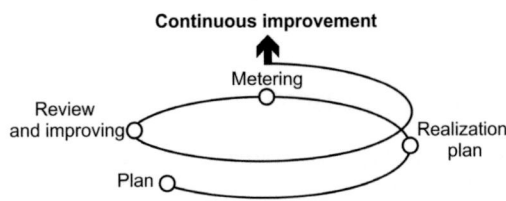

Figure 2. The cycle of continuous improvement (Energy Management Implementation Model—Short Description, 2015).

of Monitoring and Targeting be supplemented by models that are driven processes.

Deming cycle, or PDCA cycle, is also the method of gradual improvement of the processes in the management of major energy resources, carried out in an iterative implementation of four activities: P—Plan—Plan of planned improvement (plan), D—Do—implementing the plan, C— Check— verification of results of the implementation compared to the original plan, A—Act—adjustments to verification plan and its subsequent square implementation pursuing significant improvement in the use of energy sources. It is a model, by which energy management covered all aspects related to the management and fulfilling the requirements, that in the process of implementation can occur (Sheldon, Ch. and Yoxon, M. 2008).

5 PROPOSAL OF THE ACTIVITY MANAGEMENT MODEL IN AN ENERGY MANAGEMENT IN LOCAL GOVERNMENT

Each activity is described as a set of activities, gathering experience from other energy management systems, relevant literature and experiences

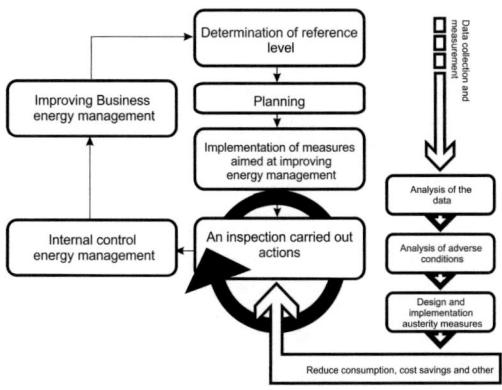

Figure 3. Model of activities management of energy management in local government.

of auditing services for energy management at home and abroad. It is understood, that the model was designed from left to right, but this does not mean that the implementation of the model in the same time. Some activities are carried out in parallel and some control activities of the individual steps are carried out concurrently. The model itself is conceived as a recurrent spiral, when the implementation phase is followed by re-understanding but enriched by experience, which is offset from the first cycle starting point. In fact, this means that each phase-change will affect the whole stage and change the overall starting position of following phase. The subsequent merger of these two methods can be presented as a final cycle model management activities that support management of important energy sources in local government (see Picture No. 3).

6 THE MEANING AND OBJECTIVE
OF THE PROPOSED MODEL
OF ENERGY MANAGEMENT
IN LOCAL GOVERNMENT

The objective of the model is to support building of energy management in public administration, defining their own priorities and timeframe. (Jašková 2015). Analysis of primary data in selected municipalities in the Prešov region identified a need to create a model of optimal practices in the use of energy. This model assumes that government leadership will constantly analyze the use of significant energy resources and their consumption on the basis of measurements. Subsequently, it will evaluate past and current use and the actual energy consumption in its premises and facilities. In terms of priorities for management of important energy resources, government leadership must:

- identify the facilities, equipment, systems, processes and personnel working for the government or on its behalf that significantly affect energy use and consumption,
- identify other important variables affecting significant energy use,
- determine the current rate of energy management, facilities, equipment, systems and processes related to identified significant energy use,
- estimate future energy use and its consumption.

An important element in terms of the present model is to create a reference level of used energies. This level must be built from the database of the initial comprehensive review of the use of the important energy sources. In terms of data processing and setting new targets, it is necessary in the management of saving energy to have data measured and compared to the reference, depending on the level of the major energy source. We must determine the time frame for achieving the goals and objectives that are in line with the government's energy policy. One of the basic conditions for fully controlled process of managing energy consumption is setting and reviewing goals and objectives. Government leadership in terms of energy use of monitored energy must take into account opportunities for improving energy management. By investigating the authorities in the Prešov region, we can conclude, that the leadership must also consider their financial and operational conditions related to the fulfillment of the main tasks of local government, technology options, including the views of stakeholders.

7 DISCUSSION AND CONCLUSION

Government leadership must consider opportunities for improving energy efficiency and operational management in the design of new, modified and renovated facilities, equipment, systems and processes that may have a material impact on the energy saving management. Results of the evaluation of energy management must be integrated into the activities related to the process of improving energy efficiency. Local government must inform the contractor about the acquisition of energy services, products and equipment that have or may have a significant impact on energy use, is partly evaluated on the basis of energy management. Local government leadership must develop and implement criteria to assess energy use and consumption during the use of products, equipment and services using energy. Energy management authorities must ensure that the key characteristics of its operations that determine energy management must be monitored, measured and analyzed at planned intervals. The results of monitoring and measuring key characteristics shall be recorded. Energy management in local government in cooperation with the management must determine the means and methods of measurement data measured, and shall examine and respond to significant deviations in energy management. The results of these activities should be maintained. Carrying out internal audits at planned intervals works to ensure, that the energy management system complies with the specified energy goals and objectives. Top management authorities must at specified intervals to reassess business energy management authorities and its effectiveness. Leadership of local government has to put emphasis on energy management in the period ahead and implement recommendations for improvement.

REFERENCES

AB Solartrip, s.r.o. [online]. [cit. 2015-04-07]. Dostupné z: http: //www.energyprukaz.cz/energeticke-rizeni-mt-monitoring-a-targeting/.

Benchmarking a schéma energetického manažmentu pre MSP—Energy Management Implementation Model—Short Description. vydané s podporou Intelligent Energy [online]. [cit. 2015-04-08]. Dostupné z: alpha.cres.gr/besss/elearning/bess/pdfs/sk/.../Tool_0.

Harčarik, M. 2011. Energetický audit a systém manažérstva hospodárenia s energiami podľa EN 16001/ISO 50001 [online]. [cit. 2015-04-07]. Dostupné z: https: //www.siea.sk/materials/files/poradenstvo/aktuality/2011/konferencia_audit_/prezentacie/03_Harcarik__Normy_upr.pdf.

Jašková, Ľ. 2015. Informačné systémy v organizáciách. [online]. [cit. 2015-04-08]. Dostupné z: http: //edi.fmph.uniba.sk/~jaskova/InformacneSystemy/tema03/tema03.html.

Sheldon, Ch. & Yoxon, M. 2008. Environmental Management Systems—A Step by Step Guide to Implementation and Maintenance. 3rd edition. New York. 280p. ISBN 978-1-84407-257-6.

STN EN ISO 50001 Systém energetického manažérstva. Požiadavky s návodom na používanie (ISO 50001: 2011).

Vlčej, J. 2013. Komparácia stavu životného prostredia v demokratických a nedemokratických štátoch z ideologického hľadiska. In: Vedecký obzor = Scientific horizont. Roč.5, č.2. ISSN 1337–9054.

Production Management and Engineering Sciences – Majerník, Daneshjo & Bosák (Eds)
© 2016 Taylor & Francis Group, London, ISBN: 978-1-138-02856-2

Corporate vision from management development in the globalization context

G. Tomek, V. Vávrová & P. Červenka
Faculty of Electrical Engineering, Czech Technical University in Prague, Prague, Czech Republic

J. Naščáková & M. Tomčíková
Faculty of Business Economics Košice, University of Economics in Bratislava, Košice, Slovak Republic

ABSTRACT: Long-term research has shown that the way to increase corporate competitiveness in a globally perceived environment is corrupted namely by factors, which inhibit development and technological progress, implementation of lean production, and last but not least satisfaction of customer's individual needs. The main problem emerges upon existence of latent and actual discrepancies between the interests of individual direct and indirect participants to the value creation process. Resolution of this problem is implemented by the authors in application of the process management, target-oriented at the customer. "Production operative management" represents in the internal value creation a chain integration of the process of planning, records-keeping, management methods and implementation of changes in the continuous process of sales management—production—procurement. Integration of the external value-creation process represents the supply chain management. Uniform database is a matter of a process of full standardisation, which currently enables flexibility upon answering customer requests.

1 INTRODUCTION

In the last ten years, we have dealt with research, which can generally be described as research of restructuring of the corporate managerial relations with regard to corporate competitive ability. That means both in terms of separate research assignments, and in terms of a wider task dealing with decision-making and management in industrial production within the scope of the Czech Technical University in Prague. Gradually, more than 350 business entities from the machine, electrotechnical, and consumption industries were addressed across the entire Czech Republic. The final project included 220 interviews held. Another source of information used was theses and dissertations, consultation with hands-on practice, participation in workshops etc.

Economic standing of a company and its objectives require that creation of benefit for the customer to be at the same time accompanied by acquiring a competitive advantage in the market.

2 ACTUAL AND LATENT CONTRADICTIONS WITHIN THE COMPANY

A theoretical starting point can be a reference to the analytical method by M.E. Porter (Porter 1993), who brought the value creation chain in theory and practice of strategic management in relation to potential assumption of competitive advantage (Magretta 2012). Company's competitive advantage is secured by company performance, which has a higher value than the competition (Gosling, Naim & Towill 2013). The value is expressed by the price, which the demanding party (buyer) is willing to pay for the performance provided. The purpose of the value-creation chain is to explain that company performance is, either directly or indirectly, a joint matter of a number of company activities, or company functions. Value creation chain analysis must thus be understood as an analysis of all company activities.

Contacts may also be based on information, contracts between parties involved or take-over of the distribution function (Woratschek, Roth & Pastowski 2002). The above-stated implies that it is necessary to leave the clearly organisational and functional attitude upon evaluating division of company activities, but—on the contrary—to consider as decisive activities those, which:

- Show a high potential of differentiation from the customer's viewpoint;
- Are significantly related to potential cost reductins.

A different attitude must be considered as there may be various causes and consequently problems upon harmonising the activities (Roberts, Kayande & Stremersch 2014).

In order to be able to analyse this issue in our research, we focused our attention on marketing as the driving force of the orientation on value-creation chain (Dorčák & Delina 2011). That is determined generally and without any further discussion by the fact that a company accepts management marketing concept in order to secure its success.

If we summarise analysis of overall environment, in which a company is operating in the beginning of the 21st century and attempt to express individual ideas in a fundamental thesis, it is possible to affirm that the decisive prerequisite for successful business activities is *company focus on creating benefit for the customer*. That means understanding what real benefit is for the customer and on the basis of that concentrating on development (construction and technology) and actual manufacture of a product (product or service) (Belz & Bieger 2004).

Someone quite aptly said that marketing is like a relay. It features numerous user interfaces with other functions. Where is the core of its sense? Namely in being able to interpret the customer's voice for the company and to take it across the entire value-creation chain upstream to the company's research and development. Despite existence of internal communication rules, there is murmur, errors, breaks, clashes and interruptions of required contacts and relations within the frame of the value creation chain of the company. Marketing function was accepted as the starting point of dispute analysis. As already stated, disputes originate as a result of existence of an interface between departments within the scope of value-creation chains (both with direct and indirect contact to customer) (Schütz 2002).

In terms of dispute analysis, activities were monitored that generally belong to the type of departments marked as: research and development, technical preparation of production, sale, production, procurement and marketing.

If we monitor the said data from the viewpoint of company size, it is obvious that representation of individual departments monitored is present at companies with over 251 employees. With the decreasing number of employees, representation of research and development and technical preparation departments of production is decreasing as well. Interesting data is provided by existence of a separate marketing department. This is where the figures range from 83.3% with companies of above 500 employees to 0–33.3% in companies with fewer than 50 employees.

In companies without a dedicated research and development department appropriate activities are secured by production technical preparation departments (25%). Often, this function is assumed by the parent company (12.5%). As regards production technical preparation department, dependency on company size is significantly reduced. A more detailed explanation regarding the issue of production technical preparation is provided by the percentage of outsourced companies. The sphere of technology is very often secured by the production department (33.3%), and research and development department (41.7%). In 50% of cases, the technological area is secured usually by the research and development department in co-operation with the production department.

The most frequent solution in companies without a sales/distribution department is merging with procurement and marketing in a single department called sales department (40%).

Procurement department is similarly often merged with sales and marketing department (57.1%). The marketing department exists basically in a smaller number of companies. Dependency on company size is apparent only with big companies, in case of small and medium companies it is approximately 1/4 of companies. The most frequent method of securing marketing activities is within the departments of sales and distribution (41.4%), further it is secured directly by company management (17.2%) or by a business partner (6.9%). An important role is played here also by the above-said merge of the departments of procurement/sales/marketing (13.8%).

A relatively significant share is represented by companies without segmentation into departments. Which for actual research of new solutions proposed plays an insignificant role, as the basis of task fulfilment analysis is the phases of the value-creation process (chain), and so the existence or non-existence of a dedicated department cannot generally negate further proposed results of solution to the clashes analysed.

Given the issue being dealt with, a response to the question about marketing department functionality is also interesting. This means a question of how it provides other company departments with information concerning the market, application of its own products, points to changes in customer requirements etc. Marketing department functionality is positively viewed by 69.6% of companies. Companies with a dedicated marketing department have better prerequisites of prosperity. The said trends that do not confirm this conclusion include only the trend of costs where the dependency is not clear. A similar conclusion applies also in terms of assessing marketing function fulfilment.

Unfortunately, very few specific answers were received regarding issues concerning insufficiencies in marketing work in a company. Specific answers were provided only by several respondents. What reasons did the respondents mention? *"Insufficient human resources/work is not a hobby to people", "Insufficient knowledge of products", "Low level of automation in work with data", "Poor organisation of marketing activities", "Insufficient exchange of information/cohesion", "Small range of customers".*

3 WHAT ARE THE SOURCES AND CAUSES OF DISPUTES?

Horizontal differentiation of tasks necessarily results in vertical differentiation. If marketing is not respected by management, then its function and mission in terms of business-making philosophy degenerates to promotion or sales of company excessive inventory. It is not reasonable to expect marketing to become a god on the Olympus of company top ma-nagers. It is rather about their mutual and equal co-operation. The worst reason for a clash of interest is following the principle of "this is mine and this is yours!" Then the clashes cerate cracks in the entire company organisation structure (Diller & Saatkamp 2002).

Other barriers in company department co-operation may be represented by issues originating from unclear or even conflict-involving determination of objectives to individual departments. Such a situation is even further complicated by insufficient professional qualifications, time pressure, and overload of both individual employees and entire departments. Insufficient personality prerequisites, application of bureaucratic attitudes, as well as lack of mutual trust often act as further complicating factors. Specifically, this shows as personal distance, egoism, space distance etc.

4 DISPUTES CAUSED BY MARKETING ITSELF

If marketing is to work as an initiator if impulses in terms of growth and innovation processes, it has to resolve its own issues as follows:

- Various ideas, which may originate between marketing and actual sales;
- Do not put blame impossibility of seeing through own concepts on negative attitude of others;
- Keep proving its contribution to creation of value in a measurable manner.

The starting point of competitive ability level is in our point of view namely attitude to creation of new products. In this sense, results are satisfactory because 63% of companies follow the potential area of demand, 22% by requirement for provision

of a higher benefit for the customer and the rest prefers use of technologies available to it.

5 BASIC PREREQUISITES FOR DISPUTE RESOLUTION

The above-stated implies that dispute and clash resolution rests on the following principles:

- application of marketing as the unifying concept of management;
- product management implementation;
- process-based management;
- communication pursuant to a uniform information basis.

Gradual interconnection of production systems inside and outside the company so that actual management is not interfered with, but mutual creative communication is enabled, briefly described as INDUSTRIE 4.0 (Industrie 4.0 2013; Takeda 2012).

5.1 *Full standardisation*

Full standardisation is understood to be its application in the following areas: in the management process, input elements of production process, further standardisation of activities and methods of conversion, standardisation of relations in consumption and use of production agents (Lyonnet & Toscano 2014). Eventually, there is also standardisation by combination upon the actual planning and management of production process and its input elements. Therefore its preparation, implementation and changes must involve (next to the departments of research and construction) the department of technology, production, procurement, sales, logistics, and marketing (Yao 2011). Losses must be defined as wasteful action, according to the Japanese attitude defined as "muda" (Takeda 2012).

We are giving some examples characterising insufficient appreciation of standardisation in corporate procedures—see Figure 1 (companies from the entire set—in %).

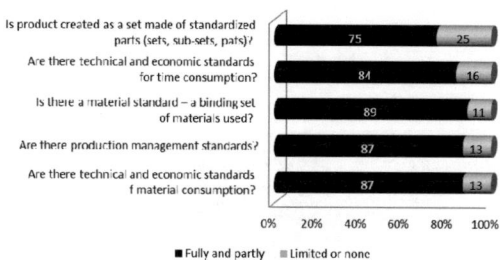

Figure 1. Standardisation within the scope of the companies reviewed.

Full standardisation must simultaneously address the above-stated issue of production process restructuring:

1. Eliminate those parts of the process, which do not comply with the technological standards accepted.
2. Secure repeated production of standard parts using the principles of economy of scale, savings on quantity, product range, speed and competence.
3. Decrease the number of different options applied to production process.
4. Integrate production process on the basis of modular organisation, if need be, using parallel production processes (Pels—Wortmann 1990).
5. Application of substitutions during production process.
6. Application of effective co-operation outside the company.

Figure 2 represents basic ideas of full standardisation as regards standardisation process elements, its basic directions, content and information, which it provides for company integrated management.

5.2 Production operative management

The decisive elements of the internal value creation chain are sales—production—procurement (Worats-chek, Roth & Pastowski 2002). The first prerequisite is co-operation upon decision-making about product range plan with the initiator, i.e. sales. Results are given in Figure 3 (companies from the entire set—in %).

Insufficiencies in co-operation within the value-creation chain then reflect in insufficient accommodation of customer requirements and then namely upon changes to their requirements and seeking alternative solutions. Results are shown in Figures 4 and 5—companies from the entire set—in %.

Figure 2. Full standardisation.

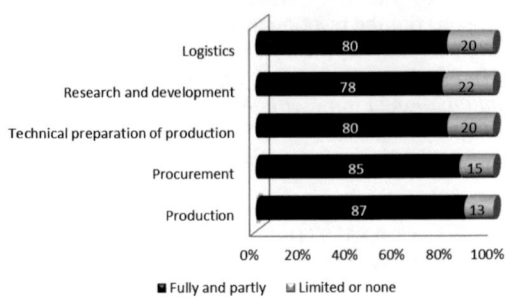

Figure 3. Co-operation upon decision-making on production plan with sales department.

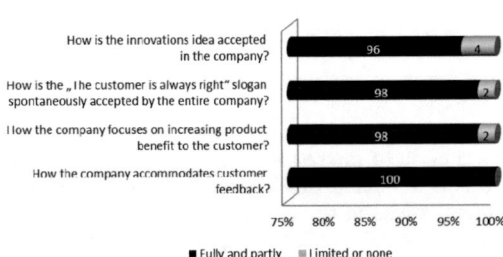

Figure 4. Relation between the company and customer requirements.

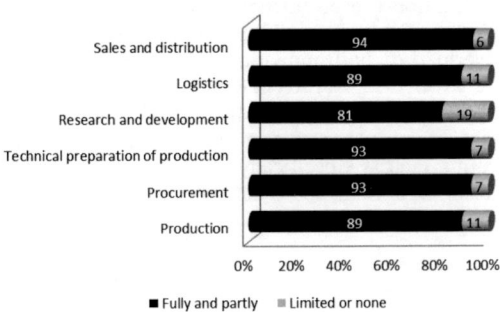

Figure 5. Accommodation of changes in requirements and seeking alternate solutions.

Figure 6 shows integration of internal value-creation chain activities, including management tools of its planning, records-keeping and management, with regard to relations both to market, and to other supporting activities of the value-creation chain.

Production operative management represents a complex tool, which within the frame of the company management system follows a common objective, common input information (standardised normative base) and resolution to mutual relations in order to liquidate friction areas between individually managed areas (departments) within the scope of the value-creation process.

Figure 6.　Production operative management.

The basis here is an operative plan, which represents a complex management tool consisting in closely linked plans of sales, production, and procurement. This is how the key players are united, i.e. those who determine what, when and using what is to be produced (Nosoohi & Nookabadi 2014).

5.3　Supply chain management

Progress towards a lean and smart company leads to multilateral support of partnership within the frame of the entire range of contractor and buyer relations (Alfalla-Luque, Medina-Lopez & Dey 2012). Again, it is possible to use the knowledge of how contractors accommodate the internal value-creation chain—see Table 1.

The extent to which certain aspects of strengthening competitive ability using supply chain management are used in an insufficient manner are then shown in Figure 7 (companies from the entire set—in %) assessing company flexibility from this point of view.

Figure 8 shows integration of external value creation chain activities, especially individual partners and their activities, with which they enter the integrated process. In relation to that, it expresses the mission of this partnership chain for the future (Baumgarten et al. 2004).

The question of a company's success, i.e. success of its value-creation chain, consists in conditions of globalisation and network and actual interconnection of companies and the issue of success of all companies participating in creation of the resulting product value (Lu, Meng & Goh 2014).

Competition between individual companies changes into competition between entire chains starting with contractors, continuing with production companies and ending by sales agents and end users (Roberts, Kayande & Stremersch 2014). That

Table 1.　Willingness of contractors upon changes to requirements.

	Companies from the entire set—in %
Fully and thoroughly	47
Only partly	47
Very little	6

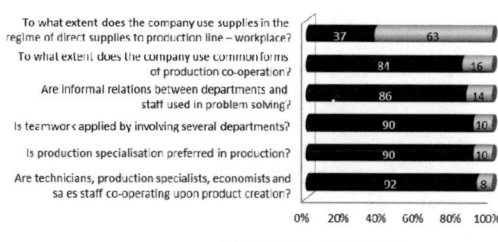

Figure 7.　Company flexibility.

Figure 8.　Supply chain management.

means that competitive ability and effectiveness of such a new value-creation chain if contributed to by a number of contractors, co-operating producers, sales agents and logistics service providers (Gallear, Ghobadian & Chen 2012; Stankovič & Bašistová, 2015). Fulfilment of the sense of this chain requires managed co-operation exceeding the company by its individual participants being co-ordinated upon creating relevant business-making processes (Gosling, Naim & Towill 2013; Groznik & Heese 2010).

The industrial base of the future is thus interconnection of scientific and research components with those of operations. There is no way this topic can be limited only to big companies. Today, it has to be dealt with by all companies provided they do not want to stay aside increasing competitive ability. Assessment of these principles is not only a question of information, but actual creation of networks, within the frame of which it is necessary to learn mutually. This is related to a number of issues to be resolved within the chain:

1. It is necessary to count on worldwide presence of the company.
2. There is a new possibility of gathering detailed information.
3. Customer interests should be analysed within a wide range of partners.
4. It is necessary to seek ways of shortening reaction time between participants to the chain.
5. Necessity of resolving cost saving issues jointly.
6. Identify with the possibility of creating a virtual company.
7. Dealing with the relation and range of activities of small companies.

6 CONCLUSION

Economic position of a company and its objectives require that creation of value for the customer be accompanied by acquiring a competitive edge. This is the only way a company can be successful in the market and keep successfully developing its business activities. The basis of value creation for both the customer and the company is the production process, i.e. those activities the company undertakes that are directly participating in product making. In a company, production factors transform into products and services, which necessitates combination, coordination and performance of various activities. Individual activities must create value in such a way that the result of one activity provides a higher value than the sum of all production factors applied. For this purpose, it is necessary that the company analyses the specifics of its production process and creates suitable organisation prerequisites so that an optimum relation is created between product and process profile.

Pull strategy application gives the internal/external chain overall concept a new sense. This principle expresses the fact that both within the company value-generating chain (internal supply chain), and within the entire supply chain (external supply chain), the entire product creation project implementation process is always "carried on" by that level (implementation) of the chain, which is closer to the customer. No intra-company activity should take place and no participant in the supplier-customer chain should make any action upstream the value-generating chain than as requested by customers (internal or external) downstream. That

means that all activities are designed to customer wishes (including individual participants in the chain acting in the role of customers here) and thus it is the customers who carry on the implementation process. Such partnerships are mutually conditional with the customer partnership.

ACKNOWLEDGEMENT

This essay originated as a result of a research project MSM 6840770038 entitled "Decision Making and Control for Manufacturing III, TU 11 Management of a company and its competitive ability".

REFERENCES

Alfalla-Luque, R., Medina-Lopez, C. & Dey, P.K. 2012. Supply chain integration framework using literature review, *Production Planning & Control*, Volume 24, Issue 8–9, p. 800–817.

Baumgarten, H., Darkow, I. & Zadek, H. 2004. *Supply Chain Steuerung und Services*. Berlin, Heidelberg: Springer Verlag.

Beckmann, H. (Hrsgb.) 2004. *Supply Chain Management*. Berlin, Heidelberg: Springer Verlag.

Belz, CH. & Bieger, T. 2004. *Customer Value*. St. Gallen: Redline Wirtschaft, Verlag Thexis.

Diller, H. & Saatkamp, J. 2002. Schwachstellen in Marketingprozessen, *Marketing ZPF/2002*.

Dorčák, P. & Delina, R. 2011. Impact of electronic marketing business solutions to economic performance, *Business magazine*, Vol. 1, Isseu 59, s. 44.

Gallear, D., Ghobadian, A. & Chen, W. 2012. Corporate responsibility, supply chain partnership and performance: An empirical examination, *International Journal Production Economics*, Volume 140, Issue 1, p. 83–91.

Gosling, J., Naim, M. & Towill, D. 2013. A supply chain flexibility framework for engineer-to-order system, *Production Planning & Control*, Volume 24, Issue 7, p. 552–566.

Groznik, A. & Heese S. 2010. Supply chain interactions due to store-brand introductions: The impact of retail competition, *European Journal of Operational Research* 203 p. 575–582.

Industrie 4.0. 2013. *Recommendations for implementing the strategic initiative Indrustrie 4.0*. Germany: National Academy of Science and Engineering, www.acatech.de

Lu, Q., Meng, F. & Goh, M. 2014. Choice of supply chain governance: Self-managing or outsourcing? *Int. J. Production Economics* 154, p. 32–38.

Lyonnet, B. & Toscano, R. 2014. Towards an adapted lean systém—a push-pull manufacturing strategy, *Production Planning & Control*, Volume 25, Issue 4, p. 346–354.

Magretta, J. 2012. *Michael Porter—Clear and understandable*. Praha: Management Press Marketing ZFP, Sonderheft 12/2002.

Majerník, M., Bosák, M., Szaryszová, P., Tarča, A., & Štofová L. 2014. *Model for assessing the environmen-*

tal performance of product systems within sustainable development. In: Geoconference on ecology, economics, education and legislation SGEM 2014: 14th international multidisciplinary scientific geoconference, Albena, Bulgaria. volume III.—Sofia: STEF92 Technology Ltd., 2014. ISBN 978-619-7105-19-3. pp. 285–292.

Nosoohi, I. & Nookabadi, A. 2014. Designing a supply contract to coordinate supplier's production, considering customer oriented production, *Computers & Industrial Engineering* 74, p. 26–36.

Pels, H.J. & Wortmann, J.C. 1990. Modular design of integrated databases in production management, *Production Planning & Control*, Volume 1. Issue 3, p. 132–146.

Porter, M.E. 1993. *Konkurenční výhoda*. Praha: Victoria Publishing.

Qrunfleh, S. & Tarafdar, M. 2014. Supply chain information systems strategy: Impacts on supply chain performance and firm performance, *Int J. Production Economics*, Volume 147, Special Issue, p. 340–350.

Roberts, J., Kayande, U. & Stremersch, S. 2014. From academic research to marketing practice: Exploring the marketing science value chain, *Intern. J. of Research in Marketing* 31, p. 127–140.

Schütz, P. 2002. Bruchstellen im Marketing. Die tausend Tode der Effizienz, A*bsatzwirtschaft*, Sonderausgabe zum Deutschen Marketing-Tag 2002.

Seebacher, G. & Winkler, H. 2014. Evaluating flexibility in discrete manufacturing based on performance and efficiency, *Int. J. Production Economics* 153, p. 340–351.

Stankovič, L. & Bašistová, A. 2015. *Introduction to the management of corporate social responsibility.* Kosice: VÚSI, 2015. 198p.

Takeda, H. 2012. *Das synchrone Produktionssystem, 7. Auflage.* München: Verlag Franz Valen.

Vahrenkamp, R. & Siepermann, CH. 2004. *Produktionsmanagement.* München: Oldenbourg Verlag.

Woratschek, H., Roth, S. & Pastowski, S. 2002. Geschäftsmodelle und Wertschöpfungskonfigurationen im Internet.

Yao, J. 2011. Supply chain scheduling optimisation in mass customisation based on dynamic profit preference and application case study, *Production Planning & Control*, Volume 22, Issue 7, p. 690–707.

Production Management and Engineering Sciences – Majerník, Daneshjo & Bosák (Eds)
© 2016 Taylor & Francis Group, London, ISBN: 978-1-138-02856-2

Error prevention at creation of concept of product under development

R. Turisová
Faculty of Mechanical Engineering, Technical University, Kosice, Slovak Republic

M. Tkáč
Faculty of Business Management, University of Economics in Bratislava with Seat in Kosice, Kosice, Slovak Republic

ABSTRACT: Errors arising in stage of creation of concept at development of new products generally have a major impact on functionality, design as well as duration or effectiveness of product development. It is therefore very important to prevent the above mentioned conceptual errors already in the early stages of product development. Presented article presents a tool the purpose of which is as far as possible to prevent the mentioned types of errors. It is a modification of classic method aimed to prevention of rise of potential errors (FMEA). Whereas, in the initial stage of product development it may be possible to use experimental procedures or experience of developers to a very limited extent, it is appropriate to approach the subjective evaluation used within the method of FMEA as the expert judgment.

1 INTRODUCTION

At development of new product or service, the stage of creation of concept of such new product is of cardinal importance. It is a stage of development in which even major interventions are generally financially undemanding. At this stage, however, there are created basic preconditions which in turn significantly affect functionality, reliability and efficiency of newly developed product. However, if an error of conceptual nature occurs which is observed by developers in later stages of product development (often in the creation of a prototype or even during production or sale), it is a very expensive matter. The effort to develop a product as well and effectively as possible, i.e. without expensive changes in the final stages of development. To this end, the method of FMEA (Failure Mode and Effect Analysis) is used in practice aimed at preventing errors when creating product concept. Classic method of FMEA (Turisová 2010) is based on a subjective evaluation of the assessors. This causes considerable problems, especially in cases when opinions of the assessors are diametrically different. As in the initial stage of product development at designing completely new, in practice untested, concepts, achievement of consensus among evaluators is often problematic, it is appropriate to approach the subjective evaluation used within the method of FMEA as the expert judgment. This approach allows to solve effectively even such cases the evaluators opinions are significantly heterogeneous. In this article, we present such a modification of the method of FMEA aimed at conceptual errors prevention that uses the Cooke model in assessing used at the expert assessment (Bedford & Cooke 2001; Cooke 1991).

1.1 Method of FMEA

The classic method of FMEA of the concept uses an ordered triplet of integral values, each ranging from 1 to 10 to determine a seriousness of the potential error, its cause as well as its effect. First, it is important to identify all possible causes of all the errors accruing in the account when the concept of products under development will have any particular proposed shape. The effort of so applied method of FMEA is to prevent possible complications as well as unnecessary limitations in further stages of product development because of conceptual error. In this case we talk about conceptual vulnerability of product under development (Yang & El-Haik 2003).

To each identified triplet—error, cause and consequence of error, there is assigned a triple of numbers characterizing (Turisová 2010):

P probability of error occurrence,
V importance of error and
Z probability of error finding or detection.

By product of mentioned number characteristics, we reach a value *M* marked also as *MR/P* (risk degree/priority) (Tkáč 2001):

$$M = P \cdot V \cdot Z \qquad (1)$$

Based on this characteristic, it is possible to arrange any potential errors, causes and consequences. The largest value represents the highest degree of risk, but numerically the lowest, i.e. first priority in terms of urgency of corrective actions taken or alternations made.

1.2 Interval FMEA

The term I-FMEA (interval FMEA) is understood as modification of FMEA to the intent that the integral ranges from 1 to 10 for the individual parameters P, V, Z will be replaced by the intervals of real numbers. It means that individual characteristics may acquire any real value in the range <1, 10>. Moreover, the evaluator will estimate above mentioned values of each parameter by means of quantiles characterizing the distribution of random values. Such an approach brings in itself gives more information about the quality of the estimation than a point estimation. Using quantile method, it is possible to estimate not only the confidence interval, but approximately also the shape of the density of distribution of random quantity. In doing so, we go out from the subjective definition of probability (Savate 1972).

Further, we will show why the point estimate in standard FMEA causes reduction in accuracy of the method. Parameter M, which is the basis for the assessment of any pair, error—cause, as we know, arises as a product of probability of occurrence P, significance of error V and the probability of detection Z. At classic FMEA, the assessor can chose only integral number from 1 to 10 for each factor. Resultant number M can acquire only some of integral values. If the frequency of occurrence of these values is processed in histogram (Figure 1), wherein the red line represents the density of Weibull distribution as the distribution of values that best characterizes the file. Median of above mentioned distribution has a value of

$\tilde{x} = 166.375$. However, if at the specific cause of error we get number M that is less than 125, it means in most cases acceptable risk which does not need any intervention. If we consider values M responding to critical state ($M > 500$) we keep at our disposal 17 different values from all 120 mutually different number values that is 15%. Provided that from 500 to 600 there are 7 possible values, from 600 to 700 there are 4 possible values, from 700 to 800 there are 3 possible values, from 800 to 900 there are 2 possible values and over 900 there is 1 possible value.

Mentioned state can be considered poor resolution of the method. Let us consider relation $M = P \cdot V \cdot Z$, but $P \in$ <1, 10>, $V \in$ <1, 10>, $Z \in$ <1, 10 > a $M \in$ <1, 10>.

Unlike point estimation, now we will understand above mentioned values as continuous from the interval from 1 to 10. Value M is therefore also a continuous value from 1 to 1000. Let us mark by $P_{0.05}$, $P_{0.5}$ and $P_{0.95}$ such values from the interval $P \in$ <1, 10> which represent respective quantiles of density of continuous distribution over the mentioned interval characterizing the estimation of the assessor and his/her conviction that real value of the objective reality is governed by density proportionality of which is determined just by above mentioned quantiles.

Figure 2 shows the density and consequently the empirical distribution function of estimation of parameters P, V, Z. On X-axis there are shown particular quantiles. The density as well as the distribution function is from the expected uniform distribution.

In Figure 3, we have the density and the distribution function of value M while for each i of the closed interval from 0 to 1, it applies that

$$M_i = P_i \cdot V_i \cdot Z_i \qquad (2)$$

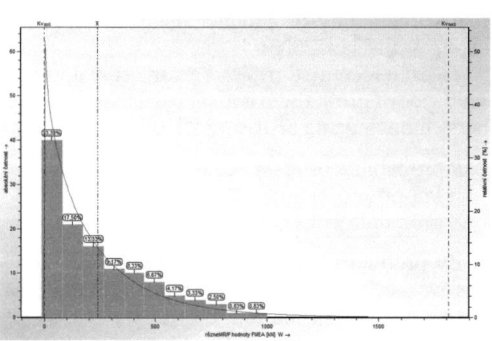

Figure 1. Histogram of different values of parameter M.

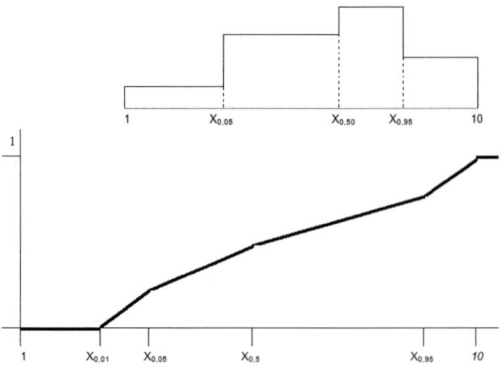

Figure 2. Density and distribution function of parameters X.

306

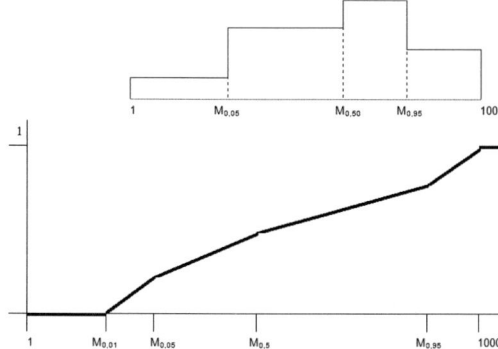

Figure 3. Density and distribution function of parameter Mi.

Table 1. Values of particular quantiles at interval FMEA.

a)				b)
Δ	p0.05	v0.05	z0.05	M0.05
$\cdot$	p0.5	v0.5	z0.5	M0.5
$\tilde{N}$	p0.95	v0.95	z0.95	M0.95

In a form of I-FMEA the assessor enters to each identified combination of error, cause of error and consequence of error respective quantiles for each characteristics of P, V, Z as shown in figure 3 (Table 1).

At I-FMEA, the assessor makes estimation for each of factors consisting of three numbers (Table 1a). By simple product of the respective quantiles, we achieve value M_i (Table 1b). We suggest to use, as a first estimation, quantile 0.05, second estimation—quantile 0.5 (median) and third estimation—quantile 0.95. The assessor may use non-integral values and therefore he/she estimates a whole interval of values which is by him/her admissible. I-FMEA has therefore significantly finer range and the assessor does not do point estimation but by means of quantiles he/she makes an estimation of distribution function. Laically, it can be interpreted so that lower limit of the estimation (Δ) expresses the lowest expected number of the respective parameter, i.e. for example the smallest expected significance of error.

The mean value ($\cdot$) corresponds to median of distribution and upper value of estimation (?) present the highest possible expected number, i.e. in case of significance the biggest expected significance. The assessor so first determines the "optimistic value", i.e. the lowest number corresponding to the smallest value. Then he/she determines the "mean" and finally the "pessimistic" estimation,

i.e. such which is by him/her as high as possible. By product of mutually corresponding quantiles of particular parameters P, V, Z we get the first iteration (some first value) of the estimation of the density of distribution function for parameter M. As it is an estimation achieved by calculation, the assessor can make any final correction of any quantile at his/her own discretion. To so modify FMEA it is possible directly to apply the Cooke model of the expert judgement.

2 COOKE MODEL OF EXPERT THE EXPERT JUDGEMENT

Cooke model (Cooke 1991) belongs to the so-called Non-Bayesian methods expert judgment. Influence of a particular assessor (expert) on final evaluation of parameter M depends on his/her performance. For assessment of performance of evaluating experts, two criteria are used. The first one is called as accuracy of estimation (calibration) and second one—informativeness. Calibration within this model is an ability of expert to achieve status when the actual value of the estimated value will lie in the middle of an interval defined by experts by means of quantile method. On the other hand, informativeness deals with the problem of the confidence interval size by means of quantile method while, of course, the interval estimation the range which is smaller is a better assessed. Calibration and informativeness are defined by means of so called calibration questions, i.e. questions to which the final assessor know the answers but the expert does not know it.

We can measure quality of calibration of the expert upon the differences between the empirical distribution of calibration variable and distribution defined by the expert, it means that the calibration is a probability characteristics of the tests of statistical hypotheses defined for each expert. Implementations can be understood as independent samples from distribution corresponding to quantiles estimated by the expert (Turisová et al. 2012).

The assessor prefers the experts whose statistical hypotheses respond to data achieved from the empirical estimation of the calibration variable distribution. More clearly but less accurately said, the assessor prefers probability estimation at the most corresponding to reality.

Let us assume that we observe a set N of such calibration variables that $s_1 N$ implementations come from the interval 0–5%, $s_2 N$ implementations are from the interval 5–50%, etc. Then the empirical density has a form $(s_1, \ldots s_4)$ and we want to measure how close it is to the hypothetic density $(p_1, \ldots, p_4) = (0.05; 0.45; 0.45; 0.05)$.

The way how to measure this closeness is provided by so called relative information in consideration of p defined by relation

$$I(s; p) = \sum_{1}^{4} s_i \log\left(\frac{s_i}{p_i}\right) \quad (3)$$

It is a non-negative value which achieves the minimum, that is 0 if and only if $s = p$. A good expert should have the empiric density $(s_1, \ldots s_4)$ close to $(p_1, \ldots, p_4)$ and so he/she should have the relative information close to 0. It is well known (Cooke, 1991), that for big N, the distribution of relative information (on range of $2N$) is well approximable χ—quadrate by distribution with three degrees of freedom

$$P(2NI(s, p) \leq x) \approx \chi_3^2(x) \quad (4)$$

where χ_3^2 is a distribution function χ—quadrate with three degrees of freedom. Calibration of the expert e is defined as probability of delivery (obtaining) of information of worse (bigger or equal) as currently obtained provided that the expert distribution is $(p_1, \ldots, p_4)$, then

$$C(e) = 1 - \chi_3^2(2NI(s, p)) \quad (5)$$

Thus, the empirical density s equal to hypothetical density p gives the best possible calibration equal to 1. However, we note that calibration is not the only way to measure the quality of expert evaluation. Other criterion is informativeness. Informativeness measurement is a basic measurement assigned to each demanded value. This basic measurement is most often uniform or log-uniform (governed by uniform or log-uniform distribution) above the inherent range of each variable. Inherent range is achieved by adding some k %, it means by extension of the smallest interval containing all quantiles and implementations. k is generally defined by the assessor, the most often value is $k = 10\%$). Distribution densities are associated with assessments of each expert for each demanded variable in such a way that:

– the densities agree with the expert quantile estimations and
– the densities are at least informative considering the base of measurement defined by quantile limits.

Informativeness is evaluated with regard to each variable and each expert by means of calculation of relative information of the expert's density for this variable considering basic measurement.

Informativeness is a positive number. Its increasing value indicates an increased relative informativeness on basic measurement.

2.1 Determining weights

For determining weight, based on performance for each individual experts, knowledge of his/her informativeness and calibration. At quantification of above mentioned weights, before start of demanding, the final assessor will define certain basic level of success α. To each expert whose calibration will be lower than level α, the weight in value of will be automatically assigned. It is important to become aware that the weight w_i of the expert i used at evaluation of expert opinions has no interpretation in a form of probability that the expert i evaluates correctly. The fact that the expert weights present in total 1 does not mean that just one expert decides correctly. It is also possible that two experts giving diametrically different answers had exactly the same proportional representation realized by weights. It is therefore not possible to consider the ability of experts to answer correctly for an exclusive feature of experts.

Valuation rule R for determining the unknown variable reaching a value of 1, …, 1000 is a function in a form of $R(p, i)$ for the probability answer p at implementation of i. The expected value for the subjective probability p, if the expert believes that the true value of the parameter M has a distribution q is

$$\sum E_q R(p/i) = \sum q_i R(p, i) \quad (6)$$

We say that the decision rule is appropriate if for all p and q is this $Eq R(p/i)$ maximized only if $q = p$.

Evaluation rule is $R(q, i) = \log q_i$. Then the expected value assigned to subjective probability p is $\sum p_i \log(q_i)$ that is known as relative information. In a model, we will use more than one calibration value and so the generalization of an idea of appropriate evaluation rule is used to provide evaluation based on a group of estimations and implementations. Let us assume that the expert believes that set of m unknown variables M_1, …, M_m acquires values of 1, …, 1000 and has distribution Q. The expected relative frequency of result i is

$$q_i = E^{\frac{\#\{M_j = 1\}}{m}} = \frac{1}{m} E\left(\sum_j {}^1 M_j = i\right) = \frac{1}{m}(M = i) \quad (7)$$

We assume that we have evaluation rule $R(p, m, s)$. If the expert defines expected relative results with frequency p on set m of variables while the observing relative output frequency is s, then the result expected by the expert is

$$E_Q(R(p, m, s)) \tag{8}$$

2.2 Approximation of expert distributions

In implementing the model, we ask experts to define limit numbers for quantiles (5%, 50% and 95%) of the parameter M for each cause of error under consideration. Since there are available 5%, 50% and 95% quantiles, we must interpolate the rest of the distribution for each variable demanded. Let $qi(e)$ be i % quantile of the expert e. Internal range will be achieved as it follows with k% exceeding coefficient. First, we will find the minimum and maximum values based on relations

$$l = \min\{q_5(e), r \mid e\}, h = \max\{q_{95}(e), r \mid e\} \tag{9}$$

where r is a value of implementation. Then i this applied that

$$q_l = l - 0,1 \times [h - l] \tag{10}$$

and likewise

$$q_h = h + 0,1 \times [h - l] \tag{11}$$

Natural (own) range is so $|q_l, q_h|$. Distribution function of the expert e is then achieved by approximation by means of linear interpolation of quantile information $(q_l; 0)$, $(q_5; 0.05)$, $(q_{50}; 0.5)$, $(q_{95}; 0.95)$, $(q_h; 1)$. It is a distribution with minimum information with regard to the uniform distribution on natural range that includes in it all expert quantiles.

Calibration, which usually controls the assignment of weights, depends only on quantiles and not on interpolation. On the other hand, informativeness depends only on quantiles and on a choice of q_l and q_h. However, the interpolation makes a difference in combined—overall assessment, i.e. at the estimation of distribution connecting all distributions of experts, so called combined expert. So it influences a limit level for all weights.

Previous procedure defines a distribution function $Fe(t)$ for each expert and each variable demanded. At the weights of experts w_e, a combined distribution function has a form of

$$\sum_e w_e F_e(t) \tag{12}$$

To determine the weights we must choose a threshold limit value of success α. For each selection of threshold limit value, the weights are changing (weights depend on α because for higher values of α, there are excluded more experts and weights are more focused on remaining experts).

Therefore also a "combined expert", which is a combination of other experts, also depends on α.

For a combined expert, we can also calculate 5%, 50% and 95% quantiles and then also calibration and informativeness. In the model, we then select the weight with respect to the expert, such that the combined expert should be given to a group of experts.

2.3 Expert interval FMEA

Expert interval FMEA (EI-FMEA) is such a modification of method of FMEA in which an interval FMEA is used as an estimation of risk and final assessment is executed by the expert evaluation in terms of the Cooke model. Calibration errors should constitute about 10–20% of all evaluated errors in the method of the FMEA. The final assessor of EI-FMEA can get the real values of the parameter M for the calibration error from the assessment of similar concepts, products, or by an exact experimental evaluation.

3 CONCLUSION

Conceptual errors that are not detected in time generally bring with themselves very expensive interventions especially in the final stages of product development. On the other hand, in the early stages of product development, i.e. in stages when their concept rises, there is a number of problems to be solved, a number of variables that need to be at least partially defined and a whole series of decisions that must be taken. Therefore a method for errors prevention in this stage can substantially increase the efficiency of product development. Classic method of FMEA implemented in conditions of concept vulnerability reduction is on the one hand very simple and easily understandable. On the other hand, it is highly sensitive from the reason of subjective assessment of the assessors. To use classic methods of expert assessment aimed to above mentioned sensitivity minimizing, we executed necessary modifications so that Cooke model of expert assessment at risk parameters evaluation within the method of FMEA could be implemented. Increased computational complexity is manifested only in the final assessment which can be simplified using appropriate statistical software. Computational complexity of the assessors—experts was not increased greatly. After completing a short training aimed to explanation of the quantile method in the estimation of distribution density, they are usually able practically to implement the present method.

ACKNOWLEDGEMENT

This contribution is the result of the project implementation VEGA No. 1/0150/15 Development of new method of implementation and verification of integrated machinery safety systems equipment systems and industry technology.

REFERENCES

Bedford, T., & Cooke, R. 2001. *Probabilistic risk analysis: foundations and methods*. Cambridge University Press.

Cooke, R.M. 1991. *Experts in uncertainty: opinion and subjective probability in science*.

Savage, L.J. 1972. *The foundations of statistics*. Courier Corporation.

Tkáč, M. 2001. *Štatistické riadenie kvality*. Bratislava. Ekonóm. 312 pp.

Turisová, R. 2010. *Štatistická analýza možných chýb a ich dôsledkov. Zlepšovanie procesov pomocou štatistických metód*, 1–20.

Turisová, R., Mihok, J. & Kádárová, J. 2012. Verification of the Risk Assessment Model through an Expert Judgment. *Quality Innovation Prosperity*, *16*(1), 37–48.

Yang, K. & El-Haik, B.S. 2003. *Design for six sigma* (pp. 184–186). New York: McGraw-Hill.

Production Management and Engineering Sciences – Majerník, Daneshjo & Bosák (Eds)
© 2016 Taylor & Francis Group, London, ISBN: 978-1-138-02856-2

The role of improvement tools in Polish production companies

M. Urbaniak

Management Faculty of University of Lodz, Lodz, Poland

ABSTRACT: The main purpose of this paper is to present the role of organizational improvement tools (like quality and environmental management systems, elements of Toyota Production System, Lean Management as well as Six Sigma) in Polish production companies. The article describes the results of an empirical study carried out in 182 enterprises operating in the Polish market. The aim of the survey was to identify the main decisive factors for the implementation of organizational improvement tools in production companies and in their support services providers. The results of the empirical study indicate that companies operating in Poland implement various organizational improvement tools. The main reason for the transfer of these tools is the requirements of the multinational concerns towards their suppliers which focus on ensuring technical quality, improvement of environmental performance, reducing costs (through an increase in the efficiency of processes) and shortening operational processes cycles.

1 INTRODUCTION

Starting from the 1980's of the last century, especially in developed countries such as the United States, Japan, United Kingdom and Germany, one can observe significant interest in implementing quality management systems—especially in the manufacturing sector. Increased competition in international markets caused large multinational companies as well as small and medium-sized local businesses to become highly interested in the implementation of Quality, Environment, Health and Safety (QEHS) management systems, as well as the introduction of others tools (like Toyota Production System or Lean Management) to improve the quality of products and processes. The dissemination of organizational standards based on the PDCA cycle in many countries of the world gave a possibility to carry out an independent assessment by independent supervisory bodies, confirmed by relevant certificates by the effective implementation of QEHS management systems. Effective implementation of QEHS management systems allows companies to control the process (define objectives, performance standards and evaluation methods) in order to achieve the expected results of processes and products. Dissemination of the QEHS management system certifications caused the formal confirmation of implementation of the requirements comprising the organizational standards in the form of appropriate certificates to no longer be a differentiating factor for competitive advantage.

2 A HOLISTIC APPROACH TO IMPROVING PROCESSES

Analysing the activities of enterprises, it can be perceived that they are often implementing integrated management systems based on organizational standards, which stem from the concept of improving the PDCA cycle (Bernardo 2014, Bernardo et al. 2012, Jankalová 2012, Jørgensen et al. 2006). In many companies, especially in global corporations, a holistic process improvement approach can be noticed, based on the implementation of QEHS management systems depending on international organizational standards (like ISO 9001, OHSAS 18001, ISO 14001) as well as on the concept of the Toyota Production System (mainly Kaizen, 5S, Total Productive Maintenance), Lean Management, and Six Sigma methodologies (Ringen et al. 2014). This holistic approach can further contribute to optimizing the use of resources and improving the effectiveness and efficiency of the processes, building relationships with stakeholder organizations (especially with customers, suppliers and employees), removing the symptoms of waste, shortening cycles of operational processes as well as reducing the negative impact of the processes carried out by organizations and their products on the environment (Idri & Zairi 2006, Koksal et al. 2011, Pool et al. 2011, Ringen et al. 2014, Sahno et al. 2015).

3 PROCESS IMPROVEMENT METHODOLOGY

According to the guidelines of international QEHS management standards, the concept of continuous improvement process should start from establishing objectives and ensuring availability of resources (such as staff qualifications, infrastructure, information, environment, relationships with suppliers and financial resources). Implementation of operational processes (like customer communication, design and development of product, purchasing, production and service provision) should be carried out in accordance with the guidelines contained in the international standards organization. Evaluation of the results of companies can be carried out by analysing the information to meet the needs and expectations of customers and other stakeholders, analysing the effectiveness and efficiency of the processes (performance matrix), audits results (internal and external), quality cost accounting, self-assessment, benchmarking and risk analysis of emerging threats (Colledani et al. 2014). Process improvement can be realized by taking corrective action (eliminating causes of occurring errors), preventive action (aiming to reduce the level of risk the appearance of non-compliance) and the introduction of ongoing improvements in operations/manufactured products (shortening the processing time, reducing costs or improving the technical parameters), as well as by the implementation of new processes or products. The implementation of these changes should be monitored and evaluated in terms of achieving its objectives, standards compliance (operational processes should be in accordance with the Standard Operating Procedures—SOP) and perceived risks of threats. The final stage of the process improvement methodology is an evaluation of compliance with the objectives as well as process optimization, focused on improving the effectiveness and efficiency of the company (Koksal et al. 2011, van Scyoc 2008). Improvement of the organization must be focused on pro-active problem solving, introducing product and process innovations that will shape the company's competitive advantage (Terziovski & Guerrero 2014).

4 THE ROLE OF PROCESS IMPROVEMENT TOOLS IN BUILDING THE RELATIONSHIP BETWEEN SUPPLY CHAIN PARTNERS

Original Equipment Manufacturers require individualized requirements for quality assurance specifications (ensuring technical quality and product safety), organizational effectiveness (time delivery), process efficiency (cost reduction) and reducing environmental impact from their current and potential suppliers, as well as having an adequate capacity to guarantee the implementation of product and process innovations. This approach is an important challenge for suppliers to make efforts related to the introduction and improvement of QEHS management systems, as well as other organizational tools such as elements of the Toyota Production System (Kaizen, 5S, TPM) Lean Management, and Six Sigma or methodologies (like DMAIC, DMADV). The implementation of such solutions engages providers to show more active involvement of employees in order to standardize operational processes and improve performance. Multinationals investing in many countries of the world, do not focus on selected purchasing sources but increasingly diversify them. These companies are looking for new sources of purchase, creating linkages with local suppliers that qualify for them. The starting point in building these relationships is a preliminary assessment of potential partners based on self-assessment questionnaires and audits at suppliers. This qualification is based on multi-criteria assessment methods. In the evaluation, the verification criteria are not only focused on the quality (safety of products and processes), timeliness of delivery, costs and the protection of the environment but also an important role is played by flexibility and ethical behavior (defined in the relevant codes of conduct for suppliers) (Azadeh et al. 2014, Dobos & Vörösmarty 2014, Morita et al. 2015, Rajesh & Ravi 2015).

5 THE RESULTS OF THE EMPIRICAL STUDY

The aim of the survey was to identify the main decisive factors for the implementation of organizational improvement tools in productions companies and in their support services providers. The empirical study was carried out from September to December 2013 through the use of a postal survey. Questionnaires were sent to 3857 companies operating in Poland. 182 questionnaires were returned (response rate at 4.7%). Companies were selected from a database of the ISO Guide 2012. Results of this study indicate that the surveyed enterprises wanting to develop their processes, implement organizational improvement tools such QEHS management systems, elements of the Toyota Production System (such as Kaizen, 5S, TPM) and concept Lean Management. The surveyed companies implementing a quality management system mainly focused on meeting customer requirements, the reduction of nonconformities (internal non-compliance and customer complaints) and the improvement of the effectiveness and efficiency of processes. Important premises for deployment of this system are the possible increase in employee involvement in solving problems related to improving the quality of products and processes

Table 1. The main reasons for implementing quality management system by the surveyed companies [%].

The main reasons for implementing environmental management system	Sector		Number of employees			Origin of capital		Target market	
	Producers N = 112	Service providers N = 70	−50 N = 45	51–250 N = 89	250– N = 48	Foreign N = 41	Polish N = 141	B2B N = 112	B2C N = 70
Reduction of nonconformities	75.21	67.21	66.67	78.65	66.67	75.21	73.05	73.21	71.43
Increase in employee engagement	77.69	57.38	62.22	80.90	60.42	77.69	73.05	73.21	67.14
Client requirements	72.73	57.38	64.44	79.78	47.92	72.73	71.63	72.32	60.00
Increase in process efficiency	71.07	57.38	60.00	73.03	60.42	71.07	68.79	69.64	61.43
Internal communication	71.07	54.10	64.44	74.16	50.00	71.07	68.09	67.86	61.43
Increase in process effectiveness	65.29	55.74	64.44	67.42	50.00	65.29	63.12	62.50	61.43
Improvement of work conditions	38.02	34.43	37.78	41.57	27.08	38.02	36.17	61.61	48.57
Shortening time of processes	42.15	16.39	22.22	39.33	33.33	42.15	34.75	36.61	28.57
Impacts on the environment	27.27	13.11	20.00	28.09	14.58	27.27	24.82	21.43	24.29

Table 2. The main reasons for implementing environmental management system by the surveyed companies [%].

The main reasons for implementing environmental management system	Sector		Number of employees			Origin of capital		Target market	
	Producers N = 55	Service providers N = 15	−50 N = 11	51–250 N = 28	250– N = 31	Foreign N = 23	Polish N = 47	B2B N = 48	B2C N = 22
Impacts on the environment	90.91	80.00	100.00	100.00	92.86	91.30	87.23	87.50	90.91
Increase in employee engagement	56.36	60.00	63.64	57.14	57.14	43.48	63.83	58.33	54.55
Client requirements	52.73	46.67	63.64	60.71	60.71	47.83	53.19	60.42	31.82
Internal communication	32.73	53.33	36.36	46.43	46.43	21.74	44.68	37.50	36.36
Reduction of nonconformities	30.91	40.00	45.45	32.14	32.14	21.74	38.30	35.42	27.27
Improvement of work conditions	27.27	40.00	45.45	21.43	21.43	34.78	27.66	29.17	31.82
Increase in process efficiency	23.64	33.33	18.18	25.00	25.00	21.74	27.66	22.92	31.82

and the improvement of internal communication (through the establishment of SOP).

A large group of companies wanting to reduce the negative impact on the environment and meet the customers' expectation implement a system solution based on the requirements of ISO 14001 management standard. The results show that the effective implementation of this system also contributes to the improvement of internal communication and safety conditions for employees.

It may also be noted that the increase in the interest of many organizations to improve safety conditions for employees cause the implementation of health and safety management system based on OHSAS

Table 3. The main reasons for implementing health and safety management system by the surveyed companies [%].

The main reasons for implementing health and safety management system	Sector		Number of employees		Origin of capital		Target market	
	Producers N = 33	Service providers N = 12	−250 N = 25	250− N = 20	Foreign N = 16	Polish N = 29	B2B N = 27	B2C N = 18
Improvement of work conditions	93.94	83.33	88.00	95.00	93.75	89.66	96.30	83.33
Increase in employee engagement	54.55	58.33	52.00	60.00	68.75	48.28	55.56	55.56
Reduction of nonconformities	24.24	50.00	28.00	35.00	18.75	37.93	29.63	33.33
Improvement of internal communication	24.24	50.00	32.00	30.00	18.75	37.93	29.63	33.33
Increase in process efficiency	21.21	41.67	32.00	20.00	25.00	27.59	25.93	27.78

Table 4. The main reasons for implementing elements Toyota Production System (Kaizen, 5S, TPM) by the surveyed companies [%].

The main reasons for implementing elements Toyota Production System	Sector		Number of employees		Origin of capital		Target market	
	Producers N = 32	Service providers N = 7	−250 N = 24	250− N = 15	Foreign N = 15	Polish N = 24	B2B N = 29	B2C N = 10
Involvement of employees	59.38	57.14	53.33	62.50	53.33	62.50	62.07	50.00
Improvement of work conditions	50.00	85.71	53.33	58.33	53.33	58.33	55.17	60.00
Shortening time of operational processes	46.88	42.86	40.00	50.00	40.00	50.00	41.38	60.00
Increase process efficiency	46.88	28.57	26.67	54.17	26.67	54.17	44.83	40.00
Reduction of nonconformities	37.50	42.86	20.00	50.00	20.00	50.00	48.28	10.00
Improvement of internal communication	28.13	28.57	6.67	41.67	6.67	41.67	31.03	20.00
Impacts on the environment	28.13	14.29	6.67	37.50	6.67	37.50	24.14	30.00

18001 standards. The results show that the effective implementation of the health and safety management system also contributes to internal communications improvement and an increase in employee engagement, which also affects the increase of their productivity and reducing nonconformities in processes and products. Details of the study in terms of cross-segmentation are presented in tables 1–5 below:

Summarizing the results of empirical study it should be noted that the most interested in implementing QEHS management systems as well as process improvement tools (such as TPS elements or Lean Management) are the medium and large size enterprises (employing more than 50 employees), manufacturing companies and firms with foreign capital offering their products in the B2B market. Enterprises with Polish capital wanting to be providers for international companies increasingly adopt these types of system solutions and organizational concepts. Effective implementation

Table 5. The main reasons for implementing elements Lean Management by the surveyed companies [%].

The main reasons for implementing elements Lean Management	Sector		Number of employees		Origin of capital		Target market	
	Producers N = 22	Service providers N = 7	−250 N = 18	250− N = 11	Foreign N = 17	Polish N = 12	B2B N = 19	B2C N = 10
Shortening time of operational processes	79.17	85.71	83.33	76.92	88.24	71.43	78.95	100.00
Increase in process efficiency	66.67	71.43	66.66	69.23	76.47	57.14	68.42	80.00
Increase employee engagement	41.67	71.43	44.00	53.85	58.82	35.71	57.89	60.00
Reduction of nonconformities	37.50	71.43	50.00	38.46	47.06	42.86	47.37	60.00
Improvement of internal communication	29.17	57.14	38.88	30.77	23.53	35.71	31.58	80.00
Improvement of work conditions	25.00	42.86	27.77	30.77	29.41	28.57	31.58	40.00

of these tools allows the local providers to meet the growing requirements of multinational corporations, which expectations are focused on providing high-quality products, timeliness and flexibility of supply (shortening cycle orders, change order ranges and volumes) reducing production and logistic costs as well as reducing environmental impact. Companies that have implemented these types of tools (like QEHS management systems, Toyota Production System or Lean Management) also recognize the many benefits, such as greater internal efficiency of communication processes and an increase in awareness and involvement of employees in the implementation of product and process innovations (for submission of new ideas and improving the quality of existing products and processes). In recent years, many countries in Central and Eastern Europe (such as the Slovak Republic, the Czech Republic, Romania, Hungary, and Poland), there is a strong increase in the interest of implementing process improvement tools such as the Toyota Production System and Lean Management. Certainly, this it related to the investments of many international companies, especially high-tech sector (particularly in the automotive industry and the electronic devices sector). These international corporations also expect their local suppliers to implement such tools to improve processes. Results of this study indicate that the implementation of these tools helps to improve internal communication and increased employee involvement in solving problems. This has a significant impact on the growth of the efficiency and effectiveness of processes and improves the technical quality of the products.

6 SUPPLIER DEVELOPMENT PROGRAMS AIMED AT IMPROVING PROCESSES AND PRODUCTS

For many companies, relationships with suppliers are not limited to establishing stringent requirements and continuous monitoring of their compliance. In order to allow suppliers to comply with these requirements, many multinational companies offer special programs for the development of their partners (Omurca 2013). The programs are based on offering a broad-based training and consulting in the field of quality development and implementation of continuous improvement tools like QEHS system solution elements of the Toyota Production System (Kaizen, 5S, Total Productive Maintenance), Lean Management and Six Sigma project for the introduction of product and process innovations (Fu et al. 2012). The special support programs are based on the principle of win-win (Blonska et al. 2013). The successful implementation of these programs allows both suppliers and customers to improve the quality of products (lower level of non-compliance, introducing product innovations, increased reliability and security), shorten cycle processes and reduce their costs (in particular with regard to operational processes such as design, customer service before and after the sale, production/services, transportation and maintenance of infrastructure) and improve mutual communication (Ahmed & Hendry 2012, Arroyo-López et al. 2012). Actions aimed at developing suppliers undoubtedly contribute to a reduction in transaction costs related to the exploration of new supply capacity, conducting audits and other forms of assessment, verification and qualification of the sources of purchase. In order to ensure the effectiveness of the supplier development program it is necessary to produce a climate of cooperation based on mutual commitment, trust and open information exchange especially in the area of performance quality (level of compliance with the requirements for the provision and improvement of products and processes) and cost of operational processes. Effectively implemented, the development programs of suppliers undoubtedly contribute to building the intellectual capital of the partners (Fu et al. 2012, Nagati & Rebolledo 2013).

7 CONCLUSIONS

Implementation of process improvement tools is strictly conditional on technical and organizational progress as well as the constantly increasing globalization in all areas of the economy and other spheres of social life (Kumar et al. 2014). These effects are particularly noticeable in the case of producers supplying in the B2B market. Recapitulating, it should be noted that many institutional buyers (especially multinationals) expect their partners in the supply chain to have efficient QESH management systems aimed at improving the efficiency and effectiveness of processes and which ensure the development of safe products. Suppliers are obliged to focus their efforts on achieving the capacity to implement continuous organizational and product innovation. This capacity is often achieved by the supplier through the implementation of organizational improvement tools, which allow them to achieve a sustainable competitive advantage.

REFERENCES

Ahmed, M. & Hendry, L. 2012. Supplier development literature review and key future research areas. *International Journal of Engineering and Technology Innovation* 2: 293–303.

Arroyo-López, P., Holmen E. & de Boer, L. 2012. How do supplier development programs affect suppliers?: Insights for suppliers, buyers and governments from an empirical study in Mexico. *Business Process Management Journal* 18: 680–707.

Azadeh, A., Gaeini, Z. & Moradi, B. 2014. Optimization of HSE in maintenance activities by integration of continuous improvement cycle and fuzzy multivariate approach: A gas refinery. *Journal of Loss Prevention in the Process Industries* 32: 415–427.

Bernardo, M. 2014. Integration of management systems as an innovation: a proposal for a new model. *Journal of Cleaner Production* 82: 132–142.

Bernardo, M., Casadesus, M., Karapetrivic, S. & Heras, I. 2012. Integration of standardized management systems: does the implementation order matter? *International Journal of Operations & Production Management* 32: 291–307.

Blonska, A., Storey, Ch., Rozemeijer, F. & de Ruyter, M.W.K. 2013. Decomposing the effect of supplier development on relationship benefits: The role of relational capital. *Industrial Marketing Management* 42: 1295–1306.

Colledani, M., Tolio, T., Fischer, A., Iung, B., Lanza, G., Schmitt, R. & Váncza, J. 2014. *CIRP Annals—Manufacturing Technology* 63: 773–796.

Dobos, I. & Vörösmarty, G. 2014. Green supplier selection and evaluation using DEA-type composite indicators. *International Journal of Production Economics* 157: 273–278.

Fu, X., Zhu, Q. & Sarkis, J. 2012. Evaluating green supplier development programs with at a telecommunications systems provider. *International Journal of Production Economics* 140: 357–367.

Idri, M.A. & Zairi, M. 2006. Sustaining TQM: a Synthesis of Literature and Proposed Research Framework. *Total Quality Management & Business Excellence* 17: 1245–1260.

Jankalová, M. 2012. Business Excellence evaluation as the reaction on changes in global business environment. *Procedia—Social and Behavioral Sciences* 62: 1056–1060.

Jørgensen, T.H., Remmen, A. & Mellado, M. 2006. Integrated management systems—three different level of integration. *Journal of Cleaner Production* 14: 713–722.

Koksal, G., Batmaz, I. & Testik, M.C. 2011. A review of data mining applications for quality improvement in manufacturing industry. *Expert Systems with Applications* 38: 13448–13467.

Kumar, D.T., Palaniappan, M., Kannan, D. & Shankar, K.M. 2014. Analyzing the CSR issues behind the supplier selection process using ISM approach. *Resources. Conservation and Recycling* 92: 268–278.

Morita, M., Machuca, J.A.D., Flynn, E.J. & de los Ríos, J.L.P. 2015. Aligning product characteristics and the supply chain process—A normative perspectives. *International Journal of Production Economics* 161: 228–241.

Nagati, H. & Rebolledo, C. 2013. Supplier development efforts: The suppliers' point of view. *Industrial Marketing Management* 42: 180–188.

Omurca, S.I. 2013. An intelligent supplier evaluation, selection and development. *Applied Soft Computing* 13: 600–607.

Pool, A., Wijngaard, J. & van der Zee, D.-J. 2011. Lean planning in the semi-process industry, a case study. *International Journal of Production Economics* 131: 194–203.

Rajesh, R. & Ravi, V. 2015. Supplier selection in resilient supply chains: a grey relational analysis approach. *Journal of Cleaner Production* 86: 343–359.

Ringen, G., Aschehoug, S., Holtskog, H. & Ingvaldsen, J. 2014. Integrating Quality and Lean into a Holistic Production System. Proceedings of the 47th CIRP Conference on Manufacturing Systems. *Procedia CIRP* 17: 242–247.

Sahno, J., Shevtshenko, E. & Zahharov, R. 2015. Framework for Continuous Improvement of Production Processes and Product Throughput. *Procedia Engineering* 100: 511–519.

Scyoc van, K. 2008. Process safety improvement—Quality and target zero. *Journal of Hazardous Materials* 159: 42–48.

Terziovski, M. & Guerrero, J.-L. 2014. ISO 9000 quality system certification and its impact on product and process innovation performance. *International Journal of Production Economics* 158: 197–207.

Production Management and Engineering Sciences – Majerník, Daneshjo & Bosák (Eds)
© 2016 Taylor & Francis Group, London, ISBN: 978-1-138-02856-2

The use of selected statistical methods as an objective tool of company's management

Š. Vilamová, A. Király & R. Kozel
Faculty of Mining and Geology, VŠB–Technical University of Ostrava, Ostrava, Czech Republic

K. Janovská & D. Papoušek
Faculty of Metallurgy and Materials Engineering, VŠB—Technical University of Ostrava, Ostrava, Czech Republic

ABSTRACT: Economic theory defines the costs of business as a cash expenditure of production factors including the public expenses that originated from production of corporate revenues. One of the most important aspects is the division of costs based on the production volume—to fixed and variable costs. There are many cases when the individual cost items are very often empirically classified as fixed or variable costs of the companies. The individual cost items often empirically classified as fixed or variable within the companies. When used properly, the method of regression and correlation analysis may become an objective tool of management process of the company and with sufficient number of observation; provable close correlation and correct interpretation of the results, the results of this method are credible and useful in practice.

1 INTRODUCTION

All Economic theory defines the costs of business as a cash expenditure of production factors including the public expenses that originated from production of corporate revenues. The costs are an important synthetic indicator of company's quality. There are many approaches in terms of their division. One of the most important aspects is the division of costs based on the production volume—to fixed and variable costs.

Division of individual cost items to purely fixed or variable depends on the type of company, however, most of the cost items (especially the overhead costs) within the company are so called combined with part being variable and part fixed. The individual cost items are often empirically classified as fixed or variable within the companies. Based on that classification the calculations are made when counting for instance the corrected plan, controlling reports, when setting the covering contribution, in case of investment ratings, creations of business plans, etc.

The problem is that this empiric division of individual cost items to fixed and variable based commonly on detailed statistics of overhead costs, is very often not reversely controlled by the use of e.g. regression analysis of time series, even though the accuracy of the allocation of cost items into fixed and variable is very important for different

economic considerations and influences the final evaluation of the development (Vilamová et al. 2015).

2 FIXED AND VARIABLE COST IN BUSINESS PRACTICE

The theory of fixed and variable costs implies that the division of costs to constant and variable is done in relation to their absolute level of production volume (Synek et al. 2011). In the businesses, however, it is very difficult to label costs "constant, fixed" and "variable" and especially the statement that results out of this sorting, saying that the fixed costs are in their full amount invariable, constant, as well as the image that the full variable costs are only the function of production volume is quite problematic. The theories e.g. (Czopek 2003), do allow for certain changes "by jump" regarding the constant costs with enormous change in production thus the need for modified configuration of production capacities (both upwards and downwards).

The most problematic costs within the companies are the costs of mixed nature, where there are some costs that are partly fixed and partly variable, and in addition to that they are influenced by other factors besides the production volume. Typical example of mixed character costs are the costs for maintenance and repairs. Basic economic consideration—the more I make, the more equipment wears out—is indeed rational, however, many repairs in enterprises are directed by safety regulations (periodic inspection and replacement of selected equipment) or unplanned factors (disturbances, accidents).

Despite all objections or comments to the theoretical distribution of costs, the classification of costs depending on the volume of production and the necessary business practice is widely used. The current way of determining the amount of fixed and variable costs in the metallurgical enterprise lies in the empirical distribution of individual cost items to fixed and variable.

The individual cost items (usually in a detailed classification) are classified as fixed or variable based on experience in a particular company and often the quotient of both parts is determined in %. This division may be quite close to reality, but the drawback may be its long-term use without an update, while the accuracy of the allocation of cost items into fixed and variable is very important for different economic considerations and influences the final evaluation of the development.

Therefore the problem lies in how to properly determine which expenses should be categorized as "fixed" and "variable" or what proportion of specific cost items have one or the other character (fixed or variable).

Given these challenges and complications associated with the distribution of cost items to fixed or variable it would be appropriate to verify this division by comparing it to another procedure. For instance, the methods of statistical analysis are objective and mathematically accurate. To verify the proportion of fixed and variable costs in practice, the most commonly used are the methods of linear regression and correlation analysis (Tošenovský et al. 2001).

3 EMPIRICAL DISTRIBUTION OF COST ITEMS

In selected metallurgical enterprise were cost items of pig iron production divided to fixed and variable based on a long time experience. The cost calculations of pig iron production for 2002–2003 were made available.

Empirical distribution of the average cost items to fixed and variable in the metallurgical enterprise in monitored period:

CC = 3 481 492 000 CZK,
FC = 205 204 000 CZK,
VC = 3 276 288 000 CZK,
CC Complete flat cost,

FC Fixed costs,
VC Variable costs.

The percentage of average fixed planned costs:

$$\%p = \left(\frac{FC}{CC}\right) \times 100\% \tag{1}$$

$$\%p = \left(\frac{205204000}{3481492000}\right) \times 100\% = 5{,}894\% \tag{2}$$

The percentage of planned fixed costs was empirically set to 5,894%. With blast furnace, where the variable costs of the charge are approximately 90% of total costs, this figure is considered to be realistic.

4 USE OF REGRESSION AND CORRELATION ANALYSIS

As already stated in the introduction, in many cases, the empirical distribution of individual cost items to fixed and variable, which is usually based on detailed statistics of overhead costs, is in practice, not verified retrospectively.

As already mentioned, the vast majority of costs in the company are so-called combined type. The relationship can be expressed by the equation:

$$y = a + b * x \tag{3}$$

when:

a fixed part of the cost,
b*x variable part of the cost,
x production,
a, b searched parameters of the regression line,
y complete flat cost.

To determine the parameters of the regression line and the correlation coefficient, the data and information from the cost calculations were used. The calculations are performed in Excel.

Calculated values of the parameters of the regression line and correlation coefficient are:

a = −720 055,
b = 4,9156,
r = 0,8753.

The result of the regression analysis for the mentioned periods is the cost function in the form:

$$y = -720055 + 4{,}9156x. \tag{4}$$

Data in the cost function are shown in thousands CZK, production in tonnes. Correlation coefficient r = 0.8753662, which means that the cost function has a high quality and functional dependence between variables is very tight.

According to the regression equation, the average value of fixed expenses during the reporting period was negative—720 055 000 *CZK*. Constant costs would be more than three times compared to empirical determination, but negative. This is in business practice impracticable. It is therefore evident that besides the dependence that the linear regression function presumes there are other factors.

Based on performed analyses and discussions with employees of the metallurgical enterprise it may be stated, that out of the factors that significantly influence the credibility of regression analysis results with variable costs is probably the most significant the impact of prices and assortment.

5 EXAMINING THE IMPACTS OF DIVIDING THE INDIVIDUAL COST ITEMS TO FIXED AND VARIABLE IN THE PROCESS OF PLANNING AND CONTROLLING

From the results, presented in the previous chapters, it is clear that verification of empirical distribution of costs into fixed and variable by the use of regression analysis resulting from raw, unadjusted data files of individual time series, cannot be used.

In the following progress of work were the costs for charge out of the calculation for individual periods converted to a single price and assortment basis in two variants:

I. on the basis I/2002
II. on the basis VIII/2003

Calculation was based on the assumption of linear and proportional cost of the charge only depends on the volume of production of pig iron. Absolute iron productions in each period were therefore multiplied by the cost of charge in *CZK/t* of iron or optionally in I/2002 and VIII/2003. The results of the calculation are stated in Table 1.

Based on thus adjusted time series of complete flat costs the regression and correlation analysis were subsequently performed depending on complete flat costs of production volumes with the following results:

The costs function when the charge converted on the basis of I/ 2002.

Calculated values of the parameters of the regression line and correlation coefficient are:

$a = 61\,538,$
$b = 4,4369,$
$r = 0,9534.$

The result of the regression analysis for the periods indicated is the cost function in the form:

$$y = 61538 + 4,4369x. \qquad (5)$$

Data in the cost function are shown in thousand CZK, production in tonnes.

Correlation coefficient r = 0.9534, which means that the cost function has a high quality and functional dependence between variables is very tight.

The cost function with the charge converted on the basis of VIII/ 2003:

Table 1. Conversion of complete flat costs for each period to a single price and assortment basis.

Quarterly periods	1	2	3	4	5	6	7	8
production [t]	721103	752224	790131	694008	771663	857703	881499	822816
Complete flat costs [thousand CZK]	3205855	3203758	3159859	3063623	3561936	3815764	3928620	3912521
of CMH: pure chargé [thousand CZK]	2857664	2779990	2812085	2565522	3144481	3353298	3535580	3396569
charge of cost for the I/2002 (at t pig iron) [thousand CZK]	2857664	2980994	3131216	2750289	3058029	3398997	3493298	3260743
Difference	0	201003	319131	184767	−86452	45699	−42282	−135826
charge for a price VIII/2003 (per ton pig iron) [thousand CZK]	2976700	3105167	3261646	2864852	3185410	3540582	3638811	3396569
Difference	119036	325176	449561	299330	40929	187284	103231	0
Complete flat costs recalculated [thousand CZK]	3324891	3528935	3609420	3362953	3602865	4003048	4031851	3912521

Calculated values of the parameters of the regression line and correlation coefficient are:

a = 61 538,
b = 4,6020,
r = 0,9565.

The result of the regression analysis for the monitored periods is the cost function in the form:

$$y = 61538000 + 4,6020x. \tag{6}$$

Data in the cost function are shown in thousand CZK, production in tonnes.

Correlation coefficient r = 0.9565, which means that the cost function has a high quality and functional dependence between variables is very tight.

Average value of annual fixed cost set empirically is 205 204 000 *CZK*. The results of the above mentioned regression analysis show that the fixed part of complete flat costs is 61 538 000 *CZK*. Calculated average annual fixed cost is the same for both options selected for recalculating the charge costs, which means that the choice of basic conversion period is not critical. The results of regression analyses therefore suggest that the items of processed costs declared in the empirical calculation as fixed, may include a certain proportion of the variable costs.

Based on the detailed analysis of the available cost calculations of pig iron production it was defined that primarily the following items belong to the calculation pattern: repairs and maintenance, overhead power. The nature of these costs does not exclude the possibility of mixed dependence on the volume of production. When doing the empirical calculations for the analysis of economic reality in the business it would be appropriate to identify and consequently separate the amount of fixed and variable parts of these costs.

6 CONCLUSION

The issue of costs and their allocation to fixed and variable parts is theoretically and in detail described in many publications that focus on costs, their classification and management, for example, (Drudy 2006), (Kutáč et al. 2014), (Skokan et al. 2013), (Vaněk et al. 2008) etc. In business practice, the individual cost items are often empirically classified as fixed or variable, and so they are calculated, for example when calculating the converted plan, in controlling reports, when setting the covering cost, in case of investments evaluation, creation of business plans etc. The problem is that the empirical distribution of individual cost items, fixed and variable, resulting in general from detailed statistics of overhead costs, is in many cases not verified retrospectively, by for instance regression analysis of time series, while the accuracy of the allocation of cost items into fixed and variable is very important for different economic considerations and influences the final evaluation of the development

The successful use of statistical analysis methods to calculate the cost function is quite difficult due to the need to eliminate some other impacts that distort the information value of the derived results. The theory of fixed and variable costs is based on a defined relationship between costs and volume of production. The amount of costs may depend on other factors then the production volume—characteristic may be e.g. seasonal influences (costs to unfreeze the ore during the landing in the winter months) and the like.

From the calculated and presented results is obvious that verification of the empirical division of costs to fixed and variable parts by the use of regression analysis that is based on the raw company data in time series is inapplicable and does not have any information value. Within the calculations the incomparability of the values of variable costs in individual periods was proved and the price and assortment influence of charge components including the effects of changes in specific consumption in many cases distorted the dependence of the charge on the cost of production volume.

REFERENCES

Czopek, K. 2003. *Fixed and variable costs: theory and practice.* Kraków: Art-Tekst, pp. 103.

Drudy, C. 2006. *Cost and Management Accounting.* London: Cengage Learning EMEA.

Kutáč, J., Janovská, K., Samolejová, A. & Besta, P. 2014. Innovation of costing system in metallurgical companies, *Metalurgija,* vol. 53, no. 2, pp. 145–288.

Skokan, K. & Pawliczek, A. 2013. Lifecycle of enterprises and its dynamics on the basis of annual turnover: an empirical study of Czech and Slovak enterprises, *In Proceedings of 22th IBIMA Conference Rome,* Norristown: IBIMA, pp. 268–280.

Synek, M., Kislingerová, E., Dvořáček, J., Dvořák, J. & Tomek, G. 2011. *Manažerská ekonomika,* 5. vydání, Praha: Grada Publishing, pp. 471.

Tošenovský, J. & Dudek, M. 2001. *Základy statistického zpracování dat,* VŠB-TU Ostrava, pp. 82.

Vaněk, M., Mikoláš, M., Růčková, H., Bartoňová, J., Kučerová, L. & Žoček, F. 2011. Analysis of Mining Companies Operating in the Czech Republic in the Sector of Non-metallic and Construction Minerals. *Gospodarka Surowcami Mineralnymi,* vol. 27, no. 4, pp. 17–32.

Vilamová, Š., Miklošik, A., Kozel, R., Samolejová, A., Piecha, M., Weiss, E. & Janovská, K. 2015. Regression analysis as an objective tool of economic management of rolling mill. *Metalurgija,* vol. 54, no. 3, pp. 594–596.

Production Management and Engineering Sciences – Majerník, Daneshjo & Bosák (Eds)
© 2016 Taylor & Francis Group, London, ISBN: 978-1-138-02856-2

Runway incursion and methods for safety performance measurement

P. Vittek, A. Lališ, S. Stojić & V. Plos
Department of Air Transport, CTU in Prague, Prague, Czech Republic

ABSTRACT: Aviation industry as a whole is detecting long-lasting safety improvement trend. Extensive legislation and internal operational rules supported the establishment of the high standard and efficient oversight for civil aviation authorities. Actual level of safety is often expressed as a ratio of the total number of accidents to realized operational cycles. However, in today's world, accidents are generally rare; hence authors integrate innovative approach to safety which enables safety performance evaluation. The approach utilizes safety performance indicators as fundamental tool in the process of evaluation. This paper focuses on Runway Incursion, one of the major phenomenons in aviation safety. It has great potential to seriously affect safety during the most critical phases of flight—during takeoff and landing. Additionally, this paper presents results of descriptive and explanatory factors analysis. It outlines their implementation into the model, which will enable future safety performance quantification of the corrective measurements preventing Runway Incursion.

1 INTRODUCTION

Use Runway Incursion is modern aviation safety phenomenon. It is often referred to as "Any occurrence at an aerodrome involving the incorrect presence of an aircraft, vehicle or person on the protected area of a surface designated for the landing and take-off of aircraft" (ICAO 2007). The protected area designates rectangular adjacent zones next to and including the runway surface. Any aircraft, vehicle or person should never enter these zones without permission granted by Air Traffic Control (ATC). Such unauthorized presence of these subjects on runway is substantially increasing risk of collision with other traffic operating on, or approaching the runway. Collisions between two aircraft or aircraft and vehicle/person can easily lead to fatal injuries and severe damage. One of the most recent cases happened in Moscow Vnukovo airport on 20th of October 2014 where departing business jet crashed after runway collision with snow-cleaning machine, killing all four occupants aboard and completely destroying the aircraft. According to available statistics, there are almost 80 cases per each million aircraft movement in Europe in year 2012 with total of 1 234 reports by 36 EU Member states (Eurocontrol 2013). In majority of reported cases there was no damage and no injuries, however, risk of collision had significantly arisen in each case. Hence these stats are still quite alarming and clear commitment to focus on this issue is expressed both by aviation authorities and EU institutions.

On the other hand, the fact that this is modern phenomenon in aviation safety suggests a lot about present level of safety. Technology itself or separate aviation components are no more of such great concern since all of them reached desired level of safety individually over last decades. There is still demand for better avoidance and alerting technology (Schönefeld 2012), but Runway Incursion is an operational issue which is quite complex in terms of causes, overlapping with many aviation parties and involving various technology. To handle such a complex issue, a lot of innovations and effort is needed to better understand why this takes place and to learn how to effectively prevent it. Obviously, there are much more of such complex events in aviation but only Runway Incursion will be described in this paper. The other are operational issues as well, however, means of understanding and mitigating risks are quite the same as this paper will introduce. The innovative approach takes advantage of modern knowledge and tools aimed at operations and organizations' management and how do they interact together. Only better understanding of the system in general can help us find reasonable solutions to these complex issues.

Additionally, safety is precondition of quality in aviation. It is not identical to quality, however, it is of much greater concern than quality itself, because the ability to effectively control safety forms inherent basis for quality product offered to customers. Hence Safety Management System (SMS) in aviation and concerned efforts on safety performance quantification may be perceived comparable to quantitative methods used in product manufacturing in other branches, such as Six Sigma method unified by International Standards Organization (ISO 2011). According to this method, if the output of production is normally distributed, it has six standard deviations between the process mean and the specification limits, practically no item will fail to meet given specifications. That is equivalent to 3.4 defective parts per million produced parts or 99.99966% percentage yield. Most of the critical aviation safety processes fall easily into the six sigma rule, but Runway Incursion with its 80 cases per one million movements in 2012 represents one of the challenges that occur regularly and that need to be handled in order to retain or improve existing level of safety in aviation.

2 PRESENT SAFETY MANAGEMENT

Safety management is nowadays combination of reactive, preventive and proactive measures (SKYbrary 2014). Reactive measures and subsequent precautions are always taken after some incident or accident happened. Safety was handled in this way from very beginning of aviation and is still present today. If there are losses of lives or costly damages, it is great motivation for respective stakeholders to find out what was the cause and prevent that particular cause to repeat again.

Over last decades, preventive measures became progressively part of safety management (SKYbrary 2015). These focus more on what is preceding accidents and incidents and apply mitigation measures in the system well before any accident or incident can take place. The ultimate stage is then proactive approach where questions such as "What else could have happened?" or "What did effectively prevent this chain from developing further?" are regularly asked and considered into safety management.

The point is not all aviation stakeholders (organizations) are of the same level with regard to safety. Some of them have more data available and take more responsibility than others, considering safety management of higher importance. Some are economically motivated and focus only on what is clearly economical for them. As a result, there are different safety management systems handling various aspects of various issues. This is the moment where both international and national aviation authorities have to become involved and unify the approach to make it effective over the whole industry.

Today, there are extensive regulations on EU level as well as many global standards, internationally accepted. They range from reporting activities

(SKYbrary 2013) to the authorities up to SMS requirements (SKYbrary 2015).

On global level, there are standards published by International Civil Aviation Organization (ICAO), in form of Annexes to the Chicago treaty from 1944 or in form of ICAO documents that specify more in detail given requirements. For ICAO member states they are compulsory and often implemented into local legislation. Safety was originally covered within more Annexes but from year 2013 it is now available as a standalone one—the very last Annex 19 (ICAO 2013). Safety Management Manual as ICAO Doc. 9859 then extends the information on safety issues to facilitate process of safety management. Consequently, ICAO member states are required to establish safety programmes and safety plans, where they specifically define safety objectives and plan actions to achieve them. They are required to report incidents and accidents to the ICAO so they can be further analyzed. Vision of global safety research and development is then regularly updated and published in Global Aviation Safety Plan (GASP) by ICAO.

At European level, the legislation and approach to safety is complementary to the global level but in many aspects it extends the requirements to certain degree (EASA 2014). With the same logic, there is European Aviation Safety Plan; incidents and accidents are required to be reported but extended to cover more occurrences than just incidents and accidents (SKYbrary 2015). SMS systems are compulsory for various organizations within EU member states, while State Safety Plans and Programmes are based on ICAO requirements and compulsory for each EU member state, according to European commission regulations No. 290/2012 and No. 965/2012. The idea is the same as on global level—to establish common approach and to engage various stakeholders into the process. Both ICAO and EU agencies have recognized the importance to interconnect all the aviation components and to create one network in terms of safety management.

3 EXPLANATORY AND DESCRIPTIVE FACTORS

To better understand how incidents and accidents happen, and to improve event analysis, there are various taxonomies. They emerged owing to huge amount of safety investigations conducted in aviation and their aim is to establish and define set of terms used to describe chain of events. Consequently, having the same incident described by two independent persons individually, there should be identical event descriptions obtained.

Some taxonomy helps to create chain of events, i.e. they allow for several additional information to be covered in the report. These might reveal then dependences, responsibilities and involvement of different entities to help better understand the chain. Chains might be assessed in terms of severity but also, having one chain repeated more frequently than the other, in terms of repeatability. All of these aim at better understanding of the complexity of chain and individual events, eventually facilitating the process of taking precautions.

The most universal today is ADREP taxonomy (The Accident/Incident Data Reporting) by ICAO. This taxonomy introduces so called "explanatory" and "descriptive" factors and as their naming may suggest, they concern detailed description of various factors involved in respective event. These are of critical importance, because they are capable of indicating root causes. Descriptive factors are supposed to describe what the background of the event situation was while the explanatory factors explain actions and activities taking place on it. The logic is that background shall constitute active barriers to stop further development of undesired actions and activities (Icao 2015).

In this way, any safety-related event in aviation can be described. The same works for Runway Incursion, which name itself, is also unified through taxonomies and recognized all over the world in the same way. Ones it occur, today it is very unlikely to happen so that after proper investigation the event couldn't be described with ADREP taxonomy. No matter if it was human factors of airport layout what allowed triggering the event; the ultimate causes are well known today. The point is, one should understand how these scenarios evolve, and build up sort of a net of barriers to break the scenario early enough. In reality, certain processes must be monitored and related data must be gathered and analyzed. To do that effectively, various tools and mathematical methods must be deployed, e.g. performance indicators and system's performance measurement. Having some irregularities or anomalies observed, proper actions must be taken in order to maintain the system's desired properties, in this case safety performance.

Concerning Runway Incursion, the event including factors may look as follows:

Figure 1 describes the way both types of factors contribute to the event chain. Basically, there are two crucial types of events: the Runway Incursion and Human error. These can be specified more in terms of ultimate responsibility assignment by other parts of ADREP taxonomy. But there will always be some kind of irregular behavior preceding the actual event of Runway Incursion. The factors contribute both to the irregular behavior

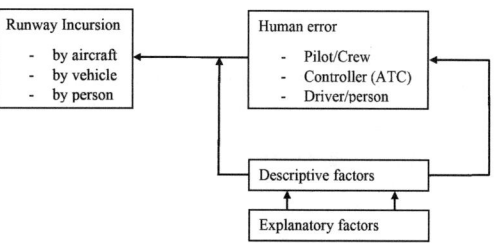

Figure 1. Runway Incursion and descriptive/explanatory factors.

occurrence as its further development. They contain the background set as well as additional chains which have impact on the Runway Incursion chain. Figure 1 simplifies the issue as much as possible, still retaining conformity with ADREP taxonomy. Obviously, both types of factors could be depicted as standalone fault trees which are simply injected between the most critical events, having then assigned responsibility. It is just a matter of taxonomy term definition that these are referred to as factors rather than standalone events, and the presumable reason for this is the root cause identification the factors include.

Typically, descriptive factors concerning Runway Incursions include, but are not limited to (Icao 2007):

a. Aerodrome as a structure—Complicated airport design
b. Runway as an entity—Insufficient spacing between parallel runways or taxiways; Taxiway being not perpendicular to active runway
c. Aerodrome/heliport marking—Inadequate markings/signage
d. Aerodrome/heliport lighting—Inadequate ground lighting
e. Aerodrome/heliport operations generally—Work in progress
f. Aerodrome/Runway obstruction—Design hindering sight and visual detection
g. The aircraft and Air Traffic Management (ATM), their systems and components—Lack of equipment/system, Equipment/system malfunction
h. Night/dark—Night time
i. Atmospheric restrictions to visibility—Weather conditions

Similarly, explanatory factors include, but are not limited to (ICAO 2007):

a. Workload task demands—Excessive workload
b. The interface between humans in relation to formal coordination—Inadequate coordination
c. Human interface-training—Inadequate training
d. Personal experience—Lack of experience

e. Experience of aerodrome—Lack of familiarization
f. Psychological-habituation—Routine behavior
g. Human knowledge acquisition factors situational awareness—Distraction
h. Liveware (human)-hardware/software interface—Human-Machine interface
i. The interface between humans in relation to phraseology—Non-standard phraseology in use
j. The interface between humans in relation to readback/hearback—Readback/hearback error
k. Psychological-misunderstanding—Misunderstanding instructions
l. Psychological-mis-recognition—Accepting clearance intended for another aircraft
m. VHF/HF frequencies—Blocked transmissions
n. The interface between humans in relation to ATC—pilot communications—Overlong or complex transmissions
o. Other defenses/warnings—System warning issued

There is lot of issues involved and ones depicted as whole fault three, these wouldn't be hierarchically ordered any more. The same factor may contribute several times to the same chain, whilst in other case even prevent it. Thus splitting figure 1 to all possible scenarios for above mentioned factors, one would get rather complex network of all possible events, barriers and links, than a hierarchically ordered fault three. This is the core complexity of the event of Runway Incursion, pointing out need to deploy more sophisticated methods to discover the true nature of the event.

4 SAFETY PERFORMANCE AND SAFETY PERFORMANCE INDICATORS

Performance Indicators (PIs) are modern tool which help assess system's performance with regard to given goal. They are universal, easy to implement though efficient and economical solution to many "soft" issues. Typically, they are being implemented when "hard" technology works fine, but where more challenging goals exist which cannot be achieved by the technology (Szabo 2014). This is the case of aviation safety today, but also other high risk industries have already implemented the PIs in domain of safety, such as nuclear power or chemical industry. Their experiences have only proved the effectiveness of safety PIs (Wano 2014; Cefic 2011).

Key processes or key parts of system relative to safety are exactly the places where safety PIs shall be established. Complete set of safety PIs is then feasible only after complex analysis, applied on the whole system. In aviation this is quite com-

plicated as the analysis includes all organizations, overlapping in many places. Hence the issue is approached through systematic breakdown of all undesired events and their possible scenarios. In each scenario for given event, comprehensive identification of root causes provides the basis for establishing safety PIs. These are then just monitoring number of specific events or states of the system against time.

Having the event of Runway Incursion described concerning all key scenarios with contributing factors, monitoring processes should be established and data sources identified. Best would be monitoring of all of the contributing factors, i.e. both descriptive and explanatory factors, and evaluate them either monthly or annually, even before they contribute to some of later-stage events. Where limitations exist, list of safety PIs may be reduced due to practical reasons. This is true for cases where there are no data available, data available are not to be relied upon, or any measurement is highly impractical.

Ones the safety PIs are established and respective statistics per time evaluated, safety performance can be quantified. This is done for each part of the system or undesired event individually, added up then to the overall figure of safety performance. The figure is relative only, i.e. non-dimensional number, to be tracked against time and subjected to subsequent trend analysis. The figure must never be taken in absolute manner because such interpretations tend to lead to meaningless conclusions.

Obviously, the overall figure cannot be obtained just by adding up numbers or total occurrences of all indicators. Not all the occurrences are of same severity and thus indicators' weight must be added. This requires prior expertise involving subject matter experts and application of multi-criteria decision methods. The most successful methodologies used are Analytical Hierarchy Process (AHP) and Analytical Network Process (ANP) invented by Professor Thomas L. Saaty. Where the chain is hierarchically ordered, AHP is more suitable and where the chain consist of a network, ANP is more effective. Both methodologies are based on pair-wise comparison of each element in cluster, and then between clusters in the super-matrix. Elements or clusters are rated by the experts from 1/9 to 9, expressing how much more or less severe one element is over the other. Number 9 stands for extremely higher severity and fraction 1/9 for extremely lower severity. Number 1 is expressing equality. Due to extensiveness of these methodologies, no detailed information will be given. Other studies on how to model runway incursion severity might be considered here (Wilke 2015).

The point is how such a scheme will look like ones the events, occurrences or system states are "weighed". For Runway Incursion by aircraft scenario, the scheme will look as follows:

On the Figure 2, evaluated network of contributing factors and critical events is depicted, and due to practical reasons for one of four lower level scenarios only. Dotted horizontal lines express non-hierarchical parts of contributing factors as they may happen before, in-between or after some event of hierarchical part, and more of them may happen simultaneously. Dotted vertical lines aim at expressing inherent weights of PIs, assigned within the cluster. Figure 2 is complementary to the figure one in terms of colors, i.e. blue are explanatory factors and green the descriptive ones.

Each of the factors and events have their assigned weight W_i and the first partial safety performance then equals

$$safety\ performance_1 = \sum_{i=1}^{n} PI_i \times W_i \qquad (1)$$

where 'n' is number of weighed events or factors, 'PI' stands for respective performance indicator. The same would apply for other scenarios so that overall Runway Incursion safety performance would be obtained as sum of all partial performances.

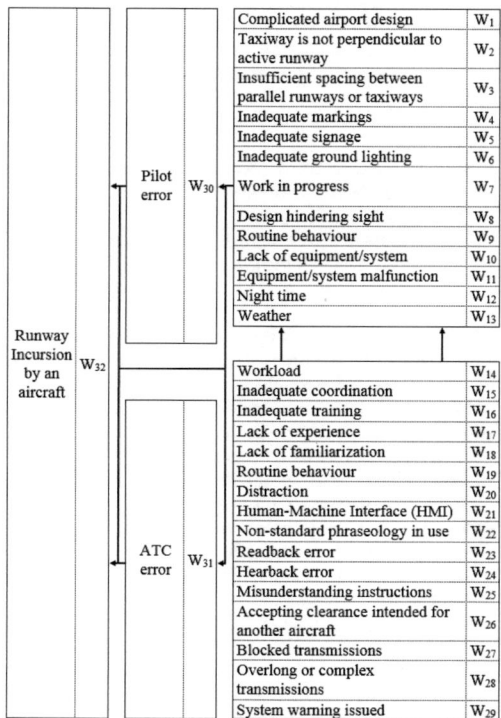

Figure 2. Runway Incursion by aircraft—weighed network.

This safety performance quantification is though not an ultimate stage of research here as there are numerous ways that allow either further development of this approach or adding several other features via applying statistical methods and modeling, such as Bayesian Belief Networks (Green 2014) which can be used also for risk predictions (Goodheart 2013).

5 CORRECTIVE MEASUREMENTS

Any corrective measurements can be taken based on quantified safety performance. As its value is not absolute, it always must be compared to other safety performance values obtained in adjacent time periods or values from the same time one year ago or else, based upon specific regularities that happen and influence the system. For instance, aviation is clearly influenced by seasonality determined by volume of traffic and thus value of safety performance would normally arise on average in summer while drop in winter. Other factors may be considered such as economic crisis or progressive increment in volume of traffic year-on-year. These and many other factors only underline the importance not to take the value of safety performance in absolute manner.

Additional valuable hint can be graphically depicted safety performance; as such representation helps better perceive relative changes in the values. Since the performance is supposed to include also occurrences which does not necessary pose imminent risk to the system and which appear in the system normally to certain degree, only more significant changes of the value shall be subjected to deeper analysis. The analysis may search for specific constellation of contributing factors or reveal irregularities that are not being tracked or recognized today. Moreover, with the benefit of having experience with specific system, after some time there might be possible to draw a line of acceptable level or acceptable region of values for the safety performance. Any extensive analysis on relevant parts of the system would be then conducted only if the figure should exceed the drawn expectations. There is huge potential for safety issues resolving which is provided with complex safety performance measurement.

The output of safety performance is primary intended for senior management, as the look on safety data it provides, is "zoomed-out". The decision process of senior management would be more facilitated and perhaps also more effective. In aviation, some measurements might be found also beneficial to aviation authorities who need to have safety oversight available, capable of tracking important changes in the system so that respective regulations and legislation would effectively fit the needs for improvements within industry.

In fact, the zoomed-out standpoint from management's perspective is significant part of six sigma quantitative method in process improvement (ISO 2011). It is covering definition and measurement phase and supporting further analysis. The management is then provided with great support for control and improvement processes which may alter the measurement when necessary, as all the phases comprise one loop.

6 CONCLUSIONS

Safety PIs and safety performance are innovative means for improving aviation safety. Their deployment on Runway Incursion phenomenon is only part of the whole, however, it enables civil aviation authorities and aviation organization to gain more control over the phenomenon. It is widely recognized that only better understanding of the issue and its contributing factors may effectively prevent the issue in the future. With help of the innovative means, this process is considerably facilitated as they bring exact logic and ultimately eliminate the need for subjective decisions or solutions to related problems. They also force all involved parties to cooperate together, tying the aviation up to a new level of cooperation on issues resolving.

The desired effect is safety-related first, i.e. aimed at preventing injuries caused by Runway Incursion evolved to collision on the runway. But it may help the organization to perform better on cost-efficiency of their operations as well. Considering the prices of various aircraft components in case of their damage after collision, or costs of non-standard situations such as go around or delay of flight caused by Runway Incursion, the PIs and safety performance measurement are very efficient means as far as costs on their establishment are concerned. Providing the organizations with better understanding how this phenomenon happens, they would be able to prevent it so that its occurrence will be far less likely and thus additional costs would be saved. On the flip side, it is difficult to precisely quantify the real savings per given timescale hence some organizations may find it inconclusive whether to implement the safety PIs and measure safety performance, or not. Therefore the involvement of aviation authorities is critical as in some cases only these can push the process forward enough.

Today, clear commitment to effectively handle Runway Incursion is made internationally. Some safety PIs are established at European level and within ICAO member states, but the system is still far from effective safety measurement or issues resolving. For academic branches or aviation research institutions this represents challenge and it is undoubtedly matter of time until the desired effectiveness and benefits are reached.

ACKNOWLEDGEMENTS

This research has been supported by the research program TA CR Alfa No TA04030465 "Research and development of progressive methods for aviation organizations' safety performance control". Authors thank all involved parties for supporting the project both financially and professionally with great commitment to improve aviation safety in Czech Republic.

REFERENCES

CEFIC—The European Chemical Industry Council, 2011. Guidance on Process Safety Performance Indicators.

EASA—European Aviation Safety Agency, 2014. SMS—Europe [online].

EUROCONTROL, 2013. Safety Regulation Commission. Annual Safety Report 2013.

Goodheart, B.J. 2013. Transportation Research Record. *Journal of the Transportation Research Board.* 2013–12–1, Vol. 2400, issue -1, p. 9–20.

Green, L.L. 2014. Development of a Bayesian Belief Network Runway Incursion Model. *14th AIAA Aviation Technology, Integration, and Operations Conference.*

ICAO—International Civil Aviation Organization, 2007. *Doc. 9870: Manual on the Prevention of Runway Incursions.*

ICAO—International Civil Aviation Organization, 2013. *Attachment D to State letter AN 8/3-13/30: SOURCES OF ANNEX 19.*

ISO—International Standards Organization, 2011. 13053–1: 2011. *Quantitative methods in process improvement—Six Sigma: Part 1: DMAIC methodology.*

Schönefeld J., Möller, D.P.F. & Leith, S. 2012. Runway incursion prevention systems: A review of runway incursion avoidance and alerting system approaches. *Progress in Aerospace Sciences*, Vol. 51, p. 31–49.

SKYbrary: The single point for aviation safety knowledge, 2013. Safety Occurrence Reporting.

SKYbrary: The single point for aviation safety knowledge, 2014. Hazard Identification.

SKYbrary: The single point for aviation safety knowledge, 2015. SMS in the Airline Industry.

Szabo, S. & Sidor, J. 2014. The performance measurement system—potentials and barriers for its implementation in healthcare facilities. *Journal of Applied Economic Sciences*. Craiova: Reprograph, Volume IX, Issue 4(30).

WANO—World Association of Nuclear Operators, 2014. *Performance Indicators.*

Wilke, S., Majumdar, A., Ochieng, W.Y. & Leith S. 2015. Modelling runway incursion severity. In *Accident Analysis*. Vol. 79, p. 88–99.

Production Management and Engineering Sciences – Majerník, Daneshjo & Bosák (Eds)
© *2016 Taylor & Francis Group, London, ISBN: 978-1-138-02856-2*

Quality management systems in competitive strategies of Polish enterprises

D. Wyrwa & B. Ziółkowski
Rzeszow University of Technology, Rzeszow, Poland

ABSTRACT: A variety of different factors can impact the competitiveness of companies, but it is well known that the biggest role in creating the stable position on the market belongs to quality. The work presents the management system as one of the instruments to control the operation and development of an organisation. The primary goal of this article is to identify the role of quality management systems in improving the competitiveness of Polish industrial companies. For that purpose the results of interviews conducted among randomly selected entrepreneurs have been used. The article also includes basic definitions and problems related to the process of forming a competitive strategy accounting for both quality and the respondents' opinions about the role of quality management systems in the competition strategies of investigated companies. The research made it possible to identify and evaluate the effects of implementation of quality management systems in enterprises.

1 INTRODUCTION

Nowadays, the majority of companies operate in a highly competitive environment. They are forced to identify the directions which allow them achieve satisfactory results. Some way to build the competitive position of a company could be maintaining and improving the quality of offered products. For this reason, quality management systems based on ISO 9000 standards started to play a significant role in business activities.

Quality management systems do not relate directly to the quality of products. They focus on the procedures which enable to ensure quality or identify possible reasons for a decrease in quality. The implementation of a quality management system compliant with ISO 9000 standard guarantees the repeatability of quality for all products offered by the company. Because of this, quality management systems are implemented and certified by companies in about 160 countries (Bednarova et al. 2010).

The article presents the results of studies realized in 280 randomly selected companies from the manufacturing sector in Poland. The studies were conducted by questionnaire interview in the first half of 2014. The identification of the role of quality management systems in creating competitive strategies was the goal of the studies.

2 COMPETITIVENESS VERSUS THE VALUE OF PRODUCTS

The value of a product is a parameter affecting the customers who make the choice of the products.

The higher the value of a product, the greater may be the interest in purchasing and satisfaction with the choice. The concept of the value of product is not homogeneous however.

The belief that the company does not achieve its success only by means of sufficient resources is becoming more common.

Recently the attention of different authors is attracted by the methods for utilization of resources as well as creation of appropriate configuration of resources.

According to M.E. Porter, the competitive advantage or lack thereof is the result of all activities carried out within the company (Porter 1998b). Thus, the competitiveness of companies is largely determined by internal processes. It is important that it is much easier for competitors to imitate products than to implement identical processes. M.E. Porter also believes that the competitive advantage of a company depends on the value which the company is able to create for its customers (Porter 1985). It is a very synthetic approach to competitiveness, and its evaluation is difficult because many factors can influence the assessment of value of a product. This assessment is usually carried out by comparison with other similar products. For this reason the appropriate shaping of all processes that affect the creation of the value of a product are of utmost importance.

According to A. Smith, in the evaluation of a product, both its value in use as well as value in exchange should be considered. The former type of value is assessed subjectively by a customer who considers the way the product is used and its significance in satisfying needs. The exchange value depends primarily on the production costs incurred to manufacture a product. According to Smith, water has a high value in use, but insignificant value in exchange. The opposite is a diamond, which with little value in use has a very high value in exchange. The final assessment of a product, and hence its price, is in most cases a combination of value in use and value in exchange (Smith 1904). When assessing the value in use of a product it is important to know that it may vary in certain conditions. The value in use of water will certainly be higher in areas where there is a deficit.

The value of a product was a subject of works by D. Ricardo. Similarly to A. Smith, D. Ricardo used the concept of value in use and value in exchange, but noted that the value in use is an essential prerequisite of value in exchange. According to D. Ricardo, it means that a product without any value in use cannot have any value in exchange (Ricardo 2011).

D. Faulkner and C. Bowman popularized an approach based on the concept of perceived value in use, which in their opinion, together with the price, is an important parameter explaining the decisions of customers during the selection of a product. The ability to increase the perceived value in use is combined with the basic competitiveness of a company (Faulkner & Bowman 1995).

The perceived value in use is determined by such factors as product design skills, the time for its launch, additional service provided, production capacity, and the level of product quality. Customers pay their attention to these factors in a specific configuration, and the task of a company is to meet their expectations (McIvor 2005).

M.E. Porter believes competitive strategy to be certain offensive or defensive actions introduced in order to: maintain one's own position in the sector, achieve high profit rates, and effectively resist competitive forces (Porter 1985). According to M.E. Porter, a competitive strategy of company aims to achieve a higher rate of profit. He proposed the most well-known classification of strategies divided into three basic types. In his opinion, the important element for the building of a competitive advantage is the ability to obtain a permanent advantage in terms of costs and guarantee comparable quality of products, which is simultaneously the essence of the strategy of cost leadership. The strategy of differentiation (embracing differentiation or highlighting the product) aims at leadership through giving the products such characteristics that will, clearly and significantly, positively distinguish them in the market, which can decide that the price becomes less important in determining the choice made by a customer. According to Porter, in both described cases it is possible to achieve a higher rate of profit. For companies which are not able to lead large-scale competition, he recommends a strategy of concentration, which covers a selected part of the market (segment) along with the selection of the aforementioned bases of competition (i.e. low costs or differentiation) (Porter 1985).

The differentiation of a product is possible through its proper design, utility features such as stability or reliability, service, or the mode of customer service. In this kind of strategy, the quality of product and service plays a significant role. The importance of quality was also emphasized by K. Obłój, who distinguished three basic competition strategies: the product delivery, competitive price and the quality strategy of the offered product (Obłój 2000).

Product quality can certainly have an impact on the creation of its image in the perception of customers and, simultaneously, the brand building. It is obvious in the case of product quality described by its attributes, which can be measured and evaluated (Leffler 1982).

Product quality can also be understood as the property reflecting the expectations of the user. In such an approach, it can be defined as the ability to use or subjectively assessed suitability to meet the expected functions (Juran & Godfrey 1993). This understanding of the notion of quality seems to correspond to the above mentioned notion of perceived value of the product.

It is natural that the customers pay attention to the perceived value of the products, which is characterized mainly by quality, and which, in combination with the price, affects the willingness to make purchases or abandon them.

3 QUALITY MANAGEMENT SYSTEMS IN COMPETITIVE POSITION BUILDING

The family of ISO 9000 standards devotes much attention to the management principles that allow improving internal processes and, consequently, facilitating operations as well as achieving economically effective results. The first of the eight principles of the ISO 9001 standard is the customer focus. The objective resulting from the ISO 9001 requirements, which correspond to this principle, is the continuous supply of products that meet customer requirements and increase customer satisfaction through the use of a system of improvements, conformity to the customer and applicable law. The ISO standard introduces the obligation to measure customer satisfaction, which at the same time is an indicator of performance in quality management system (Pacana 2014).

The companies which have implemented certified quality management systems can gain an advantage over competitors in the marketing aspect. This is a strategic argument which significantly distinguishes different enterprises. It also allows you to create a suitable image of the company in the perception of customers. The resulting benefit is the possibility to increase the confidence of customers and suppliers. ISO 9001 certificates also play an important role in the case of fulfilling environment protection rules. According to T. Hermaniuk and T. Sikora, the ISO benefits are aimed primarily at improving the performance of production and distribution processes. It is achieved by appropriate documentation of processes, which allows to identify all differences and their determinants, opens the chance for improving the system and, consequently, maintains and increases the quality offered to customers (Hermaniuk & Sikora 2010).

The implementation and certification of the quality management system, however, does not guarantee the company's success. It must be regarded as a first step to a proper prioritization and methods of achieving it.

The competitive position of the company can be achieved thanks to quality improvement methods, under several additional conditions, which primarily include: the company's ability to offer products that meet the needs of customers to a greater extent than so far, the company's activities oriented towards quality being noticed and appreciated by customers, as well as the readiness of customers to pay more for the purchase of improved products (Monachello 1996).

4 POLISH ENTREPRENEURS' OPINIONS ON THE EFFECTS OF THE IMPLEMENTATION OF QUALITY MANAGEMENT SYSTEMS

The presented study focused on the randomly selected Polish enterprises from the manufacturing sector which implemented the quality management system compliant with ISO 9001. The primary objective of the study was the identification of the role of quality management in the process of shaping the competitiveness in a company.

The research hypotheses were: H1—the quality management systems increase the confidence of customers to the company, H2—the certificate confirming the compliance of the implemented system with the standard of quality is an important element which promotes a company, H3—the lack of certified quality management system weakens the competitive position of a company.

These studies included 280 companies in total, 13.2% of which were large enterprises, 33.6% were medium-sized companies, 42.1% were small, and only 11.1% of respondents were micro-enterprises.

Among the latter group, those which implement quality management systems certified for compliance with ISO 9001 are relatively rare. The structure of the scrutinized companies was presented in table 1.

The respondents were asked about the two main motives that have determined the implementation of quality management system. The reasons for the implementation of quality management systems were presented in table 2.

Most of respondents, as many as 45.4%, said that the main reason to implement ISO 9001 was the need to meet the expectations of customers who require appropriate certificates from suppliers. For 38.9% of respondents, the desire to improve the quality of products was important. One in three respondents (32.9%) pointed out the need of improving organisation and management in the enterprise. One in four of them (27.9%) expected that the costs of activity should decrease after the implementation of the system, mainly due

Table 1. Structure of the surveyed companies.

Number of employees	Number of companies	Rate of companies
250 and more	37	13,2%
50–249	94	33,6%
10–49	118	42,1%
0–9	31	11,1%
Total	280	100,0%

Source: Own work based on the surveys.

Table 2. Reasons for the implementation of quality management systems.

Answers	Number of respondents	Rate of respondents
Meet the expectations of customers	127	45,4%
Improve product quality	109	38,9%
Better organisation and management	92	32,9%
Reduce the costs	78	27,9%
Raise the prestige of the company	63	22,5%
Improving the information flow	46	16,4%
Pressure of competitors	32	11,4%
Others	13	4,6%

to the elimination of irregularities and costs associated with complaints. Similarly numerous (22.5% of respondents) were the answers that the certificate improves the prestige of the company. Lower response frequency rates belonged to the improvement of information flow after implementation of the system (16.4%) and to the pressure of competitors who have already implemented such systems and benefit from easier access to commercial offers (11.4%). Among the responses categorized as others, the enterprises marked the raise in productivity, increase in operational revenues, easier access to tenders financed by public funds (4.6% in total).

According to the objectives of the research, the opinions of respondents on the results of implementation of quality systems were important. They were divided into internal effects, i.e. those visible within the company, and external effects, which facilitate the functioning on the market and relationships with other entities.

The most important internal effect was the arrangement of internal processes, as indicated by 72.9% of the respondents. Improving the flow of information within the enterprise also plays a big role, as 65.7% of respondents pointed out. In this case, it is important to facilitate access to information related to customer expectations and opinions. It is also combined with the ability to identify errors in the company and in the customer service system faster, which was noted by 51.1% of respondents.

The intensification of customer focus is a noticeable effect for 42.5% of the respondents. It is surprising that only one fourth of the respondents indicated that because of the quality management system the number of complaints has been reduced. Such answer was given by 28.2% of them. When analyzing the reasons for implementing the quality system, however, it is remarkable that most of the respondents felt that the customer satisfaction is of highest importance, which should be manifest in a reduced number of complaints. These results may be explained by the fact that supposedly not all unsatisfied customers submitted their complaints. It is also likely that before the implementation of the quality system in companies not all recipients' complaints were documented. 19.3% of respondents noted that a noticeable reduction in operational expenses took place in their companies. It may be a result of the increase in the quality and reduction in the number of defects during production as well as the decreased number of complaints made by customers.

It is also surprising that during the realization of the surveys some respondents pointed out the negative effects following the implementation of the quality management system. Nearly one in four respondents (23.2%) noted that after the implementation of the quality system the requirements also increased due to the need to comply with new procedures and file additional system documentation. This is similar to the effect described as development of excessive bureaucracy after introducing system documentation, which was noted by 14.6% of respondents. The compliance with the requirements of ISO 9001 actually requires a great effort from employees directed to the preparation of relevant documents. This can be especially troublesome in the initial period of system activity. The reasons for the implementation of quality systems were presented in table 3.

An important aspect of the conducted surveys was the verification of external effects that appeared after the implementation of certified quality management systems. These factors can have the greatest impact on the competitiveness of the company and build its competitive position.

Most of the respondents (85%) said that the company uses a certificate of compliance with the quality management system ISO 9001 as an important part of its marketing strategy.

In the customers' perception, the certificate can create an image of the company as an actor offering high quality products and taking care of customer satisfaction due to the applied procedures. Indeed, two out of three respondents (67.5%) noted increased confidence on the part of suppliers. It is reflected by the withdrawal of audits performed by institutional customers and increased interest in the offer from customers. A similar group of respondents (62.9%) stated that it is important to meet the expectations of cus-

Table 3. Internal effects after the implementation of a quality system.

Answers	Number of respondents	Rate of respondents
Arrangement of internal processes	204	72,9%
Improvement in the flow of information	184	65,7%
Faster identification of errors	143	51,1%
Increased customer focus	119	42,5%
Reduced number of complaints	79	28,2%
Increased load on employees	65	23,2%
Lower costs	54	19,3%
Development of excessive bureaucracy	41	14,6%

tomers, which is the result of customer attitudes, and implement procedures to test their satisfaction. More than half of the respondents (55.7%) note that owing to obtaining the certificate they can currently participate in tenders for contracts financed from public funds. The certificate is often one of the conditions for participation. 47.9% of respondents believe that the system has a direct impact on competitive advantage.

This is also the result of increased access to customers, who require the suppliers to certify their compliance with the system standard. Thus, companies that have received such certificates can define their own position as superior against those who do not have one yet. One in three respondents (33.6%) believes that, thanks to the quality management system, it is possible to enter foreign markets, which may be associated with an increase in the reliability of companies that have implemented such systems. A similar group of respondents (29.3%) noticed an increase in the number of orders after the implementation of the quality management system. According to the opinions of 11.1% of the respondents, the implementation of a quality management system had no notable external effects. As indicated by the detailed analysis of distribution of responses, the same respondents most often pointed out the negative internal effects after the implementation of the quality management system. It was probably the result of their dissatisfaction with the generated results that these respondents focused their attention on the additional effort in the company after the introduction of system documentation and responsibilities. The external effects following the implementation of a quality management system were presented in table 4.

The lack of noticeable positive effects resulting from the implementation of a quality management system in the enterprise may be caused by too high expectations on the part of a company adopting such a system. However, it is very likely that in the case of these companies the understanding of the system idea was incomplete, and perhaps the staff resistance is such that the implemented systems are not as effective as possible.

The analysis of the research results clearly shows that most enterprises can see the positive aspects of introduced quality management systems. They appreciate the importance of certification in building the company's prestige and its brand. Focus on customers and greater fulfillment of expectations generate some increase in the customers' confidence in the company, thus increasing their turnover. For the proper performance of the company, it is also important to improve internal processes and communication within the company, as well as reduce general costs. Appropriate use of the positive effects following the introduction of a

Table 4. External effects following the implementation of a quality system.

Answers	Number of respondents	Rate of respondents
The use in marketing strategy	204	72,9%
Increase in confidence of suppliers	184	65,7%
Meeting the expectations of customers	143	51,1%
The opportunity to participate in tenders	119	42,5%
Gaining a competitive advantage	79	28,2%
The possibility of entering foreign markets	65	23,2%
The increase in the number of orders	54	19,3%
Lack of significant effects	41	14,6%

quality management system can lead to the development of an effective competitive strategy.

The performed studies positively verified all the initial research hypotheses, which indicates the important role of quality management systems in building the competitive position of a company.

5 CONCLUSIONS

Competition is a phenomenon associated with most companies at a scale greater than so far. In order to meet the challenges, the enterprises are forced to constantly seek new ways to build a competitive position. For this reason they are obliged to formulate competitive strategies which allow them to better formulate the company's strengths and use them to achieve success on the market.

Companies cannot find an universal method to effectively compete on the market. It is also difficult to achieve a stable competitive advantage. For that reason it is important to introduce internally such changes that will allow a better understanding of customer needs and the use of collected knowledge in the production of goods and services. Quality management systems can significantly expand the capabilities of enterprises to establish normal relations with the market environment and the proper development of internal processes. They also facilitate obtaining current information on the problems of the company. It is important for top management, however, that they should not only focus on obtaining a certificate after meeting the requirements of the market. The correct use of information collected during the process of competitive strategy formulation is the only factor that will allow you to compete with other entities. The important issue is to try developing, e.g., an effective strategy of differentiation by increasing the durability and reliability of products and better tailoring the products to the needs of customers.

The quality management system can play an important role in the building the foundations for competitive position of enterprise mainly by optimizing internal processes and opening up to contacts with suppliers and customers. It is also important that the systems require an appropriate approach to planning the development of the company, so that you can avoid the costs associated with bad investments.

REFERENCES

Bednarova, L., Pacana, A., Gazda, A. & Gierczak, A. 2010. Zarządzanie w oparciu o system jakości zgodny z międzynarodową normą ISO 9001. *Zeszyty Nau-kowe Politechniki Rzeszowskiej, Zarządzanie i Market-ing*, 272 (17): 99–106.

Faulkner, D. & Bowman, C. 1995. *The essence of Competitive Strategy*. Hertfordshide: Prentice Hall.

Hermaniuk T. & Sikora T. 2010, Wdrażanie systemów zarządzania jakością—korzyści i trudności. *Problemy Jakości* 5: 7–11.

Juran, J.M. & Godfrey, A.B. 1993. *Yuran's Quality Handbook*, 5 edit. New York: McGraw-Hill.

Leffler, K.B. 1982. Ambiguous Changes in Product Quality, *American Economic Review* December.

McIvor, R. 2005. *The Outsourcing Process: Strategies for Evaluation and Management*. Cambringe: Cambridge University Press.

Monachello, T. 1996. *Human Resource Management in the Hospitality Industry*. New York: John Wiley.

Obłój, K. 2000. *Strategia sukcesu firmy*. Warszawa: PWE.

Pacana, A. 2014. *Synteza i doskonalenie wdrażania systemów zarządzania jakością zgodnych z ISO 9001 w małych i średnich organizacjach*. Rzeszów: Oficyna Wydawnicza Politechniki Rzeszowskiej.

Porter, M.E. 1985. *The Competitive Advantage: Creating and Sustaining Superior Performance*. New York: Free Press.

Porter, M.E. 1998. *Competitive Strategy*. New York: Free Press.

Porter, M.E. 1998. *On Competition*. Boston: Harvard Business School.

Ricardo, D. 2011. *On The Principles of Political Economy and Taxation*. Kitchener: Batoche Books.

Smith, A. 1904. *An Inquiry into the Nature and Causes of the Wealth of Nations*. London: Methuen & Co., Ltd.

Production Management and Engineering Sciences – Majerník, Daneshjo & Bosák (Eds)
© 2016 Taylor & Francis Group, London, ISBN: 978-1-138-02856-2

Quality management model in the supply chain

D. Zimon, D. Malindžák & B. Zatwarnicka-Madura
Rzeszow University of Technology, Rzeszow, Poland

D. Dysko
Technical University of Košice, Košice, Slovakia

ABSTRACT: Despite the fact that logistics has a number of instruments allowing to achieve multiple benefits in the process of supply chain management, still valid and necessary is to search for new ideas about how to solve problems in the use of procurement, transportation, inventory control, customer service and improvement of logistics processes. According to the authors the modern concepts of quality management may be helpful in this regard. The purpose of this paper is to highlight the role of quality in supply chain management, identify areas where quality and logistics support each other in pursuit of common goals and to develop a general model of quality management in the supply chain. Model presented in the article is general in nature and leaves a lot of flexibility in choosing the optimal management concepts specific to a particular supply chain.

1 INTRODUCTION

As competition in the 1990s intensified and markets became global, so did the challenges associated with getting a product and service to the right place at the right time at the lowest cost. Organizations began to realize that it is not enough to improve efficiencies within an organization, but their whole supply chain has to be made competitive. The understanding and practicing of Supply Chain Management has become an essential prerequisite for staying competitive in the global race and for enhancing profitably (Li & Lin 2006).

Despite the fact that logistics has a wide range of instruments allowing to achieve multiple benefits in the process of supply chain management it is still valid and necessary to search for new ideas how to solve problems in the use of procurement, transportation, inventory control, customer service and logistics process improvement (Malindžák 2015, Wolniak 2013) According to the authors, improving supply chain management should be based on the most advanced systems and concepts of quality management. Łunarski (2012) agrees with this point of view and points out that over the last several years in the field of quality there have been developed various systems, concepts and techniques to improve both logistic operations and production. These methods can be successfully used to solve both widely understood logistical problems and specific difficulties in the elimination of waste, minimize inventory levels, synchronization of activities or efficient use of production capacity (Malindžák & Zimon 2014).

The aim of this paper is to emphasize the role of quality in supply chain management, identify areas where quality and logistics penetrate each other in pursuit of common goals and to develop a general model of quality management in the supply chain.

2 THE ROLE OF STANDARDIZED QUALITY MANAGEMENT SYSTEMS IN LOGISTICS

Despite the fact that in recent years ISO 9001 has lost much of its popularity, it is still the most commonly-implemented quality management system in the world. This is due to the fact that the main goal of its creators was to develop an universal system addressed to any type of organization. Initially, the quality management systems in accordance with ISO 9001standards were implemented in large industrial enterprises. Today, they are more popular in all types of organizations, including those that carry out any activity in logistics. According to Urbaniak (2015) the most common organizational standard used by companies to ensure the required quality and raising its level with the growth expectations of the buyers are the guidelines contained in ISO 9001. The guidance in this standard includes criteria for the implementation of operational processes (related to product design, purchasing, production, transportation, storage and delivery of goods, installation of equipment at the customer service after the sale).

The main goal of the authors of ISO 9001 was to develop such a standard of governance that would make entrepreneurs aware of the fact that management of the company should be directed primarily at the client. In addition the main objectives of the ISO standard include:

- the introduction in the company of such a management system that allows to enter a path of sustainable development, enabling appropriate targeting, minimize losses and improve the functioning of internal processes,
- to show external customers, on the basis of objective evidence that the company does everything within its capabilities to maximize their satisfaction,
- to enable the company to creative adaptation, the general requirements of the standard, to the specifics of the business, size, sector, objectives, strategies or existing market rules.

In addition, the ISO 9001 with its requirements gives a company a guidance on the definition of the quality management system architecture, based on the process approach and the indications for its continuous improvement. The ISO 9001 with its guidelines governs such areas in a company as: the development of system documentation, management responsibility, employees' rights, methods of communication, maintenance management, acquisition and management, comprehensive implementation of the product or measurement, analysis and improvement (Zimon & Zimon 2014).

Particularly significant connection exists between the implementation of the requirements of ISO 9001 and the improvement of logistics processes and relationships with suppliers. Especially, considering the relationship with SCM is a prerequisite for many companies worldwide as they introduce and implement the ISO 9001 quality management system. In article 7.4 'Purchasing', ISO 9001 outlines the requirements related to suppliers; more specifically, it states, 'An organization should guarantee the properness of purchased products to defined requirements. The method or degree of management applied to a supplier or purchased goods should be subject to change according to next product realization or the influence of finished goods' (Lee et al. 2003).

It can be concluded that the implementation of the requirements of standardized quality management systems in enterprises in the chain of supply can significantly affect the improvement of internal processes, integration chain and establish standards permitting the production of goods or the provision of services on the repetitive level taking full account of the requirements of customers. Implementation of ISO 9001 in many areas can support the development of an efficient supply chain through:

- development and implementation of effective methods of communication with customers,

- involving customers in product design,
- coordination of the activities carried out in the supply chain and matching them to the customer's requirements,
- ensuring the optimal level of customer service,
- selection of appropriate suppliers,
- establish appropriate quality control methods,
- partnership relationships with the supply chain,
- highlighting the importance of the human factor in the implementation of quality activities,
- formation of an effective supply chain management system,
- Developing objectives, mission and vision accepted by all the links in the chain,
- increase awareness of pro-quality of management.

3 TQM IN LOGISTICS

ISO 9001 International Standards can be an excellent start to TQM, if it is interpreted in a way that encourages the company to begin the process of continual improvement and aligns its entire people toward that goal (Fonseca 2015). Generally speaking the Total Quality Management is way of thinking about goals, organizations, processes and people to ensure that the right things are done right the first time. Adopting the Total Quality Management philosophy will (Nigam 2005):

- make an organization more competitive,
- establish a new culture, which will enable growth and longevity,
- provide an working environment in which ever one can succeed,
- reduce stress, waste and friction,
- build teams, partnership & cooperation.

In opinion of Nikolaidis (2013) TQM is a system approach that works backwards and forwards along the supply chain. At the heart of TQM lie the concept of continuous improvement and "customer value". Value is a delivered to the customer during the "use process", which includes all the activities that customer go through in using a product: find, acquire, transport, use, dispose.

According to the authors between logistics and TQM there are many dependencies because the assumption TQM philosophy enhances all key processes in the company, especially those that affect the wider and multifaceted customer service. Therefore it can be assumed that the majority of TQM improves processes within the supply chain and leads to an increase in their efficiency, speed of implementation, efficiency, accuracy and flexibility (Fig. 1). In addition; it can be concluded that the TQM philosophy improving the quality of:

- relationships in the supply chain,
- logistics infrastructure,
- production logistics subsystem,
- logistics customer service,
- finished products.

Furthermore, the implementation of TQM principles in the supply chain can positively affect the development of the cells and the construction of cooperation on the principles of sustainable partnerships. Relying on test results available in the literature it can be concluded that:

- the relationships between companies applying the principles of total quality management and their suppliers are long-term and enhance mutual learning processes,
- organizations applying TQM requirements put on partnerships with various actors in the supply chain,
- there is a close relationship between quality management, supply chain management and the positive results the company achieves by cooperation,
- companies should strive to integrate the principles of TQM with the strategy of supply chain management.

With that opinion agrees Ciesielski (2008) considering that between TQM and logistics there are close relationships:

- both TQM and logistics process in the supply chain are focused on the realizations of the same objectives,
- two elements are interrelated and interdependent,
- TQM affects all factors determining the smooth functioning of the logistics.

4 PROPOSED MODEL OF QUALITY MANAGEMENT IN THE SUPPLY CHAIN

Each organization and each supply chain has its own characteristics and operates under different external conditions. With this in mind, the authors propose a general model of quality management in the supply chain (Fig. 2), which should be enhanced with industry-specific systems and supported instruments individually matched to the implemented strategy and current needs.

This model is general in nature and leaves a lot of flexibility in choosing the optimal management concepts considering the specificity a particular supply chain. Its correct implementation should include the following principles:

- The basis for the implementation of total quality management and integration is to build partnerships in the supply chain. Only cells closely cooperating with each other will be able to implement a general philosophy of quality management. Ongoing partnership between the links of the supply chain should exist before the company decide to implement a model guidelines.
- The model should take into account the purpose of missions and visions adopted in the supply chain. Should support their implementation, must be subject to them.
- Quality management in the supply chain must be continuously improved in every aspect. The strategy should take into account the ever-

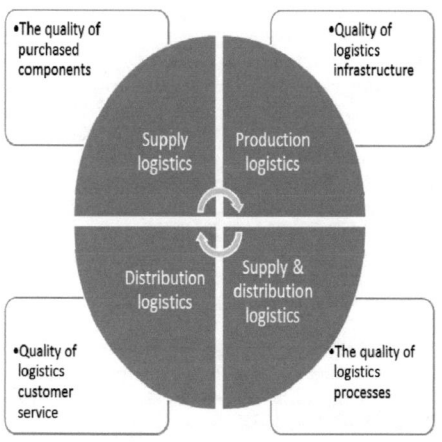

Figure 1. The impact of TQM on creation of quality in supply chain.

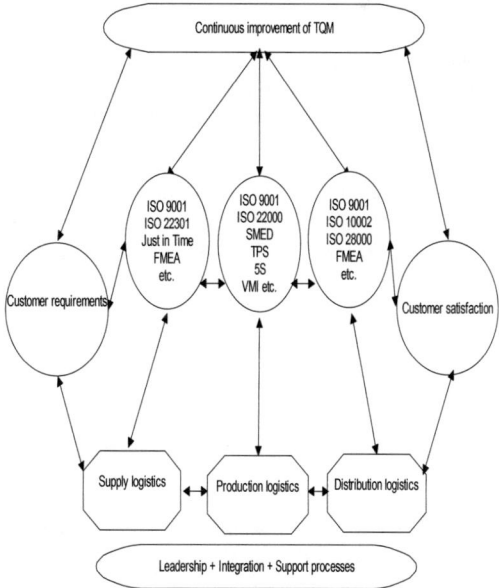

Figure 2. Model of quality management in the supply chain.

changing customer requirements and respond accordingly, through the development of new attitudes in the various pro-quality supply chain,

- The model is based on the implementation of the requirements of standardized quality management systems, industrial systems and to support their quality management instruments. Standardized quality management systems do not need to be certified, you can select only some of the items that are currently needed.
- Design, quality control and management has been created based on customer requirements and market and is supported by selected instruments, concepts and quality management systems. The idea of the model is to develop a philosophy of total quality management throughout the logistics chain.
- The premise of the model is guided by the outside (to customers) and the consolidation of the relationship between supply chain links.
- Quality control of purchased components should be almost entirely passed on to suppliers. According to the authors an external audit destroys trust, instead enter the periodic audits for the exchange of views and develop common goals.
- Development of quality management systems should be parallel with the development of technology. Develop optimal quality standards must be built on the basis of modern logistics infrastructure.
- The system should include all organizations working in the supply chain. Each cell is important and is making up the final customer satisfaction.
- Each subsystem has its own logistical tasks and different aspect of quality is the key, because each of them are addressed individually tailored concepts of quality management.
- Managing the logistics development should be balanced and take into account the needs of the customer, the environment, and to reconcile them with the objectives of enterprises (ISO 26000).

5 CONCLUSIONS

Systems supporting logistics processes and concepts of quality management seem to be necessary for each supply chain seriously thinking about the development, maintenance and customer acquisition. The individual chain links can implement standardized quality management systems, or create their own quality of their own. However, this is not recommended solution because it would lead to the development of many different concepts of quality management in the supply chain. According to the authors, due to the fact that quality management in the supply chain is complex and must be carried out according to strict schedules. It can not be based on ad hoc, piecemeal actions. Model proposed in the publication of quality management in the supply chain due to its versatility is addressed to the majority of supply chains through which should be considered only as an introduction to the development of its own concept of quality management. The model shows some links and highlights the benefits of combining quality and logistics. Next steps for the quality of the supply chain must be developed independently by the participants in the supply chain, taking into account their current, needs, aspirations and needs of the particular clients.

REFERENCES

Ciesielski, M. 2008. *Instrumenty zarządzania łańcuchami dostaw.* Warszawa: PWE.

Fonseca, L.M. 2015. From quality gurus and TQM to ISO 9001: 2015: a review of several quality paths. *International Journal for Quality Research* 9(1): 167–180.

Lee, M.S., Lee, Y.H. & Jeong, C.S. 2003. A high-quality-supplier selection model for supply chain management and ISO 9001system. *Production Planning & Control* 14(3): 225–232.

Li, S. & Lin, B. 2006. Accessing information sharing and information quality in supply chain management. *Decision support systems* 42(3): 1641–1656.

Łunarski, J. 2012. *Zarządzanie jakością w logistyce.* Rzeszow: Oficyna Wydawnicza Politechniki Rzeszowskiej.

Malindžák, D. & Zimon, D. 2014. The basic principles of the analyse for heuristic model creation in metallurgy. *METAL 2014—23rd International Conference on Metallurgy and Materials.* Brno: TANGER ltd.

Malindžák, D. 2015. The Basic Principle of Logistic Theory. *Applied Mechanics and* Materials 708: 47–52.

Nigam, S. 2005. *Total Quality Management: An Integrated Approach.* New Delhi: Excel Books.

Nikolaidis, Y. 2013. *Reverse Logistics and Quality Management Issues. Quality Management in Reverse Logistics.* London: Springer.

Urbaniak, M. 2015. The role of the continuous improvement tools of processes in building relationships in supply chain. *LogForum* 11(1): 41–44.

Wolniak, R. 2013. The assessment of significance of benefits gained from the improvement of quality management systems in Polish organizations. *Quality & Quantity* 47(1): 515–528.

Zimon, G. & Zimon, D. 2014. Influence of quality management systems on the financial capital management strategies in trading companies. *Modern Managamnet Reviov* 19(4): 123–133.

Production Management and Engineering Sciences – Majerník, Daneshjo & Bosák (Eds)
© 2016 Taylor & Francis Group, London, ISBN: 978-1-138-02856-2

The possibility of utilization of separated municipal waste as solid alternative fuel

K. Žurková, K. Kryštofová, H. Raclavská & H. Škrobánková
Centre ENET, VŠB—Technical University of Ostrava, Ostrava-Poruba, Czech Republic

ABSTRACT: The Separated Municipal Waste (SMW) produced in the town of Ostrava was used for the study on the possibility to use it for production of Solid Alternative Fuel (SAF). The content of combustible component in the separated municipal waste was the same in the summer and autumn months—68%, non-combustible proportion represented 32%. The higher heating value for dry matter of a sample from the autumn is 14.17 MJ/kg, from the summer period is 13.2 MJ/kg. The high content of glass in SAF produced from SMW represent a problem, it reached 20% in both sampled periods. Two procedures were tested for the removal of glass: the separation of waste with grain size of over 50 mm and the separation by air stream. The separation by air stream increased the higher heating to 21 MJ/kg, and the content of glass in the sample decreased by approximately 7%.

1 INTRODUCTION

Growing energy consumption and demands on environmental protection require ensuring a higher quantity of quality fuels. In order to save them, fossil fuels are replaced by alternative fuels. The advantage of the utilization of Municipal Solid Waste (MSW) as a component of the Solid Recovered Fuel (SRF) is not only a lower consumption of non-renewable energy resources, but also reduction of the amount of waste dumped in landfills. The municipal solid waste produced in the Czech Republic accounts for about 60% of municipal waste (Cenia 2013). In 2013, 3.4 million tonnes of municipal waste was produced, i.e. 320 kg/year per capita (Czech Statistical Office 2012).

SRF is defined as a solid fuel made from non-hazardous waste (or from waste not classified as hazardous), intended for energy recovery and utilization in incinerators or co-incinerator equipment, and meeting the requirements for separation and classification, as defined in CSN (Czech National Standard) EN 15359. The classification system is based on three major parameters: calorific value (economic characteristics), the chlorine content (technical characteristics) and Hg content (environmental aspect).

To ensure parameters monitored for SRF in CSN EN 15359, it is necessary to treat municipal solid waste by separating incombustible constituents and removing moisture (increased calorific value), or another pre-treatment technology for reducing the chlorine content in the fuel. The elemental composition of MSW can significantly vary among

countries, regions and cities, as a result of differences in the physical composition of MSW. The physical composition of MSW is usually dependent on the socio-economic conditions of a country, its population, the climatic conditions, and the national environmental legislation (Komilis et al. 2012).

In CSN EN 15359, fuels containing less than 0.2% of chlorine in the dry state fall within class 1, fuels containing less than 3% of chlorine fall within class 5. There is a potential for high temperature corrosion to occur on the heat transfer surface due to a high chlorine content (0.5–1.0 wt.%) in MSW (Chen et al. 2015). A high chlorine content of MSW induces a significant volume of deposit and high corrosion rates of super-heater tube. During 9 months of combustion of Solid Recovered Fuels (SRF) with the content of chlorine 1.13% up to 10 mm, thick deposit originated on super-heaters (Chen et al. 2015, Maa et al. 2010). The molten deposits with content of K, Na, Cl, or Fe increased the corrosion because liquid phase had faster chemical reactions, and it also provided an electrolyte for ionic transport or electrochemical attack. The chlorides and alkali metal compounds accelerated corrosion not only by lowering the melting temperature of the deposits, but also by attacking the metal surface with active oxidation mechanism (Lai et al. 2014). Combustion of fuels with high content of chlorine changes also the properties of energetic by-products. Approximately 60 wt.% of chloride compounds were removed in the baghouse and electrostatic precipitator, while the remaining 40 wt.% was discharged into the bottom and cyclone ash (Maa et al. 2010).

From the viewpoint of formation of deposits, glass content in municipal waste, which softens even at relatively low temperatures, may represent a problem. Glass softens within the temperature range from 500 to 1000°C, creeps at a temperature of about 900°C, and melts at a temperature above 1200°C. Man produces around 20 kg of waste glass per year, which corresponds to approximately 9% representation in municipal waste.

The aim of the presented paper is to compare the composition of municipal waste with literature data, to determine the amount of chlorine and forms of occurrence in terms of reducing its content using appropriate technology, and to assess the impact of container glass in terms of the possible formation of deposits.

2 MATERIAL AND METHODS

The representation of individual components in municipal waste was determined from waste supplied by OZO Ostrava, s.r.o., which provides municipal waste management in the city of Ostrava. For manual sorting, 120 kg of waste collected during the summer and autumn was available, and it was screened in the operating conditions on the 80 mm screen. Waste sample humidity from the summer was 53.37%, and in the autumn 34.58% (EN15414-1). For manual sorting, the sample was dried at 105°C. In addition to manual sorting and subsequent grain size analysis, which was performed according to EN 15415-1, separated components were also sorted out. Combustion heat (CSN EN15400) and ash content (CSN EN15403) was determined within the individual grain size classes. X-ray fluorescence method (Alpha Spectrometer TM of the Innow X company), heavy metals analysis was performed including the determination of chlorine. Samples for the determination of soluble chlorides from waste were prepared according to the decree no. 294/2005 Coll. on waste management in the latest version, and in the extract, chlorides were determined by ion chromatography.

3 MUNICIPAL WASTE COMPONENTS

The amount of waste produced in the Czech Republic and in the world has a slightly increasing trend. According to data from 2011, waste production per capita/year in the EU countries with the highest standard of living is more than two times higher than in the Czech Republic. Boer et al. 2010 reported the following annual MSW production: the European Union (27 countries)—503 kg/person/year, the Federal Republic of Germany—597 kg/person/year, Denmark—673 kg/person/year, Slovakia—327 kg/person/year, and Poland—238–309 kg/person/year. In 2014, the Czech Republic produced 3.11 miles of tons of municipal waste, of which Biodegradable Municipal Waste (BMW) amounted to 0.61 million tonnes.

Table 1 containing the representation of the components in municipal solid waste shows, also according to other authors, that in Europe, MSW contains significantly higher proportions of paper and cardboard, while glass content is lower, by up to more than 10%. Also in the UK, a significantly lower proportion of glass in municipal waste, 6–7%, has been reported (Burnley 2007). Differences in the proportions of various commodities in different countries are mainly due to the efficiency of waste separation at the source. The granulometric analyses of the total sample of MSW shows that the greatest amount of waste (about 50%) falls within the class 6.3–25 mm, about 20% of waste falls within the fine-grained class below 6.3 mm, and the class of 50 mm only amounts to 10% of waste.

In the summer and autumn sample of MSW in Ostrava, the proportion of BDMW was higher in autumn. Higher amounts of BDMW in the autumn sample are due to a greater proportion of green waste (Denafas et al. 2014). Besides BDMW, share of paper varied widely in the samples; unlike BDMW, it was higher in the summer sample, namely by about 10% (Fig. 1). We can assume that in the autumn months, the paper was used for heating.

The greatest amount of biodegradable waste occurs in the finest grain size class up to 25 mm. A significant difference was found in the representation of plastic in spring and autumn, in spring, the most significant share (63%) occurred in the class above 50 mm, the amount of plastic in autumn in this class dropped to ½. The largest proportion of glass occurs in spring and autumn in the fine-grained class below 25 mm (60–67%), a significant proportion was also found in the class above 50 mm, but only in summer. In addition, about 50% of paper and cardboard is found mainly in the class of 50 mm. Technologically, sorting the problem fraction on the screen is the simplest solution. The grain size distribution of the individual components of waste shows that it is easier to sort specific compounds; however, some energy parameters (calorific value) will be affected (Tab. 2). BDMW accumulates in the finest grain size classes in the same way as problematic glass. Plastics and paper mainly occur in coarse grain size class, the removal of this class would results in a significant decrease in the calorific value, but on the other hand, this would also reduce the chlorine content in the fuel.

Table 1. Comparison of the proportion of components in municipal waste [%].

	Assamoi et al. 2012	Chen et al. 2015	Ionescu et al. 2013 Europe	Leckner 2015 OECD	Ostrava Summer	Ostrava Autumn
Paper + card board	26	20.2	22	32	17.45	8.71
Food	27	48.5	24	27	38.51	53.76
Green waste	19		7			
Wood	2		4			
Plastics	8	20.4	9	11	8.61	4.05
Textile and leather	2	1.6	6		2.97	1.3
Rubber	<1					
Metals	2		3	6	2.56	2.27
Glass	5	4.1	12	7	20.87	19.84
Others (ash, non-combustible)	10	5.2	13	17	8.47	9.82

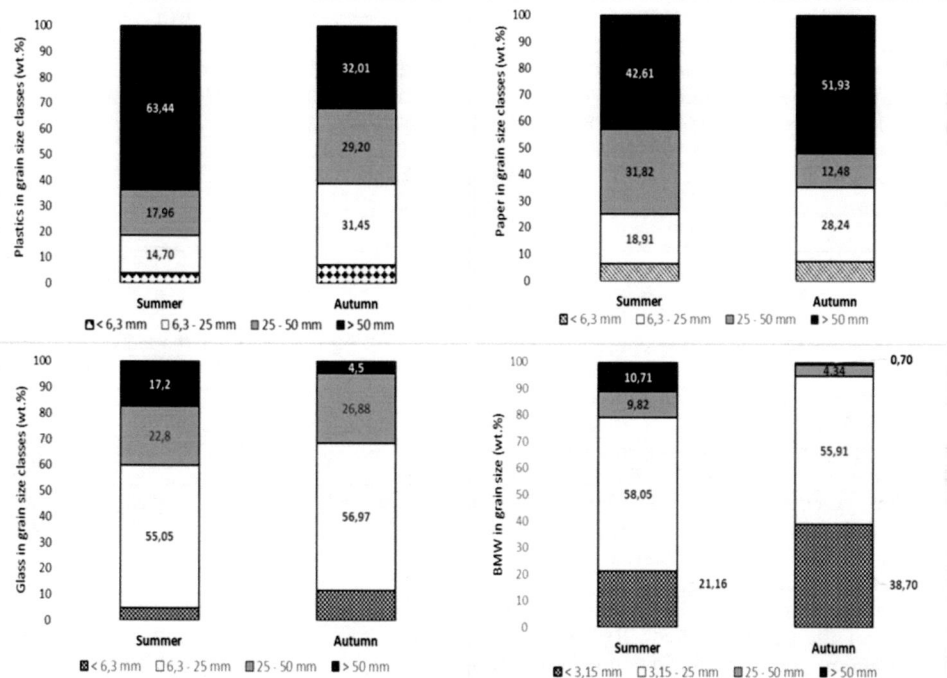

Figure 1. Representation of plastics, paper, glass and BMW in the grain-size classes.

The combustion heat value of the dry matter of the total sample is approximately 14 MJ/kg; it decreases with the particle size. The highest combustion heat value was measured in the class above 50 mm in summer, which is associated with a higher proportion of plastics in the sample. The combustion heat value of plastics in both sampling is around 24 MJ/kg. On average, the total sample contains 36% of ash. The amount of ash in the fine-grained classes increases more than two-fold (summer).

4 TECHNOLOGICAL TREATMENT OF MSW

Technological treatment of MSW was performed by sorting the waste on the 50 mm screen (sample INGEA recycling, Ltd.), and technological line was used as well: magnetic separation and air separation in OZO Ostrava, s.r.o. Both technologies aimed to achieve a reduction in the proportion of non-combustible materials and glass in MSW

Table 2.   Proximate and ultimate analysis of MSW collected in summer and autumn.

	Volatile matter	Ash	FC	C	H	N	S	O	HHV
	[%]								[kJ/kg]
MSW collected in summer									
MSW grain size >50 mm	68.38	27.52	4.11	35.60	9.11	1.20	<0.01	26.57	**15 885**
MSW grain size 25–50 mm	57.79	39.57	2.64	31.77	7.66	1.26	<0.01	19.75	**13 758**
MSW grain size 6.30–25 mm	56.23	42.43	1.35	29.36	7.79	1.56	<0.01	18.86	**13 263**
MSW grain size <6.30 mm	37.66	62.22	0.13	20.48	5.41	1.04	<0.01	10.86	**7 438**
MSW—bulk sample	62.36	34.03	3.61	31.77	5.83	1.39	0.61	26.37	**13 714**
Biodegradable waste	50.95	46.69	2.36	31.60	6.95	1.85	0.22	12.69	**10 859**
Textile	81.72	8.85	9.34	42.11	7.45	2.71	<0.01	38.88	**29 588**
Plastic	78.50	18.96	2.54	68.75	9.52	0.37	<0.01	2.41	**23 307**
Paper	78.81	22.21	1.98	37.17	8.43	0.78	<0.01	31.41	**13 411**
MSW collected in autumn									
MSW grain size >50 mm	77.59	17.95	4.45	36.26	9.38	3.31	0.01	33.10	**18 943**
MSW grain size 25–50 mm	76.55	20.16	3.28	39.04	10.05	2.96	0.01	27.78	**18 974**
MSW grain size 6.30–25 mm	59.44	38.65	1.91	33.40	6.57	3.36	0.06	17.96	**16 473**
MSW grain size <6.30 mm	42.67	57.37	0.47	22.98	4.70	2.81	0.13	12.01	**9 250**
MSW—bulk sample	59.86	38.20	1.85	29.04	4.78	2.85	0.74	25.13	**14 176**
Biodegradable waste	51.70	45.12	3.01	30.27	6.03	3.07	0.01	15.51	**11 446**
Textile	85.11	10.34	4.38	42.92	11.82	2.73	0.01	32.20	**21 904**
Plastic	86.86	7.89	5.16	69.67	9.86	1.71	0.01	10.88	**24 325**
Paper	75.81	21.81	2.23	35.97	7.34	1.98	0.01	32.91	**12 836**

Explanations: HHV—higher heating value in dry matter, FC—fixed carbon.

(Tab. 3). With air separation, the content of glass decreased by about 7%, and the content of the paper increased by about 6% and the content of plastic by about 5%. In the case of air separation, there was approximately a twofold increase in the combustion heat to 24.14 MJ/kg. The gross combustion heat in energy recovery is relatively high (Zhou et al. 2015). The ash content in the feed material is not significantly decreased. After sorting glass and ceramics, the carbon content in the sample increased on average by 15% as a result of increase of the plastics content. Removing the oversize fraction above 50 mm did not have any significant effect on the MSW composition.

The best performance in terms of combustion was observed in the sample after air separation on a line in OZO. After air separation, ash fusibility was determined for the sample. The sorted waste sample was burned in a laboratory muffle furnace at 815°C. The resulting ash represents the input for the determination of fusibility of ash according to ISO 540. The character of ash after burning the sample separated by air (spherules of glass after melting in white circle) is shown in Fig. 2.

The results of determination of fusibility of ash in Tab. 4 show that the deformation occurs in the sample after air separation at about 1100°C, while in the input sample, it is already at 1000°C; the

Table 3.   Representation of the individual compounds in MSW after technological treatment [%].

	Autumn input	Air separation OZO	Separation under 50 mm
Debris	5.51	2.89	3.90
Ceramics	4.31	1.69	4.05
Hazardous waste	0.24	0.00	0.00
Metal	2.27	1.73	2.78
Textile	1.30	2.30	1.47
Glass	19.84	12.79	17.54
Plastics	4.05	9.46	3.93
Paper	8.71	14.44	9.57
BMW	53.76	54.69	56.76
Combustion heat [MJ/kg]	**13.71**	**24.14**	**11.75**

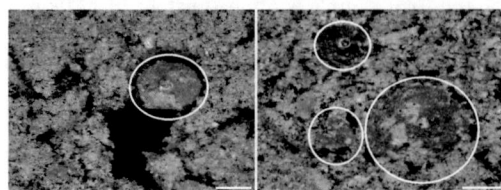

Figure 2.   Spherules of glass showing a softening and melting trend.

melting points of both ashes are identical. Jiang et al. 2009 points out the problem of sticking during incineration of hospital waste with 12% of glass (medicine bottles with high content of Na_2O up to 10% and K_2O more than 2%) (Jiang et al. 2009); they determined the softening temperature of about 830°C depending on the alkali content.

Although the deformation of the glass already manifested at lower temperatures, up to 815°C, ash melts at a higher temperature, which is also influenced by the presence of other phases in the ash. Representation of the main mineral crystalline phases in the ash from the incineration of MSW as

Table 4. Fusibility of ash [°C].

	DT	ST	HT	FT
Sample for air separation—ash	1099	1189	1213	1242
Input sample—ash	1005	1181	1200	1249

Explanations: DT—deformation temperature, ST—sphere temperature, HT—hemisphere temperature, FT—flow temperature.

Table 5. The mineralogic composition of the ash phase [%] and the melting points of minerals.

Mineral	Melting point [°C]	Separation below 50 mm	Air processing
Quartz (SiO_2)	1713	17.28	16.81
Calcite ($CaCO_3$)	1300	7.37	6.71
Microline ($KAlSi_3O_8$)	1170	6.57	7.17
Gehlenite (C_2AS)	1593	4.44	5.68
Albite ($NaAlSi_3O_8$)	1120	5.39	4.33
Muscovite 2M1 ($KAl_2(AlSi_3O_{10})(OH)_2$)	1400	4.55	4.50
Larnite (Ca_2SiO_4)	1420	10.62	9.42
Amorphous		43.78	45.38

determined by X-ray diffraction is shown in Tab. 5. Most of the identified phases in the ash from the incineration of MSW has a melting point higher than 1100°C, which significantly affects the melting point listed in Tab. 4. However, it cannot be excluded that softening glass may stick to the grids.

In addition to glass, chlorine and alkali metals may also participate in the creation deposits and in sticking. The main source of chlorine in the MSW are remains of food (0.9%) and plastics containing 0.5–6.3% of chlorine (Chen et al. 2015). Zhou et al. 2014 state similar values of chlorine; in the remains of food, they determined (0.12–2.50% of chlorine, with an average of 1.06%), wood waste (mean 0.29%), paper (0.10–0.73, mean 0.28%), textile (0.06–0.96, mean 0.36%), chlorine free plastics (0%), PVC (42.3–58.15, mean 53.53%), rubber (1.16–2.08, mean 1.68%). Polyvinyl chloride (PVC) from packaging, electrical wire insulation etc. in plastics and chloride salts (mainly NaCl) in kitchen waste are the main sources of organic and inorganic chlorine. The increase of the operating temperature from 700°C to 1000°C has more influence on the HCl formation for kitchen waste than that for PVC (Maa et al. 2010).

In a sample of municipal waste collected in the summer, the average chlorine content is 2.43%, in the autumn collection it is slightly higher—3.13%. Due to these contents, MSW falls within the class 5. according to CSN EN 15359. After the implementation of the air separation technology, the chlorine content in the fuel increased to 4.71%, which is associated with an increase in plastics in the sorted MSW. By leaching soluble chlorides, chlorine content in the fuel can be reduced by about 20–30% (Fig. 3). This reduction, however, does not allow to achieve a safe value of the chlorine content in the fuel in terms of deposit formation (Chen et al. 2015). Solving the problem of the chlorine content in the fuel is preparation of MSW with added woody biomass with low chlorine content

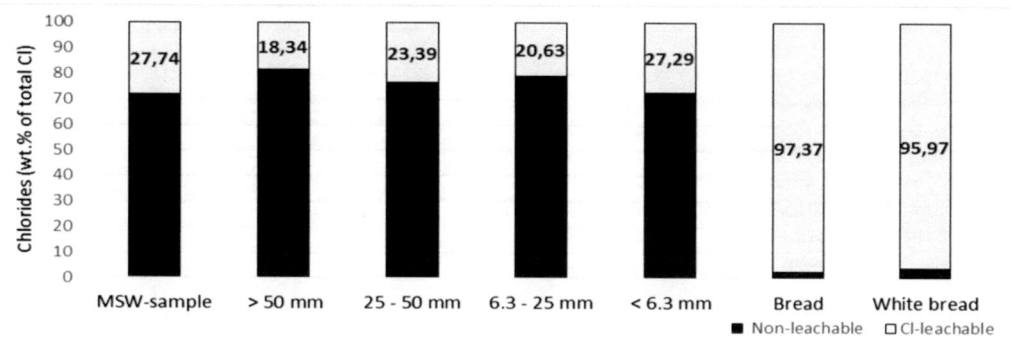

Figure 3. The leachability of chlorides from MSW in the individual grain size classes [%].

(about 80%) or sorting of plastics (PVC reduction in MSW) and increasing the PP-foil share. In fact, plastics are not a single component, since they comprise different materials such as polyethylene (PE), polypropylene (PP), polystyrene (PS) and polyvinyl chloride (PVC). Materials have various proximate and ultimate analysis results, as well as thermal kinetic parameters. For instance, the content of C in PE is as high as 85.5%, while that of PVC was only 34.24% and the Cl content of PVC is 52.21% (Zhou et al. 2014).

5 CONCLUSIONS

SRF from MSW meets the requirements of the calorific value of group 3 according to the classification system specified in CSN EN15359. Other parameters (chlorine content and mercury) rank the fuel to group 5. For improving fuel, technology of air separation was tested, as well as sorting classes above 50 mm on screens. Air separation increased the fuel efficiency and reduced the proportion of waste glass, however, the chlorine content in the fuel increased at the same time. Sorting classes above 50 mm yielded no significant effect for improving the fuel, the proportion of glass in SRF remained the same. The comparison with the results of the component content in MSW with literature data shows that the contents of waste glass in Ostrava are about 10% higher, they mostly amount to 10% in MSW. The presence of container glass with a high proportion of alkalis (Na) may also influence the formation of sticking. During microscopic study of the character of ash, thermally affected glass particles were identified. For the preparation of SRF, which would meet safe limits in terms of the chlorine content in the fuel, sorting plastics with a high chlorine content or the addition of wood waste chlorine-free can be recommended.

ACKNOWLEDGEMENTS

This paper was supported by research projects of the Ministry of Education, Youth and Sport of the Czech Republic: the Centre ENET CZ.1.05/2.1.00/03.0069 and by research projects of the Ministry of Education, Youth and Sport of the Czech Republic: SP2015/64—The interdisciplinary study of fuel behaviour.

REFERENCES

Assamoi, B. & Lawryshyn Y. 2012. The environmental comparison of landfilling vs incineration of MSW accounting for waste diversion. *Waste Management* 32: 1019–1030.

Burnley, S. J. 2007. A review of municipal solid waste composition in the United Kingdom. *Waste Management* 27: pp. 1274–1285, ISSN 0956–053X.

Chen, G., Zhang, N., Ma, W., Rotter, V.S. & Wan, Y. 2015. Investigation of chloride deposit formation in a 24 MWe waste to energy plant. *Fuel* 140: 317–327.

Den Boer, E., Jędrczak, A., Kowalski, Z., Kulczycka, J. & Szpadt, R. 2010. A review of municipal solid waste composition and quantities in Poland. *Waste Management* 30: 369–377, ISSN 0956–053X.

Denafas, G., Ruzgas, T., Martuzevicius, D., Shmarin, S., Hoffmann, M., Mykhaylenko, V., Ogorodnik, S., Romanov, M., Neguloaeva, E., Chusov, A., Turkadze, T., Bochoidze, I. & Ludwig, C. 2014. Seasonal variation of municipal solid waste generation and composition in four East European cities. *Conservation and Recycling* 89: 22–30.

Ionescu, G., Rada, E. Ch., Ragazzi, M., Mărculescu, BADEA, A. & Apostol, T. 2013. Integrated municipal solid waste scenario model using advanced pretreatment and waste to energy processes. *Energy Conversion and Management* 76: 1083–1092.

Jiang, X. G., An, C. G, Li, C. Y., Fei, Z. W., Jin, Y. Q. & Yan, J. H. 2009. Fusibility of medical glass in hospital waste incineration: Effect of glass components. *Thermochimica Acta* 491: 39–43.

Komilis, D., Evangelou, A., giannakis, G. & Lymperis, C. 2012. Revisiting the elemental composition and the calorific value of the organic fraction of municipal solid wastes. *Waste Management* 32: 372–381.

Lai, Z. Y., Ma, Q., Tang, Y. T., Li, M. D. & Ni, J. F. 2014. Deposit analysis of water-wall tubes in a municipal solid waste grate incinerator. *Applied Thermal Engineering* 66: 415–422.

Leckner, B. 2015. Process aspects in combustion and gasification Waste-to-Energy (WtE) units. *Waste Management* 37: 13–25.

Maa, W., Hoffmann, G., Schirmer, M., Chena, G. & Rotterb, V. S. 2010. Chlorine characterization and thermal behavior in MSW and RDF. *Journal of Hazardous Materials* 178: 489–498.

Mixed municipal waste. 2013. In: Multimedial yearbook of environment. Prague, Cenia (In Czech). Available: http://www.czso.cz/csu/2013edicniplan. nsf/t/F300446190/$File/w20011314.pdf (In Czech).

Production of municipal waste in selected countries in 2004–2011. 2012. The Czech Statistical Office.

Zhou, H., Meng, A. H., Long, Y. Q., Li, Q. H. & Guo, Y.. 2014. An overview of characteristics of municipal solid waste fuel in China: Physical, chemical composition and heating value. *Renewable and Sustainable Energy Reviews* 36: 107–122.

Zhou, H., Meng, A., Long, Y., Li, Q. & Zhang, Y. 2015. Classification of municipal solid waste components for thermal conversion in waste-to-energy research. *Fuel* 145: 151–157.

Scientific and technical support of sustainable production

Production Management and Engineering Sciences – Majerník, Daneshjo & Bosák (Eds)
© 2016 Taylor & Francis Group, London, ISBN: 978-1-138-02856-2

New generation of Eurocodes

I.J. Baláž & Y.P. Koleková
Slovak University of Technology, Bratislava, Slovak Republic

ABSTRACT: The special group of European standards is called Eurocodes. They are set of common European standards covering 10 main subjects in the field of structural and civil engineering, based on the requirements of the Council Directive 89/106/EEC relating to Construction Products (CPD). Ten Eurocodes have several parts, making a total of 58 parts, the last of which was published by CEN in May 2007. They consist of 10 subjects: EN 1990 Basis of Structural Design, EN 1991 Actions on Structures, EN 1992 Design of Concrete Structures, EN 1993 Design of Steel Structures, EN 1994 Design Steel and Concrete Composite Structures, EN 1995 Design of Timber structures, EN 1996 Design of Masonry Structures, EN 1997 Geotechnical Design, EN 1998 Design of Structures for Earthquake Resistance and EN 1999 Design of Aluminium Structures. Eurocodes are concerned with requirements for resistance, serviceability, durability and fire resistance of structures.

1 DEVELOPMENT OF EUROCODES

1.1 First generation of Eurocodes

The Eurocodes system was developed to facilitate the application of the first European Public Procurement Directive (1971) in the field of design rules for buildings and civil engineering works. In 1975, the Commission of the European Community decided on an action programme based on Article 95 of the Treaty of Rome. The objective of the programme was the elimination of technical obstacles to trade and the harmonisation of technical specifications for construction works. The documents, which were not really standards at the origin, were based on the works of international scientific associations and were intended to be used as a reference for the judgement of public tenders.

During 15 years, the Commission, assisted by a steering committee composed of experts from EU Member States, oversaw the development of the Eurocodes programme, which led to the publication of a first generation of European codes in the 1980s which were immediately submitted to international inquiries. After the adoption of the European Unique Act (1986) with the objective to modify and complete the Treaty of Rome, in particular to improve the decision procedures (vote at the qualified majority), new European directives were established, defining only essential requirements and the development of standards based on these requirements was transferred to appropriate standardisation bodies. One of these directives, published in 1989, dealt with construction products.

1.2 European pre-standards—ENV Eurocodes

In 1989 the special agreement between the CEN (Comité Européen de Normalisation) and the European Commission transferred the preparation and publication of the Eurocodes to the CEN, thus providing the Eurocodes with a future status of European EN standards. The agreement specified that the Eurocodes were intended to serve as reference documents to be recognised by authorities of the Member States for the following purposes:

- As a means for enabling building and civil engineering works to comply with the "essential requirements" of the construction products directive (89/106/EEC), particularly the first two—mechanical resistance and stability, and safety in case of fire;
- As a basis for specifying public construction and related engineering service contracts. At that time, this related to Council Directive 93/37/EEC of 14 June 1993 concerning the co-ordination of procedures for the award of public works contracts, and the services directive (92/50/EEC). Since the 31st of March 2004, these two directives have been replaced by a unique directive (2004/18/CE) of the Parliament and of the Council on the coordination of procedures for the award of public works contracts, public supply contracts and public service contracts;
- As a framework for drawing up harmonised technical specifications for construction products (ENs and European Technical Approvals).

From 1991 to 1999 altogether 60 parts of ENV Eurocodes (European pre-Standards),

2 Amendments and 7 Corrigenda were published by the CEN in Brussels.

From 1998 to 2004 the Slovak Standards Institute (SUTN) implemented ENV Eurocodes into national standardization system. SUTN supported the translation of ENV Eurocodes and the creation of Slovak National Application Documents (NAD) into Slovak by 160,000 Euros. 52 STN P ENV were published in Slovak language, 4 parts remained in Czech and 4 parts in English language. 59 STN P ENV Eurocodes had NAD, one was without NAD. The complete set of STN P ENV Eurocodes (4451 pages) cost consulting engineers 927 Euros.

2 EN EUROCODES

2.1 Current generation of EN Eurocodes

From 2002 to 2007 altogether 58 parts of EN Eurocodes were published by CEN in Brussels. Till today several Corrigenda and Amendments to EN Eurocodes have been published by CEN.

The benefits and opportunities of adopting the Eurocodes can be summarised as follows:

- To provide a common understanding regarding the design of structures between owners, operators and users, designers, contractors and manufacturers of construction products;
- To facilitate the exchange of construction services between European Member States;
- To facilitate the marketing and use of structural components and kits of parts in Member States;
- To be a common basis for research and development in the construction sector;
- To allow the preparation of common design aids and software;
- To increase the competitiveness of the European civil engineering firms, contractors;
- Designers and product manufacturers in their world-wide activities.

The European Commission has formally recommended the 58 structural Eurocodes as "a suitable tool" for designing construction works, checking the mechanical resistance of components and checking the stability of structures. The Recommendation 2003/887/EC of 11 December 2003 says Member States should recognise that construction works designed using Eurocodes will conform with the essential requirements of mechanical resistance and stability, safety in use and safety in case of fire.

Eurocodes are a state-of-the-art tool which will undoubtedly provide a higher level of safety than before for EU citizens. The Eurocodes constitute a coherent EU-wide framework of common calculation methods with facilities to adapt their functioning to national settings and priorities through a set of Nationally Determined Parameters (NDPs). When taking the Eurocodes on board, the Member states are expected to define the NDPs to be observed on their territory taking into account justified differences in climate, geographic conditions (e.g. seismic risk), levels of safety, or traditions regarding the way of life prevailing in their territory.

To secure the harmonization efforts, the Commission Recommendation includes a suggestion that Member-States, acting in coordination under the direction of the Commission, compare the NDPs implemented by each Member-State and assess their impact as regards the technical differences for works or parts of work. The Commission has developed and made available a structure, including appropriate databases, to collect and compare the defined NDPs, thus enabling further harmonization work.

The Recommendation 2003/887/EC underlines the necessity to support the implementation on the ground through fully integrating the Eurocodes into all relevant education and training activities on national and regional level and it indicates that continuous efforts to maintain the Eurocodes at the forefront of engineering knowledge and developments in structural design are needed, with further research on MS and EU level facilitating uptake of the latest scientific knowledge and the development of the construction market, including new materials, products and construction methods.

All EU Member States have now initiated the full adoption of the Eurocodes as national design codes including the definition of the Nationally Determined Parameters (NDPs). Due to different regulatory systems and setups the process and the time for full implementation vary depending on the Member States.

2.2 Implementation of Eurocodes in Slovakia

In the transition period SUTN published the following guidelines concerning design of structures and use of Eurocodes in Slovakia.

In the period from 30 March 2006 to 30 November 2008 three independent standard systems could be used for design of structures:

- Slovak national standards STN;
- STN P ENV Eurocodes with Slovak NAD;
- STN EN Eurocodes with Slovak NAs (only when complete required package was available).

It was forbidden to mix individual standards or their parts from different standard systems.

In the period from 1 December 2008 to 31 March 2010 two independent standard systems could be used for design of structures:

- Slovak national standards STN;
- STN EN Eurocodes with Slovak NAs (only when complete required package was available).

It was forbidden to mix individual standards or their parts from different standard systems.

Starting from 1 April 2010 STN EN Eurocodes with Slovak NAs are the valid standard system for design of structures—taking into account all Amendments and Corrigenda. The STN that were in contradiction with Eurocodes were withdrawn. Design of structures in Slovakia according to other standards (e.g. foreign standards) must be supported by an agreement in the contract and cannot have lower levels of reliability as design according to Eurocodes. The details of above guidelines were created by the Technical committee TK 111 "Application and use of Eurocodes", which was founded at SUTN in 2005.

Implementation of Eurocodes in Slovakia was supported by:

- Publishing of set of 35 informative papers in Slovak journal EUROSTAV (from 2004 till 2010). They are available for free on www.eurostav.sk.
- Publishing papers in other Slovak journals and numerous scientific papers in many national or international proceedings.
- Publishing of several proceedings containing numerical examples, which served as textbooks in courses organised by the Slovak University of Technology, the Ministry of Building and Regional Development, the SUTN, or the Slovak Chamber of Civil Engineers.
- Creation of publications: Terminology of Structural Eurocodes; English-French-German-Czech-Slovak terminological dictionary with definitions and the official Slovak national standard STN 73 001 Terminology of Eurocodes.

Education at universities is based on Eurocodes starting from 1994 to 1998. Education of foreign students at the Faculty of Civil Engineering of the Slovak University of Technology in Bratislava helps to promote Eurocodes in other countries. Before the separation of the Czech and Slovak republics, common teams of specialists prepared common Czechoslovak standard CSN 73 1401 Design of Steel Structures, the first half of which was based on its former edition from 1986 and the second half of which was based on parts of ENV Eurocodes. This standard was published later as Czech standard in 1995 and after removing errors and improving again in 1998. A similar, but not identical standard was published also in Slovakia as STN 73 1401 Design of Steel Structures in 1998. Starting from 1998 and thanks to STN 73 1401 a

lot of rules of European pre-standard ENV 1993 were used in Slovakia by consulting engineers. It is necessary to mention that Limit States Design (LSD) was used in design of steel structures in Czechoslovakia 55 years ago, and therefore consulting engineers had no problems with using new standard STN 73 1401 influenced by ENV Eurocodes.

The verification of the possibility to change Allowable Stress Design (ASD) into LSD in the design of steel structures was investigated in the former Soviet Union at the beginning of 1950s. The government of Czechoslovakia decided to verify the possible weight saving when using LSD instead of ASD in official announcement in 1960. The first verifications were done by A. Brebera in 1956 and 1957 at Research Institute of Building Production in Prague and by A. Mrazik in 1958 and 1960 at Institute of Construction and Architecture, Slovak Academy of Sciences in Bratislava. The first Czechoslovak standard STN 73 1401 Design of Steel Structures based on LSD was approved in 1966, and became valid in 1969. The first realised steel structure in Czechoslovakia designed according to ASD but recalculated and verified by LSD was the hangar at Prague airport. It was before 1966.

Design of structures according to Eurocodes comparing with design according to former Slovak standards leads in most cases to safer, heavier and more expensive structures.

The first large structure in Slovakia designed according to Slovak national standards and also according to ENV Eurocodes is the arch bridge Apollo over the river Danube in Bratislava, which was opened in 2005. The bridge holds several awards (2005 ECCS European Steel Design Award, 2006 Winner of the competition Structure of the year in Slovakia) and it has been named one of five finalists (chosen from more than 400 structures) for the 2006 Outstanding Civil Engineering Achievement Award, presented by the American Society of Civil Engineers (ASCE). It is arch bridge with main span 231 metres and with length 514.5 metres of 6 span bridge structures.

Consulting structural engineers criticise large content and complexity of Eurocodes complemented by several Amendments and Corrigenda, missing scientific backgrounds and inconsistency of some rules which came from various authors. Among the drawbacks of Eurocodes implementation there are big expenses together with paying for hard copy of Eurocodes, courses, new literature and software. On the other hand, people are satisfied having common modern European standards. Guidance paper L gives information on the "Application and use of Eurocodes" and states that one of the aims and benefits of the Eurocode programme is that it allows common design aids and software to be developed in all member states. Several handbooks, design manuals, guides, textbooks and programmes supporting design according to Eurocodes are available currently in various countries. To have an efficient system of Eurocodes it is necessary to solve the problem of availability of all national annexes at least in English language.

2.3 Permanent evaluation of EN Eurocodes

The design of buildings and engineering works is permanently evolving for several reasons.

Constructors have to take into account the evolution of societal needs or requirements, generally expressed by political authorities (at the national or international level).

Constructors have to take account of the evolution of the markets and in particular the increasing cost of raw materials to maintain as far as possible the requirement of economy of construction works. This point may have technical aspects if the manufacturing process of construction materials (in particular steel and concrete) is altered. The consequence of the more economic design is an evolution of the quality control of materials and structural components on site.

The competitiveness of construction companies depends on:

- The quality of the design;
- The quality of execution, including the quality of materials and manufactured components, and
- The ability to develop innovative solutions (innovative execution processes, use of new materials).

Of course, competitiveness implies, as well, relatively low cost.

The main developments in the construction sector in the last decades may be summarized as follows:

- Implementation of computer-aided design which facilitates the use for refined structural models and advanced calculation methods;
- Development of architectural innovation (and complexity) in the field of high-rise and monumental buildings;
- Development of new execution methods, more simple, rational and industrialized, with, as a consequence, a decrease of the manpower volume;
- Development of transportation infrastructures (motorways, railways, airports, etc.), especially in urban and suburban areas (tram, metro, liaison between various transportation means);
- Increase of the span length of footbridges and of very large bridges;
- Development of the use of new materials like glass, FRP, or very high performance self-compacting concrete.

One major aspect of standardization is to establish a clear relationship between industry, clients, public authorities, etc. from both technical and contractual points of views.

Document CEN/TC250—N 630, dated February 2006, was the final report defining the main evolution axes of the Eurocodes for the future. It gave proposals from CEN/TC 250 with regard to its role in the implementation, use and development of Eurocodes by providing on a European level:

- Maintenance. Maintenance cannot be reduced to simple corrections of mistakes and errors to be made available in accordance with the formal and rigid schedule defined in CEN internal rules and stated in document CEN/TC250—N 778. Therefore, the maintenance of the Eurocodes appears to be more and more a new technical and transparent activity aiming at the improvement of the current versions;
- Harmonization (convergence of the technical content of National Annexes). Member States sometimes complain that the present versions of the Eurocodes require too much input from regulatory bodies that could be avoided by leaving out many possibilities of choice. The comparison of NDPs is essential for the evolution of the Eurocodes. The work for this comparison by JRC is in progress. In particular, a database has been developed to facilitate the analysis of the National Annexes. The output of this work should be a proposal for a limited number of NDPs in each Eurocode Part. This proposal is the result of a detailed analysis and evaluation of the database of JRC, comparative studies and discussions among the members of the SCs—altogether a huge workload which must be mandated and funded;
- Promotion of the Eurocodes. The objectives of promotion the Eurocodes is the dissemination of the European technical culture in civil engineering outside the EU to support globally the European construction industry;
- Further developments. From a general point of view, industry is not in favour of quick and deep changes in the existing Eurocodes (their implementation and use is in a starting phase). It is in favour of improvements with the objective to clarify the existing text and rules, to simplify them as far as possible, to shorten some Eurocodes to reach a better user friendliness.

The development of new aspects is also very important: in the short term, the main subjects of interest have been identified:

- Guidelines on new materials, with a priority for structural glass, structural applications of FRP and very high performance concrete. These guidelines are intended to become additional rules to be incorporated to the existing

Eurocodes, or additional standards to the family of Eurocodes. Review and development of existing rules in the context of the wider use of recycled or re-used materials and components.

- Guidelines for the assessment of existing structures, excluding, at present, historical monuments.

2.4 EN Eurocodes—tool for advanced structural design

The Eurocodes highlight new or advanced concepts for the design of civil engineering works, the most important being:

- Durability;
- Robustness;
- Use of new materials, and
- Representation of actions.

These concepts are defined in EN 1990 Basis of structural design, and developed in the other Eurocodes.

2.5 Countries where EN Eurocodes are implemented

According to the CEN/CENELEC Internal Regulations, the national standards organizations of the following countries are bound to implement this European Standard: Austria, Belgium, Bulgaria, Cyprus, Czech Republic, Denmark, Estonia, Finland, France, Germany, Greece, Hungary, Iceland, Ireland, Italy, Latvia, Lithuania, Luxemburg, Malta, Netherlands, Norway, Poland, Portugal, Romania, Slovakia, Slovenia, Spain, Sweden, Switzerland and the United Kingdom.

EN Eurocodes are accepted in: Malaysia, Singapore, South Africa and Vietnam.

3 NEW GENERATION OF EUROCODES

3.1 Mandates M/466 EN and M/515 EN

In May 2010, the European Commission (EC), Enterprise and Industry Directorate-General, sent Programming Mandate M/466 EN to CEN concerning the Structural Eurocodes. The purpose of this mandate was to initiate the process of further evolution of the Eurocode system, incorporating both new and revised Eurocodes, and leading to the publication of the second generation of EN Eurocodes. CEN replied to this mandate in June 2011.

In December 2012, the EC sent a further Mandate M/515 EN, inviting CEN to develop a detailed standardisation work programme using the reply to mandate M/466 as a basis. This document is the reply to Mandate M/515 EN. It sets out TC 250's proposed work programme together with additional supporting information. Over 1000

experts from across Europe have been involved in the development and review of this document.

The objectives and intended impacts of the work are defined. It is widely recognised that long-term confidence in the codes requires the Eurocodes to evolve in an appropriate manner. Specifically the proposed work programme focuses on ensuring the standards remain fully up to date through embracing new methods, new materials, and new regulatory and market requirements. Furthermore, it focuses on further harmonisation and a major effort to improve the ease of use of the suite of standards for practical users.

Beneficial impacts of the work programme are presented ranging from improved efficiency and targeted extension of scope, to increased user confidence and enhanced sustainability in construction. With the European market for design services in the construction sector being approximately 75 € Billion, it is clear that even very modest efficiency savings will yield very substantial monetary benefits for public and private sector clients.

The approach to the execution of the mandate is presented in CEN/TC 250 Response to Mandate "Towards a second generation of Eurocodes" N 993 from May 2013. Guiding principles used in the development of this response are set out, together with a detailed explanation of how a series of specific issues will be dealt with. The TC 250 work programme encompasses all the requirements of M/515, supplemented by requirements established through extensive consultation with industry and other stakeholders. As such the overall work programme includes elements for which funding is sought from the EC and elements that will be wholly funded from other sources, principally industry.

The work programme is structured to comprise four overlapping phases. In Response to Mandate a complete overview of all phases is included, with further detail provided for those tasks in Phase 1 that are expected to form the basis for initial contractual discussions with the EC.

Details are provided of the organisational structure for the execution of the mandate and the means by which effective coordination will be assured. To maximise the benefit derived from the

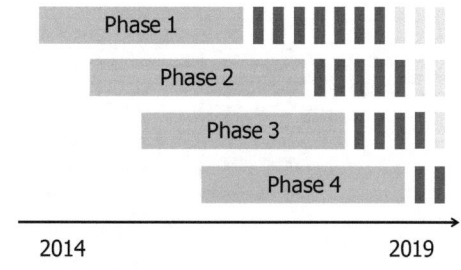

Figure 1. Indicative phasing of work.

extensive existing network of active stakeholders, the organisational structure has been based upon the current TC 250 operating model.

The TC 250 work programme is presented in detail as Annex 1. It comprises 77 discrete tasks, all of which will be undertaken under the direction of one of TC 250's existing Sub-committees, Working groups or Horizontal groups. A summary of the deliverables for each task is provided, and a mapping between the explicit requirements of M/515 and the TC 250 work programme is given.

TC 250 is committed to the successful development of the second generation of EN Eurocodes, and seeks the support of the EC in helping to make this possible.

3.2 Overview of the CEN/TC 250 work programme

The structure of the work programme has been developed to meet the requirements of M/515 and align with the operating structure of CEN/TC 250. The complete work programme has been divided into a series of tasks, under the primary leadership of an existing CEN/TC 250 Sub-committee, Working Group or Horizontal Group. Each task contains a series of sub-tasks as illustrated in Figure 2.

All tasks concerned with existing Eurocode parts include some common requirements in their scope. Standard requirements on reducing NDPs and enhancing ease of use have been included as sub-tasks 1 and 2 in all such tasks. The provision of background documents is a common requirement for all tasks. Funding for maintenance activities is not included in any tasks.

3.3 New Eurocodes and technical specifications

Working groups are preparing 2 new Eurocodes:

- WG2: Assessment and Retrofitting of Existing Structures;
- WG 3: Structural Glass, and 2 new Technical Specifications:
- WG 4: Fibre Reinforced Polymers;
- WG 5: Membrane Structures.

3.4 Working groups and project teams

The first author is member of the following 5 Working Groups:

- WG EN 1993-1-1 Design of Steel Sections and Members;
- WG EN 1993-1-3 Steel Cold-formed members and sheeting;
- WG EN 1993-1-5 Steel Stability of Plated Structural Elements;
- WG EN 1993-2 Steel Bridges;
- WG EN 1999-1-1 Update and Simplification of all parts of EN 1999 design of Aluminium Structures.

The Working Groups consists of volunteers from various countries. The number of volunteers in a group vary approximately from 10 to 20 specialists. The Working Groups have meetings two times per year. In the Summer of 2015 there is an opportunity to apply for membership in the Project Team or for being convener of the Project Team. Project Teams will be established to undertake each of the tasks. In line with TC 250 document N 250, such teams will typically have 6 members to ensure an appropriate representation from member states whilst promoting cost effectiveness.

The appointment of Project Teams that will receive funding will be undertaken in line with the Framework Partnership Agreement (FPA) 2009 rules for award of contracts, and may involve competition. The contracts for each of the Project Teams will be based on the scope of work for each task. Project teams will be responsible to their appropriate SC, WG or HG, and their responsibilities will end when they have provided a draft that the SC, WG or HG accepts as being a correct and adequate response to the contract. After the SC, WG or HG has accepted a draft from a PT as meeting its contract, the SC, WG or HG will be responsible for the finalisation of the document to Formal Vote within the CEN rules.

ACKNOWLEDGEMENT

Project No. 1/0748/13 was supported by the Slovak Grant Agency VEGA.

REFERENCES

Baláž, I.J. 2011. Eurocodes in Slovak Republic. *FEANI NEWS, The European Engineers Publication,* FEANI Brussels, Issue 09: 26–28.

CEN/TC 250—N 798. 2009. CEN/TC 250 The Eurocodes and Construction Industry. Medium-Term Strategy 2008–2013.

CEN/TC 250—N 993. 2013. CEN/TC 250 Response to Mandate M/515 EN Structural Eurocodes "Towards second generation of Eurocodes".

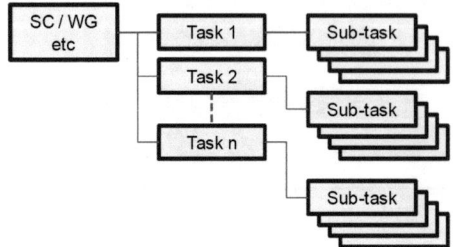

Figure 2. Illustrative structure of task and sub-tasks.

Production Management and Engineering Sciences – Majerník, Daneshjo & Bosák (Eds)
© 2016 Taylor & Francis Group, London, ISBN: 978-1-138-02856-2

Analysis of the hydraulic arm for use on a light goods vehicle

M. Blatnicky & J. Dizo
University of Zilina, Zilina, Slovak Republic

ABSTRACT: The paper deals with numerical analysis of the first part of the designed hydraulic arm. The arm is used for manipulation of load pieces up to 300 kg. Its load capacity depends on the transport vehicle to be placed on—vehicle type Pick-up. Great operability of this mechanism predetermines its wide application in construction, transport, business, etc. Locations and values of the maximum stress in the first part of the structure are calculated using the ANSYS software. These results determine the relevant data necessary for correct design and functioning of the machine. After carrying out all the analyses and calculations we will be able to determine the safe use of the machine and put it into operation. Safety is the most important requirement in any structure operation.

1 INTRODUCTION

An important role of engineers is to analyse the existing and newly-designed working processes in order to find an optimal way of executing the given operation. The best working process is generally considered the one that minimises the cost of performance, which can be achieved by mechanisation.

Mechanisation is an important means of increasing productivity, quality and production competitiveness. Successful introduction of mechanisation requires knowledge and understanding of physical dependencies of executed operations. Operations are executed by transfer of mechanical, electrical, pneumatic or hydraulic energy.

The aim is that individual working processes be as short and simple as possible, easy to learn and at the same time require minimum man-power deployment. Mechanisation significantly relieves people of hard physical work for instance in dangerous or harmful environment.

Such a means of mechanisation utilisable not only in transport is the designed hydraulic arm (Blatnicky & Dizo 2014), which significantly simplifies manipulation with great operativeness. With the requirement of minimum production costs we reach the most gainful output/costs ratio.

2 DESIGN OF ARM GEOMETRY

Every transport and handling machine consists of three main parts:

- steel structure,
- driving mechanisms, and
- other parts (e.g. cab, steps, etc.).

The aim of this work is to design a hydraulic arm steel structure to be mounted on a particular light goods vehicle (Figure 1). The whole mechanism consists of three parts of steel structure. The advantage of this solution is that the operator needs driving license only for vehicles with total mass up to 3500 kg.

In previous functional calculations (Blatnicky & Dizo 2014), force effects on the arm were detected (Figure 2) during operation due to the load with mass of 300 kg.

Consequently, analytical dimensional calculation was executed and all dimensions of the arm were designed (Blatnicky & Dizo 2014).

Designed lengths of individual parts, resulting from construction of hydraulic cylinders in Figure 2, are as follows: $c_1 = 1150$ mm, $c_2 = 1150$ mm, $c_3 = 1180$ mm, $e_1 = e_2 = e_3 = 75$ mm, $a_1 = a_2 = 210$ mm, $b_1 = 774$ mm. Calculated inner force effects of the mechanism in Figure 1 are:

Figure 1.   Platform for the designed arm.

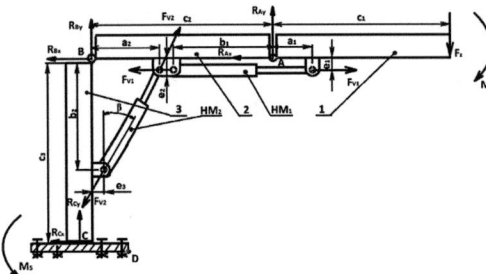

Figure 2.   Free diagram of arm.

Figure 3.   Lifting pivot arm with pick-up bed.

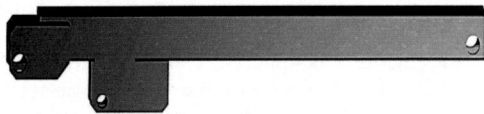

Figure 4.   3D model of the first part of the arm.

$F_z = 2943$ N, $F_{v1} = 45,126$ N, $R_{Ax} = 45,126$ N, $R_{Ay} = 2943$ N, $\beta = 10.931°$, $F_{v2} = 30,710$ N, $R_{Bx} = 5824$ N, $R_{By} = 27,210$ N, $R_{Cy} = 2943$ N.

After defining the cross-section dimensions by analytical calculation, it is possible to build a 3D model of the hydraulic arm (Figure 3) in CAD software Catia.

The whole arm consists of three parts and to simplify the problem, numerical analysis of each part of the arm is made separately.

In this paper, the analysis of the first part of the arm is carried out (the part on which the load hangs) using the FEM software ANSYS. The 3D model of the first part is shown in Figure 4.

3   BUILDING THE MODEL BY SHELL ELEMENTS

The FEM software ANSYS was used to calculate stress distribution in the structure (Benca 2005, Handrik et al. 2014, Zmindak et al. 2004). The ANSYS software allows simulating various problems of mechanics including statics, dynamics,

contact, etc. (Harusinec et al. 2007, Harusinec et al. 2008, Harusinec et al. 2011).

Firstly, it is necessary to create geometry. The arm under consideration (Figure 4) is made of normalised square seamless steel tube and therefore it is convenient to model it using shell elements (Figure 5).

When the model is created, boundary conditions can be entered into the software (Figure 6). Loading of the arm is realised by a couple of forces at the end of the arm in its most efficient position. This position is shown in Figures 2, 3. The value of the couple of forces represents the weight of the load with maximum mass of 300 kg. Gripping of the construction is by two rotational joints. One rotational joint features pin connection between the first and second part of the arm. The second joint is pin connection between the arm and the eye of the hydraulic cylinder that commands the movement of this part.

Entering the boundary conditions is followed by the model meshing. To create the mesh, we used four-node linear shell elements 181 with mesh density 5 mm. Figure 7 shows the arm after meshing.

Figure 8 shows detail of the mesh in the pin hole site. The maximum stress according to the analytical calculation is 106 MPa. This value was determined without considering the own weight of the arm. It means that the maximum stress in the arm can increase in numerical analysis when the arm own weight is taken into consideration. The chosen profile TR4 HR $90 \times 70 \times 5$ made of steel 11 423 has admissible effective stress 120 MPa within the safety factor (Balaja et al. 2014).

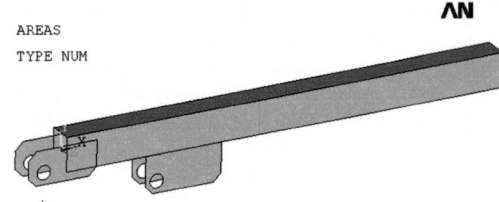

Figure 5.   Shell model of the arm.

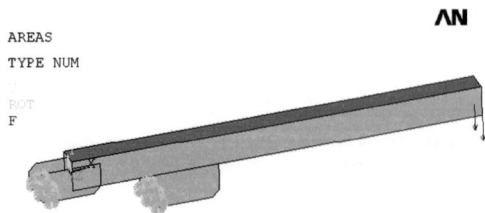

Figure 6.   Boundary conditions of the arm.

Figure 7.   Meshed model of the arm.

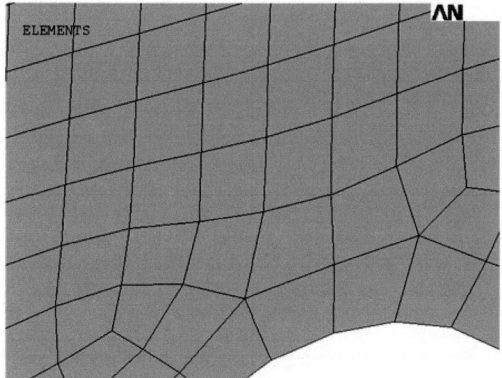

Figure 8.   Detail of the mesh used.

Meshing the model is followed by numerical calculation and then we can show the results of required quantities in a post processor.

Figure 9 shows displacement of the arm due to the load during at the machine operation. The software states maximum displacement by 1.62 mm.

The maximum admissible deflection is up to 1/600 of the entire length. If the total length of the arm is 1150 mm, then the admissible deflection is 1.92 mm, which is more than the determined value of 1.62 mm. This means that displacements in the structure are within admissible values.

The next important determined values are distribution of stress in the structure, determination of critical point in the structure and the maximum stress in the critical point of the structure. If the maximum stress value is under the specified admissible stress of 120 MPa, then it is well dimensioned and no further optimisation of the structure is needed.

Figure 10 shows distribution of effective von Mises stress in the first part of the arm. Correctness of the analytical solution was verified by numerical analysis. The maximum simulated stress value is 117.044 MPa. It is clear that the value is below the design value of 120 MPa, and therefore the conclusion is that the structure with such

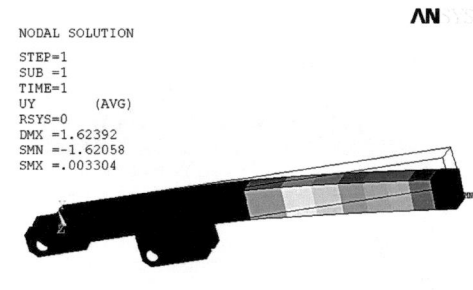

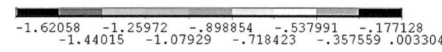

Figure 9.   Simulated displacement of the arm part.

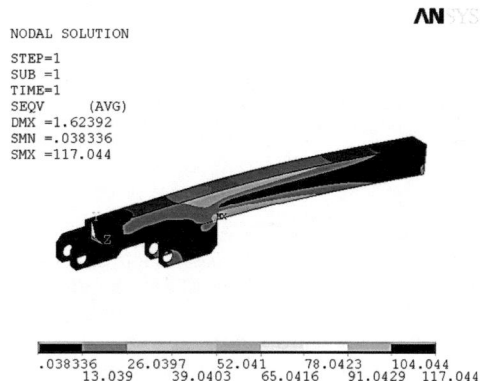

Figure 10.   Distribution of effective von Misses stress.

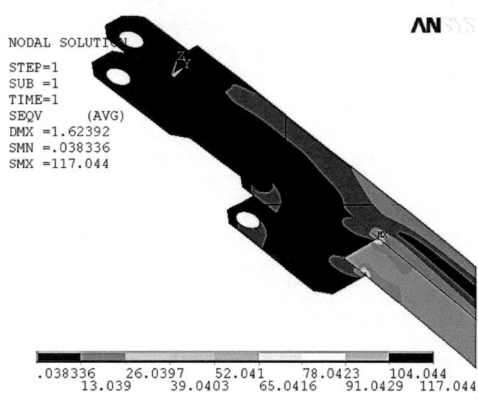

Figure 11.   Detail of the critical section in the arm.

geometry endures the load. The critical section is shown in Figure 11.

As the resultanting stress is lower than admissible stress, we can conclude that the designed

Figure 12. 3D model of the arm construction.

structure will be safe. Safety is the most important requirement posed on any structure.

4 CONCLUSIONS

As the aim of the paper was calculation of the first part of the hydraulic arm steel structure, it can be said that the aim was achieved. The designed arm part in operation endures working with the load for which it was dimensioned.

To guarantee the structure safety requires execution of numerical analysis of other parts of the arm as well (Figure 12). Numerical analysis and optimisation of all parts of the arm is followed by the next step in solving this issue—building a model of the structure in MBS software. The aim of this analysis is to simulate dynamic effects of the load in the mechanism operation.

ACKNOWLEDGEMENT

This contribution is the result of the project implementation: Modern methods of teaching of control and diagnostic systems of engine vehicles, ITMS code 26110230107, supported by the Operational Programme Education.

REFERENCES

Balaja, J., Broncek, J., Antala, J. & Sekeresova, D. 2014. Mechanical Engineering Tables. (In Slovak). Selection Standards. *Slovak Office of Standards, Metrology and Testing*, 2014. ISBN 978-80-8130-039-4.

Bathe, K.J. 1996. Finite element procedures. Prentice-Hall, Inc. New Jersey 1996. ISBN 0-13-301458-4.

Benca, S. 2005. *Computational procedures of FEM for solving linear mechanics problems*. (In Slovak). Slovak University of Technology in Bratislava, Bratislava 2005. ISBN 80-227-2404-1.

Blatnicky, M. & Dizo, J. 2014. The designed functional calculation of the hydraulic arm of a track maintenance machine to lift of piece loads. (Online). In: *Railway transport and logistics. Scientific and technical on-line journal.* (In Slovak), 2/2014, Zilina 2014. ISSN 1336-7943, Pp. 79–84.

Handrik, M., Vasko, M., Kopas, P. & Saga, M. 2014. Effective finite element solution and post-processing for wide load spectrum. In: *Communications: scientific letters of the University of Zilina*. ISSN 1335-4205. Vol. 16, no 3A (2014), Pp. 19–26.

Harusinec, J., Gerlici, J. & Lack, T. 2007. A rail/wheel contact analysis with the help of the Finite element method. (In Slovak). In: *PRORAIL 2007: XVII. International conference "Current Problems in Rail Vehicles"*, Zilina 2007, September 19th–21st, Slovak Republic, Proceedings, Scientific and technical Society at the University of Zilina. ISBN 978-80-89276-06-6, Pp. 173–180.

Harusinec, J., Gerlici, J. & Lack, T. 2011. Contact stress between railway wheel and rail head analysis with the help of the finite element method. (In Slovak). In: *PRORAIL 2011: XX. International conference "Current Problems in Rail Vehicles"*, Zilina 2011, November 21st–23rd, Slovak Republic, Proceedings, Scientific and technical Society at the University of Zilina. ISBN 978-80-89276-8, pp. 3–16.

Harusinec, J., Gerlici, J. & Lack, T. 2008. Stress conditions in the contact of rail wheel tread and rail head. (In Slovak). In: *Dynamics of rigid and deformable bodies 2008: Proceedings, VI. International conference:* Usti nad Labem, October 17th–19th. ISBN 978-80-7414-030-3, pp. 43–52.

Zmindak, M., Grajciar, I. & Nozdrovicky, J. 2004. *Modelling and Computation in Finite Element Method*. (In Slovak), Zilina, ISBN 80-968823-5, 2004.

Production Management and Engineering Sciences – Majerník, Daneshjo & Bosák (Eds)
© 2016 Taylor & Francis Group, London, ISBN: 978-1-138-02856-2

The availability of digital data for flood risk assessment in Slovakia in GIS environment

P. Blišťan & M. Zeleňáková
Technical University of Košice, Košice, Slovak Republic

M. Blišťanová
University of Security Management, Košice, Slovak Republic

ABSTRACT: Geographic Information Systems (GIS) are a group of specialized software tools capable of handling with different groups of spatial data. Geographic information systems can be effectively used to evaluate flood risks, to assess vulnerability of the area, to model the flood area, to simulate an extent of floods etc. For the purposes of flood risk assessment and flood vulnerability zone statement in GIS environment it is necessary to have good quality base input data. Most of these data is now available in digital form and are produced under the European INSPIRE directive, which was adopted by the EU to harmonize data and thus ensure interoperability. The aim of this paper is to show what kind of space data is currently available in digital format, and where and in what quality we can obtain them for the needs of computer modeling.

1 INTRODUCTION

Risk assessment is a difficult process and requires primarily high-quality input data. In assessing the risks associated with the development of flood events in the vulnerability zone and in modeling of flood situations, we are using different categories of input data (Regulation no. 313/2010 Coll.). These data are typically processed in an environment of specialized computer systems, so-called expert computer systems. This group includes computer systems and Geographic Information Systems (GIS) (Burrough 1986; STN 73 0401-3). The result of the process of data processing tools in GIS are then various synthetic maps and models (Act 7/2010; Blišťan 2007) such as:

- flood hazard maps,
- flood risk maps,
- maps of vulnerability area,
- maps of flood susceptibility of the territory,
- maps overcoming crisis situations,
- logistic models, etc.

The Slovak Republic has a responsibility to identify and analyze the flood risk by government authorities in accordance with Act no. 42/1994 Coll. on civil protection of the population and Act no. 7/2010 Coll.. The reason for the adoption of the new law on flood protection was the obligation of the Slovak Republic transposed into the legal system of European Parliament and Council Directive 2007/60/EC of 23 October 2007 on the assessment and management of flood risks. That Directive and the Act provides a comprehensive system of flood risk management.

2 INPUT INFORMATION FOR FLOOD RISK ASSESSMENT

2.1 *Location data*

To create maps of vulnerability zone, or maps of area susceptible to floods are necessary in particular spatial data—location data. These are currently obtained by conventional terrestrial surveying methods, using eg. tacheometry, Global Navigation Satellite Systems (GNSS), Terrestrial Laser Scanning (TLS), but also LIDAR technology or Unmanned Aerial Vehicles (UAV) (Cracknell & Hayes 2007; www.draganfly.com).

All major products sector have the character of geodetic and geographic information and are maintained and operated under Act no. 215/1995 Coll. on Geodesy and Cartography and the INSPIRE Directive of the European Parliament and Council Directive 2007/2/EC of 14 March 2007, which was transposed into our legislation by Act no. 3/2010 Coll. a National Infrastructure for Spatial Information (NISI). INSPIRE Directive defines 34 spatial data themes:

1. Coordinate reference systems.
2. Geographical grid systems.
3. Geographical names.

4. Administrative units.
5. Addresses.
6. Cadastral parcels.
7. Transport networks.
8. Hydrography.
9. Protected sites.
10. Altitude.
11. Land cover.
12. Orthometry.
13. Geology.
14. Statistical units.
15. Buildings.
16. Soil.
17. Land use.
18. Human health and safety.
19. Utility and governmental services.
20. Equipment for environmental monitoring.
21. Production and industrial facilities.
22. Agricultural equipment.
23. Population distribution—demography.
24. Area management/restriction/regulation zones and reporting units.
25. Natural risk zones.
26. Atmospheric conditions.
27. Meteorological geographical features.
28. Oceanographic geographical features.
29. Sea regions.
30. Bio-geographical regions.
31. Habitats and biotopes.
32. Species distribution.
33. Energy sources.
34. Mineral resources.

These data are subsequently processed by relevance to digital form in accordance with INSPIRE Directive. A large part of data, in particular location data and environmental data is already available in digital form. Location data in particular, offers Geodesy, Cartography and Cadastre Authority of the Slovak Republic and topographical institute of Colonel John Lipski resident in Banská Bystrica, which is under the administration of the Arm Forces—department of the Ministry of Defence. Other providers of spatial data in Slovakia is the Ministry of Interior and some private companies (eg. GEODIS SLOVAKIA, sro, EUROSENSE Ltd).

The basic location data needed for flood risk assessment and vulnerability assessment of area mainly include:

• Basic data base for Geographic Information System (GIS-ZB) (Figure 1).
• Digital Terrain Model (DTM) of the Slovak Republic or relief of the Slovak Republic (DMR) (Figure 2).
• Vector map of Slovakia.
• Aerial and satellite imageries.
• Cadastral maps, etc.

2.2 Statistical data

An important group of attribute data in the Slovak Republic offers the Slovak Statistical Office (SO SR), which is a central state administration of the Slovak Republic for the area of state statistics. His status is governed by Act no. 575/2001 Coll. on the organization of government activities and the organization of central state administration

Figure 1. Geoportal—map client ZB GIS.

Figure 2. Digital terrain model.

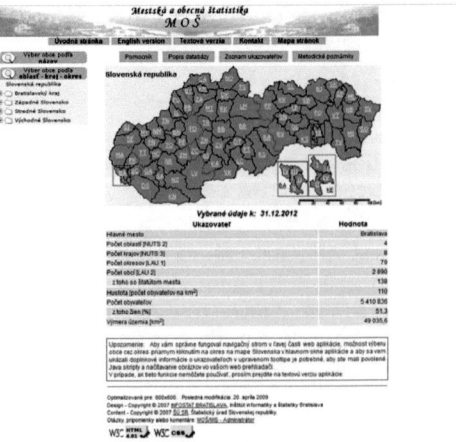

Figure 3. Sample of database of Statistical Office.

as amended. The Authority shall undertake tasks according to Act no. 540/2001 Coll. on state statistics, as amended, and functions provided by other generally binding regulations. Statistical Office of the Slovak Republic operates as a separate institution from January 1, 1993, the date of the independent Slovak Republic.

The main databases of SO SR (Figure 3) (http://slovak.statistics.sk) are:

- Slovstat—database contains time series of indicators of economic and socio-economic development in the Slovak Republic.
- RegDat—regional statistics database contains time series of indicators of economic and socio-economic development in the regions of the Slovak Republic.
- STATdat—the database contains reports (tables) for indicators of economic and socioeconomic development. The classification of individual reports is based on preserving the structure and sections, as well as on the website.
- DATAcube—includes multidimensional table (cube) for indicators of economic and socioeconomic development.
- Eurostat databases—databases are available for free directly on the Eurostat website.
- Selected tables of European Statistics—Eurostat offers on its website, among other things predefined tables.
- Register of Spatial Units (REGPJ)—is administered by the Department of comprehensive methodology of the Statistical Office of the Slovak Republic as a separate register taking into account the mutual relations of territorial codes of regions, districts, municipalities, cadastral areas and residential units of the Slovak Republic.

These data are publicly available and serve as background data on basic economic and socio-economic characteristics of the Slovak Republic.

2.3 Environmental data

The need to develop a comprehensive monitoring of the environment resulted from the resolution of the Slovak Government no. 623 from 21.12.1990. The concept of monitoring and integrated environmental information system for the Slovak Republic was approved by the Government Resolution no. 449 of 26.5.1992 and implementation of project-area monitoring system by the Government Resolution no. 620 of 7.9.1993. The principal elements of nationwide environmental monitoring of the Slovak Republic are the Subordinate Monitoring Systems (SMS), which fully assures the guarantors; methodology and coordination activities provide the Centre of SMS. Environmental monitoring currently consists of 10 SMS (www.shmu.sk):

- Air by Slovak Hydrometeorological Institute, Bratislava.
- Water by Slovak Hydrometeorological Institute, Bratislava.
- Soil by Research Institute for Soil Science and Conservation, Bratislava.
- Biota (fauna, flora) by State Nature Conservancy Banská Bystrica.
- Forests by Forest Research Institute, Zvolen.
- Geological factors by Geological Survey of Slovak Republic, Bratislava.
- Waste by Slovak Environmental Agency, Banská Bystrica.
- Food contaminants and feed by Food Research Institute, Bratislava.
- Meteorology and Climatology by Slovak Hydrometeorological Institute, Bratislava.
- Radioactivity by Slovak Hydrometeorological Institute, Bratislava.

The main aim of monitoring is to pursue a particular phenomenon or a parameter in a precisely defined temporal and spatial conditions. It serves the objective knowledge of the characteristics of the environment and assessing their changes in the monitored spatial area. Environmental monitoring provides objective information necessary for decision-making, management, supervision and scientific research area, but also the public (www.shmu.sk).

In terms of flood risk and vulnerability areas are the most important input information on rainfall, soil and land use.

2.4 Land use—Urban Atlas and Corine landcover

Urban Atlas (Figure 4) contains comprehensive information on the actual cover and utilization of the earth's surface in the large European cities. These data represent a valuable material not only for the evaluation and supervision of the current development of urban agglomerations according to the Urban Plan, but can serve as a basis for assessment of risks and opportunities

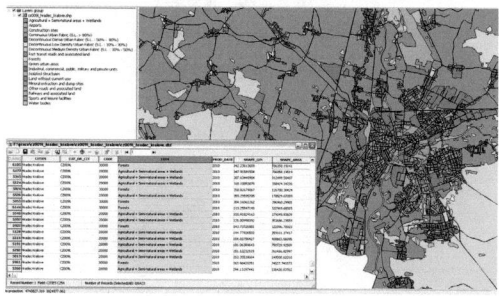

Figure 4. Urban Atlas.

in the territory, from the threat of flooding to identify new infrastructure needs. An important advantage is the homogeneity of the data. When comparing data from several time horizons will be evaluated dynamics of agglomeration and depending on the time and also to further explore the factors that affect growth. In terms of transport the Urban Atlas has importance in monitoring the construction of buildings, logistics centers and new industrial areas, which are for transport infrastructure is closely linked and their occurrence can only estimate the burden of existing transport infrastructure, but also its optimization plan, or any extension (www.czech-spaceportal.cz).

Urban Atlas project is funded by the European Commission mainly from the European Regional Development Fund (ERDF). In 2010, Atlas worked in the territory of 117 European cities with a minimum of 100 000 inhabitants. By 2011, however, the list was expanded to more than 300 cities, with a planned update every 1-3 years. One of the conditions for the inclusion of the city in the atlas was its participation in the European Urban Audit. Atlas was drawn up in scale of 1: 10,000. The basis for the creation of Atlas has become images from satellite Spot 5, with a spatial resolution of 2.5 m. Minimum size area is displayed depending on the scale 0.25 ha (www.czechspaceportal.cz).

CORINE landcover project is mapping the surface of the countries of Europe from the Landsat. The project is coordinated by the European Environment Agency (EEA). Slovak Environmental Agency (SEA) has prepared the CORINE map service in the area of Slovakia for experts and the general public (Figure 5).

The pan-European program CORINE is aimed at gathering information on the environment on the European continent. One of its activities is to create a single thematic maps category Europa's

surface at a scale of 1: 100,000. The main source of data for this database were satellite pictures that contain data from LANDSAT scanner. The smallest mapped unit has an area of 25 ha and line formation width of at least 100 m. The types of surface were set by the group of experts so as to conform to most countries and have permitted mapping and regional specificities. The whole nomenclature is conceived in a hierarchical way, and includes details of three levels that are required in all countries. At this level it is intended 44 categories marked in three number codes.

3 GIS AS A TOOL FOR THE NEEDS OF FLOOD RISK ASSESSMENT

The issue of flood risk assessment and modeling of flood situations is currently devoted considerable attention, which is predominantly based on the rapid development of information technology, enabling the generation of clear graphical outputs in a very short timeframe. Therefore a high profile tool—Geographical Information Systems (GIS) are suitable for flood risk analysis because they provide a broad spectrum of functionality within a single software package. Because the occurrence of flood events is in terms of time and in terms of localization to some extent impossible to predict, it is modeling itself inundation current priority areas of scientific studies and projects of a commercial nature.

Maps of flood protection are divided into flood hazard maps and flood risk maps. They are produced in digital form in GIS environment and displays the geographical areas in which there is a potential significant flood risk exists or in which it may be presumed that it is likely to occur. Maps of flood protection should be available to the public via the Internet (Figure 6) and EU Member States should ensure that the flood hazard maps and risk were completed by 22 December 2013.

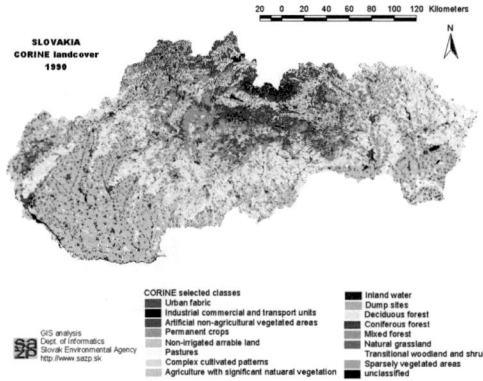

Figure 5. CORINE landcover.

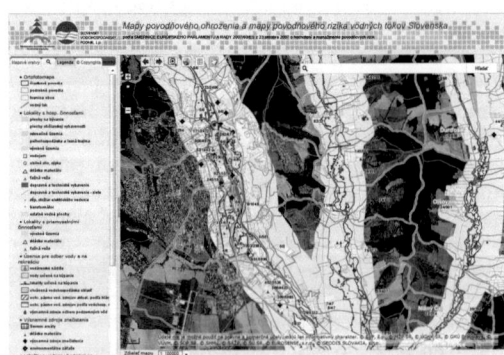

Figure 6. Portal—flood maps.

3.1 Flood hazard maps

Flood hazard maps shall cover the geographical areas which could be flooded according to the following scenarios (Zeleňáková 2009):

- floods with a low probability, or extreme event scenarios,
- floods with a medium probability (period ≥ 100 years),
- floods with a high probability.

These scenarios in flood hazard map displayed by flood lines together and also each well separately. The flood hazard map also displays the following particulars:

- flood extent (maximum flood level) shows the flood line which is the intersection of the water level flooding with the terrain,
- water depths or water level,
- the flow rate of water flow or the relevant water flow, as necessary.

Graphic design maximum level, depth and velocity of water flow as a result of the hydrodynamic modeling of floods appear as a 2D grid. This analyzes and simulations are carried out in a GIS environment using a set of specialized software (Fencík et al. 2011; Zeleňáková & Gaňová 2011).

3.2 Flood risk maps

Flood risk maps shall show the potential adverse consequences associated with flood scenarios and contains an indicative number of inhabitants potentially affected, the economic activity in the area potentially affected, installations which might cause flooding in the event of accidental pollution, and potentially affected protected areas and other information the Member State considers useful. Flood risk maps must include (Fencík et al. 2011):

- a flood line,
- indication of the estimated number of inhabitants potentially affected by flood,
- types of economic activities in the territory of the potential risk of flooding,
- sites of industrial activities,
- the position of potentially vulnerable areas for the abstraction of water for human consumption and for recreational activities,
- locations with water suitable for swimming,
- information on other significant sources of potential pollution of water after flooding and during floods,
- territories which form a national system of protected areas and the proposed European system and declaration of protected areas (Natura 2000),
- information that Ministry of Environment considers useful.

The creation of flood risk management is needed compared to the flood hazard maps, collect large amounts of data into different categories, mainly relating to the method of land use, settlement and economic activity. These data are then digitally processed by GIS tools.

4 CONCLUSIONS

Modeling the flood wave and subsequent flood risk assessment, it is important to anticipate adverse events and assess their impact on the population and the environment. For effective flood management it is necessary to have a database for logistics planning. Quality and comprehensive data is the first prerequisite for effective work of the crisis management. GIS systems offer a wide range of options for further data analysis, the results of which can be utilized in the decision-making process (Oberle & Merkel 2007). Utilization possibilities are wide, ranging from complex database, which is in digital form and accessible online at any time in the field, through classic and special spatial analysis, to the creation of outputs over to the demands of the individual components of the integrated rescue system (Kavan & Baloun 2013). We try to simulate the extent of the flood threat by modeling the flood wave. We try to thoroughly prepare for emergencies and to effectively reduce and mitigate their consequences by GIS tools using in emergency planning and management.

The need to build digital data warehouses and data sharing support under the INSPIRE Directive is an important step to streamline the process of flood risk assessment and vulnerability zone statement.

ACKNOWLEDGMENT

The contribution is written thanks to support of project VEGA 1/0609/14.

REFERENCES

A Short History of Unmanned Aerial Vehicles (UAVs). Available at http://www.draganfly.com/news/2009/03/04/a-short-history-of-unmanned-aerial-vehicles-uavs/.

Blišťan, P. 2007. Presentation of geological data in the GIS environment. *Acta Montanistica Journal*, Vol. 12, no. 3, Košice, 2007, p. 329–334. ISSN 1335-1788.

Burrough, P.A. 1986. *Principles of Geographical Information Systems for Land Resources Assessment*. Clarendon Press, Oxford.

CORINE landcover. Available at http://www.sazp.sk/slovak/structure/CEEV/RS/CLC2000/.

Cracknell, A.P. & Hayes, L. 2007. *Introduction to Remote Sensing* (2nd ed.). Taylor and Francis, London, 2007.

Databases Statistical Office. Available at http://slovak.statistics.sk.

European Parliament and Council. 2007/2/EC of 14 March 2007 establishing an Infrastructure for Spatial Information in Europe (INSPIRE). In: *Unified automated system of legal information*, the Ministry of Justice Available at http://jaspi.justice.gov.sk.

European Parliament and Council Directive 2007/60/EC of 23 October 2007 on the assessment and management of flood risks. In: *Unified automated system of legal information*, the Ministry of Justice. Available at http://jaspi.justice.gov.sk.

Fencík R., Danek, L. & Daneková, J. 2011: GIS applications and hydrodynamic modeling in flood maps. In: *GIS Ostrava*. Available at http://gis.vsb.cz/GIS_Ostrava/GIS_Ova_2011/sbornik/papers/Danekova.pdf.

Kavan, Š. & Baloun, J. 2013. *Rescue and security operations during floods in terms of water management facilities*. České Budějovice: College of European and Regional Studies, 116 p.

Monthly Maps. Available at http://www.shmu.sk/sk/?page=1610&id=

National Council of the Slovak Republic no. 215/1995 Coll. on geodesy and cartography. In: *Unified automated system of legal information*, the Ministry of Justice. Available at http://jaspi.justice.gov.sk.

National Council of the Slovak Republic no. 3/2010 Coll. a national infrastructure for spatial information. In: *Unified automated system of legal information*, the Ministry of Justice. Available at http://jaspi.justice.gov.sk.

National Council of the Slovak Republic no. 42/1994 Coll. on civil protection of the population. In: *Unified automated system of legal information*, the Ministry of Justice. Available at http: //jaspi.justice.gov.sk.

National Council of the Slovak Republic no. 540/2001 Coll. on state statistics. In: *Unified automated system of legal information*, the Ministry of Justice. Available at http://jaspi.justice.gov.sk.

National Council of the Slovak Republic no. 575/2001 Coll. on the organization of government activities and the central government. In: *Unified automated system of legal information*, the Ministry of Justice. Available at http://jaspi.justice.gov.sk.

National Council of the Slovak Republic no. 7/2010 Coll. on flood protection. In: *Unified automated system of legal information*, the Ministry of Justice. Available at http: //jaspi.justice.gov.sk.

National Geoportal. Available at www.geoportal.sk.

Oberle, P. & Merkel, U. 2007. *Urban Flood Management—Simulation Tools for Decision Makers*, Advances in Urban Flood Management. Taylor & Francis Group, London, Pp. 91–122.

Portal flood maps. [Online]. [Cited 10.3.2015]. Available at <http://mpomprsr.svp.sk/Default.aspx>.

Regulation of the Ministry of Environment no. 313/2010 Coll., Laying down details of the preliminary flood risk assessment and the review and updating. In: *Unified automated system of legal information*, the Ministry of Justice [online]. Available at http: //jaspi.justice.gov.sk.

STN 73 0401-3. 2009. *Terminology in geodesy and cartography*. Part 3: Terminology of cartography and geographic information systems. Slovak Standards Institute, Bratislava. p. 92.

The Database Urban and Municipal Statistics. Available at http://app.statistics.sk/mosmis/sk/run.html.

The Digital Terrain Model of Slovakia Available at http://www.sazp.sk/slovak/struktura/ceev/DPZ/DTM/dtm.html.

The Environmental Monitoring. Available at http://www.shmu.sk/sk/?page=218.

The Urban Atlas. Available at http://www.czechspace-portal.cz/urban-atlas /.

Zeleňáková, M. & Gaňová, L. 2011. Integrating multicriteria analysis with Geographical Information System for evaluation flood vulnerable areas. In: *11th International multidisciplinary scientific geo-conference (SGEM 2011)*, Albena, Bulgaria, pp. 433–440.

Zeleňáková, M. 2009. Preliminary flood risk assessment in the Hornád watershed. In: *River Basin Management 5*, Southampton: Wessex Institute of Technology, 15–24.

Production Management and Engineering Sciences – Majerník, Daneshjo & Bosák (Eds)
© *2016 Taylor & Francis Group, London, ISBN: 978-1-138-02856-2*

Product dependability testing of polymer composite materials

A. Čapka, L. Fojtl, S. Rusnákova & M. Žaludek
Department of Production Engineering, Faculty of Technology, Tomas Bata University, Zlín, Czech Republic

ABSTRACT: This research paper deals with a proposal of the most appropriate methodology for dependability testing of products used in transport industry. Dependability is a global concept that includes terms of availability, reliability, durability, maintainability, supportability, etc. An important part of dependability is to find limit states of studied object, which are for PCM characterized by fiber cracking and delamination. Based on these specific failures, Building-Block Approach (BBA), which allows a systematic approach is often used for evaluation of composite constructions. Dependability evaluation consists of analysis and tests. Tests are always very time-consuming and expensive, because dependability is a property of objects with the strong stochastic character and therefore it is necessary to perform numerous measurements. Conversely, the analysis itself without verification may not achieve the desired results. For every test, it is necessary to develop test plan and determine whether it is possible to use shortened or accelerated test.

1 INTRODUCTION

Polymer composites are materials, which increasingly replacing conventional construction and building materials such as steel, aluminum, concrete, etc. The replacement of traditional materials occurs because PCM can have more optimal properties than traditional materials. With the creation of a priority request to reduce the weight of motor vehicles, which is related to necessity reduce of the emission amount, PCM became basic material to fulfil this requirement.

One of the major problems of contemporary world is the quality of production and services. Countries with high quality of production and services have simultaneously a high level of population life quality. One of the most important properties belonging to the notion of quality is dependability (Malkin 2005).

Scientific approach to dependability began with introduction of electronics to aerospace and defense industry around the middle of last century (Vasiljev 1988). Dependability is very elaborately standardized in specific national and international standards. Crucial role in the dependability standardization plays International Electrotechnical Commission.

Consistent terminology is the foundation of every scientific discipline. In PCM there is no Czech terminology standard. There is standard ČSN EN 472 ISI Plastics—dictionary, in which a few terms from the field of PCM are located, but they are only in foreign languages (English, French and German). In the standard ČSN EN 4408-001—

Aerospace series—Technical drawing and in the other standards there are only a few terms from PCM.

1.1 Basic terms and definitions of PCM

Composite material—construction material composed of two or more chemically and physically different components (phases), which are not mutually dissolved and have identifiable phase interface; composite always consists of a matrix and reinforcement, where synergistic effect is always achieved.

Fiber reinforced composite—construction material with long-fiber reinforcements, where the length of fiber is much greater than its diameter; reinforcement have much higher strength than matrix.

Laminate—composite manufactured from two or more layers containing reinforcing fibers oriented in one or more directions.

Matrix—basic component (phase) of the composite, which is used to store reinforcement, transfer load on the reinforcement, ensure the product shape, and protect reinforcement against damage; matrix has usually lower strength than the reinforcement.

Polymeric Composite Material (PCM)—composite, which have the matrix (or also fibers) made of polymer.

Resin—polymer used as matrix that is by mean of hardener polymerized (cured).

Fiber—natural or synthetic product, finite or infinite length, which is an essential element of fabric or other textile structures.

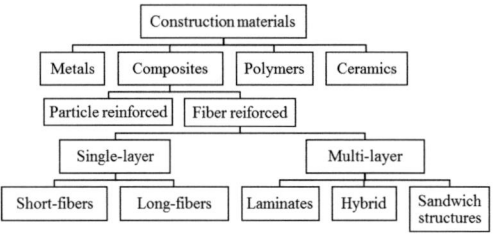

Figure 1.   Classification of composite materials.

Figure 2.   Dependability classification.

Lamina—element or set of elements in a certain level, the layer may consist of several sublayers.

Reinforcement—reinforcing component of the composite, which in fiber composites transfers most of applied load.

1.2   *Basic terms and definitions of dependability*

Standardization of technical terminology in dependability is very well elaborated, therefore individual terms are not described (except for durability).

Durability—ability of an object to fulfill a required function under given application conditions and maintenance until achievement of limit state.

1.3   *Classification of composite materials*

Figure 1 shows basic classification of composites due to their material constitution.

Composites can be divided according to different aspects, e.g. by matrix:

– Metal matrix
– Ceramic matrix
– Carbon matrix
– Polymer matrix

also by manufacturing technology:

– Hand lay-up
– Partially mechanized
– Fully mechanized and automated

or by types of individual components, form of reinforcement, structure and number of components.

2   DEPENDABILITY OF PCM PRODUCTS

2.1   *Dependability classification*

Function of any technical system can be characterized its quality. The term quality means the totality of properties that define ability of the system to fulfill specific objectives. Dependability (Figure 2) is one of the quality characteristics of technical object.

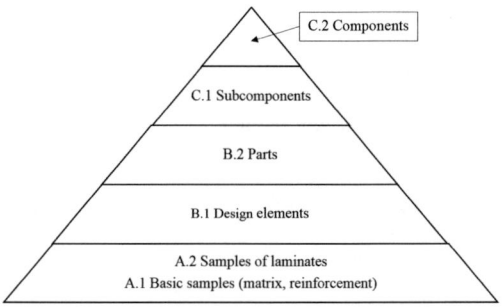

Figure 3.   Structure of BBA method.

2.2   *Building-Block Approach (BBA)*

BBA method allows to shorten and mainly cheapen tests of PCM construction. A numerous experiments have to be performed, if individual tests are carried out for constructions itself. However, these tests are very time-consuming and expensive. Furthermore, it is also impossible to perform only construction analysis using computational methods. In practice, a combination of analysis and tests is conducted. Following figure (Figure 3) depicts structure of BBA.

BBA method works analytically and experimentally at three basic levels:

– Fist level (A):
    A.1 Basic material samples; (matrix, reinforcement; data of strength and modulus).
    A.2 Lamina and laminate samples (data of strength and modulus).
– Second level (B):
    B.1 Design elements.
    B.2 Components of critical zones of construction, e.g. stiffeners, beams; (local static and fatigue loads, data of strength and stiffness, local damage, environmental impact).

- Third level (C):
  C.1 Subcomponents (subassemblies as complete wing spar).
  C.2 Components—real constructions or parts, e.g. wing; (complex tests, verification, tests until failure during both static and fatigue stresses).

2.3 *PCM limit states*

During the lifetime, every product goes through different stages. In this specific period, product properties (stages) are changed (technical, economical, ecological stages, etc.). Limit State (LS) occurs when the state variable reaches the limit value (Jones 1999). Achievement of the limit state has always probabilistic nature (probability of error-free operation).

The most frequent limit states in PCM products (sorted according to severity):

- LS related to failure of part cohesion (cohesive failure of reinforcement and matrix, cracks in matrix, fracture of the reinforcing fibers, delamination).
- LS related to the part deformation (local buckling of shell without fiber damage and delamination).
- LS related to damage of part surface (UV radiation, humidity, operating fluids, etc.).

3 ANALYSIS AND DEPENDABILITY TESTING OF PCM PRODUCTS

During all stages of product life, it is necessary to deal with product dependability. Therefore, it is necessary to develop a program of product dependability to ensure this parameter to be an essential characteristic of quality.

3.1 *Dependability analysis (analytical- modeling method)*

The specific information required for a decision about the system characteristics are obtained, examined and evaluated during dependability analysis (Trivedi et al. 1992, Gurtjahr 2000). These analysis are conducted on models and are standardized. In practice, following methods are for example used:

- Failure mode and effects analysis (FMEA), or extended FMECA to indicate critical analysis.
- Graphical methods (event and fault tree, logical block diagram, etc.).
- Marks methods.
- Simulation methods (Monte Carlo method, SBRA, method of artificial intelligence, etc.).

Dependability analysis has two basic characteristic:

- Interactive (interdependence of various stages, utilization of previously obtained empirical data, analyzes correction based on real-time operation).
- Iterative (repetition, accuracy improvement until achievement of set goal).

Dependability analysis is performed in four basic stages (as well as in all other fields):

- Functional and technical analysis (data collection, preliminary analysis).
- Qualitative analysis (objectives determination, scope, division of the system, application of one above mentioned methods, creation of dependability model, failure definitions).
- Quantitative analysis (calculation of dependability indicators, sensitivity analysis).
- Synthesis (assessment of achieved dependability level, conclusion).

3.2 *Dependability tests*

Dependability tests (Figure 4) are used for experimental determination or verification of dependability indicators. The aim of tests is to verify values of parameter distribution of corresponding quantity. These tests can be classified into following groups.

Furthermore, test plan for every dependability test have to be prepared. This plan defines methodology of progress and test end mode. Test plans are divided according to obtained data set to tests with results, which consist of:

- Complete set (n—number of specimens, U—damaged specimen is not replaced or repaired after failure and is removed from the test, τ_0—length of test).
- Set limited by number of failures (r—plan).
- Set is limited by time (t—plan).
- Progressively limited set (miscellaneous plan).

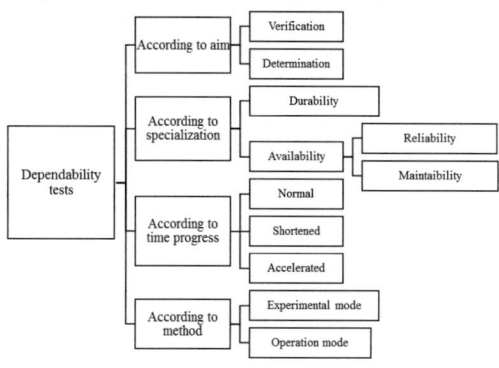

Figure 4. Classification of dependability tests.

363

Every test plan includes:

- Number of specimens.
- Information, if specimen will be removed, replaced by new, or repaired after failure.
- Method of test end (test duration or description at which failure will the test be stopped).

During the durability test of parts and subcomponents, it is time and financial advantageous to use shortened tests. These shortened tests last until the first failure and serves mainly to estimate the mean time to failure (estimation of distribution parameter to failure).

For products, where long-term trouble-free operation in a matter of decades is expected, the accelerated tests are used. There are two basic method of accelerated tests:

- Accelerated by greater use (especially for products that are not in continuous service).
- Accelerated by overload (load higher than under normal operating conditions).

The crucial dependability test of PCM products is a durability test. Especially in Aerospace Engineering with respect to safety, it is necessary to perform an extensive set of durability tests, which are given by aviation regulations.

The result of durability test is creation of damage curve, which ends by one of limit state (Harris 2003). Damage curve of PCM contains of three basic phases:

I. Phase: cracks in matrix, or/and failure of reinforcing fibers.
II. Phase: cracks linking, failures at interlaminar interface and delamination.
III. Phase: delamination grow and final failure.

4 CONCLUSION

Dependability of PCM constructions in contrast to for example steel structures are not fully investigated. It is among other things related to the fact that the mechanism of composite fatigue damage is different from metals. Methods for evaluation of dependability and also durability of PCM is still evolving. Almost totally unexplored area is dependability of PCM constructions and parts used in the field of civil transport vehicles (automobiles, rails buses, ships).

Moreover, for evaluation of PCM constructions dependability, it is necessary to take into account the modern theories of dependability, modern knowledge from materials science, technologies, mathematical modeling and statistics. It is necessary to solve dependability in all stages of product lifecycle and apply the knowledge from aerospace industry into transport industry PCM products.

ACKNOWLEDGEMENT

This study was supported by the internal grant of TBU in Zlin No. IGA/FT/2015/001 funded from the resources of specific university research.

REFERENCES

D3878-07. 2013. *Standard Terminology for Composite Material*. ASTM International.

Ehrenstein, G.W. 2009. *Polymerní kompozitní materiály*. Praha: Scientia.

Gutjahr W.J. 2000. Software dependability evaluation based on Markov usage models. Performance Evaluation 40 (4). 199–222.

Harris, B. 2003. *Fatigue in composites: science and technology of the fatigue response of fibre-reinforced plastics*. Cambridge: Woodhead Publishing.

Jones, R.M. 1999. *Mechanics of composite materials*. Philadelphia: Taylor.

Malkin, V.S. 2005. *Osnovy teorii naděžnosti i diagnostiki*. Toljati: TGU.

Trivedi, K.S., Muppala, J.K., Woolet S.P. & Haverkort, B.R. 1992. *Composite performance and dependability analysis*. Performance Evaluation 14 (3-4). 197–215.

Vasiljev, V.V. 1988. *Konstrukciji iz kompozicionnych materialov*. Moscow: Mašinostrojenije.

Production Management and Engineering Sciences – Majerník, Daneshjo & Bosák (Eds)
© 2016 Taylor & Francis Group, London, ISBN: 978-1-138-02856-2

System states and requirements of reactive maintenance

N. Daneshjo & M. Kravec
Faculty of Business Economy, University of Economics, Košice, Slovak Republic

ABSTRACT: Influenced by the requirements of global markets, the competition in high technology branches necessitates permanent operational readiness of technical machinery and equipment. Materials of high quality and reliability-oriented engineering philosophies work against system abrasion, but cannot avoid it completely. Accompanying the introduction and operation of new and existing technologies, many influencing variables need to be considered when maintenance strategies are organized or implemented. A central aspect of maintenance is the failure behavior; but as a result of complex plant structures and solid interlinking of equipment and components, rival goals grow up. Planning maintenance activities is a demanding challenge.

Effective Maintenance Management examines the role of maintenance in minimizing the risk of safety or environmental incidents, adverse publicity, and loss of profitability. In addition to discussing risk reduction tools, it explains their applicability to specific situations, thereby enabling you to select the tool that best fits your requirements.

1 INTRODUCTION

Decreasing reliability and abrasion of heavy stressed parts are indicative for increasing age of technical equipment. It is up to maintenance to constitute and reconstitute absolute operability. According to DIN EN 13306:2010 (2010) maintenance is a "combination of all technical administrative and managerial actions during the life cycle of an item intended to retain it in, or restore it to, a state in which it can perform the required function". Per definition those items are ascertainable as parts, components, devices, subsystems or systems individually describable and considerable. The main condition of maintenance is maintainability; for stated conditions of use, stated maintenance procedures and resources are defined. Maintainability influences the down time of an object substantially. (DIN 40041:1990, 1990, DIN EN 13306:2010, 2010).

The generic term of maintenance combines inspection, preventive maintenance, repair and improvement as basic procedures. (DIN 31051:2003, 2003).

An inspection is a conformity examination; items relevant characteristics are measured, observed or tested. The inspection is intended to identify and assess the actual condition of objects. (Mende 2007).

Managed by an inspection plan the activities take place. In addition to identifying the actual item condition the analysis of generated data is important to initiate future failure prevention. The following tasks are derived from inspection:

- Determining the actual state.
- Evaluating the actual state.
- Assessing the actual state.
- Initiation of further measures.

The assessment of the actual state allows monitoring the wear-out of a unit and provides an insight into its reasons. Planning necessary maintenance activities becomes possible this way.

Preventive maintenance unifies all procedures intended to reduce failure probability or functional limitations. They are carried out at defined intervals or according to stated criteria. (DIN EN 13306:2010, 2010).

Ideally, preventive maintenance activities avoid fault-induced down times of items by retarding the reduction of residual operation time. Concrete goals characterize the predetermined maintenance time frames; goals and time frames are constituted by an operation chart. The available wear-out reserve defines the intensity of preventive maintenance intervals consisting of tasks like cleaning, preserving, adjusting, lubricating, complementing and replacing parts, subsystems and systems. (Mende 2007).

Repair contains physical activities taken to restore the regular function of a faulty unit. The goal of repair activities is the compensation of the consumed wear-out reserve by initiated corrective work and item exchanges. Depending on lead time available planned or unplanned repair—the preparation is more complex compared with preventive maintenance. Only with great efforts down states are avoidable. According to DIN 31051:2003 (2003) time schedules, personnel placement and work cycles have to be created before starting repair works. Particularly the unplanned (reactive) maintenance may cause problems; here the time of system failure as well as complexity and flow of actions to initiate are unknown. (Mende 2007, DIN 31051:2003, 2003, Rasch 2000).

Improvement refers to all activities which are intended to ameliorate reliability, maintainability or the safety of parts; the original function is not changed. The traditional maintenance term gets extended by the aspect of perfecting activities.

During the construction process of an object weak spots and safety deficits may creep in. It is the task of improvement to adjust shortcomings by constructive changes; therefore, experiences derived from systems on duty are essential. By establishing a continuous improvement process the latest learning effects inure to the benefit of existing items. Thus, a permanent adjustment to the usual system performance requirements and currently applicable safety requirement has to be realized.

Now defined maintenance procedures are divided into scheduled and reactive maintenance; Table 1 provides a structured overview.

Well-designed maintenance strategies may generate synergetic effects by optimizing the combination of basic procedures; processing efforts in administrative and operative maintenance get reduced. According to the definition of maintenance the aimed target condition of an object is kept for time of usage.

Table 1. Structure of maintenance procedures and classification of mission fields.

Maintenance			
Scheduled maintenance			Reactive maintenance
Inspection	Preventive maintenance	Improvement	Repair
Determine and assess	Keep	Restore	
The actual state of an item	The target condition of an item	The target condition of an item	

2 SYSTEMS TYPICAL FAILURE BEHAVIOR

The incidence of a system failure is hard to predict; but the breakdown is guaranteed. Maintenance data analysis reveals part-specific down times which may cause a system failure. This data must be a fundamental part of planning maintenance strategies.

In figure 1 a typical failure rate $F(t)$ is drawn as a function of time. The first bringing into service happens at the moment $t0$; noticeable is the high error rate at the beginning (early failure). Mostly, quality deficits are identifiable as the cause; they were not recognized during the production process and a high error rate at the beginning is induced. After overcoming initial difficulties the error rate is falling during the steady state before it rises slightly with progressing age of the parts. (Moncrief et al. 2005).

The curve of the function $F(t)$ allows conclusions on quality management structures of part manufacturers, maintenance activities to extend the duration of life and the optimal timing of part or component replacement. In particular, the prediction of parts behavior after steady state operation is difficult.

As soon as machines, which are designed for high reliability, run into age-related wear-out the probability of an error caused breakdown becomes unpredictable; the wear limit of an object is achieved. According to DIN 31051:2003 the reserve of functional performance under specified conditions is defined as wear reserve; the reductions of wear reserve as wear. (Mende 2007)

For three components $n=1, 2, 3$ the wear reserve $An(t)$ is drawn as a function of time in figure 2. Corresponding to the technical configuration, different time courses of degradation may result; starting at $An(t0)$. The wear limit is reached at the points $An(tAwln)$; with further exceeding the components finish their functioning at $An(t) = 0$ – failure.

An unintended early termination of the system function does not need to result from a complete reduction of the wear reserve. Error-related failures may be caused by dysfunctions which interrupt the normal functioning. Such events are divided into

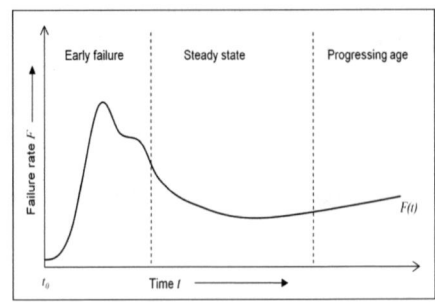

Figure 1. Typical failure rate of items.

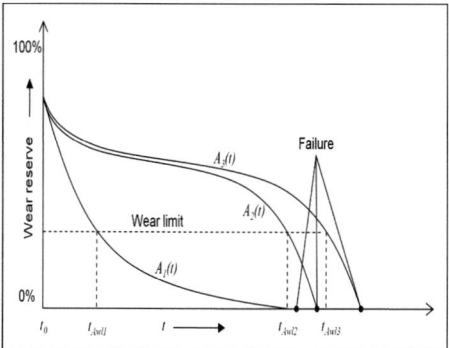

Figure 2. Component specific reduction of wear reserve.

primary and secondary failures; primary failures are caused by the unit itself and secondary failures by an error of peripheral units. (Moncrief et al. 2005).

The average operating time interval between two failures of a system is defined as the Mean Time Between Failures (MTBF). Ideally, worn components will be replaced before the expiry of the MTBF (system failure). The average time interval between two preventive component changes is defined as Mean Time Between Replacements (MTBR). (Moncrief et al. 2005).

3 SYSTEM STATES AND REQUIREMENTS OF REACTIVE MAINTENANCE

Disruptions during operation are the reason for the transition from functional to a dysfunctional system state. Further analysis leads to various substates from a functioning system to an external disabled system (secondary failure). Hereafter the two system states with their corresponding substates are presented in detail and classified according to the table 2 in an overall context.

Characteristic for the upstate of a system is the performance of a required function. The following substates are possible:

- Idle state: non-operating item during up state (system is not required).
- Operation state: fulfilling the required function.
- Standby: non-operating item during up state (system is required).
- External disabled: the item is in an upstate but there is a lack of required external resources (secondary failure). (Márquez 2007).

Systems regarded at down state are not able to keep up the required function; causal for those dysfunctions are:

1. Faults.
2. Subjects to preventive maintenance, which force the system to cease its function.

Table 2. States of an item.

Upstate				Downstate
				Disabled state
				Internal disabled state
Idle	Operation	Stand-by	External disabled	Subject to preventive maintenance	Fault

Table 3. Equipment active corrective times.

Active corrective maintenance time			
Technical delay	Trouble-shooting Repair time	Fault correction time	Check-out time

Maintenance activities reduce the systems availability. The planning of those activities needs to be well prepared; lead time is mandatory for preventive maintenance. During reactive maintenance the situation is much more complex. In case of damage necessary steps, such as scheduling, personnel approach, staging, etc. cannot be planned. Table 3 shows the course of reactive maintenance, but does not consider a possible logistic delay which can arise through administrative activities, personnel, materials and working space acquisition. (Márquez 2007).

According to systems complexity and cause of reactive maintenance the troubleshooting as well as the approach of time needed can be extended strongly. The removal of cover parts and other technical preparations require additional time before starting fault correction activities. After finishing, there is the technical follow-up; it includes the assembly, function checks and calibration work respectively.

4 MAINTENANCE MANAGEMENT

Already when designing technical facilities use times are taken as a basis; the wear reserve is the reference value. Technical measures, administrative measures and management activities ensure as parts of maintenance the operation state of a system. According to DIN EN 13006:2010 (2010) maintenance management includes: "all activities of the management that determine the maintenance objectives, strategies and responsibilities, and implementation of them by such means as maintenance planning, maintenance control, and the improvement of maintenance activities and economics".

The work is distributed to maintenance leadership with responsibility for maintenance management, and maintenance staff in an organi-

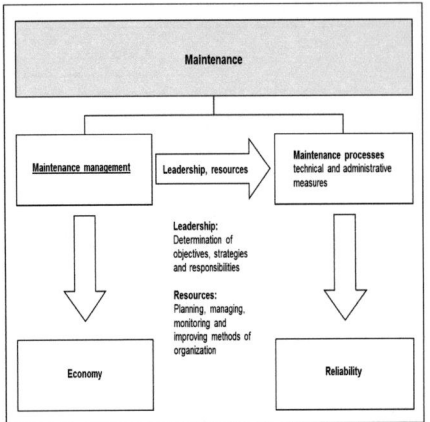

Figure 3. Management and processes in the context of maintenance.

zational structure; all together are responsible for maintenance processes. Figure 3 illustrates the separation into classical maintenance and management; reliability and economy are the core tasks of maintenance management. (DIN EN 13306:2010, 2010).

Characteristic functions of maintenance management are system procurement, resource allocation and process management. Already during the process of system procurement necessary maintenance measures should be taken into account. It is up to maintenance management to provide the required resources for reducing down state times of existing machinery. Especially in case of reactive maintenance, process management decides on the length of the down state time.

5 CONCLUSIONS

Planning and organization of maintenance activities are the fundamentals of maintenance management and various problems may affect success; especially constructing a maintenance management is highly difficult. Depending on the industrial branch and the type of investment good a variety of reasons may be causative. Referring to Márquez (2007) the most frequent reasons will be presented in more detail:

Maintenance is a complex system of activities; the lack of maintenance management models makes the cancellation of complexity difficult. It is not just repair—it is a complex methodology starting with the product design, continuing with planning preventive maintenance activities and ending with shortened down states for repair (product supporting services) in worst case. (Márquez 2007)

The diversification of potential maintenance problems is increasing more and more; administra-

tion becomes more difficult. An enormous variety of activities inhibits the implementations of standardized procedures and information support systems. Only in exceptional cases standard recipes are applicable; the demanded customized services require a maintenance management with a maximum of flexibility.

Customized products imply variants—individual maintenance measures are necessary. Building up system-related error knowledge is strongly limited. The permanent further development cuts the time for dedicated problem analysis. Additionally, ad hoc decisions and short term plans undermine maintenance management systems designed for long-term plants availability.

Staff retention and the fear of disruption by maintenance measures restrict the willingness of the top management to support. The increasing improvement of systems efficiency, productivity and complexity makes the implementation of maintenance difficult. As a result of using latest manufacturing technologies and implementing the idea of just-in-time production conditions have changed. Inventory levels have been minimized, throughput increased and the economic efficiency of plants grew as well as the pressure on maintenance. The failure of one component can bring the whole system to a standstill; huge costs are caused at efficiency-optimized systems.

REFERENCES

DIN 31051:2003. 2003. DIN 31051: Grundlagen der Instandhaltung; Berlin: Beuth Verlag GmbH.

DIN 40041:1990. 1990. DIN 40041:1990: Zuverlässigkeit—Begriffe,, Berlin: Beuth Verlag GmbH.

DIN EN 13306:2010. 2010. DIN EN 13306:2010: Maintenance terminology, Brussels: Beuth Verlag GmbH.

Knežo, D. 2014. Inverse transformation method for normal distribution and the standard numerical methods. In International Journal of Interdisciplinarity in Theory and Practice. 2 (5): 6–10.

Márquez, A.C. 2007. The Maintenance Management Framework: Models and Methods for Complex Systems Maintenance. Sevilla: Springer Verlag London LimitedSpain.

Mende, K.M. 2007. Konzeption des Modells der opportunistischen Instandhaltungsstrategie; dissertation, Dortmund: Shaker Verlag GmbH.

Moncrief, E.C., Schroder, R.M. & Reynolds, M.P. 2005. Production spare parts: optimizing the MRO inventory asset. New York: Industrial Press Inc.

Rasch, A.A. 2000. Erfolgspotential Instandhaltung: theoretische Untersuchung und Entwurf eines ganzheitlichen Instandhaltungsmanagements. Berlin: Erich Schmidt Verlag.

Rudy, V. & Lešková A. 2013. Modular systems for experimental modelling in the design process of flexible workstations. In Interdisciplinarity in theory and practice. 1(1): 28–31.

Production Management and Engineering Sciences – Majerník, Daneshjo & Bosák (Eds)
© 2016 Taylor & Francis Group, London, ISBN: 978-1-138-02856-2

The accuracy of relative navigation system

M. Džunda & N. Kotianová
Faculty of Aeronautics, Technical University of Košice, Košice, Slovak Republic

ABSTRACT: This contribution discusses the relative navigation system in an aviation communication network. It provides derived algorithms for computing the position of a flying object in a space based on data received from other users within an aviation communication network. This will enable location of a flying object without the need to complement the airborne equipment with new navigation devices. Further we discuss the errors that affect the accuracy of finding the position of such an object. Also analysed is the effect of error in measurement of the distance between the source of transmission and the object in term of the accuracy of position. This contribution is introducing an alternative option of determining the position of aircraft without using satellite navigation systems on the basis of the principles of relative navigation making use of communications systems. Such solutions are assumed to bring about remarkable economic effects in air transportation.

1 INTRODUCTION

Global navigation satellite systems are dominant in all phases of flight. They are considered primary means of future navigation and communication infrastructure. Status quo flows from the program NextGen (FAA) and the concept of ATM/CNS (Eurocontrol). The concept is intended to increase airspace capacity and safety of air traffic. (Kršák & Tobisová 2012, Pavolová & Tobisová 2013, Szabo& Ferencz & Pucihar 2013a, Szabo & Dorcák & Ferencz 2013b)

Some alternatives are being examined to determine how best to maintain the safety, security, and efficiency of the air traffic in the event of a loss of Global Navigation Satellite System PNT (Positioning, Navigation, and Timing) services. (Kurdel & Labun & Adamčík 2014, Labun & Soták & Kurdel 2012)

One of alternatives could be relative navigation system in aviation communication network. This system use a method of measuring Time of Arrival (TOA) of signals received from other users within an aviation communication network. Positioning in RelNav mode consist in measuring the distance between the Source of Transmission (SoT) and the flying object by measuring the TOA.

Relative navigation (RelNav) can be realized as follows:

1. In the relative coordinate system—in which coordinates are determined by each object and in a coordinate system determined by a single object termed as navigation controller.

2. In the absolute coordinate system—geographical coordinate system, this is bound to the planet Earth: latitudes and longitudes.

In any of the cases as above, on object of the network is marked as the Reference Station (RS), the one by which system time is determined. (Džunda 1999)

2 ALGORITHMS FOR CALCULATING THE POSITION OF A FO IN A SPACE

There are several methods of determining the position of a point in a space using matrices. Let us assume that all user of communication network is transmitting the so-called position messages or reports in short time intervals. It takes certain time for the signal to travel from one user (SoT) to the receiver of another user. This period of time is directly proportional to the distance between the receiver and the source of transmission. By obtaining the distance data from several network users, it is possible to locate the position of the receiver. Here, we present an alternative solution based on the application of Groaner's bases and the resultant of the polynomial. Let us show how to explicitly solve non-linear equations with four unknown variables using resultants of multi-polynomials, namely, by converting it into algebraic (polynomial) form and simplifying it into a linear from.

We use the equation for calculating the distance between two points in a space for formulation of algorithms.

$$\|B - A\| = d(A, B)$$
$$= \sqrt{(a_1 - b_1)^2 + (a_2 - b_2)^2 + (a_3 - b_3)^2} \quad (1)$$

By the equation (1) for calculating the distance between two points in a space the expression for the pseudo-distance d_i is obtained:

$$(d_i - b) = c.(\tau_i - \Delta t)$$
$$= \sqrt{(x - x_i)^2 + (y - y_i)^2 + (z - z_i)^2} \quad (2)$$

where i = 1, 2, 3, 4.

After adjusting the (2), the position of the receiver on the unknown FO is expressed by means of 4 equations with 4 unknown parameters: (Kotianová 2015, Awnage & Grafarend 2004).

$$(x - x_1)^2 + (y - y_1)^2 + (z - z_1)^2 - (b - d_1)^2 = 0 \quad (3)$$

$$(x - x_2)^2 + (y - y_2)^2 + (z - z_2)^2 - (b - d_2)^2 = 0 \quad (4)$$

$$(x - x_3)^2 + (y - y_3)^2 + (z - z_3)^2 - (b - d_3)^2 = 0 \quad (5)$$

$$(x - x_4)^2 + (y - y_4)^2 + (z - z_4)^2 - (b - d_4)^2 = 0 \quad (6)$$

where d_1, d_2, d_3, d_4 = measured pseudo-distances from the SoT to the receivers; $x_{1-4},\ y_{1-4},\ z_{1-4}$ = coordinates of users; $x,\ y,\ z$ = coordinates of the receiver (unknown parameter); b = shift of the time basis of the receiver converted into distance (unknown parameter).

Let us transcribe (3) up to (6) into a form with linear members on one side and non-linear ones on the other: (Kotianová 2015, Awnage & Grafarend 2004).

$$x^2 - 2xx_1 + y^2 - 2yy_1 + z^2 - 2zz_1 - b^2 + 2bd_1$$
$$+ + x_1^2 + y_1^2 + z_1^2 - d_1^2 = 0 \quad (7)$$

$$x^2 - 2xx_2 + y^2 - 2yy_2 + z^2 - 2zz_2 - b^2 + 2bd_2$$
$$+ + x_2^2 + y_2^2 + z_2^2 - d_2^2 = 0 \quad (8)$$

$$x^2 - 2xx_3 + y^2 - 2yy_3 + z^2 - 2zz_3 - b^2 + 2bd_3$$
$$+ + x_3^2 + y_3^2 + z_3^2 - d_3^2 = 0 \quad (9)$$

$$x^2 - 2xx_4 + y^2 - 2yy_4 + z^2 - 2zz_4 - b^2 + 2bd_4$$
$$+ + x_4^2 + y_4^2 + z_4^2 - d_4^2 = 0 \quad (10)$$

Subtracting (10) from (7), it holds:

$$f_{14} = x_{14}x + y_{14}y + z_{14}z + d_{41}b + e_{14} \quad (11)$$

Subtracting (10) from (8), it holds:

$$f_{24} = x_{24}x + y_{24}y + z_{24}z + d_{42}b + e_{24} \quad (12)$$

Subtracting (10) from (9), it holds:

$$f_{34} = x_{34}x + y_{34}y + z_{34}z + d_{43}b + e_{34} \quad (13)$$

where:

$$x_{14} = 2(x_1 - x_4)$$

$$y_{14} = 2(y_1 - y_4)$$

$$z_{14} = 2(z_1 - z_4)$$

$$d_{41} = 2(d_4 - d_1)$$

$$e_{14} = (d_1^2 - x_1^2 - y_1^2 - z_1^2) - (d_4^2 - x_4^2 - y_4^2 - z_4^2)$$

$$x_{24} = 2(x_2 - x_4)$$

$$y_{24} = 2(y_2 - y_4)$$

$$z_{24} = 2(z_2 - z_4)$$

$$d_{42} = 2(d_4 - d_2)$$

$$e_{24} = (d_2^2 - x_2^2 - y_2^2 - z_2^2) - (d_4^2 - x_4^2 - y_4^2 - z_4^2)$$

$$x_{34} = 2(x_3 - x_4)$$

$$y_{34} = 2(y_3 - y_4)$$

$$z_{34} = 2(z_3 - z_4)$$

$$d_{43} = 2(d_4 - d_3)$$

$$e_{34} = (d_3^2 - x_3^2 - y_3^2 - z_3^2) - (d_4^2 - x_4^2 - y_4^2 - z_4^2)$$

If variable b is considered as constant (the so-called factor of homogenisation), then out of expressions (11–13) a system of three equations with three unknown variables is obtained. We apply the method of resultants of the multi-polynomials to solve the system of linear equations.

Further by Awnage & Grafarend (2004) it holds:

$$f_1 = (x_{14}x + d_{41}b + e_{14})k + y_{14}y + z_{14}z \quad (14)$$

$$f_2 = (x_{24}x + d_{42}b + e_{24})k + y_{24}y + z_{24}z \quad (15)$$

$$f_3 = (x_{34}x + d_{43}b + e_{34})k + y_{34}y + z_{34}z \quad (16)$$

where k is turning into the factor of homogenisation. Jacobi's determinant of the coordinate x by (14), (15), (16) is expressed as:

$$J_x = \det \begin{pmatrix} \dfrac{df_1}{dy} & \dfrac{df_1}{dz} & \dfrac{df_1}{dk} \\[2mm] \dfrac{df_2}{dy} & \dfrac{df_2}{dz} & \dfrac{df_2}{dk} \\[2mm] \dfrac{df_3}{dy} & \dfrac{df_3}{dz} & \dfrac{df_3}{dk} \end{pmatrix} = \begin{pmatrix} y_{14} & z_{14} & x_{14}x + d_{41}b + e_{14} \\ y_{24} & z_{24} & x_{24}x + d_{42}b + e_{24} \\ y_{34} & z_{34} & x_{34}x + d_{43}b + e_{34} \end{pmatrix}$$

$$(17)$$

Applying (17), we obtain the value of the determinant Jx, from which the $x = g(b)$ variable is expressed as follows:

$$
\begin{aligned}
x = -(&d_{43}by_{14}z_{24} + e_{34}z_{24}y_{14} + d_{42}bz_{14}y_{34} + e_{24}z_{14}y_{34} \\
&+ d_{41}by_{24}z_{34} + e_{14}y_{24}z_{34} - d_{41}bz_{24}y_{34} - e_{14}z_{24}y_{34} \\
&- d_{43}bz_{14}y_{24} - e_{34}z_{14}y_{24} - d_{42}by_{14}z_{34} - e_{24}y_{14}z_{34}) \,/ \\
(&x_{34}z_{24}y_{14} + x_{24}z_{14}y_{34} + x_{14}y_{24}z_{34} - x_{14}y_{34}z_{24} \\
&- x_{34}z_{14}y_{24} - x_{24}y_{14}z_{34})
\end{aligned}
$$
(18)

In compliance with (14)–(16), for $y = g(b)$ it holds:

$$
f_4 = (y_{14}y + d_{41}b + e_{14})k + x_{14}x + z_{14}z
$$
(19)

$$
f_5 = (y_{24}y + d_{42}b + e_{24})k + x_{24}x + z_{24}z
$$
(20)

$$
f_6 = (y_{34}y + d_{43}b + e_{34})k + x_{34}x + z_{34}z
$$
(21)

Jacobi's determinant for the coordinate y is expressed as:

$$
\begin{aligned}
J_y = det
\begin{pmatrix}
\dfrac{df_5}{dx} & \dfrac{df_5}{dz} & \dfrac{df_5}{dk} \\[2mm]
\dfrac{df_6}{dx} & \dfrac{df_6}{dz} & \dfrac{df_6}{dk} \\[2mm]
\dfrac{df_7}{dx} & \dfrac{df_7}{dz} & \dfrac{df_7}{dk}
\end{pmatrix} \\[4mm]
= det
\begin{pmatrix}
x_{14} & z_{14} & y_{14}y + d_{41}b + e_{14} \\
x_{24} & z_{24} & x_{24}y + d_{42}b + e_{24} \\
x_{34} & z_{34} & y_{34}y + d_{43}b + e_{34}
\end{pmatrix}
\end{aligned}
$$
(22)

Applying (22), we obtain variable $y = g(b)$:

$$
\begin{aligned}
y = -(&z_{24}x_{34}d_{41}b + x_{14}z_{34}d_{42}b + x_{14}z_{34}e_{24} - z_{14}x_{34}d_{42}b \\
&- x_{14}z_{24}d_{43}b - x_{14}z_{24}e_{34} + x_{24}z_{34}d_{41}b - x_{24}z_{34}e_{14} \\
&- z_{14}x_{34}e_{24} + z_{24}x_{34}e_{14} + z_{14}x_{24}e_{34}) \,/ \\
(&x_{34}z_{24}y_{14} + x_{24}y_{34}z_{14} + {}+x_{14}y_{24}z_{34} - x_{14}y_{34}z_{24} \\
&- x_{34}z_{14}y_{24} - x_{24}y_{14}z_{34})
\end{aligned}
$$
(23)

In compliance with (14)–(16), for $z = g(b)$ it holds:

$$
f_7 = (z_{14}z + d_{41}b + e_{14})k + x_{14}x + y_{14}y
$$
(24)

$$
f_8 = (z_{24}z + d_{42}b + e_{24})k + x_{24}x + y_{24}y
$$
(25)

$$
f_9 = (z_{34}z + d_{43}b + e_{34})k + x_{34}x + y_{34}y
$$
(26)

Jacobi's determinant for coordinate z is expressed as:

$$
\begin{aligned}
J_z = det
\begin{pmatrix}
\dfrac{df_7}{dx} & \dfrac{df_7}{dy} & \dfrac{df_7}{dk} \\[2mm]
\dfrac{df_8}{dx} & \dfrac{df_8}{dy} & \dfrac{df_8}{dk} \\[2mm]
\dfrac{df_9}{dx} & \dfrac{df_9}{dy} & \dfrac{df_9}{dk}
\end{pmatrix} \\[4mm]
= det =
\begin{pmatrix}
x_{14} & y_{14} & z_{14}z + d_{41}b + e_{14} \\
x_{24} & y_{24} & z_{24}z + d_{42}b + e_{24} \\
x_{34} & y_{34} & z_{34}z + d_{43}b + e_{34}
\end{pmatrix}
\end{aligned}
$$
(27)

Applying (27), we obtain the variable $z = g(b)$:

$$
\begin{aligned}
z = (&x_{14}y_{24}d_{43}b - x_{14}y_{24}e_{34} - y_{14}x_{34}d_{42}b \\
&- y_{14}x_{34}e_{24} - x_{24}y_{34}d_{41}b - x_{24}y_{34}e_{14} \\
&+ y_{24}x_{34}d_{41}b + y_{24}x_{34}e_{14} + y_{14}x_{24}d_{43}b \\
&+ y_{14}x_{24}e_{34} + x_{14}y_{34}d_{42}b + x_{14}y_{34}e_{24}) \,/ \\
(&x_{34}z_{24}y_{14} + x_{24}y_{34}z_{14} + x_{14}y_{24}z_{34} \\
&- x_{14}y_{34}z_{24} - x_{34}z_{14}y_{24} - x_{24}y_{14}z_{34})
\end{aligned}
$$
(28)

Substituting x from expression (18), y from (23) and z from (28) into equation (7), the quadratic function for the b unknown variable is obtained:

$$
Ab^2 + Bb^2 + C = 0
$$
(29)

Solutions of the quadratic equation (29) are two roots, b^+ and b^-. Substituting the values of b^+ and b^- into the equations (18), (23),(28), we calculate the coordinates of FO (x^+, y^+, z^+) and P (x^-, y^-, z). Calculations provide two values of the FO (x^+, y^+, z^+) and $P(x^-, y^-, z^-)$ coordinates.

To arrive at a correct solution, we have to calculate the norm (the radial distance from the centre of the Earth) for re (x,y,z) b^- and (x,y,z) b^+. If the coordinates of the receiver are fed into the coordinate reference system, then the norm of the position vector will be close to the value of the radius of the Earth ($Rz = 6372.797$ km) and this solution FO (x,y,z) will be considered as correct. The norm of the position vector of the second trio will be in the space. To calculate the norm, the following expression holds:

$$
norm = \sqrt{x^2 + y^2 + z^2}
$$
(30)

Using the coordinates of four FO from Table 1, we apply derived algorithm and figure out coordinates of fifth FO and the unknown parameter b-shift of the time basis of the receiver converted into distance.

We have chosen a real situation in the air in a given time and place as by the flightradar24. Reading the coordinates of real aircraft flying over the territory of the Slovak republic was

Table 1. Coordinates of four FO and the pseudo-range.

Flight number	Coordinate x (m)	Coordinate y (m)	Coordinate z (m)	Pseudo-range D (m)
THY3HD	4014.466	1293.512	4783.064	2043.863
ETD12	4033.224	1214.930	4788.259	94076.685
WZZ86Q	4047.944	1292.268	4753.813	65443.141
UAE236	3978.388	1353.284	4796.965	55047.852

realized on 1 October 2014 at 12:20. We have randomly chosen 5 aircraft and determined their coordinates. For the purpose of verifying the derived algorithms, they were to be transformed from WGS-84 into a rectangular coordinate system.

By applying the above algorithms we calculate the two roots b^+ and b^- to finally determine the coordinates of FO (x^+, y^+, z^+) and FO (x^-, y^-, z^-):

$x^+ = 3996.978$
$y^+ = 1301.567$
$z^+ = 4793.798$
$x^- = 12169777.\ 391243594$
$y^- = 3981351.\ 2466636864$
$z^- = 14206264.77723196$

Then, using (30), we calculate the norm for the position vectors $\{X,Y,Z\}|\ b^+$ and $\{X,Y,Z\}|\ b^-$:

$\{X,Y,Z\}|\ b^+ = 6375767.33050830$ m,
$\{X,Y,Z\}|\ b^- = 19125182.519005124$ m.

Consequently, the correct solution is the position vector of $\{X, Y, Z\}|\ b^+$.

3 ANALYSIS OF THE ACCURACY OF THE RELATIVE NAVIGATION SYSTEM

Pseudo-range measurement to SoT includes some errors that affect the accuracy of finding the position of FO. In this section we will discuss the dependence of various factors affecting the accuracy of the positioning. Fundamental errors are:

– errors associated with source of transmission (clock error),
– errors associated with transmission environment (reflected wave, multipath propagation etc.),
– errors associated with receiver (clock error, receiver noise),
– errors associated with the geometric arrangement of FO.

4 ASSESSING THE IMPACT OF MEASUREMENT ERROR OF THE DISTANCE IN TERM OF THE ACCURACY OF POSITION

Positioning accuracy of FO is affected by de-synchronization of time bases of individual users in network (Figure 1). In practise the time deviation occurs between each user and source of transmission in practise. This time deviation affects the calculation of distance from source of transmission. The distance is burdened by lots of errors. Therefore, it is termed as a pseudo-distance. The time basis of the receiver is shifted by an unknown interval t, which can be converted into distance according to the relation

$$b = c.\Delta t \qquad (31)$$

where c is speed of light in vacuum = 299792458 m/s (3×108 m/s).

True distance will be equal to:

$$D_i = d_i - b \qquad (32)$$

where d_i = measured pseudo-distance; D_i = true distance; b = shift of the time basis for the receiver converted into distance.

We created program for determining and analysing the accuracy of the position of FO depending on error in measurement of the distance. We have to create to four pseudo-random number generators, because of four distances from four sources of transmission. We need four distances from four sources of transmission, therefore we have to establish to four pseudo random number generators. By way of the generators, we obtained four measurement errors in measurement of the distance and we then substituted them when calculating of position of FO. The generator generates pseudo-random measurement errors according to a standard normal distribution denoted by N(0,1).

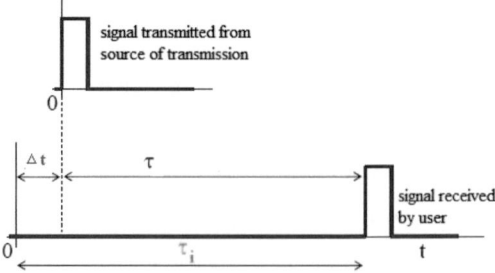

Figure 1. Time de-synchronization of users.

The parameters for a standard normal distribution are:

– mean, which characterizes the position of the distribution and is the value of the maximum density, $E(x) = \mu = 0$.
– standard deviation, variance $V(x) = \sigma^2 = 1$

In the context of the normal distribution, the random errors are often mentioned, e.g. measurement errors caused by numerous unknown and independent causes. Therefore, the normal distribution is also known as Gaussian laws which govern the distribution of some of the errors in physical an technical magnitudes. The density of probability of normal distribution is expressed as:

$$f(x) = \frac{1}{2\pi} e^{-\frac{x^2}{2}} \qquad (33)$$

Figures 2 and 3 show a normal distribution of probability of the genera numbers from our generator. The sample represents 1000 generated pseudo-numbers of measurement errors from sources of transmission 1,2,3,4. Given the extent of the contribution, we don´t introduce specific generated numbers but only graphs of their probability density function.

Generated random numbers of measurement errors will be added up to calculated pseudo-distance. Pseudo-distance including generated

measurement errors will be implemented into algorithms for determining the position of user. We will research the error in position of a flying object on the basis of different combinations and range of pseudo-range measurement errors. We verify two situations of dependence of the accuracy of finding the position of FO on distance measurement errors from SoT taking into account:

1. The randomly changing distance measurement errors from all four sources of transmissions and the impact on error in position of a flying object.
2. The distance measurement errors just from a single SoT. Other measurement errors are considered to be null.

The results of simulations carried out on a computer are shown in Figures 4 to 5.

Case No. 1: The graph in Figure 4 shows the changing generated distance measurement errors from SoT 1–SoT 4. The errors are arranged by size, so we can monitor a growing and decreasing error in position of FO.

Submission of error in position of FO does not change linear with growing/decreasing distance measurement error but it depends on size of other distance measurement errors to other SoTs. If generated measurement errors from the group of normal distribution are randomly distributed, average positioning errors of FO will be 86,31 m.

In the Table 2 we can see the statistic data of simulations carried out on a computer for Case No. 1. Minimum positioning error of FO (1.044 m) occurred when the mean value was 0.1509 m and standard deviation of 0.8246. Otherwise, the maximum positioning error of FO (373.8 m)

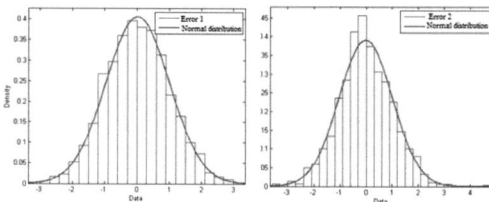

Figure 2. Probability density function of a normal distribution of random numbers generated for pseudo-range measurement errors 1,2.

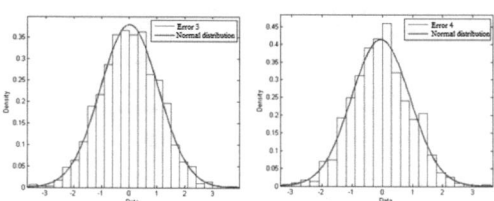

Figure 3. Probability density function of a normal distribution of random numbers generated for pseudo-range measurement.

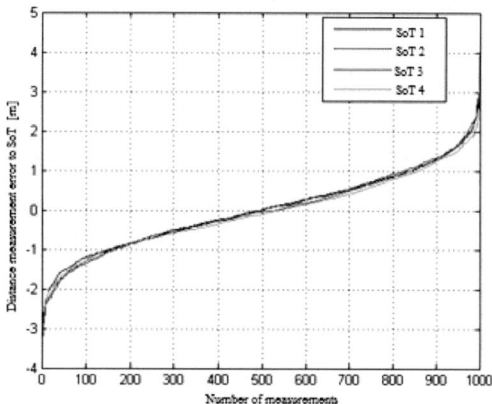

Figure 4. Generated distance measurement errors to SoT 1–SoT 4 arranged by size.

373

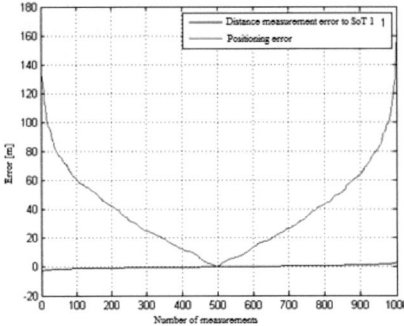

Figure 5. Positioning error of FO with distance measurement error to SoT 1.

Table 2. Statistic data of simulations carried out on a computer for Case No. 1.

Minimum positioning error of FO	1.044 m
Maximum positioning error of FO	373.8 m
Mean	86.31 m
Median	74.48 m
Standard deviation	62.74 m

Table 3. Results for positioning error of FO for Case No. 1.

Δd_1 (m)	Δd_2 (m)	Δd_3 (m)	Δd_4 (m)	$\bar{d}$ (m)	s (m)
Minimum positioning error of FO in case No. 1:					
0.84928	−1.0443	0.39931	0.39955	0.150	0.824
Maximum positioning error of FO in case No. 1:					
−0.8398	4.61381	0.23539	0.19488	1.051	2.427
Average positioning error of FO in case No. 1:					
−0.475	0.2461	1.4731	−0.2338	0.25	0.867

occurred when the mean value of distance measurement error to four SoTs was grew up to 1,051 m and the standard deviation was 2.427. The exact values are entered in Table 3.

Note that in case of the maximum value of the positioning error of FO (in Table 3) one of the distance measurement error reached extremely higher value (d_1 = 4.613819 m) compared to the average value. Therefore, we can conclude that the accuracy of finding the position of FO is affected by the difference in the size of the individual errors and their deviation from the average value. Otherwise it hold that the closer the value of measurement errors to all four SoTs to average value, the position of FO is determined with better accuracy.

Case No. 2: In the next part of article we will monitor how the positioning error of FO is changing in case of occurrence just one measurement error.

The simulation carried out on a computer shows that positioning error of FO is decreasing

Table 4. Statistic data of simulations carried out on a computer for Case No. 2.

Minimum positioning error of FO	0.062472 m
Maximum positioning error of FO	164.947567 m
Mean	39.6 m
Median	33.77 m
Standard deviation	29.54 m

Table 5. Results for positioning error of FO for Case No. 2.

Value	Δd_1 (m)	Δd_2 (m)	Δd_3 (m)	Δd_4 (m)	Positioning error (m)
Minimal	0.001243	0	0	0	0.062472
Maximal	−3.28312	0	0	0	164.947567
Average	0.787862	0	0	0	39.6

or increasing depending on change of distance measurement error to set SoT. This applies only if we are considering the existence of only one measurement error to one SoT. Other distance measurement errors are considered to be zero.

Figure 5 shows the dependence of positioning accuracy of FO on distance measurement error just to SoT 1. The specific values of errors are shown in Table 4. Minimum positioning error of FO (0.062472 m) had occurred when distance measurement error reached minimal value (0.001243 m). Otherwise, the maximum positioning error of FO (164.947567 m) had occurred in case of maximal distance measurement error (−3.283121). Complete average positioning error of FO was 39.6 m.

Table 5 shows values for measurement errors to one SoT, for which the positioning error of FO reaches maximal, minimal or average value. The same situation occurs in other three cases where we gradually generated the distance measurement errors to SoT 2, 3 and 4.

Based on the above facts, if only one distance measurement error to one SoT occurs, we can conclude that positioning accuracy of FO will depend on the size of this error. In this case the positioning accuracy of FO is better than in case No. 1, where all the four distance measurement errors were taking into account. On the basis of computer-based simulation, we can state that such errors affect largely the precise positioning of the FO in the system of relative navigation.

5 CONCLUSIONS

The submitted contribution discusses the relative navigation system as an alternative option of controlling the movement of flight object in the event of GNSS interference or breakdown. The submitted contribution is presenting algorithms for determining the positions of FO, to be potentially applicable in the design of a joint system of communication and navigation. Further, we evaluate the positioning accuracy of FO in relative navigation system, depending on distance measurement errors to individual sources of transmission. Modelling had confirmed that distance measurement errors from reference sources to flight object ranging from +4.00 m to –4.00 m, result in positioning error falling between 1.00 m and 374.00 m.

REFERENCES

Awnage, J.L. & Grafarend, E.W. 2004. *Solving algebraicc omputational problems in Geodesy and Geoinformatics,* Germany: Springer. ISBN 3-540-23425-X.

Džunda, M. 1999. *Syntéza rádionavigačných systémov.* Košice: Vojenská letecká akadémia generála M.R. Štefánika. ISBN 80-7166-029-9.

Kotianová, N. 2015. *Relatívna navigácia v komunikačnej sieti letectva*: Essay to dissertation examination. Košice: Technical University of Košice.

Kršák, B. & Tobisová, A. 2012. Prerequisites forthe implementation of information technologies in tourism small and medium size denterprises. In SGEM 2012; *Proc. Of 12th International Multidisciplinary Scientific GeoConference: Volume 3, Albena 17–23 June 2012*. Sofia: STEF92 Technology Ltd. ISSN 1314-2704.

Kurdel, P., Labun, J. & Adamčík, F. 2014. The estimation method of the characteristics of an aircraft with electromechanic analogue. *Naše more* 61 (1–2): 18–21. ISSN 0469-6255.

Labun, J., Soták, M. & Kurdel, P. 2012. Technical note innovative technique of using the radar altimeter for prediction of terrain collision threats. *The Journal of the American Helicopter Society* 57(4): 045002-1-045002-3. ISSN 0002-8711.

Pavolová, H. & Tobisová, A. 2013. The model of supplier quality management in a transport company. *Naše more* 60 (5–6): 123–126. ISSN 0469-6255.

Szabo, S., Dorcák, P. & Ferencz, V. 2013b. The significance of global market data for smart E-Procurement processes. In IDIMT, *Information Technology, Human Values, Innovation and Economy; Proc. of 21st interdisciplinary Information Management Talks, Prague 11-13 September*. Prague: Trauner. ISBN 9783990330838.

Szabo, S., Ferencz, V. & Pucihar, A. 2013a. Trust, Innovation and Prosperity. *Quality Innovation Prosperity.* 17(2): 1–8. ISSN 1335-1745.

Production Management and Engineering Sciences – Majerník, Daneshjo & Bosák (Eds)
© *2016 Taylor & Francis Group, London, ISBN: 978-1-138-02856-2*

Cross-border M&A in services and manufacturing sector within European area

J. Hečková, A. Chapčáková & E. Litavcová
Faculty of Management, University of Prešov, Prešov, Slovak Republic

A. Tarča
Faculty of Business Economy, University of Economics in Bratislava, Košice, Slovak Republic

ABSTRACT: The contribution focuses on analysis of the impact of selected predictors on the carrying out of cross-border mergers and acquisitions in the services sector and the manufacturing sector in selected countries of the European area for the period 1998–2012. For quantification of the impact of the considered predictors on the volume of cross-border mergers and acquisitions a general regression model was used. The contribution was prepared in the scope of implementing the project VEGA no. 1/0173/15 "Analytical view of aspects determined the development of cross-border mergers and acquisitions in the European area" and the project KEGA no. 032PU-4/2013 "Applying E-Learning to Teaching Economic Disciplines of the Management Study Programme and New Accredited Study Programmes at the Faculty of Management of University of Prešov in Prešov".

1 INTRODUCTION

Cross-border mergers and acquisitions are defined as an instrument for obtaining a larger share on a market. The development of cross-border mergers and acquisitions in the past ten years has copied the cyclical development in the world economy as a whole. Financial and trade liberalization in the European Union and the European Monetary Union have an impact on the conducting of cross-border mergers and acquisitions (the reallocation of capital) by increasing their profitability, because regional treaties enlarge the market and support competition through a lowering of costs for financial transactions associated with financial integration. The creation of a single market, which removed the barriers to the free movement of capital, labour, goods and services within the European Economic Community (later the European Union), and the adoption of the euro as the common currency within the European Monetary Union, the introduction of which eliminated the exchange-rate risk and supported financial integration, had an impact on the development of cross-border mergers and acquisitions. This was pointed out in the study of Petroulas (2007), for example, who investigated the impact of introducing the euro on internal flows of direct foreign investments and came to the conclusion that within the European Monetary Union flows of direct foreign investments increased by approximately 16%. The studies of other authors, such as Schiavo (2007) and Brouwer, Paap and Viaene (2008), also arrived

at a similar conclusion. An economy open to an influx of direct foreign investments can expand more rapidly than an economy which must completely rely on domestic resources. Direct foreign investments are important in that their influx not only increases domestic share capital but can also bring other benefits with it, such as access to new technologies, know-how and better managerial techniques, among others. Referring to Neary's study (2007), authors Coeurdacier, De Santis and Aviat (2009) elaborated their own study, the aim of which was to investigate the impact of the European Union and European Monetary Union on capital reallocation through cross-border mergers and acquisitions in the scope of member states of these integrational groupings. Their aim was to either confirm or refute the theoretical arguments presented by Neary in his study (2007), namely that trade liberalization and deeper integration of the European market correlates with growth in the number of cross-border mergers and acquisitions carried out. The aim was also to determine whether the European Union and European Monetary Union managed to lure capital from other parts of the world and to identify the sectors most influenced in this direction. They started from the assumption that a better understanding of capital reallocation in countries is a key for public policy makers, because the majority of countries use several methods of attracting direct foreign capital.

The aim of the contribution is to analyse the impact of selected predictors (with the emphasis on processes of financial and commercial liberalization in the European Union and the functioning of the European Monetary Union) on the carrying out of cross-border mergers and acquisitions in manufacturing sector within European area. The contribution was prepared in the scope of implementing the project VEGA no. 1/0173/15 "Analytical view of aspects determined the development of cross-border mergers and acquisitions in the European area" and the project KEGA no. 032PU-4/2013 "Applying E-Learning to Teaching Economic Disciplines of the Management Study Programme and New Accredited Study Programmes at the Faculty of Management of University of Prešov in Prešov".

The contribution is divided into two subsections in order to achieve its aim. The first subsection defines the methodological basis covering database of research and description of the applied methods for the processing of data necessary for fulfillment of the aim. The second subsection presents the results of analysis of the impact of selected predictors of cross-border mergers and acquisitions within European area.

2 MATERIALS AND METHODOLOGY OF THE CONTRIBUTION

The database which was analyzed for the purpose of this study contains 85,510 data items on mergers and acquisitions carried out in the countries of the European area and in Turkey in the period from 1998 up through 2012 (16 source countries: Belgium, Cyprus, Denmark, Finland, France, Greece, Netherlands, Luxembourg, Malta, Germany, Poland, Portugal, Austria, Spain, Italy, United Kingdom, and 25 target countries: Belgium, Bulgaria, Cyprus, Czech Republic, Denmark, Estonia, Finland, France, Greece, Netherlands, Lithuania, Latvia, Luxembourg, Hungary, Malta, Germany, Portugal, Austria, Romania, Slovakia, Slovenia, Spain, Italy, Turkey, United Kingdom). The key sources of information used are statistical data from the Zephyr, Eurostat and Freedom House databases. From the total number of records in the mentioned databases, 11,583 relate to cross-border mergers and acquisitions, 4,395 of which have the value of the volume of cross-border activities listed, and 4,285 of these also the values of other selected predictors.

For quantification of the impact of the considered predictors on the volume of cross-border mergers and acquisitions a general regression model was used, and the programs MS Excel, Statistica 12 and IBM SPSS Statistics 20 were used for processing the research results.

$M\&A_{i,j,s,t}$ indicates the total value of cross-border assets obtained through mergers and acquisitions by the source country i in the target country j in the sector s at time t. The Gross Domestic Product (GDP) of the source (i) and target (j) countries in the sector s at time t $(GDP_{j,s,t}, GDP_{i,s,t})$ can be considered as an important predictor impacting the volume cross-border mergers and acquisitions. The use of the logarithm of their product eliminates their different elasticity without affecting the overall result. *Market capitalization* represents the annual average market capitalization of the sector under consideration, which was obtained from the Zephyr database. The ratio of market capitalization *to* GDP was used as an indicator of development of the share market. It is impossible to omit even the nearness of the countries, the specificity of their culture and the nearness of their languages. The nearness of the source and target countries was quantified by the geographic distance between their capital cities, labelled as *distance_{ij}*, and the sharing of a common border of the countries was quantified with the binary variable *border_{ij}*, acquiring a value of 1 in the positive case and 0 in the negative case. The binary variable *nearness of language_{ij}*, acquiring a value of 1 in the case of the same language, otherwise 0, was considered for quantification of the impact of the nearness of the language on the volume of cross-border assets.

The goal was to estimate the weights of the considered predictors on the total value of assets purchased through mergers and acquisitions $M\&A_{ij,s,t}$ by a source country i in a target country j in sector s at time t. The regression coefficient equations are estimated in the form:

$$
\begin{aligned}
log\,(M\,\&\,A_{ij,s,t}) \\
= \beta_0 + \beta_1 log(\text{GDP}_{i,s,t}\text{GDP}_{j,s,t}) \\
+ \beta_2 log\,(market\ capitalization/GDP_{j,s,t}) \\
+ \beta_3 log\,(distance_{ij}) + \beta_4(border_{ij}) \\
+ \beta_5(nearness\ of\ language_{ij}) \\
+ \beta_6(EU_{i,t}EU_{j,t}) + \beta_7(NonEU_{i,t}EU_{j,t}) \\
+ \beta_8(EMU_{i,t}EMU_{j,t}) \\
+ \beta_9(NonEMU_{i,t}EMU_{j,t}) \\
+ \beta_{10}(civil\ liberty_{i,t}) + \beta_{10}(civil\ liberty_{j,t})
\end{aligned}
\tag{1}
$$

Another considered predictor in the given equation is the variable *civil liberty_{it}*, which was quantified using the Civil Liberties Index and assesses the quality of institutions in the source countries (i) in the individual years t. The scale of the mentioned index is from 1 (the best country) to 7 (the worst country), whereby in the analyzed database it acquires a maximum value of 3. Similarly, the variable *civil liberty_{jt}* of the target country (j) at time t acquired a maximum value of 5 in the database. Other predictors of the considered model are dummy variables relating to membership of the source and target countries in the European Union and the European Monetary Union, specifically $EU_{i,t}EU_{j,t}$ acquires a value of 1 if the source country i as well as the target country j were both members of the European Union at time

t, otherwise, a value of 0; $NonEU_{i,t}EU_{j,t}$ acquires a value of 1 if the source country (i) was not and the target country (j) was a member of the European Union at time t, otherwise a value of 0; $EMU_{i,t}EMU_{j,t}$ acquires a value of 1 if the source country i as well as the target country j were members of the European Monetary Union at time t, otherwise a value of 0; and $NonEMU_{i,t}EMU_{j,t}$ acquires a value of 1 if the source country i was not and the target country j was a member of the European Monetary Union at time t, otherwise a value of 0.

3 RESULTS

The results of the regression are presented by the total value R, R^2 and the result of an ANOVA test, by the values of the regression coefficients and the standardized regression coefficients in Table 1.

For identification of the relative importance of significant predictors, standardized β coefficients were monitored. Among the eleven variables a significant contribution to the model under consideration appeared in eight of them. The ratio of *market capitalization to GDP* in the target country contributed most positively to the significance of the model. One-percent growth in the ratio of *market capitalization* and *GDP* is associated with 0.88% growth in the volume of cross-border mergers and acquisitions. Second in line is the product of *GDP* of the source (i) and the target (j) countries in the sector s $(GDP_{ist}GDP_{jst})$, a 1% growth in which causes 0.52% growth in the volume of cross-border mergers and acquisitions. From the viewpoint of importance of standardized coefficients $EU_{it}EU_{jt}$ then follows, where membership of both countries in the EU contributes to an increase in the volume of cross-border mergers and acquisitions by 175%.

The lowering of the *quality of institutions* contributes positively to the model in the target country and negatively in the source country. If a worsening of the quality of institutions in the target country occurs, then with other predictors unchanged, the average amount of growth of the logarithm of the volume of cross-border mergers and acquisitions will be 0.42 units. If, however, under equal conditions the civil liberty$_{it}$ unit accelerates, i.e. the quality of institutions worsens in the source country, then the result is a drop in the expected value of the dependent variable by 0.40 units. Membership of both countries in the European Monetary Union with the predictor $EMU_{it}EMU_{jt}$ influences an increase in the volume of cross-border mergers and acquisitions by 32%. With the predictor $NonEU_{it}EU_{jt}$ expressing membership of only the target country in the EU, the regression coefficient achieves a value of 2.556 (very high), but the highest is the standard error, which lowers the credibility of the regression coefficient in terms of interpretation. The variable *distance of capital cities of source and target countries* contributes negatively to the model, with 1% growth resulting in a 0.15% drop in the volume of cross-border mergers and acquisitions.

Subsequently, special regression analyses were carried out in the subsets of the original database for the services sector and the manufacturing sector, during which the same model was used as in the case of the entire database. The results of the regression for the services sector, presented by the total value of R, R^2 and the result of the ANOVA test, and by the values of regression coefficients and standardized regression coefficients are presented in Table 2.

The above-presented model applied to the services sector provided a similar result as it did for the entire set. Complete agreement occurs in the sign

Table 1. Regression model with volume of cross-border mergers and acquisitions as the dependent variable.

Predictor	Coefficient β	Standard error	Standardized coefficient β
(Constant)	−.636***	.180	
$log(GDP_{ist}GDP_{jst})$	.522***	.006	.540
$log(market\ capitalization/GDP_{jst})$	.881***	.005	1.042
$log(distance_{ij})$	−.151***	.020	−.046
$Border_{ij}$	−.016	.035	−.003
$Nearness\ of\ language_{ij}$	.022	.054	.002
$EU_{it}\ EU_{jt}$	1.012***	.058	.125
$NonEU_{it}\ EU_{jt}$	2.556***	.254	.054
$EMU_{it}\ EMU_{jt}$	.281***	.036	.060
$NonEMU_{it}\ EMU_{jt}$	−.032	.035	−.006
$Civil\ liberty_{it}$	−.403***	.031	−.078
$Civil\ liberty_{jt}$	.424***	.029	.104

Note: A generalized regression model was used. Total result of the regression: R = .940, R2 = .883, ANOVA test: p = .000. Statistical significance at a 10% (resp. 5% and 1%) level of significance is labelled with * (resp. ** and ***). The number of observations is 4,285.

of the regression coefficients and also in the case of collinearity. Here, however, the relative importance of the predictor mapping geographic distance of the source and target countries accelerated, with 1% growth causing a 0.2% decline in the expected value of the volume of cross-border mergers and acquisitions. The importance of the predictor EU_{it} EU_{jt} is lower than in the entire set; the effect of membership of both countries in the EU increases the volume of cross-border mergers and acquisitions in the services sector by 135%. The predictor *nearness of language$_{ij}$*, which is not significant in the entire set, is significant in the services sector at a level of significance of 0.01. From the viewpoint

of its relative importance it is in the last position among the nine significant predictors of the model.

The results of the regression for the manufacturing sector, presented by the total value R, R^2 and the result of the ANOVA test, values of regression coefficients and standardized regression coefficients are presented in Table 3.

Application of the above-presented model on the manufacturing sector differs from the results in the entire set more than in the case of the services sector. Predictors which are in the first seven positions from the viewpoint of relative importance remained in the same order and with the same sign as in the entire set, i.e. *log(market capitalization/GDP$_{jst}$)*,

Table 2. Regression model with the volume of cross-border mergers and acquisitions for the services sector as the dependent variable.

Predictor	Coefficient β	Standard error	Standardized coefficient β
(*Constant*)	−.968***	.224	
log(GDP$_{ist}$GDP$_{jst}$)	.562***	.008	.591
log(market capitalization/GDP$_{jst}$)	.873***	.006	1.081
log(distance$_{ij}$)	−.205***	.024	−.064
Border$_{ij}$	−.058	.045	−.011
Nearness of language$_{ij}$	.253***	.065	.028
EU$_{it}$ EU$_{jt}$	.847***	.076	.099
NonEU$_{it}$ EU$_{jt}$	2.626***	.323	.055
EMU$_{it}$ EMU$_{jt}$	.286***	.044	.063
NonEMU$_{it}$ EMU$_{jt}$	−.086	.042	−.017
Civil liberty$_{it}$	−.440***	.038	−.088
Civil liberty$_{jt}$	.445***	.035	.110

Note: A generalized regression model was used. Total result of the regression: R = .939, R^2 = .882, ANOVA test: p = .000. Statistical significance at a 10% (resp. 5% and 1%) level of significance is labelled with * (resp. ** and ***). The number of observations in the service sector is 2,735.

Table 3. Regression model with the volume of cross-border mergers and acquisitions for the manufacturing sector as the dependent variable.

Predictor	Coefficient β	Standard error	Standardized coefficient β
(*Constant*)	−.805***	.295	
log(GDP$_{ist}$GDP$_{jst}$)	.498***	.010	.468
log(market capitalization/GDP$_{jst}$)	.905***	.008	.964
log(distance$_{ij}$)	−.065*	.034	−.019
Border$_{ij}$	.028	.056	.005
Nearness of language$_{ij}$	−.204**	.101	−.018
EU$_{it}$ EU$_{jt}$	1.151***	.087	.153
NonEU$_{it}$ EU$_{jt}$	2.516***	.393	.054
EMU$_{it}$ EMU$_{jt}$	.259***	.057	.054
NonEMU$_{it}$ EMU$_{jt}$	.109*	.061	.018
Civil liberty$_{it}$	−.365***	.051	−.067
Civil liberty$_{jt}$	.402***	.047	.099

Note: A generalized regression model was used. Total result of the regression: R = .947, R^2 = .897, ANOVA test: p = .000. Statistical significance at a 10% (resp. 5% and 1%) level of significance is labelled with * (resp. ** and ***). The number of observations in the manufacturing sector is 1,550.

$log(GDP_{ist} GDP_{jst})$, $EU_{it} EU_{jt}$, Civil liberty$_{jt}$, Civil liberty$_{it}$, $EMU_{it} EMU_{jt}$ and Non $EU_{it} EU_{jt}$. The ratio of *market capitalization toward GDP of the target country*, where 1% growth of the considered ratio causes 0.9% growth in the volume of cross-border mergers and acquisitions, contributes most positively to the model. Further, 1% growth of the product of *GDP* of the source (i) and target (j) countries in the sector s ($GDP_{ist} GDP_{jst}$) is associated with 0.5% growth in the mentioned volume. The effect of membership of both countries in the EU on the volume of cross-border mergers and acquisitions is in this case significant, achieving up to 216%. The 30% impact of membership of both countries in the EMU ($EMU_{it} EMU_{jt}$) is also significant. The lowering of the quality of institutions in the source country is expressed by a lowering of the volume of cross-border mergers and acquisitions, while in the target country its impact is the opposite. The relative importance of the predictor mapping the distance of the source and target countries $log(distance_{ij})$ is smaller than in the entire set, as in the services sector, which is also expressed in the level of significance equal to 0.1. Its 1% growth causes only a 0.06% drop in the volume of cross-border mergers and acquisitions. The predictor *nearness of language$_{ij}$*, which is not significant in the entire set, is significant in the manufacturing sector at a level of significance of 0.05; however, with a negative sign, in contrast to the services sector, where the significance was positive at a level of significance of 0.01. From the viewpoint of its relative importance it is one of the last positions among the ten significant predictors of the model. The predictor *NonEMU$_{it}$ EMU$_{jt}$* is significant at a level of significance of 0.1 solely in the manufacturing sector.

4 CONCLUSION

Empirical results of several foreign studies (for example Bjortvan, 2004, Brouwer, J. et al., 2008, Coeurdacier, N. et al., 2009, Neary, P., 2007 and others) indicate that European integration positively influences the development of cross-border mergers and acquisitions primarily in the manufacturing sector. We arrived at this same conclusion on the basis of our analysis, whereby integration into the EU has a crucial impact which is significantly higher in the manufacturing sector. The impact of integration to the EMU can be considered as supplementary and is the same for the manufacturing and the service sectors. We can with certainty state that institutional changes, such as the unified market of the European Union and the European Monetary Union, were initiating factors for the relocation of capital in the manufacturing sector in the entire world. The impact of adoption of the single euro currency is also a very strong determinant for the carrying out of cross-border mergers and acquisitions, primarily in this same manufacturing sector. The impact of the adoption of the euro on cross-border mergers and acquisitions was shown to be equally important in manufacturing sectors primarily from countries not belonging to the euro area to countries within the euro area. The European Monetary Union, by means of effects of unilateral financial liberalization (the lowering of marginal and fixed costs for carrying out relocation transactions), supported the relocation of capital and thus eased the carrying out of cross-border mergers and acquisitions in the European area.

REFERENCES

Anderson, D.R., Sweeney, D.J. & Williams, T.A. 1993. *Statistics for Business and Economics*. St. Paul: West Publishing Company.

Bjortvan, K. 2004. Economic integration and the profitability of cross-border mergers and acquisitions. *European Economic Review* 48(6): 1211–1226.

Brouwer, J., Paap, R. & Viaene, J.M. 2008. The trade and FDI effects of EMU enlargement. *Journal of International Money and Finance* 27(2): 188–208.

Bureau van Dijk. 2013. *Zephyr*. (zakúpené údaje z databázy Zephyr za obdobie rokov 1998–2012). http://www.bvdinfo.com/en-gb/our-products/economic-and-m-a/m-a-data/zephyr.

Couerdacier, N., De Santis, R.A. & Aviat, A. 2009. Cross-border mergers and acquisitions and European integration. *Economic Policy*, January 2009, pp. 55–106. ISSN 14680327.

European Commission: http://epp.eurostat.ec.europa.eu/portal/page/portal/eurostat/home/.

Freedom House: https://freedomhouse.org/reports#.VRlLaY4oEQk.

Hečková, J., Chapčáková, A. & Badida, P. 2014. Aktuálne problémy ohodnocovania podnikov pri fúziách a akvizíciách a ich riešenie. *Ekonomický časopis* 62(7): 743–766. http://www.infoplease.com/atlas/calculate-distance.html.

Jovanovic, B., Rousseau, P.L. 2002. The q-theory of mergers. *American Economic Review* 92(2): 198–204.

Long, N.V., Horst, R. & Stähler, F. 2007. The effects of trade liberalization on productivity and welfare: The role of firm heterogeneity, R&D and market structure. *Kiel Economics Working Paper*, No. 2007(20).

Neary, P. 2007. Cross-border mergers as instruments of comparative advantage. *Review of Economic Studies* 74(4): 1229–1257.

OECD. 2002. *Forty Years' Experience with the OECD Code of Liberalisation of Capital Movements* [online]. Paris: OECD. http://www.oecd.org/investment/investment-policy/44784048.pdf. ISBN 92-64-17612-8 No. 525632002.

Petroulas, P. 2007. The effect of the euro on foreign direct investment. *European Economic Review* 51: 1468–1491.

Schiavo, S. 2007. Common currency and FDI flows. *Oxford Economic Papers* 59(3): 536–560.

Production Management and Engineering Sciences – Majerník, Daneshjo & Bosák (Eds)
© 2016 Taylor & Francis Group, London, ISBN: 978-1-138-02856-2

CFD simulation in process of axial turbo machine design

P. Hlbocan & M. Varchola
Faculty of Mechanical Engineering STU in Bratislava, Bratislava, Slovak Republic

ABSTRACT: In the paper, it is presented an algorithm of the axial hydrodynamic turbo machine which utilizes classical methods and CFD simulation as well. CFD simulation provides a feedback during the hydraulic design process from a point of view of an obtained efficiency and other performance parameters. The classical methods utilize experimentally acquired guiding parameters in a meridional cut design. A blade cuts' design utilizes an ala-nasy theory, aerodynamic characteristics of NACA profiles and Voznesensky method. The hydraulic design algorithm was fully implemented in a design software. The design software provides a data flow from the design software environment into a CAD software and CFD software environment as well. Potential corrections of a machine's geometry are realized based on the CFD simulation due to an improvement of required performance parameters. The design software is interactively interfaced with a parameterized 3D CAD axial machine model.

1 APPROACH TO HYDRAULIC DESIGN

Several approaches to the hydraulic design of the axial impeller are accessible in literary sources and in a technical praxis as well (Vrábel 2014, Lobanof & Ross 1985, Varchola et al. 2013). The impellers' hydraulic design was dependent on approaches based on one and two dimensional methods especially before the high quality simulating software took place in the engineering praxis. Dimensions and shapes of the meridional cut are proposed by using experimental knowledge which was generalized in the form of guiding parameters when using one and two dimensional design methods. An impeller meridional velocity and impeller main dimensions are proposed based on the guiding parameters. The ala-nasy theory and well known NACA profiles' characteristic (experimentally acquired) are utilized in the blade cuts' design (Brůha et al. 1939). Voznesensky method takes into account differences between the NACA profile aerodynamic characteristic which were acquired from a single profile in a wind tunnel and characteristics of a profile placed in a blade cascade Strýček (1994).

Additional geometry modifications usually necessary for improving the performance parameters (e.g. a shift of an optimum working point, improving the efficiency, cavitation capability etc.) were realized in the past exclusively on the basis of results obtained by systematic experimental research on model machines. One and two dimensional calculation approaches provide a geometry which is basically necessary to improve. These additional improvements are necessary because

of simplifications employed in the calculation methods. All simplifications utilized in the calculation models bring systematic errors into the calculations.

Despite the mentioned deficiencies, the one and two dimensional methods are still in use especially because of their simplicity and ability to provide the explicit geometry solution of the particular hydraulic components Strýček (1994).

The geometry designed by one and two dimensional methods is called prime (initial or preliminary) geometry. Certainly also nowadays it is necessary to correct the prime geometry for the sake of the obtaining more precise solution.

Besides the classical geometry correction approach which utilizes the results of the experimental research on the model machine, it is possible to employ the numerical flow simulation or rather Computational Fluid Dynamics (CFD) tools. The performance parameters of the machine (specific energy, efficiency etc.) are assessed by means of the CFD results in the next step. An internal flow pattern analysis is utilized as well when considering the potential geometry modifications for the sake of improving hydraulic parameters of the machine. After that, the modified geometry of hydraulic components is analyzed by CFD methods and subsequent hydraulic parameters evaluation again.

It is possible to make a set of additional geometry modifications and subsequent CFD simulations for the sake of geometry improvement.

Repeated CFD analysis represents a powerful tool which provides the feedback during the hydraulic design process using one and two dimensional methods. The experimental research on the model machine is still an important element in the explained design system. The main sense of the model tests lies on a CFD results validation or realizing advanced CFD model settings.

2 PRELIMINARY HYDRAULIC DESIGN OF THE AXIAL IMPELLER

Each hydrodynamic machine is in principle reversible and it is capable to work in a pump or in a turbine operating mode (Varchola et al. 1995). Of course the performance parameters in the optimal operating point of the machine are different in the pump and in the turbine operating mode. The ratio between the nominal flow-rate or nominal specific energy of the machine in the pump and in the turbine operating mode varies from one machine to another. Despite that it is possible to observe certain tendencies which determine the ratio according to the obtained efficiencies and specific speed as well. The comparison of the characteristics (Q-Y, Q-h) of the hydrodynamic machine in the both operating modes is in the Figure 1.

Assuming it is possible to successfully assess the certain values of the ratio between the nominal parameters of the machine in the pump and in the turbine operating mode, it is possible to realize calculation of the proposal parameters of the turbine into the proposal parameters of the same machine operating in the pump mode. After that, we can successfully design the turbine's geometry by methods which are primarily intended to design a pump's geometry.

Hydraulic design of an axial impeller is relatively complex task. It consists of meridional cut design and design of several blade cuts which determine the blade shapes. The blade cuts are usually designed on cylindrical or conical surfaces.

2.1 Design of a meridional cut

The meridional cut design is realized on the basis of the hypothesis of a constant meridional velocity assumed in the impeller's passages (except the impeller with the conical hub shape). The guiding parameter K_m (2) which contains implicit generalizations of the systematic experimental research focused of improving efficiency is utilized in the meridional velocity calculation. The meridional cut main dimensions are depicted in the Figure 2. Stated generalizations employed in the meridional cut design are assumed as functions of the specific speed coefficient n_b (1).

$$n_b = n \cdot \frac{\sqrt{Q}}{Y^{0.75}} \tag{1}$$

$$K_m = 0.0688 + 0.733 \cdot n_b^{1.1} \tag{2}$$

$$c_m = K_m \cdot \sqrt{2 \cdot Y} \tag{3}$$

$$D_2 = D_A = \sqrt{\frac{4}{\pi} \cdot \frac{1.04 \cdot Q}{c_m} \cdot \frac{1}{1 - \left(\frac{D_B}{D_A}\right)^2}} \tag{4}$$

$$\frac{d_n}{D} = \frac{D_B}{D_A} \doteq 0.63 - 0.346 \cdot (n_b - 0.25) \tag{5}$$

2.2 Preliminary design of the blade cuts

Hydraulic design of the blade profiles is usually realized on several cylindrical surfaces which are spaced between the hub and the shroud of the impeller. It means that between each two surfaces the same mass flows (Fig. 2). During the calculation, experimental results obtained by means of the systematic research on model single profiles in the wind tunnel are utilized.

In principle the main effort is to design the blade profile geometry in order to achieve that a profile placed in a blade cascade provides the required circulation. Between circulation of the one blade and

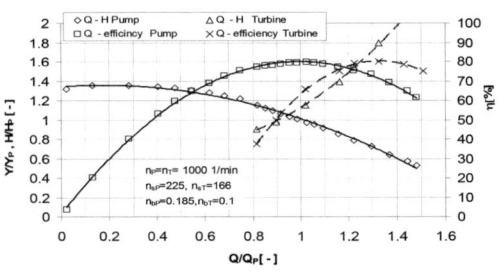

Figure 1. Performance characteristics of a hydrodynamic machine in a pump and in a turbine operating mode—Varchola & Sikhart (2010).

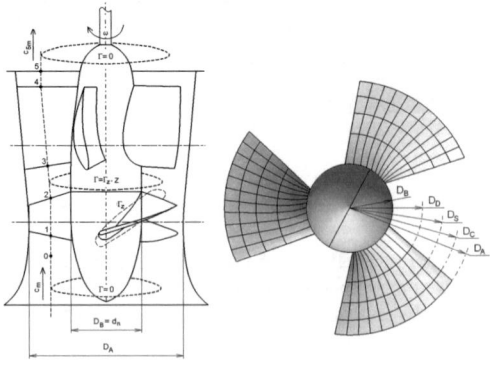

Figure 2. Meridional cut, impeller blades and cylindrical surfaces for blade profile calculation—Bielik (2010).

384

aerodynamic parameters of the particular profile are direct relations which are used in the design process. Similarly, the geometry parameters of the profile (including blade cascade lean angle) (Fig. 5) are crucial from the point of view of particular values of the aerodynamic parameters.

For the circulation of the one profile stands: $\Gamma = \Gamma_z / z$. Parameter z represents blade number. Blade number choice depends on machine's specific speeds. According to the blade count and specific speeds, it is necessary to choose suitable blade span angle of the profile ϕ_r (Fig. 5). For this purpose a nomogram might be implemented—Vrábel (2014).

Blade span angle must be chosen in the way that its value lies between upper and lower bounding curve.

In the Figures 3–4 are depicted velocity triangles, kinematic parameters and geometrical parameters of the impellers blade cascade which are significant from the point of view of attained parameters of the machine. The main task is to propose these parameters and some other geometrical parameters as well in the way that the required machine parameters would be achieved.

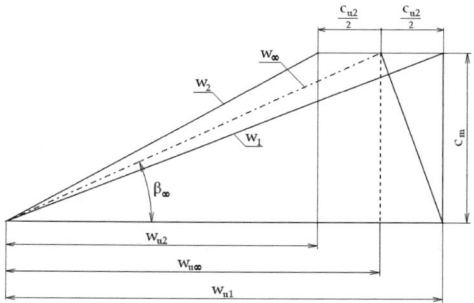

Figure 3. Velocity triangles of the axial impeller (in a pump mode)—Bielik (2010).

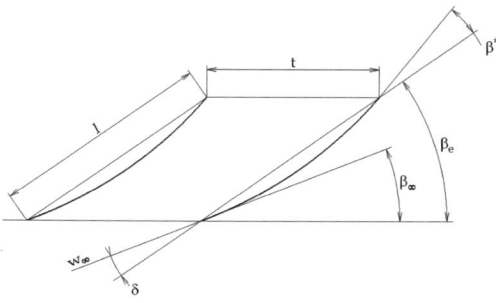

Figure 4. Blade cascade of the axial impeller (in a pump mode)—Bielik (2010).

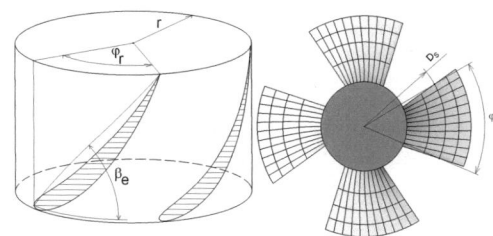

Figure 5. View on the cylindrical surface of the blade cut—Bielik (2010).

The technique is as follows. The peripheral component of mean geometrical relative velocity $w_{u\infty}$ (6) is stated at first. Then lean angle of mean geometrical relative velocity b_∞ (7) and blade cascade angle b_e (8) are stated. The angle of attack δ is considered to be zero in the first approach.

$$w_{u\infty} = u - \frac{c_{u2}}{2} = u - \frac{Y}{2\pi \cdot D \cdot n \cdot \eta_H} \quad (6)$$

$$\beta_\infty = \operatorname{arctg} \frac{c_m}{w_{u\infty}} \quad (7)$$

$$\beta_e = \beta_\infty + \delta \quad (8)$$

Then the profile chord length l (9), relative pitch t/l (10) and coefficient of the blade cascade impact ξ (11) are stated.

$$l = \frac{\phi_r \dfrac{D}{2}}{\cos \beta_e} \quad (9)$$

$$\frac{t}{l} = \frac{\pi \cdot D}{z \cdot l} \quad (10)$$

$$\xi = \frac{(c_z)_1}{(c_z)} \quad (11)$$

Coefficient of the blade cascade impact ξ is possible to assess according to the diagram in the Figure 7. The ξ value depends on relative pitch t/l and angle of cascade's inclination b_e. It is one of the most important coefficients which determines a ratio between the lift coefficients of the single profile in the wind tunnel $(c_z)_1$ (14) and in the blade cascade c_z. We have to assess this ratio. The purpose is that the experimentally assessed NACA profile values of (c_z) were acquired on the single profile in the wind tunnel. Therefore the lift coefficient of the profile in the blade cascade must be adjusted to the value of the geometrically same single profile. Besides that, we have to take into account the difference between the lift coefficients of the single profile and the profile in the blade cascade because the profile in the blade cascade is possible to consider as a profile of the infinite span. This difference is compensated by change of the blade attack angle $\Delta\delta$. The blade attack angle dependency is depicted in the Figure 7. Its value is dependent on the relative pitch t/l and the blade deflection angle β^* (13) (Fig. 6). For the mean geometrical velocity w_∞ stands (12).

$$w_\infty = \sqrt{c_m^2 + w_{u\infty}^2} \qquad (12)$$

$$\beta^* = \frac{\Gamma}{\xi . w_\infty . l} \qquad (13)$$

$$(c_z)_1 = \frac{2 \cdot \Gamma}{l \cdot w_\infty} \cdot \frac{1.6}{\xi} \qquad (14)$$

It is necessary to perform accuracy boosting of the lift coefficient $(c_z)_1$, by means of taking into account of $\Delta\delta$ value. It means that the $\Delta\delta$ value have to be taken into account when calculating angle of cascade's inclination b_e (Fig. 7). Angle b_e is calculated according to (15) and then equations (9) to (14) are calculated again.

$$\beta_e = \beta_\infty + \delta + \Delta\delta \qquad (15)$$

Then it is possible to compare the lift coefficient value $(c_z)_1$ with experimentally acquired lift coeffcient value $(c_z)_1$ of NACA profile with the

proposed chord length l, relative pitch t/l, blade attack angle δ and with a proposed maximum profile thickness t*. Then all of the parameters (16) to (21) are updated.

For the NACA profile lift coefficient (c_z) stands (16). The chosen parameters are relative maximum thickness t*/l and relative length of the maximum profile bending L*/l. Using these parameters and using calculated parameters of the maximum relative bending of the profile as well, the NACA profile topology can be calculated.

$$(c_z) = \frac{\partial c_z}{\partial \delta} \cdot (\delta_\lambda - \delta_0) \qquad (16)$$

$$\frac{\partial c_z}{\partial \delta} = 0.079 - 0.037 \cdot \frac{t^*}{l} \qquad (17)$$

$$\delta_0^\infty = -83.3 \cdot \frac{m}{l} \cdot \left(0.74 + \frac{L^*}{l}\right) \qquad (18)$$

$$\frac{m}{l} = \frac{\sin^2\left(\frac{\beta^*}{2}\right)}{\sin\beta^*} - \frac{1}{20} \cdot \text{tg}\frac{\beta^*}{5} \qquad (19)$$

$$\delta_y = \delta_i + \Delta\delta \qquad (20)$$

$$\delta_i = 57.3 \cdot \frac{(c_z)_1}{\pi \cdot \lambda} \qquad (21)$$

The attack angle has to be iterated until rough coincidence between values of c_z and $(c_z)_1$ is reached.

The geometry or the main geometrical parameters of the profile is depicted in the Figure 8. The coordinates of the pressure and suction side of the profile are marked in the Figure 8. The camber line consists of two parabolic curves which are analytically expressed through the equations (22) and (23).

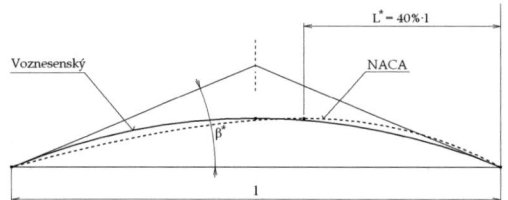

Figure 6. Camber line of the NACA profile in the comparison with the circle arc according to Voznesensky—Bielik (2010).

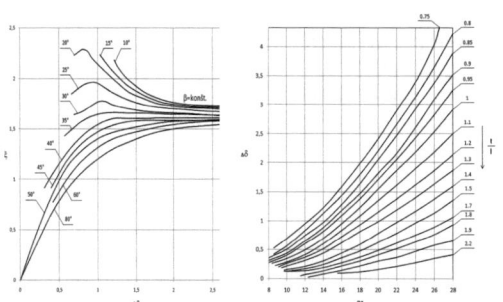

Figure 7. Dependencies of the blade cascade influence coefficient ξ and change of the blade attack angle $\Delta\delta$—Bielik (2010).

$$y_s = \frac{\frac{m}{l}}{\left(\frac{L^*}{l}\right)^2} \cdot \left[2 . \frac{L^*}{l} . x - x^2\right] \qquad (22)$$

$$y_S = \frac{\frac{m}{l}}{\left(1 - \frac{L^*}{l}\right)^2} \cdot \left[1 - 2 \cdot \frac{L^*}{l} + 2 \cdot \frac{L^*}{l} \cdot x - x^2\right]$$

$$\qquad (23)$$

$$y_t = \frac{\frac{t^*}{l}}{0.2} \cdot (0.2969 \cdot \sqrt{x} - 0.126 \cdot x$$
$$- 0.3516 \cdot x^2 + 0.2843 \cdot x^3$$
$$- 0.1015 \cdot x^4) \qquad (24)$$

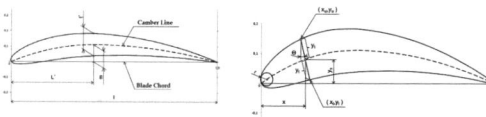

Figure 8. Main dimensions of the NACA profile and the technique of the coordinates definition of the pressure and suction profile's side—Bielik (2010).

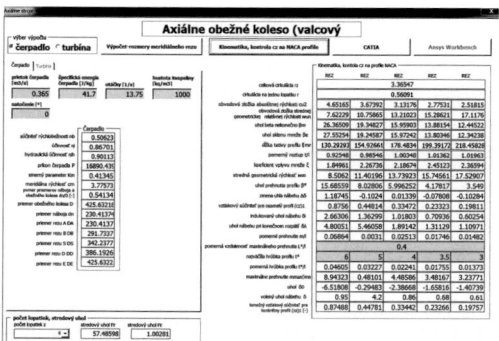

Figure 9. The example of the axial impeller geometry calculation. Q = 0,365 m³/s, Y = 41,7 J/kg, n = 825 min⁻¹. Hydraulic solution by using the own design software—Vrábel (2014).

$$x_u = x - y_t \cdot \sin\Theta \qquad (25)$$

$$y_u = y_s + y_t \cdot \cos\Theta \qquad (26)$$

$$x_l = x + y_t \cdot \sin\Theta \qquad (27)$$

$$y_l = y_s - y_t \cdot \cos\Theta \qquad (28)$$

$$\Theta = arctan\left(\frac{dy_s}{dx}\right) \qquad (29)$$

$$r = 1.1 \cdot l \cdot \left(\frac{t*}{l}\right)^2 \qquad (30)$$

3 DESIGN SOFTWARE

The algorithm presented in the previous subchapters was implemented by using an own (native) calculation procedures (developed in an environment of the VBA—Visual Basic for Application programing language). Development environment of the language VBA is an integral add-on module of the CAD software CATIA V5. In the CATIA V5 environment was created a referential parametric 3D CAD model of the hydraulic components of the axial machine. All the calculation parameters of the axial machine and parameters containing information related to proposed geometry are represented through member variables of

the design software. Integrated COM software elements ("Component Object Model") and OLE elements ("Object Linking and Embedding") provide ability to access and edit objects and parameters of CATIA V5 software using VBA scripts containing all the calculation procedures. So, one of the features of the design software is the ability to edit 3D CAD model of the hydraulic components of the axial machine directly. The generic 3D model is parameterized by means of several governing parameters. These parameters are: diameters of cylindrical surfaces (the blade profiles are defined on these surfaces), blade number and parameters of each blade profile. Blade profile parameters are: blade cascade lean angle, chord length, maximum profile thickness, maximum profile bending and the relative distance of the location with the maximum profile bending from the leading edge.

The profile shape itself (pressure and suction side) is created using special objects of the CAD model which enable to define curve courses by means of an equation with one independent variable. In this case this equation represents analytical form of the NACA profile. Parameters of the equation are t/l, t*, m/l, L*/l. The independent variable in this equation is the relative chord length.

Besides the 3D model creation using the classical CAD/CAM software (CATIA V5), the design software provides also the elements to transform the preliminary solution of machine's geometry into CFD software environment. The transformation is provided using the results of the hydraulic design into the Excel file and subsequent run of the application ANSYS WB. Data flow is provided using storing the calculated data in a special Excel file and subsequent automated run of the application ANSYS WB. Run of the ANSYS WB application is directly managed by the design software. After executing of the application ANSYS WB "journal file" (extension "wbjn") is initiated automatically.

Mentioned "journal file" is a macro written using syntax of the programing language Python. It manages execution of the referential parameterized file of the software ANSYS WB, loading the parameters from the Excel file and updating of the parameters of the referential file ANSYS WB.

Figure 10. Impeller in CATIA V5 software (in the left) and impeller blades in ANSYS BladeGen environment (in the right)—Vrábel (2014).

The results of the data flow are represented by a blade model of the axial impeller which is created using the parameters loaded from the Excel file.

We can see 3D blade model created in the ANSYS Design Modeler in the Figure 10 (its right side).

4 CFD SIMULATION

It is necessary to emphasize that for the sake of the high quality CFD calculations which are suitable to correct Q-Y and Q-h curves estimation, it is necessary to perform the CFD simulation on the computational domain which comprises an inner suction duct volume, impeller flow passages, diffuser flow passages and a volume of a discharge pipe as well. Resultant performance parameters are significantly dependent on the reciprocal interaction between the rotor and the stator. The machine's geometry is capable to influence the rotor-stator interaction significantly.

For the sake of the optimal flow-rate, specific energy and efficiency estimation, the CFD simulation on domain containing only of the one rotor flow passage is employed for several reasons. The rotational periodicity is employed in this model (Fig. 11). The reciprocal interaction between the rotor and the suction duct (or diffuser) is not taken into the account. The results of this CFD analysis are ultimately more or less different from the results obtained by CFD simulation in the complete computational domain. In principle the resultant specific energy is higher when using the reduced computational domain because there were not assumed the energy losses in the stator parts during the calculation. Obviously, the resultant hydraulic efficiency is higher as well due to the same reasons.

It is also necessary to emphasize that the hydraulic design of the diffuser must be performed in the way that no unacceptable shift of the machine's optimal flow-rate occurs due to the inconvenient rotor-stator interaction. Then, in case we want to employ CFD results performed on the reduced computational domain as the feedback to the axial impeller hydraulic design, we have to subtract loss specific energy due to the dissipation in the stator components. The hydraulic efficiency is necessary to decrease as well. The value of the loss of specific energy is not known in the first approach, so we have to make an appropriate guess at the beginning.

The computational domain scheme is depicted in the Figure 11.

The mass flow rate is a boundary condition in the inlet of the computational domain. The static pressure is the boundary condition in the outlet. Besides these two main boundary conditions, rotating speeds and spin direction are defined as boundary condition as well.

CFD calculations are performed assuming the stationary flow model. Shear Stress Transport (SST) turbulence model with an automatic near wall treatment is implemented as well. CFD calculations are realized assuming several different mass flow rates, and the values of the specific energy and the efficiency are estimated for each value of the mass flow rate. The position of a best efficiency point and rotor's performance parameters are estimated according to CFD results.

The hydraulic design of the axial pump impeller for the requested parameters $Q = 0{,}365$ m^3/s, $Y = 41{,}7$ J/kg, $n = 825$ min^{-1} was realized using the explained technique as an example. The results are summarized in the GUI shortcut of the design software depicted in the Figure 9. Blade geometry of the designed impeller is depicted in the Figure 10. CFD simulation was performed on the impeller in the next step. Its results are shown in the Table 1. We can remark that the optimum flow rate is shifted to the left and the specific energy is too high. By modification of the subsequent geometry (1° decreasing of the blade cascade lean angle), the shift of the optimum flow rate and the decrease of the specific energy were achieved. The specific energy remained still too high but it is necessary to consider neglecting the additional losses in the stator components which have significant influence on the ultimate specific energy value.

In case the impeller fulfills the required parameters, it is possible to perform CFD simulation in the complete computational domain employing advanced flow models (transient flow, two phase

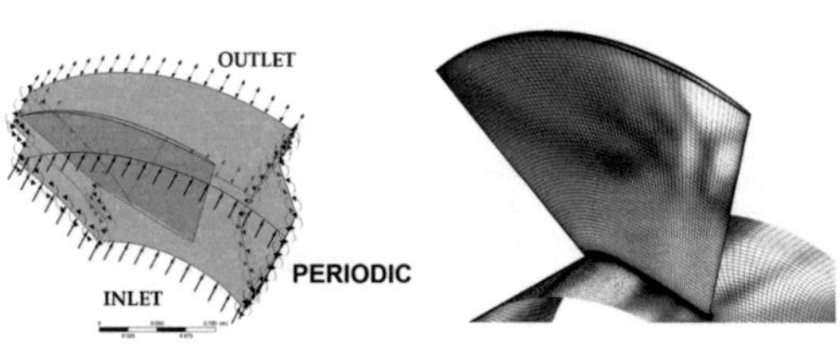

Figure 11. Computational domain and computational mesh of the axial pump impeller.

Table 1. The comparison between the performance parameters of the axial impeller—preliminary design and modified geometry (1° decreasing of the blade cascade lean angle).

	Q_n [m$_3$/s]	Y_n [J/kg]	h[%]
Preliminary design	0,39	52	89
Decrease b_e −1°	0,365	50,9	89

model etc.). Such a CFD simulation is capable to predict the performance parameters of the machine with a good precision. Using the results of the stated CFD model, it is possible to perform ultimate geometry modifications of the impeller and the diffuser as well (potentially the suction or the discharge duct).

5 CONCLUSION

The explained hydraulic solution technique of the axial pump impeller represents the automated software system which uses the procedures based on the classical approach to the impellers' hydraulic design and the tools of the advanced CFD simulation as well. The connection between the design software and CAD software (parameterized 3D CAD model) is the important element of the mentioned system. This element allows the direct integration of the hydraulic solution with the constructional processing of the machine in the CAD environment. It provides the direct connection to the CAM software. CFD simulation provides the operative tool to realize the necessary geometry modifications of the machine.

REFERENCES

Bláha, J. & Brada, K.. 1992. *Hydraulic Machines.* (in Czech *Hydraulické stroje.*) Prague: SNTL.
Brůha, O., Felber, V. & Smolař, V. 1939. *Aircraft Handbook.* (in Czech *Letecký Průvodce.*) Prague: Matice česká. http://people.clarkson.edu/~pmarzocc/AE429/The%20NACA%20airfoil%20series.pdf.

Lewis, R.I. 1996. *Turbomachinery Performance Analysis*, Oxford: Butterworth-Heinemann.
Lobanof, V.S. & Ross, R. R. 1985. *Centrifugal Pumps—Design and Application.* Houston: Gulf Publishing Company.
Nechleba, M. 1962. *Hydraulic Turbines Mechanical Design and Equipment.* (in Czech *Vodní Turbíny jejich konstrukce a príslušenství.*) Prague: SNTL.
Strýček, O. 1994. *Hydrodynamic Pumps.* (in Slovak *Hydrodynamické čerpadlá.*) Bratislava: SjF STU.
Varchola, M., Bielik, T. & Hlbocan, P. 2013. Methodology of 3D hydraulic design of a impeller of axial turbo machine. In Engineering Mechanics. *Vol. 20, No. 2. Pages 107–118.* Prague: Association for Engineering Mechanics.
Varchola M., Gančo M., Strýček O. & Sikhart R. 1995. *Research on Mixed-Flow Pump 150 BQZ in a Turbine Mode* (in Slovak *Výskum diagonálneho čerpadla 150 BQZ v turbínovej prevádzke.*) Bratislava: SjF STU.
Varchola M. & Sikhart R. 2010. Comparisson of Characteristics of Mixed-Flow Machine in Pump and in Turbine Operation Mode. (in Slovak Porovnanie charakteristík diagonálneho stroja v čerpadlovej a turbínovej prevádzke.) *HYDROTURBO 2010: International conference; Proceedings, Patince, 21.-23.92010. pages: 27–34.* Bratislava: Sjf STU
Vrábel, M. 2014. *Hydraulic Solution of Axial Machine with Output for CFD software.* (In Slovak *Hydraulický návrh axiálneho stroja s výstupom pre CFD softvér.*) Bratislava: SjF STU.

Production Management and Engineering Sciences – Majerník, Daneshjo & Bosák (Eds)
© 2016 Taylor & Francis Group, London, ISBN: 978-1-138-02856-2

Innovation and intellectual property

K. Hrazdilová Bočková & G. Sopková
Dubnica Institute of Technology, Dubnica nad Váhom, Slovak Republic

J. Gabrhel
E-learnmedia Ltd., Bratislava, Slovak Republic

ABSTRACT: The innovations bring the chances for the products and services improvements, but they involve the risks, that can lead to crisis. To realize this fact is the first step, how to predict possible problems or how to avoid them. The issue of the paper is the formulation of the risk identification in the innovation projects connected with the topic of intellectual property and mostly with the copyright upon the innovation realization, then the risk identification in the innovation projects often is neglected and in literature closer unspecified and unpublished.

1 INTRODUCTION

"Innovation is the specific tool of entrepreneurs, the means to exploit change as an opportunity to introduce a new business or service. It can be taught as a discipline, it can be learned, it can be practically used. Business people must deliberately seek sources of innovation, change and symptoms of change, suggesting the possibility of successful innovation. And they must know and apply the principles of successful innovation." (Drucker, 1993).

The ability to do business is often seen as something slightly mysterious. It is often argued that success in business requires special skills, talent, inspiration, flashes of genius. Entrepreneurship and innovation are the disciplines, however, that not only can, but must be managed. It is a systematic activity, which is the responsibility of the company's management. The growing importance of innovation is due to the fact that in today's increasingly connected and increasingly shrinking world, competition grows and shortens the life cycle of products. Therefore, at a time when some of our products are successful in the market, we already need to work on innovation that will replace them.

It is possible to find innovations that are part of a "flash of genius". However, they are unique; they cannot be taught or learned. Contrary to popular opinion, they happen very rarely. What we can learn is a purposeful, analysis-based innovation, system and hard work. Innovation is a phenomenon rather economic than social and technical. Few of technical innovations may be more important than such social innovations as newspapers or textbooks or such economic innovations as the hire purchase. Modern health care has a greater impact on health than medical advances. Management as art of combining knowledge and understanding of different people in the organization is the innovation of this century, which completely changed the world. Technology can be imported at relatively low prices. The so-called "creative imitation" can be a very successful business strategy, for although there are innovations that cause changes (e.g. the invention of the aircraft), most innovations are more prosaic: they use changes that have already occurred. And that is why intellectual property is given an ever increasing attention, because new ideas are the sources of innovation that have become the most important source of the economic growth. Approximately ¾ of the value of publicly traded companies in the US are intangible assets. Worldwide revenue from technology licenses is estimated at 100 billion US dollars and growing rapidly.

2 OBJECTIVE AND METHODOLOGY

The primary objective of this paper is to describe the modern concept of identifying risks in innovation projects with an emphasis on issues of intellectual property in the field of innovation management. This identification was made primarily on the study of literature, conference papers, scientific journals and related web sites in the period of January 2010–December 2014, taking into account topicality of the literary background. This literary research resulted in the realization of a controlled interview with the consultant and JIC fund manager, which was implemented in February 2015.

Secondarily, this article is based on the currently processed case studies on the topic of risk management of innovation projects in engineering, which form part of a project of the Internal Grant Agency of Tomas Bata University in Zlín IGA/67/FaME/10/D Comprehensive analysis of the current state of risk management in projects, the aim of which is to create a methodology for managing innovation risks, and KEGA project No. 003/DTI–4/2014 entitled Diagnostic system for identification of competencies of managers of national and international educational projects.

3 BASIC CONCEPTS

The OECD manual (CZSO © 2015) features the definition of technological innovation and the role of research and development (R&D) in the innovation process as follows: *"Technological innovations mean new products and processes and significant technological changes in products and processes. Innovation is considered implemented when placed on the market (product innovation) or used in the production process (process innovation). Innovation thus includes a number of scientific, technological, organizational, financial and commercial activities. R&D is only one of these activities and may occur at various stages of the innovation process. It acts not only as the original source of inventive ideas but also as the problem solution framework, which may be addressed at any stage of implementation."* This definition may, in our view, extend to innovations that do not have the technological basis. The summary of the activities listed in the definition implies that, aside from exceptional cases, the innovation process will involve a wide range of experts, and demands on their qualifications will vary.

In this definition and in the following text, we use the term product as a summary for goods and services.

This often involves mixing the concepts of "innovation" and "advanced technology" (high-tech). In fact, only a small portion of innovation is directly linked with advanced technologies (which usually mean computer systems, telecommunications, robotics, automation, biotechnology etc.). And many high-tech companies, by contrast, behave in a quite traditional, conservative and non-innovative manner. In accordance with Drucker (1993), we will therefore understand innovation as an ability to exploit change as an opportunity to start a new business or launch a new service or product.

Successful innovation is not a work of coincidence or a unique idea. It is a planned, controlled process. In the course of the whole process, we must realize that the later we discover a mistake that we made, the more expensive it is to correct it. Therefore it pays to focus on the early stages and try to "sort out" as many bugs as soon as possible. The first step in the process also involves most significantly creative thinking ability.

4 NEW TRENDS IN INTELLECTUAL PROPERTY

The importance of intellectual property is reinforced by several trends, as specified in (Hrazdilová Bočková, 2009).

Many experts now wonder whether the increasing number of patents does not rather hamper innovation; lawyers become part of the development teams for new technological products side by side with engineers. The accumulation of patents is often not only protecting its own inventions, but it also acts as a defence against competitors. With such a large number of patents, it is almost impossible to avoid infringement of one's rights; therefore the necessary self-defence is often based on balancing.

An example: companies in the market of Internet search engines, about which experts claim that each of them to a certain extent violates the rights of others and they all have an interest in maintaining a delicate balance, otherwise the whole complicated structure could crash.

With the number of patents, there increase R&D costs. A practical rule says that technology companies in the field of IT register nearly two patents for every million dollars invested in R&D. At the same time, the number of patents is growing even faster than the spending on R&D; it means that there are a growing number of "cheap" patents. Despite this growth, a number of economic studies mentioned in (MODERNIRIZENI © 2006), (Köppl © 2009), (Pospíšilík, 2007), (SVES © 2014) shows that only about 5% of patents brings real value and only a small fraction of them are involved in the vast majority of revenue from patents. Patents are often less effective in innovation protection than secrecy and rapid product launch. To make a business model based on selling licenses a long-term success, they must be accompanied by complementary products or services (know-how, etc.) that will help manufacturers buying licenses quickly enter the market.

After leaving Microsoft, Nathan Myhrvold founded Intellectual Ventures, which represents a new model: a combination of a venture capital fund, a law firm and R&D laboratories. The aim is—similarly to Microsoft—to break into the market with such a pervasive technology that no one can avoid being paid for it, but using different means: while Microsoft was trying to create a monopoly, which was challenged by governments, N. Myhrvold's strategy is based on the use of government—supported monopoly by means of patent granting.

In generally, according to Harris (2014), there are three possibilities for the use of intellectual property of the company:

– use in their own products without additional income from the use of IP,
– enforcement of patent claims against others,
– sale of licenses.

A new trend is the depositing of patents into the "general property" (commons). For example, IBM released 500 patents of "open source" software for the community, based on the consideration that excessive patent protection becomes an impediment to innovation. Shortly afterwards, a similar move was made by other firms— Nokia, RedHat, Computer Associates and Sun Microsystems.

A similar trend can now be observed in the field of copyright, where Creative Commons initiative was founded, under which a growing number of authors assign their work to further use with limited exclusivity (CC as Creative Commons instead of ©). In connection with the management of innovation, we can note that E. von Hippel of MIT released two of his monographs this way.

Perhaps the best example of the benefits of open standards is the internet. Open standards enabled low cost, the system is open to improvement, attracting a large community of developers and benefiting from the network effect—the more people use the system, the more its value increases for each user and attracts more new users. In favour of open standards, based on (CZSO © 2015), (MODERNIRIZENI © 2009) and (Pospíšilík, 2007), several arguments may be stated:

– growth of complexity: in an effort to maintain the competitiveness, companies add new elements to their products, but for this they often need to take advantage of innovations created elsewhere,
– convergence: multimedia applications, e.g. television programs offers over the phone,
– disaggregation and specialization, a growing modularity and interchangeability.

Special attention is paid to China and India as countries that can offer not only cheap labour. The possibility to dedicate a large number of people to a project makes it possible to shorten the development cycle or explore alternative approaches, which would not be possible with fewer people. It is not the case of just the ability to do the same thing cheaper, but also that something can be done, which is not possible in different conditions.

5 NEW RISK OF INTELLECTUAL PROPERTY PROTECTION

In the period of one of the worst recessions mankind has known, the protection of sensitive information and intellectual property is more important than ever. The only leak of sensitive data can cause irreparable financial harm, damage the reputation of the company, shrink the stock price or destroy the trust of customers. Surveys (Jirásek, 2004), (Pospíšilík, 2007), (TYDEN©2014) show the following:

– More and more digital information, such as intellectual property or sensitive customer data is transferred between companies and continents, and this causes their evasion.
– The global economic crisis can cause an actual storm in the information security area. Companies are cutting costs on security and staff numbers. This creates an ideal opportunity for cybercrime.
– Individuals and groups in some countries represent a growing threat to sensitive data. The perception of geopolitical differences in corporate policies for working with data becomes a reality. Particularly China, Pakistan and Russia are considered problematic regions.
– Cyber thieves are no longer satisfied with conventional attacks and credit card or personal accounts data theft. They develop targeted theft of intellectual property. Why spend money on their own research and development, when the results can just be stolen?

Most intellectual property today is still physically located in North America and Western Europe. Central and Eastern Europe, especially countries disposing of adequate technological infrastructure and skilled labour force, increasingly serve the Western European companies to store their data.

A large amount of data is now stored also in India, China, South Korea and Japan. In China, a large amount of production is concentrated and therefore the companies must make available their intellectual property, even though they are aware of the risks. For the selling of electronics in the Chinese market, it is necessary to register the product, and the certificate is granted on the condition of providing detailed documentation. Companies are afraid that this technique serves Chinese government to gain access to their intellectual property.

The problem of leakage of sensitive information is even more serious than the "raw data" shows. By far not every incident is reported. Half of the respondents are worried that reputational damage as a result of a data breach will cause more damage than the direct economic losses or penalties for failure to comply with legislation.

Companies today are increasingly worried that intellectual property theft may be committed by leaving employees. People are trying to increase their value in the labour market and steal data they think the future employer, most often the competitor of their current employer, would be able to appreciate. While the theft of corporate data for direct financial gain took place already in the past, the aforementioned motivation is typical of the current economic crisis (MODERNIRIZENI © 2009).

6 COPING WITH TECHNOLOGICAL INNOVATION RISK

Innovation activities of any organization are always accompanied by a number of business risks. Innovation is in fact focused on the future and it is uncertain—nobody can predict it in detail. It is exactly the incomplete knowledge of the future that is a source of innovative business risks. The main risk category of innovative entrepreneurship can be identified through an analysis of the value chain (set of internal processes), used by organizations to prepare the innovation entry to the target market, and are detail cited and described in (Hrazdilová Bočková, 2009a).

The truth is that many innovative projects are unsuccessful for example, because the company's management is not able to find the way to effective commercialization of the project and the company itself. Successful innovation depends not only on the outcome of scientific and technological development, but especially on the successful exploitation of new business opportunities, and therefore it is no exception when innovation is accompanied by a change in the current business concept.

In the creation of innovative projects, organizations must therefore manage to avoid getting stuck in the trap of technological development. Many companies get caught in this trap for fear of falling behind. Therefore they intensely, though unprofitably, invest in the use of technological developments in their innovation. It happens that these innovations, however, fail to address the customers who are either not ready or the innovation does not bring any value to them and does not meet needs. Large losses due to ill-considered investments under the pressure of technological development therefore lead the company to failure to change its model of business. On the other hand, the use of technological developments and research is very important. No company, creating innovative project cannot afford to ignore technological development and believe to maintain high business performance and earning power. Nevertheless, mismanagement of new technology is not

the cause of a decline of business performance, but neither its excellent management leads to mastering its sharp rise.

The cause of failure is also the fact that in their conceiving, the company issues from historical data acquired through marketing research, instead of focusing on the circumstances of the use of the innovation by the customer. This leads to the fact that the innovative project and innovation itself do not address the potential customer with the necessary power.

Innovative projects are focused on the future and management of their processing is provided with limited rationality of the decision-making project managers. The results are therefore affected by a number of risks which both the project manager and the team members must be able to handle.

The technological risk assessment applies a structured approach, which is based on practical methodology of finding correct answers to questions, as stated in (Hrazdilová Bočková, 2009b).

– How can the selection of technical solutions of new products/services influence the success of innovative business of the organization?
– Can a progressive technology of innovative solutions be a source of business risk?

In seeking answers to the above questions, representatives of the organization's management must be guided by the principles presenting a conceptual core of four basic steps of effective prevention of technological risks methodology at various stages of preparation of launching the innovation to the target market. First, the creators of innovative solutions must understand the business potential of the new technology, which the organization wants to use when creating the offered new products/services. Along with this, the creators of the innovative plan must accurately identify the target group of (existing and new) customers.

In the next stage of the procedure, the managers responsible for preparing the innovation launch on the identified target market segment must predict with sufficient accuracy how the new technology will affect the dynamics of its development. It is an estimate of how it changes the user habits of potential customers and also the structure of the value chain. This will enable the organization to anticipate the changes that new technology and innovative solutions based on it will cause in the field of business.

A possible solution is to use a disciplined, systematic process in the company that allows a balanced plan for innovation, both large and small, in terms of risk and potential impact on the growth and profitability of the company. One of the tools that enable organizations to apply a balanced approach to innovation is a risk matrix, which allows you to graphically express the risk of the entire innovation portfolio. This matrix is described in (Hrazdilová Bočková, 2009b).

Known products intended for known markets are placed in the lower left part of the matrix. Such innovations have a small probability of failure. Conversely entirely new products for markets in which the company is not involved yet are placed in the upper right quadrant of the matrix. Such innovations are burdened with high probability of failure.

Once the innovative risk matrix is prepared for all projects, there usually show two things: that the company has more innovative projects than it is able to effectively manage, and that the risk distribution between the small and big innovation is unbalanced. Usually it is found that most of the innovative projects are located in the lower left quadrant of the matrix, and only a small part of the innovative projects are aimed at the upper right quadrant. Such an imbalance is unhealthy. Many small projects responding to customer requirements and vendors represent a steady stream of incremental innovations that deplete the financial resources available for innovation. On the other hand, financial experts who use the standard financial instruments such as the cash flow analysis are usually averse to innovation projects promising high returns in the distant future and are polluted with substantial risk.

7 FIGHT AGAINST PIRACY

Imitation of products and know-how theft is a widespread phenomenon against which firms must defend themselves. The number of cases of confiscation of counterfeit goods and pirated copies on the external border of the European Union has increased, according to statistics from the European Commission, from 37,000 in 2013 to 43,000 in 2014. So far, confiscation and patent disputes stood at the forefront, while the comprehensive management of legal and substantive measures is missing in most enterprises (ZK—INOVACE © 2014).

In patent applications, typically, only technologies are protected. Often, however, it does not take into account whether the patent also protects the customer benefit and substantial competitive advantage. At the same time, only the integrated protection of patents, trademarks and design entry forms offers options to ensure an almost monopolistic advantage over the competition.

For example, STIHL company, a global manufacturer of small forest and agricultural mechanization, in addition to patents, has tried to protect a combination of colours on their prod-

ucts (orange—gray). This colour is the hallmark of her chainsaw and pirates often mimic it. Given the numerous plagiarism from China, the company now applies to protect also the design, because it is easier to make a claim in the event of a dispute.

In the event of a detected violation of the rights, the damage is reflected in different ways. This includes in particular the damage caused by:

- loss of sales due to falling demand, stagnation royalties, loss of market share, and the outflow of customers or price fight with dumping prices,
- damage to good reputation,
- claims under the warranty and product liability.

A timely identification of threats and violations of rights can prevent costly consequences. It is therefore necessary to systematically identify violations of property rights. For example, the Swiss manufacturer of machines checks for this reason, especially the Chinese market with patents and utility models. First, his scanning identifies yet unknown competitors. At the same time before the fair, in which the company will participate, the research cases of infringement of property rights and prepare patent documentation so that patent infringement at the fair could be proposed as soon as possible for interim measures. This process reveals machines whose components infringe proprietary rights and therefore they can be removed from the fair on the very first day (INOVACE © 2014).

However, businesses often reveal an alternative trade with counterfeiters only on the basis of a strongly declining demand and out flux of their clients to the competitors. Their name and address is usually difficult to determine, but it is possible to identify by test purchases in cooperation with local agencies. At the same time, users can obtain evidence.

Direct legal protection forms a good basis to proceed against the infringer. Offensive strategy of protecting rights and disclosure of successful lawsuits also has a deterrent effect. STIHL company builds an image in the media of offensive action against pirates that draws attention to the destruction of plagiarism and deployment of detectives (INOVACE © 2014).

In addition to the legal safeguards, subjective measures (technological, marketing and human) represent—particularly in China—an effective supplement to counteraction against the different forms of imitation products and theft of know-how. To protect the product or its packaging, both visible and invisible technology can be used. For example, Danfoss, a Danish manufacturer of mechanical and electronic components, replaced the two-dimensional bar code with a three-dimensional one and so checks its logistics chain. Visible technologies include holograph, artificial lines (line structures that deform when copied), tactile detection (e.g. low and high embossing), security labels, seals etc. Security inks, micropigments, nanotechnology or chemical or electronic calibration, such as Radio Frequency Identification (RFID), on the contrary, provide invisible protection. Using RFID and other automatic identification technologies, products can not only be recognized, but it also allows continuous monitoring within the chain, in which the product moves. The principle of a *black box* protects modular products by producing technologically demanding components in safe countries) (Tesařová, 2014).

Intensive contact with customers, suppliers' loyalty, proactive work in associations, lobbying and strengthening relationships with local authorities contribute to better protection. For example, GE Philips actively cultivates relationships with customs authorities. Because the quality of counterfeits is still growing, they offer training to be able to distinguish the genuine product from an imitation. LINDE company refuse to introduce several devices that are easily replicable to the Chinese market. Global knowledge management program allows continuous documenting and improvement of the know-how of engineers located throughout the world

8 CONCLUSION

All risks (as long as they are met) have negative impacts on the economy of the organization caused by the causal relationships between the yield and the influence of other perspectives of innovative entrepreneurship. The company management must therefore take measures that will lead to the elimination of potential innovation risks, or at least if their origin cannot be effectively prevented, to minimize their negative impacts on the internal environment of the organization. These measures have the character of risks prevention, the fulfilment of which can affect the organization, or preparation of mobilization procedures that serve to minimize the negative impact of the risks, the origination of which the organization's control.

Considering the high levels of piracy and multilayered nature of the hazards, controls of protective measures and continuous innovation process are crucial for success. When checking these, effect and costs are opposing each other. From the assessment of quality or success of protection it is possible to derive future protection measures. Costs are difficult to quantify in non—legal measures, but they can be estimated. These procedures allow the company to introduce the continuous innovation process as protection against imitation of products.

Prevention and reaction form a cycle, which is supported by both positive and negative experience of the company itself and examples of other companies and industries. Finally, there is a need of active management of intellectual property as a competitive advantage. Fast prevention and response to these dangers alleviates or reduces the impact of piracy and theft of know-how.

REFERENCES

Drucker, P. F. 1993. *Inovace a podnikavost: praxe a principy.* Praha: Management Press.

Harris, T. 2014. What innovation can do to your business during a crisis. *Enterprise innovation* Retrieved from http://www.enterpriseinnovation.net/content/what-innovation-can-do-your-business-during-crisis?page=0%2C0.

Hrazdilová Bočková, K. 2009a. *Řízení inovací.* Zlín: FaME UTB ve Zlíně.

Hrazdilová Bočková, K. 2009b. *Firemní inovační politika I.* Zlín: FaME UTB ve Zlíně.

Hrazdilová Bočková, K. & Škoda, M. 2014. Study of culture of project oriented society. *International Journal of Information Technology and Business Management.* Vol. 29.

Inovační aktivity podniků v České republice. 2015. Praha: Český statistický úřad. Retrieved from http://www.czso.cz/ csu/redakce.nsf/i/inovacni_aktivity_podniku_v_cr/$File/inovace.pdf.

Inovační infrastruktura Zlínského kraje. 2014. Retrived from http://www.zk-inovace.cz/.

INOVACE.CZ—proměňte nápady na peníze. 2014. JIC, zájmové sdružení právnických osob. Retrieved from http://www.inovace.cz/pro-podnikatele/inovacni-management/jak-a-kde-inovovat/.

Jak na inovační rizika. 2009. *Moderní řízení.* Retrieved from http://modernirizeni.ihned.cz/ c4-10000605-39393190-600000_d-jak-na-inovacni-rizika.

Jirásek, J. A. 2004. Nesnesitelná lehkost řízení. *Moderní řízení* Praha: Ekonomia a.s. Volume XXXVIII. No. 7. pp. 11–13.

Komplexnost a spolehlivost v plánování inovací. 2006. *Moderní řízení.* Retrieved from http://modernirizeni.ihned.cz./c4-10065460-17978480-600000_d-komplexnost-a-spolehlivost-v-planovani-inovaci.

Köppl, D. 2009. Inovace v době krize platí dvakrát tolik. *Marketing a media.* Retrieved from http://mam.ihned.cz./c4-10010830-36537610-100000_d-inovace-v-dobe-krize-plati-dvakrat-tolik.

Pospíšilík, K. 2007. Inovace a strach. *Inovace.* Retrieved from http://www.inovace.cz/for-business/veda-vyzkum/podpora-inovaci/clanek/inovace-a-strach/.

Průzkum: Krize nezabrání investovat do inovací. 2014. *Týden.* Retrieved from http://www.tyden.cz/rubriky/byznys/cesko/pruzkum-krize-nezabrani-investovat-do-inovaci_136840.html.

Rizika inovačních aktivit. 2014. Retrieved from www.svses.cz/skola/akce/konf/inovace06/pps/pitra3.pps.

Škoda, M. & Hrazdilová Bočková, K. 2014. Project Team: Chimera of Today's Project Management. *International Journal of Management: Theory and Application (IREMAN).* Volume 2, No. 4. pp. 123–130.

Tesařová, H. 2014. Inovační management. *Inovace.* Retrieved from http://www.inovace.cz/for-business/manazerske-dovednosti/clanek/inovacni-management/.

Production Management and Engineering Sciences – Majerník, Daneshjo & Bosák (Eds)
© *2016 Taylor & Francis Group, London, ISBN: 978-1-138-02856-2*

Evaluation the effectiveness of rehabilitation treatment using the tools of statistics

G. Ižaríková & J. Halčinová
Faculty of Mechanical Engineering, Technical University of Košice, Košice, Slovak Republic

P. Hermel
REHAMED, s.r.o., Košice, Slovak Republic

ABSTRACT: This paper is concerned with analyzing the effectiveness of rehabilitation treatment by method of neural mobilization at basic diagnosis of radicular syndrome in the lumbar spine. Health monitoring and monitored parameters in patients were assessed and compared before the start of rehabilitation, during rehabilitation and after completion of rehabilitation treatments. The aim of the paper is to highlight the significance and the importance of special therapy method using neural mobilization focusing on monitoring effect of neural mobilization on subjective parameters (pain). Gathered information has been described by using basic statistical characteristics and their changes were evaluated graphically. Difference of monitored parameters in patients before the rehabilitation, during and after completion of rehabilitation treatments was evaluated by analysis of variance. In this paper, has been showed the pain mitigation after the rehabilitation.

1 INTRODUCTION

1.1 *Mobilization of the nerve tissue*

The article is focusing on the valuation of therapy using neural mobilization by monitoring the effect of neural mobilization on subjective parameter— pain in patients suffering from essential diagnostic radicular syndrome in the lumbar spine. The obtained results show the neural mobilization, which by using comprehensive method allows more efficient improved and speed up the process of treatment and rehabilitation to improve the healing of the patient's problems. Neurodynamic disorders can be found in connection with pathological processes in periradicular and perineural space, manifesting the root symptomatology. For the purpose of practice is important to determine changes in neural disorders mobility and differentiated nature of the symptoms. The pain comes from damaged or regenerating afferent nerve fibers. For the purpose of comprehensive approach to therapy is appropriate to use the classification of spinal disorders by Quebec Task Force of spinal disorders (De Rosa 1992).

- back pain without radiation,
- back pain with radiation into the proximal part of DK,
- back pain with radiation to the distal portion DK,
- DK pain greater than back pain,
- back pain with radiation and neurological symptoms,
- postoperative conditions (6 months),
- chronic pain syndrome.

When centralization fault occurs at the spinal level to integrate signals from the periphery in general can be signs of damage to the peripheral nervous system, divide to: irritative (presenting the different qualities of pain and related antalgic reaction of presenting the tension changes surrounding tissues and changing mobility related segments) and paretic (presenting the reversible or irreversible reduction in muscle tone, muscle weakness, numbness and reducing reflexes).

Techniques for mobilization of the nerve tissue composed of passive or active movement aimed to restoring the ability of the nervous system to tolerate normal pressure, friction and tensile forces associated with daily and sporting activities. These therapeutic movements can have a positive effect on symptoms (Merskey 1994).

To assess changes in neural dynamics and share neural pain clinical condition is used in testing the elasticity of the nervous tissue tension movements—neurodynamic tests. Neurodynamic tests in patients exhibit considerable individual variation. The main goal of diagnosis is the ability to interpret neurodynamic tests and their ability to reproduce, analyse and understand input and output information to find the key point of tissue dysfunction. For analysis, we used the results of the research, which was conducted in the same rehabilitation facilities. Standardized tests were used, which have been adapted to the objectives and tasks of research under precisely defined principles so that they obtain real data. For data collection was chosen individual approach (questionnaire), for the purpose of better understanding of the disease process and careful clinical examination at baseline diagnosis of radicular syndrome in the lumbar spine. After obtaining the information approval, patients were guaranteed anonymity of data and measurement results. Pharmacotherapy for patients was not excluded during testing, as opposed to the ethical principles of research. Patients were applied neural mobilization techniques with classical rehabilitation. The intensity of pain was assessed by visual analogue scale the 10-point scale (VAS). VAS scale ranges from 0 up to 10. The highest value indicates the unbearable pain.

The scale of pain:

- 0—no pain,
- 1-2-3—discomfort, little pain,
- 4-5-6—moderate pain,
- 7-8—servere pain,
- 9-10—unbearable pain.

This simple division helps treating, patients and physicians both in assessment and also in evaluating the effectiveness of nursing interventions and effectiveness of analgesics.

2 CHARACTERICS OF THE RESPONDENTS

2.1 Classification of respondents

After the evaluation questionnaires and analysis of this information for further analysis, we used the response 159 respondents—patients who were executed study of the effectiveness of rehabilitation treatment of neural mobilization with the addition of classic physiotherapy outpatients with primary diagnosis of radicular syndrome in the lumbar spine. Assessment and comparison were done of the difference health and endpoints in patients before rehabilitation, rehabilitation during and after completion of rehabilitation treatments. Study group consisted of 94 women and 65 men, the age structure is given in Table 1. and Figure 1 and Figure 2. Patients aged over 50 years were excluded from the research. It is likely that most patients were aged 36 to 40 years—69 respondents, representing 43.4% of the whole set. In this group, 39 were women,

Table 1. Structure of the study group.

| Age | Gender | | Σ |
	Female	Male	
Less than 35 years	27	14	40
36–40 years	39	30	69
41–50 years	28	21	49
Σ	94	65	159

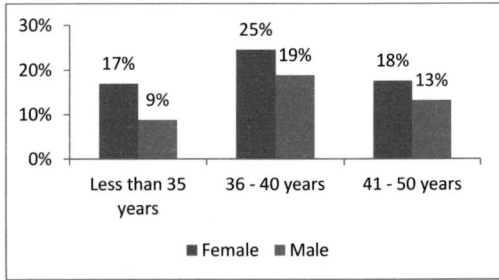

Figure 1. Distribution of respondents by gender.

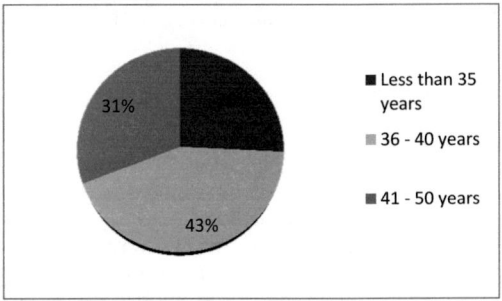

Figure 2. Distribution of respondents by the age.

accounting for 24.53% of all respondents, 41.49% of all women respectively 56.52% of all 36–40 annual respondents (patients). The largest group were 30 men, representing 18.87% of all respondents, 46.15% of all men respectively 43.48% of all 36–40 annual respondents (patients).

3 EVALUATE THE IMPACT OF THERAPY ON PAIN INTENSITY USING STATISTICS

Table 2 contains data on the number of patients with spinal disorders during rehabilitation treatment. During treatment changes occurred in spinal disorders in these patients, changes in development can be seen in the graphs.

In the graph shown in Figure 3. observe patients without pain, number of patients with back pain without radiation increased, the number of patients with back pain with radiation into the

Table 2. Spinal disorders during rehabilitation treatment.

	After therapy	During therapy	Before start of therapy
Without pain	12	0	0
Back pain without radiation	82	27	40
Back pain with radiation into the proximal part of DK	53	64	93
Back pain with radiation to the distal portion DK	12	55	25
DK pain greater than back pain	0	13	1

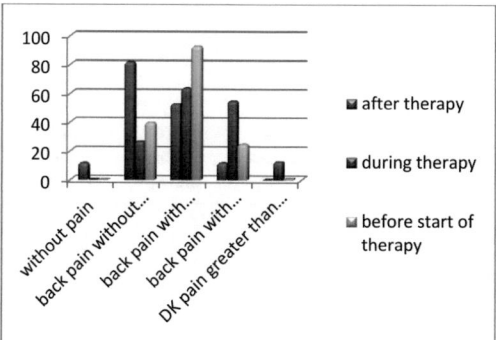

Figure 3. The changes in development of spinal disorders.

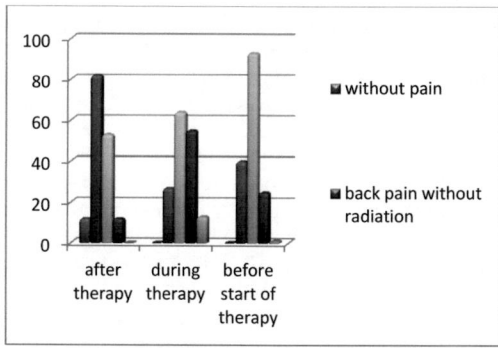

Figure 4. The changes in development of spinal disorders by phases of treatment.

proximal part of DK and pain with radiation to the distal portion DK and DK pain greater tha back pain decreased. The graph shown in Figure 4. shows those spinal disorders dominated during the different phases of rehabilitation treatment.

In this contribution we consider comparing the intensity of pain (as assessed by VAS), indicated by the patients in research before rehabilitation, rehabilitation during and after completion of rehabilitation treatments. The values obtained are presented in Table absolute numbers and shown in Figure 5. Before starting therapy lowest value quoted in pain intensity was 4-moderate pain by only one respondent stated, most of them stated value of 6-moderate pain (53 respondents). During therapy, the lowest value quoted in pain intensity decreased to 1-little pain (2 respondents) and the most frequently reported pain intensity value was 3-little pain (69 respondents). After the therapy has respondents reported pain intensity values only from 0-no pain to 3-little pain, most to 73 of them have a value of 2- little pain and 23 indicated 0-no pain. And from said values it is clear that lowering the level of pain, is confirmed by the statistical characteristic (tendency and variability). Comparison of the values of pain intensity before rehabilitation, rehabilitation during and after completion of rehabilitation treatments using descriptors, their point and interval estimates in Table 3.

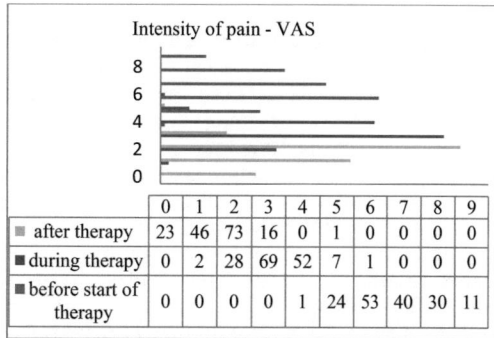

Intensity of pain - VAS

	0	1	2	3	4	5	6	7	8	9
after therapy	23	46	73	16	0	1	0	0	0	0
during therapy	0	2	28	69	52	7	1	0	0	0
before start of therapy	0	0	0	0	1	24	53	40	30	11

Figure 5. The values the intensity of pain according to the scale (as assessed by VAS).

Table 3. Statistical characteristic of the value the intensity of pain.

Statistical characteristic	Initial medical conditions, before start of therapy	Medical conditions during therapy	Medical conditions after therapy
Average	6.67	3.23	1.54
Modus	6	3	2
Median	7	3	2
Sample standard deviation	1.16	0.86	0.90
Koeficient of variation	0.17	0.26	0.59
95% IS for μ	(6.49–6.86)	(3.10–3.37)	(1.40–1.68)

399

In this analysis is important the selection of appropriate indicators and selection of appropriate statistical tools, enabling comparisons. To measure differences can be used several statistical tools and methods such as cluster analysis, arithmetic mean, standard deviation, coefficient of variation. In this paper are used the characteristics of tendency and variability and confidence intervals. From the graphical tools are used columns and scatterplots and boxplot. Mathematical calculations and analysis software used MS Excel and Statistica.

Mean, median, and mode are three kinds of averages. The arithmetic mean (average) is the most commonly used feature of position, it is a quotient of the sum of all the values, character and their number.

The median is the "middle" value in the list of numbers. To find the median, your numbers have to be listed in numerical order. The mode is the value that appears most often in a set of data. The calculated values of the means of signaling a fact that before starting, pain intensity VAS scale provided in excess of that during and at the end of therapy. The arithmetic mean diameter is reduced by the five grade.

Sample standard deviation belongs to the overall variability characteristics, can be interpreted as the average difference between the values and the average of the marks ignored. The greater the variability, the larger the deviation.

The coefficient of variation is a relative measure of dispersion derived from the sample standard deviation (the ratio of the sample standard deviation and the average expressed as a percentage). It is used to compare the variability between datasets with different diameters.

The values of the resulted footprint can be evaluated using the confidence intervals. The confidence interval contains the true value of the estimated parameter of the population with a predetermined probability. The value of 95% confidence interval for the before initiating therapy is 6.49–6.86 and obtained by the after therapy is 1.40–1.68. It is clear that the confidence interval calculated from the value of the before initiating enters into higher values than the interval calculated from the value of the after therapy.

From the calculated values shows that the influence of therapy decreased pain intensity, which was assessed by visual analogue scale at 10 point scale. Achieved values are graphically represented by the boxplots. (Figure 6.) In descriptive statistics boxplot is a convenient way of graphically depicting groups of numerical data through their quartiles. The boxplot generally shows mean, median, 25th and 75th percentiles, and outliers (Table 4.). A standard box plot is composed of the median, upper hinge, lower hinge, upper adjacent value,

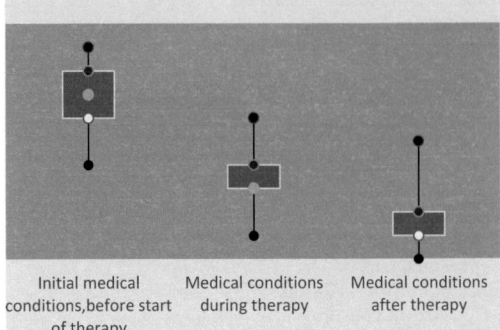

Initial medical conditions,before start of therapy   Medical conditions during therapy   Medical conditions after therapy

Figure 6. Boxplot for the intensity of pain medical conditions, before start of therapy during therapy and after therapy.

Table 4. The parameters in the boxplots.

Statistical characteristic	Initial medical conditions, before start of therapy	Medical conditions during therapy	Medical conditions after therapy
Lower quartile	6	3	1
Min	4	1	0
Median	7	3	2
Max	9	6	5
Upper quartile	8	4	2

lower adjacent value, outside values, and far out values. Box plots may also have lines extending vertically from the boxes indicating variability outside the upper and lower quartiles. Boxplots can be drawn either horizontally or vertically.

The mean values the intensity of pain was tested using Analysis of Variance. ANOVA is a technique that allows you to compare the mean values of several k independent basic files. Its purpose is to detect whether the differences of mean values for these files are statistically significant or only incidental. At the significance level of the test of the null hypothesis, which holds that all mean values essential files are the same compared to the alternative hypothesis, which argues that at least one pair of mean values are significantly different from each other. The decision on acceptance or rejection of the null hypothesis, we conducted using *p-value*—significance level.

The decision on acceptance or rejecting the null hypothesis is true:

- if $p < \alpha$ it is then the null hypothesis is rejected,
- if $p > \alpha$ it is then not reject the null hypothesis.

Value $\alpha = 0,05$ respectively $\alpha = 0,01$. The decision on acceptance or rejection of the null hypoth-

Table 5. ANOVA.

Source of variation	SS	df	MS	F	P-value	F crit
Between Groups	2174,9	2	1087,4	1118,4	3,26E-180	3,015
Within Groups	460,9	474	0,9723			
Total	2635,8	476				

esis can be made by comparing the test criterion (F) with the Fisher distribution quantiles (F crit).

From Table 5. that the group have different diameters in the sample are too large to be random, are thus statistically significant ($p<\alpha$, $F>F$crit.). The null hypothesis of equal mean values rejecting ANOVA, different diameters can be identified on Figure 6., it means that after the rehabilitation treatment is expected pain relief.

4 CONCLUSION

The aim of this paper was to highlight the importance and relevance of the selected methods of special neural diagnosis and therapy focusing on the dynamics of neural activation in patients suffering from primary diagnosis of radicular syndrome in the lumbar spine. Endpoint was pain intensity, assessed by visual analogue scale at 10-point scale (VAS). Change in the intensity of pain in patients before rehabilitation, rehabilitation during and after completion of rehabilitation treatments were statistically evaluated based on the results obtained. Patients after completion of ther-

apy (after conversion neurodynamic mobilization and classical rehabilitation) reported lower values when assessing pain intensity, pain intensity differences averages are statistically significant.

ACKNOWLEDGMENT

The article has been supported by the research grant No. ITMS 26220220185 of the project: University Medical Science and Technology Park in Košice (MediPark).

REFERENCES

Amird d Aczel. 1989. Complete Business Statistcs. Irwin. Boston.
Butler, D. & Gifford, L. 1989. The concept of adverse mechanical tension in the nervous system. Part 1: Testing for dural tension'. *Physiotherapy 75.*
Butler, D. 1991. Mobilization of the nervous system. London. Churchill Livingstone. *Mobilisation des Nervensystems.*
De Rosa, C.P. & Porterfield, J.A. 1992. A physical therapy model for treatment of low back pain. *Phys. Ther. 72.*
Maitland, G. 1985. The slump test: Examination and treatment. *Australian Journal of Physiotherapy 31.*
Merskey, H. & Bogduk, N. 1994. Classification of chronic pain: Descriptions of chronic pain syndromes and definitions of pain terms. 2nd ed. Seattle, International Association for the Study of Pain, WA: IASP Press.
Zgodavová, K., Živčak, J., Hudák, R., & Čurilla, E. 2013. Quality in healthcare, Košice, Faculty of Mechanical Engineering TU in Košice.
Zvárová J. 2007. Základy statistiky pro biomedicínské obory. Praha, Karolinum.

Production Management and Engineering Sciences – Majerník, Daneshjo & Bosák (Eds)
© 2016 Taylor & Francis Group, London, ISBN: 978-1-138-02856-2

Probabilistic tools supporting risk analysis of investment projects

J. Janekova, J. Kováč & P. Malega
Faculty of Mechanical Engineering, Technical University of Kosice, Kosice, Slovak Republic

ABSTRACT: In theory and practice, a variety of tools for risk analysis of investment projects are being utilized. The aim of this paper is to clarify the nature of the selected instruments and specify their features, strengths and weaknesses. The emphasis is put on probabilistic tools, specifically scenarios and Monte Carlo simulation. Monte Carlo simulation utilization is demonstrated on a specific example of economic practice. In conclusion the contribution deals with problems related to the implementation of probabilistic tools into economical practice.

1 INTRODUCTION

Knowledge of economic practice points to lack of integration of risk and uncertainty into the investment decision. This is confirmed by results of the survey in 2013, which was conducted in 62 small and medium enterprises in Slovakia. As a result it emerged from the survey that companies risk the analysis of investment projects either not implement at all (65%) or only in a limited extent (35%), the most often intuitively and they express the risk projects in words (Belanova 2014). As a result, wrong decisions arise in companies that especially for large-scale investment projects may threaten their prosperity and financial stability. To prevent such situations, it is necessary that companies implement evaluation of the effectiveness of investment projects comprehensively, including of consistently taking into account risk and uncertainties. This means that assessment of investment projects using only traditional approaches in today's dynamic time is insufficient. Into the investment decision-making is necessary to include probabilistic approaches that will increase its quality thereby they will allow to assess the investment project comprehensively, thus also in terms of strict respect of risk and uncertainty.

2 TRADITIONAL APPROACHES TO INVESTMENT DECISION MAKING

The traditional approach to evaluation of investment projects is based on financial criteria that the risk and uncertainty associated with the project either do not respected at all or only indirectly. Non-respecting of risk is associated with static criteria such as profitability and payback period of the project. Indirect integration of risk and uncer-

tainty is associated with dynamic criteria such as net present value, index of present value, internal rate of return or discounted payback period. In this case, compliance of risk is implemented through a risk premium, which forms a part of the discount rate of the project.

The following facts can be considered as shortcomings of this approach. This is a single-scenario approach because cash flows of the investment project under consideration are based on a single, usually the most likely development of internal and external factors affecting cash receipts and cash expenditures of the project during its economic lifetime. Risk and uncertainty are taken into account only non-formalized as another aspect of the evaluation of investment projects. The shortcomings of this approach highlight the optimism of managers who often underestimate the probability of an unfavourable development of individual risk factors affecting the results of the evaluated projects.

Shortcomings of the traditional approach to the evaluation of investment projects can be somewhat weaken by a sensitivity analysis, which identifies effects of isolated changes of individual risk factors (such as production volume, sales prices of products, required rate of return, etc.) for the selected financial criterion of the project (such as net present value, index of present value, etc.), whereby all other factors remain on predicted values.

3 PROBABILISTIC APPROACHES IN INVESTMENT DECISION MAKING

The substantial increase in quality of investment decision making in terms of respect of risk and uncertainty can be provided by probabilistic approaches. Their significant representatives are scenarios and Monte Carlo simulations.

3.1 Scenarios

In most cases, scenarios are understood as internally consistent pictures of the future, based on a certain set of interrelated key risk factors (Fotr & Soucek 2011). Schoemaker states that the starting point of their creation shall be what is known about future developments, i.e. trends and specifications of what are unknown, thus key uncertainties. Each scenario is thus based on a link between certain trends and uncertainties.

In practice, two forms of scenarios are used. In the first case are qualitative scenarios that provide a narrative description of fundamentally different possibilities of the future development of the business environment. They have mainly macro-economic nature. In the second case are scenarios of quantitative nature. These scenarios allow different future situations to characterize quantitatively by some combination of values of risk factors. These scenarios have primarily microeconomic nature. Each scenario records a different future development, or status of the business environment. It is therefore clear that also values of financial criterion at various scenarios will be more or less different.

Process of creation of quantitative scenarios takes place in three stages:

Selection of risk factors for development of scenarios. When developing scenarios it is necessary to limit the number of risk factors to two, maximum three major factors, so that the number of scenarios generated does not exceed the acceptable level (five, maximum ten scenarios). The starting point for identifying key risk factors is a sensitivity analysis. In principle, the smaller the number of risk factors, the simpler the scenarios creation.

Determination of risk factors. Their determination depends on whether it is the discrete factors (theoretically take a finite number of values) or continuous (theoretically take an infinite number of values). A more complex situation occurs with continuous factors that need to be replaced with discrete one, by means of two to four values that represent these linear factors.

Creation of scenarios and determination of their probabilities. Set of scenarios can be displayed using a table or decision tree. Decision trees are usually used to show scenarios in case of two or more risk factors. They deal with the probability of occurrence of consecutive events. Probability of a whole sequence of events (i.e. the common probability) is found by multiplying the individual probabilities following each other. This common probability is then used in the calculation of the specific criteria in various branches of the decision tree (Fotr & Kislingerova 2009, Scholleova 2009).

When creating scenarios it is extended also an approach which results usually as three scenarios, namely optimistic, most likely, and pessimistic scenario. In this case, for each key risk factor is necessary to estimate its optimistic, most likely and pessimistic values and each scenario to expressed as a combination of the corresponding values of risk factors. However, also in this case it is necessary to require that the values of risk factors have been consistent in each scenario (Hnilica & Fotr 2009). The advantage of this approach lies in the fact that in some cases it can be easily concluded that the economic benefits and drawbacks of the project (for example, if the net present value of the project is positive even in pessimistic variant or negative in optimistic variant). The drawback of this approach is the ambiguous perception of estimates of optimistic and pessimistic values of risk factors.

In any case scenarios are a suitable tool for risk analysis of investment projects. Their advantage is in the simplicity and visual presentation of the project results (results of operations, cash flows, financial criteria, etc.) for probable future situations represented by different scenarios.

3.2 Monte Carlo simulation

Monte Carlo simulation belongs to the most important simulation models. It is used when there are more risk factors, usually of continuous nature. Its essence lies in generation of a large number of scenarios (hundreds to ten-thousands) and calculation of criteria for each scenario.

It is carried out in several steps. The first step is to develop a mathematical model of the object of the risk analysis and process the program in the selected spreadsheet. The second step is to determine the key risk factors. It means such factors that are highly uncertain and the results of the simulation are very sensitive to their changes. The third step is to determine the probability distribution of key risk factors, including their statistical dependencies. In the last step is carried out the simulation itself, the results of which have graphical form (distribution of probabilities of selected financial criteria) as well as numerical form (in the form of risk features of variance, standard deviation, variation coefficient, etc.).

The main reason for using a Monte Carlo simulation is the quantification of the probability distribution for the overall project risk. On the basis of this distribution can be stated the expected value of project risks and how likely this value will be in the range that interests us. Thus, it allows a deeper understanding of risk site of evaluated objects, leading to better decision making in the choice of variants of a company development, acceptance or rejection of investment variants and projects, etc.

This method have also some drawbacks. These include high labour intensity and complexity especially when determining the probability distribution of risk factors and respect of their dependence. The greatest deficiency is considered the fact that the key risk factors, that influence the most the results of the risk analysis are often based on an assessment of the present and the past, are unpredictable. This can lead at the simulation to so called tunnelling effect and thus decrease the sensitivity of the search of new risk factors (Kopecky & Trkovsky 2011).

4 IMPLEMENTATION OF MONTE CARLO SIMULATION FOR A SPECIFIC INVESTMENT PROJECT

Risk analysis based on Monte Carlo simulation is applied on the project, which is aimed to expand the production of sanitary products. This is a development project, namely the acquisition of a new production line for women's sanitary napkins production. Its aim is to ensure an increased production volume and by selection of a suitable production equipment to create preconditions for flexible respond to changes in market requirements.

Economic evaluation of the investment project is processed by a financial model created in MS Excel. The financial model includes:

Input data for determination of cash flows of the project and monitored financial criteria (such as investment costs, production volume, selling price, material consumption, energy consumption, repair and maintenance costs, personnel costs etc.).

Cash flows of the project which take into account the construction period and the operation period of the production facility. The construction period of the project is set at two months. The economic lifetime of the project is identified with the depreciation period of production equipment, i.e. estimated for a period of six years.

Financial criteria for evaluation of the economic efficiency of the project are net present value, index of present value, internal rate of return and discounted payback period.

Financial criterion for risk analysis is the net present value.

4.1 Preparation of Monte Carlo simulation

Risk analysis carried out by Monte Carlo simulation is processed in the Crystal Ball system, which is an extension of MS Excel.

The output parameter is NPV. Its base value (base case) is provided by traditional approaches,

thus on basis of the most probable input variables (investment costs, expected production volume, sales prices, material costs etc.) that affect individual income and expenditure parts of cash flow of the project. Thus determined NPV value received a value of 5,046,759 EUR.

Furthermore, it is assessed the uncertainty of input variables that affect the economic performance of the project. In this case the expert evaluation is being utilized. As risk factors of the project are determined production volume, selling price, material costs and investment costs.

To display the uncertainty of risk factors a triangular probability distribution is used. The symmetrical one is selected for risk factor of production volume and asymmetrical one for risk factors of investment costs, selling price and material costs. The probability distribution of selected risk factors of the investment project is illustrated in Figure 1.

4.2 Monte Carlo simulation results

Primary outputs of the Monte Carlo simulation are probability distribution of NPV (Fig. 1) and statistical characteristics of this distribution (Tab. 1).

Based on the values of statistical characteristics it is possible to make the following conclusions. Average value (mean) of the Net Present Value is by EUR 213,145 less than the NPV calculated by traditional approaches. With 95% confidence the positive NPV of the project is expected in the range from EUR 3,335,975 to EUR 6,452,748. The probability distribution of NPV is approximately symmetrical, a little skewness has a negative value, indicating that it is slightly deflected to the left, i.e. towards lower values of NPV. This is documented on Figure 2, which indicates that only with the probability of about 31%, the project would reach NPV greater than shown on the figure and with the probability of about 69% it would reach lower value of NPV. Given that the probability distribution of the NPV is approximately symmetrical, characteristics of variability in the form of standard deviation, variance and coefficient of variation they represent good measure of risk of the project in relation to NPV. Kurtosis of the distribution corresponds to the ideal normal distribution.

The third significant output of the simulation is a graph of the NPV sensitivity. It provides information about contribution of selected risk factors to the overall risk of the project in relation to the NPV, in graphical and numerical form.

It is shown in Figure 3 that two risk factors contribute most to the risk, investment costs and selling price. The most important risk factor is

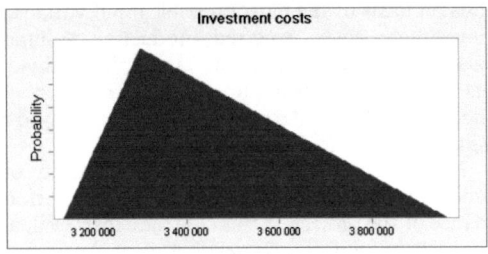

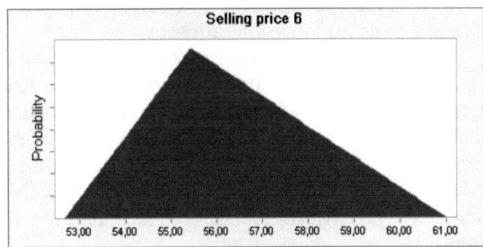

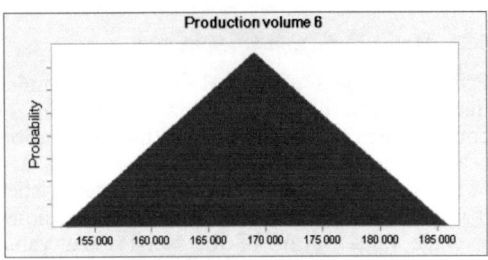

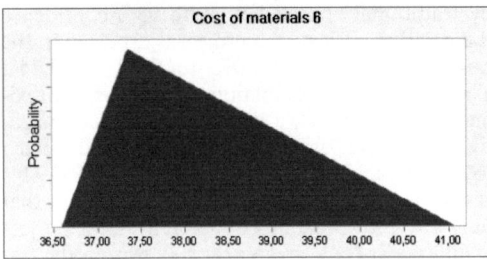

Figure 1. Probability distribution of selected risk factors of the project.

Table 1. Forecast of NPV.

Statistics	Forecast values
Trials	10,000
Mean	4,833,614
Median	4,838,904
Mode	–
Standard Deviation	432,996
Variance	187,485,905,521
Skewness	−0.0370
Kurtosis	3.00
Coefficient of Variability	0.0896
Minimum	3,335,975
Maximum	6,452,748
Range Width	3,116,773
Mean Std. Error	4,330

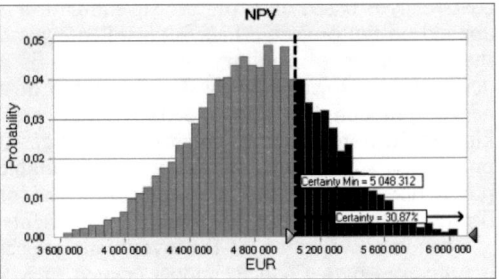

Figure 2. Probability distribution of NPV.

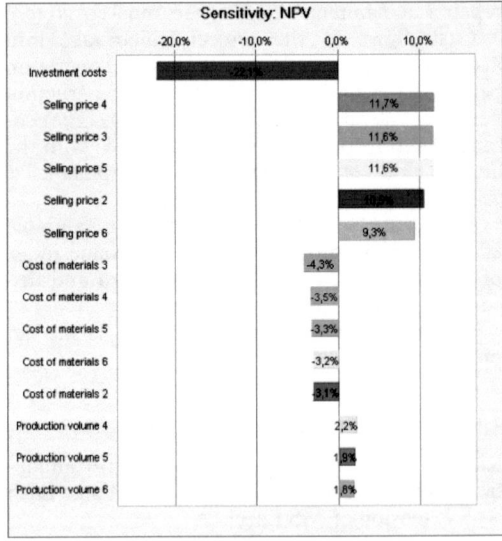

Figure 3. Sensitivity of NPV chart.

the investment costs that contribute to the risk of the project by 22.1%. Another important factor is the selling price in year 4, whose contribution to the risk is 11.7%. It is necessary to take into consideration also the fact that the selling price is a risk factor monitored in time series. Thus, its cumulative contribution to the risk of the project for the entire period of operation is up to 54.7%, whereby the highest contribution is in the fourth year of the project operation. The amount of contribution of risk factors material costs and production volume circulates by respective years in range from 1.8% to 4.3%. In both cases, these are also risk factors followed in time series. Attention should be paid to material costs for which the cumulative contribute to the risk of the project amounts to 17.4%. In order to reduce the risk of this project it is appropriate to focus attention on the first three risk factors.

5 CONCLUSION

Finally, it should by emphasized that Monte Carlo simulation as a tool for risk analysis improves the quality of decision making at risk and uncertainty thereby that leads to a deeper knowledge of the risk aspect of the object being analysed. Its application is not a one-time issue. Simulation should be repeated at detected changes in the development of the analysed risk factors of the project, or when new risk factors have been identified. The application of this probabilistic tool is particularly suitable in strategic decision making. However, in order that these probability instruments have been successfully implemented into a business practice it requires support of senior management, change of thinking, change of the traditional work style, education and training in this specific field and suitable software support.

ACKNOWLEDGEMENT

The article was elaborated within the project VEGA 1/0879/13 Agile, adapting to market business systems with highly flexible corporate structure.

REFERENCES

Belanova, K. 2014. Empirical Research of the Investment Projects´ Evaluation from the Aspects of Risks and Uncertainty in Small and Medium-Sized Enterprise in Slovakia. In. *Financial markets*, No. 3, p 12.

Fotr, J. & Kislingerova, E. 2009. Integration of risk and uncertainty in the investment decision-making and evaluation. In *Political Economy*, No. 6, pp 801–826.

Fotr, J. & Soucek, I. 2011. Investment decision-making and project management. How to develop, finance and evaluate projects, manage their risk and create project portfolio. Praha: Grada Publishing, a.s.

Hnilica, J. & Fotr, J. 2009. Applied Risk Analysis in financial management and investment decisions. 1. edition Praha: Grada Publishing, a.s

Kelliber, CH.F. & Mahoney, L.S. 2000. Using Monte Carlo Simulation to Improve Long-Term Investment Decisions, In *The Appraisal Journal*, No. 1, pp. 44–56.

Kislingerova, E. et al. 2010. *Management finance*. 3. ed., Praha: C.H. Beck.

Korecky, M. & Trkovsky, V. 2011. Management of risks of a project with the intention of projects in industrial. Praha: Grada Publishing, a.s.

Mangla, S.K., Kumar, P. & Barua, M.K. 2014. Monte Carlo Simulation Based Approach to Manage Risk in Operational Networks in Green Supply Chain. In *Procedia Engineering97*, pp 2186–2194.

Schoemaker, J. 2002. Profiting from Uncertainty. Strategies for Succeeding No Matter What the Future Brings. The Free Press, New York.

Szabo, L., Varcholova, T. & Dubovicka, L. 2005. *Management of risk*. Bratislava: Ekonom.

Schoellova, H. 2009. Investment controlling. How to evaluate investment intentions and manage company investments. Praha: Grada Publishing, a.s.

Zmeskal, Z., Dluhosova, D. & Tichy, T. 2013. *Financial models. Concepts*, methods, *applications*. Praha: Ekopress.

Production Management and Engineering Sciences – Majerník, Daneshjo & Bosák (Eds)
© 2016 Taylor & Francis Group, London, ISBN: 978-1-138-02856-2

Tools for organizational changes managing in companies with high qualified employees

Z. Jurkasová, M. Cehlár & S. Khouri
Institute of Earth Resources, Košice, Slovak Republic

ABSTRACT: The main goal of article is to describe tools for managing organizational changes. Mainly tool InLook system, which bring many advantages in this field. Described tool is perspective for enterprise management and employees, mainly in manufacturing corporations, like mining companies, where should be high qualified employees. Mainly this days, which are specific by constantly changes and uncertainty, as for employers as for employees, as for enterprise. At these days, when the implementation of organizational changes represent often only way how to solve enterprise problems, which are at crisis. The main reason of this tool using is that lot of enterprises do not know when somebody is not at work, who is responsible for the work of his colleague, what negative influences work quality. Managers also can not dismiss these experts even if there are excess, because they know that, if necessary, they are no longer available in labor market.

1 INTRODUCTION

1.1 *Effective management of organizational changes*

Still lasting global economic and financial crisis affected all parts of business. This crisis also affects acceptation of legislative rules, which describes relations between employers and employees. This is the main reason why the most affected place is employment market. Enterprises do not have contracts, they have to save money, dismiss people, also review opportunities. That is the way how to worsen of demand at employment market and also why the Labour Office has more and more unemployed.

Based on this argument some theoretical models show that demand is slack, enterprises try to introduce new management processes, also reorganize its own processes of production, which mainly lead to organizational changes implementation. Through this changes enterprises are due to effectively and on time react to still lasting changes at market and also be facing to competition.

The main problem is that organizational changes bring together also stress, uncertainty, which are the main reasons why are people resistance to them. Only very good prepared projects of enterprise transformation, which are support by management and other key people have the chance for success. They also have to be prepared during the right culture of couching and managing.

By this comprehensive view can be maximized and effectively used mineral deposits in mining enterprises. Effectively use in this case, "by setting" the part of mining and processing of raw material, which is governed by the law and in advance define regularities like location and method of using bearings. Also economic factors like size of the investment, the investment structure, methods of financing investments, timing of investments and timetable mining, extraction size. Consequently, due to the changing external and internal conditions in the company it is required to effectively respond and adapt "internal structure" undertaking with respect to the emerging situation. It is also possible by organizational and personal changes, which are in this kind of companies very complicated.

This comprehensive view of the enterprise in terms of personnel and financial management, provided the use of innovative, supportive and available tools requires the achievement of the main objective of the article, which is mention for an effective management of organizational change by new available tools.

2 TOOLS FOR EFFECTIVE MANAGEMENT OF ORGANIZATIONAL CHANGES

2.1 *Conventional usages of tools*

Many companies, as was mentioned previously are exposed to unprecedented impact of changes, such as globalization, competition, privatization, changes in legislation, splitting and merging, reduc-

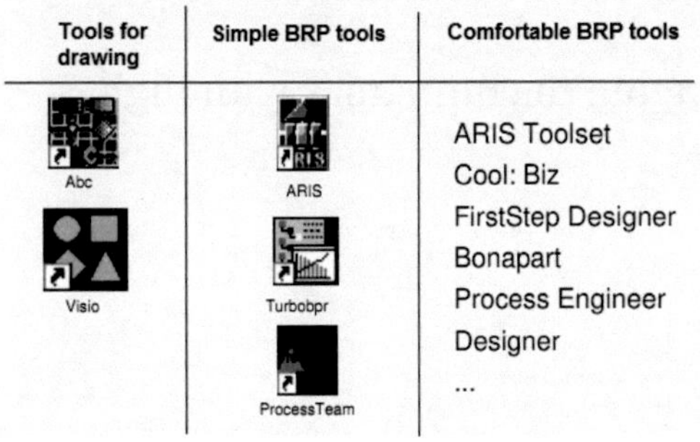

Tools for drawing	Simple BRP tools	Comfortable BRP tools
Abc	ARIS	ARIS Toolset
Visio	Turbobpr	Cool: Biz
	ProcessTeam	FirstStep Designer
		Bonapart
		Process Engineer
		Designer
		...

Figure 1. Tools for re-engineering business processes.

tion in personnel, changes in the product portfolio and many others. The reasons for changes can be difficult. Companies can be at difficulties, or they expected or wanted to be among the best on the market, or they want to be prepared for any possible unforeseen difficulties. Especially in the current technological progress rapidly changing times, when it is possible managing and monitoring changes in companies by using different available software-supported tools. They allow you to use not only monitoring of organizational changes, as well as to simplify and speed up the decision-making process and subsequently introducing the necessary changes.

In today's market there are a large number of tools for re-engineering business processes, which can be divided into simple drawing tools, simple and comfortable BRP tools (see. Fig. 1):

COOL: Biz is product that provides for their users comprehensive support for the development of business models. However, it has only static modeling options.

ARIS allows modeling of processes, organization, data, functions and create relationships between these objects, which are used to manage the substantial changes in the company. It also includes facilities for monitoring and evaluating the costs of business processes. It also supports ISO 9000 certification and implementation of process oriented standards such as Oracle, Baan and SAP (Krajčíová 2015).

2.2 InLook system®

As from the English name of the application implies "InLook", it means integrated inside look at the organization. This application is an analytical tool of personnel management, organizational change management, which by analyzing the current state of the organizational structure with respect to the activities enables to efficiently process data relating to the organization and its management.

This way allows to model different whether current or expected, but also possible future states of the enterprise and through individual modules to get different views of the modeled situation. This allows get a better idea and flexible response to the changing environment.

In this way, the user can select from a number of modules and therefore from the general usability of the application. This way user also chooses from the views to the company by the change managing (cost management, people, processes and others). This data can also be exported for further use in the management. At the same time the application is a tool of organizational controlling, in which are processed data of management of organizational change and knowledge into a coherent structured database of activities of the organization. InLook system is also very useful for creating organizational policy, for planning and deployment of human resources, training and compensation of employees, also for evaluation of their job performance. It identifies and describes in detail the effectiveness of organizational units and also the activities of individual employees. It allows creating at the time working database with high level of details as:

Using applications are strictly designed different activities, competencies described departments, locations and functional roles. They are also defined their ability, knowledge, skills and qualifications;

It analyzes the contents of the actual fund employee time;

Determining the actual occupancy of staff;

Allows connected documents necessary knowledge to functional sites and processes;

It describes the current situation of the company, allows to model desired, future state;

It eliminates duplication of activity and fill up missing activities;

Determine the optimal number of employees per activity (Inlook system®).

In terms of organization, this tool provides using mainly for:

– on the creation of a structured record data in a central database, collection, data editing by protected access through the Internet,
– it improves communication within the special technical terminology,
– it allows of the structured hierarchical views of the database information and overview in real time,
– it allows to users to define and to expand their own items,
– it implements results of process changes to personal agency with to the time change takes effect,

410

– improves the scope of organizational units, organizational structure as well as the activities of employees,
– it permits to employees to allocate required capabilities, to optimize the human resources in the context of process and strategic management (Inlook system®).

Besides upper described available software supported tools, it is also further possible use many traditional management approaches that are equally effective tool for managing the process of change. In the context of their use is the most important effective communication, skills and leadership styles of managers, but mainly target setting, in this case based on the change we want to achieve.

2.3   *Working with InLook system*

For the InLook system tool running is important to the overall interconnection of components, subsystems such as named Organizational structure, Types of working places, Catalogs, Objects, Processes. Within each part it is necessary to fill a specific database, input data for the management and operation of the enterprise, then the logical connection of different parts and mainly showed at a time. In this way, there is a "live" database, showing data over time, with the possibility of effective evaluation, according to user needs, respectively enterprise.

As a basis for guidance in the application InLook System it can be regarded division of the window into four parts. At the left side of the screen it may be selected from the basic module of Inlook System subsystems of applications, named Organizational structure, Catalogs, Types of working place, Processes, Objects (shown on the left side under each other) and the right side is a further description of the individual subsystems within the application, which are different Tabs (see Fig. 2).

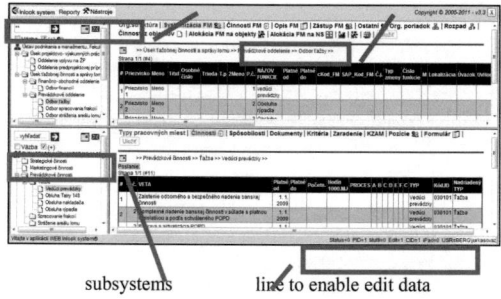
subsystems   line to enable edit data

Figure 2.   User environment of application InLook system.

By the first steps in working with the application it is necessary to fill the data in the individual subsystems. On the left side of the screen you can select only one of the selected subsystems, which are displayed in a tree structure. The first step is subsystem Organizational structure, on the right side is his closer specification in which we operate by clicking on the Tabs. The next steps are similar to a filling of the other subsystems and their tabs containing information to the relevant parts. Here it is important to note that the input data can be obtained from internal company documents, organizational structure, process maps, diagrams of mining, from management consulting, but also from employees of enterprises, as well as in the earlier stages during the assessment of the possible future scope and operation of mining enterprise, that is processed from various reports and studies.

In this way it is possible to describe the various subsystems within the tree structure of modules and consequently also different tabs. Then they must be mutually and logically interconnected the "binding" in dependence on part of evaluating each of processes, areas or part of enterprise.

The last part of working with the application allows users to use one of the basic modules of InLook System—InLook Reports. Like all other modules Reports are displayed on the left side of the working window in a tree structure. There is allowed select from fifteen possible options for the use of these resulting reports. These are consequently important for their further using and as a basis for the design or subsequent acceptance of change.

3   EXAMPLE OF SWITCHING MEN IN CASES OF ABSENCE

3.1   *Example from mining company*

At first we can model example where one of the employees of long-term sick leave. In this case, it is necessary for him to seek compensation, either in-house resources—rearranging the individual changes, or seeking new external staff who are capable of adequate crowd.

Selection of staff is an important step especially since high demands are placed on workers and their specific properties. For his reason, it is a great advantage of the use of database applications InLook System, specific subsystem Catalogs in which are precisely specified jobs with the necessary particulars to the cast.

Assuming that it is necessary in order to safeguard the operation because of the long-term sick leave to ensure the representation of a particular job on a certain date in the application it is possible

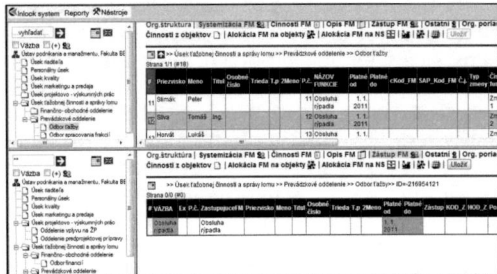

Figure 3. User environment of application InLook system.

Table 1. The necessary training for the staff position of operation of construction equipment.

Kind of training	Costs €
Basic course of operation of construction equipment	183,00
Extension of card of level	120,00
Update training (once/5 years)	60,00
Periodic retraining (one/2 years)	25,00
Issue of license	33,00
Safety trainings of employers	25,00
Safety training of managers	30,00
Together	476,00

to see through the partition Organizational structure, which is necessary to show that the lower and the upper working window, where in the first step at the bottom right of the screen dump data on staff who will represent in the *Tab* "Other". Then click around on the *Tab* "FM Crowd" at the top right of the window, I set in the *Tab* "Systemization FM" where I find a job of particular employee who is represented by a double click it switches to the lower windows to create a representation of a specific time (see Fig. 3). Consequently, setting the highest position within the Organizational structure and on the Tab "FM Crowd" you can see who is who within the company represents.

Modeling this condition is especially important because in this way it is possible to assess the adequacy of the crowd to the individual capabilities, which is represented employee and therefore to what extent the representation adequate and sufficient. Important perspective is the fact that the mineral extraction sector as mentioned places of great emphasis on a high degree of specialization, which is contained in the relevant laws, regulations defining the essential training and obtained certificates. It is therefore necessary to look also for the costs of training staff and the update training, without which the given position cannot perform its function.

As example of this is position Operation of construction equipment. Since the construction machine can operate only person with a valid engineer construction machinery (Tab "Competences Documents"), so the machine can be allocated to an employee who owns such authorization. It is one of the fundamental obligations of employers to organize training on occupational health and safety for its employees. Both of the obligations provided for by law no. 124/2006 Coll. OSH (Igmar, 2014).

Consequently, it is necessary to calculate the cost of training, retraining, or renewal of licenses (see Table 1). If we assume that the costs in this connection per employee is about 476€/year for a total of 23 employees in the field of mining and

processing of fractions, the implementation of job requires completion of each training costs can amount to around 10,948€/year after cannot be considered negligible amount.

4 CONCLUSIONS

As in mentioned in article, also as we can see from the special example from mining industry, area of managing organizational changes is very specific and for all enterprises important. From the mining enterprises it is also known that they are sometimes seasonal industries depend on what they are mining and processed. So they know that they have more than 10% employees which are not needed, but when they get some bigger order, they have to have this kind of qualified staff. When management dismiss this kind of employees, they also have to know that the same experts have not find. Si this kind of enterprises cannot save costs from human resources, mainly because competition is very huge. That is the fact why tools for effectively managing of organizational changes important.

REFERENCES

Igmar školiace stredisko. [online]. [cited 7.9.2014]. Available on internet: <http: //www.igmar.sk/druhy-skoleni/vysokozdvizne-voziky-kurz-vzv.html>.
Inlook system®—systém riadenia znalostí. [online]. [cited 7.9.2014]. Available on internet: <http: //www.inlook. eu/products/>.
Krajčíová, M. 2015. Nástroje BPR na slovenskom a českom trhu. [online]. [cited 5.2.2015]. Available on internet:<http: //www.krajciova.sk/dokumenty / clanky/c1_ nastrojebpr.pdf>.
Legel, R. & Krajčíová, M. 2014. Nástroje na reinžiniering podnikových procesov. [online]. [cited 7.12.2014]. Available on internet: <http: //www.krajciova.sk/_ dokumenty/clanky/c2_clanoknastrojebpr2.pdf>.

Production Management and Engineering Sciences – Majerník, Daneshjo & Bosák (Eds)
© 2016 Taylor & Francis Group, London, ISBN: 978-1-138-02856-2

Unmanned aerial vehicles vs. safety

R. Klír
Faculty of Aeronautics, The Technical University of Košice, Košice, Slovak Republic

ABSTRACT: The introductory part of the article is dealing with the issue of classifying and defining aerial equipment without pilots on board. It provides a short overview of ways these vehicles are applied, including a short overview of their development. Next part is devoted to the development of research, manufacturing and engagement of unmanned aerial vehicles. The author is pointing out the dynamics of the development in this field for the period of several years. Further part provides a brief review of the conditions, which are to be observed when operating them. The final part is focused on the issue of the prepared concept of the European Agency for Safety of Aviation (EASA) with regard to the unmanned aerial vehicles and their operation.

1 INTRODUCTION

The term drone has its origin in English expression used to designate the Dynamic, Remotely Operated Navigation Equipment, forming the acronym of DRONE. As it is commonplace in Anglo-Saxon countries, the current form of the word should be searched for elsewhere, not only in an abbreviation marking a certain group of flying platforms. Simultaneously, the English word drone marks member of the flying insect too. Similarly such the representative of the living world as the technical equipment termed of drone has been recently on the margin of the aviation community. Technical and technological progress, mainly miniaturization and electronisation contributed to the rise of such small flying objects importance.

Over the recent years, drones have become fixed part not only of the armed forces but civilian life as well. Advantages of unmanned aerial vehicles can also be utilized by various companies and private persons as well. Modern technologies, particularly their availability, enable a wide spectrum of the population legally acquires and makes use of these devices. They are small in size, controllable and with low noise signature. Most importantly, they can be easily adapted to the realization of many activities. The extent of activities in which drones can be currently applied is limited only by one's phantasy and the availability of the technology that is to fulfil the goals. It creates problems in the issues of safety. Practical application can involve both leisure activities with families or friends but also illegal actions threatening privacy of other persons, security of various objects or infrastructure. It is the availability of drones and applicable technology which forces our society to resolve legislative issues regarding their use. Currently, there are a great number of norms and rules that regulate utilization of technics and technology, yet there is no country having tackled satisfactorily the legislative issue of unmanned aerial vehicles.

2 CLASSIFICATION AND CHARACTERIZATION OF UNMANNED AERIAL VEHICLES

Currently, there are several classifications unmanned aerial vehicle. Among the most frequently used are by their way of control during flight and their elementary use.

By way of control ability, drones are divided into:

– Remotely Piloted Aviation Systems (RPAS) are equipment controlled by a pilot from a certain distance. It means that the flight is controlled by the pilot, who is not aboard of the aircraft. Such equipment must be approved by appropriate authority responsible of safety in the airspace where flights of such vehicle take place.
– Unmanned Aerial Systems (UAS) are automatically controlled flying objects, performing mission on the basis of an exactly defined time-schedule pre-programmed before the flight. Officially, operation of such platforms is not allowed by ICAO or by EU standards.

By way of application, drones can be classified for:

– Military use;
– Civilian use.

2.1 Military use

Military use of UAV is not a phenomenon of this era. Active use of such pilotless flying objects is known from WWII era when the Nazi Germany was bombing the territories of its military counterparts using the flying bombs called V-1 (13/JUN/1944-29/MAR/1945). After the war, this concept has been adopted by the armed forces of France, USSR and the USA. As a matter of course, the development has not stopped and many other countries started to build and successfully applied unmanned systems for military purposes, for example state of Israel or the former republic of Czechoslovakia. Currently, the technological development has reached a stage, when the devices developed for military purposes are no longer used only for passive missions of monitoring, surveillance or air-support but they can by useful in performing active roles in military engagements. It means that they can carry various weapon systems, which can deliver direct hits to the opposing forces.

2.2 Civilian use

Civilian use of UAV has been used mostly on the level of theoretical assumptions as a result of a strict hierarchy of security and technical or technological scarcity. UAV were represented only by small models, mostly miniaturized versions or models of their real pattern. Predominantly, they were radio-controlled models, which due to financial demandingness, were available only to a small group of fans of flying.

Development of technology, miniaturization or further progress in electronics, unmanned aerial vehicles became more affordable to wider and wider circle of potential users, with the ever increasing number of number types and ways of new applications. Some companies, not only private ones launched new areas of use for drones, e.g. fulfilling tasks of delivering small-sized parcels or inspecting technological infrastructure, in heavily accessible areas such as oil drill rigs or towers in the sea. In view of the current status in technological development, UAV are currently present on a wide scale of civil applications, e.g. in a heavily soiled or contaminated areas or environments dangerous for human beings.

Nowadays, a lot of European countries use civilian drones for safety checks and inspections of various infrastructural items such as railway lines, water dams or the electric supply networks. In most countries civil drones are used to resolve task of crisis management, namely monitoring of the flooded areas, surveillance and support of fire fighting in mountainous terrain or in a larger complex of buildings.

Development of drones using in the future will be affected by development particularly on the field of power units and miniaturization. Currently used engines will be replaced by smaller and more powerful turbine engines; transition to electrical engines will eliminate environmental pollution. Miniaturization will again enable making smaller and smaller drones down to the size of insects etc.

3 CURRENT STATUS

Over the recent period of time, the topic of drones has increased on the dynamism of its development. There already exist drones, with Maximum Take-Off Weight (MTOW) from some grams to tens tons. The airspeed of some of them exceeded 1000 km/h and the maximum time of operations is between several minutes to orders of months. Among them are systems making use of the various principles of flying, such as rotary wings, fixed wings or lighter-than-air systems.

3.1 Manufacturing and application

Development in the field of manufacturing and application for UAV is growing steadily over the recent period of time. According to some sources of information, their manufacturing and selling over the recent decade made up roughly 10% of commercial activities related to manufacturing and selling aviation equipment. The estimates speak of a gross turnover of cca 15 billions of Euros per year. In 2013 research and development of UAV, including those produced for military purposes, used up globally a sum of 5.2 billions of USD and by 2023 it is assumed to grow up to 11.6 billions of USD.

In 2013, only in Europe, there were 566 official manufacturers who offer more than 1708 different types of UAV. On global scales, only in 2013 their number grew up to about 471 companies, from which 176 were only in Europe. Estimates reveal that by 2050, only in Europe, their manufacturing will employ more than 150 thousand people. Apart from European manufacturers, production of UAV involves countries such as the USA, Israel; where the output was mostly focused on military application and numbers of production in these two countries positions them in the forefront of the world production. The large potential for manufacturing is increased in countries such as Brazil, China, India or the Russian Federation.

Growth is recorded not only by products, but also by geometric series growth a number of the operators. In the EASA member countries (28+4) there are more than 2500 operators registered in using UAV with MTOW below 150 kg. For example, only in France, their number grew up, from Dec/2012 to Feb/2014 from 86 to 431 operators. Similar trend is present in other countries, especially in the United Kingdom or Sweden, where annual rate of those obtaining licences is more than 200 new operators.

3.2 *Legislative*

Legislative environment of UAV has reacted on development of situation in the manufacturing and application, but many areas of it are dealt with more comprehensive meaning and it can be divided to three basic levels:

- Global;
- Particular;
- National.

Global level of resolving the issue of UAV is mostly based on the legislation of the International Civil Aviation Organization (ICAO), which is issuing documents of standardization. The ICAO member states are accepting these rules for aviation transport activities but those rules are not prescriptive but declaratory only. It means that these documents are not a directive what to do, but they are recommendation what to adhere to prevent violation of safety in air transportation. Therefore, the ICAO emphasizes the fact that drones are not flying in their separate airspace, but in a joined airspace along with piloted flying objects. Logically, their use should be subjected to the same rules, which ensure the required level of safety and the ICAO transferring responsibility for control and the directives issued in this regard by national authorities. The most important legislative norm in this field is the document Circ 328,— Unmanned Aircraft Systems (UAS), issued by the ICAO in 2011.

Particular level of the legislative environment with the competence applicable to Slovak Republic is solved mostly at the level of the European Union. Some countries, such as Sweden, France, Denmark, Italy, Germany, Great Britain or the Czech Republic have adopted their own legislations for UAV. However, they are reacting only to lightweight, radio-controlled flying platforms. These countries have adopted their legislative norms mostly for the reason to avoid to issuing each licence to operators of UAV separately. The main concern regarding this legislation is in that it is applicable only on country and cannot be used for international activities. No of this legislation provides area for inevitable cooperation or joint actions in the field of safety, protection of privacy, etc.

The European Commission (EC) is preparing a legislative framework for the operation of UAV, which would be valid in all member states and would ensure that drone will fly safely and they will be no longer pose any threats. It is assumed by EC, that such legislation will enable issuance of European licences enabling operation of UAV. At a wider level, this topic was dealt with experts, representatives of governments and manufacturers of UAV. Based on the results of these negotiations by the end of year 2015 the EASA will suggest recommendations regarding use of pilotless flying objects which could enter into force throughout the EU in 2016.

New standards of the EU should cover several important areas related to the operation of UAV. Among the most decisive ones are mainly:

- Strict rules of licensing valid throughout the European Union;
- Strict checks of privacy and data protection;
- Reviews regarding implementation of safety;
- Clear framework for responsibility and insurance;
- Higher efficiency of research, development and support to new branch of the aviation industry.

The national level of the legislation environment is the exclusive competence of each country. Based on the ICAO recommendations, each country is held responsible for ensuring safety at a required level within its sovereign airspace and the national legislation has to be adapted to this necessity. The problem of UAV use has been expanded even to areas, which have not been taken into consideration by national authorities during creating of Air Law of their own country. Currently, some countries still have managed this problem only partially or not at all. The majority of them solve problems by referring to their current Air Law. That situation brings problem which need to be solved immediately and the ways how to solve it can be very different.

For example, Germany is applying the articles of its current legislation and violation of the rules of the air applicable by UAV are sanctioned and fined. In Belgium, in present time, using of UAV is prohibited, with the exception matters licensed by the Directorate General for Air Transportation (DGTA). Nowadays Belgium is preparing a law, which will handle issues related to drone separately.

This proposal, however, met fierce critique, mostly from the potential operators because this document is suggesting limitations of flight level for UAV to 60 meters altitude only, which by experts, would limit the development and use of some types of such platforms, mostly those serving scientific purposes.

In present time, in Slovak Republic, there is no separate legislative norm dealing exclusively with the operation of drones. The currently applicable legislation has been adopted earlier and there are applied two parts of legislation, national and international (EU).

International legislation which is applicable in Slovakia is dealing with mainly pilotless aircraft with MTOW exceeding 150 kg and consisting of:

- Directives of the European Parliament and Council (EC) No. 216/2008
- Directive of the European Parliament and Council (EC) No. 785/2004.

In case of UAV with maximum MTOW up to 150 kg operation is legal only in compliance with Law of the National Council of the Slovak Republic No. 143/1998. By § 7, sect. 2 of the this law, flights in the airspace can be performed only under the conditions with due regard to flight safety and the conditions are laid down by the Civil Aviation Authority (Transport Authority of Slovak Republic) following the agreement with the Ministry of Defence of the Slovak Republic. All requirements for performing flights in the whole airspace of the Slovak Republic must meet requirements defined in Air Law for the aircrew, airworthiness of aircraft, conducting air transportation and aerial works. In all these cases, there is a requirement that aircraft can be operated in compliance with the conditions and requirements valid for the use of the airspace and aircraft capable of flying without a pilot on board (regardless of their weight) and airworthy must be strict separated from other aircraft.

4 CONDITIONS OF OPERATIONS

Using any kind of flying systems, including drones, is always regulated by the national legislation based on internationally accepted rules and recommendations of the ICAO. This should be performed so as to prevent violation of air traffic safety. Therefore, rules of air traffic and flying of UAV are strictly defined. The permission for their operations is always issued by the national authority authorized with competence to make decision on the basis of appropriate legislative norms. The rules do not apply only to copters, but to everything that flies without a pilot such as platforms, balloons, airships, paragliders, hovercrafts and kites.

In abroad it is common practice to see fly Radio-Controlled (RC) models on special airports for modellers only. In some countries it is strictly prohibited to perform flight of UAV through FPV (*First Person View*), i.e. flight with controlled model based on a view from a cockpit, such as it is in real aircraft, not by view at model from the ground. Similarly, aerial photography and filming is prohibited in many countries, or it is possible only on a special licence issued for a single, concrete flight. In these countries, violation of such rules is either sanctioned by fees, at best, or even sanctioned by prison, at worst.

Flights of UAV must be realized with emphasis on fulfilling some prescriptions and limitations. In the Slovak Republic they are determined by the Transport Authorities, for example:

- Operator of any aircraft capable of flying without pilot is:
  ∘ held responsible for the technical status of the aircraft and its airworthiness;
  ∘ held responsible for the preparation and performing the flight;
  ∘ knowledgeable with the state of weather prior to the actual use of the airspace.
- An UAV is prohibited to fly:
  ∘ to the distance exceeding visual contact of the operator of the UAV or distance 1000 m, which is reached first;
  ∘ to the minimum distance 50 m from any people, buildings, ships or ground vehicles except for the equipment or the persons needed for the take-off and landing;
  ∘ in the controlled airspace or segregated areas, including areas or zones sensitive to noise of aircraft operation, dangerous, prohibited, restricted areas, etc.;
  ∘ in the vicinity of all airfields of general aviation (airports for sports flying) including heliports not less than 5500 m from the reference point, without the permission and coordination of the operator or the administrator of the airfield.
- Flying an UAV must be performed so as to:
  ∘ prevent threats to other aircraft, persons or property on the ground;
  ∘ ensure protection of the environment against noise and emissions from UAV.
- Flying an UAV is recommended at a distance larger than 1500 m from densely built spaces or crowds on an open space, if the UAV is powered by a petrol engine and its MTOW is exceed 7 kg;
- An UAV cannot be used to transport persons, dangerous loads, materials or equipment, which could pose threats or endanger health of persons, property or environment;
- Dropping objects from the UAV during flight and performing flights by night is prohibited;
- The navigation lights have to be on during flight, if they are part of the equipment;
- Flight with UAV can be performed under VMC (Visual Meteorological Conditions) in uncontrolled airspace "G" while the flight level from ground cannot be higher than 330 ft./100 m AGL (Above Ground Level) and the operator have to maintain visual contact with the UAV;
- Flights on UAV can be performed with respect to (not to fly over) densely built areas, crowds

and areas or protected zones (e.g. water reservoirs, communication lines).

These rules are valid also in case when the UAV is operated only as a model of an aircraft, without commercial purposes, (sport, and competition or leisure activities). In the Slovak Republic, when negotiation with the Transport Authority, the decisive criterion is the MTOW of the UAV. When exceeding the weight limit of equipment (20 kg) it is necessary to register it in CAA and in compliance with the Directives of the European Parliament and Council (EC) No. 785/2004 it must be insured.

The person controlling the drone must obtain the permission for safe operation and possess the license certificate of airworthiness. Prerequisites of obtaining the licence is passing theoretical and practical test at the Transport Authority.

5 THE EUROPEAN AGENCY FOR SAFETY OF AVIATION CONCEPT OF SAFETY

The EASA has developed the concept of operation UAV relating to safety in the airspaces of the countries of the European Union, aimed to limit the risks resulting from use of UAV. Based on the wide spectrum of applying civilian UAV and the basic types of UAV there were defined three categories:

– Open category.
– Specific category;
– Certified category.

5.1 *Open category*

Open category should be pre-designed for operation of those UAV, which will remain under the control and supervision of police. It is a principle similar to the one applied to for registering and controlling cars. Activities listed in this category could be performed even without specific certification from the national Civil Aviation Authorities, or other responsible authorities. This measure should be applicable not only for the use of UAV, but for the operators as well. Included should be all small and medium sized activities, which can be realized separately from the standard air traffic, i.e. in the uncontrolled airspace. UAV will have to be used in compliance with the following limitations:

– A flying UAV must be under visual control of the operator to the maximal distance of 500 m;
– Maximal height must not exceed 150 m AGL or water surface;
– The flight must not be performed in the vicinity of specifically designated areas such as airports, restricted areas owing to ensuring safety of objects within it or for the purpose of protecting the environment.

Similarly, they will also have to comply with the measures adopted for protection of persons present in the immediate vicinity, where such flights are planned. In view of the fact that any time during flight and for any reasons operator can fail control over UAV, flights over crowd are prohibited generally. Excluded from this category can be all activities realized by flying toys, which mass do not exceed 500 g and those that can be controlled by persons younger than 14 years of age.

5.2 *Specific category*

Specific category will include activities with higher risks. It should be applicable for flights of UAV such as overflight persons or flights, when other flying object may be in their vicinity. This category will include all activities, which do not meet the requirements for the open category and where the necessary risk will have to be reduced by way of introducing operational limitations or ensuring of higher levels of capabilities of the UAV or operator.

Authorization of activities in this category should be performed exclusively by the National Civil Authorities or a specialized organization, which, however, would require changes in the organizational structure of the EASA and the related legislation.

The operator performing the activity in this category should be able assess the safety risks, identify measures to reduce them, which will have to be considered and approved by the National Aviation Authority. Assessment of safety risks on the part of the Civil Aviation Authorities by the concept of the EASA should not be necessary, if the operator is approved (certified) for performing such activities and is licenced to approve his own assessment of the safety risk. In case when the activities realized by UAV are performed in an airspace with standard air traffic, where other piloted aircraft will be in the air, the certified operator have to submit his own risk assessment to the provider of air traffic control services for approval.

Minimal level of assessing the maintenance of UAV should be based on the decision about safety risks and should be defined via adequate industrial standards.

Risks concerning the operator should be handled by defining the conditions for specific training for obtaining the EASA licence.

5.3 *Certified category*

Certified category should include all activities, where the level of safety risk to the operation of UAV achieved the level considered as standard to the normal air transportation. Such activities, as well as the UAV used to their purpose, will have to be certified in a way standard for normal operation of piloted aircraft. It means that not only the performance of the flight, but also the qualification of the operator or maintenance of the UAV will have to meet the same requirements and certification measures as it is standard for pilots, technicians, mechanics or aircraft. Thus, this category is comparable to standard air traffic operation.

Final definition of the differences between the certified and specific category is currently open and should be based mostly on the type of the engine, activities types or on the level of UAV autonomy.

There are several reasons for defining of the certified category, for example, noise, pollution, or other factors affecting the environment as a result of flying the UAV. The basic types of certificates, which should be issued for each UAV, could be e.g. Certificate of environmental protection, individual Certificate of Airworthiness or Certificate of noise level limitations.

Similarly, the process of certification should be applied to the radio, board equipment and other components responsible for controlling the UAV or providing not only radio contact but safety of the activities performed. When awarding of certificates, it will be necessary to take into consideration the configuration of the UAV, namely if it is a fixed-wing, rotary-wing or lighten-than-air system. Unification of the process of certifying the drones into an independent category would currently require changes in the directives on the establishment of the EASA.

6 CONCLUSION

Despite the efforts of responsible nations, international organizations, institution and authorities, it can rise situation, which will not respect appropriate legislative standards. Firstly, it is necessary to keep an eye on fact that the legislative norms issued by the ICAO are only declaratory and violations cannot be sanctioned. However, many of the legislative norms issued by the EASA are prescriptive and must be respected and violators should be sanctioned. It applies for both intentional but also unintentional interference into air traffic safety.

The most critical problem consists of the fact that the legislative norms of both the ICAO and EASA are covering exclusively civil air traffic, i.e. only use of UAV for civil purposes.

REFERENCES

Digitálny svet pod lupou: Nemci pokutujú za používanie dronov bez povolenia. 2014. [online]. [cit. 2015-03-20].

Digitálny svet pod lupou: Používatelia dronov sa na Slovensku musia registrova'?, nemôžu lieta'? nsad l'ud'mi ani budovami. 2014. [online]. [cit. 2015-03-20].

European Aviation Safety Agency. 2014. Concept of Operations for Drones: A risk based approach to regulation of unmanned aircraft. Brussels.

European Commission. 2014. Communication from the Commission to the European Parliament and the Council: A New era for aviation—Opening the aviation market to the civil use of remotely piloted aircraft systems in a safe and sustainable manner. Brussels.

European Commission. 2014. Remotely piloted aviation systems (RPAS): Frequently Asked Questions. Brussels.

Fözö, L., Andoga, R., Madarász, L., Kolesár, J. & Judièák, J. 2015. Description of an intelligent small turbo-compressor engine with variable exhaust nozzle. Source of the Document SAMI 2015—IEEE 13th International Symposium on Applied Machine Intelligence and Informatics, Proceedings: 157–160

International Civil Aviation Organization. 2011. Unmanned Aircraft Systems (UAS). Montreal.

Koblen, I., Szabo, S. & Krnáèová, K. 2013. Selected information on European union research and development programmes and projects focused on reducing emissions from air transport. Naše more, International Journal of Maritime Science & Technology, Vol. 60 No. 5–6.: 113–122.

Tlaèová agentúra Slovenskej republiky. 2015. Belgicko zvažuje zákon o bezpilotných lietadlách [online]. [cit. 2015-03-19].

V-1 flying bomb. 2015 [online]. [cit. 2015-03-25]. www.letectvo.nsat.sk

Production Management and Engineering Sciences – Majerník, Daneshjo & Bosák (Eds)
© 2016 Taylor & Francis Group, London, ISBN: 978-1-138-02856-2

Application of forecasting methods in aviation

J. Kolesár & M. Petruf
Department of Air Transport Management, LF TUKE, Košice, Slovak Republic

R. Andoga
Department of Avionics, LF TUKE, Košice, Slovak Republic

ABSTRACT: The aim of the article is to present the possibilities of application of forecasting methods in computation of prediction of the number of aviation passengers in a certain time frame. These methods can be also used for computation of forecasts during peak times during a day, week or year. The methodology results from estimation and precision of demands in logistic processes of future trends in supply and sales of goods and services on the market of transport aviation. The content of the article also includes a brief description of application of certain statistical and predictive methods in the area of creation of a general plan, economic, technical and organizational development of airports and airport companies. Possible fluctuations due to political, social, economic or weather factors are not considered in computation of forecasts of the number of air transported passengers.

1 INTRODUCTION

Only these companies that can efficiently and quickly satisfy the needs of their customers can achieve success on the aviation transport market. This relates not only air transport companies, but also to airports and airport companies, which need to adapt to the needs of transport companies. Highly professional approach and the supply of qualitative logistic services form the base of economic growth in aviation transport. This can result in increase of safety, efficiency and also the volume of the supplied transport services. This is a long term and difficult process, base of which is formed not only by qualitative employees and modern technological base, but it also demands a thorough planning process. The creation of prognostic plans, analytic studies and the ability to adapt to actual needs of the market and changes in legislative, technical, technological, safety and organizational structures all belong into this area (Johnson 1999, Dorčak et al. 2014).

The base for strategic, tactic and operative decisions of the accepted measures is also represented by forecasting of demand for air transportation. It is about planning of new flight lines and destinations, forecasting of the amount of the transported passengers, fulfillment of seating capacities or tonnage of transported goods and so on. Air companies as well as airports; have to be personally technically and organizationally prepared for it. Possible scenarios of future development and the creation of a forecasting plan has to come

from relevant and real materials, basis of which is formed by reliable information, consistency, results of analysis, simulations and modeling (Formánek 2004, Szabo et al. 2013).

Helpful tools in the process of demand and sales forecasting in air transportation can be certain mathematic forecasting/prognostic methods, by use of which demand for air transportation can be forecasted and evaluated in a time frame of several years (Endrizalová et al. 2014).

2 SIMPLE MOVING AVERAGE METHOD

It is a relatively simple method of a time series, base of which is represented by statistical data. Their obtaining and evaluation by solution of a specific decision making problem like the prognosis of the air transported passengers for the upcoming year is not trivial. Statistic data, which are at disposal have to be put into a statistic order, while the data have the character not only as section data, but also time series data expected in the following periods.

The method can also be used to forecast a number of air passengers in a certain period or to predict the number of passengers (tons of cargo) for a certain flight route, or seat utilization of a certain aircraft.

The formula for estimation of the forecast for demand of air transport by passengers (cargo, utilization of a route, line) for t+1 time period $P_{t,t+1}$ in t-th period has the following form:

$$P_{t,t+1} = \frac{\sum_{i=0}^{n-1} S_{t-i}}{n} \qquad (1)$$

where S_{t-i} is the number of transported passengers (tonnage of cargo, utilization of a line) during the previous n periods.

This method is however complicated to use for predictions in case of seasonal fluctuations in air transport, for example during one year, week or day. Such forecasts computed by the method of moving averages do not diverge from the total mean when looking at a larger time span (e.g. 5 and more years) (Král 2001a).

3 THE NAIVE MODEL

The naïve model represents a hypothesis of relation between two values of the same variable (seating utilization of the chosen aircraft type for a certain route), which we observe in several consequent periods. This type of model can also be used for comparison of precision with other forecasting models, for example the regression models (Garaj 1993, Socha et al. 2014).

In one of the naïve models, we start from the assumption that the predicted value of a variable in the following period Y_{n+1} is best represented by a value from some common period Y_n. At the same time we assume that the error of the created formula is a stochastic element of the model with zero mean value and a constant dispersion. This model can include seasonal and trend fluctuations like the number of passengers traveling during a peak. The formula of the model is the following:

$$Y_{n+1} = Y_{n-3} \frac{Y_n - Y_n - 4}{4} \qquad (2)$$

The average value for the last four terms can be added to the term value of a certain variable (like a single year), thus adding the estimation of trend into the model.

4 WEIGHTED MOVING AVERAGE METHOD

By using the method, average values of a certain variable from a certain chosen number of empirical values from a time series can be computed. The computed average value is added to the mean period of the moving part of the time series (Král 2001b).

During the calculation of prediction of air transport passengers (tons of cargo, seating utilization), the basis is formed by real weighted demands for passengers transport in real time and the transported passengers during a certain time period in the past.

We start from the formula:

$$S_{t+1-n} = P_{t,t+1} \qquad (3)$$

After inclusion:

$$P_{t+1,t+2} = P_{t,t+1} + \frac{1}{n}(S_{t+1} - S_{t+1-n}) \qquad (4)$$

It holds that:

$$P_{t+1,t+2} = P_{t,t+1} + \frac{S_{t+1} - P_{t,t+1}}{n} \qquad (5)$$

If we include for $\frac{1}{n} = \alpha$ into the formula we obtain

$$P_{t+1,t+2} = (1-\alpha)P_{t,t+1} + \alpha S_{t+1} \qquad (6)$$

while it holds, that the coefficient $\alpha = \frac{1}{n}$, where n is the number of time periods.

The calculation using the weighted moving averages can be combined with algorithmic processing. The selection of the correction factor in algorithmic calculation is quite problematic, this is the value which will need to be added or subtracted to obtain realistic results. If the correction factor equals 1, it means that the estimation of prediction of passengers in the following time periods is based on real transported passengers in the last (real) period. As the value of the correction factor gets smaller than 1, the prediction will be closer to the simple weighted average and will be less susceptible to actual changes in passengers numbers in the closest upcoming period (Wessling 2003).

The correct selection of α solves the compromise between elimination of random fluctuations and adjustment towards the actual state of the transported passengers, while the coefficient does not have to be constant. Methods, which can help to adjust the correction factor 1n are called adaptive methods (Gros. 2010).

5 THE COMBINATION OF WEIGHTED MOVING AVERAGES AND SEASONALITY

To improve the resulting value of the prediction (estimation) in numbers of potentially air transported passengers during a certain period, it is necessary to correct the predictions. This is mainly relevant in computation of seasonal fluctuations in the demand. Analyses unambiguously show that the density of air transportation and number of passengers is higher in summer months compared to winter months. This is mainly caused by climatic factors and vacations during summer periods, which is more attractive to tourism. Seasonal fluctuations in air transport are evident in peaks during a day or a week. The density of air transport in the morning and hours before lunch is several times higher than evening and night flights.

The algorithm of computation of the weighted averages and seasonality is done as follows:

1. the method of moving averages is applied describing the whole development trend of the researched time series,
2. seasonal indexes based on moving averages are computed describing the seasonal fluctuations according to the formula:

$$I_s = \frac{S_t}{\bar{S}_t}, \tag{7}$$

where S_t is weighted average for the t-th period.

Correction of the forecast for seasonal fluctuations of air transportations during a year (week, day) is then computed as

$$P_{t,t+1} = I.S_{t+1} \tag{8}$$

6 EXPONENTIAL SMOOTHING

This method represents an application of the weighted moving average method. In forecasting of the number of air transported passengers with application of such smoothing, the coefficients $\alpha \in \langle 0,1 \rangle$ are computed. The values of the coefficients are derived from exponential division. Higher importance in this case have the most current (actual) statistical data about the number of dispatched passengers during a certain time period.

The formula for computation of the prediction from n historical data by application of the base relation for weighted average the equation for the exponential smoothing has the following shape:

$$P_{t+1} = \alpha S_t + \alpha(1-\alpha)S_{t-1} + \alpha(1-\alpha)2S_{t-2}$$
$$+ \cdots + \alpha(1-\alpha)^{t-n+1}S_{t-n+1} \tag{9}$$

or

$$P_{t+1} = \alpha S_t + (1-\alpha)[\alpha S_{t-1} + \alpha(1-\alpha)S_{t-2}$$
$$+ \cdots + \alpha(1-\alpha)^{t-n+2}S_{t-n+1}] \tag{10}$$

It is also possible to apply the Holt method in exponential smoothing, if the demand for air

transport achieves a certain value level and trend element. This method however does not have any element of seasonality. It is the case of double exponential smoothing (Král 2001c).

The algorithm of the computation is as follows:

1. equalization of the series of known values from the third period
2. the first value for the second period is $P'_{2,1} = S_2$
3. the first difference will be $d'_1 = S_2 - S_1$
4. the equalized value for the third period then equals:

$$P'_{3,2} = (1-\alpha)(P'_{2,1} + d'_1) + \alpha S_3 \tag{11}$$

$$d'_2 = (1-\beta)d_1 + \beta(P'_{3,2} - P'_{2,1}) \tag{12}$$

To adjust the series with seasonal fluctuations in air transport Holt-Winters method can be used, however it has to be sufficiently long to catch the seasonality. Moreover, this method also uses time indexes, which are computed as follows:

$$I_t = \frac{S_r}{\sum_{i-1}^{s} S_i/S} \tag{13}$$

where s is the length of the seasonal series.

The time series can be further equalized including differences and time indexes according to the following formulas:

$$P'_{T+1,T} = (1-\alpha)(P'_{T,T-1} + d'_{T-1}) + \alpha \frac{S_t}{I_{T-s}} \tag{14}$$

$$d' = (1-\beta)d'_{T-1} + \beta(P'_{T+1,T} - P'_{T-1,T}) \tag{15}$$

$$I_{T+1} = \gamma \frac{S_T}{P'_{T+1,T}} + (1-\gamma)I_{T-s} \tag{16}$$

5. Prediction for the k-th period is then:

$$P_{T+i} = (P'_{T-1,T-2} + id'_{T-1})I_{T-s} \tag{17}$$

The methodology of the computation is as follows:

1. computation of seasonal indexes in every seasonal series,
2. computation of means of seasonal indexes for every period in a seasonal series,
3. estimation of the first difference as a subtract of average needs for the last two seasons divided by the seasons' length,
4. estimation of the equalized centered moving average for the last period of the last season as an average of the last season plus the multiple of the estimated mean difference,

5. computation of the equalized moving averages and the computation of prediction for the next season.

One of the applicable methods in prognostic planning of demand in air transportation can also be the Brown exponential equalization, which includes automatic weighting of all previous data, where the weights are decreasing exponentially over time. This method can set a prediction for one period ahead. The method is suitable only for time series with a constant trend.

7 DECOMPOSITION MODELS OF TIME SERIES

These models have an important place among the prognostic models. Except the equalization according the means, analysis is used to estimate trends or seasonal fluctuations.

These models distinguish three elements of time series:

1. trends,
2. seasonal elements,
3. stochastic elements.

The task in decomposition of time series is to quantify their trend and seasonal elements. The process of quantification of these elements is called decomposition of time series

The models of decomposition are divided into the following categories:

1. additive models
2. multiplicative models.

Additive decomposition can be defined as:

$$D = T + S + E \qquad (18)$$

The multiplicative decomposition can be defined as:

$$D = T.S.E \qquad (19)$$

where

D = time series of historical data,
T = the trend element,
S = the seasonal element,
E = the stochastic element.

The additive model can be applied where the course of the trend of demand is quite stable. If the trend shows some considerable growth or decline it is more suitable to use the multiplicative model. The information gathered by the decomposition can be used in evaluation of the present development of demand and its future behavior.

If the trend is to be quantified, it is necessary to use moving averages for setting the trend and further use the knowledge to set the seasonal factors (Petö 2001a).

Decomposition of time series consists of the following algorithmic steps:

1. search of the trend, which represents the main tendency of the relatively long period of the analyzed demand described by the methodology known as centering the trend (mean of two moving averages),
2. finding of the seasonal element by the afore mentioned formula for additive decomposition where the stochastic element is neglected: $S = D - T$,
3. creation of the cleaned time series from seasonal influences. The result is represented by trend elements including only random elements. This means that $D - S = T + E$, where the left side represents the seasonally cleaned series.

The task can be considered as finished only when we see that the seasonal element is not changing, this means that the forecast is done by the trend. If it is not the case, causal models can be used. One of the models can be represented by a simple linear regression, which represents a relation between two parameters in a general shape as $y = f(x)$, where y is the dependent variable and x is the independent variable.

The precision of the regressive analysis is largely dependant on selection of the correct data that will be used in regressive analysis. In case where several variables exists, multiple level regression has to be applied (Král 2001d).

8 EVALUATION OF THE FORECAST

In order to evaluate precision of the forecast the methodology of computation of mean absolute deviation or mean absolute error can be used. The mean absolute error is a basic tool to evaluate error created by the forecasting model in relation to historic data of the time series

The formula for computation can be expressed by the relation:

$$MAD = \frac{\sum_{t=1}^{n} |D_t - F_t|}{n}, \qquad (20)$$

where,

D_t = the real demand in a time period t,
F_t = forecast of the demand in a time t,
n = number of the used time periods.

The absolute value is needed to ignore the direction of the deviation. This characteristics is used when we want to express the error of the forecasting in the same units as the previous statistical series.

8.1 *Mean square error*

Mean square error uses the same variables as MAD for computation. This characteristics penalizes large deviations in the forecast because the error is squared. Based on the square error, we prefer the model, which shows only slight values of error in front of a model which shows only very small values of errors and one extreme forecast error (Petö 2001b).

Mean square error can be computed as:

$$MSE = \frac{\sum_{t=1}^{n} |D_t - F_t|^2}{n} \quad (21)$$

8.2 *Mean forecast error*

It happens only seldom that the forecasted value and the real demand are the same. However, the difference between the mean forecast value (scope of several time intervals) and the average value of the real demand should be the smallest possible (Petö 2001c). To evaluate this demand, the following computation of the average error of the prognosis can be done:

$$MFE = \frac{\sum_{t=1}^{n} |D_t - F_t|}{n} \quad (22)$$

8.3 *Mean absolute percentage error*

The last important characteristics is an addition to improve the understandability in evaluation of forecasts. Even though this characteristics is named as absolute it denotes relative error. The formula to calculate this error can be expressed as follows:

$$MAPE = \frac{100}{n} \sum_{t=1}^{n} \left| \frac{D_t - F_t}{D_t} \right| \quad (23)$$

9 CONCLUSION

The forecasting of demand and supply in air transport belongs to important analytic studies in creation of future flight plans, logistic planning, financial, organizational, operational and personal elements. Forecasting of air transport is generally a combination of analysis of historic data, intuition, experiences of analytics and prognostics of air and airport companies.

ACKNOWLEDGEMENT

This work is supported by the project KEGA 014TUKE-4/2015—Digitalization, virtualization and testing of a small turbojet engine and its parts using stands for modern applied education. This work was co-funded by the Slovak Research and Development Agency under the contract No. DO7RP-0023-11.

REFERENCES

Dorčak, P., Pollak, F. & Szabo, S. 2014. *Analysis of the possibilities of improving an online reputation of public institutions*. In: IDIMT 2014. Sept. 10–12. Poděbrady: IDIMT Networking Societies—Cooperation and Conflict 22nd Interdisciplinary Information and Management Talks. 275–281.

Endrizalová, E. & Němec V. 2014. *Demand for Air Travel.* In: MAD—Magazine of Aviation Development. Vol. 2, No. 12. Praha.

Formánek, T. 2004. *Systém On Line: Demand planning; Cesta k úspěšnému supply chain managementu* Dostupné na: URL: http://www.systemonline.cz/clanky/demand-planning.htm.

Garaj, V. 1993. *Úvod do ekonometrického modelovania.* Bratislava: Ekonomická univerzita.

Gros, I. 2010. *Řízení dodavatelských řetězců.* Interný učebný text. Přerov: VŠLG.

Johnson, J.C. 1999. *Contemporary Logistics.* 7th Edition, Upper Saddle River, Prentice Hall.

Král, J. 2001. *Podniková logistika.* 1.vyd. Žilina: Vydavateľstvo ŽU.

Petö, Z. 2001. *Metódy odhadu poptávky.* [Diplomová práca] Přerov: VŠLG.

Socha, V., Kutilek, P., Stefek, A., Socha, L., Schlenker, J., Hana, K. & Szabo, S. 2014. *Evaluation of relationship between the activity of upper limb and the piloting precision.* In: Proceedings of the 16th International Conference on Mechatronics—Mechatronika 2014. IEEE. 405–410.

Szabo, S., Dorčák, P. & Ferencz, V. 2013. *The Significance of Global Market Data for Smart e—Procurement Processes.* 2013 In. IDIMT—2013. Sept. 11–13. Prague: IDIMT—Interdisciplinary Informatin and Management Talks. 217–224.

Wessling, H. 2003. *Aktivní vztah k zákazníkům pomocí CRM: Strategie, praktické příklady a scénáře.* 1.vyd. Praha: Grada Publishing a.s. 191–192.

Production Management and Engineering Sciences – Majerník, Daneshjo & Bosák (Eds)
© 2016 Taylor & Francis Group, London, ISBN: 978-1-138-02856-2

Investigations on thermophysical parameters of polyurethane filled with SWCN

I. Kopal & K. Kováč

Faculty of Industry Technologies in Puchov, Alexander Dubček University of Trencin, Trencin, Slovak Republic

ABSTRACT: The work is focused on the analysis of temperature dependence of thermophysical parameters of polyurethane filled with different amounts of single-walled carbon nanotubes having a high purity. The regression models of experimental data have shown that temperature dependence of thermal diffusivity can be described with cubic polynomial, temperature dependence of specific heat with bilinear model and temperature dependence of thermal conductivity with two third degree polynomials. The temperature dependence of thermal conductivity and thermal diffusivity showed an inflection point corresponding to the temperature of secondary relaxation transition occurred in all tested materials, while temperature dependence of specific heat indicated a secondary relaxation transition at the sharp change in the slope of bilinear regression confirmed by dynamic mechanical analysis. The experimental results also showed that embedding carbon nanotubes into polyurethane increases the thermo-physical parameters in accordance with a first-order exponential model over the entire monitored temperature range.

1 INTRODUCTION

Polyurethanes are a kind of unique plastic materials used in a wide range of practical applications. They offer elasticity of rubbers combined with toughness and durability of metals. Being composed of soft and hard segments, polyurethanes represent one of the most versatile construction materials in recent times because of their properties can be widely modified by selecting appropriate raw materials, catalysts and auxiliary compounds, by employing various production methods or by employing various methods for further processing and for shaping the final products (Randal & Lee 2003). These polymers are perfect materials for using as a strength and elastic matrix in nanocomposites (Prisacariu 2011).

Nanocomposites are composite materials consisting from two or more different components, but at least one of them must be in the form of particles with a size from 1 up to 100 nanometers which are uniformly dispersed in an inert matrix. In the most cases these nanoparticles are made of an active substance with interesting magnetic, electric or another properties. The reason of using active substance in the form of nanoparticles are their qualitatively different properties compared to matrix caused by a monodomain structure of nanoparticles, their high ratio of surface atoms to interior atoms, impossibility of nanoparticles

interactions and more other effects which are not yet examined (Drobny 2007).

Carbon nanotubes offer exciting opportunities for fabricating new composites due to amazing chemical and physical properties, such as excellent tensile strength, superior thermal and electrical properties and many others (Chen et al. 2012, Slobodian et al. 2012). One of the unique properties of carbon nanotubes is the possibility to make its only one atom layer thick. So this is the reason, why carbon nanotubes can be about 50,000-times thinner than a human hair. Nowadays, scientists try to make them in the large quantities with a very high degree of purity (with very little or no material defects) what leads to more possibilities of usage (Biercuk et al. 2002, Ghosh et al. 2010).

Nanocomposites based on polyurethane matrix with carbon nanotubes filler are characterized by very unique mechanical, surface and multifunctional properties and strong interactions with matrix resulting from the nano-scale microstructure and extremely large interfacial area. These are the main reasons why they have stimulated the great interest in recent years (Wang et al. 2011).

In this paper we present temperature-dependent thermophysical parameters of hardened polyurethane matrix filled with different amounts of single-walled carbon nanotubes with 90% purity in the temperature range from −25°C up to +75°C.

2 MATERIALS AND METHODS

2.1 Materials

In the present work, four sets of samples of hardened polyurethane filled with different amounts of single-walled carbon nanotubes with purity of 90 percents were investigated. The Nanocyl™ carbon nanotubes were stirred at normal pressure and at laboratory temperature by an ultrasound mixer WELDER in the reaction Axson technologies polyol component PX 522/HT POLYOL during about two minutes. After ten minutes of the vigorous agitation by the ultrasound in a vacuum chamber for degassing the mixture before its polymerisation, the Axson technologies PX 521-522 HT ISO ISOCYNATE was adding and the subsequent curing process was carried out for 4 hours at 80°C and 16 hours at 100°C. From the cured material four sample sets with 0 wt %, 0.5 wt %, 1 wt % and 1.5 wt % amount of single-walled carbon nanotubes were produced. Each sample was pressed by the displacement speed of 2 mm/min to a value of force 100 kN. The pressure modulus and the coefficient of compressibility were in the range of tension from 20 to 60 MPa.

2.2 Methods

The temperature dependence of thermal conductivity, thermal diffusivity, as well as specific heat capacity of polyurethane nanocomposites under investigation were analyzed in the temperature range from −25 to +75°C by professional Thermophysical Transient Tester 1.02, developed at hoc at Institute of Physics of Slovak Academy of Science, using a pulse transient method which is one of the most effective transient methods for investigation of thermophysical parameters of solids.

In the pulse transient method, one side of specimen surface is subjected to short stimulation by heat pulse. The temperature response in the form of temperature function (Kubièár 1990).

$$T(h,t) = \frac{Q}{c_p \rho \sqrt{\pi \alpha t}} \exp\left(\frac{h^2}{4\alpha t}\right) \qquad (1)$$

is registered on the opposite side of specimen surface. Variables T, h, t, Q, ρ, c_p and α represent surface temperature, specimen thickness, time, amount of heat generated by a planar heat impulse source per unit area, density, specific heat capacity at constant pressure and thermal diffusivity of specimen material, respectively, whereas

$$Q = RI^2 t_0, \qquad (2)$$

where R is the electrical resistance of the heat source and I is the electric current applied at time t_0.

The unknown thermophysical parameters c_p, α as well as λ are calculated from the transient temperature response (1) by using the one point method according to expressions

$$c_p = \frac{Q}{T_m h \rho \sqrt{2\pi e}}, \qquad (3)$$

$$\alpha = \frac{h^2}{2t_m} \qquad (4)$$

and

$$\lambda = \frac{Qh}{2T_m t_m \sqrt{2\pi e}}, \qquad (5)$$

where T_m is the maximum temperature response of the thermally stimulated specimen reached at the time t_m and e is the Euler's number work (Štofanik et al. 2007).

In the present work, the thermal stimulation of all tested specimens was provided by a 13 seconds wide current pulse at driving voltage of 1.4 V and a supply current of 1 A. The temperature response of specimens was measured during 420 seconds in a vacuum chamber at 0.1 Pa pressure. The time of specimens temperature stabilization at each test temperature was two hours, approximately. The experimental setup as well as experimental procedure was described in detail in our earlier work (Kopal et al. 2011).

3 RESULTS AND DISCUSSION

The temperature dependences of thermal conductivity, thermal diffusivity as well as specific heat capacity at constant pressure for all tested nanocomposite specimens in the temperature range from −25°C up to +75°C, include with their appropriate regression models (solid lines), are presented in Figure 1–3, respectively.

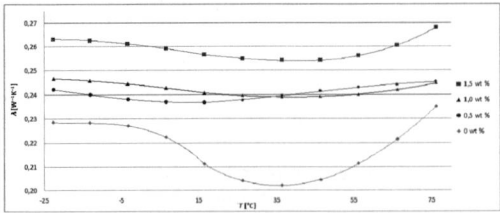

Figure 1. Temperature dependence of thermal conductivity for pure polyurethane and nanocomposites with 0.5 wt %, 1 wt % and 1.5 wt % amount of single-walled carbon nanotubes.

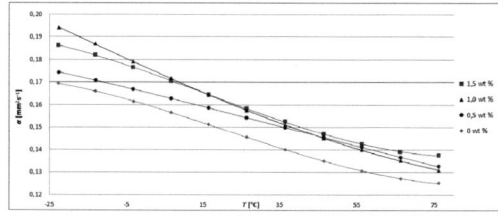

Figure 2. Temperature dependence of thermal diffusivity for pure polyurethane and nanocomposites with 0.5 wt %, 1 wt % and 1.5 wt % amount of single-walled carbon nanotubes.

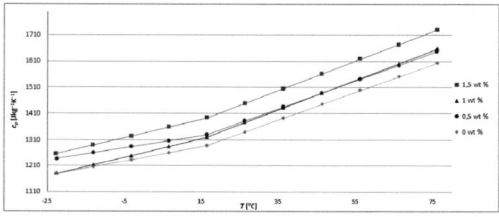

Figure 3. Dependence of thermal conductivity on carbon nanothubes content in polyurethane matrix at temperature of 25°C.

The particular regression models along with their coefficients of correlation R^2 with experimental data are summarized in Tables 1–3.

From the Figure 1, it is apparent that the temperature dependence of thermal conductivity for pure polyurethane matrix as well as for polyurethane nanocomposites with 0.5 wt %, 1 wt % and 1.5 wt % amount of single-walled carbon nanotubes can be described with bicubic polynomials (Tab. 1) having a decreasing trend in the first part and an increasing trend in its second part. The temperature dependence of thermal diffusivity for all tested specimens, presented in Figure 2, can be described with the cubic regression models (Tab. 2) having a decreasing trend and temperature dependence of specific heat capacity (Fig. 3) can be described with bilinear models (Tab. 3) having an increasing trend in both parts.

The rising content of carbon nanotubes inside the polyurethane matrix induces the apparent grow of all of three thermophysical parameters under investigation. However, this grow shows ambiguous character for different amount of nanotubes in polyurethane based composite at various temperatures (probably due to insufficient dispersion of non functionalized carbon nanotubes inside polyurethane matrix done by ultrasound mixing) and in the case of temperature dependence of thermal conductivity they have also a non-monotonic character.

Table 1. Regression models for temperature dependence of thermal conductivity.

CNT* wt %	Model	R^2
0	$\lambda_1 = -5\text{E-}07T^3 - 2\text{E-}05T^2 - 0.0004T + 0.226$	1
	$\lambda_2 = -8\text{E-}08T^3 + 3\text{E-}05T^2 - 0.0019T + 0.2348$	1
0.5	$\lambda_1 = -1\text{E-}08T^3 + 4\text{E-}06T^2 - 0.0001T + 0.2372$	1
	$\lambda_2 = 9\text{E-}09T^3 - 2\text{E-}06T^2 + 0.0003T + 0.2317$	0.9999
1	$\lambda_1 = 1\text{E-}08T^3 - 2\text{E-}06T^2 - 0.0002T + 0.2439$	0.9999
	$\lambda_2 = y = 1\text{E-}08T^3 + 2\text{E-}06T^2 - 0.0002T + 0.2439$	0.9997
1.5	$\lambda_1 = 3\text{E-}08T^3 - 3\text{E-}06T^2 - 0.0002T + 0.2605$	1
	$\lambda_2 = 1\text{E-}07T^3 - 6\text{E-}06T^2 - 3\text{E-}05T + 0.2582$	0.9999

Table 2. Regression models for temperature dependence of thermal diffusivity.

CNT* wt %	Model	R^2
0	$\alpha = 5\text{E-}08T^3 - 3\text{E-}06T^2 - 0.0005T + 0.1629$	1
0.5	$\alpha = 6\text{E-}09T^3 - 7\text{E-}07T^2 - 0.0004T + 0.1655$	1
1	$\alpha = 1\text{E-}08T^3 + 1\text{E-}06T^2 - 0.0008 + 0.1765$	1
1.5	$\alpha = 5\text{E-}08T^3 - 2\text{E-}06T^2 - 0.0006T + 0.1746$	1

* Carbon nanotubes.

Table 3. Regression models for temperature dependence of specific heat capacity.

CNT wt %*	Model	R^2
0	$c_{p1} = 2.7807T + 1241.9$	0.9998
	$c_{p2} = 5.2988T + 1200.2$	1
0.5	$c_{p1} = 2.3505T + 1290.4$	0.9999
	$c_{p2} = 5.3366T + 1241.4$	1
1	$c_{p1} = 3.5879T + 1260.6$	0.9999
	$c_{p2} = 5.6439T + 1226.8$	1
1.5	$c_{p1} = 3.5159T + 1335.8$	1
	$c_{p2} = 5.639T + 1300$	1

* Weight percent.

The inflection points of functional dependencies $\lambda(T)$ and $\alpha(T)$ as well as the point of the sharp change in the slope of $c_p(T)$ at temperature 16.5°C, approximately, indicate a secondary relaxation

427

transition, occurred in all tested nanocomposite materials, confirmed by a dynamic mechanical analysis method (Kováė & Kopal 2013). At the same time, the amount of carbon nanotubes has no practical influence on the observed secondary relaxation transition temperature.

The above mentioned regression models of experimental data showed a good agreement with physical models based on a quantum interpretation of thermal transport phenomena in solids above Debye temperature, presented in our earlier work, in the form of

$$\lambda_1(T) = AT + B \qquad (6)$$

and

$$\lambda_2(T) = \left(-CT - \frac{D}{T} + E\right)^{-1} \qquad (7)$$

for thermal conductivity,

$$\alpha_{1,2}(T) = \mp A_{1,2}T^{\pm B_{1,2}} \pm C_{1,2} \qquad (8)$$

for thermal diffusivity and

$$c_{p_{1,2}}(T) = A_{1,2}T + B_{1,2} \qquad (9)$$

for specific heat capacity at constant pressure, where A, B, C, D and E represent positive constants, different for particular thermophysical parameters, dependent on carbon nanotubes content inside the polyurethane matrix and they have to be determined from relevant experimental data. Indexes 1 and 2 correspond to the temperatures before and after registered relaxation transition, respectively.

The dependencies of analyzed thermophysical parameters λ, α and c_p on the amount of carbon nanotubes inside the polyurethane matrix along with their first-order exponential regression models at temperature 25°C are presented in Figures 4–6, respectively. The first terms of these regression models have a very interesting practical importance by reason that they can predict a maximum possible value of relevant thermophysical parameter which can be reached by the aid of carbon nanotubes into the polyurethane matrix at given temperature. The regression models of these functional dependencies for all others temperatures from monitored temperature interval -25°C to +75°C show analogical first-order exponential forms which can be expressed by the term of

$$p(w) = p_{max} - (p_{max} - p_0)e^{-kw}, \qquad (10)$$

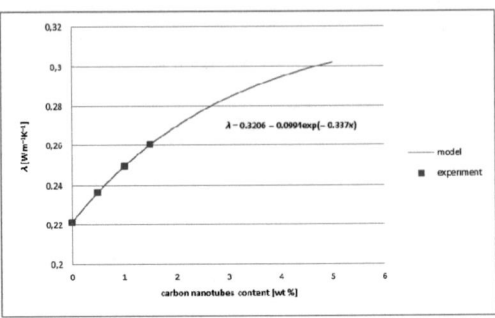

Figure 4. Dependence of thermal diffusivity on carbon nanothubes content in polyurethane matrix at temperature of 25°C.

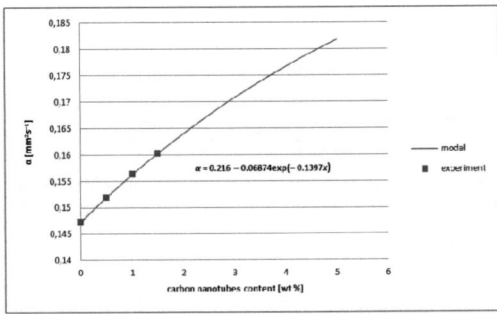

Figure 5. Dependence of specific heat on carbon nanothubes content in polyurethane matrix at temperature of 25°C.

where p and p_{max} are thermophysical parameter and its limit value at given temperature, p_0 is the value of this thermophysical parameter for pure polyurethane, w is the amount of carbon nanotubes inside polyurethane matrix and k represents the amount of nanotubes which rises the nanocomposite thermophysical parameter p to 37% of its value for pure polyurethane.

4 CONCLUSIONS

The temperature-dependent thermophysical parameters of hardened polyurethane matrix filled with different amounts of single-walled carbon nanotubes with 90% purity in the temperature range from −25°C up to +75°C were investigated in this work by using the pulse transient method. It has been shown that the temperature dependence of thermal conductivity for pure polyurethane as well as for polyurethane nanocomposites with 0.5 wt %, 1 wt % and 1.5 wt % amount of single-walled carbon nanotubes can be described with bicubic polynomials having the decreasing trend in the first part and the increasing trend in second part, the temperature dependence of thermal diffusivity for all tested specimens can be described with the cubic regression models having the decreasing trend and the temperature dependence of specific heat capacity with bilinear models having the increasing trend in both parts. The identified regression models showed good agreement with physical models based on quantum interpretation of thermal transport phenomena in the solid materials above Debye temperature. The experimental results have shown that inflection points of functional dependencies $\lambda(T)$ and $\alpha(T)$ as well as the point of the sharp change in the slope of $c_p(T)$ at temperature 16.5°C, approximately, indicate the secondary relaxation transition, occurred in all tested nanocomposite materials, confirmed by dynamic mechanical analysis, and that the amount of carbon nanotubes has no practical influence on the observed secondary relaxation transition temperature. Moreover, the first-order exponential model for dependence of thermophysical parameters under investigation on amount of carbon nanotubes in polyurethane matrix has been created.

ACKNOWLEDGEMENT

The authors gratefully acknowledge the contribution of the Scientific Grant Agency of the Slovak Republic under grant VEGA 1/0385/14 and KEGA 007 TNuAD-4/2013.

REFERENCES

Biercuk, M., Llaguno, M.C, Radosavljevic, J.K, Hyun, J.K., Johnson, A.T. & Fisher, J.E. 2002. Carbon nanotube composites for thermal management. *Applied Physics Letters* 80: 2767–2769.

Chen, S., Wang, Q. & Wang, T. 2012. Damping, thermal, and mechanical properties of carbon nanotubes modified castor oil-based polyurethane/epoxy interpenetrating polymer network composites. *Materials and Design* 38: 47–52.

Drobny, J.G. 2007. *Handbook of Thermoplastic Elastomers*. New York: Elsevier.

Ghosh, S., Bachilo, S.M. & Weisman, R.B. 2010. Advanced sorting of single-walled carbon nanotubes by nonlinear density-gradient ultracentrifugation. *Nature Nanotechnology* 5: 443–450.

Kopal, I., Bakošova D. & Šišáková, J. 2011. Temperature Dependency of Thermal Properties of Polyurethane. *Hutnické listy* 7: 69–72.

Kováč, K. & Kopal, I. 2013. Detection of primary and secondary relaxation transitions in hardened polyurethane by two different methods. *Hutnické listy* 7: 35–37.

Kubičár, L' 1990. Pulse Method of Measuring Basic Thermophysical Parameters. *Comprehensive Analytical Chemistry* 12: 350.

Prisacariu, C. 2011. *Polyurethane elastomers*. Wien: Springer-Verlag.

Randal, D. & Lee, S. 2003. *The polyurethane book*. New York: Willey.

Slobodian, P., Riha, P. & Saha, P.A. 2012. Highly-deformable composite composed of an entangled network of electrically-conductive carbon-nanotubes embedded in elastic polyurethane. *Carbon* 50(10): 3446–3453.

Štofanik, V., Markovič I, Boháč, V., Dieška, P. & Kubičár, L'. 2007. RT-Lab—the Equipment for Measuring Thermophysical Propertis by Transient Methods. *Measurement Science Review* 7(1): 15–18.

Wang, T.L, Yu, Ch.Ch, Yang, Ch.H., Shieh, Y.T., Tsay, Y.Z. & Wang, N.F. 2011. Preparation, Characterization, and Properties of Polyurethane-Grafted Multi-walled Carbon Nanotubes and Derived Polyurethane Nanocomposites. *Journal of Nanomaterials* 2011: 9–18.

Production Management and Engineering Sciences – Majerník, Daneshjo & Bosák (Eds)
© 2016 Taylor & Francis Group, London, ISBN: 978-1-138-02856-2

Selected aspects of modeling of movements of flying objects

N. Kotianová, T. Vaispacher & D. Draxler
Faculty of Aeronautics, Technical University of Košice, Košice, Slovak Republic

ABSTRACT: The contribution is dealing with the selected aspects of modeling the movements of flying objects. Using the derived algorithms we model the movement of flying object in an airspace. The position of the flying object is evaluated in a geocentric coordinate system with the center of the Earth's center of gravity. Presented further are transformational relations between the individual coordinate systems, which are used in the process of simulation. The given models are demonstrating the movement of flying objects in the airspace. Simulation of the flying objects is realized in the Matlab program environment. The developed algorithms can be used when modeling a system of relative navigation.

1 INTRODUCTION

Modelling and simulation process can be used in many scientific disciplines. They form important part of work of mathematicians, technicians and engineers working in various industries. By modelling we can conduct experiments out of real system without interfering with its operation. It allows us to get an idea of system operation before it is proceeding to its production. (Džunda et.al 2011, Džunda & Cséfalvay 2013)

Modelling is an effective tool by way of which we obtain new information about the structure of the system and its behaviour under changed conditions. (Džunda 2009a, b)

Our contribution is devoted to selected aspects of mathematical modelling the movement of the Flying Object (FO). Experiment on the real system is replaced by simulation of developed mathematical models on the computer. Such a mathematical simulation is called mathematical experimentation. Models as well as subsequent simulation are implemented in MATLAB program environment.

2 PRINCIPLES OF MATHEMATICAL MODELLING AND SIMULATION

Modelling is a research method, the basis of which is in replacing the object of research by its model. Modelling is aimed at obtaining information on the system subjected to research. The concept modelling is often wrongly confused with simulation, whereas one cannot accept them identical. One could say that the modelling process precedes the simulation. Modelling process means making models on which the experiments are conducted subsequently. (Rozenberg et al. 2014)

The mathematical model is an abstract model that uses mathematical notation to describe the system behaviour. The mathematical model must be goal-oriented, there is no universal model. Mathematical models can be analytical, statistical and statistical-analytical.

3 THE PRINCIPLE OF DEVELOPING THE MODEL OF THE MOTION OF A FLYING OBJECT

The design of simulation model is based on mathematical description of current idea and on the assumed system and its movement. (Socha et.al 2015) Consider the case when modelling scene is three-dimensional and we want to create three-dimensional image in ECEF Cartesian coordinate system (ECEF—Earth-Centred, Earth-Fixed). It is a three-dimensional Coordinate System (CS) with origin at the Earth's centre of gravity. The X-axis is identical with prime meridian. The y-axis is directed from the west to the east. The Y-axis has a north-south direction. It is a rectangular coordinate system. (Bréda et al. 2006).

As the initial coordinates of FO we have chosen point in geodetic coordinate system specified by latitude, longitude and altitude. For each part of the movement of FO we use local East, North, Up (ENU) Cartesian coordinates.

Simulation of movement of flying object consists of three phases (Figure 1):

1. Linear movement (time duration 300 s)
2. Turning No. 1 (time duration 100 s, angular increment 1 rad/s)
3. Turning No. 2 (time duration 100 s, angular increment 1 rad/s).

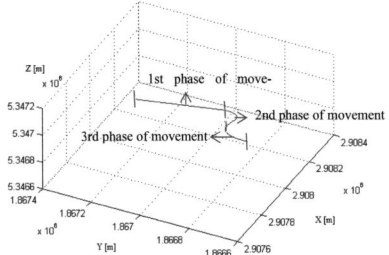

Figure 1.   Model of movement of flying object.

1st phase of movement:

```
for t = 1: 300;
x(t) = 25 + t;
y(t) = 0 + t;
z(t) = 1700;
end;
```

2phase of movement: Turning No. 1

```
for i = 1: 100
x(t+i) = s((t+i)−1) + 2.2*r * cos(ink(i) + beta);
y(t+i) = w((t+i)−1) + 2.2*r * sin(ink(i) + beta);
z(t+i) = 1700;
end;
```

3rd phase of movement: Turning No. 2

```
for k = 1: 100;
x(k+t+i) = s((k+t+i)−1)−(4.2*r) * cos(ink(k) +
(beta+0.1));
y(k+t+i) = w((k+t+i)−1) + (4.2*r) * sin(ink(k) +
(beta+0.1));
z(k+t+i) = 1700;
end;
```

where:

ink—step vector for generating motion of the same length
beta—angular increment
$r = 1$ is radius (dimensionless)

The initial coordinates of our point are entered into the geodetic CS, while by modelling only local motion is considered. Thus, the resultant coordinates of movement of FO are not linked with WGS-84. Therefore, it is necessary the model of local movement of FO convert to geodetic CS (LLH). Geodetic coordinate system is used on most maps and it is also used for data processing in autonomous navigation and air-navigation equipment. (Bréda & Soták & Váci 2006).

Geodetic coordinate system determines the position of a point on the surface of the ellipsoid. The coordinates are latitude, longitude and ellipsoidal height, for which it holds:

$$H = h + \varsigma \qquad (1)$$

where h is an altitude and ς is the height of a geoid or of a quasi-geoid.

Transformation process of the coordinates of resulting movement of FO with start in the entered into the geodetic CS is divided into two steps:

1. Transformation of initial geodetic coordinates of FO into geocentric CS (ECEF),
2. Transformation of local movement into geocentric CS (ECEF).

For coordinate transformation, we used the following MATLAB functions:

• Function for transformation of initial coordination from LLH into ECEF: llh2xyz

The function is used to convert geographic coordinates (latitude, longitude and altitude of WGS-84) into the rectangular geocentric coordinates X, Y, Z. The latitude and longitude are in radians and the altitude is in metres.

• Function for coordinate transformation of local movement of FO from ENU into ECEF: enu2xyz

The ENU is a local topocentric coordinate system. This system has its beginning in the referential point. The u-axis (up) is given with the direction of the normal to the ellipsoid in the initial point, while the n and a—axes are in the horizontal plane and perpendicular to the v-axis, which intersects the topocentre. The horizontal axes are in the plane of the geodetic meridian. The n-axis runs in northward direction and the e-axis to the east. The topocentric coordinate system is also helpful for applications where the area being mapped is sufficiently small to allow the curvature of the earth to be ignored, thereby rendering projections unnecessary. Thereby, in the defined system, we are evaluating measurements and the individual points shifts in relation to the reference points. (Volker 2009)

• Function for coordinate transformation final model of movement of FO from ECEF into LLH: xyz2llh

Using these functions, we can determine the real position of our FO on the map, because outputs are latitude, longitude and altitude related to reference system WGS-84.

Our task is to determine and display position of FO in rectangular ECEF coordinate system. Therefore, it is necessary to transform the coordinates of the final model of movement of FO into the latitude, longitude and altitude.

• Function for transformation final model of movement of FO form LLH into ECEF: xyz = llh2xyz(lat, long, h)

After coordinate transformation processes we can proceed to visualize the modelled movement of FO. We work with the geocentric coordinate system ECEF. Position of FO is determined by the coordinates x, y, z (in meters) at every second. The CS origin is located at the Earth's centre of gravity. The developed models can be modified by the required input parameters, namely: initial position of FO, time duration of motion or some motion of its part, the trajectory of motion. Next chapter is devoted to some cases modifying the initial parameters of models and their visualization.

4 THE APPLICATIONS OF THE MOTION MODEL OF FO

We apply the above algorithms for developing models of motion of FO. By changing the input parameters we will show more options of FO motion in space. Such models can be applied to any point in the space defined by latitude, longitude and altitude.

Case No. 1: We will use five models of FO with identical trajectories but with a height difference of 300 m (1000 ft) which is the minimum allowable vertical separation between aircraft for flights below FL 290 (8850 m). All motion models of FO labelled as A, B, C, D, E are shown in Fig. 2. The initial coordinates for each motion model of FO in a geodetic coordinate system related to WGS-84 are registered in Table 1.

By applying coordinate transformation processes described in chapter 2, we obtain the coordinates of the initial points of the FO in the geocentric coordinate system (Table 2).

In figure No. 2 an No. 3 we can see the modelled motion of five flying objects flying along an the identical track with different height. The motion involves a straight-and-level flight with time duration 300 s (5 min) and two turns (right, left) with time durations of 100 s (1,67 min) each. Radius of turn is r = 1, angular increment beta = 1 rad.

Case No. 2: By changing the input parameters, we show several cases of movement of FO in space. Fig. 4 shows model of movement of FO, which consists of consist of straight-and-level motion and one right-hand turn with six times greater radius than in case No. 1. The initial coordinates

Table 2. The geographic coordinates of initial points in ECEF.

FO	X (m)	Y (m)	Z (m)
A	2.90732795e+06	1.86634914e+06	5.34582022e+06
B	2.90738249e+06	1.86621593e+06	5.34631227e+06
C	2.90749676e+06	1.86649128e+06	5.34674921e+06
D	2.90774021e+06	1.86662900e+06	5.34716410e+06
E	2.90798069e+06	1.86677392e+06	5.34745926e+06

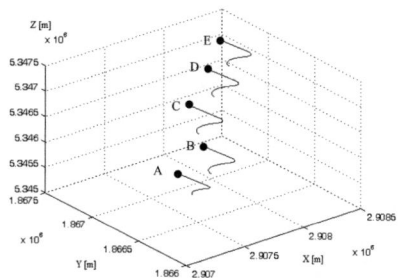

Figure 2. Models of movement of flying objects No. 1.

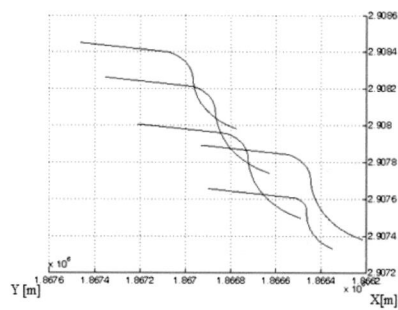

Figure 3. Models of movement of flying objects No. 1 (top view).

(Tab. No. 1, 2) and angle increment (beta = 1 rad) remained unchanged.

Case No. 3 in: Model of movement of FO on Fig. 5 shows the movement with time duration 900 s. (15 min.) of which 11,67 min. is made up of a straight-and-level flight. The initial geodetic coordinates of FO are [1,1,1000]. For local topocentric coordinates in model of FO it holds:

1st phase of movement: Straight-and-level motion

for t = 1: 700;
x(t) = 270 + t;
y(t) = 100 + t;
z(t) = 1000;
end

Table 1. The geographic coordinates of initial points in LLH.

Flying object	Latitude (rad)	Longitude (rad)	Altitude (m)
A	1	1	1000
B	1	1	1300
C	1	1	1600
D	1	1	1900
E	1	1	2100

433

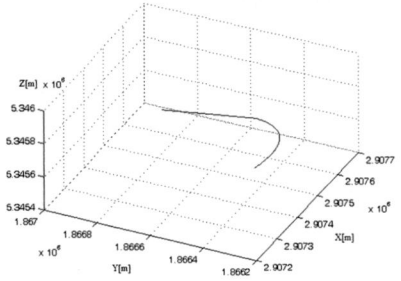

Figure 4.   Model of movement of flying object No. 2.

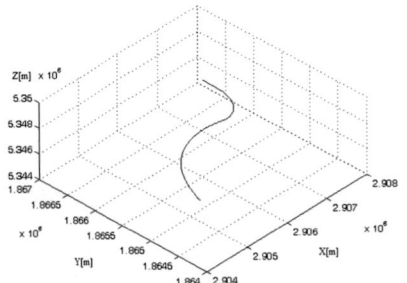

Figure 5.   Model of movement of flying object No. 3.

2phase of movement: Right-hand turn

```
for i = 1: 100;
x(t+i) = x((t+i)−1) + r * cos(ink(i) + beta);
y(t+i) = y((t+i)−1) + r * sin(ink(i) + beta);
z(t+i) = 1000;
end
```

3rd phase of movement: Left-hand turn

```
for k = 1: 100;
x(k+t+i) = x((k+t+i)−1)−(3*r) * cos(ink(k) + beta);
y(k+t+i) = y((k+t+i)−1) + (3*r) * sin(ink(k) + beta);
z(k+t+i) = 1000;
end
```

where:

beta = 1, r = 12 (dimensionless)

5   CONCLUSIONS

The paper deals with aspects of modelling the movement of flying object in space in the MATLAB program environment. These models can be modified and used for the modelling of movement of any flying object. Its initial coordinates are defined in the geodetic coordinate system related to WGS-84. Further we present functions by use of which it is possible to transform of FO coordinates from geodetic to geocentric coordinate system.

The model of movement of FO described in this article was developed to verify the principle of relative navigation system. In this case, we assume that users (FO) communicate with one another (transmission and reception of signals). Based on measuring the time of arrival of these signals from at least four other FO, the user can calculate its own position. Model of movement of FO is used to obtain information about the user's position in space. We abstracted the forces exerted on the FO during the flight. For the purposes of simulation of relative navigation system, these forces acting on the FO can be neglected.

REFERENCES

Bréda, R., Soták, M. & Váci, Ľ. 2006. INS alignment algorithm in simulation environment. In ASIS, *Advanced Simulation of Systems*; *Proc. XXVIIIth International Autumn Colloquium, Vranov, 12–14. September 2006*. Ostrava: MARQ. ISBN 8086840263.

Džunda, M., Humeòanský, V., Draxler, D., Csefalvay, Z. & Bajusz, P. 2009a. The impact of windmills on the operation of radar systems. In Marine Navigation and Safety of Sea Transportation; *Proc. of international conference TRANSNAV 2009, Gdynia 17–19 June 2009*. London: CRC Press Taylor & Francis Group. Balkema ISBN 9780415804790.

Džunda, M., Humenansky, V., Draxler, D. & Bajusz, P. 2009b. Possibilites of eliminating radar signal clutters from windmills. In ICMT 2009; *Proc. of the International Conference on Military Technologies 2009. Brno, 5–6 May 2009*. Brno: University of Defence. ISBN 9788072316489.

Džunda, M., Žák, P., Noswitz, M. & Draxler, D. 2011. Modern billing solutions for aircraft inflight entertainment. In AEIT 2011, *Aerospace Engineering and Information Technology; Proc. of International conf., Beijing, 5–6 May 2011*. Beijing: University of Science and Technology, International Industrial Electronics Publisher (Center) ISBN 978-988-19116-3-6.

Džunda, M. & Cséfalvay, Z. 2013. Selected methods of ultrawide radar signal processing. In Marine Navigation and Safety of Sea Transportation: Advances in Marine Navigation; *Proc. of international conference TRANSNAV 2013, Gdynia, 19–21 June 2013*. London: CRC Press Taylor & Francis Group. Balkema. ISBN 978-1-138-00106-0.

Rozenberg, R., Szabo, S. & Šebešèáková, I. 2014. Comparison of FSC and LCC and their market share in aviation. *International Review of Aerospace Engineering (IREASE)*. 7 (5): 149–154. ISSN 1973-7459.

Socha, V., Schlenker, J., Kaľavksy, P., Kutilek, P., Socha, L., Szabo, S. & Smrcka, P. 2015. Effect of the change of flight, navigation and motor data visualization on psychophysiological state of pilots. In Applied Machine Intelligence and Informatics (SAMI); *Proc. of IEEE 13th International Symp. Herľany, 22–24 January 2015*, Slovakia.—Danvers: IEEE.

Volker, J. *Understanding coordinate systems, datums and transformations in Australia*. 2009 <http://www.lpi.nsw.gov.au/__data/assets/pdf_file/0006/129408/2009_Janssen_SSC2009_coords_datums_transformations_in_Australia.pdf> (viewed 18 May 2015).

Production Management and Engineering Sciences – Majerník, Daneshjo & Bosák (Eds)
© 2016 Taylor & Francis Group, London, ISBN: 978-1-138-02856-2

Indicators of green public procurement for sustainable production

A. Kottner
University of Central Europe, Skalica, Slovak Republic

L. Štofová, P. Szaryszová & Ľ. Lešková
*Faculty of Business Economics with Seat in Košice, University of Economics in Bratislava, Košice,
Slovak Republic*

ABSTRACT: Sustainable, thus balanced, social, economic, environmental and institutional, socio-economic development is now coordinated globally and implemented in the form of standardized procedures and preventive tools. The authors analyzed post within the green public procurement of goods, services and construction works according to international and European approaches from the aspect of its implementation in Slovakia. They submit the methodology and criteria for its implementation in connection to the standardized indicators of sustainable production, consumption and green growth from the level of OECD. Authors identify and formalize intervention and feedback between green growth indicators and criteria for the green public procurement, including barriers of such a processability.

1 INTRODUCTION

Sustainability of tenders is an issue dealing with problems linked to procurement of goods and services. This procurement takes into account social, economical and environmental aspects of development and its influence on society and economy. Green Public Procurement (GPP) is one of the voluntary instruments of international environmental policy.

GPP is supported by a number of international development policies and strategies, which proves it is an important instrument contributing to the achievement of economic and environmental objectives. The application of green public procurement supports sustainable use of natural resources, achieving changes in behavior towards sustainable production and consumption, and also encourages environmental innovation and green growth.

The process quality of GPP and its indicators within evaluation criteria must be based on globally standardized and used indicators for monitoring of sustainable development, sustainable production, consumption and green growth by OECD. Authors within the essay introduce the methodology of GPP, applying globally monitored indicators of sustainable production and consumption of green socio-economic growth at present.

2 GREEN GROWTH AND SUSTAINABLE DEVELOPMENT

To achieve the strategic objectives of sustainable development there have been formulated several concepts and models in the past. Examples include green economy, which has been drafted by UNEP and green growth, formulated from the level of OECD. These concepts do not replace the original strategy for Sustainable Development (SD), they are just instruments to achieve objectives of the original strategy.

According to the OECD, green growth is based on already existing initiatives oriented at sustainable development in several countries. It focuses on identification of cleaner sources for economic growth, including scoping the possibilities to develop new green industries, technologies and jobs, while it also controls the structural changes associated with the transition to greener economy (OECD, 2010).

In 2011, European Commission defined that green growth is part of the strategy Europe 2020, which is oriented on innovation, education, employment, social inclusion, energy or climate changes.

The strategy further aims to overcome the crisis and to prepare the European economy for the next decades within the intentions of sustainable development and low-carbon economy.

The main initiatives are focused mostly on efficient use of natural resources and green growth in industrial policy (European Commission 2011).

Green growth is a key tool for those countries that seek to restore economic growth, create new economic opportunities and jobs while ensuring a high standard of living. The objective of green growth is to achieve long-term growth while maintaining environmental quality and the most efficient use of natural resources and raw materials (Kanianská 2013).

Green growth is not a substitute for sustainable development, but rather we should consider it as its part.

Its scope is narrower and it aims to promote the necessary conditions for innovations, investment and competitiveness. These conditions yield the new sources of economic growth in accordance with the flexible ecosystems (OECD 2011).

Green growth is the path not only to promote economic growth and development but also way to protect natural resources, so we can continue to use environmental services and sources which our prosperity depends on. Therefore it is important to support investments and innovations that will be basis for sustainable growth and enable creation of new economic opportunities as well (enviroportal.sk).

2.1 Green growth strategy

OECD green growth strategy uses a broad base of analyzes and political plans. The aim to achieve social and environmental growth and economic recovery is monitored while strategy development.

The strategy of green growth by OECD in 2010 is "the first comprehensive strategy for economic growth in the post-crisis period, which takes into account environmental aspects. Its core is economy and its characteristic feature is the link between economy, taxes, innovation, knowledge economy, labor market, business environment and environmental aspects and impacts" (OECD Green Growth Strategy 2011).

"In general, green growth is defined as growth which does not increase environmental impacts, but rather stabilizes and gradually reduces environmental impacts per unit of growth. The green growth defines separation of economic growth from growth of environmental impacts" (Filčák 2012).

The aim of OECD green growth strategy is mainly to construct a bridge between policy makers in the economic, employment, finance, trade and the environment area. In one hand the green growth tries to minimize the consequences of distribution for disadvantaged population groups and in the other hand to manage negative economic consequences for companies in the current

level of incentives—to maintain higher economic performance.

"The strategies of green growth must encourage eco-behavior of companies and consumers, facilitating the smooth and fair jobs relocation, capital relocation and technology relocation towards greener activities and to provide adequate incentives and support for eco-innovation" (Filčák 2012).

The green growth strategies use various economic, politic and market actions and interventions on state level (Fig. 1) by support of investments and innovations into environmentally suitable technologies and their implementation. Using these methods they want to ensure environmental protection and effective use of natural resources at the highest possible level. Due to climate change, one of the green growth variants received quite considerable attention: so-called low-carbon economy.

The objective pursued by green growth strategy development is to reach economic recovery, environmental and social sustainable growth.

Fulfillment of this objective is subject to politic instruments necessary for transition to effective economy with specification of resources. Resources are specified to monitor green growth progress in form of indicators (Majerník et al. 2011, Bosák & Olexová 2013.).

2.2 Green growth indicators

Indicators can be characterized as measurable variables that provide information on developments and trends of processes and phenomena in quantitative and qualitative terms and monitor the environmental performance of the company. Green growth indicators help us in tracking whether the

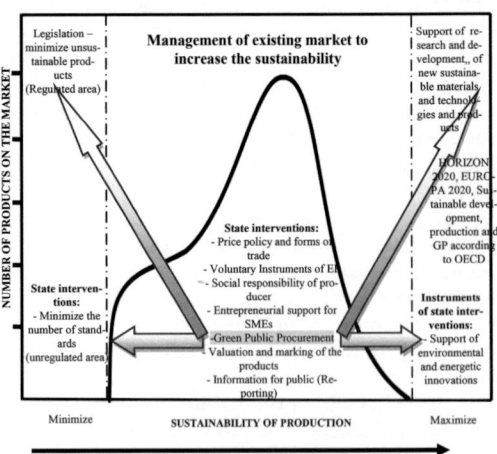

Figure 1. Key government interventions to support the sustainability of production and the status of GPP—system approach.

undertaking is deteriorating or improving their environmental performance and whether it is consistent with current legislation, which is relevant for the activities carried out.

The Slovak Republic as a member state of the OECD belongs to those states which have developed their own national set of indicators of green growth. The Netherlands was the first country of the OECD, which introduced its own indicators, later followed by Czech Republic and South Korea. Methodology and selection of indicators SR is based on a set of indicators that have been proposed by OECD (Guštafíková et al. 2014).

In terms of green growth, indicators are used for characterization of the baseline situation in Slovakia and should help as a benchmark for a comprehensive assessment of the direction of Slovak economy in the future, when considering the next steps (Tab. 1).

The cycle of economic activity is a natural starting point for the definition of indicators of sustainable green growth sector of production where the outputs (goods and services) are produced by business inputs. In several economic models such inputs include, eg. services, capital and intermediate inputs which are used directly in the production.

The trend towards green growth requires an assessment of the quality and composition of growth, starting from the one who ultimately makes the most of this growth. In the area of economic activities, natural starting point for defining indicators or green growth indicators is an area of production. Production system is an input-output system.

The green growth indicators are always limited in some areas, same as other assessment methodologies and indicators. In particular in international comparisons they must be interpreted according to the specifics of each country.

All indicators, however, should reflect the political importance of green growth, analytical validity and measurable quality of the underlying data. In 2011 OECD proposed 5 main groups of indicators (Fig. 2).

The proposed indicators should capture major trends and the importance of green growth and point to ways of development eliminating environmental damage and risks that could ultimately become an obstacle to economic growth. Reparation of environmental damage is yet often more expensive than the cost of prevention. The purpose of the indicators and the assessment itself is to find ways of development providing opportunities for economic growth and prosperity.

Today, standardized indicators shall in particular:

– *Ensure even coverage of the important elements of green growth with an emphasis on those that are of common interest to the members of OECD and partner countries*
– *To guarantee transparency and clarity for users in terms of the importance of values and their changes over time*

Table 1. Groups of indicators and evaluated areas for green growth.

Group of indicators	Assessed area
1—Indicators of environmental performance and resource productivity	Productivity of carbon and energy Productivity of sources: materials, water Multifactor Productivity
2—Indicators of stocks of natural capital	Stocks of renewable sources: water, forests, fish stocks of non-renewable raw materials: rocks, minerals Biodiversity and Ecosystems
3—Indicators of environmental quality of life	Environmental health and risks Environmental services and equipment
4—Indicators of policy response and economic actions	Technology and innovation Environmental goods and services International financial flows Prices and taxes Skills and training Governance and Management
Socio-economic aspects of growth	Economic growth and structure Labor productivity and trade Labor market, education and income Socio-demographic trends

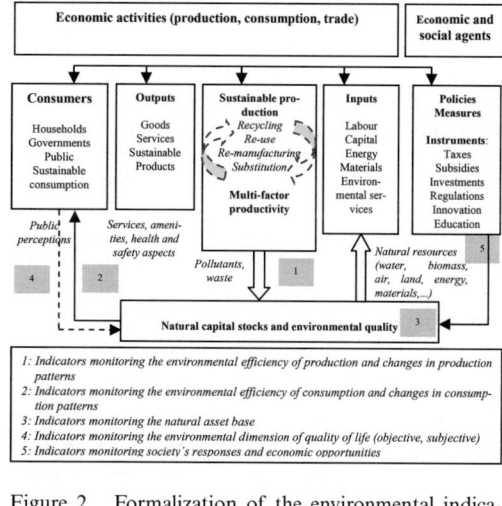

Figure 2. Formalization of the environmental indicators of sustainable production, consumption and green growth.

– *Serve as a core for comparison between other countries*
– *Adapt to diverse national differences and analyze disparities.*

Indicators should mainly reflect (Majerník et al. 2011):

– *The current situation (eg. For the last year)*
– *Changes in trends and intensities (linking "green" and "growth", ie. Since 1991)*
– *Changes in composition of environmental ingredients and indicators (eg. By sector and by source of emissions)*
– *The total value, if necessary, to inform about overall pressure on natural resources and the environment.*

These indicators should be significantly linked to the objectives and tasks to be useful. They should involve employees and other stakeholders to provide comprehensive and production adjusted measure of performance and should be based on an appropriate system for monitoring various data (Majerník et al. 2013).

Their measurability should be based on data that are of good quality, reasonably available and regularly updated.

3 BACKROUND AND STATUS OF GREEN PUBLIC PROCUREMENT

The emphasis on green public procurement at global level has already been placed into plans of preparation for the World Summit on Sustainable Development in Johannesburg in August/September 2002. OECD pays attention to this issue for a long period. In 2002, its recommendations were taken into account about the impact of public procurement on the environment and there was an attempt to contribute to favor greener products in the public sector and subsequently also in private contracts (draft National Action Plan for Green Public Procurement in the Slovak Republic for 2007–2010).

The potential of GPP was for the first time presented in detail in the EC Report for ER and the European Parliament-Integrated Product Policy, where the European Commission recommends EU Member States to process publicly available plans for greener procurement. Renewed EU Sustainable Development Strategy was adopted in June 2006, one of its key strategies is creation the models of sustainable production and consumption. In June 2008, the European Commission developed a report to the European Parliament and the Council, the European Economic and Social Committee and the Committee on Regional Action Plan on Sustain-

able Consumption and Production and Sustainable Industrial Policy. This report provides an integrated set of measures to achieve sustainable production and consumption for current state of improved sustainable production of the European economy.

The objectives of the national action plan for green public procurement in Slovakia (GPP NAP II 2012):

– *The strategic objective of the national action plan is to increase the share of promotion of GPP in Slovakia at the level of the key organs of state administration to 65 % and on the level of self-governing regions and cities by 50 % by 2015.*
– *Partial objectives:*
  1. Build awareness among contracting authorities and contracting entities on the issue of GPP in Slovakia.
  2. Strengthen the implementation of the environmental performance of public contracts.
  3. Value of the level of application of GPP in Slovakia in relation to the Commission's requirements.

Tools GPP NAP II to the partial objective 1:

– *To organize regular educational activities for contracting authorities and contracting entities*
– *Update and publish more exact GPP methodological guide for contracting authorities and contracting entities*
– *To provide timely information to the contracting authorities and contracting entities through the SEA website*
– *To organize regular meetings of the Working Group for GPP.*

Tools GPP NAP II to the partial Objective 2:

– *To prepare and issue a standardized tender documentation for the 3-5 most frequently procured product groups*
– *To promote administration of the environmental characteristics through GPP Hepldesk*
– *To provide information on available green products on the Slovak market*
– *Continue to implement the Green Procurement project.*

Tools GPP NAP II for Partial aim 3:

– *To monitor the GPP issue in the EU to cooperate with the Commission and EU Member States*
– *Carry out regular annual monitoring and evaluation of the GPP level in Slovakia.*

Despite the fact that GPP is a voluntary instrument of environmental policy, it plays a key role in the pursuit of resource-efficient economy (Majerník et al. 2014).

Public authorities can provide industry with incentives to develop environmentally friendly technologies and products through promotion and use of GPP. In some sectors, the contracting authorities have a large market share (e.g. Public transport and construction, health and education), therefore their decision has a significant impact. Purchasing power of these authorities when choosing environmentally-friendly products, services and works, can significantly contribute to sustainable production and consumption.

3.1 Characteristics of Green Public Procurement

GPP is characterized as an voluntary instrument, what means that Member States and public authorities themselves may set the extent of GPP implementation (SAŽP 2015).

In pursuing the efforts of the European Union to be more successful from the dimension of resource management, GPP is still a key factor despite the fact it is only a voluntary instrument. This solution helps to support sustainable product demand. These products would encounter difficulties when reaching the market. GPP is also stimulation for eco-innovation (Commission of the European Communities 2008).

GPP needs clear and verifiable criteria for products from environment due to success cover of the whole process. Most European countries have already developed national criteria. It is important that these standards are consistent between Member States with each other and relevant to the sustainability policies. This kind of behavior reinforces the single market and ensures what is good for the EU is also good for the environment (Green Public Procuremen, 2013).

Starting concept of green public procurement is focused on the existence of clear, verifiable, reasonable and ambitious environmental criteria for products. These criteria are based on life cycle data and the scientific knowledge base. The Commission in its Communication "Public procurement for a better environment" introduced a proposal to establish a procedure to set common criteria for green public procurement.

The criteria used by the Member States should be the same in order not to undermine the single market and neither to reduce EU-wide competition. Similar criteria reduce the administrative burden for economic operators and for public administrations carrying out the GPP. The common criteria for GPP represent a significant benefit for companies operating in several Member States. They are also beneficial for SMEs, which have limited ability to cope with different procedures for awarding public contracts.

The European Commission has developed more than 20 common criteria for GPP till today. The development lasted since 2008. Implementation

of GPP through multi-criteria analysis, selected criteria (European Commission 2014):

– *Scope for improvement of the environment*
– *Public spending*
– *The potential impact on suppliers*
– *The potential for an example of private and corporate customers*
– *Political sensitivity*
– *The existence of relevant and easy-to-use criteria*
– *The availability of market and economic efficiency.*

The criteria are regularly updated and are based on data from a database, on existing criteria of Ecolabel and on information collected from stakeholders from industry, civil society and Member States. The database uses available scientific information and data, adopts a life-cycle approach and engages in discussions involved parties who deal about issues of consensus.

3.2 Procedures of implementing Green Public Procurement in organization

Two informal practices of implementation are used at present (SAŽP, 2013):

1. Model Management GPP in the organization.
2. Quick start by implementing GPP (QuickStart).

GPP Management Model is a guide for organization which realizes public procurement for gradual integration of green procurement in GPP. Organization also integrates this approach into whole management system. This model offers simple, flexible and integrated approach for the systematic implementation of GPP within the organization based on the Deming cycle of continuous improvement "Plan-Do-Check-Act" (Majerník et al. 2015).

GPP management model has been developed to ensure its efficient and effective use. If the organization will manage the proposed measures, it will be able to provide a chronological improvement of its environmental performance.

GPP management model in order to exhibit features of systematic approach and became effective for the organization, it is necessary (SAŽP 2013):

– *Ensure management support and management*
– *Establishing clear responsibilities and duties for those involved*
– *Monitoring benefits, achievements and problems*
– *Ensuring continual improvement.*

The first step in the implementation of GPP model management is training, which consists of:

– *Ensuring analysis of planned public procurement,*
– *An analysis of the costs of GPP in public procurement*

– *Market research*
– *The choice of product groups for GPP.*

The second step is setting goals which consist of the following elements:

– *Political commitment*
– *Operational objectives*
– *Specific objectives directly related to the procurement process.*

The third step is the creation, implementation and operation of the action plan. It includes:

– *Actions and activities of the Action Plan*
– *Assignment of responsibilities and powers.*

The fourth step is the evaluation and monitoring of the results and benefits of GPP. This step provides assessment of achieved and attained goals. This fourth step also provides identification of encountered problems during the GPP implementation and seeking solutions to remedy problems and improve the GPP implementation in the future (Boďová & Kovárová, 2008).

4 INTEGRATION OF SUSTAINABLE PRODUCTION, CONSUMPTION AND GREEN GROWTH INDICATORS INTO THE GREEN PUBLIC PROCUREMENT PROCESS

Requirements for environmental characteristics corresponding to the monitored of indicators of sustainable production, consumption and green growth must be incorporated into public procurement procedures in such a manner to maintain conformity with European and national legislation as well as the following principles:

– *The principle of equal treatment*
– *The principle of non-discrimination of candidates or tenderers*
– *The principle of transparency*
– *The principle of economy and efficiency.*

The legislative regulations on public procurement define the possibilities of incorporation of environmental characteristics in the award conditions. Individual processes of proposed methodic approach of indicator integration into GPP are pictured in flowchart. This process can be also performed via standardized green growth indicators derived from indicators of sustainable production and consumption as follows:

– *Define the subject of the contract under the GPP*
– *Formulation of conditions for participation in GPP*
– *Sizing of the award criteria*
– *Specification of the conditions for the contract realization*

– *Monitoring and continuous improvement of the procurement process and results*
– *Reporting the level of GPP in comparison with public procurement (PP).*

An important decisive step is to precisely size criteria for tender evaluation within GPP and integration of indicators of sustainable production, consumption and green growth.

In case that these take into account standardized indicators for sustainable development according to OECD, it is possible to continue smoothly in the GPP process. Subsequently, conditions for tender

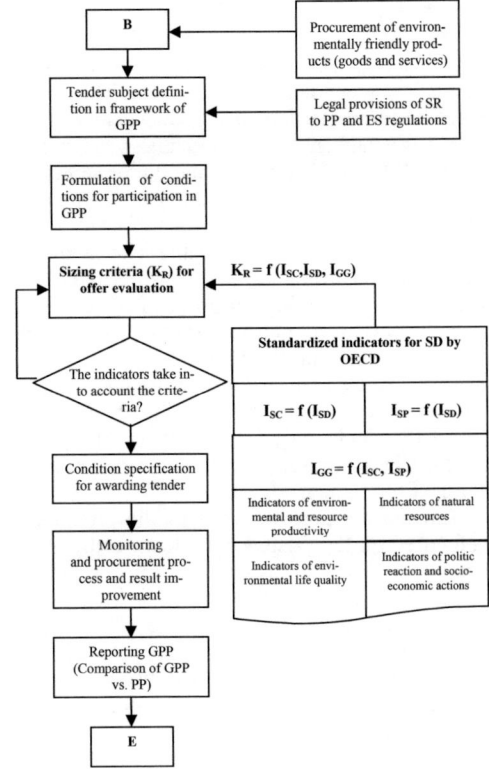

Legend:
  SP - Sustainable Production
  SC - Sustainable Consumption
  SD - Sustainable Development
  GG - Green Growth
  I_{SP} - Indicator of Sustainable Production
  I_{SC} - Indicator of Sustainable Consumption
  I_{SD} - Indicator of Sustainable Development
  I_{GG} – Indicator of Green Growth
  K_D - Criterion of Development

Figure 3.   Processability of GPP to provide sustainable production, consumption and green growth.

realization can be specified and GPP monitoring and reporting can be done.

Sustainable development in terms of GPP methodology is thus possible to achieve by indicator monitoring of sustainable production, consumption and green growth. On the basis of practical experience and our knowledge it is possible to identify following barriers to wider implementation of GPP:

- *A limited number of products assessed in terms of environmental aspects, impacts and risks*
- *The absence of more exact methods for calculation of life-cycle costs of products and relevant costs of environmentally friendly products*
- *Low awareness of the benefits of environmentally friendly products and sustainable development goals*
- *Lack of information about possibilities and forms of integration requirements for environmental characteristics in tenders*
- *Lack of communication and exchange of information about best practices between institutions and organizations at regional and local levels.*

These obstacles are encountered with increasing number of activities and new initiatives, thus preventing better GPP prosperity on the national level. Following the recommendations of the European Commission and support of the GPP implementation, many Member States adopted national action plans for GPP or implemented information systems or national guides.

5 CONCLUSIONS

GPP is an instrument of environmental friendly behavior of organizations that are funded by public resources through the implementation and realization of system-oriented measures for environmental operation in frames of sustainable development.

Benefit of so formulated GPP is reduction of environmental impact through the promotion of eco-innovation, influencing trends in production and consumption, in particular when creating demand for environmentally friendly products, what leads to supply escalation of such products. From an economic point of view GPP ensures rationalization of costs and what is important, it has a very positive educational impact on citizens, office goers, external companies, and also on other offices, businesses and institutions (Štofová et al., 2014).

Restrictions occur with increasing amount of activities and new stimulations, thus preventing more prosperous GPP at national level. Following the recommendations of the European Commission

to support the implementation of GPP, many of Member States adopted national action plans for GPP or implemented information systems or national GPP guides. Increasing the efficiency of green procurement and purchasing is from our point of view determined mainly by changes in approach to the issue. Raising awareness and implementation of more exact methods and procedures causes changes in the implementation process as well.

ACKNOWLEDGEMENTS

This contribution is a result of the Project for young teachers, researchers and PhD students, no. I-15-110-00, 2015: Methodology for implementation of Integrated Management for small and medium-sized enterprises in SR and EU levels.

REFERENCES

Boďová, E. & Kovárová, P. 2008. *Príručka pre verejných obstarávateľov a obstarávateľov.* Banská Bystrica: VKÚ, a.s., 2008. 80 p. ISBN 978-80-88850-85-4.

Bosák, M. & Olexová, C. 2013. Recent environmental trends and innovations in the Slovak small and middle enterprises, In. *GeoConference on ecology, economics, education and legislation SGEM 2013, 16–22 June, 2013*, Albena, Vol. 2., 247–254.

Commission of the European communities. 2008. *Public procurement for a better environment* [online]. Brusel, 2008. [cit. 2014.11.04]. Available at: <http://eurlex. europa.eu/LexUriServ/LexUriServ.do?uri = COM: 2008: 0400: FIN: EN: PDF>. Drawn from the Sustainable Products and Materials Progress Report. Available at: <http://www.defra.gov.uk/environment/ business/pdf/prod-materials-report0708.pdf>.

ENV/EPOC/SE. 2010. *Monitoring Progress towards Green Growth ministerial Report on Green Growth Indicators.* Draft outline and measurement framework. 23–25. November 2010. Paris, OECD Conference Centre.

European Commission. 2014. What is GPP. [online]. 2014. [cit. 2014.11.04]. Available at: <http://ec.europa. eu/environment/gpp/what_en.htm>.

Filčák, R. 2012. *Spoločnosť trhu a environmentálna politika: aktéri a konflikty.* 1. vyd. Bratislava: VEDA: vydavateľstvo Slovenskej akadémie vied. 2012. pp. 302. ISBN 978-80-224-1216-2.

Guštafíková, T. et al. 2014. *Vybrané indikátory zeleného rastu v Slovenskej republike.* [online]. Banská Bystrica. 2014. [cit. 2015.02.01]. Available at: <http:// www.oecd.org/greengrowth/Green%20Growth% 20Indicators%20 in%20the%20Slovak%20Republic. pdf>. ISBN: 978-80-89503-35-3.

Informácia o implementácii Národného akčného plánu pre zelené verejné obstarávanie v SR na roky 2011 až 2015 za rok 2012. Na podnet Uznesenia vlády Slovenskej republiky č. 22/2012, č. materiálu: UV-16754/2013.

Kanianská, J. 2013. *Zelený rast a zelená ekonomika Zelené iniciatívy zamerané na dosiahnutie cieľov trvalo udržateľného rozvoja.* [online]. Banská Bystrica, 2013.

[2014 08.12]. Available at: <https://www.enviroportal. sk/uploads/files/Zeleny%20rast/ZREANALYZA. pdf>.

Majerník, M., Bosák, M., Szaryszová, P., Tarča, A. & Štofová, L. 2014. Model for assessing the environmental performance of product systems within sustainable development, In *Geoconference on ecology, economics, education and legislation* SGEM 2014: 14th international multidisciplinary scientific geoconference, 17–26 June 2014, Albena, Bulgaria. volume III.—Sofia: STEF92 Technology Ltd., 2014.—ISBN 978-619-7105-19-3, pp. 285–292.646. doi: 10.1108/02656710610672461.

Majerník, M., Szaryszová, P., Bosák, M., Štofová, L. & Kabdi, K. 2015. Integrated management of environmental—safety and technical risks of plant producing automobiles and automobile components. In *Communications: scientific letters of the University of Žilina.*—Žilina: University of Žilina, 2015. - ISSN 1335–4205.—Vol. 17, no. 1 (2015), pp. 28–33.

Majerník, M., Panková Juríková, J., Bašistová, A. & Halagová, L. 2011. Manažment a hodnotenie udržateľnej spotreby, produkcie a zeleného rastu na Slovensku, In *Spoločenská zodpovednosť—súčasť environmentálnej a firemnej kultúr: interaktívna konferencia o spoločenskej zodpovednosti organizácií*, 2011, Banská Bystrica: Fakulta prírodných vied UMB v Banskej Bystrici, 2011. ISBN 978-80-557-0135-6.

Majerník, M., Mihok, J., Tkáč, M., Bosák, M., Szaryszová, P. & Tarča, A. 2013. *Environmentálne manažérstvo v integrovanom systéme.* Recenzenti: Janko Hodolič, Ondrej Hronec. 1. vyd. [Košice: Podnikovohospodárska fakulta EU], 2013. 302 p. [15,1 AH]. ISBN 978-80-971555-1-3.

OECD, 2010. Predbežná správa k Stratégii zeleného rastu: Implementácia nášho záväzku pre trvale udržateľnú budúcnosť, [online]. [cit.2014-09-10]. Available at: <http://www.oecd.org/greengrowth/45638609.pdf>.

OECD GREEN GROWTH STRATEGY. 2011[online] .[cit.2014–09–10] Available at: <http://www.oecd.org/greengrowth/48012345.pdf>. ISBN: 978-92-64-094970.

OECD Green Growth Studies. Green Growth Indicators 2014. [online]. [cit.2015-04-10] Available at: <http://www.oecd-ilibrary.org/docserver/download/9713101e.pdf?expires = 1434972010&id = id&accname = oid020161&checksum = 79DB4C3454E3CF0 A9362 A036 A657E397>.

SAŽP. 2013. *Metodická príručka pre verejných obstarávateľov a obstarávateľov.* Centrum odpadového hospodárstva a environmentálneho manažérstva. Bratislava.

SAŽP. 2015. GPP v praxi. [online]. 2015. [cit. 2014.11.04]. Available at: <http://www.sazp.sk/public/index/go.php?id = 2201>.

Štofová, L., Szaryszová, P., Majerník, M. 2014. Proaktívna podniková stratégia environmentálneho manažérstva v dodávateľsko-odberateľskom reťazci. In *Zelená energia—Environment—Udržateľný rozvoj: zborník príspevkov z 1. medzinárodnej vedeckej konferencie*: Poprad, 23 januára 2014 / Editor: Peter Adamišin; Recenzenti: Ondrej Hronec, Martin Bosák.—Prešov: Prešovská univerzita v Prešove, 2014.—ISBN 978-80-555-1170-2.—pp. 94–100.

Zelené verejné obstarávanie. 2013. *Metodická príručka pre verejných obstarávateľov a obstarávateľov.* Slovenská agentúra životného prostredia, Centrum odpadového hospodárstva a environmentálneho manažérstva. 2013.

Zelený rast. 2014. In *Enviroportal.sk.* [online]. [cit. 2014–11–12]. Available at: <http://www.enviroportal.sk/environmentalne-temy/starostlivost-o-zp/zeleny-rast>.

Production Management and Engineering Sciences – Majerník, Daneshjo & Bosák (Eds)
© 2016 Taylor & Francis Group, London, ISBN: 978-1-138-02856-2

Future directions and gaps of the global supply chain risk modelling

J. Kováč, J. Kadarova & L. Kalafusova
Mechanical Engineering, Technical University of Kosice, Kosice, Slovak Republic

ABSTRACT: The paper discusses the current gaps and future directions of global supply chain defined by the literature. The most used research methods in the field of supply chain were identified by their quantitative occurrence in the literature. According to the review provided by various companies, the future directions of supply chain risk management were defined and pointed in the map. In the first part of the paper the influence factors of failure are discussed according to the literature review. In the second part of the paper, current research gaps in the Supply Chain Risk Modeling (SCRM) are specified and further analyzed according to the information from literature reviews. The third part is dedicated to potential future directions of global SCRM. Developing the future supply chain require time, insight and input, so it is necessary to start with the collaboration between them to become more prosper in this dynamic environment.

1 SUPPLY CHAIN MANAGEMENT

1.1 Current supply chain

Supply Chain Risk Management (SCRM) is one aspect of Enterprise Risk Management (ERM) which starts with the risk identification. Many different definitions of risk in a Supply Chain (SC) context are provided in literature. According to (Jüttner, Peck, & Christopher, 2003) Supply Chain Risk (SCR) can be defined as "the variation in the distribution of possible supply chain outcomes, their likelihoods, and their subjective values". This definition highlights two risk dimensions, namely: (1) impact and (2) likelihood of occurrence. The supply chain may be divided into four dimensions: (1) traditional, (2) lean, (3) agile and (4) leagile supply chain (Faisal, Banwet, & Shankar, 2006). Chopra & Sodhi (2004) classified supply chain risks as disruptions, delays, systems, forecast, intellectual property, receivables, inventory, and capacity. According to (Sinha, Whitman, & Malzahn, 2004) the most important supply chain risk areas are standards, supplier, technology, and practice. Finch (2004) mentioned three levels of risk coverage, namely: (1) application level; (2) organizational level; and (3) inter-organisational level. Norrman & Jansson (2004) categorized all types of SC risk into three categories: (1) operational accidents; (2) operational catastrophes; and (3) strategy uncertainty. According to (Tang C., 2006) there are two kinds of risks in SC: (1) operational risk; and (2) disruption risk. Kleindorfer & Saad (2005) divided risk in two categories as: (1) risks arising from the problems of coordinating supply chain and demand; (2) risks arising from disruptions to natural activities. Companies usually manage supply chain risks at the strategic (long term) or at the tactical (medium term) level (Tang & Tomlin 2008).

1.2 Literature review

Managing risks in a global supply chain requires lot of information sharing, close relationships and trust on the partners, alignment of incentives and knowledge about risks (Faisal, Banwet, & Shankar, 2006).

The literature suggests that proactive approaches at supply chain level should be implemented in order to effectively manage disruptions (Colicchia & Strozzi, 2012). Research community has developed numerous supply chain design models. To confront the challenges of a complex and unstable competitive environment and to gain long-term benefits, it is necessary to include resilience and robustness considerations into supply chain design (Colicchia & Strozzi, 2012). Dynamic supply chain models under uncertainty are needed, as well as tools able to consider interactions characterizing supply chain factors and risk sources. Nevertheless, very complex interactions among supply chain factors make the supply chain inherently vulnerable to disruptions. That is the reason why only the most relative risk factors of SC should be considered. This creates a challenge for managers to choose the most relevant risk factors in their models. Many researchers analyzed supply chain risks through different influence factors pointed in below (Table 1):

We believe that empirically grounded research supports the development of quantitative modeling using empirical studies and conceptual models. WEF (2011) presented triggers of global supply chain disruptions with the top external disruptions highlighted in the Table 2. According to their research the initial event may results in a cascading disruption or failure across regions or industries.

Modeling the critical point of failure within a Supply Chain Network is un-attempted and demands further research (Finch, 2004). We believe that disruption impact evaluation in terms of cost, duration, and service will provide better transparency for effective SCRM. According to (Musa, 2012) the percentage of publications by types of research methods were analyzed. In the literature review process were identified four types of research approaches (Figure 1).

According to (Colicchia & Strozzi, 2012) it is necessary to analyze how key concepts and performance measures from other disciplines can be incorporated into SCR models. Also there is a need to investigate—through empirically based research—how companies assess their supply chain risk exposure and how they develop mitigation capabilities in collaboration with their supply chain partners.

Table 1. Factors of failure pointed in the literature.

	Influence	Description	Previous research
Delivery	Capacity	Supplier is near or at full capacity	(Lee, Padmanabhan, & Wang, 1997); (Chopra & Sodhi, 2004); (Norrman & Jansson, 2004)
	Cycle time	Unreliable cycle times of suppliers	(Zsidisin, 2006)
	Material availability	Unreliable raw material sources	(Norrman & Jansson, 2004) (Noordewier, John, & Nevin, 1990); (Hillman, 2006)
	Logistic	Unreliable logistics infrastructure of suppliers	(Chopra & Sodhi, 2004); (Hillman, 2006)
	Natural disaster	Suppliers are located in areas prone to natural/political disruptions	(Chopra & Sodhi, 2004) (Zsidisin , 2006) (Norrman & Jansson, 2004)
Cost	Market strength	Suppliers are to strength or to weak to dictate the price	(Kraljic, 1983)
	Currency	Suppliers common currency is volatile	(Hillman, 2006)
	Financial	Supplier is financially weak	(Chopra & Sodhi, 2004); (Babich, 2004)
	Cost management	Poor management skills of suppliers	(Hillman, 2006); (Zsidisin, 2006)
Quality	Quality system	Substandard control quality methods of suppliers	(Hillman, 2006); (Zsidisin, 2006)
	Legal standards	Supplier is unaware/unconcerned with legal/environmental standards	(Hillman, 2006); (Huang, A.; Yen, D. C.; Chou, D. C.; Xu, Y.;, 2003)
Flexibility	R&D	Poor product development methods of suppliers	(Droge, Jayaram, & Vickery, 2004); (Noordewier, John, & Nevin, 1990); (Laseter & Ramdas, 2002)
	Flexibility	Suppliers have processes which don't allow significant changes in volume	(Chopra & Sodhi, 2004); (Noordewier, John, & Nevin, 1990), (Norrman & Jansson, 2004); (Zsidisin , 2006)
Confidence	Management	Lack of clear management vision or expertise	(Hillman, 2006)
	Market characteristics	The market in which the supplier operates is volatile	(Chopra & Sodhi, 2004); (Noordewier, John, & Nevin, 1990)
	Information	Unreliable information system of suppliers	(Chopra & Sodhi, 2004); (Hillman, 2006)
	Product type	The supplier may not be able to handle the complexity or sensitivity of product requirements	(Chopra & Sodhi, 2004); (Hillman, 2006)
	Relationship	It is difficult to manage the relationships with suppliers (communication barriers)	(Noordewier, John, & Nevin, 1990); (Hillman, 2006)

Table 2. Triggers of supply chain disruptions (WEF, 2011).

Environmental	Geopolitical
Natural disasters	Conflict and political
Extreme weather	unrest
Pandemic	Export/import
	restrictions
	Terrorism
	Corruption
	Illicit trade and organized crime
	Maritime piracy
	Nuclear/biological/chemical weapons
Economic	Technological
Sudden demand shocks	Information and
Extreme volatility	communications
in commodity prices	disruptions
Border delays	Transport
Currency fluctuations	infrastructure failures
Global energy shortages	
Ownership/investment	
restrictions	
Shortage of labor	

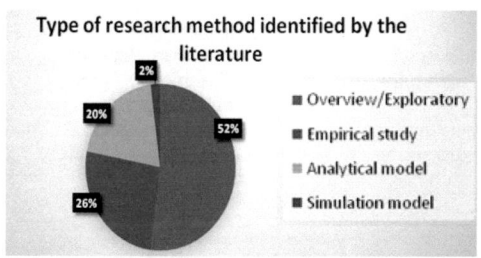

Figure 1. Percentage of publications by the type of research methods in the field of SCRM.

2 GAPS AND FUTURE DIRECTIONS OF SCRM

2.1 Current gaps of SCRM

Using literature review, main research gaps in the field of SCRM were determined. According to the research, the holistic SCRM approach is lacking in the current literature. Williams, Lueg, & LeMay (2008) claims that studying dyadic relationships are extremely important for the holistic understanding of SCRM. This holistic "system of systems" approach is expected to bring fresh thinking for current SC problems (Mingers & White, 2010). There is a need to develop novel quantitative as well as qualitative methods to quantify the risk propagation and create an effective quantitative modeling approach to solve SCRM problems. There are many quantitative tools, like mathematical programming and simulation models (Rao & Goldsby, 2009) or graph theory (Colicchia & Strozzi, 2012), which are not developed to solve SCRM problems. According to (Khan, Christopher, & Burnes, 2008) it is necessary to develop well-grounded quantitative models which could consider other interdisciplinary research approaches. Quantitative tools and theories which could be used to the dynamic behavior of the complex SC system might be: (1) System Dynamics using causal loops diagrams and flow diagrams, (2) Real Option Valuation, and (3) Simulation modeling. According to (Jüttner, Peck, & Christopher, 2003) there is an increasing importance to identify risk drivers and their potential impact to SC processes. Bryson, Millar and Mobolurin (2002) forced the need for creating an effective recovery system to mitigate the effect of disasters. Only some authors like (Khan & Burnes, 2007) and (Natarajarathinam, Capar, & Narayanan, 2009) examined the effects of risk disruption and recovery planning. For these reason, the robust contingency/recovery planning strategies for potential future disruption are needed. We believe that create an appropriate risk recovery model will require proactive planning combining appropriate information and human intervention. The global compliance standards and environmental thinking become more important in recent years. This environmental awareness force logistics activities to be more effective. Also there is a pressure to use natural resources and recycle. Many companies are aware of this recent trend and take into consideration the interrelationship between risks and sustainability perspective to become more successful in the current competitive global market. Companies need to be able to monitor real time logistics networks. Low information and communication technology make the visibility of SC performance almost insignificant. According to (Tang C., 2006) and (Rao & Goldsby, 2009) current technologies such Radio Frequency Identification (RFID), Enterprise Resource Planning (ERP) or General Packet Radio Service (GPRS) will become important information tools for management of supply chain risks in terms to improve SC performance. Nevertheless, the recent technology used to analyze risk management demands needs to be investigated more extensively. Network information communication and sharing avoids defaults and generate trust between shareholders. Nevertheless, long-terms contracts for disruption are still lacking in the literature. Only few research papers were focused on contracts relating to price and demand fluctuations.

Other additional gaps which might be considered are the lack of understanding of the nature of complexity among many supply chain researchers and the low effectiveness of risk management.

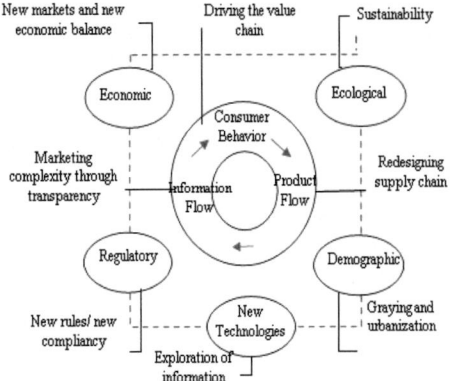

Figure 2. Future supply chain proposed by (CapGemini, 2008).

2.2 Future directions of SCRM

Each supply chain should follow 5 steps of improvement provided by (Logisitics, 2010), namely: (1) Analyze the entire supply chain, (2) Connect Electronically, (3) Anticipate Issues, (4) Set up Alternative Plans, (5) Manage by Exception.

The most important aspect of future supply chain were proposed by CapGemini (2008) and were pointed in the Figure 2.

2.3 Case study

Global supply chain suffers many breakdowns in recent years. Many of them were related to natural disasters, e.g. Japan Tsunami (2011), Hurricane Katrina in US (2005), Indian Ocean Tsunami (2004), etc. These natural disasters caused cascading effect of disruptions pointed in Table 3.

According to our study, future directions of global supply chain should be oriented on scenario planning, quantitative risk metrics and data/information sharing. The most important directions of Global Supply Chain are also pointed in Figure 3.

The importance of the selected areas was measured by their key words occurrence in the literature. The importance of individual areas was numerically and graphically classified.

3 CONCLUSIONS

The common challenge for the future of global supply chain might be the (1) collaboration between consumers and shoppers which should be informed about the sustainability impact of their shopping choices, (2) collaboration and standardization between retailers in order to be profitable and achieve sustainable growth, (3) collaboration between manufacturers and suppliers with the aim

Table 3. Cascading effect of disruptions.

Cause	Effect
Physical/ infrastructure disruption	All major ports were closed with intense effect on the global logistic services. The spilling effect of the earthquake and tsunami led to a nuclear power plant breakdown in Fukushima. Nuclear crisis and electricity shutdown forced companies to shut down their plants.
Financial disruption	After the Japan Tsunami, three sectors, namely: (1) automotive; (2) steel industry; (3) electronics, were identified by global supply chain analysts as most affected business areas (hike in prices of all affected products).
Technology disruption	Impact on stability and integrity of e-services, loss of intellectual property.
Information disruption	Massive impact on information security (lack of communication due to a complete shutdown of the supply network).
Human resource disruption	Loss of lives. Remove of employees (affect productivity).
Social and ecological disruption	Corruption, organized crimes, global imbalance, etc.

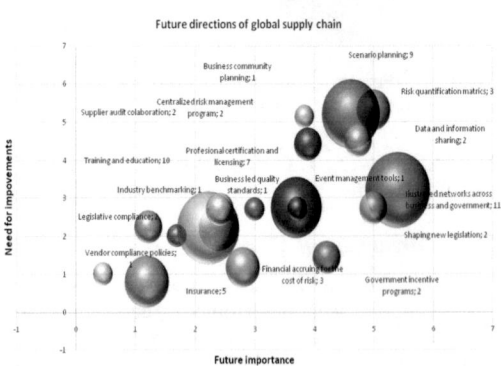

Figure 3. Potential future directions of global supply chain identified by the literature.

to manufacture, market and supply the products that consumers need in a cost-efficient manner, (4) facilitate the distribution process from supplier to consumer using longer-term contracts with logistics service providers. Developing the future supply chain require time, insight and input from a wide range of industry "players" (e.g. retailers, consumer products, manufacturers, industry standards organizations, technology companies, etc.) so it is necessary to start with the collaboration between them to become more prosper in this dynamic environment.

This contribution is the result of the projects implementation: Project VEGA 1/0669/13 Proactive crisis management of industrial enterprises based on the concept of controlling.

REFERENCES

Babich, V. (2004). *Vulnerable Options in Supply Chains: Effects of Supplier Competition.* Michigan: *Department of Industrial and Operations Engineering, University of Michigan. Dostupné na Internete: http://www. realoptions.org/papers2004/BabichVulnerable.pdf

Bryson, K.M., Millar, H., Joseph, A., Mobolurin, A. (2002). Using formal MS/OR modeling to support disaster recovery planning. (Elsevier, Ed.) *European Journal of Operational Research, 141,* 679–688.

CapGemini. (2008). Future Supply Chain 2016: Serving customers in a sutable way. Dostupné na Internete: http://www.slideshare.net/BCLadd/future-supply-chain-2016?related = 2

Chopra, S., & Sodhi, M.S. (2004). Managing risk to avoid supply-chain breakdown. (Fall, Ed.) *MIT Sloan Manage, 46,* 53–61.

Colicchia, C., & Strozzi, F. (2012). Supply chain risk management: a new methodology for a systematic literature review. (Emerald, Ed.) *Supply Chain Management: An International Journal, 17*(4), 403–418.

Droge, C., Jayaram, J., & Vickery, S.K. (2004). The effects of internal versus external integration practices on time-based performance and overall firm performance. (Elsevier, Ed.) *Journal of Operations Management, 22,* 557–573. doi: 10.1016/j.jom.2004.08.001

Faisal, M.N., Banwet, K., & Shankar, R. (2006). Mapping supply chains on risk and customer sensitivity dimensions. (Emerald, Ed.) *Industrial Management & Data Systems, 106*(6), 878–895. Dostupné na Internete: http://www.emeraldinsight.com/doi/pdfplus/10.1108/02635570610671533

Finch, P. (2004). Case study: Supply chain risk management. (Emerald, Ed.) *Supply Chain Management: An International Journal, 9*(2), 183–196.

Hillman, M. (2006). Strategies for managing supply chain risk. *Supply Chain Management Review*(10), 11–13.

Huang, A., Yen, D.C., Chou, D.C., Xu, Y. (2003). Corporate Applications Integration: Challenges, Opportunities, and Implementation Strategies. *Journal of Business and Management* (9), 137–145.

Jüttner, U., Peck, H., & Christopher, M. (2003). Supply Chain Risk Management: Outlining an Agenda for Future Research. *International Journal of Logistics: Research & Applications, 6*(4), 197–210.

Kern, D., Moser, R., Hartmann, E., & Moder, M. (2012). Supply risk management: model development and empirical analysis. (Emerald, Ed.) *International Journal of Physical Distribution & Logistics Management, 42*(1), 60–82.

Khan, O., & Burnes, B. (2007). Risk and supply chain management: creating a research agenda. (Emerald, Ed.) *The International Journal of Logistics Management, 18*(2), 197–216.

Khan, O., Christopher, M., & Burnes, B. (2008). The impact of product design on supply chain risk: a case study. (Emerald, Ed.) *International Journal of Physical Distribution & Logistics Management, 38*(5), 412–432.

Kleindorfer, P.R., & Saad, G.H. (2005). Managing Disruption Risks in Supply Chains. (POMS, Ed.) *PRODUCTION AND OPERATIONS MANAGEMENT,* 1–16. Dostupné na Internete: http://opim.wharton.upenn.edu/risk/downloads/05–08-PK.pdf

Kraljic, P. (1983). Purchasing must become supply management. *Harvard Business Review, 61,* 109–117.

Laseter, T.M., & Ramdas, K. (2002). Product Types and Supplier Roles in Product Development: An Exploratory Analysis. *IEEE TRANSACTIONS ON ENGINEERING MANAGEMENT, 2,* 107–118. Dostupné na Internete: http://strategy.sauder.ubc.ca/nakamura/iar515p/laseter_product_types.pdf

Lee, H.L., Padmanabhan, V., & Wang, S. (1997). Information distortion in a supply chain: The bullwhip effect. *Management Science, 43,* 546–558. Dostupné na Internete: http://www.ie.bilkent.edu.tr/~ie571/lee%20et%20al%20(1997),%20 ms.pdf

Logisitics. (2010). How to Enable a Proactive Supply Chain: 5 Steps to Supply Chain Empowermen. (Thomas, Ed.) NY. Dostupné na Internete: http://www.inboundlogistics.com/cms/article/how-to-enable-a-proactive-supply-chain/

Mingers, J., & White, L. (2010). A review of the recent contribution of systems thinking to operational research and management science. (Elsevier, Ed.) *European Journal of Operational Research, 207*(2010), 1147–1161. doi: 10.1016/j.ejor.2009.12.019

Musa, S.N. (2012). *Supply Chain Risk Management: Identification, Evaluation and Mitigation Techniques.* Linköping, SWEDEN: Linköping University. Dostupné na Internete: http://liu.diva-portal.org/smash/get/diva2:535627/FULLTEXT01.pdf

Natarajarathinam, M., Capar, I., & Narayanan, A. (2009). Managing supply chains in times of crisis: a review of literature and insights. (Emerald, Ed.) *International Journal of Physical Distribution & Logistics Management, 39*(7), 535–573. doi: 10.1108/09600030910996251

Noordewier, T.G., John, G., & Nevin, J.R. (1990). Performance outcomes of purchasing arrangement in industrial buyer-vendor relationships. (A. Associacion, Ed.) *Journal of Marketing, 54*(4), 80–93. Dostupné na Internete: http://part-timemba.carlsonschool.umn.edu/marketinginstitute/research/documents/John_PerformanceOutcomesofPurchasing_1990.pdf

Norrman, A., & Jansson, U. (2004). Ericsson's proactive supply chain risk management. (E. Publishing, Ed.) *34*(5), 434–456. doi: 10.1108/09600030410545463

Rao, S., & Goldsby, T.J. (2009). Supply chain risks: a review and typology. (Emerald, Ed.) *The International Journal of Logistics Management, 20*(1), 97–123.

Sinha, P.R., Whitman, L.E., & Malzahn, D. (2004). Methodology to mitigate supplier risk in an aerospace supply chain. (Emerald, Ed.) *Supply Chain Management: An International Journal, 9*(2), 154–168.

Tang, C.S. (2006). Perspectives in supply chain risk management. (Elsevier, Ed.) *Int. J. Production Economics, 103*(2006), 451–488. doi: 10.1016/j.ijpe.2005.12.006

Tang, C., & Tomlin, B. (2008). The power of flexibility for mitigating supply chain risks. (Elsevier, Ed.) *Int. J. Production Economics, 116*(2008), 12–27. doi: 10.1016/j.ijpe.2008.07.008

WEF. (2011). New Models for Addressing Supply Chain and Transport Risk: An Initiative of the Risk Response Network In collaboration with Accenture. Geneva, Switzerland.

Williams, Z., Lueg, J.E., & LeMay, S.A. (2008). Supply chain security: an overview and research agenda. (Emerald, Ed.) *The International Journal of Logistic Management, 19*(2), 254–281. doi: 10.1108/09574090810895988.

Zsidisin, G.A. (2006). Managerial Perceptions of Supply Risk. *Journal of Supply Chain Management: A Global Review of Purchasing and Supply Copyrigh, 39*(4), 14–26. Dostupné na Internete: http://onlinelibrary.wiley.com/doi/10.1111/j.1745–493X.2003.tb00146.x/pdf

Application of current methods and techniques in the enterprise, case study

A. Krištanová
Ekologické služby-Anna Krištanová, s.r.o., Petrovce, Slovak Republic

L. Stankovič
Faculty of Business Economy with Seat in Košice, University of Economics in Bratislava, Košice, Slovak Republic

ABSTRACT: Changing dynamics of the external environment brings out unexpected opportunities and risks and threats. It all opens up new challenges and new requirements to address them. To the forefront of practice in enterprises shall receive methods and techniques such as: new concepts in strategic management, change management, re-engineering, project management, quality management, benchmarking, marketing development, information management, information engineering company, lean management, increase efficiency and effectiveness, stress the added value of new trends in the development and use of human resources, changes in the approach to the management of people, and so on. The paper deals with the introduction and implementation of selected new trends in the business. The case study points to their application, advantages and disadvantages in selected enterprise.

1 INTRODUCTION

While the 80 years of the 20th century were marked by particular effort to respond effectively to the Japanese call, the following years were strongly marked by the advent of globalization and its consequences. In developed countries, market saturation slows the growth needs; it raises the need to move from "market vendors" to "market customers." Increased pressure on competitiveness, the pressure of the need to produce cheaply and efficiently, quickly innovate, provide a high level of service. It develops information technology and the opportunities for their use. The globalization of markets brings new challenges and new opportunities, but on the other hand, conflicts and threats. It strengthens the impact of multinational companies especially when the fall of socialism expands into new countries. New conditions mean, amongst others:

- that the world will be in the near future super competitive global market, where the information technology will have a crucial role;
- the beginning of the new economy and new technologies, but also the development of political and economic relations will be blurred boundaries of countries, cultures and time. This also can bring except the positive growth also contradictory tendencies;

- a global network economy creates new competitive space, the competitive advantage will belong to those, who are able to develop and utilize human capital (knowledge, readiness, creativity, ability to execute) and who will be able to seek new opportunities and to respond to them.

All this is reflected in the new trends of management theory and practice. New trends will bring new demands on employees and the managers.

2 APPLICABILITY OF INTERNATIONAL KNOWLEDGE AND EXPERIENCE IN SLOVAKIA

The possibility of applying management theories and practice of foreign managers in Slovak enterprises and in non-profit organizations and public administration is determined by several factors. It should be taken into account in particular geographical, temporal, economic and socio-cultural factor (Bosák & Olexová 2013, Moeller & Kuľka 2012). It is not enough just to have a deep knowledge of what has worked elsewhere, but it must be respected the character and structure of the economy, certain peculiarities that needs to be suppressed, but also the comparative advantages that should be used, and some facts that are likely to change soon. In terms of time to be understood that what effective 10 years ago was in the US in presence already effectively might be less.

When we assessing the capabilities and limitations of the application of foreign knowledge and experience, we should respected in particular the following principles:

- know the essence of new approaches, theories and experiences, as well as its conditions,
- accept the prior results of their application in the geographical characteristics in Central Europe and particularly in Slovakia,
- take into account the demographic, cultural and other specifics influencing the needs, values, people in an organization and how its conduct,
- use its own positive experiences of the effective functioning of foreign and domestic enterprises in Slovakia.

The practice confirms that the globalization process and the development of information technology explore current trends in management and learn from the best. Applying new theories and downloading experience abroad will, however, be a long time to respect the socio-economic and cultural-historical peculiarities of particular country.

2.1 Strategic management

In the mid-seventies there were some changes in the environment that made it impossible to predict the development of the external environment. Unexpectedly the oil and energy crisis happened that practically from one day to change relatively predictable environment to turbulent environment changing and unpredictable. As was reported by H. Mintzberg, strategic planning fell into a deep crisis of such magnitude that even his biggest supporters began to doubt his prospects. Strategies developed on the premise of capability to predict the development environment has become obsolete and unworkable, and did not offer any strategic planning methods and techniques to predict developments in the new character of the environment. The reversal came in 1978, when the top experts in the field of strategic planning at a scientific conference in Pittsburgh in Pennsylvania (USA) have concluded, that the way out of the crisis is not necessary to look at improving the centerpieces of planning, but changed view on management and decided to establish a new field of science "*strategic management*" (Papula & Papulová 2010).

2.2 Development of the theory and practice of strategic management

How can an organization get into a situation as it is today? Why organizations produce these particular products or services? Why it is aimed at a particular part of the market? All these questions relate to different but interrelated aspects of the organization and these aspects converge and affect the efficiency of the organization. Planning and decisions about products, the production itself, the target market, etc. represents a binding decision which can be called strategic. The ways in which these decisions are being made are subject to strategic management (Bowman 1996).

Strategic Management was established as a product of the search for new methods of corporate governance in the constantly complicating the business environment conditions. The accelerating rate of change, developing into discontinuities, steep increase investment risk, increasing capital intensity, the formation of strategic surprises, intense domestic and global competition, unpredictability behavior of competitors, enormous growth demands of customers for quality products and service, terms of delivery and many other factors it was not possible to master traditional methods of management. Strategic management procedures do not ensure quick and easy success as instructions for variants to be further creatively to guess (Pearce & Robinson 2000).

Strategic management cannot be regarded only as a fad in the development of management theory. It is especially actual response to the current conditions and emerging development trends of the market economy.

2.3 Management as a process

Strategic management in the new sense is primarily *process*. An impact on current management presents our next Figure 1.

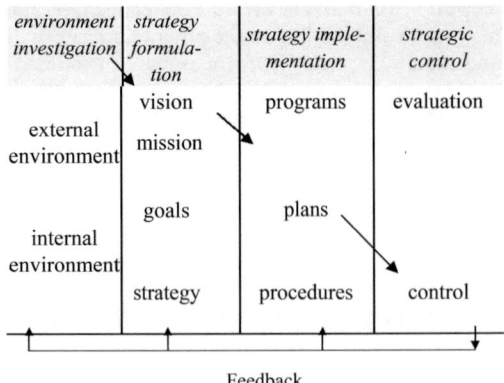

Figure 1. Impact on current management.

450

2.3.1 Strategic management as a process

Strategic management is a process whose main attributes are continuity, compactness and inner harmony. J. Papula in its work graphically shows a simple sequential model as a process consisting of four basic phases: "exploration of the environment, strategy formulation, strategy implementation, evaluation and control (Bowman 1996)."

A. Exploration of the environment

The starting phase of the process of strategic management is phase of exploration of the environment (external and internal environment). Compared to the classic approach of strategic planning is the phase that is not focused solely on making a case study related to the development strategy, the support and additional analyzes related to the process of strategy. Its role is to continuously monitor the environment and provide information about ongoing changes.

B. Strategy formulation

Phase of the strategy formulation consists of identifying the strategic direction of the organization by the vision, mission and strategic objectives and the completion of the strategy. In the case of diversified organizations is a developed strategy for all its hierarchical structure, from the general strategy of the corporation as a whole, through competitive strategies of strategic business units, to the functional strategies for different functional areas (marketing, research and development, operation, logistics, HR, IT, finance etc.).

C. Implementation of the strategy

Relatively new phase in its initial strategic planning process is the implementation phase of the strategy. Its primary task is to develop a strategy to basic development programs and projects (organizational, innovation, restructuring, investment, motivation, information, etc.), which are developing specific tasks and procedures for each priority area strategy emphasized. Implementation of programs and projects is then incorporated into the strategic plans and supported the establishment of standard practices and procedures.

D. Strategic control

Independent and the most recent phase of strategic management phase of strategic is control. This phase is a form of feedback related to all the previous phases of strategic management. Feedback form can initiate changes in the implementation of strategies (changes of projects, programs, plans, procedures) can further initiate changes in the formulation of the strategy (change in strategy, strategic goals, mission and vision) and, finally may require broadening and deepening of external and internal investi-gations environment in the context of the ongoing and anticipated changes in the environment.

Each of these phases and their individual steps are interested in the whole process of strategic management important position and plays a vital role. Management theory elaborated and offers to all of these phases a number of methods and techniques that allow the process of effective strategic management process that supports not only the survival but also the success of the organization in turbulent changing environment (Papula 2004).

3 ANALYSIS OF THE STRATEGIC MANAGEMENT OF SCHOOLS SUITABLE FOR THE ENTERPRISE OF THE 21ST CENTURY

P. Sakál in his book Strategic Management in Practice of manager identified based on systemic analysis of the different profiles of schools:

School of design—the central idea is the consistency of internal and external possibilities of the company as a basis for strategic decision-making. Often we used here the SWOT-analysis. Representatives of this school are A.D. Chandler, F. Selznik and K.R. Andrews.

School of planning—strategic management process examines how process stages formalized management activities (tasks or determine the identification of the objective, external and internal audits, evaluations elaborated strategy, detailed development of strategic plans and tactical planning and control. Thanks to this school is realized, formalized strategic management process (Sakál 2007). This school represents H.L. Ansoff, P. Lorange, and G.A. Steiner.

School of borrowing—each sector has a number of key strategies. Strategic management therefore needs to analyzed it, and choose from accepted the best for the industry and the circumstances. As models here we used for example BCG and GE matrix. Representatives are M.E. Porter, D.E. Schendel and K.J. Hatten.

The business school—the process of drawing up strategies treated as a mental process of top leading organizations based on its strategic foresight. H. Mintzberg and O. Collins are its representatives.

Cognitive school—strategic management process examines how the process of learning and knowledge looks like. In this school there are two directions—objectivist (based on the assumption of objectivity and process knowledge of the world) and subjectivist (assuming that personality itself interprets and creates the world). This school was the first to draw attention to the possibility of future change in strategic decision-making on an active role of the organization in relation to its future external environment (Sakál 2007). The representatives of this school are H.A. Simon, S. Makridakis, I.M. Duhaim and C.R. Schwnk.

School of learning—this school is based on the fact that strategic management is also an effective method of governance based on changes not only control tool modification in the company. As representatives we can mention L. Rapierr, C.E. Lindblom and J.B. Quinn.

School of power—access to treatment strategy is as the quantitative process through which policy is formed organization; the basic efforts must be concentrated on policy processing and production-economic activities of the organization. Representatives are A. McMillan, J. Sarrazin, and A.M. Pettigrew.

School of culture—based on the formation of internal organizational culture, development ideology for the whole company, the motivation of each employee are to create a strong competitive with the company. The main representatives of this school are S.P. Feldman, J.B. Barney, M. Firsirotu and R. Rieger.

School configurations—the main attention is paid to implementation of the strategy, theory and practice of transitional processes in the organization. It examines the transformation of the organization from one stable state to another with the changed business conditions (Sakál 2007). Representatives are D. Miller and A. Chandler.

4 THE APPLICATION OF STRATEGIC MANAGEMENT IN PRACTICE— EXAMINATION OF THE EXTERNAL BUSINESS ENVIRONMENT—CASE STUDY

Presentation of the Centre and companies

Jasna in Low Tatras is classified in the highest category of tourism centers of international importance and among the Slovak centers belongs to him first place. It is located in beautiful surroundings on the northern and southern slopes of Chopok in the Low Tatras National Park. Tatry Mountain Resorts, Inc. operates five hotel facilities and one chat, which provides accommodation, catering and additional services, as well as operates a personal mountain transport facilities. The summer season is the biggest attraction of tourism in the countryside Low Tatras.

4.1 *Examination of the external environment centers*

4.1.1 *Political and legal environment*
In the group of political and legal factors that influence the choice of strategies pursued by the Centre include in particular the following: political stability, international relations, legislative support of tourism at home and abroad, and the state policy in tourism.

4.1.1.1 Political stability
Political stability in the country is at present seems to present a transition from a centrally planned economy to a market presupposes a progressive strengthening it.

4.1.1.2 Foreign relations
Another important factor for influencing the ski resort of political and legal environment for foreign relations of the Slovak Republic in connection with a so-called are foreign relations. The Schengen area and the visa are required for citizens of other countries. Ski resort, whose significant part of revenue, comes from foreign tourists, is dependent on international relations and barriers posed to potential customers abroad. Slovak relations can account for that face to watch as part of the EU's external relations as a whole.

4.1.1.3 State support
From a legislative point of view, central and local government have several options how they affects tourism. It can tend to hinder the development of tourism in the form of tax subsidy or by asking a number of conditions for investors in the region. The positive effect is expected from direct subsidies for entrepreneurs and allowances of domestic tourists in the form of tourist vouchers. Direct subsidies for businesses in the Slovak Republic on the authority of the Ministry of Economy is granted by state-owned institution SACR (Slovak Tourist Board) as a non-repayable contributions under "Priority Axis 3—Tourism" 4 Operational Programme "Competitiveness and Economic Growth". Possibility to use the help from state is a big plus for the entrepreneur and that opportunity is good to use when it is considering an investment in improving services or infrastructure resort.

4.1.2 *Economic environment*
Economic factors are by its nature close to management and it is easier to link them with business practice, with the impact on business. The most important of these factors are following: economic cycles and state of the economy, foreign exchange markets, the purchasing power of the population and standard of living, interest rates and the situation on the capital markets.

4.1.2.1 GDP
Probably the hardest part is to assess the situation in which the economy is; partly it's because of its complexity, but also because of its dynamic nature and constant changes. But economics offers a number of indicators that in this difficult task

can help. Real gross domestic product is measured at constant prices, i.e. prices of a previous period. Real GDP reflects only the changes in production; it's not taking into account changes in the price level.

4.1.2.2 Inflation
European Central Bank defines inflation as a general rise in prices of goods and services, not just to the increase in prices of individual items. Inflation is thus associated with a decrease in the purchasing power of monetary units, i.e. that means for one euro we can buy less than in the past.

4.1.2.3 Unemployment
Unemployment within the district in which the center Jasna Low Tatras is, copied unemployment nationwide trend. For the enterprise, the rising of unemployment means that, the company can find in market highly skilled people. Increasing labor supply also means that is small pressure on wages.

4.1.2.4 Foreign exchange markets
From 1st of January 2009, the Slovak Republic joined the European union. This has largely affected the segment of tourism in Slovakia, especially because half to two-thirds of clients of Jasna in Low Tatras consist of tourists from neighboring countries. The currencies of neighboring countries against the Euro have been recently affected by major emerging economic crisis. The weakening of currencies resulted in a decrease in purchasing power of neighboring currencies, causing significant loss of tourists from neighboring countries

4.1.2.5 Purchasing power
Private consumption is one of the items in the calculation of GDP. These costs can be divided into three basic groups: 1.expenditure used on such durables, for example furnished apartment, buying a car, 2. expenditure on non-durables such as purchase of food, clothing, 3. expenditure on services e.g. rent or hotel services. For the company expected growth in household expenditure means, that it can expect increasing sales.

The situation on the capital markets and interest rates.

The last factor in the field of economic environment, which deserves special attention, is the situation on the capital markets and changes in interest rates. The company is participating in the equity markets and their development will affect the flow of funds to the company. Interest rates are also strongly linked to investment, in the case of the ski resort and have an impact on planned and potential investments.

4.1.3 *Social and cultural environment*
Among the trends in society, which could affect the observed resort or area of tourism globally, maybe

pick a few examples: the role of lifestyle, fashion behavior, the emphasis on healthy living, trends in sports, and the expansion of the Internet and other means of communication.

Threat for the Centre as a provider of physically demanding activities is population aging. Although the ent demographic prospects of young people in developed European countries will decrease in number rather than accrue, there will always be an important and interesting group for Mark vapor-collection. Young people in their search for self are often subject to the latest fashion trends. For ski resorts have great benefits trendiness snowboarding, sports, and also a lifestyle that many young people that their personable and able to attract large numbers on the slopes. The use of this increasingly strong trend, it is a good opportunity for the ski resort.

4.1.4 *Technical and technological environment*
In terms of technical equipment in the ski resort are especially important cableways, gondolas and chair lifts, also snowmaking technology but also the technology tickets. Snowmaking technology is particularly important in areas with a lack of snow, a problem many Slovak ski resorts, not excluding clear; despite its altitude is a major advantage over other.

4.1.5 *Environmental environment*
In terms of the Environmental for the needs of this paper, we identified two important factors affecting the ski resort. The first is the protection of nature in the area, other environmental requirements of customers. We cannot forget global warming and the consequent lack of snow cover. Nature protection has great influence over the decisions of the entrepreneur, preventing interference with the environment, which is more valuable than the potential profit from his conversion. Although the legislation restricts some activities to designated areas, activities can bring about profit and raises, for privacy is often at the same time as their reason for visiting tourists. Ski resort Jasna Low Tatras is situated in the territory of a third level of nature protection, particularly in the Low Tatras National Park (NAPANT).

5 CONCLUSION

The possibility of applying management theories and practice of foreign managers in Slovak enterprises and in non-profit organizations and public administration is determined by several factors. The paper deals with the introduction and implementation of selected new trends in the business, and this case study points to application of ties factors, advantages and disadvantages in Jasna in Low Tatras. It should be taken into account in particular geographical, political, economic, technical and technological, environmental and socio-cultural factor. We can therefore conclude that the impact of these factors on the enterprise is essential.

REFERENCES

Bosák, M. & Olexová, C. 2013. Recent environmental trends and innovations in the Slovak small and middle enterprises, In. *GeoConference on ecology, economics, education and legislation SGEM 2013, 16–22 June, 2013*, Albena, Vol. 2., 247–254.

Bowman C. 1996. *Strategický manažment*. Praha: Grada, 1996. 147 s.

Moeller, J. & Kuľka, J. 2012. Medium-sized businesses and logistics under the sign of globalization. *Transport & Logistics*, 12 (23): 1–5.

Papula J. & Papulová Z. 2010. *Strategické myslenie manažérov*. Bratislava: Kartprint, 2010. 302 s.

Papula J. 2004. *Vývoj teórie strategického manažmentu pod vplyvom meniaceho sa prostredia*. Bratislava: Kartprint, 2004. 268 s.

Pearce, J.A. & Robinson, R.B. 2000. *Strategic management: Formulation, implementation, and control*. Irwin/McGraw-Hill.

Sakál P. & Podskľan A. 2004. Strategický manažment. Bratislava: STU, 2004. 256 s.

Sakál, P. et al. 2007. *Strategický manažment v praxi manažéra*. STU v Trnave 2007, 160 s.

Production Management and Engineering Sciences – Majerník, Daneshjo & Bosák (Eds)
© 2016 Taylor & Francis Group, London, ISBN: 978-1-138-02856-2

Standardization in Slovakia in the historical context and its European framework

J. Krivosudská, M. Holecek & Z. Schreier
Slovak Office of Standards, Metrology and Testing, Bratislava, Slovak Republic

J. Markovič
Slovak Legal Metrology, Banska Bystrica, Slovak Republic

ABSTRACT: The article describes the national system of standardization in Slovakia and its links with the European and international standardization structures. Within the article will be described history of standardization and recalled the basic principles of "good standardization practices" and their current application at the national level. Following will be described binding technical standards on legislation, references to standards in legislation and the New Approach principles. With varying structure of the EU single market is increased the role of standardization, metrology and conformity assessment, which are essential elements of liberalization, simplification and openness in the European market as well as the global environment. In this context is the significant contribution of standardization in support of increasing the competitiveness and innovation potential. In the section of the article devoted to trends contributed to sustainable economic development are described priorities disclosed intentions of European, partly also international standardization organizations.

1 INTRODUCTION

If we want to characterize what can be in the succinct the sense of standardization, it would be enough to draw a single word—understanding. Standardization from the very beginning of its existence—regardless of where you choose to count—directly or in directly has served and serves just for this purpose—understanding partners in the art. Acting as agents themselves own: an expression of the rules appearing in the vast majority consensus of the parties. Gradual development influenced the development of industrial production and the continued globalization, the scope of creation and use of technical standards shifted from the original corporate and national standards to international and regional level.

2 EUROPEAN AND INTERNATIONAL CONTEXT OF STANDARDIZATION

Standardization is a prerequisite for ensuring a conducive environment for market development. Standards have always been an integral part of the market system and have played a key role in increasing the wealth of nations. Standards tend to increase competition and allow lower costs of production and sales, which benefits the economy

as a whole. Technical standards reduce diversity, ensure interoperability, maintain quality and provide the necessary information. In order to ensure their own competitiveness on world markets today's participants of the market environment are fully aware of the need to create a clearly defined respect for internationally agreed recommendations serving as industry standards for products and services. Compatibility of products and services then contribute to healthy competition between producers, ensuring diversity and quality of the users and form the basis for innovation and streamlining production.

Intensive efforts to standardization on the international level have appeared around the mid-twenties of the last century. In the field of electrical engineering since 1906 operated the International Electrotechnical Commission (IEC), but the general level of standardization organizations was lacked. In 1928 in Prague was established International Federation of National standardizing Associations (ISA), in particular thanks to the initiative of European national organizations and mainly the Czechoslovak Society Standardization (CSN). In 1942 due the demise of ISA because of WW2 international standardization completed. Its function after the war took the International Organization for Standardization (ISO), founded in Geneva in 1947.

Standardization process, based on an international consensus ensures that the final standards for products and services represent the collective knowledge and experience of all stakeholders—industry, governments, research institutes, testing laboratories and consumer organizations. The value of international standards is today recognized across industry sectors without residue. It is not simply to enforce the creation and use of technical standards as widely as possible, but primarily to replace the national technical regulations with the international standards wherever possible. Lack of united standards in different countries is a cause of technical barriers to international trade. To remove these barriers was necessary to agree on globally accepted standards, which led to the establishment of an international structure of standardization bodies. The current international standardization system is formed by three most important organizations:

1. ISO—International Organization for Standardization,
2. IEC—International Electrotechnical Commission,
3. ITU—International Telecommunication Union.

Which have a direct link to the European standardization system composed also of three organizations:

1. CEN—European Committee for standardization,
2. CENELEC—European Committee for electrotechnical standardization,
3. ETSI—European Telecommunications Standards Institute.

The main reasons for global standardization are:

1. Global progress in the liberalization of the market—in today's market economy where market is more strength leads to the fact that is possible to select from various kinds of offers coming from all over the world. For the competitiveness of these offers is necessary, that there would be clearly defined internationally applicable recommendations, serving as industry standards for products and services, approved by market processes participants.
2. Interpenetration of the various sectors of industry—today no industry is not completely independent of components, manufacturing processes etc. created in another sector. Examples are products or technologies environmentally friendly.
3. Build a global communications systems—in this rapidly developing field must be used technologies quickly standardized. Compatible products then contribute to healthy competition between producers, it guarantees the quality for the user and form the basis for innovation and streamlining production.

4. Emerging technologies—the development of new technology is necessary to build on already operational prototypes. The importance of standardization is in the development of terminology and databases, gathering the necessary information.
5. The economy of developing countries—it is known that to use industry standards is an important condition for the successful development of the economy. Therefore, developing countries need build standardization infrastructure to improve their productivity and competitiveness in world markets.

3 THE SYSTEM OF SLOVAK STANDARDS (STN) AND THEIR LEGAL CHARACTER

The basis of the STN became valid the library of Czechoslovak standards (CSN) approved up to 31.12.1992 in accordance with Act No. 142/1991 Coll. about Czechoslovak technical standards in the Agreement between the Czech Government and the Government of Slovak Republic on cooperation in the field of standardization, metrology and testing and related activities. Standardization system in Slovak Republic was thus CSN transferred to the STN. STN abbreviation has been recognized internationally in September 1993 with the right of use from 1.1.1994. By legislation was the abbreviation of the Slovak Standard (STN) notified by amendment to Section 11 of the Act No. 142/1991 Coll.

STN is therefore protected designation of Slovak standards issued by Slovak Office of Standards, Metrology and Testing. It is protected by law solely the word indication of Slovak Standard. Next to the word indication STN the designation continues within six-digit number, where the first two digits are separated by space and means a category of standards (00–99 gives a broader economic field). The third and fourth digit indicates the group and subgroup of standard and the last two digits are serial numbers of the standard. Adapted (harmonized) European standards are referred to their original indication, preceded by the abbreviation STN added. The standard than may be labeled for example 'STN EN 12899-1', 'EN ISO 9001', 'STN IEC 61713', 'ETS 300 976' and the like. Technical standard is usually also assigned by classification symbols as a traditional six-designation by category of STN. According to European standards, it is introducing the practice of identifying year of issue next to the colon, for example ISO 9000:2001. The Act introduces concepts such technical regulation—regulation containing technical requirements for the product; Slovak standard—a standard adopted by the procedure under Act No. 264/1999 Coll. and reported in the Journal of the Slovak Office of Standards, Metrology and Testing; standard publication—the term is not precisely defined, it includes in particular the standards adopted in a similar way in other states or supranational institutions; technical document—another document containing the technical requirements for the product, which is not a technical regulation or technical standards.

4 NATIONAL SYSTEM OF STANDARDIZATION AND ITS LINKS TO THE EUROPEAN AND INTERNATIONAL STANDARDIZATION STRUCTURES

Essential tool to ensure the free movement of goods within the EU are:

1. the principle of mutual recognition of national regulations on products (non-harmonized area),
2. prevent new barriers to trade (information on draft national technical regulations and programs of new standards development),
3. technical harmonization (regulated at the Community level of the utmost importance in terms of public interest).

4.1 The "New approach"

Since 1985, when it was conceived in the EEC "New Approach to technical harmonization" it were, according to this philosophy, drawn up and adopted a series of Directives. They already contain only basic, general requirements for product safety. On the task of finding solutions to specific technical requirements in the form of European Standards (EN) harmonized with these directives have undertaken to European standards organizations on the basis of the "Framework Convention" concluded with the Commission in 1984. Harmonized standards have become a supporting component of EC legislation. Directives regulate the legal form of how to use European standards and they are with them (harmonized or mandated) firmly tied. In both the cited sections BK are discussed in detail and measures in the field of standardization, conformity assessment and market surveillance, without which the adoption of Directives into the national legal system had no practical effect. The supervisory task of compliance rules with generally formulated base is transferred to the independent "notified bodies", respectively the producers themselves. The basic principles of the "New Approach":

1. Legal harmonization is limited to the adoption of basic, mostly safety requirements which the product must comply with the placing on the market. These requirements are determined so to fulfil a high level of safety of each product.
2. For the development of technical specifications (i.e. an exemplary model to meet the essential requirements) are responsible relevant standardization organizations. These technical specifications (harmonized standards) are not mandatory and shall keep the status of voluntary standards. State authorities are obliged to recognize that products that have been manufactured in conformity with harmonized standards meet the essential requirements set out in the Directives.
3. The New Approach addresses the broad groups of products and horizontal risks. The standards attributed an important role and the authorities empowered to assess compliance with the requirements of the Directives (Notified Bodies) can be private organizations. Directives define the essential requirements that products must meet, but do not have the ability to meet their specific ways.
4. Different Directives covers variety of risks. It may be that the products are subject to several Directives. It is the result of the consistency of the structure of the Directives, based on the risk that a product may pose.

5 MANDATORY STATUS OF STANDARDS

The obligation to follow the non-binding provisions of technical standards may arise in the following cases:

1. If it is established by generally binding legislation (Act),
2. If the setting of specific legislation is laid down by reference to the designated STN,
3. May arise on the basis of a decision issued by the state administration body in administrative proceedings (e.g. Building permit),
4. Agreement between two or more entities (e.g. supplier—consumer contract).

6 RESEARCH AND INNOVATION

Standardization can help bridge the gap between research, innovation and the market for example in codification and dissemination of relevant research, development and innovation activities. In this way standards can support the process of bringing new ideas and technologies.

European standardization contributes significantly to achieving the goals of the "Innovation Union" the initiative launched by the European Commission in the framework of the Europe

2020. Standardization is particularly important in the context of the program "Horizon 2020"—the multiannual program of the European Union, which supports research, development and innovation projects.

CEN and CENELEC work closely with the European Commission and other partners in promoting the idea of "integrated approach", which means that any reference to standardization are addressed during the planning and implementation of research and innovation activities.

CEN and CENELEC prepared and has been maintained Helpdesk "Research", through which provide advice and support to institutions responsible for the preparation and implementation of research and innovation projects. At the national level the members of CEN and CENELEC create a network of the National Research, Development and Innovation Correspondents (RDI-COR) composed of 31 national contact points in 26 countries.

CEN and CENELEC activities are related to research and innovation coordinated by working group for Standardization, Innovation and Research (STAIR). These include activities aimed at promoting closer collaboration between researchers and the standardization experts at the national level (e.g. project "Bridge the Gap") and other activities at European level, including the "Bridging Platform" (such as the initiative "STAIR-EMPIR") which are focused on specific sectors or technologies.

As part of its ongoing efforts to promote cooperation between the scientific community and standardization, CEN and CENELEC cooperates with the Joint Research Centre (JRC) of the European Commission and the European Association of Research and Technology Organizations (EARTO). This includes the organization of joint seminars on topics of common interest, which are held annually.

6.1 Project "Bridge to Gap"

Bridging the gap (Bridgit) is the name of a two-year project, which started in 2013. The project partners include 9 alongside its national members of CEN and CENELEC with the European Commission and EFTA. The project aims to highlight the different ways in which standardization encourage innovation and strengthen the links between standardization, research and innovation communities.

Various examples of good practice shared with stakeholders. They were organized trainings and awareness-raising events to raise awareness of technical standardization, including several national events and on a larger scale the European conference "Technical standards: Your Bridge to innovation", held in Brussels on 30 October 2014.

The "Bridging the Gap" ends in March 2015 and its output will be a series of publications and other products. This will include brochures, videos and other teaching materials aimed at the discretion of managers and employees of national standards bodies, to help them establish and develop relationships with innovative and research communities in their countries. Specific recommendations will be provided to members of CEN and CENELEC in terms of how to organize seminars and educational events.

Specific publications will be created for the National Contact Points (NCPs), which are responsible for providing practical information and assistance to research and innovation staff, particularly on all aspects of participation in the "Horizon 2020". There will also be a separate brochure focused on the research and innovation community especially designed for managers in research institutes and companies which carry out research, development and innovation activities. The brochure will aim to explain all attributes of technical standardization and should help those who may have little or no prior knowledge of standardization.

6.2 STAIR-EMPIR

CEN and CENELEC work together with EURAMET—European Association of National Metrology Institutes joining in metrology research with standardization activities. This cooperation is currently supported by adoption of a new initiative called "STAIR-EMPIR".

Metrology as the science of measurement is the cornerstone of our industrial society and has an impact on almost all aspects of daily life: precision industrial products, the reliability of medical devices, environmental monitoring, and many more.

In 2010, CEN and CENELEC signed a cooperation agreement with the EURAMET, which brings together the national metrology institutes of 37 European countries. The three partners working to promote closer links between standardization and metrology, for example in sharing information on metrological needs identified in addition to standardization work (and vice versa).

STAIR-EMPIR brings together stakeholders from CEN, CENELEC and EURAMET to identify areas where metrology research may contribute to standardization activities in accordance with the specific needs of CEN and CENELEC. It will also provide a common forum for sharing knowledge between standardization and metrology community to discuss strategic issues of common concern.

The first joint meeting STAIR-EMPIR held 6 November 2014 at the headquarters of

CEN-CENELEC in Brussels. This initial meeting was an opportunity for stakeholders to learn more about EMPIRE program and the possibilities of how this program can support standardization. Particular attention was given to strategic research agenda in priority themes: energy, environment and health.

7 THE IMPORTANCE OF EDUCATION OF STANDARDIZATION

The discipline of standardization passed in Slovak Republic for last about 20 years an interesting way of change. That most crucial was in the very beginning the change of legal force of the standards, binding institute of standards was repealed and replaced by institute non-binding and voluntary. The conditions of membership in the European standardization organizations changed national standardization system and processed the creation of new (European) standards. These rapid adjustment of the entire system is reflected not only in the underestimation of the importance of standards by the newly forming management structures in industrial production, but also led to some stagnation on the part of the administrator role of development of standards—national standards body. Technical work that forms the basis for creating standards has been withdrawn in many fields and has become many times more work of enthusiasm than ordinary part of the process on an industrial scale.

For a proper understanding of the new system it was necessary to introduce a system of education. By European standardization organizations have been organized seminars, workshops, etc. which aim was understanding and mastering so called "good standardization practice". These training activities were organized primarily for employees of the national standardization body and selected representatives of the government, but engineering community representatives, technical staff and representatives of associations and professional associations or businesses were invited as processors of standardization work. Educational events of this type were focused on the interpretation of the administrative process of creating European standards, but they also aimed to show how is possible to be involved in technical work, how to access the text of the draft new standards and how to join projects linked to the development of standards.

Changing the legal force of the standards resulted in the loss of (social) value of standards mainly by the new management structures in industrial production. Despite the effort to provide training programs for managers explaining the meaning, purpose and benefits of standardization, the necessary changes, which would lead to the provision of suitable sources of financing creation of technical standards, avoid. Through several exceptions, mainly in the construction, some sectors of electrical engineering, adequate support for the creation of technical standards lacked. The government provided (and provides) a contribution to development, but it is a primary contribution to harmonized/mandated European standards, which does not fully cover the costs of development and therefore it is always necessary to seek the necessary resources.

Step by step the importance of standardization perceived more and more serious. It is associated with the business development in the European market, but also outside the EU. As I mentioned in the previous sections of the content of educational activities in terms of familiarity with the meaning and process of development standards, which certainly is constantly needed for decision-makers, the current interest becomes much more significant for the sector directly involved in the design and production process needs. The interest is more and more in terms to have orientation in content and requirements or recommendations of standards to handle the entire width of one standard reference for others. This need presents challenges to capture corporate orders and preparing the content of the educational process "tailor made". The programs of standardization work on national, European or international level give the possibility of planning costs of participation on development of selected standards in their financial budgets, including the future financial benefits of the company due to the readiness of the entire system of the requirements of standards issued.

REFERENCES

Czechoslovak Republic. 1991. Act No. 142/1991 Coll. on Czechoslovak Technical Standards, as altered and amended. Prague: Collection of Laws of Czechoslovak Republic.

Slovak Republic. 1999. Act No. 264/1999 Coll. on technical requirements for products and on conformity assessment and on change and amendment of some acts as amended. Bratislava: Collection of Laws of Slovak Republic.

Production Management and Engineering Sciences – Majerník, Daneshjo & Bosák (Eds)
© *2016 Taylor & Francis Group, London, ISBN: 978-1-138-02856-2*

Black Carbon as an indicator of air pollution in industrial region

M. Kucbel, B. Sýkorová, H. Raclavská & K. Raclavský
ENET—Energy Units for Utilization of Non Traditional Energy Sources, Ostrava, Czech Republic

ABSTRACT: Concentration of Black Carbon (BC) in $PM_{2.5}$ was measured continuously at an urban site in Ostrava (the Czech Republic, Moravian-Silesian Region) during the period from August 2014 to December 2014. The monthly average BC concentrations ranged from 1.32 to 6.77 $\mu g/m^3$ and the average of $PM_{2.5}$ concentrations varied from 19.7 to 65 $\mu g/m^3$. The BC mass concentration formed 8.33% of the total $PM_{2.5}$ mass. It was found that BC does not only have significant seasonal variations, with the highest concentrations in the winter season, but also diurnal variations with two peaks formed by morning and evening increased values. Daily average mass concentrations of BC reached the level of 3.98 $\mu g/m^3$, whereas the average morning concentration was 3.69 $\mu g/m^3$ and evening 4.48 $\mu g/m^3$. The influence of selected meteorological parameters on the BC concentrations was observed.

1 INTRODUCTION

Black Carbon (BC) forms an important component of dust particles in the atmosphere. Black Carbon is released into the air during incomplete combustion. The predominant BC sources in urbanised regions (particularly in the Northern Hemisphere) include combustion of fossil fuels as well as biomass, and transport (Cao et al. 2009, Guo et al. 2015, Sahu et al. 2011, Saha & Despiau 2009, Monks 2012). The combustion of biomass can be the prevailing source of BC in tropical regions (prevalently in the Southern Hemisphere). It was estimated that 38% of BC originates from combustion of fossil fuels, approximately 20% from biofuel combustion and about 42% from biomass combustion (Sahu et al. 2011) Black Carbon is the main component of $PM_{2.5}$ related to emissions from transport and local heating of solid and liquid fuels (Monks 2012). Sillanpää et al. (2005) in their study reported that average BC contribution to $PM_{2.5}$ in European atmosphere is 8% and 3% in $PM_{2.5-10}$. BC is the main constituent of anthropogenic atmospheric aerosols, which has the distinctively different optical properties in comparison with other present components (Saha & Despiau 2009). BC is a stable (inert) aerosol, water insoluble, with a short life cycle—days to weeks (Costabile et al. 2015, Brasseur et al. 2015). BC strongly absorbs visible, near UV and IR radiation due to its graphite structure. Aerosols with this high absorption can exert a strong influence on regional climatic conditions, visibility decreasing, radiation balance of Earth or hydrological cycle (Cao et al. 2009, Guo et al. 2015, Sahu et al. 2011, Zhang et al. 2009). Information on long-term variation of BC concentration is of great importance, it can help to estimate the influence of BC emissions on regional and global climate.

BC has very often distinctive diurnal variation with two peaks of the highest concentrations during morning rush hours (from 6 to 9 a.m.) and the evening peak (from 7 to 9 p.m.) which are partially related to the transport sources and also the lifestyle of inhabitants in the region (Feng et al. 2014, Wang et al. 2011). During the winter season, concentrations of BC in the morning and evening hours are higher than those in the remaining part of the year. It is caused mostly by the presence of permanent temperature inversion in the lower level of the surface boundary layer of the atmosphere. BC concentrations do not exhibit noticeable variation during all the year in the hours around the middle of the day due to the efficient particle dispersion and mixing in the boundary layers of the atmosphere (Srivastava et al. 2012). Other explanation for the evening peak is connected with the shallowness of the night boundary layer (Kumar 2011). The morning peak is related to the BC load from local anthropic activities and fumigation effect in the boundary layer of the atmosphere immediately after the sunrise (Stull 1989). It was also found that concentrations during working days are higher than those measured on weekends (Feng et al. 2014, Pereira et al. 2012, Cheng et al. 2014). In addition, Wang et al. (2011) described the morning peak of BC concentration measured on weekends in Rochester, USA (in the period from 1 August to 31 December 2009) which was substantially lower than values measured during working days. The evening peak was higher on weekends than during working days.

The meteorological parameters considerably influence the concentration of BC which decreases exponentially with wind velocity (Wang et al. 2011, Begum et al. 2012), the similar relationships exist also for rainfall (Ramachandran & Rajesh 2007, Babu & Moorthy 2002) and temperature (Tiwari et al. 2013).

The goal of this article is to evaluate a significance of diurnal and seasonal variation in BC concentration from $PM_{2.5}$ and to utilize this information for identification of BC proportion formed by combustion products in $PM_{2.5}$. The BC concentration measurement was performed by the instrument Aethalometer during five months of 2014 in the city of Ostrava.

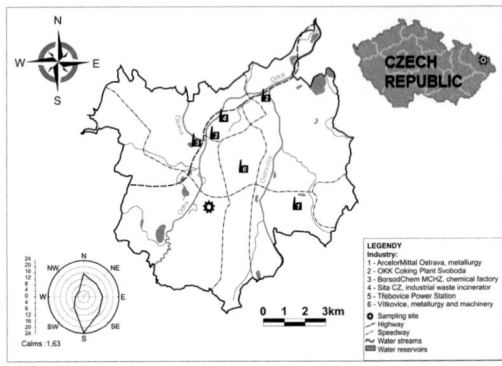

Figure 1. The location of sampling site at the territory of the city of Ostrava (the Czech Republic).

2 THE DESCRIPTION OF SAMPLING SITE AND METHODOLOGY OF SAMPLING

2.1 Sampling site

The sampling of Black Carbon and particles $PM_{2.5}$ was performed during five months from August to December 2014 in the city part Ostrava-Zabreh (Fig. 1) 49°47'23"N, 18°13'48"E. The sampling site is located in the area of blocks of flats (urban residential area) with a frequented road nearby. Two important industrial sources of pollution are located in the maximum distance of 7 km (ArcelorMittal metallurgy complex and Vitkovice Steel).

Ostrava conurbation has considerable environmental load caused by industrial activity, particularly metallurgy. The limit values for PM_{10} and $PM_{2.5}$ concentrations are exceeded several times in the long-term within the whole territory of the city. The limit value of annual average for $PM_{2.5}$ concentration (25 µg/m³ according to the Act on Air Protection 201/2012 Coll.) was exceeded by 6 µg/m³ in 2014 (CHMI 2015).

2.2 Methodology of sampling

The measurement of BC concentrations was performed by portable multiwave instrument Aethalometer, $PM_{2.5}$ by DustTrak aerosol monitor, model 8535. Meteorological parameters (temperature, wind direction and velocity, humidity and precipitation) for the time of measurement were measured by the Czech Hydrometeorological Institute (Ostrava) at the automatic climatological station at Ostrava—Poruba. The determination of BC and $PM_{2.5}$ was performed in the morning (7–8 a.m.) and the evening (7–8 p.m.). The average of concentrations measured in 5-minute interval was used for further evaluation. The average of morning and evening concentrations was used as a representative value for the particular day. The measurement

of BC concentrations was performed by Aethalometer AE-42 (Magee Scientific) at the wavelengths 370, 470, 520, 590, 660, 880 and 950 nm. Air flow in the instrument was 2l/min. The Aethalometer measures attenuation of light emitted from LED source at wavelengths in the range from 370 to 950 nm. Attenuation of light beam is caused by BC particles deposited at a quartz filter tape.

3 RESULTS AND DISCUSSION

3.1 Seasonal variation of BC and $PM_{2.5}$ concentration

The daily average BC concentration in the monitored period was 3.98 µg/m³, the morning mean value was 3.69 µg/m³ and the evening mean value was 4.48 µg/m³. The daily average BC concentrations were divided into five classes in the range from <1 µg/m³ to >9 µg/m³. The frequency distribution of BC concentrations revealed that 13.1% of BC concentrations were lower than 1 µg/m³, the highest frequency 43.1% was found in the class 1–3 µg/m³, the class 3–6 µg/m³ contained 28.8% of values, frequency in the class 6–9 µg/m³ was 5.9% and in the class >9 µg/m³ 9.2%.

The month average variation of the daily BC and $PM_{2.5}$ concentrations (Fig. 2) shows a gradual step-wise increase from August (1.32 ± 0.58 µg/m³ BC, 19.7 ± 8.9 µg/m³ $PM_{2.5}$) to December (6.77 ± 6.68 µg/m³ BC, 65.0 ± 61.86 µg/m³). The geochemical background of BC concentrations was determined by the measurement in the Moravian-Silesian Beskydy Mts., in the vicinity of Celadna, Mt. Magura. The morning value of BC concentration was 0.232 ± 0.056 µg/m³. The presence of BC in the atmosphere is influenced by the long-distance transport. The industrial and transport

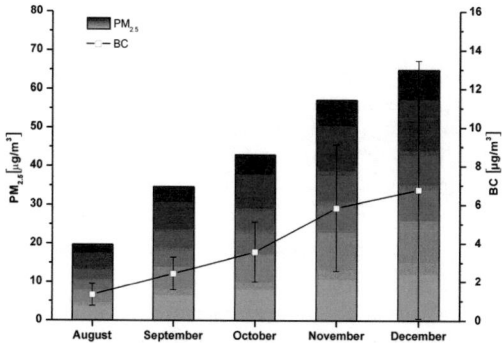

Figure 2. Monthly mean BC concentrations [μg/m³] with standard deviations and PM$_{2.5}$ concentrations [μg/m³].

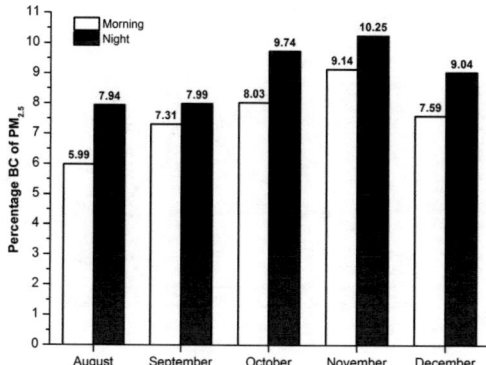

Figure 3. Monthly mean percentages of BC in the morning and evening concentrations in PM$_{2.5}$.

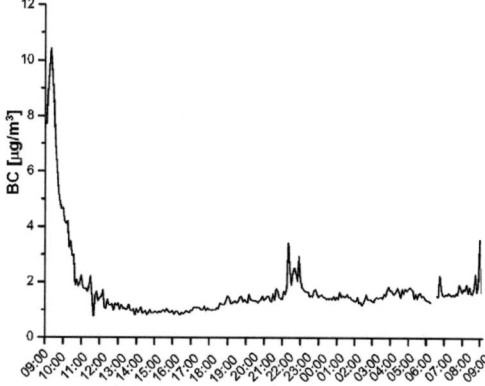

Figure 4. The course of diurnal variation of BC concentration during 19 September–20 September 2014 in Ostrava—Zabreh.

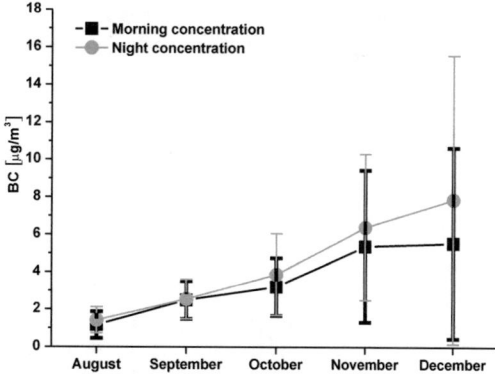

Figure 5. Monthly mean morning and evening BC concentrations with their standard deviations.

load in Ostrava manifests approximately six-fold increase of BC concentration in August. The percentage of BC in PM$_{2.5}$ for individual days varied in the range from 2.51% to 25.06%. The increase of BC concentrations during the heating season is apparent in Figure 3, where the average percentages of BC in PM$_{2.5}$ for individual months are illustrated, separately for the morning and evening measurement. The lowest BC contributions were determined in August both for the morning (5.99%) and for evening (7.94%) measurement. The highest BC percentage in PM$_{2.5}$ was recorded in November for both parts of the day (9.14% and 10.25% respectively).

3.2 Diurnal variation of BC

Twenty-four-hour measurement of BC concentrations was performed during 19–20 September 2014. The highest BC concentrations were recorded in the morning from 9 to 10 a.m. (Fig. 4).

This high concentration is related to the morning rush hours. The BC concentrations gradually decreased with increasing temperature during the day and from 11 a.m. to 5 p.m. fluctuated reaching only 1.5 μg/m³. The BC concentrations again increased from 6 p.m. with maximum at 10 p.m. This process is typical in most cases, sometimes reaching maximum values around midnight. The variation of BC concentration is also influenced by meteorological parameters. The course of BC concentrations is not always similar in the morning and evening. Some days, BC concentrations are higher in the morning, lower in the evening and vice versa.

The average morning and evening BC concentrations were calculated for individual months (Fig. 5). The morning average concentrations were lower than evening averages. This difference has the minimum value during August and September when the morning and evening BC concentrations are influenced mainly by vehicles. The evening BC concentrations increase from October with the beginning of the heating season and the increase of local heating contribution. The maximum value of the morning BC concentration 23.13 µg/m³ was recorded on 9 December, and the minimum value of 0.395 µg/m³ on 23 December. The maximum value of the evening BC concentration 27.39 µg/m³ was reached on 4 December, and the minimum value of 0.45 µg/m³ was measured on 9 August.

Fluctuation in BC concentrations is illustrated in Figure 6 for the summer season (August and September) and the winter season (November and December). The highest BC concentrations in the summer season were measured on Fridays (the mean value 2.31 µg/m³ and proportion of BC in PM$_{2.5}$ 7.42%), the lowest BC concentrations were measured on Sundays (1.28 µg/m³ with BC proportion of 3.41% in PM$_{2.5}$). The maximum morning BC concentrations in the winter season were determined on Mondays (6.11 µg/m³ and BC propor-

tion in PM$_{2.5}$ 8.56%), the maximum evening value of BC concentrations was determined for Saturdays (9.42 µg/m³ with BC proportion of 9.59% in PM$_{2.5}$). The average morning BC concentration on working days in the winter season was approximately 2.8 times higher compared with the summer period, in the evening, it was 3.2 times higher.

3.3 The influence of meteorological parameters on BC

An influence of selected meteorological parameters on BC concentration was studied by correlation analysis. Figure 7 displays changes of monthly

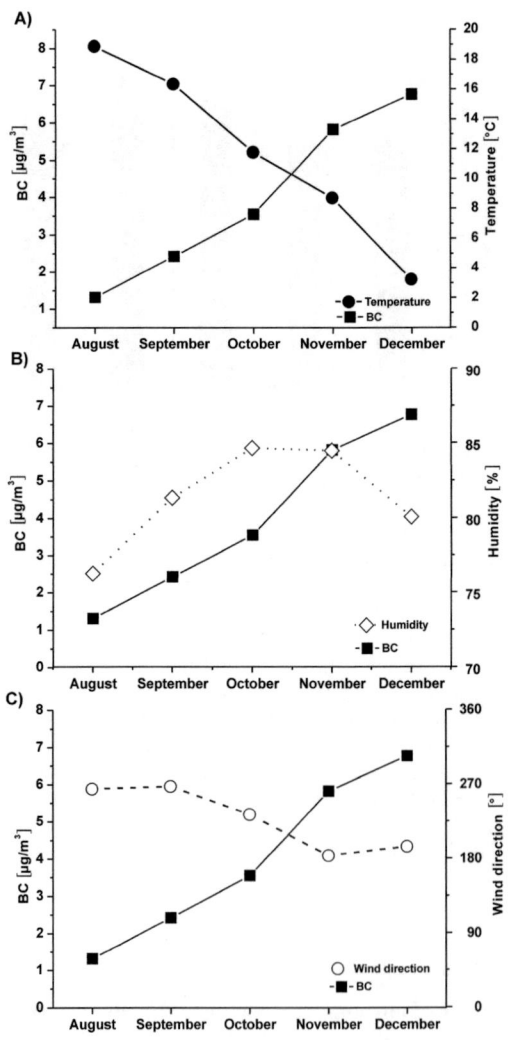

Figure 7. Monthly variation of selected meteorological parameters.

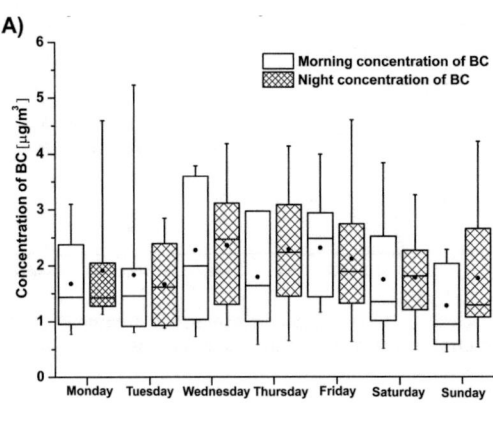

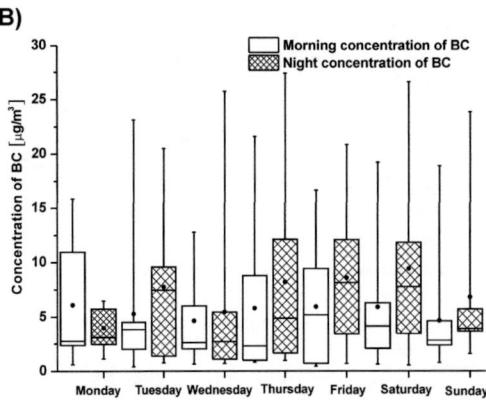

Figure 6. The weekly mean BC concentrations for the morning and evening: A—August–September, B—November–December.

average values of selected meteorological parameters—daily mean temperature, daily mean relative humidity, evening wind direction together with changes in monthly average BC concentration.

Temperature is a highly important parameter, which influences the total BC concentrations. The highest average temperature is in August 18.9°C and the lowest in December 3.2°C. The total BC concentrations increase with the decrease of temperature. This is also proved by inversion correlation coefficients $r_s = -0.42*$ between temperature and BC concentration in the morning ($\alpha < 0.05$, the same significance level was used for all correlation coefficients). The correlation coefficient for temperature and BC concentration in the evening was $r_s = -0.56*$, the all-day value for temperature and BC concentration was $r_s = -0.57*$. The relative humidity ($r_s = 0.50*$) is another very important parameter which influences total all-day BC concentrations in the atmosphere. The average monthly value of relative humidity varied from 76.2% (August) to 84.6% (October).

By means of correlation analysis of the whole set, it was possible to prove neither the relationship between BC concentration and wind velocity ($r_s = -0.01$) nor wind direction ($r_s = -0.14$). ANOVA single factor analysis of variance (at the level of significance $\alpha < 0.05$) was applied for significance determination of wind direction and velocity on BC concentration in the atmosphere. Bonferroni *post hoc* test (at the 0.05 level) was used for multiparameter comparison. This method also proved with probability of 95% that the wind speed does not have the statistically significant influence on BC concentration in Ostrava-Zabreh.

The wind velocity most often measured in the studied territory was lower than 1 m/s (frequency 42%) with average BC concentration of 4.65 μg/m³. The second highest frequency was reached by the wind speed in the interval 1–2 m/s (33.3%), during which the average BC concentration of 4.83 μg/m³ was measured. The wind velocity in the interval 2–3 m/s (frequency 15.7%) was connected with the value of average BC concentration somewhat lower—3.56 μg/m³. The interval of the wind speed 3–4 m/s (4.4%) is corresponding to the average BC concentration of 3.38 μg/m³. The lowest average BC concentration of 2 μg/m³ was measured for wind velocity higher than 4 m/s (4.4%). Figure 8 shows average BC concentrations and PM$_{2.5}$ values in relation to the wind direction. The wind directions were divided into four categories, 90° each. The north-eastern wind direction—from Poland (13.2%) is the only direction with proven correlation dependence of BC concentration, $r_s = 0.49*$. The highest average concentration of BC—9.89 μg/m³ and PM$_{2.5}$ even 91.5 μg/m³ were measured during air flow from this direction. No significant

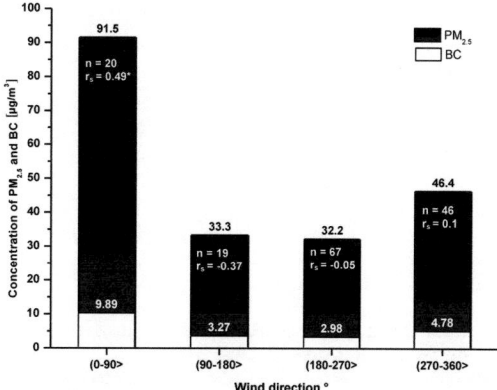

Figure 8. The dependence between the BC concentration and wind direction.

relationship was found for any other wind direction. The highest frequency had SW wind direction (44.1%) and at the same time, the lowest BC concentrations 2.98 μg/m³ were measured for this direction. The analysis of variance was used for the verification of the wind direction influence. It was proved with statistical probability of 95% that the wind direction has a significant influence on the BC concentration in the atmosphere. *Post hoc* test proved that NE wind direction (0–90°) has the most important influence on the BC concentration.

4 CONCLUSIONS

The daily average BC concentrations in Ostrava-Zabreh in the monitored period reached the value of 3.98 μg/m³. The BC concentrations in December were 5.1 times higher in comparison with August. The contribution of vehicles to the BC concentration in the heating season can thus be estimated as 20% of total concentration. During the monitored period, the morning BC concentrations were always lower (the average value for the entire monitored period 3.69 μg/m³) in comparison with evening values (4.48 μg/m³). The percentage of BC in the class PM$_{2.5}$ determined in Europe (8%) was comparable (6–8%) in Ostrava-Zabreh in the summer season, while in the winter season it increased to 9–10.5%. The highest contribution of BC in the PM$_{2.5}$ was determined during NE wind direction.

ACKNOWLEDGEMENTS

This paper was supported by research projects of the Ministry of Education, Youth and Sport of the Czech Republic: The National Programme for Sustainability LO1404—TUCENET and SP2015/64—The interdisciplinary study of fuel behaviour.

REFERENCES

Babu, S.S. & Moorthy, K.K. 2002. Aerosol black carbon over a tropical coastal station in India. *Geophysical Research Letters* 29(23): 13-1–13-4.

Begum, B.A., Hossain, A., Nahar, N., Markwitz, A. & Hopke, P.K. 2012. Organic and black carbon in $PM_{2.5}$ at an urban site at Dhaka, Bangladesh. *Aerosol and Air Quality Research* 12: 1062–1072.

Brasseur, O., Declerck, P., Heene, B. & Vander-Straeten, P. 2015. Modelling Black Carbon concentrations in two busy street canyons in Brussels using CANSBC. *Atmospheric Environment* 101: 72–81.

Cao, J.-J., Zhu, CH-S., Chow, J.C., Watson, J.G., Han, Y.-M., Wang, G. Shen, Z. & An, Z-S. 2009. Black carbon relationships with emissions and meteorology in Xi'an, China. *Atmospheric Research* 94(2): 194–202.

Cheng, Y-H., Liao, CH-W., Liu, Z-S., Tsai, CH-J. & Hsi, H-CH. 2014. A size-segregation method for monitoring the diurnal characteristics of atmospheric black carbon size distribution at urban traffic sites. *Atmospheric Environment* 90: 78–86.

CHMI (Czech Hydrometeorological Institute) 2015. *Preliminary Assessment of Air Quality and Dispersion Conditions in 2014, Annual report.* Prague: CHMI (In Czech).

Costabile, F., Angelini, F., Barnaba, F. & Gobbi, G.P. 2015. Partitioning of Black Carbon between ultrafine and fine particle modes in an urban airport vs. urban background environment. *Atmospheric Environment* 102: 136–144.

Feng, J., Zhong, M., Xu, B., Du, Y., Wu, M., Wang, H. & Changhong, Ch. 2014. Concentrations, seasonal and diurnal variations of black carbon in $PM_{2.5}$ in Shanghai, China. *Atmospheric Research* 147–148: 1–9.

Guo, Q., Hu, M., Guo, S., Wu, Z., Hu, W., Peng, J., Hu, W., Wu, Y., Yuan, B., Zhang, Q. & Song, Y. 2015. The identification of source regions of black carbon at a receptor site off the eastern coast of China. *Atmospheric Environment* 100: 78–84.

Kumar, K.R. 2011. Characterization of aerosol black carbon over a tropical semi-arid region of Anantapur, India. *Atmospheric Research* 100: 12–27.

Monks, P. 2012. *Air Quality Expert Group—Fine Particulate Matter ($PM_{2.5}$) in the United Kingdom.*

Pereira, S.N., Wagner, F. & Silva, A.M. 2012. Long term black carbon measurements in the southwestern Iberia Peninsula. *Atmospheric Environment* 57: 63–71.

Ramachandran, S. & Rajesh, T.A. 2007. Black carbon aerosol mass concentrations over Ahmedabad, an urban location in western India: comparison with urban sites in Asia, Europe, Canada, and the United States. *Journal of Geophysical Research* 112: D06211.

Saha, A. & Despiau, S. 2009. Seasonal and diurnal variations of black carbon aerosols over a Mediterranean coastal zone. *Atmospheric Research* 92(1): 27–41.

Sahu, L.K., Kondo, Y., Miyazaki, Y., Pongkiatkul, P. & Kim Oanh, N. T. 2011. Seasonal and diurnal variations of black carbon and organic carbon aerosols in Bangkok. *Journal of Geophysical Research* 116: D15302.

Sillanpää, M., Frey, A., Hillamo, R., Pennanen, A. & Salonen, R. O. 2005. Organic, elemental and inorganic carbon in particulate matter of six urban environments in Europe. *Atmospheric Chemistry and Physics* 5: 2869–2879.

Srivastava, A.K., Singh, S., Pant, P. & Dumka, U.C. 2012. Characteristics of black carbon over Delhi and Manora Peak a comparative study. *Atmospheric Science Letters* 13: 223–230.

Stull, R.B. 1989. *An Introduction to Boundary Layer Meteorology.* Dordrecht: Kluwer Academic Publishers.

Tiwari, S., Srivastava, A.K., Bisht, D.S., Parmit A, P., Srivastava, M.K. & Attri, S.D. 2013. Diurnal and seasonal variations of black carbon and $PM_{2.5}$ over New Delhi, India: Influence of meteorology. *Atmospheric Research* 125–126: 50–62.

Wang, Y., Hopke, P.K., Rattigan, O.V. & Zhu, Y. 2011. Characteristic of ambient black carbon and wood burning particles in two urban areas. *Journal of Environmental Monitoring* 13: 1919–1926.

Zhang, X., Rao, R., Huang, Y., Mao, M., Berg, M.J. & Sun, W. 2015. Black carbon aerosols in urban central China. *Journal of Quantitative Spectroscopy and Radiative Transfer* 150: 3–11.

Production Management and Engineering Sciences – Majerník, Daneshjo & Bosák (Eds)
© 2016 Taylor & Francis Group, London, ISBN: 978-1-138-02856-2

Environmental analysis of hospital waste and carcasses treatment and disposal

J. Ladomerský

Faculty of Natural Sciences, University of Matej Bel in Banská Bystrica, Banská Bystrica, Slovak Republic

E. Hroncová

Faculty of Ecology and Environmental Sciences, Technical University in Zvolen, Zvolen, Slovak Republic

ABSTRACT: The submitted paper analyses, on an actual case, the possibility of building a facility for the disposal of biological waste by incineration. The strategic objective was evaluated by the Environmental Impact Assessment (EIA) methodology. Based on the preliminary research, three solution variants have been chosen, besides the zero variant. The first variant was a combination of a small incineration facility and a line for sterilization of potentially infectious hospital waste. The second variant was a construction of a complex small incineration plant of veterinary, hospital and municipal waste with an effective flue gas cleaning system. An in the third variant, it was proposed to use the existing capacity of municipal waste incineration plant and post-adjust the technology for the disposal of this type of waste. The results of the analyses have shown, that the variant promoted by the potential investor is unsuitable. A different solution procedure has been recommended.

1 INTRODUCTION

Health care is very important, even if it produces different waste, e.g. sharp objects, infectious material that pose a risk to human health and the environment. The waste that is produced in healthcare facilities can be divided into two categories, namely hazardous and other waste. Most waste, i.e. 75–90%, is of the other waste category, character of the municipal waste, which presents no risk (Karamouz et al. 2007; Chaerul et al. 2008). The remaining 10–15% is hazardous waste or special waste, as defined by the World Health Organization (WHO) (Pruss et al. 1999) and the US Environmental Protection Agency (US EPA 1986). Therefore, the issue should be paid more attention to and dealt with comprehensively, from the collection, sorting, storage, transport to the disposal of waste.

In view of the potential risk of infection, similar to the medical waste is veterinary waste. Similarly, every farm produces various type of waste. This waste must be disposed of in an appropriate way, so as to minimalize the probability of soil and water contamination.

The disposal of infectious medical waste and carcasses can be carried out in large continuous incineration plants. For smaller units, with less capacity, discontinuous incineration plants with single dosing batch system are chosen. In case of biological waste, a very serious issue is that of infectivity of these wastes and the possibility of the spread of infection in case of a transport of these wastes at greater distances (e.g. to rendering plants). From this perspective, it is desirable and beneficial to build small incineration units allowing waste disposal at the place of their origin. The pre-requisite is a suitable and verified incineration facility and compliance especially with the prescribed incineration temperatures, as well as an appropriate way of collection and handling of the wastes, including urgent incineration without extended periods of storage.

For the disposal of infectious medical waste, non-combustion technologies can also be used. In terms of the process of decontamination of infectious waste, the use low-thermal and chemical processes against radiological and biological processes, is preferred in practice. Before using any of the processes other than incineration technologies, waste is mechanically treated, while a reduction in its volume occurs. Before the introduction of non-incineration technologies, it is necessary to perform the so called waste audit, to identify accurately the nature of the wastes that are generated, as well as their quantity and location of establishment. When choosing a suitable non-incineration facility, it is necessary to know the technology description, given by the manufacturer. Most manufacturers indicate, what kind of waste can be disposed of at their facility.

The aim of the presented paper is to assess the plan for the disposal of biological waste in 3 solution variants, besides the zero variant.

2 WASTE DISPOSAL TECHNOLOGIES OF CARCASSES AND HOSPITAL WASTE

2.1 *Combustion technologies*

There are several combustion technologies on the market, which are appropriate for the given project type. We have selected four types of combustion installations of the required capacity and complying with the requirements of air protection:

- From the point of view of the construction, the incineration plant represents a batch-type facility. It is discontinuous and is intended for a long term day time operation in two or three cycles but also for occasional use.
- A mobile device for burning carcasses and selected hospital waste.
- An incineration device is mobile, serving for the incineration of biological waste of animal provenience.
- An incinerator of animal waste and hazardous waste generated at livestock production. A modern, proven and safe facility, the operation of which is strictly controlled and subject to strict regulations and controls. Emissions from the waste incineration plants are generally very low and are lower than those of other energy sources burning coal or other fuels. They burn in particular organic waste, waste from food production and industry. To a lesser extent, the incinerators are used for burning also hazardous waste. An incineration plants of animal waste and hazardous waste are designed primarily for meat production, small and medium-sized slaughterhouses, veterinary clinics (cremation of animals), for farms rearing poultry, pigs, cattle, sheep, etc. They are suitable wherever it is necessary to dispose of rendering materials, dangerous animal waste, or veterinary sanitation.

2.2 *Non-incineration technology*

Non-incineration technologies can be, in terms of the process of decontamination of infectious waste, divided into: low-thermal technologies, chemical processes, radiation processes and biological processes. In practice, especially the low-thermal technologies and chemical processes are used.

The low-thermal technologies of waste decontamination are based on the principle of hot steam or dry heat utilization. Chemical processes are based on the principle of the action of chemical agents.

For the given type of project, especially the low-thermal systems of waste disinfection based on the activity of micro-waves, were assessed:

- A continuous sterilization and waste treatment device with a patented microwave generator with the power of 12 kW and a patented four-shaft shredder with the power of 37 kW in four rotors with automatic dosing and monitoring the process of waste treatment. The capacity of the device is up to 175 kg.h^{-1}.
- Further equipment is based on the combination of waste shredding and microwave radiation. The capacity of waste treatment is 100–181 kg.h^{-1}. It is available in a stationary, as well as mobile, version. It is not necessary to connect the device to the sewage system, because no liquid waste is being generated. The process is monitored by sensors and micro-processors, which ensure the proper temperature during the whole disinfection procedure. Temperature— 95–100°C; disinfection time-min. 30 minutes.
- A microwave unit designed for a small amount of waste that can be installed in the vicinity of the waste source. It is based on microwave irradiation and saturated steam effect on the disinfection and sterilization of medical waste, depending on the selected program and temperature. The waste is placed into a vapour-permeable sack, so that it can penetrate through to the waste. The operator opens the lid and inserts the sack with the waste into the disinfectant chamber (1 sack per cycle). Waste surface is exposed to the steam, whilst microwave radiation heats the waste from inside, destroying the micro-organisms in such way. The temperature in the disinfection chamber is 121°C but it can reach up to 134°C, if necessary. It is possible to adjust the duration of the microwave treatment. Usually, the disinfection takes 10 to 30 minutes, depending on the temperature selected. Capacity: up to 35 kg.h^{-1}.

3 MATERIAL AND METHODOLOGY

3.1 *Disposable wastes*

Potential types of disposable waste in the proposed waste disposal facilities under the Ministry of Environment Decree no. 284/2001 Coll., and European Waste Catalogue are in Tab. 1.

When choosing the best treatment and disposal technology we chose, opposed to the usual procedure of Multi Criteria Decision Analysis (Dursun et al, 2011), the elimination method by pair comparison of the technical solution of individual

Table 1. Potentially disposable wastes according to the European Waste Catalogue.

Type of waste*	Specification type of waste	H and O wastes**
18 01 03	Wastes whose collection and disposal is subject to special requirements in order to prevent infection (clinical waste—autoclaved, clinical waste, used stoma bags, bags—stoma (used), stoma bags (used), dressings—soiled, soiled dressings, soiled swabs, swabs soiled, hospital—clinical waste, clinical waste n/o/s, infectious materials—clinical)	H
18 01 04	Wastes whose collection and disposal is not subject to special requirements in order to prevent infection (for example dressings, plaster casts, linen, disposable clothing, diapers)	O
18 02 02	Wastes whose collection and disposal is subject to special requirements in order to prevent infection (clinical waste, infected animal parts, animal bedding—soiled, animal carcasses, animal faeces, animal tissue—infectious, carcasses, needles (clinical), excrement—animal, manure—animal, swabs—soiled, healthcare risk waste, paper towels (used)	H
18 02 03	Wastes whose collection and disposal is not subject to special requirements in order to prevent infection (clinical waste, animal bedding—soiled, animal carcasses, animal faeces, animal tissue—non-infectious, autoclaved clinical waste, carcasses, excrement—animal, manure—animal, swabs—soiled, paper towels (used), paper wipes—contaminated, hospital)	H

*six figure code of waste.
**Hazardous (H) and Other (O) wastes.

devices. In view of the difficult access to some sources of waste, mobile devices were excluded from the analysis. In addition to the zero variant, a simpler variant (1) with a combination of a discontinuous combustion device and a continuous sterilization device, a variant (2) with a complete small incinerator of veterinary, medical and municipal waste with efficient combustion gas cleaning and a variant (3) with modified existing waste incinerator, present is in the area, were chosen.

3.2 Zero variant

The zero variant represents the state, at which the activity would not be realized. In case of not realizing the incineration plant, or even a sterilizing device, the location would continue to be used as before.

Veterinary wastes and selected hospital waste would be disposed of scattered in different ways, in some cases with longer transit routes and longer times from the waste occurrence to the time of their disposal. The transport of these wastes in small quantities would not be economical, so their longer collection would be necessary. According to the preliminary investigation, up to 2,000 tons of such waste is generated at the given site per year, for which an interest for a different than the present way of disposal has been shown.

3.3 Variant no. 1

Includes the installation of the combustion device and the installation of the microwave line for the sterilization of potentially infectious hospital and eventually veterinary waste, on the existing premises of the selected area.

There is a large number of incineration device producers for the incineration of animal carcasses and crematoria facilities for animals of small capacities that suit most operators of agricultural farms and the like.

Following the comparison of the devices from different manufacturers, we recommended device, which has primary combustion chamber of $1 \ m^3$ volume and a single batch is up to 500 kg. The capacity of the device is up to 50 $kg.h^{-1}$. These incineration devices do not require a building permit, since their construction is not firmly fixed to the ground. This is a two-stage incineration device, which should comply with all the requirements of environmental protection. The device has the CE DEFRA Certificate, which confirms its compliance with the requirements of safety, construction and environment.

Besides this, the sterilization device, which would be operated simultaneously, in cooperation with the incineration device, has been recommended. This device serves for the sterilization of potentially infectious hospital wastes. It is regarded the best available technique for the disposal of hospital waste, as among other things: it does not negatively affect the environment, does not produce emissions, neither does it use water nor generate waste water. The capacity of the device is up to 175 $kg.h^{-1}$. The device is suitable for the following ways of waste treatment: treatment of medical waste, disposal of clinical waste and hospital wastes.

3.4 Variant no. 2

In this variant, we have been recommended one complex small incineration plant of veterinary, hospital and municipal waste, with an effective flue gas cleaning. The facility consists of two chambers, a pyrolysis chamber in the first stage. The capacity is up to 500 kg.h^{-1} for solid wastes, resp. up to 1,000 kg.h^{-1} for liquid wastes. The whole procedure is controlled automatically. The facility can dispose of: human wastes, microbiological and biotechnological wastes, discarded medicines, smears of plaster, cotton soaked in blood, bedding, waste from pathological laboratories, blood banks, animal wastes. The advantages of this type of incineration plants is: combustion efficiency of at least 99%, the flue gas residence time in the secondary combustion chamber of at least 1.5 seconds, the temperature in the primary combustion chamber in the range of 750 to 850°C. And in the secondary combustion chamber of 1,000 to 1,100°C.

Can be upgraded with absorbers (Venturi, shower or filler, the dosage of alkali), continuous emission monitoring system, automatic dosage, automatic ash removal, system using the heat.

3.5 Variant no. 3

In this we recommended to use the existing capacity of municipal waste incineration plant and post-adjust the technology for the disposal of this type of waste.

The strategic objective and the individual recommended variants were assessed in terms of their environmental impact. As part of the EIA process, the direct and indirect impacts of the proposal on public health, environment, natural resources, property and cultural monuments, have been assessed. Not only the operational side of the proposal has been assessed but also its implementation and disposal.

4 RESULTS AND DISCUSSION

Hospital waste management plays a very important role in the urban environment (Dohare et al. 2013; Shareefdeen 2012). The same significance is attached to the environmentally suitable disposal of carcasses (Morrow et al. 2000; Gwyther et al. 2011). Pursuant to the Act no. 24/2006 Coll., on Environmental Impact Assessment, Annex No. 8, the operation of the device for waste disposal by technical means, included in Chapter 11, Agriculture, Section 5, rendering plants and animal sanitation facilities with a capacity of less than 10 tons/day in Part B Inquiry Proceedings.

The proposed activities of medical waste treatment fall under the list of activities in Annex 8 into Chapter 9, "Infrastructure", Part 7—"Disposal Facilities for Incineration of Hazardous Waste and Facilities for the Treatment, Processing and Recovery of Hazardous Waste" without limit, are subject to obligatory assessment.

According to the applicable categorisation of the small and medium pollution sources listed in Annex 1 to the Decree of the Ministry of the Environment no. 410/2012 Coll., as amended by Decree no. 270/2014 Coll., these combustion facilities are categorized as follows:

5 WASTE MANAGEMENT

5.1 Veterinary sanitation (incineration) device with a projected aggregate processing capacity up to 10 t per day

5.1.1 Medium pollution source

In terms of Part E of Annex 7 of the Ministry of Environment Decree no. 410/2012 Coll., as amended and the issued "Information on the Requirements for the Air Protection at the Incineration of Carcasses in Combustion Facilities with a Capacity up to 50 kg/h with a Capacity of more than 50 kg/h, including those of 10 t/day" (published by the Department of Air Protection and Climate Changes of the Ministry of Environment of the SR on February 1, 2008) the veterinary sanitation facilities of this category are further divided into low capacity (capacity of burned animals up to 50 kg per hour) and high capacity (capacity of 50 kg per hour to 10 tons per day), while they burn exclusively these animal by-products:

• whole bodies of dead pet animals, laboratory animals and animals from holdings of poultry and lagomorphs.

In the operation of the incinerator would incur gaseous residues, the composition of which is generally dependent on the composition and type of waste input and the perfection of thermal decomposition in the combustion, but especially in the post-combustion part.

Processed animal tissues are mainly proteins, that is high-molecular substances composed of amino acids. In terms of the elemental composition, they are substances made up of carbon, oxygen, hydrogen and nitrogen. The content of the other elements—sulphur, phosphorus, halogen (chlorine, fluorine), and some metals is much smaller. From this composition, can be deduced the composition of the waste gases from the incineration of dead animals, since carbon, oxygen, hydrogen and nitrogen is released from the combustion chamber in the form of gases—in case of perfect combustion, and sufficient oxygen in the form of the combustion end products, such as carbon dioxide, water, nitrogen or nitrogen oxides—and also a small amount of chlorine, is converted to HCl (the presence of fluorine is not expected—the content in animal tissues, as in the human body, is insignificant). From metals, relevant is calcium, which remains in the ashes in the form of calcium oxide.

The composition of the waste gases would depend to a great extent from the incineration conditions. The suppliers of the device declare compliance with a temperature of min. 850°C in the post-combustion chamber, in which the organic matter is fully decomposed and there is a precondition of the conversion into the final products stated above. With respect to these conditions, polluting substances can be determined as follows: solid material, sulphur dioxide, nitrogen oxides, carbon monoxide, organic substances as total carbon and HCl. Other polluting substances shall be, based on a qualified assessment, unimportant.

The pre-condition of a satisfactory composition of emissions is the combustion of only the carcasses of dead animals and the absence of other materials such as plastics, textiles (cloth), contaminated wood and the like containing halogens (chlorine or fluoride).

Location of the facility according to paragraph 1.2.2 of Annex 7, point E of the Decree no. 410/2012 Coll., of the Ministry of Environment of the SR. With respect to local disposition conditions and the direction of prevailing winds, the combustion facility should, if possible, be placed in the greatest distance from other objects, especially administrative and residential, and from publicly available areas, such as public roads and the like.

The technical requirements and the operating conditions are listed in point 1.3 of Annex 7, section E of the Decree no. 410/2012 Coll., of the Ministry of Environment of the SR:

1.3.1 Facility with MTP <0.3 MW must be equipped with low-emission burners.

1.3.2 The incinerator of carcasses cannot burn covers, eventually packaging of dead animals containing chlorine, fluorine, metals or impregnating substances, such as tar and gum like asphalt or waste wood, rags and the like.

1.3.3 Fuels for burning carcasses of dead animals. In incineration plants for carcasses only natural gas, liquefied petroleum gas, biogas, heating gas oil, recovered fuel oil and diesel fuel, under a special regulation, can be burnt.

1.3.4 Incineration requirements. The temperature required for combustion and the residence time is determined by special regulation.

1.3.5 Reducing the production of odorous substances. To reduce the production of odorous substances, it is especially necessary.

a. equip and operate the facility with a secondary post-combustion chamber with a secondary burner or other restriction of odorous substances,
b. operate the facility so as to reach, as soon as possible, the operating temperature of incineration and a perfect combustion of organic material,
c. store the odorous materials in sealed containers and closed spaces.

The above mentioned Decree no. 410/2012 Coll., of the Ministry of Environment of the SR, assigns the incinerators of carcasses emission limits stated in Tab. 2.

Implementation of one of the variants 1, 2 or 3, in comparison with the zero option, would facilitate the organization of the management of hospital and veterinary waste at the site of their origin, accelerate and facilitate their transport to disposal site and shorten the time from waste generation to its disposal. This would reduce the cost of waste disposal for the producers of hospital and veterinary waste.

Based on the composition of the bodies of animals, farm and domestic animals, there are no concerns about the emergence of major air pollutants and substances that threaten human health. Therefore, on the following combustion plants, such strict requirements, as e.g. on the incinerators of municipal waste or hazardous waste, are not imposed.

Table 2. Emission limits—for new facilities.

The conditions of the validity of the EL	Standard State Conditions – TZL, SO_2, NO_x and CO: dry gas, O_{2ref}: 11 % of the volume – TOC: humid gas, O_{2ref}: 11 % of the volume
MTP [MW]	Conversion to O_{2ref} shall be performed only if the actual content of O_2 is >11 % of the volume.

	Emission limit [mg/m³]				
	TZL	SO_2	NO_x	CO	TOC
<0.3 MW	100[1]	500[2]	[3]	[3]	10
≥0.3 MW	100[1]	500[2]	850	250	10

[1] For facilities with capacity <50 kg/h, the TZL emission limit is not applicable.
[2] Applicable for low calorific gases, such as biogas, and others. For the rest of fuels, the emission limit for SO_2 is not applied.
[3] Emission limits for NO_x and CO shall not apply in case of combustion exclusively in the area of the concerned holding, slaughterhouse and poultry house, where the death or killing and processing of the animals occurs; emission requirements apply according to current engineering standards for the burner or combustion device for the respective fuel.

The most important factor is the possible odour, nitrogen oxides (from the oxidation of proteins), or solid polluting substances, in case of improper construction of the incinerator or mismanagement of the combustion process. Incineration facility of a smaller capacity, up to 50 kg · h^{-1} is generally a batch-type and two-stage one. Such a device is not firmly fixed to the ground and therefore its installation does not require a building permit, but of course, the device must obtain the consent of the respective local environment office.

The basic condition for minimizing emissions at an appropriate incineration device is do not overfill the combustion furnace and automatically control the temperature in the combustion furnace—the primary and secondary combustion chamber. Ovens are equipped with a start-fire burner and a finishing burner using either natural gas, propane, biogas or fuel oil or diesel. A natural requirement is the accurateness of the hygienisation and the complete burning of waste, sufficient incineration time and minimizing the mass of the resulting ash.

Decision, which variant to choose, was based on an assessment of the broader context of waste incineration on the site and on the basis of the variability of potential waste disposal and economic costs. Comparison of the alternatives for the construction of the facilities according to variants 1–3 and zero variant can be done verbally, comparing only the relevant impacts.

Impact on the air: The analysed state indicates that requirements to minimize emissions are attainable by all three variants, as for a thorough combustion of the waste, it is sufficient to comply with the basic conditions for complete combustion and no exhaust gas cleaning system is necessary. The impact of the incineration facility on the pollution in the given area will be minimal.

Impact on the health of the population: The impact of the incineration plant on the state of health of the population is related to its impact on the air and will be practically zero. The variants 1 to 3 will, in comparison with the zero variant, represent a reduction of the risk resulting from infectious waste. However, a fear of some residents that a further waste incineration plant will be built in the respective area and the same weight will be attributed to its effects as to those of the existing municipal waste incineration plant in the region, may arise.

Impact on the management of waste: Variants 1 to 3 represent, compared with the zero option, improved conditions and increased variability of the disposal of hospital waste and veterinary wastes, as well as increased operability and cost reduction.

Variant 2 in its complex form, however, presents a higher level of incinerator, since the combustion plant is equipped with a wet flue gas cleaning system and other systems that are necessary for the waste incineration plant. The choice of variant 2 would enable the incineration of a far wider range of waste as it is possible according to variant 1. At the same time, such an incinerator is much more costly—in terms of investments and also operationally.

6 CONCLUSIONS

Currently, biological hospital waste and veterinary biological waste is collected in the appropriate boxes and until its transport for disposal, it is stored on the sites of its origin. Subsequently, the waste is transported to several places of disposal by an authorized organization.

The above mentioned waste cannot be prevented and due to its potential infectivity, nor recovered. The assessed activity is proposed in the given area because it forms a natural center of several sources of hospital and veterinary waste.

An incineration facility of biological hospital waste and veterinary waste is an indispensable technique for the disposal of these wastes, as well as some other less hazardous wastes.

There are many manufacturers and suppliers of combustion facilities for the designated waste types. Important are their operating parameters, approval certificates, proofs of their reliability and compliance with the emission limits. Several manufacturers of these facilities fulfil these parameters. The devices are characterized by economic, semi-automatic or full automatic operation, easy maintenance and easy manipulation.

Besides the incineration device, an automatic sterilisation device is proposed in variant 1. It is a patented continuous sterilization and disposal device, which is regarded as the best accessible technology. This device is considered a BAT especially because no negative impact on the environment has been detected at operating it. There is a list of wastes, which can be safely treated in this device.

In variant 2, practically, a small waste incinerator with intensive cleaning of flue gases that would allow a much wider range of incineration of waste as it is possible according to variant 1, is being proposed.

Variant 1 allows for greater variability in the methods of waste disposal, in terms of costs and time—there is a chance of a simultaneous operation of the combustion equipment and the sterilization equipment.

Variant 2 represents the most costly means of operation for the disposal of a wider range of different types of waste, which is not necessary, since in the area, there is a large municipal waste incinerator.

Variant 3 proposes to use the existing capacity of a waste incineration plant and post-adjust the technology for the disposal of this type of waste.

For the above stated reasons, variant 3 has been proposed for implementation, which is economically more feasible than variants 1 and 2. Consistency of style is very important.

ACKNOWLEDGEMENTS

This work was supported by the Slovak Research and Development Agency under contract No. APVV-0353-11. This research was also supported by the Cultural and Educational Grant Agency of the Ministry of Education, Science, Research and Sport of the Slovak Republic project no. KEGA 035UMB-4/2015.

REFERENCES

Act from 14th December, 2005 no. 24/2006 Coll. on the assessment of environmental impacts as amended. *The Ministry of Environment of the SR.*

Chaerul, M., Tanaka, M. & Shekdar, A.V. 2008. A system dynamics approach for hospital waste management. *Waste Management* 28(2): 442–449.

Decree from 11th June, 2001 no. 284/2001 Coll. waste catalogue as amended. The Ministry of Environment of the SR.

Decree from 30th November, 2012 no. 410/2012 Coll. implementing the clean air act as amended. *The Ministry of Environment of the SR.*

Dohare, S., Garg, V. & Sarkar, B.K. 2013. A study of hospital waste management status in health facilities of an urban area. *International Journal of Pharma and Bio Sciences* 4(1): (B) 1107–1112.

European Waste Catalogue and Hazardous Waste List. Valid from 1 January 2002. *Environmental Protection Agency.*

Gwyther, L., Williams A.P., Golyshin, P.N., Edwards-Jones, G. & Jones, D.L. 2011. The environmental and biosecurity characteristics of livestock carcass disposal methods: A review. *Waste Management* 31(4): 767–778.

Karamouz, M., Zahraie, B., Kerachian, R., Jaafarzadeh, N. & Mahjouri, N. 2007. Developing a master plan for hospital solid waste management: A case study. *Waste Management* 27(5): 626–638.

Morrow, W.E.M., Ferket, P.R. & Middleton, T. 2000. Alternative methods of carcass disposal. *The Pig Journal* 46: 1–8.

Pruss, A., Giroult, G. & Rushbrook, P. 1999. *Safe management of waste from health-care activities.* Geneva: WHO.

Shareefdeen, Z.M. 2012. Medical Waste Management and Control. *Journal of Environmental Protection* 3: 1652–1628.

US EPA 1986. *Guide for Infectious Waste Management,* 1986. EPA/530-SW-86-014, Washington, DC, May 1986.

Production Management and Engineering Sciences – Majerník, Daneshjo & Bosák (Eds)
© 2016 Taylor & Francis Group, London, ISBN: 978-1-138-02856-2

Influence of selected digitization methods on final accuracy of 3D model

M. Mantič, J. Kuľka, J. Krajňák & M. Kopas
Faculty of Mechanical Engineering, Technical University of Košice, Košice, Slovak Republic

M. Schneider
Technogym E.E. Ltd., Malý Krtíš, Slovak Republic

ABSTRACT: This paper deals with research of chosen methods for digitization of machine parts in order to compare the creation possibilities of a parametric model with regard to accuracy and promptness of digitization processes. There are compared three various contact methods developed for acquirement of the measured data and one contact-less scanning method. The scanning of dimensional data using the 3D scanner, together with the scanning by means of the coordinate measuring machine and the classic measuring with the caliper are three compared contact methods. The following operation was modeling of the machine part in a parametric CAD system. The x-ray scanner demonstrated application possibility of the contact-less digitization method. This paper also presents other methods applied for data contact scanning using a process for scanning of the cross-sections, longitudinal-sections and point clouds. All the described procedures are compared with a reference CAD-model created by means of the coordinate measuring equipment.

1 INTRODUCTION

At the present time technologies of reverse engineering are becoming a reality. This trend leads to the search methods for data obtaining and processing as quickly and simply as possible to create a parametric model with the required accuracy. Before the digitization process is necessary to choose an appropriate digitalization device but also a method for digitizing process with respect to the possibility of following processing of scanned data. The ideal solution is to integrate the quick scanning and data processing with sufficient with costs as low as possible.

2 DIGITALIZATION OF A COMPONENT

For experimental purposes there was chosen a simple cylindrical component (Figure 1). Digitization of component was performed with four different methods. The first one was the classical method, where using a caliper there were measured the required dimensions and then in consequence there was 3D model created. In the second case there was used the touch scanner to create a 3D model. By means of the scanner and various methods there were created 3D model of component. In the third one there was applied a contactless

scanning by means of the x-ray scanner, and in the fourth case there was created a 3D model by means of data from the coordinate measuring machine.

2.1 The creation of a model using a hand-held measuring instrument

The 3D model of component was created on the base of measured dimensions by means of a hand-held caliper Mitutoyo with measurement accuracy to a hundredth of a millimeter. The measurements of individual dimensions were repeated several times and the resulting values were determined on the base of statistic processing of measured data. Based on

Figure 1. Digitalized component.

these data there was created 3D model of component in the CAD system. The process of creating for 3D models is possible only for simple shape components with an exact defined geometry. The output of this measurement is not positions of points in the space but only length dimensions. During the creating of model there has been observed both time of measurement itself, time of measured data processing as well as the time of creating of the model.

2.2 A component digitalized with a touch scanner 3D Creator

A touch scanner 3D Creator from the company Innovation Group, Inc. enables to directly scan and so to obtain profile curves as well as coordinates of individual points in space as a point cloud. A workspace of 3D Creator varies in dependence on a used sensor from 300 mm³ to 1.5 m³. The accuracy is from 0.01 mm to 0.2 mm (3D Creator Innovative system for digitalization and measurement 2010, 3D Creator Industrial system 2012).

By means of the scanner 3D Creator there can be the component digitized with three different methods. The first one was a scanning of transversal cross-sections of component (Figure 2), the second one was a scanning of longitudinal cross-sections (Figure 3) and the third one was a scanning of point cloud (Figure 4).

A used 3D touch scanner cooperates directly with CAD system and therefore it can be applied as a supply tool for modeling of component. By this means it is possible to immediately create a parametric model without an additional conditioning of scanned data. This fact considerably

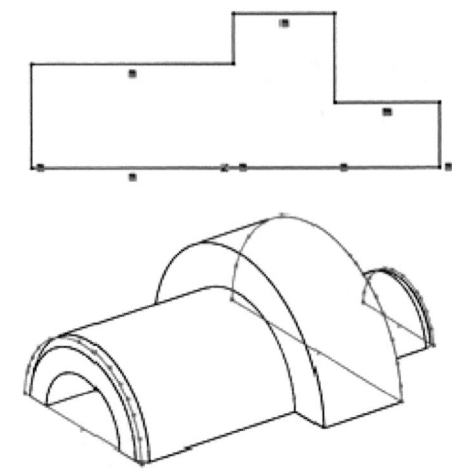

Figure 3. Model of component created from transversal cross-sections.

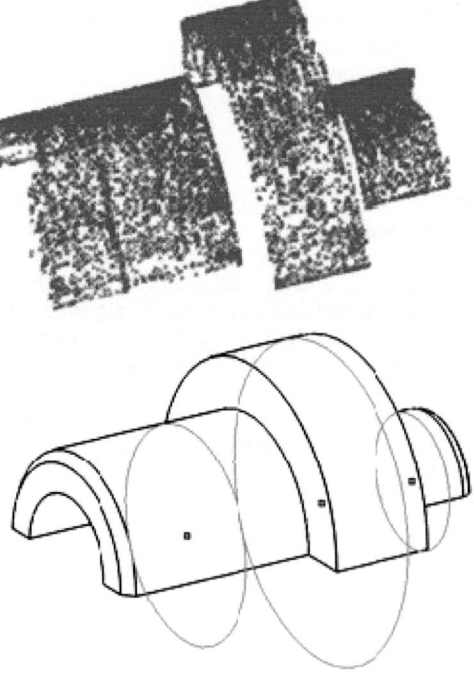

Figure 4. Model of component created from a scanned point cloud using the device 3D Creator.

affects the time for an obtaining of parametric model predominately in the application of profile curves (Figure 2 and Figure 3).

The third method to create a parametric model of the shaft from the cloud of scanned points

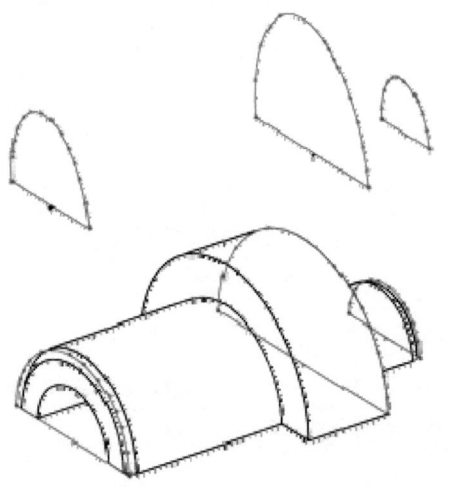

Figure 2. Model of component created from transversal cross-sections.

(Figure 4) is more time-consuming than the previous two methods but is much more effective at creating a component with demanding shapes. After scanning the point cloud was corrected and redundant interference data were removed. It was subsequently developed polygonal network with minimal modifications to keep maximum accuracy. After creating the polygonal network there was an extraction of surfaces, in our case it was the cylindrical surface. It is possible to build a parametric model of component from originated cross-sections (Figure 4).

2.3 The creation of a model using a contactless scanning technology

In this case, there was used of x-ray scanning device Metrotom 1500 for digitalization. The output of this device is a component scanned as a cloud of points. The scanning inaccuracy of the machine MPE_E is indicated in µm and is given by the equation (1), (Métrologie Industrielle 2012).

$$MPE_E = 9 + \frac{L}{50} \tag{1}$$

where L is the largest parameter of a scanned component in v mm. The value of MPEE is 9 µm, so 0.009 mm for the component, which we scanned.

The procedure for processing of point cloud and creating of the parametric model is analogous to the procedure of processing a point cloud from 3D Creator device. In this case, however, we have a more complex and denser point cloud (Figure 5). It is necessary this point cloud to adjust, to remove interference data off and to create a polygonal model. Then, if it is needed, this polygonal model can be modified and smoothed. Profile curves of the cylindrical component can be extracted from thus obtained polygonal model. On the base of these curves there is built a resulting parametric model of component (Figure 5).

2.4 A component digitalized by means of the coordinate measuring machine Contura G2

The Coordinate Measuring Machine (CMM) Contura G2 is a portal type for high-speed scanning. CMM are mostly used with contact sensors. High accuracy and very good results in repeatability are advantages of contact systems (Katuch et al. 2012). The output of this device is a data set with the coordinates of points on which the measurement (scanning) was taken place. This text file is then necessary to convert and create cloud of points from it. The equation (2) indicates the maximum inaccuracy MPE_E of this device in µm (Zeiss

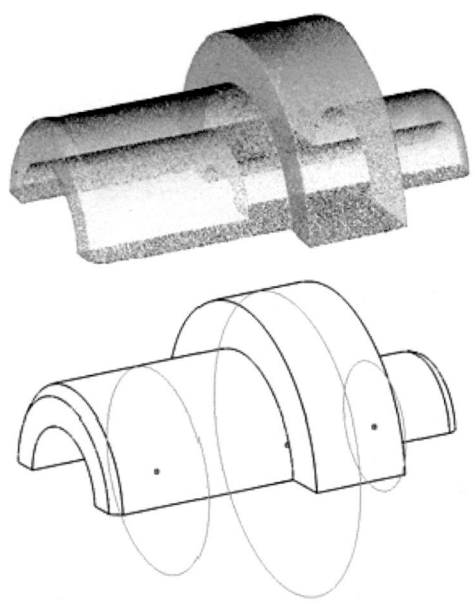

Figure 5. Model of component created from a scanned point cloud using the x-ray scanner Metrotom 1500.

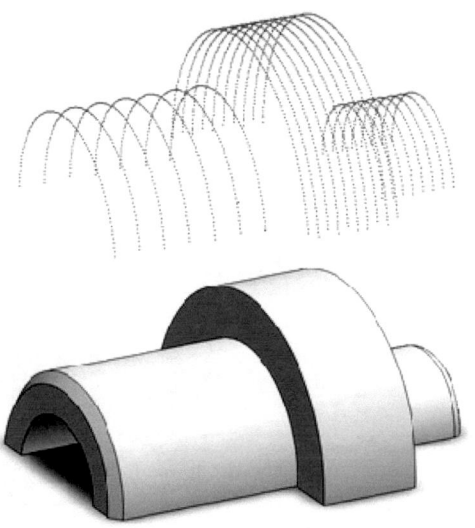

Figure 6. Model of component created from a scanned point cloud using the device Contura G2.

Contura G2 Coordinate Measuring Machine 2012) according to ISO 10360–2.

$$MPE_E = 1,8 + \frac{L}{300} \tag{2}$$

where L is a measured length in mm. An inaccuracy of MPE_E is 0.0018 mm in this case.

The converted point cloud (Figure 6) was used for the creation of polygonal model while the technique of its creation was the same as in previous cases. The result is a developed three-dimensional model of component.

3 CONSIDERATION OF DIGITALIZATION TECHNIQUES, TIME OF 3D MODEL CREATION AND PRICES OF INSTRUMENTS

The creation of the final CAD model is conditioned by certain techniques, which are different at individual digitization methods. In Table 1 there are the individual operations, required to create the model.

Based on the operations which are necessary to create the resulting CAD model (according to Table 1) there was developed a time-consuming for the creation of CAD parametric model of the components according to different digitization methodologies (Figure 7). Because each digitization device needs some time for preparation and calibration, this time was excluded from the analysis. Only the effective time for a scanning and for various operations, which are necessary to create the final 3D parametric model of components, was taken into account.

In Figure 7 there is showed that the longest digitization time has got, x-ray device. However, it is necessary to specify that scan time of the device Metrotom results from the operational principle. That is why the time will be the same for any complex component. It does not apply to other methods. To apply

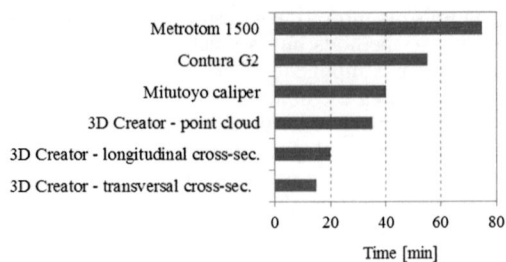

Figure 7. Time dependence of component digitization.

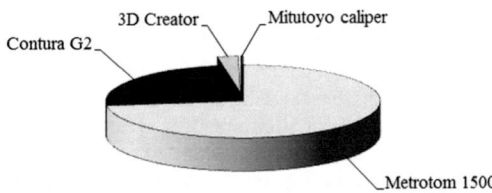

Figure 8. Comparison of used instrument prices.

these methods, the more complex the surface, it will scan time longer and the limits will be significantly reflected in the shape of the scanned surface.

A price of suitable digitalization device is another equally significant factor in its choosing. The devices, which we used, are considerably different prices, (Figure 8).

The price comparison shows that it is very important to consider a digitization device investment. It is necessary to accentuate on the shape complexity and sophistry of scanned components, required accuracy and workspace dimensions.

4 OPTION OF A REFERENCE MODEL

For the comparison of individual digitization methods and creation of 3D parametric model there was necessary to determine the reference model, which will have related, identified dimensional tolerances. On the base of accuracy for the individual scanning devices, there was chosen the reference model generated from data of Contura G2 instrument by Carl Zeiss company because this instrument reaches the smallest inaccuracy, $MPE_E = 0.0018$ mm.

5 COMPARISON OF THE DIGITALIZED MODELS

In Figure 9 there is the performed comparison analysis of shape accuracy for individual 3D parametric models with respect to the reference model.

Table 1 The techniques of component digitalization

Devices and digitalization method	Scanning or measurement	Data processing	Data transfer	Polygonal model	Creation of CAD model
Mitutoyo caliper hand-held measuring instrument	•	•			•
3D Creator transversal cross-sections	•				•
3D Creator longitudinal cross-sections	•				•
3D Creator point cloud	•			•	•
Contura G2 CMM	•		•	•	•
Metrotom 1500 contactless scanning	•		•	•	•

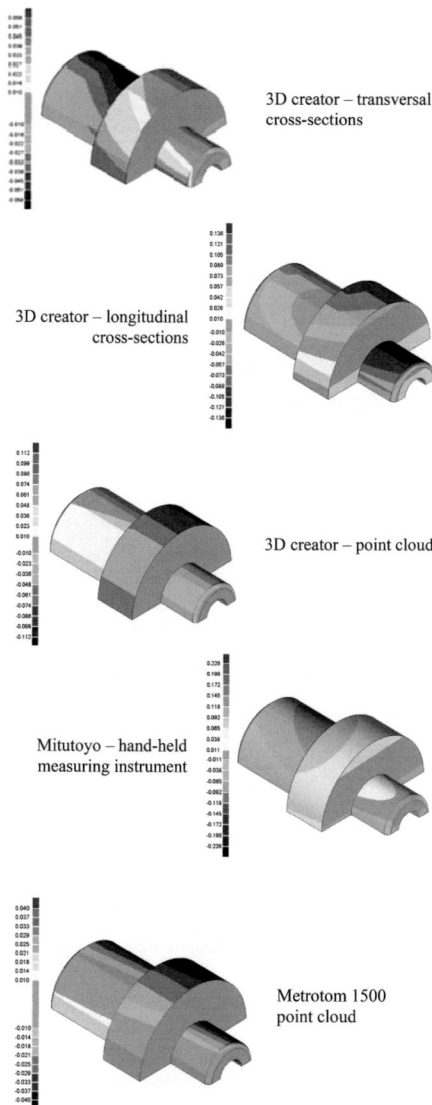

3D creator – transversal cross-sections

3D creator – longitudinal cross-sections

3D creator – point cloud

Mitutoyo – hand-held measuring instrument

Metrotom 1500 point cloud

Figure 9. Comparison of dimensional divergences from the reference model.

The digitized models were compared on the base of the value of standard deviation (3) for a created model with respect to the reference model.

$$\sigma = \sqrt{\frac{1}{n} \sum_{i=1}^{n} \left(x_i - \bar{x} \right)^2} \qquad (3)$$

where n-is a number of scanned points.

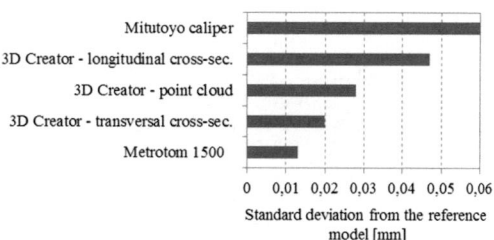

Figure 10. Comparison of accuracy for the digitalized models with the reference model.

x_i-is a deviation of a scanned point from the reference model.

$\bar{x}$ -is a mean arithmetic value of deviations for scanned points from the reference model.

The data of Figure 9 were processed and compiled in the graph of Figure 10. The comparison shows that the lowest value of a deviation from the reference model has got a 3D model generated from the data of Metrotom 1500 instrument. At using the 3D Creator the applied scanning methodology has also got a significant influence on the accuracy of the resulting CAD model what is demonstrated by the results of three different digitization methods.

6 CONCLUSION

In conclusion, except a suitable choice of digitalization device as well as a choice of appropriate digitization methodology is the considerable influence of the final time and the accuracy of the resulting CAD model. Right choice of scanning device is conditioned predominately by a shape, by a size and by a complexity of components. An important factor is the price of scanning device, which is given by its precision, its scanning technology and workspace dimensions. Generally, the simpler a shape of component, they are also less demands for a digitization device. It is necessary to take into account all the above aspects and so to obtain the resulting CAD parametric model with sufficient accuracy and following to obtain an optimal scanning efficiency ratio for the required accuracy of the model taking into account the reasonable cost of device.

ACKNOWLEDGEMENT

This paper is the result of the project implementation: VEGA 1/0198/15 and KEGA 021TUKE-4/2015

479

REFERENCES

3D Creator Industrial system [online] 2012, available on web pages: *<http://www.boulderinnovators.com>*.

3D Creator Innovative system for digitalization and measurement, [online] 2010, available on web pages: *<http://www.solidvision.cz>*.

Katuch, P., Dovica, M., Slosarčík, S. & Kováč, J.:. 2012. Comparision of contact and contactless measuring methods for form evaluation. *Procedia Engineering* vol. 48: 273–279.

Métrologie Industrielle—Metrotom 1500 [online] 2012, available on web pages: *<http://www.zeiss.fr>*.

Zeiss Contura G2 Coordinate Measuring Machine [online] 2012, available on web pages: *<http://www.gcmfg.com>*.

Production Management and Engineering Sciences – Majerník, Daneshjo & Bosák (Eds)
© 2016 Taylor & Francis Group, London, ISBN: 978-1-138-02856-2

Evaluation of the new fire retardants on wood by proposed testing method

P. Mitrenga & M. Vandlíčková
Faculty of Security Engineering, University of Zilina, Zilina, Slovak Repulic

M. Dušková
Institute of Macromolecular Chemistry of Academy of Sciences of the Czech Republic, Prague, Czech Republic

ABSTRACT: Currently, wood is often used as a building material. The wood is flammable and in case of the fire can cause extensive damages. You can be protected by fire retardants, which to certain extent limit the emergence and spread of fire. The Czech Academy of Sciences is also dealing with development of fire retardants for wood. Such retardants are being tested. Study deals with testing and evaluating of some new retardants developed by above herein scientific institution. Testing was carried out according to methods proposed by us. It is being briefly described in the paper.

1   INTRODUCTION

According to the EU testing standards, all materials may be classified into classes based on their reaction to fire. However, if we want to evaluate some flammable materials and compare them with each other, it is necessary to test with different testing methods. Such methods should enable comparison of similar combustible materials and allow detection of even minor differences during burning. We have compiled a method for wood fire retardants testing. Tests are executed on wood, in order to maintain the conditions of use. The method is based on the knowledge of the methods used so far for testing a variety of flammable materials. The testing will keep track of weight loss out of which relative speed of burning will be calculated. Based on herein speed, fire retardants for wood will be compared. Retardants being compared and evaluated are currently developed by The Czech Academy of Sciences. Herein retardants will be compared with the samples of non-retardants and with samples of commercial fire retardants. Considering "know how" of Academy of Sciences, provisional designations with no details on composition will be listed.

2   FIRE RETARDANTS FOR WOOD AND TESTING

Fire retardants present an effective way to protect wood against fire. These are substances preventing easy ignition and combustion because of their physical, chemical or combined characteristics. Various transfer and reaction of materials induced by heat are suppressed. By means of different physical and chemical mechanisms the thermal decomposition of wood, ignition and burning are affected. The principle is based on influence of heat generation and heat transfer from the reaction zone of combustion process, which ultimately will result in the cessation of the process of combustion (Osvaldová 2005).

In the retardation process, it is necessary to influence activities which are causing the cessation of burning. It is possible to influence the rate of heat generation and heat drainage from the reaction zone, creation of flammable gases, etc. Based on accompanying activities, we can divide fire retardants into several groups. The first group consists of retardants releasing non-combustible gases within flame heat range, when flammable gases are released as a result of wood decomposition. The second group consists of retardants, which cool down the heat source by heat accumulation from such a source. The third group consists of so-called intumescing fire retardants. These are substances that swell during burning and create protective layer of foam on the wood's surface which prevents it from further burning (Osvald 1997).

Because of constant development of new fire retardants testing is necessary. If we want to know which retardant is really effective and optimal in comparison with others retardants, series of tests need to be performed. In the EU countries, there is unified classification system created in order to involve material into class based on its response to fire. The classification into the following classes shall be carried out on the basis of sophisticated standardized tests. In addition to the classification tests of products' reaction to fire, however, there are many other tests, which allow evaluating the materials from firefighting point, primarily for scientific purposes and the comparison of individual products. The wood is specific matter, it is not homogenous and test is more complicated than with other homogenous materials. It is therefore necessary to create an appropriate testing method for fire retardants applied on wood, which will evaluate, compare and determine appropriate parameters and testing conditions (EN 13501-1+A1, 2010).

3 TESTING METHOD OF FIRE RETARDANTS FOR WOOD

This method is not standardized; it is not aimed to classify materials into classes based on reaction to fire nor comparison of different materials. This method is exclusively suitable for comparison and evaluation of fire retardants, which are tested on the wood. It allows continuous measurement of weight loss and subsequent calculation of relative burning speed. So we can better compare and find out how retardants affect combustion of wood.

The testing device consists of a gas pump, gas flow regulator, burner, burner, and burner's holder of the sample, precision scale and computer. Equipment scheme—Figure 1.

Gas bomb is filled with propane-butane gas. The rack is resting on the scales. Burner is placed under the sample's holder by means of burner's holder the way not to fit on the scales, and thus not affecting ongoing test. Scale weights in grams with accuracy of two decimal places and being connected to the computer. By means of installed software, scale records weight in preset interval.

The test samples are of pine wood of size $50 \times 100 \times 10$ mm. Humidity of samples should be about 12%. Fire retardants shall be applied

to samples by coating of both sides and edges. After retardants are being applied, the samples shall be dried during sufficient time before experiment launch; unless fire retardant will dry up and acquire its functionality.

Principle of test lies in exposure of flame directly to the tested sample. The flame height is set at 5 cm. The distance of the port block from the center of the sample should be approximately 4 cm. Flame operates for 5 min. Weight is recorded every 10 seconds during the test. There are 5 samples tested out of each type of retardant. For comparison we have tested 5 samples of non-treated retardant as well.

We observe and record weight loss of the sample under the effect of flame. Relative weight loss is calculated according to the relation (1) (Osvald 2009).

$$\delta_m(\tau) = \frac{\Delta m}{m(\tau)} \cdot 100 = \frac{m(\tau) - m(\tau + \Delta \tau)}{m(\tau)} \cdot 100 \, [\%] \quad (1)$$

where $\delta_m(\tau)$ = the relative weight loss at a time (τ) [%]; $m(\tau)$ = the mass/weight of the sample at the time (τ) [g]; $m(\tau + \Delta \tau)$ = the mass weight of the sample at the time ($\tau + \Delta \tau$) [g]; Δm = difference in weight [g].

Relative speed of burning is calculated according to the formula (2) (Zachar 2008).

$$v_r = \frac{\delta_m}{\Delta \tau} [\% \cdot s^{-1}] \quad (2)$$

where $\delta_m(\tau)$ = the relative weight loss at a time (τ) [%]; $\Delta \tau$ = the time interval at which masses to be deducted.

We have elaborated charts based on the data calculated from the relative burning speed, in which we can observe changes of particular fire retardants on wood behavior.

4 EXPERIMENT-FIRE RETARDANTS TESTING

Testing will cover results being compared and evaluated due to newly developed fire retardants by The Czech Academy of Sciences. Testing will be held according to compiled method. Some of the testing parameters are different though. We have modified them in our method after the actual experiment. We have found some deficiencies being implemented into changes made later on. Original parameters of the experiment represent:

– Sample's sizes: $40 \times 100 \times 10$ mm,
– One-sided application of fire retardants only.

Other parameters remain preserved. The reason why we decided to change herein parameters after the experiment will be clarified by testing results.

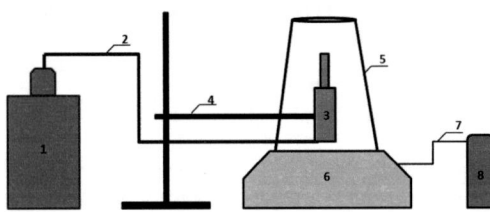

Figure 1. Testing equipment scheme. Description of the image: 1-gas bomb, 2-gas supply to the burner, 3-burner, 4-burner's holder, 5-sample's holder, 6-scale, 7-connection of scale with the computer, 8-computer.

4.1 Utilized fire retardants on wood

Newly developed fire retardants for wood by The Czech Academy of Sciences are tested. We have been supplied by Academy with samples and materials. Production and composition of the fire retardants belong to "know-how" of herein institution; therefore, confidential information will not be published per request. Only corporate labeling is used. Certain selected commercial fire retardants will be tested in order to compare them with the new ones.

Following fire retardants have been used for the experiment under labels:

– FR COM-Falun Red commercial,
– PS SDT-Plamostop standard,
– XP 6-original designation by The Czech Academy of Sciences,
– R 07-original designation from The Czech Academy of Sciences.

4.2 Results of the experiment

Results are processed via diagrams. Please note Figures 2–6 relative burning speed of the samples. One chart always shows burning speeds of five samples with the same retardants' cover.

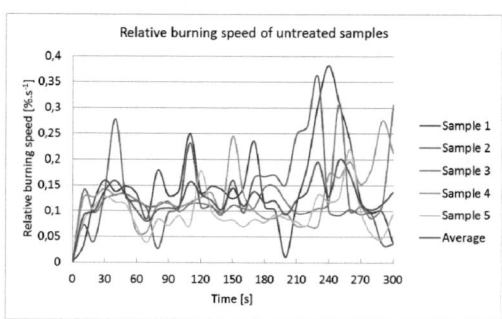

Figure 2. Chart presenting relative burning speed of non-retardant samples.

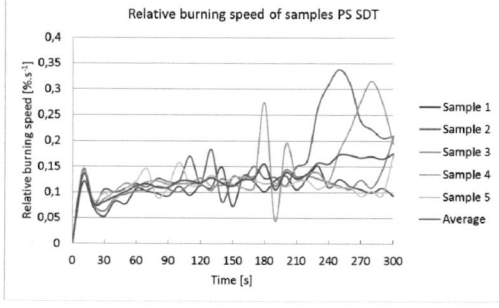

Figure 3. Chart presenting relative burning speed of samples with retardant PS SDT.

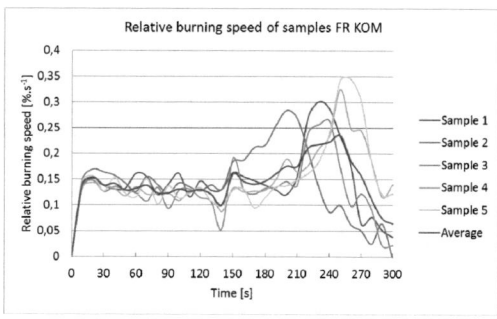

Figure 4. Chart presenting relative burning speed of the samples with retardant FR KOM.

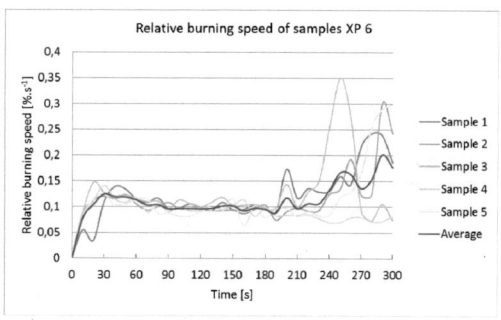

Figure 5. Chart representing relative burning speed of the samples with retardant XP 6.

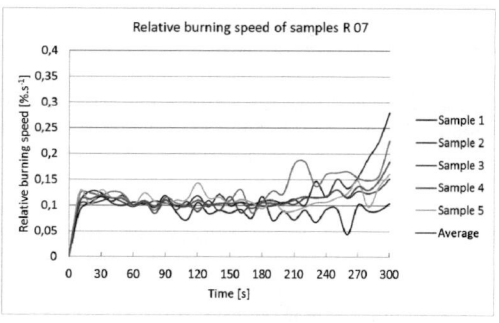

Figure 6. Chart representing relative burning speed of the samples with retardant R 07.

Please note that Figure 2 shows that non- retardant samples have higher dispersion rate of burning as retardant ones. It is partly the evidence of certain effect of retardant on wood. Figure shows that non- retardant samples have also greater burning speed since the very beginning. Wood without retardant covering is gradually burning, while different characteristics of wood, such as its uneven structure, different density, etc., cause uneven burning of the material. For retardant samples we found more uniform burning speed with lower refraction. Retardants partially reduce impact of these factors, for example by clogging vessels in the wood and preventing oxygen access, creating a protective layer on the wood that prevents rapid release of combustible gases. In addition non-combustible gases can be released, which may dilute flammable atmosphere upon the surface of samples and thus limit burning speed.

Please note that samples containing retardant application on Figures 3–6, approximately after the third minute of testing (180 seconds), have significant difference in burning speed. The scattering rate of burning after this time is quite significant, clearly visible on Figure 5 with samples covered by XP 6 retardant. This phenomenon is caused by the burning of side surfaces and also top area missing fire retardant. During the experiment, we have noticed that most of the samples began to burn from the top starting from about the third minute. It is the cause of the significant increase in the burning testing method. Adjustments are based on the dimensions of the samples being extended from the original width of 40 mm to 50 mm in the testing method. Increase of the samples´ width should partly prevent fire penetration through the edges of the samples up to the top side. Another very important adjustment consists of the way fire retardants are being applied. During the experiment, fire retardants were applied only on one side of the samples, which caused burning through the edges up to the top of the samples in the process of testing. In the early stages when the flame was drawn to the retardant portion of the sample, the speed of burning was lower exactly because of applied flame retardants. However, after a few minutes (after about 3 minutes), material exposed to the heat source began to overheat, more combustible gases began to build up which caused side burning of the samples. Whereas such samples had no fire retardant applied, flame could also spread to the top side of the sample. This is the main cause of major increase in the burning speed. At the stage when material burns from the edges and the top, it is biased to evaluate and compare the materials further. For this reason, further adjustment was designed, and that is the application of fire retardants on both sides and all edges of the sample. This should prevent the fire penetration into the material and samples can be objectively assessed during the entire period of testing that lasts for 5 min.

Based on the above listed issues occurring during the experiment, we evaluated tested fire retardants only within the first three minutes of testing, when lateral edges have not yet been burned. In order to ensure transparent comparison, we record and publish chart of average burning speeds shown in Figure 7, where we can observe the changes in effects of individual fire retardants on wood.

Figure 7 displays the progress of the relative burning speeds. We got them by averaging the samples from the same set and with the same type of burning retardant. In the initial stage there are immediate increases in the burning speeds of all the samples. During the first 10 seconds the burning speed has reached value of 0.08 to 0.14 %.s^{-1}, at retardant samples as well. It causes rapid increase in temperature due to heat source. Suddenly, there is a thermal degradation of wood and release of combustible gases. Retardant samples may also face partial burning of retardant itself. It can be observed especially in samples with retardant FR COM and PS SDT. Retardants begin to operate under the influence of heat, for example, they release non-combustible gases and dilute flammable atmosphere above the wooden surface. This limits combustion. It can also be the reason for increased burning speed of retardant samples at the beginning of the test.

In the following seconds we can observe leaping progress of the burning speed; especially at non-retardant samples. Figure 7 presents burning speed progress that is literally leaping. When burning the wood creates carbonic layer. It is causing downturn in burning by creating a barrier that limits the access of oxygen to the wood below that layer, restricts formation of flammable gases above the surface and partly forms the insulating layer. Non-retardant samples, however, burn faster than retardant ones and have internal distortions in the form, for example, fine cracks, which again cause an increase in the speed of burning. Non-retardant samples also face faster fire penetration into the edges. As far as wood has no barrier against the fire such as retardant samples do, also inner layers of wood are burning. All these knowledge is the cause for step over process of non-retardant samples burning speed.

After approximately one minute, burning speed of retardant samples is stabilized. Samples applied

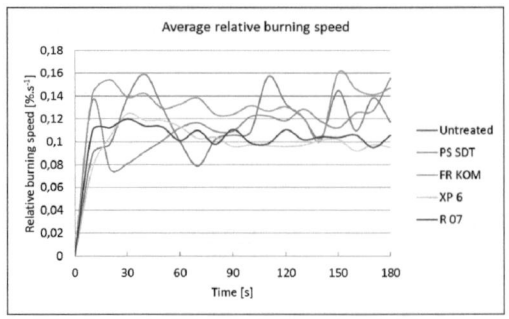

Figure 7. Chart on average relative burning speeds.

with retardants XP 6 and R 07 had the smoothest progress. At the same time these samples had the lowest burning rate almost throughout the whole period of the test. Slightly worse results were reached by retardant PS SDT with gradually increasing burning speed. In this case intumescing paint creates foam on the surface and it begins to burn slowly. However, wood could be left intacted from the inside.

Follows from the foregoing, our experiment selected the best samples covered with retardant XP 6 and R 07 by The Czech Academy of Science. These samples have not exceeded beyond the speed burning limit of 0.13 %.s^{-1}. They ranged about the value of 0.1 %.s^{-1} approximately after one and a half minute until the end of the experiment. All other samples have exceeded the limit of 0.15 %.s^{-1}, with an average value of about 0.13 %.s^{-1}.

5 CONCLUSION

Testing of fire retardants is a long-term and constantly evolving process. There are many testing methods, but the most suitable one needs to be selected in order to fulfill specific purposes. We have compiled a methodology that serves as a base for fire retardants on wood comparison via observation of relative burning speed development at the time. Precisely, mass loss during testing methods was evaluative criterion in the past. It is well readable and redeemable parameter as far as during thermal degradation of wood, substances are released from wood in particular in the form of gases. Monitored burning speed is based on the mass loss. However, fire retardants shall be tested and evaluated by several methods in order to receive more comprehensive information about their effect on the burning wood, respectively, when they are exposed to a heat source. Other appropriate parameters for the evaluation of fire retardants could be, for example, speed of flame spread along the surface, tracking of char layer after exposure to a heat source, keeping track of times of ignition and spontaneous combustion of samples and also toxicity of gases produced by the retardant's combustion should be monitored. Our testing method does not provide a comprehensive assessment of fire retardants on wood, but considerably proves achieved effect. The results of experiment show that newly developed fire retardants by The Czech Academy of Sciences were the best. Those are fire retardants labeled as XP 6 and R 07.

REFERENCES

Osvald, A. 1997. *Assessment of fire safety materials and wooden or wood-based products.* (University textbook), Zvolen: Technical University in Zvolen, 1997, 104 p. ISBN 80-228-0595-5.

Osvald, A. 2003. *Comprehensive protection of spruce.* In: Proceedings of the International Scientific Conference "Fire Protection 2003". Ostrava: VŠB—TU Ostrava, 2003. p. 312–321.

Osvald, A., Krajčovičová, J., Mitterová, I. & Orémusová, E. 2009. *Evaluation of materials and structures for the needs of fire protection.* Zvolen: Technical University in Zvolen, 2009. 355 p. ISBN 978-80-228-2039-4.

Osvaldová. L. 2005. *Fire retardants.* Arpos, 18–19, 2005, p. 18–21, ISSN 1335–5910.

STN EN 13501-1+A1. 2010. *Fire classification of construction products and building elements. Part. 1: Classification using test data from reaction to fire tests.*

Zachar, M. 2008. *Mutual comparison of some fire-technical properties of beech, spruce and poplar wood.* Building Materials, 2008, n. 4, volume 4, p. 14–18.

Production Management and Engineering Sciences – Majerník, Daneshjo & Bosák (Eds)
© 2016 Taylor & Francis Group, London, ISBN: 978-1-138-02856-2

Characteristics of wood particulates as a function of safety parameters

E. Mráčková
Faculty of Wood and Sciences and Technologies, Technical University in Zvolen, Zvolen, Slovak Republic

V. Milanko
School of Professional Higher Education, Novi Sad, Serbia

ABSTRACT: This article deals with the issue of stress and flowability characteristics of wood particulates, formed as waste during wood processing. Wood dust together with sawdust and wood chips are exhausted and stored in silos, where vault is created from dust by deposition, and thus causing a real risk of explosion. We carried out experiments with the selected samples of dust particles from the wood species such as the Norway spruce (Picea abies), the European beech (Fagus sylvatica) and the third composite sample, the mixture of powder from five wood species, to determine the characteristics of particulates. We tested selected samples with Jenike shear machine and with RST-01 tester. All selected samples were demonstrating the flowability parameters and the stress values ranging 4 < ffc < 10, classified as smoothly flowing wood particulates. According the values of those parameters were taken precautions to avoid the risk of explosion.

1 INTRODUCTION

1.1 *The particulates*

The mechanical behaviour of particulates reflects their structure. In general, their behaviour can be described through various interactions among particles (Brazda & Zegzulka 2011).

The flow characteristics of particulates are influenced with the distribution of particle size, shape and chemical composition, humidity and temperature.

The issue of flow characteristics is predominantly used in the sphere of the construction of equipment if material is required to flow from bins freely. Material does not flow smoothly if it is consolidated during transport or storing (Gondek et al. 2014).

In such cases it is necessary to use metal tools, which can represent some source of initiation and explosion (Lepik et al. 2014).

The purpose of the article is to know the characteristics of wood particulates; i.e. loose wood material which properties are not constant, but are influenced with many factors. With respect to storing, it is necessary to determine flow characteristics such as the angle of external friction σ_w (Zegzulka 2003).

Graphical specifications, Mohr's circle with coordinates $\sigma - \tau$ are used to determine maximum stress σ_1 and minimum main stress σ_2.

It is not possible to measure the characteristics directly, but it is possible to determine them indirectly using the shear test numerically or graphically (Schulze 2009). This determination of characteristics using the shear tests was specified by A. Jenike (Schulze 1995).

2 METHODS AND MATERIAL

If some loose material moves on a slippery surface, several force interactions occur. It is necessary to overcome the friction that corresponds with the normal load of the individual particles of material coming into contact with the slippery surface.

2.1 *Wood composition*

Many experiments were carried out with wood particulates such as the Norway spruce *(Picea abies)*, the European beech *(Fagus sylvatica)*, the composite mixture of the wood dust of the Norway spruce *(Picea abies)*, the European beech *(Fagus sylvatica)*, *(Fagus sylvatica Atropurpurea)*, the English oak *(Quercus robur)* and the black poplar *(Populus Nigra)*. Wood is a natural material of organic origin, composed of cellulose, hemicellulose, lignin, bitumen, lipids, wax, and so on. The content of cellulose is 30%, hemicellulose 20–35% and lignin 15–35% from the overall chemical composition of wood. Accompanying substances, though represent small percentage, influence burning despite the fact that it is not a compact wood material, but a loose wood material. The percentage of the content of constituents and the presence of accompanying substances depends on a specific wood. Therefore, there is a different resistance of individual wood species against mechanical decomposition as well as thermal decomposition (Očkajová et al. 2014).

2.2 Method of the measurement of shear characteristics

Open used the so called Jenike shear machine, also called the translational shear machine, to measure shear characteristics. There is also another type, which is also used, the rotary shear machine (Kovshov et al. 2015). The rotary shear machine has two geometric arrangements; i.e. torsional shear machine and ring shear machine (Tureková et al. 2014).

2.2.1 Jenike shear machine and accessories

The main part of Jenike shear machine is a shear cell (Schulze 2008). We used steel (stainless) and aluminium (sheet metal) to measure the external angle of wood dust.

The external angle of friction is very important for the movement of loose material on a slippery surface (for example the stainless wall of bin). The determination of the value of external angle of the friction for dust particles and the origination of slippery layer are very important for the design of conveyors, bins, silos and other equipment (Mračková et al. 2015).

2.2.2 Rotary shear machine

We used Schulze ring rotary machine RST-01.pc. to measure the internal angle of wood dust. This machine measures the flowability characteristics of dust and loose material (Schulze 1995). The sophisticated software designed for his purpose calculates friction of the particles of measured substance. Individual measurements are shown in the graph with Mohr's circles of stress.

3 RESULTS OF MEASUREMENTS

3.1 Text and indenting evaluation of the measurements of Jenike shear machine

We measured the external angle of the friction of dust particles from the wood of the Norway spruce *(Picea abies)*, the European beech *(Fagus sylvatica)* and the third sample of the composite mixture from the Norway spruce *(Picea abies)*, the European beech *(Fagus sylvatica)*, *(Fagus sylvatica*

'Atropurpurea'), the English oak *(Quercus robur)* and the black poplar *(Populus nigra)* with Jenike shear machine. We evaluated the external angle of friction for every wood dust individually with respect to two kinds of surfaces; i.e. stainless steel surface and metal sheet surface.

They are graphically presented according to the kinds of wood dust from the Norway spruce *(Picea abies)*, the European beech *(Fagus sylvatica)* and the third sample of the composite mixture of dust from the Norway spruce *(Picea abies)*, the European beech *(Fagus sylvatica)*, *(Fagus sylvatica 'Atropurpurea')*, the English oak *(Quercus robur)* and the black poplar *(Populus nigra)* in Figure 1 and Figure 2. Table 1 and Table 2 summarizes all the obtained results of shear and normal stress on stainless tray and metal tray and Table 3 shows the angles of external friction and the coefficients of friction according to the kinds of trays for the individual kinds of wood dust.

The comparison of the results of all three samples on the stainless tray and the sheet tray shows that the composite mixture has the smallest angle of external friction; i.e. the value 12.72 with the

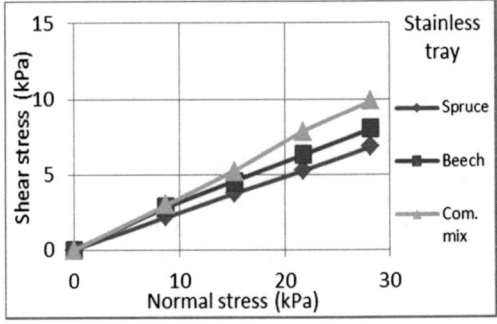

Figure 1.   Stress of three samples on the stainless tray.

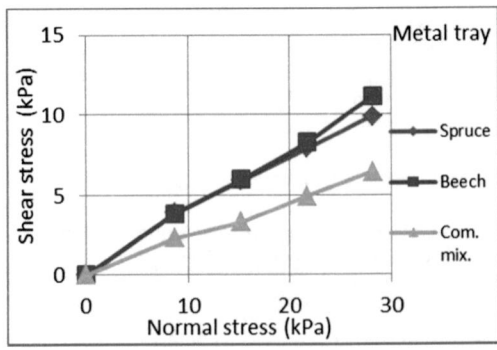

Figure 2.   Stress of three samples on the sheet metal tray.

Table 1. The measured values of stress for the individual kinds of wood dust on stainless tray.

Stress/wood dust	Stainless tray			
Normal σ_I (kPa)	8.729	15.222	21.715	28.208
Shear τ_I (kPa) spruce	2.208	3.765	5.267	6.834
Shear τ_I (kPa) beech	2.858	4.565	6.291	8.010
Shear τ_I (kPa) composite mixture	3.012	5.218	7.831	9.876

Table 2. The measured values of stress for the individual kinds of wood dust on metal tray.

Stress/wood dust	Metal tray			
Normal σ_I (kPa)	8.729	15.222	21.715	28.208
Shear τ_I (kPa) spruce	3.886	5.895	7.933	9.891
Shear τ_I (kPa) beech	3.843	5.987	8.277	11.152
Shear τ_I (kPa) composite mixture	2.301	3.311	4.953	6.456

Table 3. The angles of external friction and the coefficient of friction according to the kinds of trays for the individual kinds of wood dust.

Dust	Measurement of parameters		Kinds of trays	
			Stainless	Sheet
Spruce	Angle of external friction	φ_0 (°)	13.69	19.20
	Coefficient of friction	(–)	0.24	0.35
Beech	Angle of external friction	φ_0 (°)	16.16	21.22
	Coefficient of friction	(–)	0.29	0.39
Composite mixture	Angle of external friction	φ_0 (°)	12.72	19.13
	Coefficient of friction	(–)	0.23	0.35

coefficient of friction 0.23, but 19.13 and the coefficient of friction 0.35 on the sheet tray.

3.2 Determination of flowability curves

Many The data presented in subchapter 3.1 were obtained based on the evaluation of measured data with Jenike shear machine and after their processing using MS Excel. The obtained measured values of force for relevant shear and thrust were recalculated for shear stress Table 3 and Table 4. The flowability curves of material for a relevant preshear normal stress were determined from the calculated values σ and τ. These curves are shown in Figures 1–2 with preshear 10 KPa.

Table 4. Flowability parameters for the individual kinds of wood dusts.

Material		Spruce	Beech	Composite mixture
Presh level of stress	σ_p (kPa)	10	10	10
Angle of internal stress	φ_1 (°)	44.9	45.1	44
Effect angle of friction	φ_e (°)	49	49	49
Yield point	σ_c (kPa)	50636	51154	49308
Main normal stress	σ_l (kPa)	10133	7739	9811
Flowability	ff_c	4.99	6.61	5.02

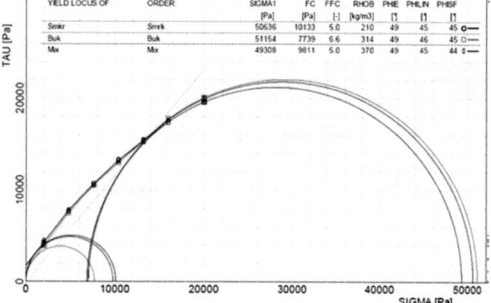

Figure 3. The comparisons of results for three kinds of wood dust with Mohr's circle.

We determined the values of yield point σ_c, main normal stress σ_1, the internal angle of friction ϕ_1 and the external angle of friction ϕ_e using geometry software, and the results are shown in Table 3. Mohr's circles are drafted in accordance with the above mentioned data Figure 3.

We evaluated the flowability of all three kinds of wood dust according to the classification of the flowability of particulates σ_1, σ_c (with the values shown in Table 4, the value of flowability ff_c, which are within the interval for dust, the Norway spruce *(Picea abies)* 4.99, the European beech *(Fagus sylvatica)* 6.61 and the third sample of composite mixture from the Norway spruce *(Picea abies)*, the European beech *(Fagus sylvatica)*, *(Fagus sylvatica 'Atropurpurea')*, the English oak *(Quercus robur)* and the black poplar *(Populus nigra)* 5.02 smoothly flowing. This internal is $4 < ff_c < 10$ for smoothly flowing material. The value 4 determines the limit of cohesion, and our measurement proves Figure 4. that the lowest value is obtained for the Norway spruce *(Picea abies)* in Figure 4.

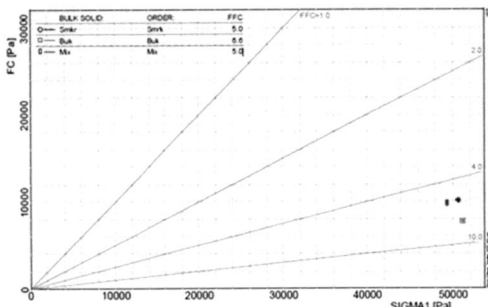

Figure 4. The resulting curves of flowability for all three kinds of wood particulates.

4 CONCLUSION

Based on the experiment, we determined the physical parameters of flowability for wood dust using Jenikeho shear machine and Schulze ring shear machine. We carried out standard tests for the wood dust of the Norway spruce *(Picea abies)*, the European beech *(Fagus sylvatica)* and the third sample of the composite mixture from five wood species—the Norway spruce *(Picea abies)*, the European beech *(Fagus sylvatica)*, *(Fagus sylvatica Atropurpurea)*, the English oak *(Quercus robur)* and the black poplar *(Populus Nigra.)* Based on the curves of the flowability of all three wood particulates we conclude that with respect to classification, all three samples can be included in the group of smoothly flowing wood particulates.

However, the most favourable results can be attributed to the wood dust of the English oak *(Fagus sylvatica)*. The English oak *(Fagus sylvatica)* belong to the group of diffuse-porous wood species. The density of wood in absolutely dry state is 490–680–880 kg/m³. After processing, waste in the form of dust originates, and thus during storing, it creates conditions for the smooth flowability of wood particulates in bins and silos. With respect to the selected tested samples, the evaluation of shear characteristics with Mohr's circles proved the most favourable flowability parameters and the values of stress for the wood dust of the Norway spruce *(Picea abies)*.

There is a constant explosive concentration of wood particulates in silos and bins, and therefore it is necessary to take precautions for storing wood dust in the silos; i.e. to ensure the optimal alleviation of explosion—with membranes, in pipelines, to ensure timely detection of flying sparks and their extinguishing with water mist, to clean and empty the equipment, and thus to eliminate the risks of explosion (Horváth et al. 2014).

With respect to the fact that the influencing of the behaviour of particulates has a considerable impact on storing, handling and the precise dosing of materials in practice, we consider necessary to develop research and knowledge about this issue not only from the point of view of safety, but also for the above-mentioned reasons (Grladinović et al. 2007).

REFERENCES

Brazda, R. & Zegzulka, J. 2011. Wall pressure issues in the aeration of bulk material silos. *Powder Technology* 206 (3): 252–258.
Gondek, H., Neruda, J. & Pokorný, J. 2014. The Dynamics of Impacts Tools the Loading Boom Bucket Wheel Excavators. *Applied Mechanics and Materials* AMM.683.213: 213–218.
Grladinović, T., Oblak, L. & Hitka, M. 2007. Production management information system in wood processing and furniture manufacturing. *Drvna industria* 58(3): 141–146.
Horváth, J., Balog, K. & Scarafilo, D. 2014. Hazards of Explosibility Dust from Wood Pellets. *Advanced materials research* Trans Tech Publication Switzerland AMR.1001.324: 324–329.
Kovshov, S., Nikulin, A., Kovshov, V. & Mračková, E. 2015. Application of equipment for aerological researching of characteristics of wood dust. *Acta Facultatis Xylologiae*. 57(1): 111–118.
Lepik, P., Mynarz, M., Serafin, J. & Bernatik, A. 2014. Explosion limits of industrial spirit and their affecting by temperature *Process Safety Progress* 33(4): 380–384.
Mračková, E., Hitka, M. & Sedmák, R. 2014. Changes of Anthropometric Characteristics of the Adult Population in Slovakia and their Influence on Material Sources and Work Safety. *Advanced Materials Research* Trans Tech Publication Switzerland AMR.1001.401: 401–406.
Očkajová, A., Stebila, J., Rybakowski, M., Rogozinski, T., Krišták, Ľ. & Ľuptáková, J. 2014. The Granularity of Dust Particles when Sanding Wood and Wood-Based Materials. *Advanced materials research* Trans Tech Publication Switzerland AMR.1001.432: 432–437.
Schulze, D. 1995. Zur Fließfähigkeit von Schüttgütern: Definition und Meßverfahren. *Chemical engineering* 67(1): 60–68.
Schulze, D. 2008. Powders and bulk solids: behavior, characterization, storage and flow. Berlin: Springer.
Schulze, D. 2009. Pulver und Schüttgüter. Berlin: Springer.
Tureková, I., Depešová, J. & Bagalová, T. 2014. Machinery risk analysis application in the system of employee training. *Applied Mechanics and Materials* AMM.635–637.439: 439–442.
Zegzulka, J. 2003. Granular States of Material Aggregation-A Comparison of Ideal Bulk Material with Ideal Fluid and Ideal Solid Matter. *Bulk Solids Handling* 23(3): 162–167.

Production Management and Engineering Sciences – Majerník, Daneshjo & Bosák (Eds)
© 2016 Taylor & Francis Group, London, ISBN: 978-1-138-02856-2

Economic of non-metallic materials in the Slovak Republic in the European context

M. Muchová, L. Domaracká, Ľ. Štrba & B. Kršák
Faculty of Mining, Ecology, Institute of Earth Resources, Technical University of Košice, Košice, Slovak Republic

J. Adamkovič
Faculty of Business Economy with Seat in Košice, University of Economics in Bratislava, Košice, Slovak Republic

ABSTRACT: Mineral resources of Slovak Republic provide a rather wide range of minerals. A significant portion of the quantity of resources, mining, processing, as well as price of minerals and their export share is represented by non-metallic materials which significantly contribute to the gross domestic product. Some non-metallic materials exploited in Slovak Republic manage to hold their stable position not only on the Slovak market, but also on the European market. The submitted article deals with the economic importance of the selected non-metallic materials in Slovak Republic and in Europe. It describes the most perspective non-metallic materials, their resources, methods of use, while focusing also on the economic importance of non-metallic materials in Slovak Republic, compared to the European Union countries (hereinafter referred to as "EU"). It analyses the resources of silicates, carbonates, evaporates in construction materials and compares their production, import and export with EU countries.

1 INTRODUCTION

In the last year, the mining of minerals represented 0.5% of the gross domestic product of the Slovak Republic. In 2012, this percentage was 0.46% and the corresponding value was € 330.3 mill in common prices. The most important proportion consists of the mining of non-metallic, construction, and energetic materials (Baláž et al. 2013). In the article, we analyse the situation regarding the selected non-metallic materials in the Slovak Republic, while considering the excavated volume, import and export, their economic importance, compared to EU countries.

2 SELECTED NON-METALLIC MATERIALS ON THE SLOVAK TERRITORY

According to Kraus (2008), non-metallic materials in Slovak Republic can be divided, depending on their potential, into the following groups:

SILICATES: bentonite, molten basalt, kaolin, talc, pearlite, zeolite

CARBONATES: dolomite, magnesite, limestone

EVAPORITES: rock salt, plaster stone, anhydrite,

CONSTRUCTION MATERIALS: silica sand, building stone, gravel sand, and brick materials which indicates their wide-range use in various industries and a stable position on the market.

2.1 Silicate materials

Silicate materials represent important raw-material potential for Slovak Republic, which is shown in Figure 1., presenting their production volume, import and export from/to the territory of Slovak Republic.

2.1.1 Bentonite

Two bentonite deposits are currently being exploited (Stará Kremnička-Jelšový potok I and Kopernica). One of the deposits of the highest quality, from the European as well as the global prospective, is the bentonite deposit Stará Kremnička-Jelšový potok I, as it does not contain the opal ingredient or any other clay minerals. It is suitable for the most demanding application, including the underground deposits of high-level radioactive waste. The entire added value—without the adequate finalisation at the deposit applying adequate treatment methods—is intended for export from Slovak Republic. Out of the total reserves registered in Slovak Republic (42,531 kt), this type represents only about 2,000 kt, which,

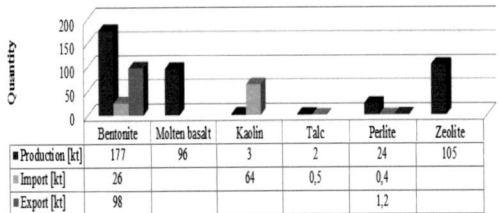

SILICATE MATERIALS

	Bentonite	Molten basalt	Kaolin	Talc	Perlite	Zeolite
■ Production [kt]	177	96	3	2	24	105
■ Import [kt]	26		64	0,5	0,4	
■ Export [kt]	98				1,2	

Figure 1. Silicate materials—production, import and export in 2012.

with current production of 100 kt/year, limits the service life of this unique and exceptional deposit to less than 20 years. The Slovak Republic should focus on possible application of lower-quality mineralogical and technological types containing kaolinite and opal CT, on the issues related to nitrification, as well as modernisation of treatment capacities. Otherwise it can happen that after the complete exploitation of the deposit Stará Kremnička-Jelšový potok I, its production will be significantly reduced. Consumption of bentonite in Slovak Republic is financed from the country's own sources and a rather large proportion of the production is exported to Poland, Austria, and Germany. In 2012, the value of this exported commodity represented EUR 10.3 mill.

2.1.2 Molten basalt

Due to its large thermal insulation potential, molten basalt is becoming the important environment-friendly raw-material in the industry and the building industry. Registered basalt deposits include: Husiná, Konrádovce, Tekovská Breznica—Brehy, and Bulhary. In 2012, basalt production represented the volume of 96 kt, covering the domestic consumption. As for the processing of molten basalt in other countries, new technologies based on nanotechnologies and geopolymers are developed.

2.1.3 Kaolin

Kaolin deposits in Slovak Republic (in the area of Poltár and Rudník in the Košice Basin) are of low quality. This material is best used in its natural condition or in form of metakaolin as an ingredient in the Portland cement. Consumption of kaolin in Slovak Republic is covered by the import from the Czech Republic, Ukraine, and Germany.

2.1.4 Talc

The only and the most important talc deposit is the Gemerská Poloma deposit which has the size and quality ranking among the deposits of the European significance. This can be perceived as a chal-

lenge for the domestic research base for expanding to the European market, which would stop exporting talc from China to the European market. Talc consumption in Slovak Republic is low, which is reflected also in the quantity of 0.5 kt of the imported commodity.

2.1.5 Pearlite

The annual pearlite production volume represents 24 kt and is carried out at the only exploited deposit in Slovak Republic, in the area near Lehôtka pod Brehmi. Another pearlite deposit in Slovak Republic is near Jastraba, which has not been opened so far, despite better qualitative parameters than the Lehôtka pod Brehmi deposit, being exploited for dozens of years. The above specified amount of exploited materials covers the domestic consumption and certain portion of the production is exported to Poland and the Czech Republic. Pearlite is becoming less demanded material also in the building industry, due to competing insulation materials made of molten basalt and with much efficient marketing and promotion of these materials. The most serious reason for this situation is that the development of new pearlite applications (filling, filters, cosmetics, glass industry, etc.) has ceased in 1980s. A solution of this situation lies in the research, focused on new pearlite applications.

2.1.6 Zeolites

Economically important zeolite deposits are located in the East-Slovakia Lowland, at the deposits of Nižný Hrabovec, Majerovce, Kučín and near Bartošova Lehôtka. In 2012, 105 kt of zeolite were excavated, which covered the domestic consumption. Natural zeolites are currently used mainly in agriculture—approximately 70 to 80% of the current global production. In the future, it will be necessary to develop applications facilitating further increase in the existing production (Baláž et al. 2013, Kraus 2008).

2.2 Carbonate materials

Carbonates represent the most important raw-material potential of Slovak Republic, generating the highest profit from export (Figure 2.). In 2012, it was approximately EUR 311.7 mill; out of this amount, the value of exported limestone, lime, and cement was EUR 227 mil; the value of exported magnesite was EUR 77.5 mil, and for dolomite it was EUR 7.2 mil. This potential has also the largest drawbacks, as the export assortment is traditional, very poorly diversified, and showing only a very low added value, which results in the need for innovation and modernisation in the field of assortment treatment and financing (Baláž et al. 2013, Kraus 2008).

CARBONATE MATERIALS

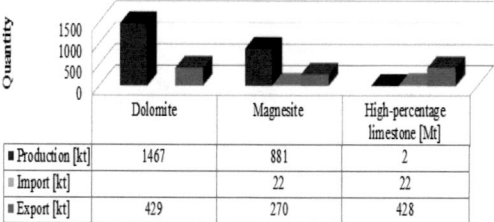

	Dolomite	Magnesite	High-percentage limestone [Mt]
■ Production [kt]	1467	881	2
■ Import [kt]		22	22
■ Export [kt]	429	270	428

Figure 2. Carbonate materials—production, import and export in 2012.

2.2.1 Dolomite
In Slovak Republic, there are dolomite deposits of very high quality. The most important deposits are located in areas near Strážov Highlands (Mníchova Lehota, Trenčianske Mitice, Malé Kršteňany, Šuja), in Považsky Inovec (Hubivá), in Veľka Fatra (Rakša) and Mala Fatra (Kraľovany). In the Eastern Slovak Republic, there are well-known deposits in the covering series of Čierna hora (Družstevná pri Hornáde—Malá Vieska). It is the highest-quality Slovak raw material located on 5% of the surface. Dolomite production in 2012 represented 1.67 kt; out of which the export represented 429 kt in the amount of EUR 7.2 mil. Over the time, dolomite consumption in the building industry decreased and it is now more and more frequently used as an additive in the ceramic industry, in the production of mineral fibre from molten basalt, as well as inorganic fertilisers (Baláž et al. 2013).

2.2.2 Magnesite
In 2012, 881 kt of magnesite were excavated in Slovak Republic, which sufficiently covered the domestic consumption. The most important deposits in Slovak Republic include Jelšava—Dúbravský masív, Lubeník, and Hnúšťa—Mútnik. Curretly, almost entire domestic production of magnesite is used for the so-called sintered magnesia. Majority of magnesite excavated in Slovak Republic is intended for export. The value of such exported commodity represented in 2012 the value of EUR 77.5 mil. The largest global producers of natural magnesite include countries such as China, Russia, Turkey.

2.2.3 Limestone
Limestone deposits are located in Slovak Republic in almost all geological objects. The most important is the high-percentage limestone, production of which achieved in 2012 the amount of 2 Mt, and other limestone types, which were exploited in the amount of 4.9 Mt, and this production covers the domestic consumption. Limestone exploited in Slovak Republic is exported to neighbouring countries—Poland, Czech Republic, Austria, and Hungary. Limestone was, and still is, used only in traditional industries: metallurgy, chemical industry, rubber industry, food industry, glass industry, and ceramic industry. In the last years, mining and processing of fine-grinded Ca carbonates has been very intensively developing. It is mostly used as the filling to paper and plastics. The most important processing procedure is grinding. Foreign companies which are focused on the exploitation of this material suitable for ultra-fine grinding (significantly limits the application of kaolin as the filling to paper) have not arrived to Slovak Republic yet to operate here. In our conditions, the limestone is traditionally used especially for the production of the Portland cement. Foreign companies already established in Slovak Republic are currently not interested in such innovation. High-quality limestone of the Wetterstein type is still "leaving" the country for a very unfavourable price, compared to the presented innovation, especially for the production of cement, which significantly increasing each year (Kraus 2008, Kraus 2011).

2.3 Evaporites
Deposits of rock salt, plaster stone, and anhydrite, according to Kraus (Kraus, 2008), belong to the group of evaporites. Production and import of plaster stone with anhydrite has been rather balanced lately (Figure 3).

2.3.1 Rock salt (halite)
It is a sedimentary rock composed mostly or exclusively of sodium chloride NaCl. It can be of white, orange, yellow, reddish, or even blue colour, but most frequently it is colourless and forms massive, granular, or compact masses. We distinguish two sedimentary genetic types of deposits of halite: fossil bedded deposits, salt stems, and recent deposits (formed by sea water evaporation). Rock salt is used especially in the chemical industry in

EVAPORITES

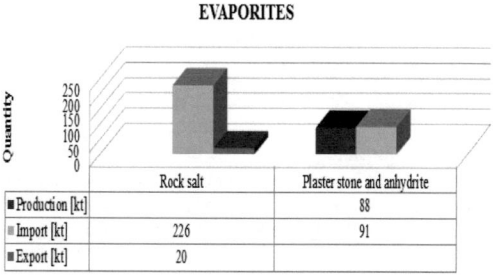

	Rock salt	Plaster stone and anhydrite
■ Production [kt]		88
■ Import [kt]	226	91
■ Export [kt]	20	

Figure 3. Evaporites—production, import and export in 2012.

the production of chlorine, soda, some inorganic salts, in the food industry, as a conservation agent, for winter road gritting, etc. From the economic point of view, important deposits of rock salt are located in the northern part of Prešov Basin and the northern part of Trebišov Basin. Both formations represent chemogenic evaporites. Since 2012, in Slovak Republic there are no organizations mining the rock salt. In 2012, rock salt in Slovak Republic was imported especially from Austria and Romania in the total quantity of 20 kt and the value of EUR 23.4 mil (Baláž et al. 2013).

2.3.2 Plaster stone and anhydrite

Deposits of plaster stone and anhydrite in Slovak Republik are associated to lagune-sea sediments of the early Permian and Late Triassic of Gemericum and Silicicum. Occurrence of evaporites is known also in early Triassic of the klippen belt and in Neogene. The largest complexes of anhydrite and plaster stone protrude on the northern edge of the Spiš-Gemer Ore Mountains, where plaster stone and anhydrite objects can be found.

Excavated deposits are in Spišska Nova Ves (2 deposits). Other deposits are not being exploited: Gemerská Hôrka, Markušovce, Gemerská Ves, Mlynky—Biele Vody, Matejovce nad Hornádom.

Domestic exploitation of plaster stone and anhydrite covers almost 50% of domestic consumption and the other half of the consumption is covered by the import from Austria, Hungary, and the Czech Republic.

The main tasks in the field of more efficient use of deposits of evaporites include:

• Reduction of plaster stone import by opening new deposits in Permian-Triassic period of Gemericum, Meliaticum, or Silicum;
• Completion of the construction of the plant processing salt brine from Zbudza in the salt works in Prešov
• Renewal and focusing the research on the innovation of the use of plaster stone and anhydrite especially in the building industry (Baláž et al. 2013, Kraus 2008).

2.4 Building materials

They include deposits of building stone, gravel sand, silica sand, and brick materials. Their production completely covers the current consumption (Figure 4). The primary issue is now the lack of research coordination to support more efficient use in the field of finalisation, diversification, treatment and innovation. Every real benefit in this field brings an incomparable economic effect, with high volume of exploited materials, compared to all other presented materials (Kraus 2008, Puzder et al. 2014).

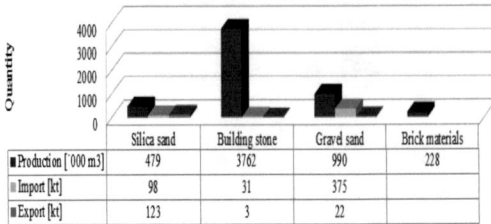

CONSTRUCTION MATERIALS

	Silica sand	Building stone	Gravel sand	Brick materials
■ Production [′000 m3]	479	3762	990	228
■ Import [kt]	98	31	375	
■ Export [kt]	123	3	22	

Figure 4. Constructional materials—production, import and export in 2012.

According to Kraus (Kraus 2008), to reach this, with regard to constructional materials, the following measures are required:

– implement non-conventional procedures in the evaluation of quality of constructional materials by amending existing technical standards, for example in case of andesite, basalt and others;
– focus the research and development on the production of cellular sandwich constructional elements with high thermal-insulation and sound-insulation parameters;
– develop the compound eco-cement, special types of cement and light synthetic stone material of the liapor, siopor and other types;
– develop the application of puzolan additives to Portland cement based on available silicate materials (metakaolin and zeolites);
– renew and interconnect the research base in the field of building industry on the cross-section Slovak level, while engaging the existing capacities from universities, the Slovak Academy of Science, and private companies.

3 POSITION OF NON-METALLIC MATERIALS OF SLOVAK REPUBLIC IN THE EU

Experts classify the non-metallic minerals by the national importance, including for example magnesites, dolomites; and by the regional importance and mass demand, such as gravel sand, building stone, or brick materials. Based on analyses of the selected non-metallic materials, we can state that in comparison to other countries, they are used only partially.

Table 1 shows the position of Slovak Republic, in terms of production volume of the selected non-metallic materials, compared to EU countries.

Mining of the selected non-metallic materials is economically important not only within the Slovak Republic, but also in the EU, which is reflected in the positions achieved by Slovak Republic, in

Table 1. Position of the Slovak Republic in the EU, by the production volume of non-metallic materials.

Non-metallic material	Position	Exploited volume [metr. t]
Bentonite	5.	158,400
Kaolin	12.	46,000
Magnesite	1.	1,196,600
Talc	7.	7,000
Pearlite	3.	23,000
Plaster stone and anhydrite	10.	143,000

terms of quality of the excavated materials. The data [ea] indicate that Slovak Republic holds the most important position in the mining of magnesite, compared to other EU countries, and kaolin is the least excavated material, which is mainly influenced by the quality of deposits located on our territory.

In order to ensure the economic importance of non-metallic materials and their future applications, it is necessary to take into consideration the position, genesis, and mineral composition of the mother rock, occurrence and form of bonds of harmful ingredients, etc.

It is also important to consider the quality of materials, the nature thereof, and profitability of treatment, which is related to the finishing and diversification. From the economic point of view, it is crucial to know the size and grade of resource verification, quantity of production, potential market, estimated price, profit, and the risk of the invested capital.

It is necessary that Slovak Republic and companies operating here strive for efficiency of application and replacement of minerals, which would ensure certain form of sustainable exploitation and consumption and efficient management not only of non-metallic materials of Slovak Republic but also EU.

Therefore, it is required to do the following, with regard to non-metallic materials (Kraus 2008, Blišťan et al. 2012):

• Eliminate the negative impact of Fe and Ti oxides in minerals with the aim to become aware of the form of their bond and then increase the rate of their application by more efficient treatment. This particularly applies to silicate (talc, bentonite, kaolin and silica sand) and carbonate materials (limestone, dolomite and magnesite).
• Significantly innovate traditional use of constructional materials (building stone, gravel sand, silica sand, and brick clay) and the related materials (pearlite, metakaolin, molten basalt, and plaster stone). Within the processing of these

materials, energy consumption and contamination by carbon dioxide should be reduced.
• Significantly diversify traditional use of the most important materials into unconventional segments by their application to the domestic market and export, and thus reduce the risk rate and the disadvantage of the current situation, regarding particularly magnesite—almost exclusively focused on refractory materials—limestone with excessively single-direction focus on the production of cement or lime, but also varied rate of urgency in case of dolomite, talc, bentonite, zeolite a plaster stone.
• Prepare a wide-extent concept of production of sorbents based on bentonite, zeolite, pearlite, diatomite, limestone, dolomite and compound products for the treatment of utility water and drinking water, oil accidents, and their application to sealing barriers, community waste and highly toxic waste dumps, as well as cleaning of waste air and gas drying.

It is also necessary to focus on the production of sorbents able to bind (Kraus 2011):

– Toxic and heavy metals, inorganic contaminants and organic pollutants from the water environment;
– Soluble nitro-compounds, phosphates, amines, and NH4 ions from the waste water;
– Detergents, herbicides, pesticides, toxic and heavy elements from the contaminated soil.

As for sorbents, special attention should be paid to bentonite or other natural additives for the planned national deposit of high-level radioactive waste, where it is necessary to consider their

– Long-term stability and behaviour when loaded (temperature and solutions from the surrounding environment);
– Contact with a container with high-level radioactive waste;
– Technology for the production of bentonite blocks.
  • To focus on the use of the waste produced mainly in form of mud, light ashes, and small grainy fractions after processing applying treatment procedures, especially in case of magnesite, limestone, dolomite, and building stone. To develop and implement new procedures especially based on thermal treatment and chemical decomposition focused on the production of synthetic minerals from natural materials or synthetically produced materials with the aim to multiply their added value. As for minerals, they can include limestone, magnesite, talc, zeolite, smectite, kaolinite, graphite, silica sand (Kraus 2008, Hlavňová et al. 2014).

4 CONCLUSION

A significant proportion of minerals in Slovak Republic, in terms of resources, exploitation, processing, as well and price or their share in export, is represented by non-metallic materials.

The change in the use of the existing potential of non-metallic materials in Slovak Republic in its economic development can be brought by applying four basic trends:

- Maximum finalisation within the processing, while increasing their added value
- Increase in the diversification during the application thereof in various industries
- Reduction of the environmental and energy load during the exploitation and processing
- Implementation of the latest treatment methods

The development of unconventional forms of application of the most important, and for Slovak Republic traditional materials, together with a more realistic evaluation of their material potential, should be regarded as the main priority for the upcoming years.

It is necessary to realize the special position of the selected non-metallic materials on the domestic and foreign markets. To emphasise the importance of the support of exploitation and processing of non-metallic materials not only from the political point of view, but also from the point of view of support of business environment.

REFERENCES

Baláž, P. & Kúšik, D. 2013. Mineral Resources of Slovak Republic 2013. State geological Institute of Dionýz Štúr. Spišská Nová Ves—Bratislava 2013: 134.

Blišťan, P., Blišťanová, M., Molokáč, M. & Hvizdák, L. 2012. Renewable energy sources and risk management. In: SGEM 2012: 12th International Multidisciplinary Scientific GeoConference: conference proceedings: Volume 4: 17–23 June, 2012, Albena, Bulgaria. Sofia: STEF92 Technology Ltd.

Hlavňová, B., Pavolová, H., Bakalár, T. & Pavol, M. 2014. Marketing Strategy of Building Stones Mining Industry in Slovakia. In: SGEM 2014: 14th International Multidiscilinary Scientific Geoconference: GeoConference on Ecology, Economics, Education and Legislation: conference proceedings: Volume 3: 17–26, June, 2014, Albena, Bulgaria. - Sofia: STEF92 Technology.

Kraus, I. 2008. New trends and possibilities of industrial minerals in Slovakia. Mineralia Slovaca, 40 (2008): 175–182, available online at: http://www.geology.sk/doc/min_slov/ms_2008_3_4/13_MS_3_4_08_Kraus.pdf.

Kraus, I. 2011. Mineral resources and human on the example of environmental sources from Slovakia Geosciences for everyone, 2011: 6, available online at: http://www.fyzickageografia.sk/geovedy/texty/kraus.pdf

Puzder, M., Pavol, M., Podolský, D. & Lechan, P. 2014. Mining gravel from the water construction Liptovská Mara In: Scientific seminar of PhD students 2014. Economy of earth sources and exploration and protection of earth resources: 7th of November 2014, Košice. Košice: TU, 2014: 67–71.

Reichl, C., Schatz, M. &, Zsak. G. 2013. World-mining data. International Organizing Committee for the World Mining Congresses. Vienna, 2013, Volume 28: 255., available online at: http://ec.europa.eu/eip/raw-materials/en/system/files/ged/20%20WMD2013.pdf

Production Management and Engineering Sciences – Majerník, Daneshjo & Bosák (Eds)
© 2016 Taylor & Francis Group, London, ISBN: 978-1-138-02856-2

Automatization of subsidiary activities and its influence on costs

D. Onofrejová & D. Šimšík

Technical University of Košice, Košice, Slovak Republic

ABSTRACT: According to economy analysts, the productivity growth reached its maximum in the last decades of the 20th century. On the contrary, there are opinions on new opportunities to improve productivity, using outputs of Internet Revolution: smart devices, smart grids and smart decision-making and control tools. The first innovative wave was called Industrial revolution, next wave was considered and known as Internet revolution, and the latest innovative wave can be technically named as "Industrial Internet", and is expected to bring significant changes to the companies and the global economy.

The paper is focused on introducing features and improvements to the industrial working environment after it's equipped with smart control elements.

1 INTRODUCTION

1.1 A chances for productivity rebound emerged

Historically, an increase in productivity had a slow cadence, until the period before about two hundred years ago, when the Industrial Revolution had begun. Innovative importance of the industrial revolution consists in replacing human and animal muscle power with plant. The Industrial Revolution proceeded over several waves and brought inventions as the steam engine, the internal combustion engine and electricity.

Productivity and economic growth has increased dramatically. Seventies of the last century experienced stagnation in productivity growth (Figure 1). Next change in innovation followed later, and was presented with the advent of computers and the global Internet, which depended on the break-throughs in information storage and computing and communication technologies. The impact of these innovations on productivity was significant, but it seems that after ten years, around 2005, this change has lost its "momentum". Some analysts believe that the productivity growth exhausted the potential benefits of a single industrial revolution.

Conversely, there are opinions and beliefs on new opportunities to improve productivity, using outputs of Internet Revolution: smart devices, smart grids and smart decision-making and control. This innovative wave can be technically named "Industrial Internet", and is expected to bring significant changes to the companies and the global economy.

Options for machines are still not yet depleted, there is potential for increasing the efficiency of the systems level. Computing, information and communication systems can now support a variety of devices and provide monitoring and systems analysis. Price apparatus, respectively for sensors was reduced substantially, enabling the deployment and the monitoring of the state of the machines in a wide range. Constantly, the production tends to increase and physical machines can be equipped with a digital intelligence. Remote Storage data sets vast amounts of data (Big Data) and advanced analysis tools have been improved substantially, and are affordable.

These changes in its deployment to machines in service and the network jointly create new opportunities. The sharp decline in the prices of devices and sensors is also related to the impact of cloud computing services that enable the collection and analysis of much larger amounts of data at a lower cost than it was ever before possible. The best cases in this area can mention applications as remote

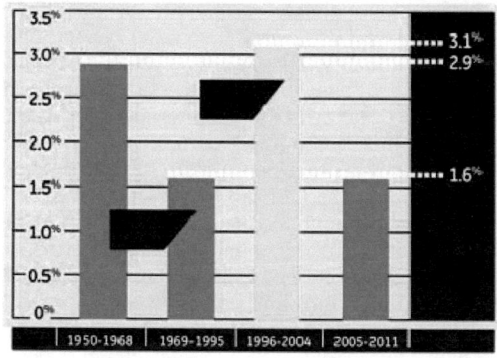

Figure 1. The productivity decline and rebound in the United States.

monitoring and control of industrial production plants, energy distribution and personalized and mobile medical care.

1.2 *Industrial internet*

The global industrial system is huge. Even a small increase in efficiency at the level of individual industries, and subsequently summing up such gain within the whole economic system could pose significant benefits. Industrial Internet could become part of the global economy in the coming years.

Industrial Internet unsealed an access to variety of benefits for the industrial economy. Smart measuring devices allow optimization of individual machines, leading to their better performance, lower cost and higher reliability. Optimized machine can operate at peak performance, and cost of operation and maintenance are minimized. Through the intelligent network, the optimization of interconnected machines can be performed.

Some companies have realized the benefits and overcome the challenges and problems associated with the acquisition and processing of data streams. It is now expected that with the continuous cost reduction of the sensors and data processing, more extensive linking of the systems and subsystems through smart devices at the level of the products would be seen in practice.

Noticeable changes in the enterprise system performance were achieved by widespread deployment of software applications and solutions for business process management, whose task was to increase the efficiency of enterprises in terms of their organizational functioning, Goldman et al. (1995). The benefits have subsequently sensed mainly in better monitoring and coordination of work, better coordination of supply chains management, improved quality, compliance with laws and regulations or the sale and production, even within a geographically distant operation plants and production lines (Evans & Annunziata 2012; Gerer 2013).

Optimization at the whole integrated system allows workers to achieve improvements in efficiency and cost reduction, higher of those which can be achieved by optimizing individual machines. Through to the intelligent software modules, running at the level of machines and systems, will be feasible to implement intelligent decisions and gain much more benefits. The continuous learning unseals new opportunities for the design of new products and services, what leads to a productivity spiral consistently yielding more and better products and services with higher efficiency and lower costs.

Technical objects in the industry sector are adapted to the needs of each customer. Therefore, the benefits of Industrial Internet will vary and impact various its various features in each business sectors differently. Nevertheless, there are common areas such as risk reduction, higher efficiency of a fuel consumption, higher employee productivity and lower costs in the each business sector (Evans & Annunziata 2012; Gerer 2013).

As a fundamental influence in the wave of the Industrial Internet is attributed to the area of development in smart systems, the next section will focus on the framework for smart system integrations.

2 FRAMEWORK FOR SMART SYSTEM INTEGRATION

2.1 *Smart integrated system management*

Smart integrated system is managed by the main control unit represents superior system for end devices managed for monitoring, communication of services for other end devices or among each other. Communications between the devices is mediated via a set of rules or protocols. Connectivity between devices on the network can be realized either through cable or a wireless transmitter/receiver. Smart buildings are built on open and standard communications networks. Building system integration takes place at physical, network and application levels. Integrated systems share resources, Sinopoli (2010). System integration involves bringing the building systems together both physically and functionally, on the contrary of interfaced systems, which share data but function as single device alone.

The physical dimension refers to the cabling, space, cable pathways, power, environmental controls, and infrastructure support. It also touches on common use of open protocols by the systems. The functional dimension refers to an interoperation capability; this means integrated systems provide functionality that cannot be provided by any single system, as it integrates the functionality of all single devices into one. International Standards Organization's (ISO) defines standards for development of the Open System Interconnection (OSI) model. The OSI model presents seven layers of network architecture (the flow of information within an open communications network), with each layer defined for a different portion of the communications link across the network. This framework model has proven its stability and functionality and should serve as the reference point for network integration (Figure 2). The model is straightforward. A network device or administra-

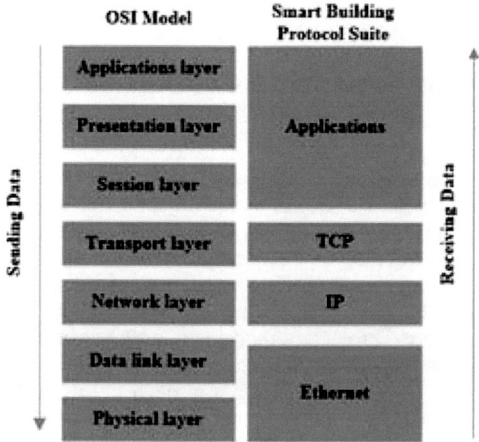

| Sending Data | | Receiving Data |

Figure 2. ISO network model layers.

tor creates and initiates the transmission of data at the top layer (the application layer), which moves from the highest layer to the lowest layer (physical layer) to communicate the data to another network device or user.

At the receiving device the data travel from the lowest layer to the highest layer to complete the communication. When the data packet is initially sent each layer takes the data of the preceding layers and adds its own information or header to the data Sinopoli (2010).

2.2 Intelligent system equipped with smart elements

Intelligent installation comprises of a bus where devices (modules), which communicate with each other, are attached. In most cases, communication between modules connected on the bus tends to be controlled via Main Controller (called Master). Operation and monitoring via Personal Computer (PC), mobile devices (tablet, mobile phones), either continuously establishes extension module (mostly directly connected to the controller), or it directly the controller. To be more specific, another trends as decentralized systems as KNX, DALI, LonWorks are used in automatized systems, which operates without controller (Networked Control System—NCS).

The intelligent automation systems offer solutions for those customers who consider and aim for safe, convenient, cost saving, smart environment and can control and manage following events:
• Lighting (and Dimming)
• Shades, Blinds (outdoor/indoor), Windows (open/close), Unlocking/locking the doors (elec-

trical lock), Garage doors and gates opening/closing, etc.
• Heating, Cooling (Air Conditioning), Fan-coil Units
• Measure: Temperature, Humidity (Dew Point), Light Intensity, Concentration of Gases, etc.
• Detect: Motion, Respond to closure of any switch
• Optional device controlled by contactor or power relays
• Any device, which can be operated via communication standard RS 232, RS485, Ethernet, Infrared port.

Similar solutions are implemented also in area of domotic automation (Onofrejova, Simsik & Onofrej 2014).

The more devices will be connected to a wireless network (Internet of Things), the more features will smart environment incorporate. The most common, centrally managed technologies in today's home automation are referred to:

• automatic door locks and security systems: they can be controlled by a smartphone or other electronic devices;
• temperature control and ventilation;
• devices that monitor energy consumption;
• entertainment systems;
• intelligent lighting control;
• smart appliances;
• detection systems for vehicles;
• systems for monitoring plants and animals (Onofrejova & Simsik 2015).

A common part of an automated smart homes are also systems mediating video and audio communication from room to room, or transmitting notifications about the status or activities of home on smartphone or other tele-user facility, upon the occurrence of events causing disruption of normal operation of the household (for example burglary), Simsik et al. (2012).

Building automation has been around for a several decades but installation of wireless technologies and integration of wired and wireless systems is propelling the market forward. There is now a great demand for energy efficient buildings, high-tech devices and enhanced security systems that are now a central component of BAS (Building Automation System). Wireless technology has revolutionized BAS. In addition, evolving technologies and expanding markets are affecting each type of control. For example, newly developed dimming systems and sensors are in especially high demand.

2.3 The importance of good systems design

The Human-Machine System (HMS) represents a system which integrates the function of human operator or group of operators and a machine (International Federation of Automatic Control, 2009). The HMS has three basic components: a human, a machine and an interface. The interface represents the means by which an operator interacts with the machine; it can be a smartphone, a robot, a touchpad, a vehicle, etc. Interaction between operator and machine has a goal to perform a specified task. Interaction is the way of human communication with the machine. HMSs require operators, capable of problem solving, especially unstructured problems for which rules do not currently exist. However, humans have a limited capacity to remember things—around seven items (± two) (Miller 1956; Wang et al. 2009; Lewis et al. 2010; Hou, Banbury & Lou 2015). With limited memory, easily distracted by background noise, humans find hard to resume tasks that have been interrupted. Humans can also have difficulty with detecting subtle changes in their environment due to limited signal detection and vigilance capabilities Wickens et al. (2013), and are overwhelmed by amount of information as volume of information, knowledge spectrums have expanded and information distribution has accelerated. Information overload can compromise attention, making it difficult to for human or machine to perceive, sort and interpret relevant information and to choose proper response within the time constrains of the situation Hou, Banbury & Lou (2015). The definition of automation is the use of machines and technology to make processes run fully on their own without manpower or with a slight interference of the human.

Intelligent interface is a mediator for commands transfer from human to the machine, in order to initiate expected action from the machine, which will perform all necessary action to adapt and fulfil desired goal. Such system does not process data from external environment. Designing the intelligent systems and using smart solutions requires caution as bad design can be annoying, perplexing and distracting for the user. Such circumstances may cause wrong decision making while evaluating information from the system, or cause time wasting while searching for error or cause displaying improper error message, etc. Challenge for intelligent designing involves Intelligent Adaptive System (IAS). In an IAS, the machine component not only provides feedback to its human partner, but also provides feed forward intelligent assistance based on the needs of the operator in a given situation communicates with and process information from external environment (world).

3 ECONOMICS OF SMART SYSTEM INSTALLATION

3.1 Savings of smart system installation

Considering expenses on the overall project for smart installation, there is a need for one initial decision-making that has to be evaluated: whether the installation project of the object (exterior/interior of building, house) involves contractor, or subcontractors, as well. In case of subcontractors share, infrastructure cost, consisting of both cable and cable pathways, has a considerably lower cost for labor. Project management and engineering costs are typically reduced as well. The net result is that the smart building approach is less costly than installing systems separately. Removing the redundancy saves capital construction monies (Shioshansi 2012; Sinopoli 2010). Figure 3 shows estimated infrastructure cost integrating two companies into one, compared to traditional approach contracting and subcontracting more than one company for smart installation services from the view of Asset and Facility Management, involving Owners and Property Manager.

First costs for integrated systems (including management hardware and software, network upgrades, web services, and devices) were 56% less than non-integrated systems. Annual costs for changes, alterations, and upgrades after a system's warranty period (including service contracts, additions and remodeling, software upgrades and reserves for systems replacement) in integrated systems were 32% less than non-integrated systems. Annual operating and maintenance costs (e.g., staff, training, IT support and management reporting) for an integrated system are 82% less than a non-integrated system. Integrated systems saved 10% in utility costs (including integrated lighting and HVAC, improved load factor, coordinated supply and demand strategies) as compared

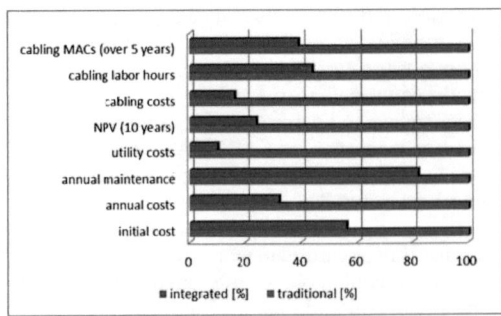

Figure 3. Reduced costs integrated vs. traditional systems (added value of smart integrated systems).

to energy costs for non-integrated systems. The NPV of the life-cycle costs of an integrated system (10 years with a discount rate of 9%) was 24% less than a non-integrated system.

Sinopoli (2010) further claims, that from the view of Concept Design and Construction, involving Real Estate Developing Organization, Funding Institutions, Architects, Engineers and Construction company savings in overhead costs are, as follows:

- Cabling: 25–40 % of labor cost, 12–20 % of the overall cost of the cable installation,
- Cable pathways: potential costs savings range from 15–60 %, for example cabling MACs—Media Access Control sublayer that manages transmission of data between two devices.
- Project management: approximately 30 % of project management for the systems is eliminated by consolidating the systems and cable installation,
- Equipment: integration of the systems results in less hardware, less space, reduction in software licenses,
- Training: standard browser and GUI (Graphical User Interface) interfaces—less training of personnel on system management tools and platforms,
- Schedule compression and time to commission: integrated systems take less time to install, less time to configure,
- Power: potential power and cooling reduction.

Operational savings using integrated smart building system are feasible to achieve and have influence on following activities:

- Energy savings due to increased building efficiency,
- Standardized infrastructure allows easier implementation changes in the building automation system controls and devices during operational life of the building,
- Improved emergency evacuation due to system integration of devices as fire alarm, video surveillance, access control, lighting control, HVAC (heating, ventilation, air-conditioning), telephone and elevator monitoring,
- Reduced training costs,
- Employees Productivity enhancement,
- Integration with additional business systems is easier,
- And others.

Building operating costs are rising, what is the reason for more owners and operators to search for new solutions: Overall energy consumption for buildings refers up to 40% of total energy consumption worldwide. While searching for substitute ecological energy sources is one solution, energy-efficient automation for new and existing buildings is gaining in importance to permanently reduce these energy costs. Considering the total costs for a functional building, only 10% of the costs are used on the actual construction of the building. And estimates for operating costs correspond with number 75% of the costs. Therefore, recent approaches support the strategy including aspects relevant to low-energy and maintenance costs into the design concept phase.

4 CONCLUSION

Industrial Internet has a potential to enhance productivity and by optimistic experts it is considered as the next wave of innovation. Assumptions say according to Evans & Entuziata (2012), that the potential benefits are sizeable, just a 1% gain in fuel efficiency over fifteen years would yield $30 billion in savings in aviation and $66 billion in the power generation industry, while a 1% efficiency gain would yield $63 billion in the healthcare industry and $27 billion in the rail industry. Obviously, the industrial internet has the potential to impact a much wider range of industries, as well as services.

Industrial internet can be understood as a network that contains intelligent machines, software analytics and people. The declining cost of tools is beginning to enable a much wider use of sensors in single machines and networks. Software analytics can evaluate the enormous amount of data obtained in order to optimize the performance of individual machines, and networks.

Small improvement in efficiency of the integrated system has possibility to bring great results. By installing the intelligent devices in every production plant or household, every vivid building that requires operating and maintenance care upon the requirements of the end-users and unit itself that creates a chance to enhance global effectiveness. Savings in costs, improving safety and the whole management in business buildings will enforce their competitiveness and even comfort for habitants. Households will become more convenient, comfortable and cost saving, especially regarding to energy consumption, which is one of the program question of the European Union future projects. Though, late trends refer that applications of smart systems, thus creating smart environments is expecting to have growing tendency in building and home automation.

ACKNOWLEDGEMENT

The research, leading to these results, has received funding from the scientific project VEGA 1/0879/13 Agile, adapting to market business systems with highly flexible corporate structure, and also this work has been supported by the Slovak Grant Agency VEGA contract No. 1/0911/14 "Implementation of wireless technologies into the design of new products and services to protect human health".

REFERENCES

Evans, P.C. & Annunziata, M. 2012. Industrial Internet: Pushing the Boundaries of Minds and Machines, *General Electric Report*, 2012, pp. 37.

Gerer, A. 2013. Priemyselný internet: posúvanie hraníc mysle a strojov, *ATP journal*, 2013–2014, Vol. XX–XXI.

Goldman, S., Nagel, R. & Preiss, K. 1995. *Agile Competitors and Virtual Organizations: Strategies for Enriching the Customer*, 1995, pp. 414, Van Nostrand Reinhold, NY, USA.

Hoffmann, T. 2008. The Convergence of Building Automation Systems and Information Technology: Global Standards Leads to Cost Reduction And Increased Functionality, *Johnson Controls, Inc.*, 2008, pp. 1–8.

Hou, M., Banbury, S. & Burns, C. 2015. *Intelligent Adaptive Systems. An Interaction-Centered Design Perspective*, 2015, pp. 293, CRC Press, Boca Raton, Florida, USA.

Kaufmann, M. 1998. *Intelligent User Interfaces. An Introduction, RUIU*, 1998, pp. 1–13, San Francisco, USA.

Onofrejova, D. & Simsik, D. 2015. Aktuálne trendy v automatizácii domácností a tvorbe inteligentného prostredia. *Automatizácia a riadenie v teórii a praxi: ARTEP 2015: workshop odborníkov z univerzít, vysokých škôl a praxe v oblasti automatizácie a riadenia*: 11.–13. február 2015, Stará Lesná. Košice: TU, 2015. (name in Slovak language).

Onofrejova, D. 2014. Process Innovation Models in Service Providing Companies. *Transfer inovácií*. č. 29. 2014, s. 259–263.

Onofrejova, D., Onofrej, P. & Simsik, D. 2014. Model of Production Environment Controlled With Intelligent Systems, *Procedia Engineering: Modelling of Mechanical and Mechatronic Systems MMaMS 2014*: 25th–27th November 2014, High Tatras, Slovakia. No. 96, pp. 330–337.

Palensky, P. & Dietrich, D. 2011. Demand Side Management: Demand Response, Intelligent Energy Systems, and Smart Loads, *IEEE Transactions on Industrial Informatics*, Vol. 7, No. 3, pp. 381–388.

Shioshansi, F.P. 2012. *Smart Grid. Integrating Renewable, Distributed & Efficient Energy*, 2012, pp. 501, Elsevier, Netherlands.

Simsik, D., Galajdova, A., Bujnak, J. & Onofrejova, D. 2012. Rozvoj technológií pre inkluzívnu a bezpečnú spoločnosť. *Bezpečné Slovensko a Európska únia: zborník príspevkov z 6. medzinárodnej vedeckej konferencie*: 15.–16. november 2012, Košice.—Košice: Vysoká škola bezpečnostného manažérstva v Košiciach, 2012. (name in Slovak).

Sinopoli, J. 2010. *Smart Building Systems for Architects, Owners and Builders*, 2010, pp. 213, Elsevier, Netherlands.

United States Department of Labor, Bureau of Labor Statistics, *Labor Productivity and Costs Database, Annual Data*, November 2012.

Wang, F.Y. & Liu, D. 2008. *Networked Control Systems. Theory and applications*, 2008, pp. 344, Springer-Verlag London, Ltd.

Production Management and Engineering Sciences – Majerník, Daneshjo & Bosák (Eds)
© 2016 Taylor & Francis Group, London, ISBN: 978-1-138-02856-2

New methods in the evaluation of flammability properties

A. Osvald & L. Makovická Osvaldová

Faculty of Security Engineering, University of Zilina, Žilina, Slovak Republic

ABSTRACT: In paper, we address the historical comparison methods with current methods for the assessment of flammability characteristics for materials an especially for wood, wood components and wooden buildings. Nowadays in EU brings harmonization in evaluated of standards into each European country and try to make one concept of evaluated the flammability properties. In each European country to the one standard level which will be used by evaluation of materials regarding flammability. In our article we focused mainly on improving the evaluation methods in terms of flammability characteristics of using materials at building industry. In the article we present examples of different assessment methods at their own test methods in terms of fire prevention. On the base of old compared of materials by STN, BS and DIN methods for testing materials on fire and new methods of evaluating the flammability properties regarding EU standards before and after starting the flash over.

1 INTRODUCTION

Beside the study of flammability there are ideas to customise the materials so (mainly natural and flammable materials) they are less flammable and not so easy to ignite. The need to customise the materials occurs also in legislative regulations, e.g. Emperor Joseph II. regulation from 1788 sets in its regulation to side wooden frames with reed and then apply coating in order to prevent fire spreading.

In the 18th century patents for retardation material customisation were granted to: J. Wild (1735) for a protective tool against fire made of alum and borax; Gay Lucas (1781) for a solution from inorganic salts; Fuchs (1820) he recommended the use of sodium silicate to prevent fire spreading. In the 19th century ammonium salts of phosphoric acid were used (Osvald 1997).

In those times the origins of material testing and evaluation for the need of anti-fire security are being born and we can talk about material evaluation and its customisation. The beginning of testing methods starts. We are going to mention only those which found the widest scope.

2 HISTORY OF EVALUATION OF FLAMMABILITY PROPERTIES

All older methods were based on the principle of constant conditions of thermal loading which should simulate the fire conditions. The evaluating criteria were the change of any physical quantity, the most frequent one, the loss of weight which presented the burnt material part given the test conditions. To mention just the most used ones, Genal-Kopytkovsky method, Truax-Harrison method via lath chimney and Schlyter method.

The Genal-Kopytkovsky (Xu et al. 2010) could have been declared as a flammability and ignitability method. Material with the thickness up to 5 mm was tested, the samples had the dimensions 100×200 mm. Onto the material which was placed under a 45° angle to the horizontal diameter the fire had effect which was created by burning 2 ml of ethanol. The bowl with alcohol was placed 25 mm under the middle of the sample. The evaluating criteria were the loss of weight, time needed for sample ignition and the time for spontaneous burning after alcohol burnt out. The testing method scheme according to Genal-Kopytkovsky is on Figure 1. From this method another method was developed and it was anchored in norms and used in the 70 s and 80 s of the 20th century under the name ČSN 73 0853.

Other methods which were used in the past were 2 similar methods, the lath chimney method and the Truax-Harrison method. For the lath method testing, spruce laned laths with the dimensions $100 \times 20 \times 40$ mm were used. The wood had given moisture 10–12%, and volume density 450–550 kg.m^{-3}. A chimney was built from the laths with the cross section 260×360 mm with 12 laths in number, whereas the distance between the laths was given as well 33–38 mm. A gas burner was used for the test and it acted on the lath chimney from bottom. The duration of gas burner acting was set to 15 min. During the test the weight is deducted every minute. It was possible due to

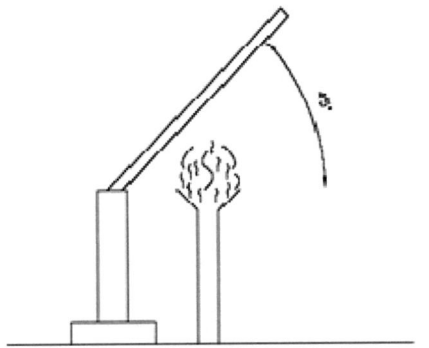

Figure 1. The testing method scheme according to Genal-Kopytkovsky.

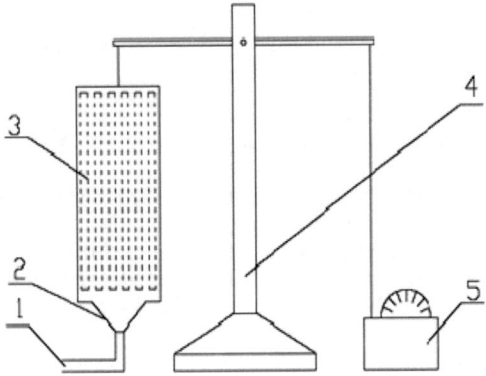

Figure 2. The lath method scheme with the lath chimney. 1—gas inlet, 2—gas burner, 3—lath chimney, 4—mounting structure with weighing possibility, 5—measuring device.

a tool, because the chimney was hung on an arm that was balanced before the test. In the upper part of the tool the thermoelement temperature was measured. The flame of the gas burner was extinguished after the prescribed time and the observation of the weight loss and temperature continued for the next 20 minutes. The course of weight and temperature loss was represented in the chart. The lath method scheme with the lath chimney is on the Figure 2.

Similar method to the lath chimney method was the Truax-Harrison method. It was anchored in the norm ČSN 49 0608 (Osvald 1997). It was mainly used for the evaluation of retardation material. The device consisted of metal tube with holes on both sides. The bottom part of the tube had a funnel shape and it was prolonged by netting basket with longitudinal hatch to light the gas burner. The tube itself hangs on scales. Under the tube is the burner.

The distance between the burner and the experimental element was 25 mm. The experimental element was $1000 \times 10 \times 20$ mm. The given moisture was 12% and weight 100–300 g. Prior to the test the gas burner was lit and the luminous flame was set by gas consumption in order to reach constant temperature in the top part of the tube to 120°C.

When the constant temperature on the top part of the tube was reached the burning burner was moved away into the tube and inside was placed experimental element and the scale was set so that the pointer was stabilised in zero position. Then the burning gas burner was again placed into the experimental device so that the axis of the experimental element was in the axis of the burner tube. From this moment on the weight loss is being observed in 30 second intervals. Simultaneously the increase of temperature in the top part of the burning tube is being observed. After for 4 min. the gas burner was removed and the weight and temperature loss in the top part of the tube are being observed until the absolute burn out (stabilisation of weight) or break-up of the experimental element. The test evaluation consisted of:

– the main weight loss (g),
– the time of burning out (min.),
– the weight loss during burning out (g),
– the time of heating (min.),
– the weight loss during heating (g),
– flammability level (general number that indicates the area size specified by curve of weight loss calculated according to the formula 1.1)

$$SH = \int_0^6 dt..\Delta m \qquad (1)$$

where SH—flammability level (-); Δm—weight loss (%/min).

The test method scheme according to Truax-Harrison (ČSN 49 0608) (STN 73 0862 b: 1986) is on the Figure 3.

The last historical and interesting method is the Schlyter method. Its origins are in Sweden. The sample was 300×800 mm, the width was 25 mm. It consisted of 2 boards (see Figure 4). The boards were placed in a parallel way after each other that the front wall was 120 mm higher than the back one. Both sample parts are connected via iron or wooden board. The distance between both boards was 50 mm. Onto the back wall effected the gas burner with flame length of 200 mm in given interval 15 min.

The evaluating criteria were:

– the flame cannot spread into a fire
– the front wall cannot burn

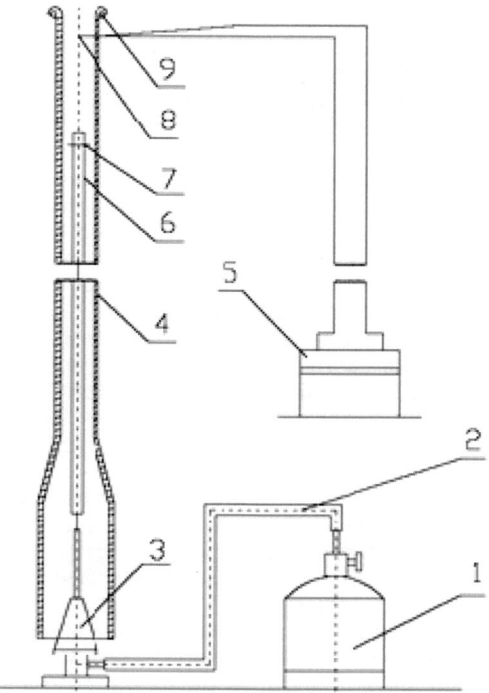

Figure 3. Scheme of the test method according to Truax-Harrison. 1—gas bottle, 2—hose, 3—gas burner, 4—metal tube, 5—measuring and recording equipment, 6—test specimens, 7—fit test specimen, thermocouple, 8, 9—hinge on scales.

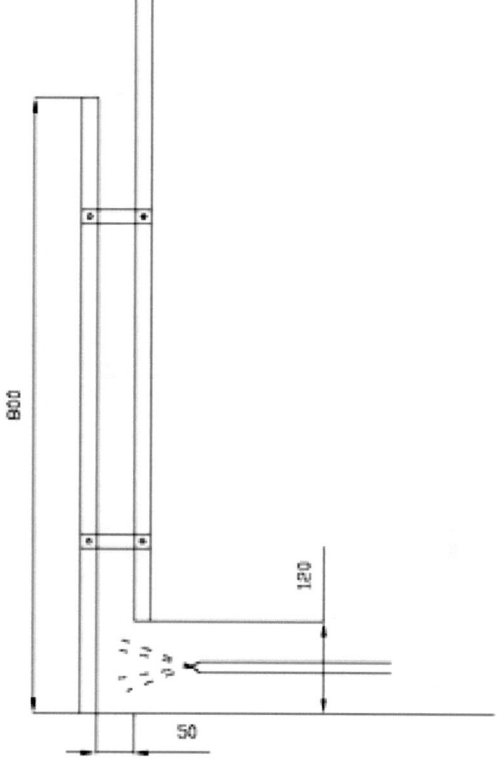

Figure 4. The test method scheme according to Schlyter.

– after flame removal the fire must stop within 30 seconds.

The test method scheme according to Schlyter is on Figure 4.

On these methods it is possible to see the testing methods development which had the biggest boom in the 1970s of the 20th century, when due to new materials—plastic, foam (foam rubber) material in bolstering, starts the "change" of fire. In those times the fire peaked at 15 min. by 800°C, nowadays it peaks at 3 minutes and by 1300°C.

3 TESTING METHODS FROM 1970S

Before the introduction of testing methods according to the EU each state had own regulations. BS 476, NS 3906, UNI 7678, NBN S 21-205, ... neither it is possible to name them all nor it is necessary. Comparison of results among these norms was not possible even though it was a frequent demand from the customers. When looking at picture number 5 we can see the differences among

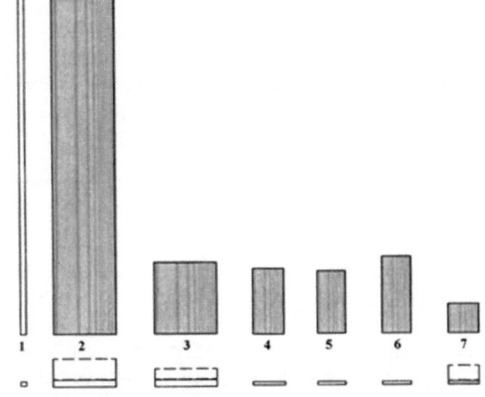

Figure 5. Shape and dimensions of experimental elements used by tests for the anti-fire protection. 1—Truax—Harrison, 2—DIN 4102/A2-B1, 3—STN 73 0862, 4—STN 73 0862/B 5, 6—DIN 4102/B2, 7—Con. Cal.

Table 1. The testing conditions for chosen testing methods.

Methods	Sample size (mm)		
	Length	Widths	Height
1/ Truax—Harrison	1000	20	10
2/ STN 73 0862	220	195	10–40
3/ STN 73 0862/B	200	100	10
4/ DIN 4102/A1	40	40	40
5/ DIN 4102/A2-B1	1000	190	10–60
6/ DIN 4102/B2	240/190	90	10–10
7/ Con. Cal	100	100	1–100

Source of heat	Duration of action of the heat source (min)	The main evaluated criteria
1/ flame	5	loss of weight
2/ flame and radiation	20 (by mode)	difference between temperature
3/ flame	10	loss of weight
4/ radiation	20	change of flame shape
5/ flame	10+10	temperature and rest
6/ flame	10	rat of flame spread
7/ radiation	(by mode)	more information

experimental elements, in their shape and dimensions. Table specifies the testing conditions, concerning the type of thermal source, length of effect and criteria for test evaluation. Which test was more reliable? And besides, the testing principle was different. For example the method STN 73 0862 (STN 73 0862 b: 1986) (was modified from BS 476.6 (BS 476.6: 1968)). To evaluate the material it needed just one device and one testing method and based on the calculated value the materials were divided into 5 flammability classes and for example the DIN 4102 (DIN 4102: 1970) needed 3 testing methods for 5 flammability classes (Figure 5).

4 CONCLUSION

There is no question whether to test or not to test. Definitely we can say yes. The question however is how to test. It is clear that each test whether the most simple or the most complicated one should as thoroughly as possible simulate the fire conditions in order to reveal the material behaviour by fire. Poor testing can cause poor material evaluation. The non-objective evaluation can be in positive as well as negative sense, i.e. on one hand it can apply material which does not have so good fire-technical

qualities and on the other hand it does not apply good quality material which was however evaluated by the test as poor.

It is important to set correct evaluating test criteria. The most simple and also the fairest evaluating criteria was the weight loss. It is given in grams or percentage. Material which was burning clearly declined. The materials were tested given standard conditions, standard source by standard time of the thermal source effect. By applying materials with moisture content, plasterboard, wood or by applying a retarder this evaluating criteria stopped to be objective. The weight before and after the test was measured. The weight loss was determined but it was not determined what was the share of the weight change caused by material drying, burning out of retarder and own material burning. In this case for example well retarded materials (visually not seen any marks of fire) were incorrectly categorised because the measured weight loss caused by the loss of retarder, or its functioning during the test. When going into extreme in such case also a table of ice is flammable.

Another problem in the testing methods was the samples' choice, mainly their size. There are number of methods based on thermogravimetry that work with milligram samples however only materials that are homogeneous (based on their chemical and physical properties) can be used. The influence of these properties cannot be reviewed (Osvald 1997).

Due to this reason wide range of laboratory testing methods has been designed with constant testing conditions, they had certain way of evaluation and its results were used many years. With scientific re-evaluation of technical design or mathematical interpretation of its results it was found out that they cannot objectively evaluate some materials. To enhance them more efficient would be certain restriction of material testing or applications effect definition on the results (Gašpercová 2010; Xu et al. 2010; Vandelvelde 2003).

Because of the critic and evaluation there has been a shift towards more quality testing methods and mainly their harmonisation, in respect of testing conditions and results application. The EU has helped a lot. To evaluate building material it is necessary to use several types of testing methods and not just one which was used to evaluate all materials and they were categorised into certain classes. The more method philosophy and strictly set criteria bring better results (Vandelvelde 2003).

The basic division can be made from various aspects. One aspect is the evaluation according to the fire phase i.e. methods that evaluate materials from first two fire phases, i.e. the possibility of inflammation—initiation and fire spread. Important it is not to evaluate the materials themselves but the whole construction to fire resistance that represents the third fire phase.

The second aspect for method division is to divide them according to what they evaluate, that is the testing of solid substance and consequently arise methods for evaluation of building material, plastic material, textile etc. We could add to these methods also special methods that evaluate solid substance however in the form of e.g. upholstered furniture, cables, clothes and other products, as well as waste such as dust firms whether in settled dust form or whirled up dust form. There are special tests for flammable liquids. The physical property of these materials (solid, liquid) does not allow to apply one test for given material (Sundström 2003).

The last possible division of testing methods is the division according to the sample size that are being used for the test. We know the division into analytical methods where small samples are used and its weight ranges from milligrams to grams; laboratory tests where the samples from the weight point of view have some kilograms and size-wise from some centimetres up to some centimetres. By these samples it is possible to take into consideration the influence of physical properties, the influence of various customisations or also errors for the final value of the conducted tests. The last group of this division are the out-sized tests.

Recently the fire tests have been combined with other tests, where not only the basic physical processes of material burning are observed but also accompaniments such as smoke where the intensity and smoke toxicity is observed (Balog & Kvarčák 1999).

It is up to the customers to give the testing organ the correct requirement for the material evaluation, so that this evaluation is complex and the results contribute to the correct application of fire resistant materials.

REFERENCES

Balog, K. & Kvarčák, M. 1999. *Dymanic of fire*. Ed. SPBI, Ostrava.
BS 476.6: 1968: Fire propagation test.
DIN 4102: 1970: Brandverhlaten von Baustoffen und Bauteilen.
Gašpercová, S. 2010. Spôsoby zvyšovania protipožiarnej odolnosti drevených konštrukcií. AlumnilPress (ed), *Environmentálne a bezpečnostné aspekty požiarov a havárií: MTF STU Trnava.—Trnava 2010.*
Osvald, A. 1997. *Assessment of fire safety materials and products of wood and wood-based. (textbooks)*, Zvolen: Technical University in Zvolen, 1997, 104 p.
STN 73 0862 b: 1986: Determination of flammability degree of building materials.
STN 73 0862: 1982: Determination of flammability degree of building materials.
Sundström, B. 2003. The Room Fire. *SP—Swedish National Testind and Research Institute,* 2003 (Polyvision videofilm).
Vandelvelde, P. 2003. Classification of fire performance. *EGOLF (ed.) Fire safe products in construction,* Luxembourg: 1999, s.1–12.
Xu, Q., Que, X., Cao, L., Jiang, Y. & Jin, C. 2010. Study of heat flux distribution of wood crib fires: bench scale tests, Thermal Science, Vol. 14, No. 1, pp. 283–290.

Production Management and Engineering Sciences – Majerník, Daneshjo & Bosák (Eds)
© 2016 Taylor & Francis Group, London, ISBN: 978-1-138-02856-2

Modelling ammonia pipeline leakage for the proposed pipeline change

H. Pacaiova, Z. Kotianova & T. Brestovic

Faculty of Mechanical Engineering, Technical University of Košice, Kosice, Slovak Republic

ABSTRACT: The paper describes the principles of modelling a leakage of liquid ammonia from the piping technology system. The methodology responds to the customer's requirement—to verify the acceptability of societal risk related to changing the ammonia pipeline with no impact on technological inputs and outputs. The ammonia treatment technology is used to store liquid NH_3, in order to produce protective atmosphere (HN_x) by cleavage of liquid ammonia into a gas product ($N_2 - H_2$). This product is used in the metallurgical industry to form a passivation coating that protects galvanized strips against corrosion. The amount and the properties of stored ammonia are factors influencing the categorization of the operation among plants with the possibility of a major industrial accident. Therefore, for any plan regarding technological changes, it is necessary to consider all aspects that may have an impact on the acceptability value of societal and individual risk the company poses.

1 CHARACTERISTICS OF HAZARDOUS SUBSTANCES

Under normal conditions, anhydrous ammonia is a colourless, strongly smelling, poisonous gas with a characteristic pungent odour. When released, liquid ammonia quickly passes into the gas phase forming a cold fog. Given the density of the vapor relative to the air, its tendency to stick to the ground is low. However, the decrease of the ambient air temperature during the ammonia leak cools its vapors and increases the specific gravity of ammonia vapor compared to the air. The vapor then spreads over the earth's surface and heats gradually. When heated, its original behavior returns and it lifts upwards. Thus, ammonia vapors may reach a much greater distance from the leak point than might be expected from its specific gravity. The dispersion and expansion of the leaked ammonia depends on the source size, extent of damage, build-up area and terrain, and also the vertical stability of the atmosphere, direction, speed, stability and force of wind and humidity. A sudden pipe breakage results in the leakage of liquid ammonia into the atmosphere. It evaporates from the opening due to the severe subcooling upon receipt of the evaporation heat from the environment of approximately 1250 kJ · kg^{-1}. On this basis, it is possible to calculate the ammonia emission as a single-phase medium.

2 PROPERTIES OF AMMONIA

In an environment with a concentration range of 300 to 500 ppm without any problems, while a 30-minute stay at a concentration of 2500 ppm is extremely dangerous and the concentration exceeding 5,000 ppm causes death quickly. According to probit functions (CPR18E, 1999), the concentration of 2,508 ppm at a 30-minute exposure is a boundary value of 1% mortality rate.

TLV (Threshold Limit Value) concentration levels set by legislation are prescribed with regard to a long-term exposure of an organism (Kotianová, 2009). Taking into account the rise of emergencies or accidents, especially their limited duration, and the need to assess individual risk, TLVs do not allow a realistic assessment of the acute toxicity of hazardous chemicals (Bernatík 2005). Therefore, in practice, further concentration limits are applied, which make emergency planning possible within acceptable limits of compromise between the acute effects and the necessity of implementation of measures to rescue lives and health of people at risk, environment and property (Oravec 2011).

In industrial practice, two concentration levels are commonly used in the assessment of the effects of toxic dispersion:

– LC50 for the quantification of societal risk,
– concentration with mortality rate below 1%.

Their values are calculated by a probit function in the form below (CPR E18, 1999):

$$Pr = a + b \times \ln(C^n \times t) \qquad (1)$$

where Pr = the probit value; a, b, n = constants characterizing the toxicity of a chemical substance; C = concentration [mg · m^{-3}] ; and t = exposure time [min].

The exposure time is limited to a maximum of 30 minutes. Probit is a function of toxic dose D for humans, which, provided that the concentration level is constant throughout the duration of the exposure, can be expressed as follows:

$$D = C^n \times t \qquad (2)$$

where D = toxic dose for humans [(mg · m$^{-3})^2$.min].

3 REQUIREMENTS FOR TECHNOLOGICAL CHANGES AND RISK ASSESSMENT

The technology used for storage, transport and processing of ammonia is, due to its dangerous properties, subject to a number of legislative requirements relating to occupational safety and health (Directive 89/391/EEC), and protecting the quality of the environment and human health (SEVESO III).

The assessment of the effects of substituting the pipeline of inner diameter DN80 with DN20 (employed to carry 30 tons of ammonia from the tank to the ammonia cleavage technology) on the reliability and safety of operation required modelling of ammonia leakage from distribution pipelines for the purpose of calculating the maximum zone of contact in the event of a pipeline rupture (see Figure 1). It was necessary to assess whether the proposed diameter DN20 would be sufficient

and how the changes affect the value of acceptable societal risk in case of pipeline rupture.

The pipeline was designed to ensure the operation's requirement about the maximum quantity of the transported liquid ammonia (LNH$_3$) in the pipeline, which was 3,000 kg/day, or 0.03472 kg/s.

The pipe inlet is at the ground level ($h_1 = 0$ m) behind pipes and shut off valves for refilling storage containers from transport tanks, whose diameter is kept at DN80 to ensure the required kinetics when refilling the tanks with liquid ammonia. The pipe outlet is at the altitude of 1.1 m above the ground, assuming the entry and exit to the pipe to be at the same elevations. Due to the low flow velocity and little pressure decrease at the end of the pipe, the corresponding calculations (Pacaiova 2010) confirmed that the use of pipes with inner diameter of DN20 is satisfactory for the operation.

3.1 Modelling a ruptured pipeline leakage

The complete proposed route for DN20 comprises 20 elbows, and its length is 258 meters. For the purposes of modelling ammonia leakage caused by a total rupture of the pipe, 3 points were chosen. Points A and C represent shutoff valves (inlet / outlet) and point B is midway between two U-shaped compensators, see Figure 2.

Model calculation of ammonia leakage in point A

– partial rupture
– total rupture.

The size of the partial rupture considered was 3.14×10^{-5} m^2 (0.1 cross sectional area of the pipe) at 1 m above ground level (3 m from the pipe inlet). At the given distance from the inlet, ammonia has passed through two elbows.

Model calculation of ammonia leakage in point B

– partial rupture,
– total rupture.

The rupture size considered was 3.14×10^{-4} m^2, which represents a complete rupture of the piping at the height of 6 m, in the middle of the pipe bridge (70 meters from the pipe inlet). At the

Figure 1. Ammonia piping prior to the change (Pacaiova 2010).

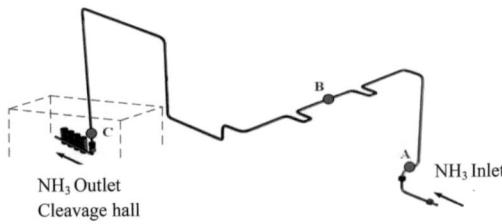

Figure 2. Layout of the model calculation points of ammonia leakage (Pacaiova 2010).

given distance from the inlet, ammonia has passed through 9 elbows.

Model calculation of ammonia leakage in point C

– partial rupture
– total rupture.

The size of the partial rupture considered was 3.14×10^{-5} m^2 (0.1 cross sectional area of the pipe) at the height of 2 m in the cleavage hall (222 meters from the pipe inlet). At the given distance from the beginning, ammonia has passed through 19 elbows.

Parameters: LNH$_3$ overpressure at the inlet is 800 kPa (pressure being highest in summer). The density of liquid ammonia pertaining to given pressure and temperature of saturated vapours is 612 kg · m^{-3}.

3.2 Simulation of emergency condition

The emergency condition simulation assumed low roughness of the inner surface of the pipe (the pipe is not corroded and is in the same condition as newly produced) k = 0.04 mm, since the maximum flow rate was considered at the leakage. Then the relative roughness of the proposed diameter (CPR16E, 1992) is:

$$k_r = \frac{k}{d} = \frac{0.00004}{0.020} = 0.002 \qquad (3)$$

Bernoulli equation for the considered pipe can be expressed as follows:

$$p_1 + \frac{v_x^2}{2}\rho = p_{atm} + \frac{v_y^2}{2}\rho + h_y\rho g + \lambda \frac{l_y}{d} \cdot \frac{v_x^2}{2} \cdot \rho$$
$$+ k \cdot \zeta_k \cdot \frac{v_x^2}{2} \cdot \rho + \zeta_v \cdot \frac{v_x^2}{2} \cdot \rho \qquad (4)$$

where p_1 = the absolute pressure at the inlet [Pa]; p_{atm} = the atmospheric pressure [Pa], normal pressure of 101,325 Pa is considered; v_x = the flow speed in the pipe after the rupture [m · s^{-1}]; v_y = the flow speed at the point of rupture [m · s^{-1}]; h_y = the elevation of the ammonia leak point [m]; l_y = the distance from the pipe inlet to the leak point [m]; and k = the number of elbows between the pipe inlet and the leak point.

This equation can be modified by applying the continuity equation to formulate an equation for calculating the leakage velocity in the following form:

$$v_y = \sqrt{\frac{P_p - h_y \cdot \rho \cdot g}{\frac{\rho}{2} + \left(\frac{4 \cdot S}{\pi \cdot d^2}\right)^2 \cdot \frac{\rho}{2}\left(\lambda \cdot \frac{l_y}{d} + k \cdot \zeta_k + \zeta_v - 1\right)}} \qquad (5)$$

where p_p = the overpressure at the pipeline inlet (Pa).

The leakage velocity depends on the linear loss coefficient, which depends on the flow rate. When using selected relations for the calculation of λ (e.g. Moody's relation), it is possible to get a transcendental equation, therefore, iterative calculation is selected with a starting value $\lambda = 0.036$.

A continuity equation is used to calculate the speed inside the pipeline during the ammonia leakage:

$$v_x = \frac{4 \cdot S}{\pi \cdot d^2} \cdot v_y \qquad (6)$$

The mass flow rate at the opening can then be calculated according to the equation:

$$Q_m = \mu \cdot v_y \cdot S \cdot \rho \qquad (7)$$

where Q_m = the mass flow rate of ammonia from the leak point [kg · s^{-1}]; μ = the flow coefficient ($\mu = 0.62$ for a sharp-edged opening; $\mu = 1$ for a total rupture); S is the size of the opening area after rupture [m^2].

Case study A—total rupture
Model calculation of ammonia leak in point A— total rupture: if the size of rupture under consideration is 3.14×10^{-4} m^2, it represents a complete rupture of the pipe at the elevation of 1 m above ground level (3 m from the pipe inlet). At the given distance from the inlet, ammonia has passed through two elbows. Overpressure of LNH$_3$ at the pipe inlet is 800 kPa (pressure being highest in summer). The density of liquid ammonia pertaining to the given pressure and temperature of saturated vapours is 612 kg · m^{-3}.

For a total rupture it holds that $v_x = v_y$.

The discharge velocity in the first iteration can be calculated as:

$$v_y = \sqrt{\frac{800\,000 - 1 \cdot 612 \cdot 9.81}{\frac{612}{2} \cdot \left(0.036 \cdot \frac{3}{0.02} + 2 \cdot 0.092 + 0.2\right)}}$$

$$= 21.18 \; m \cdot s^{-1} \qquad (8)$$

According to Moody's diagram (Moody, 1944), the coefficient of linear loss corresponding to the calculated speed v_y, is:

$$\lambda = f(R_e = 1851737, k = 0.002), \lambda = 0.0235. \qquad (9)$$

Then the rate of discharge in the second iteration can be calculated as:

$$v_y = \sqrt{\frac{800\,000 - 1 \cdot 612 \cdot 9.81}{\frac{612}{2} \cdot \left(0.0235 \cdot \frac{3}{0.02} + 2 \cdot 0.092 + 0.2\right)}}$$

$$= 25.76 \; m \cdot s^{-1} \qquad (10)$$

Table 1. Modelling the opening size and the amount of ammonia leaked from a pipe DN20.

Leak point	Opening size [m²]	Pipeline height [m]	Distance from the inlet [m]	Mass flow rate [kg · s⁻¹]
A Partial rupture	$3.14 \cdot 10^{-5}$ (10% of section)	1	3	0.598
Total rupture	$3.14 \cdot 10^{-4}$			4.95
B Partial rupture	$3.14 \cdot 10^{-5}$ (10% of section)	6	70	0.439
Total rupture	$3.14 \cdot 10^{-4}$			1.05
C Partial rupture	$3.14 \cdot 10^{-5}$ (10% of section)	2	222	0.315
Total rupture	$3.14 \cdot 10^{-4}$			0.596

According to Moody's diagram, the coefficient of linear loss corresponding to the calculated speed v_y, is:

$$\lambda = f(R_e = 2252000, k = 0.002), \lambda = 0.0235 \quad (11)$$

Given the unchanging value of the linear loss coefficient, the iterative calculation may be finished and the values of the second iteration considered final.

The mass flow rate from the total rupture in point A is:

$$Q_m = \mu \cdot v_y \cdot S \cdot \rho = 1 \cdot 25.76 \cdot 3.14.10^{-4} \cdot 612$$
$$= 4.95 kg \cdot s^{-1} \quad (12)$$

The summary of various sizes of openings and leaked quantities of liquid ammonia in the observed points A, B, C is presented in Table 1.

4 CALCULATION OF THE FREQUENCY OF NEGATIVE EVENTS

The management of major industrial risk should be one of the most important preoccupations for operators (Tixier, 2002). Risk assessment uses common analytical methods; however, it is necessary to put emphasis on the collection and utilization of the most reliable information from the plant (Pacaiova, 2009).

Modelling of hazardous properties within the risk assessment on the pipeline route, i.e. calculation of the frequency of negative events

accompanying the ammonia leak, was carried out by means of ETA (Event Tree Analysis). The following figure describes the frequency of events at the complete rupture of the pipe (according Table 3.6 Frequency of LOCs for pipes—nominal diameter <75 mm; CPR E 18E, 1999), Figure 3.

Figure 4 describes the frequency of events in the partial pipeline rupture (leakage), which considers an opening of diameter equivalent to 10% of DN.

For a 10% pipe leak, it is recommended to consider a frequency 5 times higher than the frequency in the catastrophic scenario, i.e. total rupture (CPR 18E, 1999; CPR 16, 1992), Figure 5.

The resulting scenario frequency in the considered leak point—toxic dispersion was calculated by multiplying the frequency and pipe length in meters per year (e.g. leak in A: 8.13 ×10E-7 × 73m/year), Table 4.

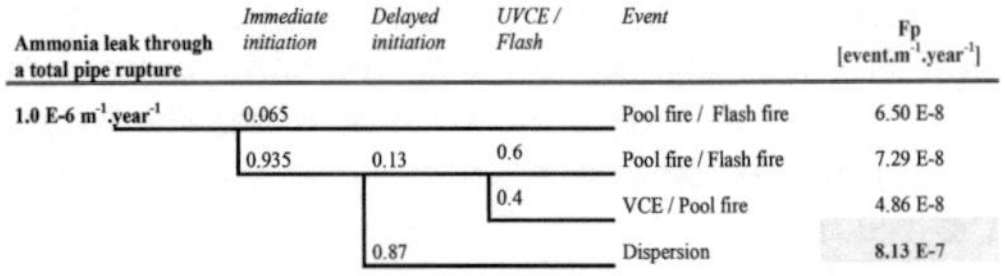

Ammonia leak through a total pipe rupture	Immediate initiation	Delayed initiation	UVCE / Flash	Event	Fp [event.m⁻¹.year⁻¹]
1.0 E-6 m⁻¹.year⁻¹	0.065			Pool fire / Flash fire	6.50 E-8
	0.935	0.13	0.6	Pool fire / Flash fire	7.29 E-8
			0.4	VCE / Pool fire	4.86 E-8
		0.87		Dispersion	8.13 E-7

Figure 3. ETA for negative events in the transport of ammonia by a pipeline—total rupture of the pipe.

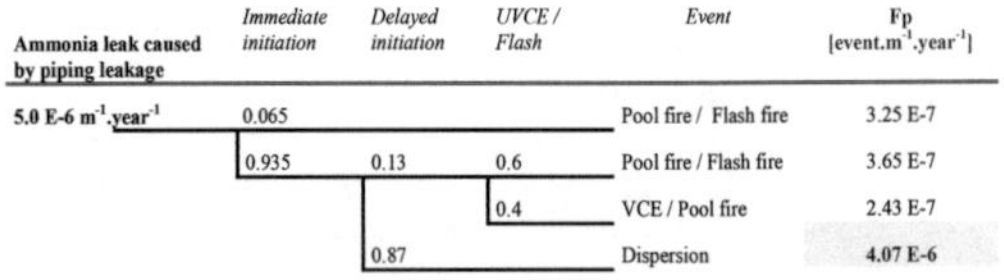

Ammonia leak caused by piping leakage	Immediate initiation	Delayed initiation	UVCE / Flash	Event	Fp [event.m⁻¹.year⁻¹]
5.0 E-6 m⁻¹.year⁻¹	0.065			Pool fire / Flash fire	3.25 E-7
	0.935	0.13	0.6	Pool fire / Flash fire	3.65 E-7
			0.4	VCE / Pool fire	2.43 E-7
		0.87		Dispersion	4.07 E-6

Figure 4. ETA for negative events in the transport of ammonia by piping—pipeline leakage (an opening of diameter equivalent to 10% of DN.

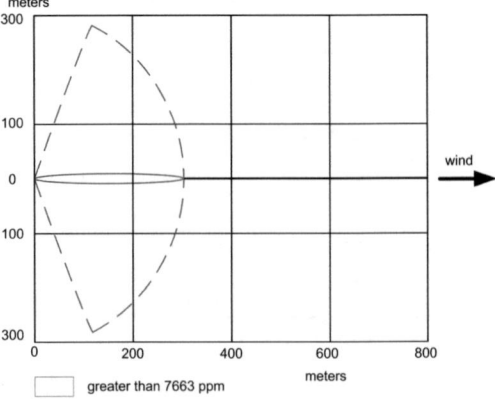

greater than 7663 ppm

Figure 5. Results of program ALOHA for ammonia dispersion DN20, threat zone for LC50 304 m.

5 MODELLING AN OUTDOOR DISPERSION

Specifying the area affected by dispersion of leaked ammonia was conducted using ALOHA ver. 5.4.4 utilizing a Gaussian dispersion model according to the equation:

$$c(x, y, z) = \frac{q}{2\Pi u_a \sigma_y \sigma_z} \cdot \exp\left(-\frac{y^2}{2\sigma_y^2}\right) \cdot \exp\left(-\frac{(h-z)^2}{2\sigma_y^2}\right) \quad (13)$$

where $c(x, y, z)$ = the concentration in the point (x, y, z) [kg · m^{-3}]; q = the substance mass flow (leakage velocity) [kg · s^{-1}]; u_a = the air velocity (wind speed) [m^{-1}]; σ_y, σ_z = the dispersion characteristics [m]; h = is the cloud height [m], and z = the vertical coordinate [m].

With regard to the above classification of blocking systems and ammonia toxicity properties, we evaluated:

– LC$_{50}$ for a 30-minute exposure (continuous cloud from a 30-minute leak)
– LC$_{50}$ for a 10-minute exposure (short-term cloud from a 10- or 2-minute leak).

The following tables (Table 2 and Table 3) describe the zones affected by leakage of substance from the considered pipe rupture points (points A and B)—*worst case scenarios*.

Table 2. Toxic dispersion for a 30-minute leakage from pipes DN20 in the leakage point A; total rupture.

LEAKAGE SOURCE (point A)	
Source height	1 m
Leakage velocity	4.95 kg/s
Leakage time	30 min.
Total amount of leaked substance	8.910 kg
AFFECTED ZONE	
IDLH 300 ppm	2.4 km
LC50 7663 ppm	346 m

Table 3. Toxic dispersion for a 30-minute leakage from pipes DN20 in the leakage point B; total rupture.

LEAKAGE SOURCE (point B)	
Source height	6 m
Leakage velocity	1.05 kg/s
Leakage time	30 min.
Total amount of leaked substance	1.890 kg
AFFECTED ZONE	
IDLH 300 ppm	692 m
LC50 7663 ppm	100 m

Table 4. Frequency of toxic dispersion, pipe length, fu and leak duration.

Source name	Frequency of toxic dispersion f_u [m^{-1} · year^{-1}]	Pipe length [m]	f_u [year^{-1}]	Leak duration [min.]
pipe DN20; A—leak after a total pipe rupture at +1 m, at a distance of 3 m from the pipe inlet	8.13E-07	73	5.935E-05	30 10 2
pipe DN20; A—the leak through a hole in the pipe with a diameter of 10% DN at +1 m, at a distance of 3 m from the pipe inlet	4.07E-06	73	2.971E-04	30 10 2

Modelling of the dispersion following a pipe leakage was carried out for individual cases under meteorological conditions characterized by atmospheric stability class F with wind speed of 1.2 m/s (worst-case dispersion scenario).

Modelling toxic dispersion for DN20 (see Figure 5).

Tables 4 and 5 show an example of the calculation of societal risk (CPR 18E, 1999) for the highest mass flow rate in point A of the pipeline for stability class F.

6 COMPARISON OF SOCIETAL RISKS FOR PIPING DN20 AND DN80

When applying the equations (4), (6) and (7) modelling a leak from DN80, it can be concluded that the biggest leak will occur in point A, similarly to the case of pipe DN20, while the leak from DN80 is approximately 31.5 times greater than that from DN20.

It is possible to express the value of acceptable societal risk (leak point A, leak duration 30 min.) by applying the FTA and ETA quantitative methods for DN80 (see Table 6).

7 CONCLUSION

The requested study into the changes in ammonia plant pipeline dimensions served the company managers for decision making, as their main priority is the reliability and safety of operation.

Table 5. Assessment of the acceptability of societal risk in point A.

Source name	Leak duration [min.]	N* [persons]	F_{pr} [event · year^{-1}]	N_{act}* [os.]	Risk [−]
pipe DN20; A—leak after a total pipe rupture at +1 m, at a distance of 3 m from the pipe inlet	30 10 2	3 2 1	1.111E-04 2.500E-04 1.000E-03	4 4 4	A A A
pipe DN20; A—the leak through a hole in the pipe with a diameter of 10% DN at +1 m, at a distance of 3 m from the pipe inlet	30 10 2	1 1 1	1.000E-03 1.000E-03 1.000E-03	1 1 1	A A A

* N—number of affected persons; N_{act}—accepted number of affeted persons, depending on the scenario frequency: A—acceptable.

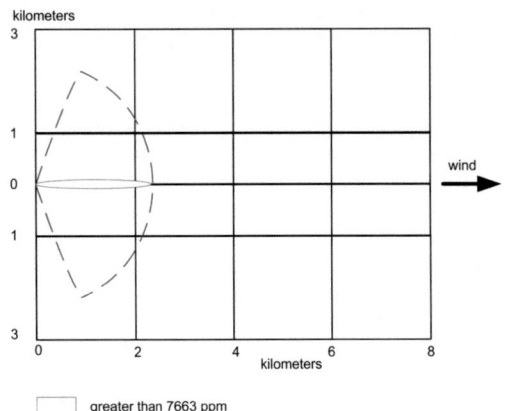

kilometers

greater than 7663 ppm

Figure 6. Results of program ALOHA for ammonia dispersion DN80, threat zone for LC50 2.4 km.

The highest mass flow rate was calculated in case of a total rupture in point A, located 3 m from the pipeline inlet, where, at the given distance from the inlet, ammonia has passed through two

Table 6. Comparison of acceptability of societal risk for DN80 and the proposed pipe diameter DN20 in point A.

Source name	Pipe DN20	Pipe DN80
Frequency of tox. dispersion f_u [m^{-1} · year^{-1}]	8.13E-07	2.44E-07
Pipe length [m]	73	73
f_u [y^{-1}]	5.935E-05	1.78-05
N [persons]	4	213
F_{pr} [events · year^{-1}]	6.25E-05	2.2E-08
Risk	acceptable	unacceptable

elbows. On the basis of the assessment of acceptability of societal risk (given the number of people affected), it was obvious that the value of the considered leakage from a pipe of diameter DN20 does not exceed the acceptable value, as opposed to DN80 (Table 6). This study confirmed that, in terms of operational efficiency, the reconstruction of pipelines substituting DN80 by DN20 does not affect ammonia flow capacity for the cleavage technology nor increase the value of societal risk expressed for the previous piping system solution. On the contrary, relative to the density of employees in service, the considered change will increase the operation safety by reducing the societal risks to an acceptable level.

ACKNOWLEDGMENT

The research has been supported and obtained financial fund from APVV-0337-11 Research into new and newly emerging risks of industrial technologies within integrated safety as a precondition for management of sustainable development.

REFERENCES

Balog, K. & Zapletalová-Bártlová, I. 1998. *Základy toxikologie*. 1.vyd. SPBI.Ostrava,.s.107. ISBN 8086111296.
Bártlová, I. & Pešák, M. 2003. *Analýza nebezpečí a prevence prumyslových havárií*. Ostrava. ISBN 80-86634-30-2.
Bernatík, A. 2005. Hodnocení rizik závažných havárií vybraných technologických zařízení obsahujících amoniak. In. *Sborník vědeckých prací VŠB—TUO*. Řada bezpečnostního inženýrství. VŠB-TU Ostrava. FBI.
Folwarczny, L. & Pokorný, J. 2006. *Evakuace osob*. 1. vyd. Ostrava. s.125. 2006. ISBN 80-86634-92-2.
Green Book. 1992. Methods for the determination of possible damage—(*Green Book CPR 16E*) First edition. Voorburrg. ISBN 90-5307052-4.
Grencik, J., Pacaiova, H., Legat, V., Dravecky, G., Hrubec, J. & Zvolensky, J. 2013. Maintaince

management, Synergie of theory and prax. press Kosice: *Slovak Maintenance Society*. p. 630. ISBN 978-80-89522-03-3.

Jusková, Z. 2008. Proposed assessment method for companies handling below-the-threshold quantities of hazardous substances. *Magazine of the European Agency for Safety and Health at Work*. No. 11. pp. 38–41. ISSN 1608-4144.

Kotianová, Z. 2009. Sources with below-the-threshold quantities of hazardous substances In: Kolloquien zum Qualitätsmanagement. No. 3. pp. 45–47. ISSN 1611-6267.

Marhold, J. 1980. *Přehled průmyslové toxikologie—Anorganické látky*. AVICENUM. Praha. p. 528.

Moody, L.F. 1944. Friction factors for pipe flow, *Transactions of the ASME 66* (8): pp. 671–684.

Oravec, M. 2011. *Manžérstvo priemyslených havárií*. ICV TU Košice. ISBN 978-80-553-0727-5.

Pacaiova, H., Oravec, M., Glatz, J. & Caky, L. 2010. *Experimental industry study for Steel Company*, TU Kosice.

Pacaiova, H., Sinay, J. & Glatz, J. 2009. *Safety and risk of technical systems*. 1. press. Košice: TU. SjF. p. 246. ISBN 978-80-553-0180-8.

Purple Book. 1999. *Guidelines for Quantitative Risk Assessment* (Purple Book CPR 18E). Committee for the Prevention of Disasters, The Hague. ISBN 9012087961.

Sanders, R.E. 1999. *Chemical process safety—learning from case histories*. Butterworth-Heinemann. ISBN 0-7506-7022-3.

Tixier, J., Dusserre, G., Salvi, O. & Gaston, D. 2002. Review of 62 risk analysis methodologies of industrial plants. *Journal of Loss prevention in the Process Industries*. 15: pp. 291–303.

Production Management and Engineering Sciences – Majerník, Daneshjo & Bosák (Eds)
© 2016 Taylor & Francis Group, London, ISBN: 978-1-138-02856-2

Application of the reverse engineering in the manufacturing process

J. Pacana, A. Pacana & A. Woźny
Rzeszow University of Technology, Rzeszów, Poland

L. Bednárová
The University of Economics in Bratislava, Košice, Slovak Republic

ABSTRACT: In the study, the process of reconstruction of a bevel gear has been investigated with the use of the Roland MDX-40 mill. For the reconstruction of the model an additional measuring head, mounted optionally instead of a tool holder, was applied. The results of scanning obtained as a cloud of points served as a benchmark for creating a complete spatial virtual model of the analysed detail. On this basis, the processing programs were prepared and the prototype of the gear was created. The whole process was efficient due to the fact that all actions were carried out on one machine.

1 INTRODUCTION

Emergency situations, in which component parts of a machine are damaged, happen frequently during the production process. In the standard procedure, which is usually time-consuming, the service is provided by ordering new components which are then assembled. However, due to the quick development of IT as well as of Numerical Control devices (NC), the process may be considerably accelerated.

The process, called the Reverse Engineering (Budzik & Pająk 2010, Slota et al. 2010), whose significant development happened simultaneously with the creation of the first copy of milling machines, is herein applicable. Thanks to that process, it is possible to measure precisely the original component and then create its exact copy on a single tool. The new element, which is the reproduction of the original part, may be immediately installed in the damaged device, re-establishing its complete efficiency. The reverse engineering is particularly beneficial for the flow of the production process as, thanks to such proceedings, time of unnecessary standstills may be shortened and financial resources may be saved.

The reverse engineering is a particularly useful process for copying the elements of a complex surface geometry (Krabowski 2008). It enables the production of a digital copy of an existing physical object, i.e. digitisation and, after processing results, a virtual model. In many cases, making a CAD model of complex elements only on the basis of simple linear measurements or two dimensional documentation would be hampered, if not even impossible. Digital data indispensable for the creation of a CAD model are obtained by contact and non-contact methods. Non-contact methods, which use e.g. optical scanners, laser scanners or X-ray computed tomography, are quicker in obtaining the results (Mantič et al. 2012). However, they are not always accurate. Contact methods, which use Coordinate-Measuring Machines (CMMs), are much more precise, but the preparation and measurement are time-consuming (Grzesik et al. 2008). An alternative for the CMM machines is a Computerized Numerical Control machine (CNC) with an installed measuring head. This connection enables the application of one device in carrying out geometrical measurements as well as the creation of a model. This solution is a significant characteristic of the reverse engineering process which, in a single process, combines both, measurement and production. Another advantage is coherent and compatible software supporting one device, which additionally accelerates the process of reconstructing a product. The coordinate milling machine Roland MDX-40, with an additional measuring head, is an example of the device which has the above mentioned features. The article presents the reconstruction process of a damaged toothed gear of a bevel drive carried out on this machine.

2 PRESENTATION OF THE ROLAND MDX-40 DEVICE FUNCTIONALITIES

The coordinate milling machine Roland MDX-40 (Figure 1) is a three-axis device designed for creating high-quality prototypes of complex geometry. The possibility of application various modules

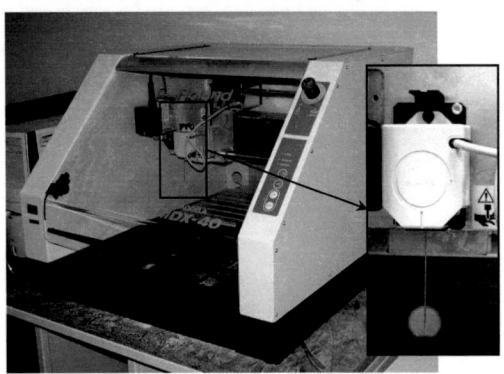

Figure 1. Coordinate milling machine Roland MDX-40, general view of the device, with the scanning head ZSC-1.

significantly increases the functionality of the machine, which allows processing by milling, scanning with the use of measuring module (ZSC-1) and carrying out engraving operations. The process of scanning is conducted on the milling machine with the contact method, with the help of the head (Fig. 1). For the configuration of the machine and for executing the scanning process particular software, Dr. Picza3D, is required. It allows to choose a method, scope and parameters of scanning as well as to trace the course of the process and save the obtained data. 3-D digitisation is based on a digital reproduction of a three-dimensional detail and its result is visible as sequentially provided points reflecting the surface of a physical model. Those points create clouds of points with the determined coordinates in the space of the scanner unit. The scanning result may be exported into the external software such as TXT, DXF, IGES, VRML files.

It is possible to enter any modifications to the model, to prepare processing software and to produce a physical replica of the component in any CAD/CAM software. On the basis of the model obtained in the process of digitisation, it is possible to create the analysis of geometry of the particular component as well as to perform measurements with the application of the finite-element method (Markowski et al. 2010; Pacana et al. 2014). After making the final virtual model in the process of scanning, a natural step is to reverse directly the process and to produce its copy on the same device. Without establishing any modifications of the model, the process requires only disassembly of the measuring head on the machine and the generation of the processing software for the particular component. ModelaPlayer 4 software, dedicated to the Roland MDX-40 machine, is used to define the productive process. This programme enables to define the processing parameters, to select appropriate tools and to trace the processing results in the form of simulation on the computer screen. When the complete process of the reverse engineering is conducted on the same device, the transition from the measurement stage into ModelaPlayer 4 software is possible with only one click after the scanning process is accomplished. Moreover, it is possible to design the processing software on the basis of the models for other milling machines, or even to prepare the processing actions for models imported from any external CAD programme.

3 THE PROCESS OF REVERSE ENGINEERING FOR A GEAR

In order to demonstrate the functionalities of Roland MDX-40 machine in the widest scope possible, three processes, different in terms of the range and type of obtained results, have been presented. Three different measurements of the toothed gear of the bevel drive (Fig. 2) have been carried out with the application of the measurement equipment and software described above. First, an initial scanning of the complete gear was executed, with rather low resolution. After that, precise digitisation of a single tooth was performed in order to reconstruct a complete gear in a later stage. The last step was to carry out very accurate measurement of the tooth outline.

In order to set the scanning parameters manually, the first process of digitisation of the whole gear was made in "Make setting and scanning" mode. Because of the large dimensions of the gear, it was important to precisely limit the measuring scope and the resolution so that the scanning process would not last too long. The aim of this process was only to determine initially the shape of the gear and to designate basic geometrical parameters.

Figure 2. Scanning of a toothed gear of the bevel drive.

The leap between subsequent measuring points in horizontal axis X, Y was set every 0,4 [mm]. Restrictions on the vertical axis Z were read with the use of the automatic sensor at the highest point of the measured gear as well as on the measuring table of the machine. The process of scanning proceeded without any difficulties and the obtained point map of a part of the gear (Fig. 3) was saved in .dxf format in order to use it to create a CAD model. The final result of scanning in Dr. Picza3D software was presented in Figure 4. Despite low accuracy of the measurement, the digitisation time was quite long, about 2,5 hours.

The digitisation of a single tooth was also conducted with the flat scanning method, with scanning parameters set manually. Parameters of measuring leap in X, Y axis were set every 0,1 [mm], hence much more accurate mapping of the tooth shape was obtained (Figure 5). The obtained measurement results reflect the real shape of a tooth precisely enough so that it can be reproduced by milling on the Roland MDX-40. The time required for conducting the measurement with this method was 30 minutes.

The third process of scanning was the accurate check of the tooth outline. The measurements were carried out by using the same methods as in two other processes. However, the scanning range was limited to the minimal acceptable width for X axis. What is more, the lowest measuring leap value, i.e. 0,04 [mm], possible for this measuring equipment, was set.

The result of the scanning process is a precise tooth outline in cross-section presented in Figure 6. Thanks to this measurement, the disturbances of the tooth outline on the active side of the tooth may be observed. Pattern wear is clearly visible in the form of a hollow on the right side of the tooth outline.

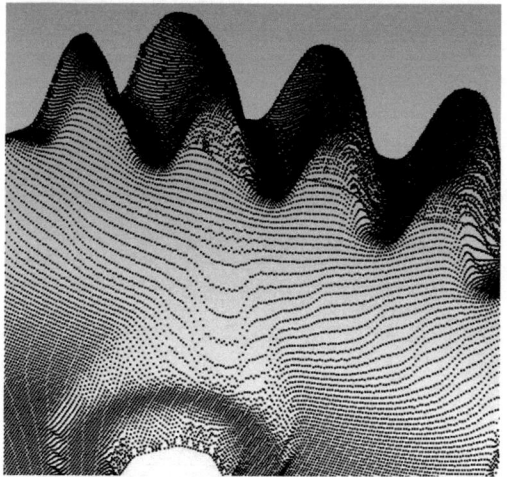

Figure 3. The result of scanning in the form of a point cloud of the gear (a fragment).

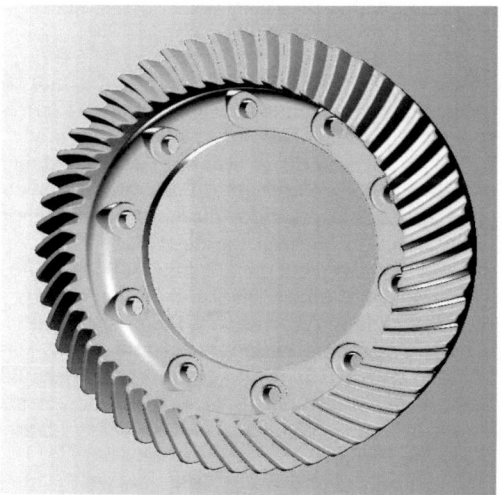

Figure 4. Final effect of gear digitisation in Dr. Picza3D software.

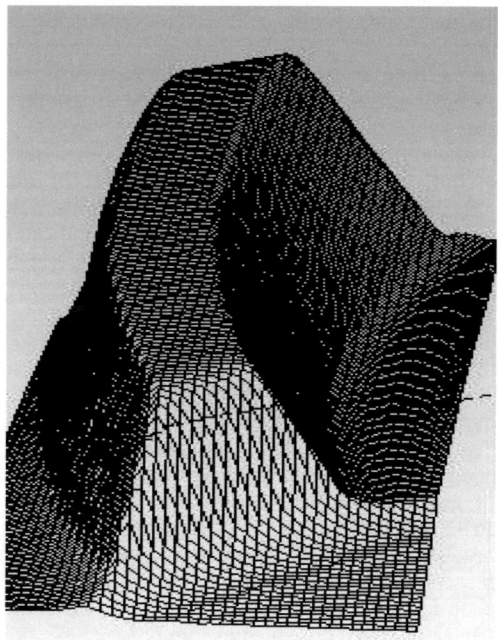

Figure 5. The result of scanning of a single tooth.

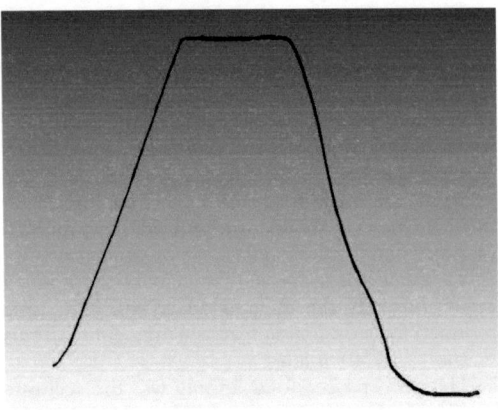

Figure 6. Tooth outline obtained in the scanning process of high resolution with the use of ZSC-1 head.

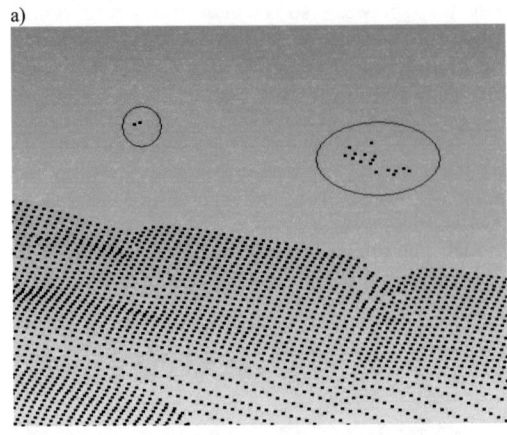

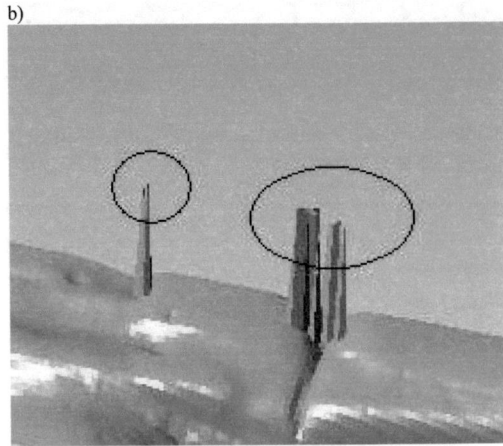

Figure 7. View of the scanned surface with exemplary errors presented as: a) cloud of points, b) shaded view.

During the digitization process, the obtained coordinates of points are presented in the window view of Dr. Picza3D software as a map, a cloud of points or as a shaded view. The results of scanning are loaded in rows which are updated on the computer screen in stages for subsequent reading series. Thanks to this, the results of scanning may be traced on-line. The exemplary visualisation of the points read during the process is presented in Figure 7.

During measurements with high accuracy, with the use of touch probe, random errors appear occasionally and they do not describe the measured geometry in an evident manner. In Figures 7 and 8 similar situations has been presented for another object measured on Roland MDX-40 device.

Such errors did not appear for the toothed gear of the bevel drive and this functionality of the software is important in the process of digitisation and worth being described. Red colour indicates noises, i.e. points which were read by the device in a wrong way, in places where the surface is smooth. The faulty points assume maximal value of height, established for the measurement, and they substitute points which should appear for the same coordinates in the model. The noises may be reduced by the software and, in order to do it, one should highlight the area of points which contain errors by applying e.g. the function "Area Rectangle". Next, by applying the function "Remove Noise", the errors will be reduced on the basis of algorithms installed in Dr. Picza3D software. An exemplary process of the random error reduction, before and after deleting the improperly scanned points in the model, has been presented in Figure 8.

This procedure is very quick and convenient for small amount of faulty points. In the case of larger areas of faulty geometry, this method may not be applied as the obtained surface might not correspond to the real shape of the measured model. If, during the process of scanning, larger number of faulty readings appears (of the unknown origin), one should interrupt the process and check the connection between the device and the computer. The measuring process may be re-launched again, however with a slightly modified accuracy of measurement. Alternatively, the previous measuring process may be continued and, when finished, the area (where the errors appear) should be re-scanned. The results of both measurements should be connected in any CAD software, removing surface errors.

The analyzed gear needs to be replaced so that the accurate functioning of the device is possible. A new gear may be created on the basis of the measurements of the correct model or a used

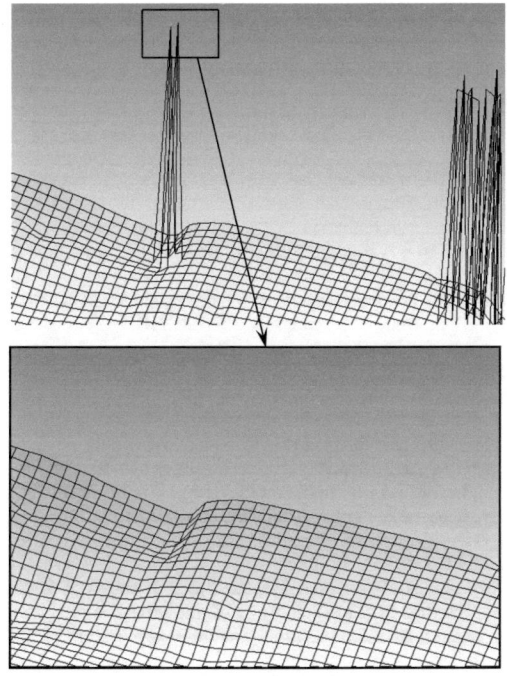

Figure 8. The process of correcting the errors of the scanned surface in Dr. Picza3D software.

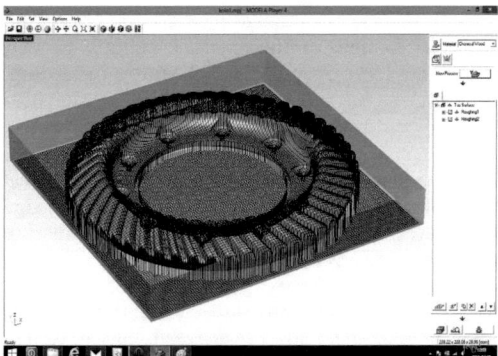

Figure 9. The view of ModelaPlayer4 software window during the process of defining gear processing.

Figure 10. The prototype of the gear produced on Roland MDX-40 device.

model may be modified on the basis of the already measured geometrical parameters. The method of the precise measurement of details on Roland MDX-40 machine enables to verify the quality of components after processing or to compare models after their modifications.

The models obtained in the process of scanning the gear were also used to produce its prototype. The physical model was created from a modelling board; therefore it was not completely functional. However, it enabled to assess the accuracy and quality of the conducted process of the reverse engineering. The model of the gear, used to create the prototype, was prepared on the basis of the measuring results of a single tooth whose outline was corrected to the point where it didn't assume its previous use.

The definition of the production process, the development of the processing software and the configuration of device parameters were made in ModelaPlayer4 software. The window view of the software, with the fragment of the gear and with tracks of tool, has been shown in Figure 9.

After verifying the prepared processing operation with the use of computer simulation, the process of creation a physical model of the gear by milling was initiated. The final result of subtractive

manufacturing in the form of the prototype of the gear has been presented in Figure 10.

The model was reconstructed on MDX-40 device with very high accuracy. In the visual assessment the prototype seems to be very good, but its precise measurements will be conducted with the use of another professional coordinate measuring machine. The obtained results enable to assess the quality of the complex detail reconstruction process that was carried out (Zielecki et al. 2013).

The time of creating the prototype was 2,5 hours and was determined by the material applied, the capabilities of the machine and tools that the lab was equipped with

4 SUMMARY

The above study depicts the functionalities of applying the coordinate milling machine Roland MDX-40 with measuring module ZSC-1 as the tool in the process of reverse engineering. Due to the use of machines available in majority of manufacturing companies it is possible to create independently, for the current needs, even complex elements of the device. Such a situation may happen in case of damage to a component being a crucial element of the machine being a part of the manufacturing process. Schematic procedure, i.e. ordering a damaged component at the producer's company and waiting for an exchange, would result in considerable delays in production and financial loses. However, the current technical possibilities, along with the engineers' knowledge permit for more efficient conduct of the missing component production process.

By applying already known procedures from reverse engineering process a damaged element may be scanned and transformed into a virtual model. Subsequently, due to the computer CAD/CAM software and numerical control machine tools, it is possible to create a physical model being a reconstruction of the damaged initial model.

The above study traced the reconstruction process on the example of a bevel gear's cogwheel. For the model recreation an additional measuring head, installed optionally instead of a tool holder of the milling machine, was applied. In the scanning process and during creation of the described component, specialised software installed on the machine was applied. The scanning process was conducted for three diverse steps, different in scope and measuring resolution. The scanning results received as points cloud served as a benchmark for creating a complete 3-D virtual model of the analysed detail. On its basis, processing software was prepared and a cogwheel prototype was created. The complete process ran smoothly because of the fact that all actions were carried out on the one device. The whole task, together with scanning results processing and preparing models did not take much time and could be realised during one shift. If the appropriate technical possibilities exist in the workplace, application of the described scheme of procedure is very advisable, in order to minimize the standstills in the production process

ACKNOWLEDGMENTS

Financial support of Structural Funds in the Operational Programme—Innovative Economy (IE OP) financed from the European Regional Development Fund—Project "Modern material technologies in aerospace industry", Nr POIG.01.01.02–00–015/08–00 is gratefully acknowledged.

REFERENCES

Budzik, G. & Pająk, D. 2010. Metody inżynierii odwrotnej, Stal, *Metale & Nowe Technologie,* vol. 11/12: 66–67.

Grzesik, W., Niesłony, P. & Bartoszuk, M. 2008. *Programowanie obrabiarek.* NC/CNC. Warszawa: Wydawnictwo Naukowo-Techniczne.

Krabowski, K. 2008. *Podstawy rekonstrukcji elementów maszyn i innych obiektów w procesach wytwarzania.* Kraków: Wydawnictwo Politechniki Krakowskiej.

Majerník, M., Szaryszová, P., Bosák, M., Štofová, L. & Kabdi, K. 2015. Integrated management of environmental—safety and technical risks of plant producing automobiles and automobile components, *Communications: scientific letters of the University of Žilina.* vol 17 (1): 28–33.

Mantič, M., Schneider, M., Kuľka, J. & Bigoš, P. 2012. Relation between accuracy of digitalized model and selection of 3D-scanning technology. In: *38. mezinárodní konference kateder dopravních, manipulačních, stavebních a zemědělských strojů,* ZU Plzeň.

Markowski, T., Budzik, G. & Pacana, J. 2010. Kryteria doboru modelu numerycznego do obliczeń wytrzymałościowych walcowej przekładni zębatej metodą MES, *Modelowanie Inżynierskie,* vol. 39: 135–142.

Pacana, J., Pacana, A. & Bednárová, L. 2014. Strength calculations of dual-power path gearing with fem, *Acta Mechanica Slovaca,* vol. 18(2): 14–19.

Slota, J., Mantič, M. & Gajdoš, I. 2010. *Rapid Prototyping a Reverse Engineering v strojárstve,* Technical University of Košice.

Tomčíková, M. &, Živčák, P. 2012. *The practical contribution of information systems in times of globalization.* In: Management 2012, Prešov: Bookman, 2012. pp. 253–256, ISBN 978–80–89568–38–3.

Zielecki, W., Pawlus, P., Perłowski, R. & Dzierwa, A. 2013. Surface topography effect on strength of lap adhesive joints after mechanical pre-treatment. *Archives of Civil and Mechanical Engineering,* vol. 13: 175–185.

Production Management and Engineering Sciences – Majerník, Daneshjo & Bosák (Eds)
© 2016 Taylor & Francis Group, London, ISBN: 978-1-138-02856-2

Material flow in logistics management

M. Pružinský & B. Mihalčová
Faculty of Business Economics with Seat in Kosice, Economic University in Bratislava, Kosice, Slovak Republic

ABSTRACT: Performance requirements and production efficiency are the starting solution for material flow logistics management. The development of logistics business is directly related to the requirement to reduce its cost. This trend is based on the fact that logistics costs assume all of the activities (i.e. marketing, production as well as distribution logistics). The aim of this paper is to examine the material flow of the enterprise. Material flow includes purchasing, inventory control of raw materials and finished goods, the receipt and storage of materials, production planning, and transport. At present, it represents fully developed technical systems that efficiently and effectively cooperate with production systems. We performed recalculation material consumptions from the largest supplier point of view. Based on the analyzes' results and calculations we suggest possible solutions in optimization of material flow within the enterprise.

1 INTRODUCTION

Logistics management is a set of planning, implementation, and management of effective, powerful flow and storage of goods, services and related information from point of origin to point of consumption, which aims to satisfy customer requirements. Business logistics deals with all movement-storage activities that allow the flow of material, products and services from the point of origin of acquisition of raw materials to the point of final consumption, as well as information flows, which show materials, products and services to move to meet the needs of customers at a reasonable cost. The logistics system according Viestová & Štofilová (2002) identifies three main subsystems. The distinction according to whose activities deal with:

- Material subsystem, which ensures the implementation of material flow,
- Management subsystem, which is carried out planning, management and control,
- Information subsystem.

The logistics process objects according to Lambert, Stock & Ellram (2005) are transformed from the initial to the final state, while changing at least one of their parameters without undue there was a change in its properties. System quantity may be e.g. time, place, quantity, kind. By some studies the existing manufacturing capability of machines use is only of 30 to 40%. The other sources say that technological processes spend only 5% of time needed to manufacturing. The rest of time is spending by manipulation, transport and storage.

The flexible manufacturing system contains some CNC machine tools supported by industrial robot for material handling. This system is designed to manufacture a group of similar work pieces. The system is characterized by its internal material and information flow. The manufacturing process represents a complex dynamical process included technological, manipulation, and control operations.

2 MANAGEMENT OF MATERIAL FLOW

The management of material flow is the entire logistics process important though that is not directly concerned by the final customer. Kolesár & Petruf (2011) underline the decisions taken in the management subsystem of the logistics process directly affect the level of customer service provided; the company's ability to compete with other companies and the level of sales and profit that the company is able to reach at the market. Materials managers perform functions purchasing, inventory control of raw materials and finished goods, receiving and storage of materials, production planning and transport. In Figure 1 we visualized the main objectives and tasks of materials management. We examined an efficiency of material flow according to different stages of the production process. Effective flow requires advanced material manufacturing process progressively organized as the shortest transport routes. Parts of the management in the area of materials are such anticipation of material requirements, identifying sources and sourcing, delivery of material and introduction to the company and monitoring of the material as current assets.

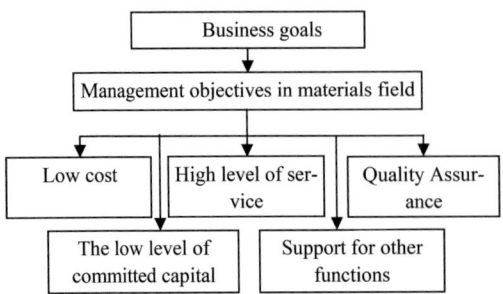

Figure 1. The objectives of integrated management, materials.

Quantifiable parameters of material flow are: the amount of material—Q; path—s and time—t. The amount of material is expressed by the following variables:

- In the case of bulk materials is expressed in units of mass or volume [kg], [t], [m³].
- In the case of lump materials shall be expressed in handling units [m].

The path/track is expressed by the length of the track route of the movement of material in the space between the source (supplier) and consumer (customer). It is expressed in units of length [m, km]. Time is expressed by the duration of some work (operation process) or the date when to start an activity or end (time of submission of vehicle for loading, delivery time inventory etc.). Continuous handling process is the time (time-consuming) to pass the path between the supplier and the customer. Time intermittent handling process is the sum of the periods of movement cycles of which the handling process has. Established units of time are second [s], or hour [h]. In some cases it should be day, week, month, and year.

The goal of efficient material flow solutions is economical movement of passive elements—materials, raw materials, intermediate products, security products through active elements such as: transportation, handling and storage systems. For the basis of efficiency is considered straightforward and simple course, as well as low frequency material flow, which allows the formation of larger, comprehensive volume handling units, which is treated as a single piece using mechanization straightforward and simple course depends on the optimal spatial distribution of company, and the objects of production and warehouse space.

According to Chromjaková (2009) the solution to the question "What?" and "Why?" defines the correct material to be transported. The questions "Where?"; "When?" and "Why?" are dealing with the identification of the necessary handling and

transport activities. Questions "How?"; "Who?" and "Why?" define the correct method of transport to be used. The questions "What system?" and "Why?" are pointing to the draft resolution material flow.

An integral part of the management of material flows are areas of:

- Purchasing and Procurement,
- Management of production,
- Transportation of material towards the enterprise and the enterprise, storage,
- Management Information System (MIS)
- Planning and Inventory management,
- Waste disposal.

2.1 The use of management information system for the management of material flows

Materials managers need to efficiently manage material flows towards the enterprise and the enterprise direct access to the corporate information system. This approach mainly affects the demand forecasts produced production, supplier lists and data on them, information on prices, production plans, routing and transport planning. Materials managers turn this system complements the data on stocks of material supply plans, fixed-price purchases in stores and vendor information.

Inventory management in the broadest sense represents obtaining and maintaining an optimal amount and type of physical resources needed to implement the strategic plan. Movement and flow of goods and materials are key elements in inventory management. To maintain a certain stock levels are several reasons:

- Decoupling between supply and demand,
- Security (protection) against the uncertainty due to suppliers, to cover unexpected demand, etc.
- The expected demand, for example due to the increase in sales, due to seasonality or advertising.
- Providing services to customers, for example the cyclical stocks of finished products or in the form of contingency reserves for unexpected demand.

Supply strategy has to respect the principles of rational supply:

1. Buy only what you need,
2. Buy only what is necessary, because the market economy is characterized by a rather excess of supply over demand and a higher number of suppliers in the market,
3. Consider the costs associated with acquiring the inventories and according to them to decide when, and how much to buy.

3 DATA AND METHODS

The basic objects of investigation in the paper were the material flows. We analysed stocks and consumption of individual materials. To meet the objective we had to pay attention to the theoretical basis of logistics, inventory management and production management. Based on the findings, we analyzed the situation of the company in the management of material flows. Acquired knowledge we interpret and seek potential opportunities to streamline processes. Necessary data and information was sought in the documentation and materials provided by employees. These were: consumption of materials, sizing each state material, the actual state of materials in stock, real supply of materials from suppliers, the methods used in materials management.

We investigated an optimization of material flow solutions within the enterprise by using analysis and synthesis (e.g. data on deliveries by each supplier, together with the capacity dispositions in warehouses and enterprise data of consumed materials on the basis of the documents and calculations). We compared the consumption of different types of materials. We also assessed the formulation of proposals to improve the management of material flows different variants Radio Frequency Identification (RFID) products. On the basis of this evaluation, we specify a valid option, which should be implemented within the enterprise.

4 RESULTS AND DISCUSSION

For production planning and managing we use in general two basic systems: type PULL, PUSH system type. In this paper we are not dealing approachable PULL. Just-In-Time (JIT) systems came widespread from the base of KANBAN. These systems link purchasing, production and logistics. The primary objective is to minimize the JIT inventory, improve product quality, maximize production efficiency and provide the optimum level of customer service. The concept is useful if an enterprise has frequent deliveries, produces in small series, and supplies small quantities. At the same time contributes to quality assurance in production, motivation of staff, eliminating losses and maintaining long-term strategic line. Studied enterprise was established in 1993 as a subsidiary of transnational corporations. In 1998 it was bought by the largest German customers. The production program of the plant is assembling of electric motors for washing machines, dryers and dishwashers. The plant produces annually more than seven million engines consumption class A and AA. More than 80% of the motors are fitted

to these products. All production is exported. The quality requirements for silent run of these demand high technological standards in acoustics and non-contact and optical methods. The plant has a special laboratory to perform tests of engines and complete products.

4.1 The manufacturing process and the course material flow in the enterprise

Enterprise uses PULL principle of production (production according to the needs) type supermarket, which we basically visualized in Figure 2. Supermarkets are materially struts between business departments in areas where there is no possibility of a smooth flow of materials. The supermarkets are preparing exactly the minimum and maximum quantities of raw materials, components and assemblies for the present fill up storage process. This is the principle "fill up producing" according to given rules only what is wanted. When determining the minimum amount of new parts are made preliminary for storage process, unless reached the maximum state/amount. The process is similar to the real supermarket.

Another type of production in the enterprise is KANBAN. This is continuity rather examined the type of production "Supermarket". KANBAN is free technology stocks, supported by a self-regulatory control circuits consist of supplying and taking the article. These articles are associated with direct chain. Their relations are determined by the

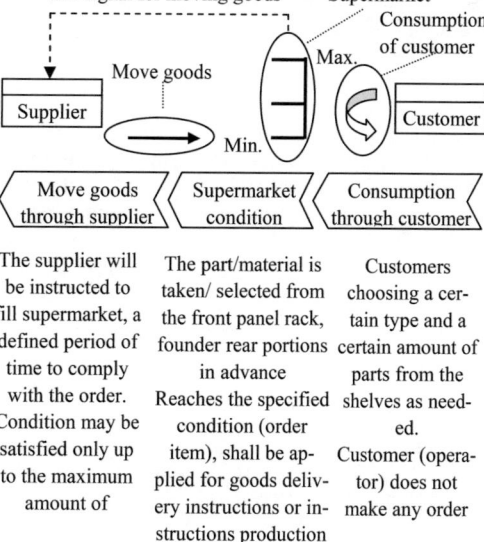

Figure 2. Production as necessary (pull-principle-Supermarket).

principle PULL. The ordered and delivered quantity corresponds exactly to the capacity of the transport vehicle (or multiple). Emmett (2008) stress the quality of the material supplied is guaranteed by supplying entities. The technology is suitable in terms of a uniform consumption without major fluctuations and changes in product. This leads to continuity processes, high productivity and efficiency.

The main supplier of metal pressed sheet needed for the components in the enterprise is a neighbouring company. The company's major supplier of metal is US Steel Košice. Enterprise in order to optimize and control of manufacturing processes uses SAP R/3 Enterprise system. The information system was purchased for the needs of production optimization. It's a good tool to support business processes in the areas of production and minimize the costs of operation and maintenance.

Currently, the system supports managing of some production lines, allowing streamline manufacturing processes to reduce costs and increase the quality of production. The system was upgraded by modules of production management, quality control, maintenance, and human resources module.

4.2 Production calculated according to consumption based on shipping claims

The benefits of this system for the enterprise would be (i.e. optimization of production processes, improved transparency of data, accurate records of production, material and commodity stocks, improving the supply system production lines, promotion of quality management, support for electricity exchange data with trading partners, and reduction of the workforce in the administration).

For the design condition of starting materials of stator and rotor plates for different engines, we examined under a scheme attributes of processes shown in Figure 3, which are the following activities such as: A—information, reaction times, B—time blockade, C—rest period, D—time production, E—tolerance, F—transit time, G—trading time.

We offer an example of the stator plates to universal motors. Needed sizing stator plates delivered

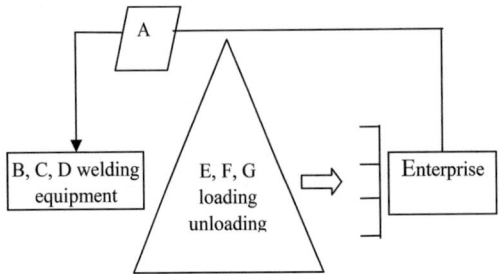

Figure 3. Basic attributes of the process flow sheet stator.

by the information on the stator core for universal motor:

a. pallet unit contains—280 pieces,
b. daily consumption is—15210 pieces,
c. safety factor—1.5.

Based on the opinion of times re-acquisition, which are:

A—60 minutes,
B—0 minutes,
C—70 minutes, and
D—84 minutes (calculated from the daily capacity of 13,000 pieces with 3 welding equipment. On one welding equipment welding accounted for 1 piece stator plate average time of 0.3 min/piece, we can multiply × 280 pieces per pallet),
E, F, G—120 minutes (loading and unloading).

Total time needed is 334 minutes.
The highest need for installation of 5070 pieces (all one line at a functional change).

- Time of re-acquisition/installation cycle = 334 minutes/min = 0.09 p = 3,711 (13.2 palettes).
- Determine the minimum condition:
- Minimum status = 3711 pieces/280 pieces for each pallet × 1.5 = 19.8 pallet (20 pallets = 5600 pieces)
- Determination of maximum state:
- The maximum state = 5600 pieces + (3 × 5070 pieces) = 20 810 pieces = 74 pallets.
- Determine the average needs:
- (minimum-maximum state condition)/2 = (5600 + 20810)/2 = 13 205 pieces = 47 pallets.

We also address the sizing of the state for the UM rotor plates (when counting to 65 mm rotor plate). We provide the basic information about the metal rotors.

d. pallet unit—900 kg (average),
e. daily consumption—16 890 pieces,
f. safety factor—1.5.

Re-acquisition time for one moulding machine
A—60 minutes
B—100 minutes (replacement disc)
C—0 minutes
D—187.5 minutes on 1 pallet (108,433 pieces per pallet, 8.3 g per sheet)
E, F, G—120 minutes (loading and unloading + transport).
Total time needed is: 467.5 minutes.

4.3 Detection of state production assembly over the re-acquisition

The highest need for installation of 5,630 rotors, for what we need 563,000 pieces of sheets, represents 4672.9 kg/all lines at 1 change from 1 pallet 1125/at 0.9 kg per piece.

- Time of re-procurement/installation cycle = 467.5 minutes/0.08 = 5 min 843.7 piece determine the minimum condition:
- The minimum condition 4 = 850.3 kg/900 pieces for each pallet × 1.5 = 8.08 pallets (8 pallets corresponds to 7.200 kg).
- Determination of maximum state:
- The maximum state = 7200 kg + (3 × 4 672.9 piece) = 21 218 kg = 5.23 = 24 pallets.
- Determine the average needs:
- (minimum-maximum state condition)/2 = (7200 + 21 218.7)/2 = 14 209.3 kg; 16 pallets; state/daily requirement = 8 636.4/14 018.7 = 1 day.

We also calculated the dimensioning of the state for the stator plates and rotor assemblies T20 motor, the stator and rotor plates GV motor and 1 BA 57. The calculated value of the stocks in the supermarket sheets for stators and rotors manufactured engines we present in Table 1.

We stated in Table 1 number of pallets for each type of product that the firm uses in production and the total number of machines deployed for half the production lines, which means that the maximum number of lines is 156 pallets.

Actual stock, where each line is used to 100% (no preparation time) is expressed in Table 2.

In Table 2 we introduced the actual condition of pallets needed for each change on the day and store for the weekend. It follows that the maximum number of machines on the line calculated for one day is 129 pallets and store for two days this weekend is 258 pallets.

The analysis proved that the company produces a regular supply of materials. Some orders with a yearly interval, other semi-annually and more per month. We calculated the estimated consumption of materials in pallets and subsequently converted its proposal the consumption of materials.

4.4 The results of the analysis of material flow

On the basis of X, Y, and Z analysis of the company each material was evaluated as follows:

- The materials which are most often consumed (stator and rotor assemblies—individual marking materials),
- The materials to use less and
- Materials to consume the least, only if the necessary requirement (carbons).

The amounts of materials are calculated according to the items of the pallets, which are placed in stock and leaving for the consumption. The materials that are consumed daily should be kept in the warehouse so that they are always on the same place in the same amount and the supply exceeded the daily consumption.

4.5 The proposal for building a consignment store

The consignment stores, we propose to place the daily production materials consumption. We calculated that it would be 258 pieces of pallets, allowing 100% of the production lines in all three shifts.

Table 1. Total stock at the supermarket, according to calculations of the enterprise.

| | Motor UM | | Motor T20 | | Motor GV | | Motor 1 BA 57 | | |
	Sta*	Rot*	Sta*	Rot*	Sta*	Rot*	Sta*	Rot*	Pallets totally
Min	20	8	3	4	5	11			51
Max	74	24	13	7	17	17	3	0,5	156
Average	47	16	8	6	11	14			102

* Sta, resp. Rot mean: Stator, resp. Rotor.

Table 2. The actual state at 100% utilization.

Field	Stocks to change unit	The need for day [pc]	Pallets for change	Pallets per day	Stock (the need for the weekend—2 days)
UM Stator	5.070	15.210	16	48	95
UM Rotor	4.673	14.019	5	15	30
T20 Stator	1.280	3.840	4	12	24
T20 Rotor	1.200	3.600	1	3	6
GV Stator	2.450	7.350	12	36	72
GV Rotor	4.000	12.000	2	6	13
BA 57 Stator			3	8	15
BA 57 Rotor			1	2	3
Together			43	129	258

To create a consignment store for a supplier in our company it is necessary to conclude the consignment respectively development of Raci matrix. Economic benefits of creating solutions consignment store will bring the enterprise the following advantages:

- reduce the cost of supplies,
- regular supply provided by the supplier,
- continuity of production,
- the exact location of the material,
- saving money (because the material is in consignment stores owned supplier, resp. Supply undertaking, it assumes responsibility for our business at a time when the material gets on the production line).

A further rationalization is the use of Radio Frequency Identification (RFID) in the enterprise. RFID (Radio Frequency Identification) is a new generation of AIDC (Automatic Identification Data Capture) technology. It offers the advantages that they provide other AIDC systems (e.g. barcode). RFID system is operating in the radio frequency electromagnetic waves. This is a contactless identification, which is used for data transfer and storage. Most of the supply comes into the company by trucks from main suppliers and other vendors. After passing the entrance gate and the adoption of the necessary documentation to the material is interpreted trucks forklifts and follow the correctness of the delivery of these materials. This control is performed by authorized staff. The download process is lengthy and laborious. The solution is in the extension of the functioning of SAP3 Company and supplier of software for cooperation with stationary, portable and built-readers for cars. Supplier marked pallets or individual components and materials chips and shipped to our company. The premises for unloading materials in the production hall built enterprise should counter scanner or RFDI gate. After passing through the gates (scanner) happen to load data from individual chips pallets.

5 CONCLUSION

Based on analyzes of available data and calculations, we have proposed two options for improving the current situation—creating a consignment store and the introduction of RFID into service within the company in cooperation with contractors undertaking the capacities of their enterprise information and management systems SAP3. Both proposals call for the creation of conditions, but thanks to the use of compatible systems of corporate governance and suppliers. No need for high costs. Both methods assume cooperation in the establishment of supplier relations. Use of RFID devices will be higher initial investment costs, which are in a relatively short time, while the volume of production returned.

ACKNOWLEDGEMENTS

The paper conducts research in area of Sustainable development of higher education in the fields of management in the frame of VEGA project No. VEGA č.1/0708/14.

REFERENCES

Chromjaková, F. 2009. *Projektovanie materiálového toku.* Electronic source [online]. 2009. [Cited 19.2.2015]. Retrieved from: <http://www.ipaslovakia.sk/slovnik_view.aspx?id_s=137>.

Emmett, S. 2008. *Řízení zásob.* Brno: Computer Press, 2008. 298 s.

Kolesár, J. & Petruf, M. 2011. *Logistické obslužné procesy v leteckej doprave.* Košice: Technická univerzita v Košiciach. 284 s.

Lambert, D.M., Stock, J.R. & Ellram, L.M. 2005. *Logistika.* Brno: CK Books, 2005. 589 s.

Viestová, K. & Štofilová, J. 2002. *Distribučné systémy a logistika.* Bratislava: Vydavateľstvo EKONÓM, 2002.

Production Management and Engineering Sciences – Majerník, Daneshjo & Bosák (Eds)
© 2016 Taylor & Francis Group, London, ISBN: 978-1-138-02856-2

Metrology in the Slovak Republic and objectives of its development

Z. Schreier & J. Krivosudská
Slovak Office of Standards, Metrology and Testing, Bratislava, Slovak Republic

J. Markovič
Slovak Legal Metrology, n.o., Banská Bystrica, Slovak Republic

ABSTRACT: The article presents the national metrology system in the Slovak Republic, laid down by the applicable legislation of the Slovak Republic and by other legal documents specifying its scope in the field of the state administration, public services and entrepreneurial environment. The individual entities involved in the field of metrology are described, including their mutual relations and activities. Besides that, the principle duties and responsibilities of the particular subjects involved in the placing the measuring instruments on the market and their putting into operation or use, as well as duties and responsibilities concerning the metrological control and market surveillance are described. The next part of the article deals with the aims relating to the development of the metrology and the national metrology system of the Slovak Republic in relation with the measuring instruments that are making use of the latest results of science and research and of the state-of-the-art technologies.

1 INTRODUCTION

In general meaning, the metrology is the science about measurement; we may define it as a science dealing with the measurement methods and measuring instruments. Of course, the science itself is not sufficient, the science must be applicable and used in practice. In the field of metrology, this requirement is fulfilled: the science is applied in the industrial metrology, legal metrology and, naturally, the binding force of the metrology is given by the metrology legislation. In the every state, it is the regulatory subject, i.e. the respective state body that is responsible for the metrology legislation and its practical application.

The metrology deals with measurements, units of measurement, methods and means of measurement, principles of the measurement results processing, in order to assure the uniformity and correctness of measurements in various fields of the human activity. The legal metrology concerns the measurements that have an impact on the transparency of the business transactions, consumers' protection, health, safety and environment protection and law enforcement. The legal metrology covers all the activities stipulated in the legal rules that are related with the measurements, units of measurement, measuring instruments and measuring methods. These activities are performed either by the state institutes, or by the bodies that were entrusted with these duties by the state. Their aim is to assure the credibility of the measurement results in the state-regulated sphere.

2 NATIONAL METROLOGY SYSTEM

The entities that are involved in the metrology and metrology-related activities are namely the state administration bodies acting within the scope of their assigned powers, the bodies performing verification and calibration of the measuring instruments and official measurements, the subjects manufacturing, repairing and installing the measuring instruments, accredited testing and calibration laboratories and certification bodies, research and educational organizations, inspection bodies, but also the users of the measuring instruments, as well as the system of the legal and technical rules specifying their status, and the complete set of technical means and equipment—that all constitutes the National Metrology System (NMS).

At present, the Slovak NMS is constituted by the Act No. 142/2000 Coll. on metrology (Law on metrology) governing the principles of assuring the uniformity and correctness of measurements, including the metrology infrastructure.

2.1 Slovak Office of Standards, Metrology and Testing

Slovak Office of Standards, Metrology and Testing (SOSMT) are the central state administration body in the Slovak Republic (SR) for the field of the technical standardization, metrology, quality, conformity assessment and accreditation of the bodies performing the conformity assessments. For these spheres of action, SOSMT develops the state policy concept, carries out the methodical activities and supervises the fulfilment of tasks in the field of normalization, metrology, quality, conformity assessment and accreditation of the bodies performing the conformity assessments. The SOSMT sphere of authority is governed also by Law on metrology and by the Act No. 264/1999 Coll. on technical requirements for products and on conformity assessment (Act on the technical requirements for products and on the conformity assessment). Within the system of the international relations, SOSMT fulfils the role of the National Metrological Authority (NMA).

Department of metrology is the professional body of SOSMT, responsible for the state policy application in the field of metrology. When performing its competency tasks, the Department of metrology observes, in particular, the Law on metrology, Ordinance of the SR government No. 399/1999 Coll. specifying details on the technical requirements for non-automatic weighing instruments (Ordinance of the SR government on non-automatic weighing instruments), Ordinance of the SR government No. 294/2005 Coll. on measuring instruments (Ordinance of the SR government on measuring instruments), Decree No. 206/2000 Coll. on legal units of measurement, Decree No. 207/2000 Coll. on e-marked pre-packages, Decree No. 210/2000 Coll. on measuring instruments and metrological control and Decree No. 419/2013 Coll. on pre-packages.

2.2   Slovak Institute of Metrology

Slovak Institute of Metrology (SMU) is the state allowance organization that preferentially arranges the activities in the field of the fundamental metrology, i.e. research, development and preservation of the national standards, transfer of the values from the national standards to the respective standards according to the SR economy requirements, so that they may be considered as the basis for validation of the measurements both on the national and international level. SMU carries out the type approvals and verifications of the legally controlled measuring instruments, calibrations of other measuring instruments, official measurements. It provides all the services on the highest metrological level in the SR. Within the international context, SMU fulfils the role of the national metrological institution—National Metrological Institute (NMI).

2.3   Slovak Legal Metrology, n.o.

Slovak Legal Metrology, n.o. (SLM) is the non-profit organization providing activities related with measurements, as specified by the legal rules

(by the requirements on the measuring instruments and on the methods of their testing) in order to assure the required level of confidence in the results of measurements in the sphere regulated by law. In accordance with the Law on metrology, SLM is the organization designated by SOSMT for the metrological control of the measuring instruments and for other activities resulting out of the Law on metrology. SLM represents the basic pillar of the consumers protection in the fields of measurements related with payments, if the field of health, property and environment protection, as well as in other public fields, where conflict of interests on the measurement result may arise, or where the incorrect measurement results might harm the interests of the natural and legal persons, or those of the society.

2.4   Conformity assessment

SOSMT authorized and notified SMU and SLM for the conformity assessment in accordance with the Law on technical requirements for products and on conformity assessment within the scope of the Ordinance of government on the non-automatic weighing instruments and of the Ordinance of government on the measuring instruments; so they are entitled to provide the respective metrological services both for the Slovak and foreign manufacturers of the measuring instruments when placing them on the EU common market.

2.5   Slovak Metrological Inspectorate

Slovak Metrological Inspectorate (SMI) is the budget-financed organization performing the state metrological surveillance (the surveillance on observing the duties stipulated by Law on metrology for the state administration bodies, for manufacturers and importers of the legally controlled measuring instruments, service organizations, users of the legally controlled measuring instruments, packing plants and importers of the pre-packaged products). It fulfils also the role of the surveillance body in the sphere of measuring instruments placed on the market as specified by the Law on technical requirements for products and on conformity assessment. On the international level, SMI fulfils the task of the authority for the sphere of the metrological surveillance—Metrological Supervision Authority (MSA).

In addition, all bodies involved in any metrological, or with the metrology related activities belong also to the national metrology system, i.e. the bodies that carry out the verifications of the legally controlled measuring instruments and the official measurements—authorized bodies, the bodies performing the calibrations of the measuring instruments—calibration laboratories, subjects repairing and installing the legally controlled measuring instruments and packers or importers of the pre-packaged products—registered bodies, manufacturers of the measuring instruments, accredited calibration laboratories and certification bodies, research and educational entities, and metrological organizations and associations.

2.6 *International co-operation*

Since the metrology is not isolated or bordered and its principles are applicable both in the Europe and worldwide, the international cooperation in this sphere is important. The metrological legislation is divided into national legislation—governing the national requirements on the measurement and measuring instruments, and the legal rules transposed from the common European legislation; that means the transposition of the directives applicable within the European region into the national legislation—at present, they are transposed most often into an approximation ordinance of the SR government. In order to prevent the creation of the technical barriers to trade and hampering the free movement of goods and services within the common European region and even worldwide, the national metrological legislation must be developed and codified with regard to these requirements.

The Slovak national metrology system is linked to the international metrological organizations, taking an active part in their activities. The most significant representation of the SR is its membership in Metric Convention, the representative of which is International Office for Weights and Measures (BIPM).

In the field of the legal metrology the most important is the membership of the SR in International Organization of Legal metrology (OIML), where the member of the International Committee for Legal Metrology (CIML) is the president of Slovak Office for Standards, Metrology and Testing (SOSMT) and in addition, the OIML secretariat in the SR is managed by SLM. Within OIML, Slovakia administrates the Secretariat of the Technical Committee TC-4-Standards and standard equipment. In addition, Slovakia is represented in several technical committees, where its representatives have a status of their members or observers.

Within the European framework, Slovakia is the active member of WELMEC (European Cooperation in Legal Metrology), where it takes part in the meetings of WELMEC Committee and is represented by the representative of SOSMT department of metrology. The Slovak representatives attend also the meetings of WELMEC working groups.

Slovak Institute of Metrology as NMI is represented also in the European Association of National Metrology Institutes (EURAMET) that coordinates the co-operation of NMIs in Europe in the fields of the metrological research, traceability of measurements on the SI units, international acceptance of the national standards and of the respective Calibration and Measurement Competences (CMC) of their members.

Slovak Institute of Metrology is also the member of the Euro-Asian regional metrological organization (COOMET). The member countries of COOMET are co-operating particularly in the field of the comparison measurements of the national standards. In COOMET, SMU is represented in the Technical committee for quantities of length and angle, in the Technical committee for flow rate, in the Technical committee of the forum for quality, and it governs the COOMET national secretariat.

2.7 *Science and research*

Metrology, as a branch of science is related also with the research in this field. It concerns the basic research in the field of standardization, measurements methods and applied research in the sphere of the metrological control. SMU as National Metrological Institution (NMI) is intensively involved in the metrological research, mainly as the participant in the European Metrology Research Programme (EMRP). Since 2010, SMU was engaged in 17 projects of EMRP.

Since 2008, SLM is competent to carry out research and development in the field of metrology in accordance with Act No. 172/2005 Coll. on organization of state support for research and development. First of all, SLM deals with the applied research in the field of new systems for calibration and verification of the measuring instruments and for implementation of the automation and information technologies for their calibrations and verifications.

2.8 *Objectives of developing the metrology and the national metrology system of the SR*

Nowadays it is evident, that the application of new tendencies in the metrology requires inevitably also further changes in the organizational arrangement of the metrology infrastructure. In order to assure an effective co-ordination of tasks within the national metrology system for future, it is necessary to delegate a part of NMI competencies to SOSMT as to the National Metrological Authority (NMA). This step would create a space for SMU enabling so SMU to focus its endeavors (in co-operation with SLM) to the solution of scientific and research metrological tasks and challenges more intensively.

The main tendencies in question:

1. Application of the metrology in relation to the competency spheres of the individual branches in the SR, i.e. the assurance of the legal rules mutual interrelation, namely of the legal rules in the field of legal metrology with those under the auspices of various ministries.
2. The individual measuring instruments will be replaced in turn by measuring systems interlinked in the networks. They will provide the comprehensive functions, combining various kinds of measurements, and being capable to evaluate lot of measured data (intelligent measuring instruments). The particular constituents of these systems need not be complex, but the sensors, modules and data processing systems will communicate mutually.
3. The measuring instruments and systems will be able to fulfil the tasks that, up to now, could be performed exclusively by the metrological or other specialized subjects; it is the question of autocalibrations and adaptation on the environmental or measurement conditions. The measuring instruments and systems of future may be even able to behave in a relatively intelligent fraudulent manner—there is therefore a danger that such ability could be misused in the business relations, where the measurements for the purposes of price determination are carried out by the legally controlled measuring instruments liable to the periodical metrological control.
4. The work in the field of the legal metrology will be globalized. The requirements on the elimination of the trade barriers will result in the necessity to support the international harmonization, mutual confidence and acceptance among individual bodies and entities.
5. The role of the international organizations (especially that of OIML) will increase and it will be more supported. These organizations will have to respond to the requirements of the legal metrology bodies, bodies of metrology supervision, market surveillance bodies and those of conformity assessment. These organizations will require the uniform conformity assessment procedures for new products and for their subsequent metrological control.
6. It can be presumed that EU will promote regulative functions within the metrology, especially in the fields of public interest, health and safety, public order, environment protection and consumer protection.
7. The increased number of requirements on the metrology means that SMU and SLM will have to be able to react on these requirements, consequently, this situation will ask for the highly qualified staff and the adequate technical equipment.

The goals of development in the field of metrology are based on the principal elements of the current global measurement system (the system of the national regulations in the field of the legal metrology, the unified system of the technical standards in the non-harmonized sphere, the acceptance of the results of measurements based on the traceability to the national standards realizing the units of measurement in accordance with the International system of units SI and the harmonized requirements on the competence of testing and calibration laboratories and certification bodies, as well as the need for both the European and international co-operation and for the interconnection of metrology, research and development). The assurance of the adequate level of enforcing the observance of the obligations set up by the metrological rules with the aim to protect the people—the consumers, is another non-negligible task. The ongoing development of the metrology system is one of the measures that may enhance the economy competitiveness and support the entrepreneurs when developing their production and business.

3 SUMMARY

Finally, we can state that the Slovak Republic operates a functioning national metrology system, accepted also abroad that provides the comprehensive metrological services both for the national economy and people when protecting their rights as of consumers, as well as for the products export and import activities. Naturally, every system requires upkeeping and developing with regard to the use of the latest scientific and technical knowledge and to the use of the state-of-the art technologies. Therefore, it is inevitable to support the scientific and applied research in the field of metrology, to develop the international co-operation in this sphere, where the information concerning the measurements and measuring instruments are exchanged, and the common recommendations for the metrological legislation are prepared. And what is the most important requirement: it is the necessity to assure the education in the metrology field on all the levels of the education system in order to have the skilled professionals to the disposal in this field in future.

REFERENCES

Slovak Office of Standards, Metrology and Testing. 2000. Decree No. 206/2000 Coll. on legal units of measurement. Bratislava: Collection of Laws of Slovak Republic.

Slovak Office of Standards, Metrology and Testing. 2000. Decree No. 207/2000 Coll. on e-marked pre-packages as later amended. Bratislava: Collection of Laws of Slovak Republic.

Slovak Office of Standards, Metrology and Testing. 2000. Decree No. 210/2000 Coll. on measuring instruments and metrological control as later amended. Bratislava: Collection of Laws of Slovak Republic.

Slovak Office of Standards, Metrology and Testing. 2013. Decree No. 419/2013 Coll. on pre-packages. Bratislava: Collection of Laws of Slovak Republic.

Slovak Republic. 1999. Act No. 264/1999 Coll. on technical requirements for products and on conformity assessment and on change and amendment of some acts as amended. Bratislava: Collection of Laws of Slovak Republic.

Slovak Republic. 2000. Act No. 142/2000 Coll. on metrology, and on amendment and supplement of some acts, as later amended. Bratislava: Collection of Laws of Slovak Republic.

Slovak Republic. 2005. Act No. 172/2005 Coll. on organization of state support for research and development as later amended. Bratislava: Collection of Laws of Slovak Republic.

Slovak Republic. 1999. Governmental Ordinance No. 399/1999 Coll. specifying details on the technical requirements for non-automatic weighing instruments as amended. Bratislava: Collection of Laws of Slovak Republic.

Slovak Republic. 2005. Governmental Ordinance No. 294/2005 Coll. on measuring instruments. Bratislava: Collection of Laws of Slovak Republic.

Production Management and Engineering Sciences – Majerník, Daneshjo & Bosák (Eds)
© 2016 Taylor & Francis Group, London, ISBN: 978-1-138-02856-2

Assessment of wood fire protection effectiveness using blocking gel Firesorb

B. Štefanický & P. Poledňák
Faculty of Safety Engineering, VŠB—Technical University of Ostrava, Ostrava, Czech Republic

P. Rantúch & K. Balog
Faculty of Materials Science and Technology, Slovak University of Technology in Bratislava, Trnava, Slovak Republic

ABSTRACT: In the article, a procedure for assessing a protective layer of fire protection/blocking gel Firesorb with concentrations of 1%, 2% and 3% in protection of wood specimens exposed to heat fluxes is described. The purpose of testing was to assess the influence of heat flux on ignition time, the determination of both heat release rate and heat release in the course of exposure of the specimens to heat, specific mass loss rates of unprotected and protected wood specimens and total smoke release.

1 INTRODUCTION

The assessment of a protective layer of fire protection/blocking gel Firesorb applied to specimens of common spruce (Picea abies L.) was carried out on the basis of values measured with the use of a cone calorimeter, in accordance with the testing procedure according to ISO 5660-1: 2002 (ISO 5660-1: 2002). In the course of testing, specimens were subject to thermal radiation of a heat flux density of $50 \, kW.m^{-2}$. Cone calorimeters belong to the best and most commonly used devices in the area of fire testing. At present, they are mainly used for the determination of heat release rate (henceforth referred to as HRR), Mass Loss Rate (MLR), CO yield, Total Heat Release (THR) and calorific values of materials (Martinka et al. 2013).

2 FIRESORB MO

Firesorb concentrate consists of a superabsorbent polymer that increases its volume when being in contact with water and produces a heat resistant gel forming an excellent fire barrier (Firesorb 2010; Chromek et al. 2010). On a burning material, the Firesorb gel has, thanks to its high density, a uniform cooling effect, which reduces the rate of evaporation. This cooling effect is crucial to fires of very high temperatures to which the application of a stream of water would be ineffective owing to almost immediate water evaporation.

The gel impedes access of oxygen to the burning material and has a smothering effect on it as well (Firesorb 2010; Firesorb 2013). An aqueous solution of the fire protection/blocking Firesorb is studied intensely especially for the purpose of use both in suppression of fires involving solid flammable materials and as fire retardant for forest fires and construction protection (Gimenéz et al. 2013, Grand 2013, Burke 2002, FP Innovations 2013).

3 EXPERIMENT DESCRIPTION

Aqueous solutions with 1 vol%, 2 vol% and 3 vol% of fire protection/blocking gel Firesorb were prepared in 1000 ml volumetric flasks one hour before experimental measurements (Firesorb MO, batch number 4500275632). Common spruce (Picea abies L.) specimens had a size of 100 mm × 100 mm × 20 mm, average density of specimens was $460.26 \pm 4.15 \, kg \cdot m^{-3}$; absolute moisture content of specimens was 6.7%. The test specimens were dipped in the prepared gel for 1 minute.

The time to ignition, heat release rate, specific mass loss rate and total smoke release were determined with the use of a cone calorimeter in accordance with the testing procedure according to ISO 5660-1: 2002 (ISO 5660-1: 2002), at a heat flux density of $50 \, kW \cdot m^{-2}$. For each concentration, six measurements were made and average values are taken as resulting values.

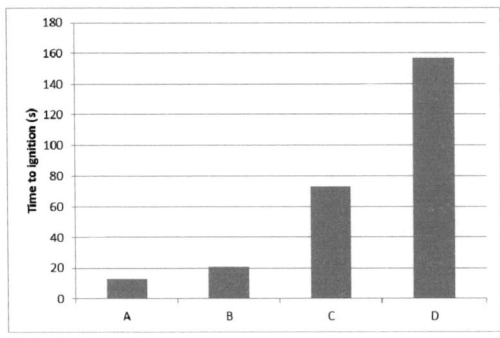

Explanatory notes:
A – spruce wood specimen
B – spruce wood specimen protected with 1 vol% aqueous solution of Firesorb
C – spruce wood specimen protected with 2 vol% aqueous solution of Firesorb
D – spruce wood specimen protected with 3 vol% aqueous solution of Firesorb

Figure 1.   Times to ignition of test specimens.

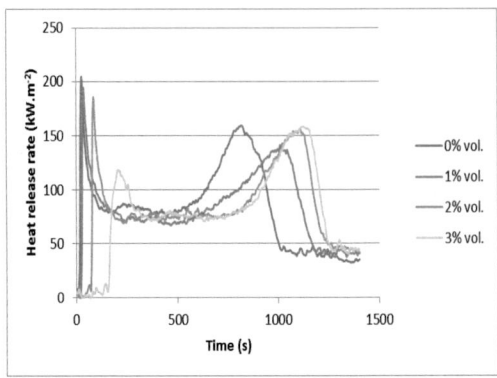

Explanatory notes:
A – spruce wood specimen
B – spruce wood specimen protected with 1 vol% aqueous solution of Firesorb
C – spruce wood specimen protected with 2 vol% aqueous solution of Firesorb
D – spruce wood specimen protected with 3 vol% aqueous solution of Firesorb

Figure 2.   Rates of heat release from test specimens.

4   RESULTS AND DISCUSSION

The times to ignition of test specimens of spruce (Picea abies L.) are illustrated in Figure 1. The time to ignition of the specimen of unprotected spruce wood was 13 s, that of the specimen protected with a 1 vol% aqueous solution of Firesorb was 21 s and that of the specimen protected with a 2 vol% aqueous solution of Firesorb was 78 s. The specimen protected with a 3 vol% aqueous solution of the fire protection/blocking gel was ignited (and burned further) 157 s after the beginning of the test.

The fire protection/blocking gel in a concentration of 3 vol% provided more than an eleven-fold increase in protection level in comparison with the unprotected specimen. If comparing the specimen protected with the gel in a concentration of 1 vol% and that in a concentration of 3 vol%, the time of protection (time to ignition) increases more than six-fold; if comparing the specimen protected with the gel in a concentration of 2 vol% and that in a concentration of 3 vol%, the time of protection is more than double. This fact can be explained by the density of the gel at individual concentrations and by the thickness of the applied protective layer (gel with a higher concentration has a higher density and can be applied more easily and more thickly).

The rates of heat release from the test specimens are illustrated in Figure 2. The rate of heat release from the specimen unprotected with the gel and that from the specimen protected with the gel in a concentration of 1 vol% show a small difference. The maximum heat release (first peak) corresponds to the time shortly after specimen ignition, when the whole surface of the test specimens burns. A rather significant time lag in the increase in heat release rate is obvious in the case of the specimen protected with the gel in a concentration of 2 vol% (in contrast to the specimen unprotected with the gel, the time lag is 45 s).

The first peak of the curve of time dependence of HRR for the specimen protected with the gel in a concentration of 3 vol% is, in contrast to the other specimens, considerably shifted (max HRR at time of 205 s, max HRR of 119 kW · m^{-2}). Here, cooling effects of the gel fire extinguishing agent manifest themselves fully in specimen protection. During the burning of the test specimen, a charred layer, which slows further warming and thermal decomposition of the specimen, is gradually formed on its surface. Both the problem of charred layer formation on the surface of wood and wood materials in the course of fire and its influence on further thermal decomposition are described in detail in (Kačíková 2009). The second peak represents the maximum rate of heat release from the test specimens after total gel degradation and after top charred layer deterioration (glowing of the specimen). The comparison of the specimens shows that the use of the fire extinguishing agent Firesorb in a concentration of 1 vol% affects the time to ignition and heat release rate only minimally. With increasing concentration of the applied gel, the times to ignition and the rates of heat release from the test specimens increase evidently. This fact may be attributed to the cooling effect of the gel with a higher concentration and thickness of the gel layer. A comparison of values of maximum heat release rate in the phase of flameless burning of the char residue (charred layer on the surface of the specimen) after complete degradation of the gel shows that the heat release rate at this flameless burning (second peak) exhibits minimum differences between individual specimens (with the exception of the specimen protected with the gel in a concentration of 1 vol%, which can be explained by the fact that the 1 vol% aqueous solution of Firesorb provided lower level of protection, the protective layer was degraded more rapidly, the second peak occurred earlier. The lower maximum heat release rate can be explained by the shorter time of protection by the 1 vol% gel and thus also by the shorter time of action of thermal radiation and the shorter time of warming the specimen). The difference lies in the time of maximum rate. The growing concentration of Firesorb affected considerably the increasing time of maximum heat release rate.

The total heat releases in the course of the experiment from individual specimens are given in Figure 3.

When comparing the specific mass loss rates of the specimens, as shown in Figure 4, we can state that with increasing concentration of the fire protection/blocking gel applied to the specimens, the mass loss rate increases as well. This fact can be explained by mechanism of action of the gel cooling effect, when water is gradually released as a result of thermal radiation. After gel degradation, the curves of dependence of mass loss on time for individual specimens are approximately identical. A comparison of values of specific mass loss rate in the phase of flameless burning of the char residue (glowing) also shows merely small differences. The time difference between the maximum mass loss rates of the specimens can be attributed to the concentration of the applied gel (density and thickness).

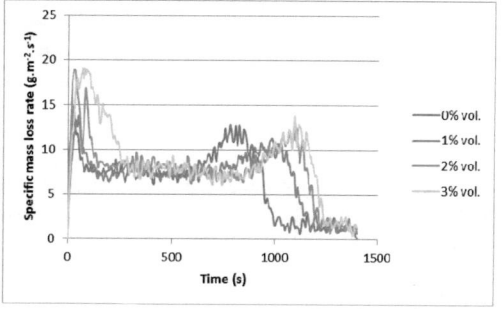

Explanatory notes:
A – spruce wood specimen
B – spruce wood specimen protected with 1 vol% aqueous solution of Firesorb
C – spruce wood specimen protected with 2 vol% aqueous solution of Firesorb
D – spruce wood specimen protected with 3 vol% aqueous solution of Firesorb

Figure 4.   Specific mass loss rates of test specimens.

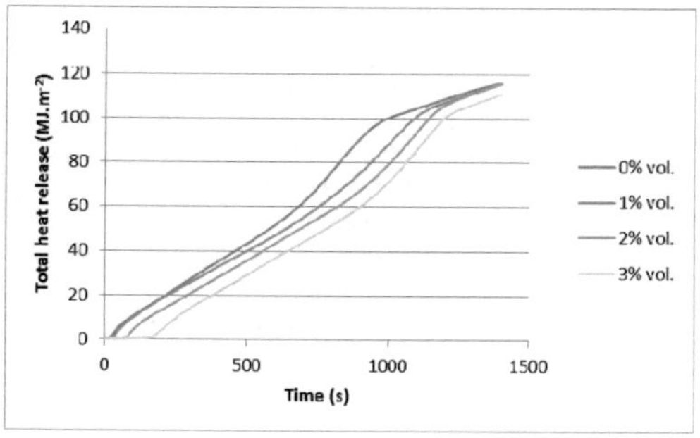

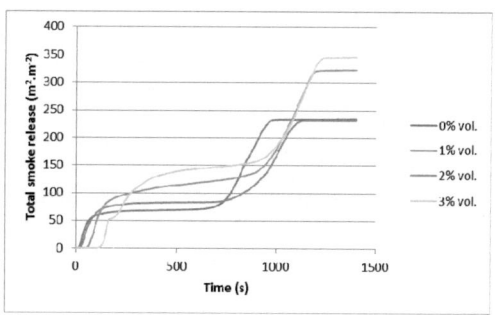

Explanatory notes:
A – spruce wood specimen
B – spruce wood specimen protected with 1 vol% aqueous solution of Firesorb
C – spruce wood specimen protected with 2 vol% aqueous solution of Firesorb
D – spruce wood specimen protected with 3 vol% aqueous solution of Firesorb

Figure 3.   Total heat releases from test specimens.

Explanatory notes:
A – spruce wood specimen
B – spruce wood specimen protected with 1 vol% aqueous solution of Firesorb
C – spruce wood specimen protected with 2 vol% aqueous solution of Firesorb
D – spruce wood specimen protected with 3 vol% aqueous solution of Firesorb

Figure 5.   Total smoke releases from test specimens.

In Figure 5 the dependence of total smoke release on time is illustrated. As already mentioned above, the gel density and the thickness of the applied layer increase with increasing gel concentration. When exposed to the heat flux, water is released from the applied gel until the time of gel degradation and initiation of burning of the specimen. The released water vapour is captured by a smoke meter in the extraction system and taken as released smoke. The considerable difference between the smoke releases from individual specimens in the initial phase of the experiment can be regarded as action of the smothering effect of the gel. In the course of burning of the specimens, a charred layer is gradually formed on their surface and the layer slows further warming and thermal degradation of the specimens. In this phase, the smoke releases from individual specimens have only a slightly rising character. A considerable increase in the smoke release occurs in the phase of flameless burning after top charred layer deterioration (glowing of the specimen).

5 CONCLUSION

In the submitted article, the effects of the protective function of a new fire protection/blocking gel Firesorb on protection of wood constructions were illustrated. Whereas the 1 vol% and 2 vol% aqueous solutions of Firesorb are, according to the producer [3], designed primarily for the suppression of fires involving flammable solids, the 3 vol% aqueous solution of Firesorb is designed for the protection of constructions. The advantages of 2 vol% aqueous solution of Firesorb were confirmed by experimental tests concerning a comparison of the fire extinguishing efficiency of this fire protection/blocking gel and that of water (Štefanický & Balog 2014).

The protective function of the mentioned fire protection/blocking gel Firesorb was tested with the use of a cone calorimeter. Presented data were acquired at the density of heat flux of $50 \, kW \cdot m^{-2}$. It follows from the obtained data that the application of 1 vol% aqueous solution of Firesorb does not affect substantially the time of protection of constructions. The 2 vol% aqueous solution of Firesorb increases considerably the time to ignition; the cooling and smothering effects grow. In the course of testing, the 3 vol% aqueous solution of fire protection/blocking gel Firesorb exhibited excellent protective properties with very good cooling and smothering effects. We attribute this fact to the density and thickness of the applied gel layer.

REFERENCES

Burke, K. 2002. Evaluating the need to add a water based fire retardant polymer gel to orange county fire rescue departments arsenal for fire protection. 2002. 32p. [online]. URL: http://www.usfa.fema.gov/pdf/efop/efo35307.pdf [cited 2013–07–13].

Chromek, I., Benedik, V., Šmigura, M. & Hlaváč, P. 2010. Ochrana materiálov na báze dreva pred ohňom gélovými prípravkami (Fire Protection of Wood-Based Materials with Gel Preparations). In: *ACTA FACULTATIS XYLOLOGIAE ZVOLEN 52(2)*. Zvolen: Technická univerzita vo Zvolene. 2010. pp. 81–90. ISSN 1336−3824.

Firesorb—publicity CD [CD-ROM]. Krefeld (SRN): Evonik Industries, 2010-. [cited 2015–02–06].

Firesorb® Bluswater besparend, Gel-vormend Brandblusadditief. [online]. URL: http://www.nater.nl/index.php?p=27 [cited 2013–03–08].

Firesorb More extinguishing power for water. [online]. URL: http://www.fireretardant101.info/uploads/5/4/4/9/5449109/firesorb_mo_info.pdf [cited 2013–03–08].

FPInnovations, Home. [online]. URL: http://www.fpinnovations.ca/Pages/home.aspx#.UoDJLGVxzIE [cited 2013–08–25].

Gimenéz, A. et al. 2013. Long-term forest fire retardants: a review of quality, effectiveness, applications and environmental considerations. In: *International Journal of Wildland Fire*. 2004. No.13. pp. 1–15. URL: http://www.publish.csir.au/journals/ijwf [cited 2013–04–12].

Grand, A.F. 2013. A*n Investigation of the Effectiveness of Fire Resistant Durable Agents on Residential Siding Using an ICAL-Based Testing Protocol (NIST GCR-00-792)*. 2000. 152p. [online]. URL: http://fire.nist.gov/bfrlpubs/fire00/art027.html [cited 2013–06–23].

ISO 5660–1: 2002: Heat release, smoke production and mass loss rate—Part 1: Heat release rate (cone calorimeter method).

Kačíková, D. & Makovická-Osvaldová, L. 2009. Rýchlosť odhorievania dreva z rôznych častí stromu vybraných ihličnatých drevín (Mass Burning Rate of Various Parts of Trees of Selected Conifers). In *Acta Facultatis Xylologiae*. 2009, vol. 51(1). pp. 27–32.

Martinka, J., Balog, K. & Chrbet, T. 2013. Nové trendy posudzovania požiarneho rizika lignocelulózových materiálov kónickým kalorimetrom (New Trends in Assessment of Fire Risk of Lignocellulosic Materials with the Use of Cone Calorimeter). In II. Medzinárodná vedecká konferencia *Advances in Fire & Safety Engineering 2013*. Žilina: Fakulta špeciálneho inžinierstva ŽU v Žiline, 2013, pp. 57–64.

Štefanický, B. & Balog, K. 2014. Experimentálne porovnanie hasiacich vlastností protipožiarneho gélu FIRESORB® s koncentráciou 2 % a vody (Experimental Comparison of Fire Extinguishing Properties of Fire Protection Gel FIRESORB® with a Concentration of 2% and Those of Water). In Zborník príspevkov z konferencie s medzinárodnou účasťou *Rozvoj teorie bezpečnostních rizik a tvorba krízových scenárov*, APZ Bratislava, pp. 260–267. ISBN 978-80-8054-604-5.

Production Management and Engineering Sciences – Majerník, Daneshjo & Bosák (Eds)
© 2016 Taylor & Francis Group, London, ISBN: 978-1-138-02856-2

Development trends in forestry mechanization at the TU Zvolen

V. Štollmann
Technical University in Zvolen, Zvolen, Slovak Republic

Š. Ilčík
ILČÍK, Banská Bystrica, Slovak Republic

ABSTRACT: The paper describes a new approach towards the construction of machines. According to this new conception, the machines are called RELAZ devices. The name RELAZ was derived from Rekuperačné lanové zariadenie (Recuperative Cable Device). At present, the name RELAZ has a wider meaning and designates the machinery whose technical design is based on the use of a traction accumulator of energy, which enables precise power economy of this machinery. The advantages of this new approach towards machine construction become evident mainly in economic and ecological spheres. Several patents, utility models and prizes have been obtained. The paper presents the latest results of research work carried out in this field at the Technical University in Zvolen between the years 2006–2015.

1 RELAZ DEVICES

Recuperative cable devices are part of special RELAZ devices, which is world-new energy saving and ecologically friendly technical designs of cable devices. They are equipped with a traction accumulator of energy and work as hybrid devices. The traction accumulator stores energy, which was previously wasted inside a device or in its surroundings. This accumulated energy is repeatedly transformed into useful work. The energy system of the RELAZ devices, which employs a traction accumulator of energy, has universal use, and can be effectively used in any machines and devices. The conceptual technical designs of RELAZ devices are industrially and legally protected by means of several patents, utility models, and home and foreign patent applications (Štollmann, V. & Ilčík, Š. & Suchomel, J. 2010).

2 RECUPERATIVE CABLE DEVICES

2.1 Principle of operation of recuperative cable devices

Recuperative cable devices operate on the principle of energy saving and ecologically clean propulsion, using mountain energy—potential energy of trees growing in mountain areas. The basic principle of recuperative cable devices is the accumulation of gravitational energy of a cable carriage and load obtained during the phase of gravitational skidding down the slope. The energy is then used to drive the cable device. The basic principle of recuperative cable devices is the fact that when the carriage with load (Figure 1) moves down the slope, the surplus energy is used to charge the traction accumulator of energy, which is then used as a source of energy for moving the empty carriage (Figure 2) along the skyline back up the slope.

The basic operational principle of a recuperative cable device is demonstrated on the comparison of the amount of mechanical work in the phase of gravitational skidding of a cable carriage with load down the slope, and in the phase of pulling up the empty carriage back up the slope.

In this part of the solution, we only present the basic calculation without regard to its completeness, where we take into account the amount of the resulting force which causes gravitational skidding of the carriage with load down the slope, and the

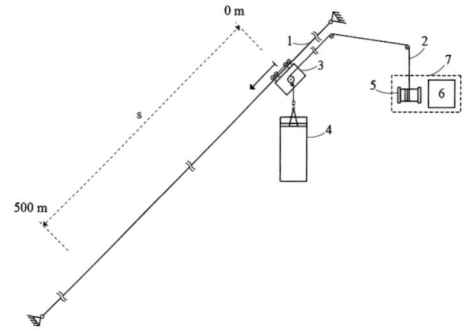

Figure 1.   Phase of gravitational skidding of a cable carriage down the slope.

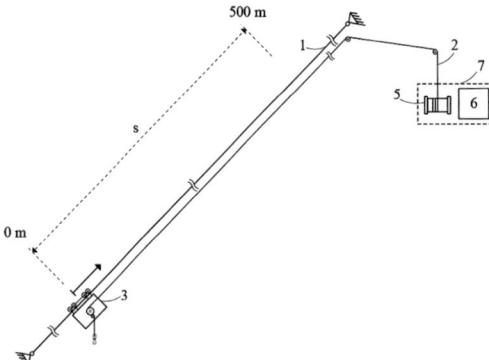

Figure 2. Phase of pulling an empty cable carriage back up the slope.

amount of force which is necessary for pulling the empty carriage up the slope, the angle of gravitational skidding, the path of gravitational skidding or pulling up. In our calculation we will also take into account the traction coefficient of friction $f = 0,007$ (Dukeľskij 1966).

2.2 Calculation of work in the phase of gravitational skidding of a cable carriage with load down the slope

The work of a cable carriage with load A_{vn} in the phase of gravitational skidding of a cable carriage with load down the slope (Example No. 1) will be calculated using Formula 1.

Example No. 1
Assuming that:

- mass of the carriage with load mvn = 1600 kg
- gravitational acceleration g = 9.81 m · s⁻²
- traction coefficient of friction f = 0.007
- angle of gravitational skidding α = 30°
- path of gravitational skidding s = 500

$$A_{vn} = [m_{vn}.g.(\sin\alpha - f.\cos\alpha)].s \qquad (1)$$

$$A_{vn} = [1600.9,81.(\sin 30° - 0,007.\cos 30°)].500$$
$$A_{vn} = 7752,84.500 = 3876420J = 3876,42kJ$$

The amount of work A_{vn} in the phase of gravitational skidding of a cable carriage with load is 3876420 J = 3876.42 kJ. This energy is obtained during gravitational skidding of a carriage with load down the slope and is accumulated in the traction energy accumulator of the recuperative cable device.

2.3 Calculation of work needed for pulling the empty cable carriage back up the slope

The work of the empty cable carriage A_v in the phase of pulling the empty cable carriage back up

the slope (Example No. 2) will be calculated using Formula 2.

Example No. 2
Let us assume that:

- mass of the carriage mv = 300 kg
- gravitational acceleration g = 9.81 m · s⁻²
- traction coefficient of friction f = 0.007
- angle of gravitational skidding α = 30°
- path of pulling up s = 500 m

$$A_v = [m_v.g.(\sin\alpha + f.\cos\alpha)].s \qquad (2)$$

$$A_v = [300.9,81.(\sin 30° + 0,007.\cos 30°)].500 =$$
$$A_v = 1489,34.500 = 744670J = 744,67kJ$$

The amount of work A_v in the phase of pulling the empty carriage back up the slope is 744670 J = 744.67 kJ. This energy is taken from the traction accumulator of the recuperative cable device and is used for pulling the empty cable carriage up the slope.

2.4 Difference between works

The above examples show that the amount of work of a cable carriage with load A_{vn} in the phase of gravitational skidding of a cable carriage with load down the slope makes 3876.42 kJ, and the amount of work of an empty carriage A_v in the phase of pulling the empty carriage back up the slope is 744.67 kJ. The difference in work ΔA is calculated using Formula 3.

$$\Delta A = A_{vn} - A_v \qquad (3)$$

$$\Delta A = 3876,42 - 744,67 = 3131,75kJ$$

The difference in the amount of work ΔA makes 3131.75 kJ. This simplified model example shows that the energy accumulated in the traction accumulator of the recuperative cable device is sufficient to pull the empty cable carriage back up the slope. Moreover, the difference in work represents the energy accumulated in the traction accumulator that the recuperative cable device can use for other working activities or covering its other energy demands.

3 ENERGY SYSTEM OF RELAZ DEVICES

The energy system of RELAZ devices with a traction accumulator of energy can be effectively used in any machines and devices which use the energy from the traction accumulator for their propulsion. For this reason, RELAZ devices are at present characterized as energy saving special devices with a power accumulator of energy, and

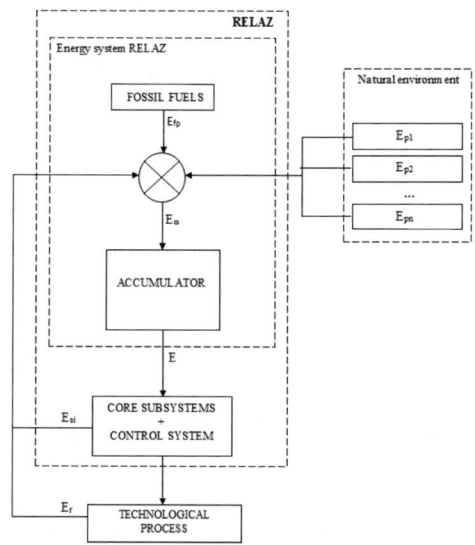

Figure 3.   Energy system of RELAZ devices.

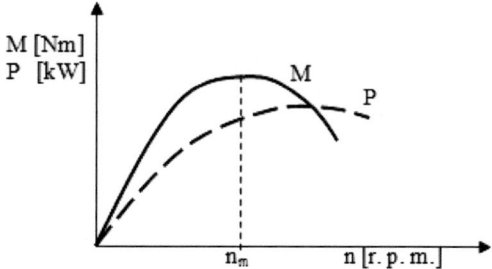

Figure 4.   Speed map of a combustion engine.

are primarily oriented towards the utilization of alternative sources of energy.

The research ground for RELAZ is a new approach towards the construction of machines, primarily their energy system. Characteristic features of this new approach in the sphere of RELAZ devices are the emphasis on systematism and precise energy management during operation. An important element of RELAZ energy system is a traction accumulator of energy, which can continuously store energy, and from which energy can be continuously taken. The basic operation principle of RELAZ devices is, in a simplified way, shown in Figure 3.

The traction accumulator of energy system of RELAZ devices stores: E_{fp}—energy obtained from fossil fuels; E_{si}—inner dissipated energy of the machine; E_r—recovered energy, that is the energy which is generated during technological process; E_{p1}—solar energy, E_{p2}—wind energy, E_{pn}—energy from biomass, etc., that is free energy from natural resources, and is an additional source of energy which from a long-term point of view can contribute to large savings of fossil fuels. The amount of energy stored in the accumulator is calculated using Formula 4.

$$E_{\omega} = E_{fp} + E_{si} + E_r + E_{p1} + \cdots + E_{pn} \qquad (4)$$

On the basis of the relationship between energy E_{ω}, which enters the accumulator and energy E, which is taken from the accumulator (Figure 3.), three basic operating modes can occur during the

operation of RELAZ devices; operating mode *RP1*, operating mode *RP2*, and operating mode *RP3*. In operating mode *RP1* the energy from alternative sources is sufficient to maintain the accumulator in a charged condition; if there is surplus energy, the accumulator is charged. Energy from fossil fuels is not needed. In operating mode *RP2* the accumulator is maintained in a charged condition by energy from fossil fuels and at the same time from alternative energy resources. In operating mode *PR3* the RELAZ device covers its energy demands from fossil fuels, however, more efficiently in optimal working conditions of the combustion engine. The effectiveness of RELAZ devices in operating mode *RP3* can be calculated using Formula 5.

$$\eta = \frac{A_u}{E_{\omega}} = \frac{A_u}{E_{fp} + E_{si} + E_r + E_{p1} + \cdots + E_{pn}} = \frac{A_u}{E_{fp}} \quad (5)$$

If a conventional machine with a combustion engine is equipped with a RELAZ-type energy system, fuel consumption will decrease approximately η 2.5–4.0 times due to the increase in effectiveness. This phenomenon can be explained by the fact that RELAZ energy system working in operating mode *RP3* uses the combustion engine for charging the accumulator and not for the propulsion of the device. The engine can then operate at the optimal point of its speed map, at its normal speed n_m, i.e. s with minimum fuel consumption.

Figure 4 shows a speed map of a combustion engine, rated speed n_m, engine torque M, and the performance P of the engine.

In working modes *PR1* and *PR2*, effectiveness η with respect to energy E_{fp} drawn from fossil fuels substantially increases due to savings from the use of alternative energy resources $E_{si}, E_r, E_{p1}, E_{p2}, E_{pn}$.

The principal difference between RELAZ devices and conventional machines lies in the fact that RELAZ devices contain an accumulator of energy, which enables precise energy management during operation. The accumulator stores energy which was previously wasted in the inside of the machine or its surroundings. This accumulated energy is repeatedly transformed into useful work.

The presented conceptual design of the energy system of RELAZ devices provides effective work of recuperative cable devices even in flat areas or in conditions with any angles of slope. Moreover, the energy system of RELAZ devices can be effectively used with any machines and devices. It is a universal ecological concept of working machines, which can also be used in combination with combustion engines.

4 RESEARCH AND DEVELOPMENT IN THE SPHERE OF RELAZ DEVICES

Research work in the sphere of energy saving and ecologically friendly special recuperative cable devices began at the Technical University in Zvolen in the year 2006 within the framework of grant VEGA 1/3526/06. In the years 2010–2012, applied research in the field of recuperative cable devices followed within the framework of the project RELAZ I Applied Research and Development of Special Cable Devices—Special Cable Carriage (ITMS project code 26220220036), and the project RELAZ II Applied Research and Development of Special Cable Devices—Special Flywheel (ITMS project code 26220220035). These projects were co-financed by the ministry of Education, Science, Research and Sports of the Slovak Republic— from structural funds of the EU, the European Regional Development Fund—operational programme Research and Development. A functional model of a special cable carriage was made within the framework of the project RELAZ I. The special cable carriage allows recuperative cable devices to effectively utilize gravitational energy of a skidded carriage with load down the slope, because the skidded load is in full horizontal suspension, or in full vertical suspension, and that is why it does not touch the ground. The special cable carriage enables work in overlapping time and thus increases work productivity. Another advantage of the special cable carriage is an ecological way of skidding loads in full horizontal and full vertical suspension. A functional model of a special flywheel was made within the framework of the project RELAZ II. The special flywheel serves as a traction accumulator for recuperative cable devices.

In the working period 2013–2015, the research of recuperative cable devices has been going on within the framework of project VEGA 1/0931/13, entitled Fundamental Research of New Principles of Cable Carriages for the System of RELAZ Devices. At present, the problem of RELAZ devices is being implemented into educational process under the support from the Ministry of Education, Science, Research and Sports within the framework of the project KEGA 011TU Z-4/2015, entitled New Forms and Methods of Education in Forestry Mechanization.

Research and development of recuperative cable devices is so perspective that the Technical University in Zvolen was awarded Ján Bahýľ Prize 2010 for extraordinarily valuable, industrially and legally protected technical solution in the sphere of special cable devices—Patent SK 287441 Mechanical Recuperative Cable Device with a Flywheel "Gyrocableway". The Technical University in Zvolen is the first university in the world which began research and development of energy saving and ecologically friendly special recuperative devices.

On the occasion of the award ceremony for Ján Bahýľ Prize 2014 for extraordinarily valuable, industrially and legally protected Slovak solution, the originators of the design of recuperative cable devices were presented "Certificate of Merit" by the chairman of the Industrial Property Office of the Slovak Republic for their patent SK 288179, entitled Recuperative Cable Device with Fuel Cells.

In connection with the research of recuperative cable devices, the Technical University in Zvolen co-operates with Izhevsk State Technical University of M. Kalašnikov (IžGTU), Udmurt Republic, Izhevsk, Russian Federation. Specifically, it is the co-operation with the Department of Mechatronic Systems of the Faculty of Quality Management, Izhevsk State Technical University of M. Kalašnikov. The Technical University in Zvolen also successfully co-operates with North-Eastern Federal University (SVFU) in Yakutsk, Yakutia, with which they signed a contract on co-operation in October 2012.

5 CONCLUSION

RELAZ devices significantly contribute to the protection of the environment, and support exploitation of alternative energy resources. The potential for the application of these energy saving and environmentally friendly devices has world-wide significance and substantiation. RELAZ devices can be used for transportation of various loads in forest management, agriculture, building industry and in other branches of industry where it is necessary to transport loads effectively.

An industrial advantage of recuperative cable devices, compared to conventional cable devices, is that these devices use part of surplus kinetic and potential energy of a skidded cable carriage with load, which in a standard case is wasted in the brake system of cable devices.

The knowledge presented in this article is part of the solution of the project VEGA No. 1/0931/13, entitled Fundamental Research of New Principles of Cable Carriages for the System of RELAZ devices.

We would like to express our special thanks for the support of our research activities and the realization of project RELAZ I Applied Research and Development of Special Cable Devices—Special Cable Carriage (ITMS project code 26220220036), and the project RELAZ II Applied Research and Development of Special Cable Devices—Special Flywheel (ITMS project code 26220220035) to the Ministry of Education, Science, Research and Sports, the Agency of the Ministry of Education, Science, Research and Sports for structural funds of the EU, the European Fund for Regional Development, and co-financing from the EU resources.

REFERENCES

Božek, P. 2007. Virtual Control Systems. *Strategic Management in Manager's Practice.* pp. 612–618, Trnava: Tripsoft.

Dukeľskij, A.I. 1966. Forest Cableways and Cable Cranes. 481. *Mašinostrojenie.* Moskva.

Fujinaka, K. 1998. *Carrying Device.* Publication Number 10250986, IPC: B66C 21/0, Japan Patent Office, Japan.

Ilčík, Š. 2009. Research of Operating Principles of Recuperative Cable Devices, *Dissertation thesis,* Technical University in Zvolen, pp. 242.

Ilčík, Š., Štollmann, V. & Suchomel, J. 2010. *Thermoelectric Recuperative Cable Device.* Patent No. 287413, Int. Cl.: B66D 1/00, B66C 21/00, B61B 7/00, A01G 23/00, PP 80–2007, 08.06.2007, Industrial Property Office of the Slovak Republic, Slovak Republic.

Ilčík, Š., Štollmann, V. & Suchomel, J. 2012. *Method of Independent Drums Control in the System of Cableway Carriages in Superimposed Time: World Intellectual Property Organization (WIPO).* Patent Application, International Application Number: PCT/SK2011/050018, International Filing Date: 15. November 2011, International Publication Number: WO 2012/067591 A1.

Ilčík, Š., Štollmann, V. & Suchomel, J. 2013. *Method of Independent Drums Control in the System of Cableway Carriages in Superimposed Time: United States Patent and Trademark Office.* Patent Application, Int. Cl.: B61B 7/02, U.S. Cl.: CPC—B61B 7/02, USPC – 104/112. USA. Pub. No.: US 2013/0263756 A1.

Harashima, T. & Kawakami, T. 2007. *Guide Pulley.* Publication Number 2007031126, IPC: B66D 1/54, B66D 1/36, Applicant: Shin Caterpillar Mitsubishi LTD, Japan Patent Office, Japan.

Štollmann, V., Ilčík, Š., Suchomel, J. & Šmál, P. 2012. *Recuperative cableway system with fuel cells. World Intellectual Property Organization (WIPO).* Patent Application. Number: PCT/SK2011/050021, International Filing Date: 28. November 2011, International Publication Number: WO 2012/074494 A1.

Štollmann, V., Ilčík, Š., Suchomel, J. & Šmál, P. 2013. *Recuperative Cableway System with Fuel Cells: United States Patent and Trademark Office.* Patent Application. Int. Cl.: H02J 7/34. U.S. Cl.: CPC—H02J 7/34. USPC—307/154. USA. Pub. No.: US 2013/0241315 A1.

Štollmann, V. & Ilčík, Š. 2009. Recuperative Cable Devices, *Scientific Paper,* Technical University in Zvolen, Zvolen, pp. 76.

Štollmann, V. & Suchomel, J. 2009. *Recuperative Cable Device for Timber Harvesting.* Patent No. 286944. Int. Cl.: B66C21/00, B66B7/00, A01G23/00, PP 53–2006. Industrial Property Office of the Slovak Republic, Slovak Republic.

Štollmann, V., Ilčík, Š. & Suchomel, J. 2010. *Mechanical Recuperative Cable Device with a Flywheel—Gyrocableway.* Patent No. 287441. Int. Cl.: B66D 1/00, B66C 21/00, B61B 7/00, A01G 23/00, PP 0108–2007. Industrial Property Office of the Slovak Republic, Slovak Republic.

Štollmann, V., Ilčík, Š. & Suchomel, J. 2010. *Hydraulic Recuperative Cable Device.* Patent No. 287411. Int. Cl.: B66D 1/00, B66C 21/00, B61B 7/00, A01G 23/00, PP 103–2007. Industrial Property Office of the Slovak Republic, Slovak Republic.

Štollmann, V., Ilčík, Š. & Suchomel, J. 2010. *Pneumatic Recuperative Cable Device.* Patent No. 287412, Int. Cl.: B66D 1/00, B66C 21/00, B61B 7/00, A01G 23/00, PP 109–2007. Industrial Property Office of the Slovak Republic, Slovak Republic.

Štollmann, V. & Ilčík, Š. 2007. Gyro-cableways—Theory of Flywheels. *Final Report of partial research task No. 1/3523/06,* Technical University in Zvolen, SIG. SLDK VSL 275, Zvolen, pp. 15.

Walters, V. 2000. *Shiftable tail-block logging skyline.* Publication Number 6145679, IPC: B66C 21/00, Applicant: Walters Victor, United States Patent and Trademark Office.

Wang, J. & Niu, Z. 2013. *Curve transporting ropeway system.* Publication Number: 102079308, IPC: B61B 10/02, B61B 7/00. Applicant: Henan Xudelong Electric Power Equipment Co. Ltd., State Intellectual Property Office of the People's Republic of China.

Yang Q. 2007. *Antislip device for manned ropeway main driving wheel steel cable.* Publication Number: 101007534, IPC: B61B 12/12. State Intellectual Property Office of the People's Republic of China.

Yamamoto, H. 2002. *Endless Wire Rope.* Publication Number: 3979777, IPC: D07B 1/10, D07B 1/18, Applicant: Tokyo Seiko Co., LTD, Japan Patent Office, Japan.

Production Management and Engineering Sciences – Majerník, Daneshjo & Bosák (Eds)
© 2016 Taylor & Francis Group, London, ISBN: 978-1-138-02856-2

Analysis of productivity in enterprises automotive production

Z. Tekulová & M. Králik
Faculty of Mechanical Engineering, Slovak Technical University, Bratislava, Slovak Republic

Z. Chodasová
Institute of Management, Slovak University of Technology, Bratislava, Slovak Republic

ABSTRACT: Present is represented by a constant pressure on the competitiveness of enterprises. Increasing of overall productivity of the company is one of the factors enabling the creation of added value and achieving long-term growth of the company. Apart from introducing new business systems, where the accents of maximum utilization and long term and circulate property and the reduction of inventories dominate, the firm must establish an internal reporting system that allows tracking of productivity and its effective management. The paper analyzes the productivity of the three major carmakers operating in Slovakia. The aim of analysis is a complex total view of the overall productivity, as well as the decomposition of productivity and factors as the impact of current assets, of the labor force and current assets represented by material and its effectiveness. The result of the analysis is to compare the impact of various factors on the overall productivity.

1 CAR PRODUCTION IN SLOVAKIA

The automotive industry is a key industry and economic pillar in several countries of Central and Eastern Europe. This sector is one of the main sources of FDI in the region over the past 20 years. Car manufacturers use educated, productive and relatively cheap labor and a quality connection to Western European markets as well as a convenient location to export eastwards. Slovakia is currently one of the important centers of global automotive industry, and produces the highest number of cars per capita in the world. This position is mainly due to the presence of modern plant three automotive companies: Volkswagen Slovakia, as (Bratislava); PCA Slovakia llc. (Peugeot Citroen Trnava) and KIA Motors Slovakia llc. (Žilina), and global supply undertakings. Development of the automotive industry in Slovakia and the integration between global automotive centers began in the 90's, when a German car company Volkswagen AG launched a factory for car production near Bratislava. This important step will help drive the Slovak economy after the demise of former Community CMEA markets and the end of military production. By accessing automotive company Volkswagen Slovakia began to build the supply chain, which attracted further investments into the country. Slovakia follows a new path in the development of industrial production in the automotive and mechanical engineering. The second major wave of investment in the automotive industry in Slovakia

was realized in 2003–2005 thanks to the arrival of two other car manufacturers PSA Peugeot Citroen and KIA Motors which rolled out their manufacturing and assembling plant in Trnava (PSA) and Žilina (KIA). Car production in Slovakia after the start-up of production lines has risen so dramatically that Slovakia has been ranked among the first 20 world producers.

In 2013, production in Slovak production assembly automobile plants reached 987,724 units of passenger cars, which is an increase compared to 2009 by 82%, i.e. about 446,120 units of cars more. This means that in the Slovak Republic were produced 182 cars per 1,000 inhabitants and the country's the largest producer of cars per capita in the world. So in a country with just over 5 million. people is thus produced more cars than in Italy and half of the entire production of the UK where the population exceeds 60 million residents.

For comparison we add the data about the number of automobiles produced in neighboring countries in 2013: the Czech Republic in 2013 produced 1,128,473 automobiles, Hungary 220000 cars, 475000 cars Poland, Austria produced 146,566 cars. Overall, the world produced 65638451 passenger cars. The automotive industry is a crucial industry and driving force of the development of the Slovak economy. Its share in total sales in industrial production in 2013 represents 27% with the sales volume of almost 17 billion €. As shown in the graph no. 2, the share of individual car manufacturers in this issue is important. The graph

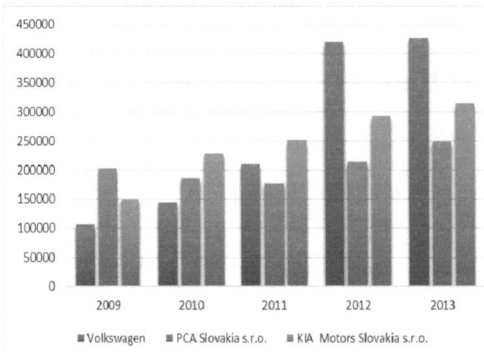

Figure 1.   The number of cars produced.

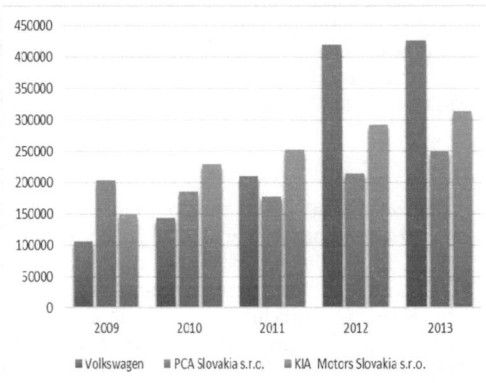

Figure 2.   Revenues in billion €.

also points to developments in the financial crisis and the onset of recovery measures in firms.

Since 2010 Volkswagen Slovakia llc. significantly took a position of leadership, as evidenced not only in the growing trend of production of the number of cars, but also steadily increasing trend in the number of production cars. Since 2012 it has been reaching 2.23 multiple production in 2009, as well as stable growth returns. Currently, Volkswagen Slovakia llc., reaches the proportion of the turnover in the sector in the range of 48–52%, so its position in the industry achieves dominance compared with other companies.

2   PRODUCTIVITY AND ANALYSIS

The object of measuring productivity of the production system is in general production system—subsystem, which is defined by an input unit, the

output of the production and the production process. Generally we can define the productivity as the ratio between the value of output and the value of production inputs. (Tokarčíková 2011). General definition remains the same, whether they are working place, production systems, enterprise, national economy or political system. Inputs can be human resources, capital, material, energy, information and so on. Outputs can be, for example, goods, services, income, profit, added value and alike. Productivity concerns all enterprises—manufacturing and non-manufacturing because it represents the efficiency of inputs and outputs-products or services. The aim of productivity is formulation of the efficiency with which factors of production are used in their transition. Productivity is a measure of the efficiency of the transformation process.

According to M. Synek, we can construct a number of different formulas depending on what units are used for inputs and outputs (Synek 2008). Outputs can be measured as subsistence units (kg, m, t, l, etc.), working units (the share of man-hours on hours worked), or in monetary terms, which we must cleanse of the price effects. Inputs are most frequently measured as hourly, daily, monthly or yearly productivity.

Labour Productivity (LP) expressed by the following share:

$$LP = Output / Input \qquad (1)$$

Output: Variables are defined in the numerator in specialist publications: value added, revenues, income, profit or loss, gain, benefit.

Input: Average number of employees, the number of working hours (accurate indicator) costs is generally defined as inputs.

Productivity indicator is a relative indicator. In its numerator may often be used the volume of production, sales or revenues, or added value. The added value is that portion of the value which entrepreneur adds with his/her work to the original entrance value (Ďurišová & Jacková 2007).

In financial terms, the added value is the sum of the profit margin of the entrepreneur, sales of own products and services, activation and changes in interior organizational supply lowered by the production consumption, which presents the consumption of materials, energy and services.

Labour productivity expressed by the LP = Added Value/Personnel, as the ratio of value added and number of employees put the social efficiency of the economic process at the company level. As the number of employees is the independent unit on prices and the value of the indicator of labor productivity measured by added value is dependent on the price changes and therefore the real growth of labor productivity is achieved only if the growth rate of productivity growth is higher than inflation.

Labour productivity, expressed as follows PPP = PH/ON, is the ratio of value added to personnel costs, the pointer is not related to the number of workers. This indicator shows how many euros value-added produce one euro of staff costs. It should have a value greater than 1, if the value is less than one, the company suffered losses. If the value is less than 0, the company suffers a loss that exceeds the size of personnel costs. We put analysis of selected entities and productivity results calculated from this relationship is expressed in graphs. Results calculated data productivity can be summarized as follows:

- The positive growth trend, more pronounced in companies with lower growth in revenues
- There are significant differences between firms in more than 50% in the amount of revenues and added value to one euro wage costs
- A necessity of a detailed analysis of the determinants of productivity and decomposition scope

Each measuring productivity in the business sector talks about how inputs translate into outputs. Each measuring productivity must focus on satisfying customer demand. The measurement is based on available data. Measurements must be logically justified and understood by the person who performed the measurement. The measured values are subsequently analyzed by management according to their preferences so as to ensure sustainable growth of measured productivity.

Given the sources of data used in the decomposition of labor productivity is not the number of employees as a value in the financial statements and thus the relevance of the data reduces as a result of obtaining values from other sources and also the rate of changing data has not been validated with the pace of inflation, we will use a modified indicator of labor productivity expressed the value-added of labor to personal cost (Kucharčiková 2011).

Labour productivity, expressed as follows LP = Added Value/Personnel expenses, is the ratio of value added to personnel costs, the pointer is not related to the number of workers. This indicator shows how many euros value-added produce one euro of staff costs. It should have a value greater than 1, If the value is less than one, the company suffered losses. If the value is less than 0, the company suffes losses that exceeds the size of personnel costs. We put analysis of selected entities and productivity results calculated from this relationship is expressed in graphs. Results calculated data productivity can be summarized as follows:

Productivity is being increased in the case of:

- Achievement with lower inputs and equal outputs,
- Higher outputs with the same inputs,
- Lower inputs and outputs above,
- Reduces the outputs but the inputs will decrease proportionally more,
- Increasing outputs and inputs but inputs are increased proportionately less.

The role of labor productivity analysis is to identify the factors that crucially affect its level. Labor productivity affects all three factors of production:

- Living work
- Work equipment
- Working Materials

Labor productivity levels depend on the presence of all three factors of the reproductive process. Therefore, the synthetic indicator of labor productivity broken down into analytical indicators which allow characterize the productivity of certain selected perspectives: the use of machinery and equipment (ME), use of the material (CM), compliance with performance standards and so on. Impact all three factors expresses model:

$$AV / PE = AV / CM * CM / ME * ME / PE \qquad (2)$$

where:
AV- Added value
PE- Personal expenses
CM- Consumption of material
ME- Machinery and equipment

From the relationship it shows that worker productivity can thus influence:

- Using of material (material efficiency AV/CM)
- Using of machinery and equipment (CM/ME)
- Rooms with workers (ME/PE) Used personnel costs of machinery and equipment.

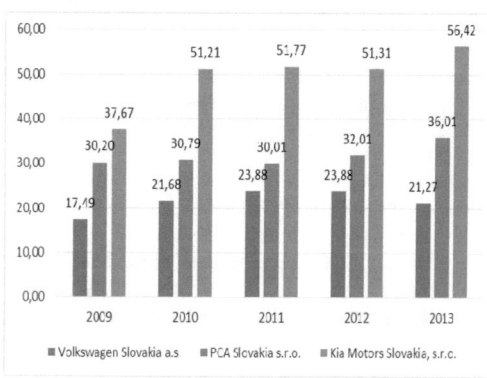

Figure 3.   The share of revenues on labor costs €.

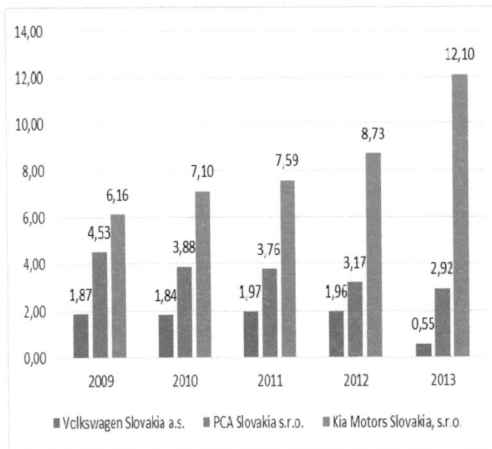

Figure 4.   The share of value added on labor costs €.

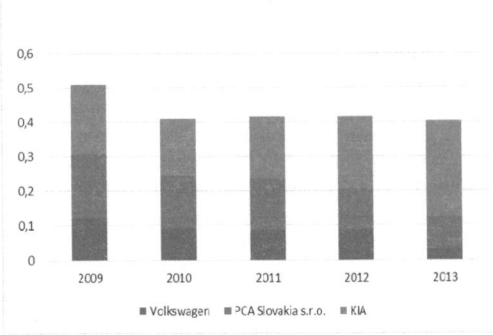

Figure 5.   Using of material.

Material efficiency of the added value is expressed by the following graph. The graph shows that the greatest value added per unit of consumed material has the company KIA, efficiency is more or less balanced in all years, the smallest material efficiency has Volkswagen and material efficiency is declining. Therefore what has significant impact on productivity? Let's continue further in analyzing productivity.

The second analyzed factor of production is an indicator of the use of machinery and equipment. Using of machinery and equipment is the best in the first surveyed company. This ist he proof that it has the best equipment and devices. They are currently operating at Volkswagen Slovakia llc., which are automated to 85%. This ranks the company among the companies with the highest degree of automation production.

The last analyzed partial indicators are the facility of workers by machinery and equipment. It is not the surprising result of the analysis of this indicator because as second indicator said analysis of direct human labor in the first surveyed company is in the range of 15–30%, which is different in compared companies as a consequence from the automation of production process.

Summary analysis of the impact of production factors points at a significant difference in the producing factors in the researched companies. Although the company Volkswagen Slovakia came as the company with the lowest labor productivity from this analysis. Currently the main factor on productivity is the efficiency of machinery and equipment, the lowest it he using of material, which can confirm that production factors have significant impact that can be influenced directly by company; by its efficiency of machinery and equipment and employees facilities with machinery and equipment.

The other two companies have pursued similar output monitoring indicators, where the highest share in the total productivity of workers has just the facility of workers of devices and equipment.

This fact testifies of the significant share of people working on the overall productivity and lower degree of automation of the production process. This situation is demanding on staff, its quantity and qualifications, which is favorable for the environment of company and state of the establishment and the state from the perspective of policy of employment. However in terms of efficiency in the company it is required to maintain the global trends of automation and reducing the needs of personnel and orientation on the effectiveness of other factors of production.

3   CONCLUSION

Labor productivity is the additional use of partial productivity. Labor productivity is expressed as the ratio between output and input, as the input consists of live work.

For the analysis of labor productivity it is necessary to define the indicators and units of expressing the volume of production, the time period for which productivity is detected (annual, monthly, daily, hourly) and the number and category of workers whose productivity was surveyed. A labor productivity level depends on the presence of all three factors of the reproductive process. Therefore, we spread synthetic indicator of labor into analytical indicators that allow characterize the productivity of labour from certain selected perspectives: the use of Machinery and Equipment (ME), using of the material (CM), filling performance standards and so on. Tracing of sample of 543 companies operating in the machinery industry, we found influence of the individual subfactors on productivity. A detailed analysis of these elements allows understanding of the attributes of

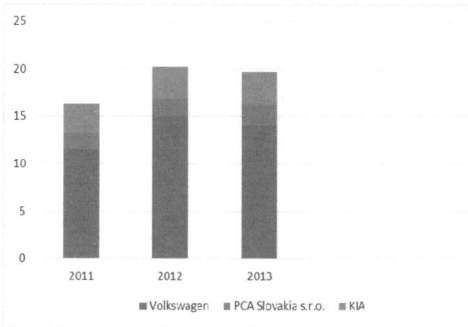

Figure 6.   Using of machinery and equipment.

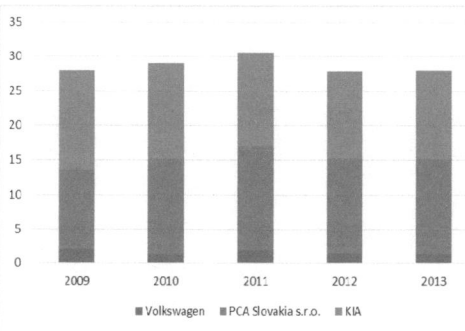

Figure 7.   Rooms with workers.

labor productivity and the possibilities for long term maintenance and increasing.

ACKNOWLEDGEMENTS

The research presented in this paper is an outcome of the project No. APVV-0857-12 "Tools durability research of progressive compacting machine design and development of adaptive control for compaction process" funded by the Slovak Research and Development Agency.

REFERENCES

Chajdiak, J. 2011. *Ekonomika firmy.* Bratislava. STATIS, 2011.
Chodasová, Z. 2005. Asset of Modern Management Methods in Development Process, *Vadyba management*, Univerzity Vilnius Nr. 1 (9): 13–20.
Chodasová, Z. 2012. *Podnikový controlling—nástroj manažmentu.* Bratislava: STATIS.
Chodasová, Z. & Tekulová, Z. 2014. Monitoring of indicators of competitiveness to improve the business strategy. *Kontrolling na malych i srednich predprijatijach: sbornik naučnych trudov 4. meždunarodnogo kongressa po kontrollingu. Praga,* Česká republika, 25 aprelja 2014. NP, Moskva: Obedinenie kontrollerov.
Ďurišová, M. 2012. Process model of managerial functions applied in business performance. *Economics and management spectrum.* No. 1(2)/ VI. Žilina Faculty of Operation and Economics of Transport and Communications, University of Žilina: 2012.
Ďurišová, M. & Jacková, A. 2007. *Podnikové financie.* Žilina: Vydavateľstvo EDIS.
Hilmar, J.V. 2008. *Controlling a new management tool.* Praha: Profess.
Horváth & Partnes. 2004. *Nová koncepce controllingu.* Praha: Profess Consuling.
Klečka, M. 2008. *Ekonomika a management.* http: // www.ekonomikaamanagement.cz/cz/clanek-produktivita-a-jeji-mereni-nove-pristupy.html. [Online] 2008. [Dátum: 1. 8 2014.]
Klečka, J. & Matějka, M. 2004. *Nové podnikové systémy: Materiály ke cvičením.* Praha: Nakladatelství Oeconomica, 2004.
Kucharčiková, A. 2011. *Efektivní výroba.* Brno: Computer Press, 2011.
Ručková, P. 2011. *Finanční analýza—metody, ukazatele, využití v praxi.* Praha: GRADA Publishing.
Tekulová, Z. & Chodasová, Z. 2012. Controlling the capital structure in a company. *Scientific proceedings 2012: Faculty of Mechanical Engineering, STU in Bratislava.* Bratislava: ed. Slovak University of Technology.
Tokarčíková, E. 2011. *Influence of social networking for enterprise's activities.* Periodica polytechnica Office. Social and Management Sciences 19 (1), Hungary: 37–4.

Production Management and Engineering Sciences – Majerník, Daneshjo & Bosák (Eds)
© 2016 Taylor & Francis Group, London, ISBN: 978-1-138-02856-2

Spectral graph partitioning in security analysis related to electrical network

D. Valek & R. Scurek
Faculty of Security Engineering, Technical University of Ostrava, Ostrava, Czech Republic

A. Veľas
Faculty of Security Engineering, University of Zilina, Zilina, Slovak Republic

ABSTRACT: This paper describes utilization of Spectral graph partitioning in security analysis. Electrical network is perceived as a system of nodes which together form a graph (network). We use very simple and understandable examples using Fiedler's theory. Within application of this theory, the most important power lines of the entire power grid are selected. Facilities related to power lines are logically ordered and considered by author's modified analysis. The used analysis has been improved and optimized for hazards related to illegal acts. Each facility of electrical network is connected with possible kind of attack and each of these devices is gradually evaluated by five coefficients. These coefficients take values from 1 to 10 points. On the coefficient basis, the final level of risk was assessed. Finally, the most risky facilities are selected and security measures are set up accordingly.

1 INTRODUCTION

1.1 *Electricity importance*

Within last fifty years, significant diversification of technical infrastructure from telegraph to internet was realized. From wide point of view, the electric network, traffic network and communication network created the foundation for all prosperous companies. Usually these networks are based on a big amount of heterogeneous components characterized by complex dependences and relationships among them. To ensure its structural integrity, effectivity and reliability of network, its illegal acts, security and protection against terrorism has to be taken into account.

Within the last two decades the electricity consumption extremely increased. It was caused mainly by social-economic changes in Eastern Europe. Directly proportional to the energy consumption, nuclear power plants were newly built and expanded as well as power plants producing electricity from renewable sources (Kampová & Loveček 2013.).

In Europe, after the disaster in Fukushima, nuclear power plants evoked discussions about safety of nuclear power plants and consequences of the radioactive substances leak. If there is a reduction in production of electricity from nuclear power plants in Europe, there would be a significant shortfall in electric energy production. This shortfall should be covered mainly by the electricity produced by renewable energy sources and fossil fuels plants. If this happens, it will have a very serious impact on energy provision. If there is a massive production of electricity by renewable energy sources, there would be many more opportunities to single impacts to the transmission system.

These unpredictable impacts are mainly caused by renewable resources, especially by wind power plants. Wind power plants work on the opposite principle than conventional power plants because they supply electricity only in case of blowing wind.

During these large fluctuations of electricity transfers, the important thing is having a robust electricity system which has to be resistant to damage of its single components such as electrical wiring, transformers or power plants. Due to this, it is necessary to have all single components protected and secured against the attacks of various criminal or terrorist organizations. These attacks could cause a blackout which could have very severe consequences to human lives and country economy. Therefore, it will be increasingly important to deal with the security measures of devices across our electricity network. The utilization of the graph theory which is described in this article can help the selection of the most important components.

2 ELECTRICITY SYSTEM

All facilities which provide electricity from the production to a final customer are the assets, decommissioning of which can result in a threat to the electricity supply to the final consumer. Among the main devices in the electricity system we can mention electrical grid, electrical stations and electrical wiring.

2.1 Electrical grid

The facilities considered in the analysis are parts of the electrical grid. Electrical grid is a system of interconnected devices which are used for electricity transmission, transformation and distribution. Further devices are used for metering and control and other for the security system and the power plants.

The transmission system is a system of devices used to transmit electrical energy with a voltage of 400 kV and 220 kV from the manufacturers to the power nodes. Distribution system transmits electrical energy with a voltage of 110 kV and 22 kV from the transmission system to customers. The customers in this context are cities, factories and households.

2.2 Electrical wiring

Voltage is transmitted by different types of wiring. Usage of a specific wiring type depends on many factors such as quantity of the transmitted voltage, quantity of the transmitted electric current, voltage drop and so on. Wiring is most often outdoors but can also be cabled on the ground or under the ground but this solution is more expensive.

Outdoor wiring has to be resistant to weather changes, extreme weather conditions, humidity and it must have sufficient mechanical solidity against intentional damage. Cable wiring is used in residential areas, industrial areas and in the buildings. Outdoor wiring is carried by electrical pylons. Construction of these electrical pylons can be made of wood, reinforced concrete, steel or aluminium alloy. There exist many types of electrical pylons and the difference is mainly it their design of construction. Existing types of pylons are shown in Fig. 1.

Electrical pylons are designed from the construction point of view to resist extreme weather conditions and wind power. Instead of main cantilevered pylons, the grid consists of reinforcement pylons to ensure stability in case of wires break because without the main cantilevered pylons would not stay in the right position. These reinforcement pylons are made of special steel which can resists the climatic exposure.

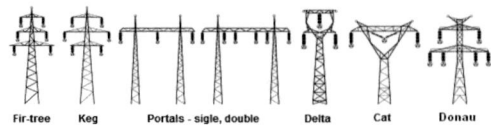

Figure 1. Types of pylon constructions.

2.3 Electrical stations

Electrical stations belong to electrical grid and are divided into transformations, switch stations and substations. Big parts of these devices are created by the substations which can be a single building or a bounded area. These substations take care of the in-put and output electricity flow and they are consisted of conductors, insulators and switch, security or control devices. In the buildings are mainly situated substations with voltage up to 35 kV and in the outdoor areas are situated substations with very high voltage over 52 kV.

3 UTILIZATION OF GRAPH THEORY

Electrical wiring, telecommunication, transport infrastructure or others engineering network form and system of nodes generate together graphs of different shapes with many degrees of complexity. Many networks are designed on the basis of landscape relief.

3.1 Graphs characteristics

Using graphs, we can present a set of objects illustrating the interdependence of various elements. Objects are assigned as vertices (power plants, transformers etc.) and their connections are called edges (transmission network, distribution network). Graph can be basically represented by a simple model of a real network which emphasizes topological properties of objects and neglects their geometric properties.

Graphs can be divided as a directed and undirected. Undirected graph is defined as $G = (V, E)$ where V indicates vertices and E indicates edges. In case of undirected graph we do not consider the order of vertices which is used in case of electrical network analysis (Ochodková 2009).

3.2 Graph theory

Considering a connected graph G, it is possible to divide it into two smaller graphs having roughly the same number of edges and vertices and one shared edge. There are many methods to achieve this. One is the Spectral graph Partitioning. This method involves the use of Laplacian matrix and divides vertices of a connected graph G into two subgraphs by the use of Laplacian matrix eigenvectors (Slininger 2013).

3.3 Adjacency matrix

Graph can be represented by the adjacency matrix. It is defined as G, A(G) = (ai, j)

$$a_{i,j} = \begin{cases} 1, (i,j) \in E \\ 0, (i,j) \notin E. \end{cases} \quad (1)$$

where ai, j = adjacency matrix, i = row, j = column, G = graph

This means that the adjacency matrix A represented by the graph G has for i-th row and j-th column value equal to 1 in case there is an edge between node i and j. Otherwise there is assigned value 0 for this position.

3.4 Degree matrix

Other matrix defined by the graph G is degree matrix D(G) = (di, j)

$$d_{i,j} = \begin{cases} d(i), & i = j \\ 0, & i \neq j. \end{cases} \quad (2)$$

were di, j = degree matrix, i = row, j = column, G = graph

Degree matrix is a diagonal matrix, which provides us information about degrees of each vertex—number of edges entering or exiting from the concrete vertex. This matrix is used together with the adjacency matrix to create a Laplacian matrix.

3.5 Laplacian matrix

Laplacian matrix is another way how to present a graph. Matrix L(G) defined by graph G.

$$L(G) = D(G) - A(G). \quad (3)$$

where G = graph, L = Laplacian matrix, A = Adjacency matrix.

Laplacian matrix is the difference between the degree matrix and the adjacency matrix.

3.6 Spectral graph partitioning

In Spectral graph partitioning is based on a simple principle. For the graph G defined by vertices and edges G = (V, E) which has to be dividend, Laplacian matrix L(G) is calculated. Using the spectral partitioning of Laplacian matrix, the eigenvalues

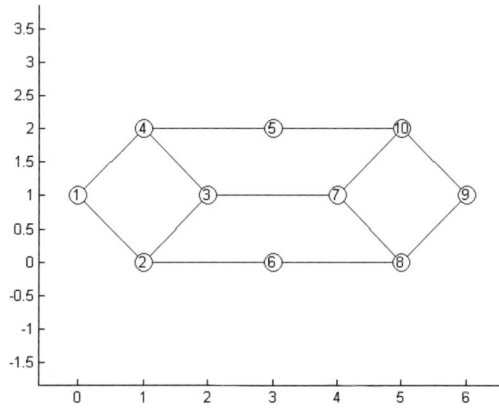

Figure 2.   Example of simple graph.

and eigenvectors are calculated. Such eigenvector which is related with the second smallest eigenvalue provides us with required graph partitioning. This vector is named a Fiedler's vector (Byrtusová 2015).

4   ALGORITHM FOR FINDING THE MOST IMPORTANT ELECTRICAL WIRINGS

Input parameter of the algorithm to calculate the most important electrical wirings is adjacency matrix. Values of single rows and columns of the adjacency matrix are the inputs given by the user according to the power lines map. These values are uploaded into the Excel and then exported to a computer program.

The computer program then calculates the degree matrix and Laplacian matrix from adjacency matrix followed by the spectral graph partitioning and finding of the required Fiedler's vector (Leitner & Míka 2012).

Distribution of the final graph is determined by the sign related to the Fiedler's vector. Vertices with positive numbers are assigned to the first part of the graph and vertices with negative numbers are assigned to the second part of the graph. The line which connects both parts of graph is then selected as the most important power line (Aven 2010).

5   EXAMPLE OF SPECTRAL GRAPH PARTITIONING

To illustrate the above theory, a very simple example of the spectral graph partitioning is shown. The given graph is a simple network of ten vertices and eleven edges. To each of the vertices we assigned coordinates to display them in the program according to its potential deployment in the territory.

553

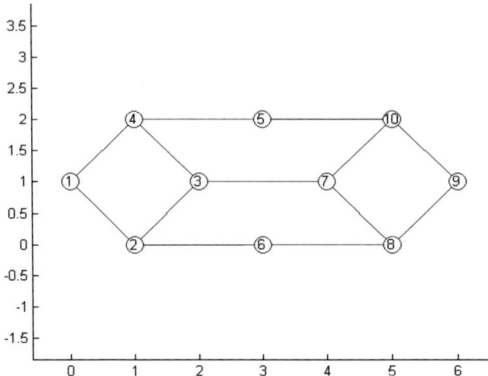

Figure 3.   Selection of the smallest number of edges.

Table 1. Adjacency matrix.

	j=1	j=2	j=3	j=4	j=5	j=6	j=7	j=8	j=9	j=10
i=1	0	1	0	1	0	0	0	0	0	0
i=2	1	0	1	0	0	1	0	0	0	0
i=3	0	1	0	1	0	0	1	0	0	0
i=4	1	0	1	0	1	0	0	0	0	0
i=5	0	0	0	1	0	0	0	0	0	1
i=6	0	1	0	0	0	0	0	1	0	0
i=7	0	0	1	0	0	0	0	1	0	1
i=8	0	0	0	0	0	1	1	0	1	0
i=9	0	0	0	0	0	0	0	1	0	1
i=10	0	0	0	0	1	0	1	0	1	0

Table 2.   Laplacian matrix and related Fiedler's vector

										Fiedler's vector	Vertices number
2	-1	0	-1	0	0	0	0	0	0	0.5117	1
-1	3	-1	0	0	-1	0	0	0	0	0.3162	2
0	-1	3	-1	0	0	-1	0	0	0	0.1954	3
-1	0	-1	2	-1	0	0	0	0	0	0.3162	4
0	0	0	-1	2	0	0	0	0	-1	-0.0000	5
0	-1	0	0	0	2	0	-1	0	0	0.0000	6
0	0	-1	0	0	0	3	-1	0	-1	-0.1954	7
0	0	0	0	0	-1	-1	3	-1	0	-0.3162	8
0	0	0	0	0	0	0	-1	2	-1	-0.5117	9
0	0	0	0	-1	0	-1	0	-1	3	-0.3162	10

According to picture above (Fig. 1) is designed adjacency matrix where single rows and columns are valued on the basis of existence the edges between single vertices.

From the adjacency matrix (Table 1), according to the defined equation, we can get the Laplacian matrix and the Fiedler's vector related to the second smallest eigenvalue.

The Fiedler's vector divides the network into two approximately equal parts (Table 2). Positive values of vertices are the part of the first portion

and negative values of vertices are the part of the second portion.

Red marked edges displayed in the Figure 3 represent the smallest amount of edges between both network parts. If we bring this situation into the electrical grid environment, we talk about electrical lines which failure means a blackout in the largest possible area.

The graph theory and its spectral graph partitioning are able to select the most important power lines.

Then we can use conventional risk analysis only on the selected components of transmission and distribution network (Válek 2012).

6   ANALYSIS OF AN ATTACK ON THE ELECTRICITY SYSTEM

Considering the vulnerability of single objects in the electricity system, modified method FMEA can be used. This method was modified by the author to be more suitable for anticipation of illegal acts. Further in the article, an analysis named by shortcut FMEAIA (Failure Model and Effect Analysis of Illegal Acts) will be represented (Sharma & Kumar & Kumar 1984).

Classic FMEA method is most commonly used in the automotive industry to search and evaluate potential defects in processes and products (Process Quality Management 2014). Level of risk is determined by the multiplication of subjectively evaluated coefficients which are Occurrence "O", Severity "S", Detection "D" and in addition, in FMEAIA analysis is Appeal "A" and Accessibility "AA". Level of risk is calculated according to formula:

$$R = \frac{O \cdot S \cdot D \cdot A}{AA} \qquad (4)$$

Values of each coefficient may vary from 1 to 10 according to bellow mentioned table:

In the evaluation process of the electricity system, we should start with the selection of the most important places provided by the Spectral graph partitioning method described above. Selected devices placed in the calculated territory should be considered in the analysis according to above described method of FMEAIA (Máca & Leitner 2002).

Table 3.   Classification of each coefficient.

	O	S	D	A	AA
Evaluation 1	Low probability	Low severity	Very easily detectable	Not very appealing to attack	Easily accessible
Evaluation 10	Very high probability	Very high severity	Almost impossibly detectable	Very appealing to attack	Almost inaccessible

Table 4.   FMEA

Purpose of device	Type of device	Specific device	Kind of attack	O	S	D	A	AA	R
Production	Thermal	Conveyor belt	Overload	1	2	2	1	3	1,3
			Blockage	2	2	2	1	3	2,7
		Cooling tower	Bomb attack	2	6	1	6	7	10,3
			Air attack	1	6	1	6	7	5,1
		Boiler	Diversionary activities	3	5	3	4	6	30
	Nuclear	Cooling tower	Bomb attack	2	9	1	8	9	16
			Air attack	1	9	1	8	8	9
		Reactor	Diversionary activities	3	9	3	9	9	81
		Pump	Pump blockage	2	7	2	3	9	9,3
	Water	Turbine	Blockage	2	4	2	7	7	16
		Electric generator	Removal of wires	4	4	4	4	8	32
		Water feeder	Feeder blockage	3	3	4	6	7	30,9
		Drainage canal	Canal backfilling	3	3	3	6	8	20,3
	Wind	Rotor	Collision with aircraft	2	1	6	1	2	6
				1	1	8	1	1	8
		Tower	Placing the bomb	3	1	8	2	3	16
		Electrical connection	Cutting the wires						
		Control system	Disposal of computer equipment	3	1	8	2	3	16
	Solar	Photovoltaic panels	Damage by stones	6	1	8	1	2	24
		Wire jumpers	Cutting the wires	5	1	7	1	2	17,5
		Substation	Placing the bomb	2	1	7	1	2	7
			Removal of wires	5	1	7	2	3	23,3
Wiring	Transmission system	Outside wiring	Air attack	1	5	7	5	8	21,9
			Pressure load	3	7	6	4	9	56
		Underground wiring	Cutting the wires	4	7	5	4	7	80
		Steel pylon	Fusion	3	7	6	6	4	189
			Incision	4	7	7	7	3	457
			Undermining of foundation	3	7	4	6	5	101
		Reinforced pylon	Placing the bomb	2	7	8	6	3	224
			Fusion	3	8	6	7	4	252
			Incision	4	8	7	7	3	523
			Placing the bomb	3	8	8	6	3	384
	Distribution system	Outside wiring	Damage by constr. machine	4	5	3	2	2	60
		Underground wiring	Cutting the wires		4	3	4	6	40
		Wooden pylon	Ignition	5	4	2	5	2	100
			Incision	5	4	3	4	2	96
			Dent	4	4	3	4	2	72
			Undermining	3	4	2	3	3	16
			Intentional breakage	2	4	3	2	4	18
		Steel pylon	Fusion	3	5	5	5	4	125
			Incision	4	5	5	5	3	125
			Undermining	3	5	4	4	4	40
			Flection	2	5	4	4	5	32
				2					

Example of the analysis procedure according to FMEAIA method can be used as follows:

1. Distribution from the system into devices in terms of production and in terms of the electricity wiring.
2. Distribution from the devices into buildings—pylons, wiring and further devices which are fixed to the ground (for example wind power plants, photovoltaic power plant etc.).
3. Distribution into the further smaller technical devices.

Example of evaluation of several selected devices is shown in the Table 4.

6.1 Spectral graph partitioning

The procedure of risk evaluation is followed by the phase of security measures proposal for the devices evaluated as the highest risk. The security measures should be efficient, economical and easily feasible. In general, the principle of ALARA should be considered in this case because this principle takes into account the value of the protected object and the value of the devices which protect this object.

For minimization of the risk represented by the incision and fusion of a reinforced and steel pylon in the transmission system, it is possible to apply measures which physically prevent contact with a single pylon or make the activity of an attacker more uncomfortable and time consuming. Among the suitable security measures we can classify:

- Usage of hardened steel for the lower part of pylon construction.
- Concreting of the pylon foundations up to the height of 2 meters above the ground level.
- Definition of the perimeter around the pylon by the fence or barbed wire.

The proposed security measures are suitable to be implemented mainly in terms of reinforced pylons which are included in the grid of the cantilevered pylons.

Distance between cantilevered pylons of Donau type can be 500 meters in suitable terrain.

In case of a bomb attack, similar security measures can be applied like in the case of the incision and fusion of the pylon, however, in case of the bomb attack, it depends on the level of energy which is released within the explosion.

7 CONCLUSIONS

The article described facilities which are a part of the electricity system. The components used for the transmission and distribution of electric energy are connected to this network.

This network can be understood as vertices connected together with edges. In this case, the method of graph spectral partitioning was used to make the selection of the most important power lines.

After this selection, we are able to consider components included in the calculated territory with conventional risk analysis. Applied risk analysis was modified by the author using FMEA analysis to create more suitable analysis for the illegal acts. Two new coefficients were applied to this modified analysis—appeal and accessibility.

Due to this modification, the FMEAIA analysis is more suitable for illegal acts consideration. The result from the analysis shows that the most risky devices in terms of illegal acts are mainly the devices for transmission of electric energy which have much better accessibility then the devices for electricity production. Due to this conclusion, it is necessary to select the most important devices in specific terrain. Security measures to reduce the risk of attack to acceptable level should be applied to these devices.

REFERENCES

Aven, T. 2010. *Misconceptions of Risks*. Stavanger: John Wiley and Sons.
Byrtusová, A. 2015. *The security environment and factors influencing security environment*. In Beli-anum: Security forum 2015.
ISO 31000: 2010. *Risk management—Principles and guidelines*.
Kampová, K. & Loveček, T. 2013. *Managing of security in organization*. Verejná správa a regionálny rozvoj: ekonómia a manažment: 105–110.
Leitner, B. & Míka, V. 2012. *Reliability and safety of technical means in critical infrastructure with respect to fatigue damage processes*. Logistics and transport: 103–110.
Máca, J. & Leitner, B. 2002. *Operational analysis*. Zilina: University in Zilina.
Ochodková, E. 2009. *Graph algorithms*, Technical university of Ostrava.
Process Quality Management. 2014. *FMEA—Failure Mode and Effect Analysis*, available at: www.pqm.cz/NVCSS/fmeacs.html.
Sharma, R., Kumar D. & Kumar, P. 1984. *Systematic failure mode effect analysis (FMEA) using fuzzy linguistic modeling*. International Journal of Quality & Reliability Management. 22(9): 986.
Slininger, B. 2013. *Fiedler's Theory of Spectral Graph Partitioning*, University of California.
Válek, D. 2012 *Analysis of the possibility of disruption the electricity*.

Production Management and Engineering Sciences – Majerník, Daneshjo & Bosák (Eds)
© 2016 Taylor & Francis Group, London, ISBN: 978-1-138-02856-2

The sustainability management of the means of production

Š. Valenčík & J. Kováč
Faculty of Mechanical Engineering, Technical University of Kosice, Kosice, Slovak Republic

ABSTRACT: This contribution analyses in detail the challenges and issues associated with sustainability management in the production plant. The first issue is focused on not preferring the financial results to the maintenance from the short term perspective as it is done at the expense of renewal aimed at longer period of time, i.e. the sustainability of its infrastructure. It is associated with the ability to predict long-term effects on different maintenance strategies. The second problem is a challenge for managers to optimize tactical organization of intervention operations considering the maintenance cost and the cost of downtime. The sustainability strategy is based on defining factors and methodology of optimal sustainability. It aims to assess the cost of downtime, whether planned or unplanned, pointing to the broader context associated with disruption of the production process.

1 INTRODUCTION

Production is radically affected worldwide by complex economic, led by socio-political and technological dynamics. In the production management and application of production processes, machines and systems there is increasing pressure on cost and maintainability, as a result, there is support for the identification and quantification of the inputs and outputs of the standalone subjects of the production company for the purpose of transparency of economic and financial flows, see Jovane, Yoshikawa, Alting, Boër, Westkamper, Williams, Tseng, Seliger & Paci (2008). On the other hand, it is a general requirement to revitalize production (improving employment, increase productivity), which puts pressure on financial resources (internal, external) provided with funds and leads to increased demands on the availability of the property (utilization of production base) see Valečík (2012).

Based on mentioned above, following two problems are picked up to discuss. The first comes from sustainable management of the means of production through policy maintenance or renewal (renewal of the means of production) aimed at a longer time period. The second problem affects the maintenance organization, including the organization of working periods (typology maintenance work on the production base), in the context of increased utilization of the production plant.

2 PROBLEMS OF MAINTAINABILITY

2.1 *Policy of maintenance and renewal*

Strong pressure on the financial results of production often leads to the temptation to immediately restrict investments in favour of reducing the cost of renewal, i.e. at the expense of sustainability of property, see Putallaz & Rivier (2003). Chronic shortage of funds needed to regenerate a steady reduction in the value of assets. In substance the production base built on the old elemental base is characterized by its small residual service life, so the value of the assets is consequently low, resulting in loss of control over the means of production what is explained in Figure 1.

Imagine the means of production with stable asset value (corresponding to the value of the property at the time of the average life-time of its components), whose condition meets the requirements of the sustainability of operation (state A). Maintenance managers decide to reduce the funds defined for the reconstruction of production base. The value of assets (production base) is permanently reduced despite the fact that the maintenance requirements for care of the means of production are respected (condition B).

In the coming years, it is consequently difficult to maintain the quality and condition of the production base, without significant increasing the amount defined for maintenance (condition C). This further reduces the value of the production base and production quality, the amount of funds provided for the maintenance of production infrastructure assets is not sufficient to maintain the state of the infrastructure (condition D). The system becomes unstable, requires a lot of maintenance what is in the nature a temporary solution which is very expensive and has its limits and results in a reduction in production capacity (state E). In this case, there are two solutions: either to interrupt the operation of the production department or to make an enormous investment in renewal.

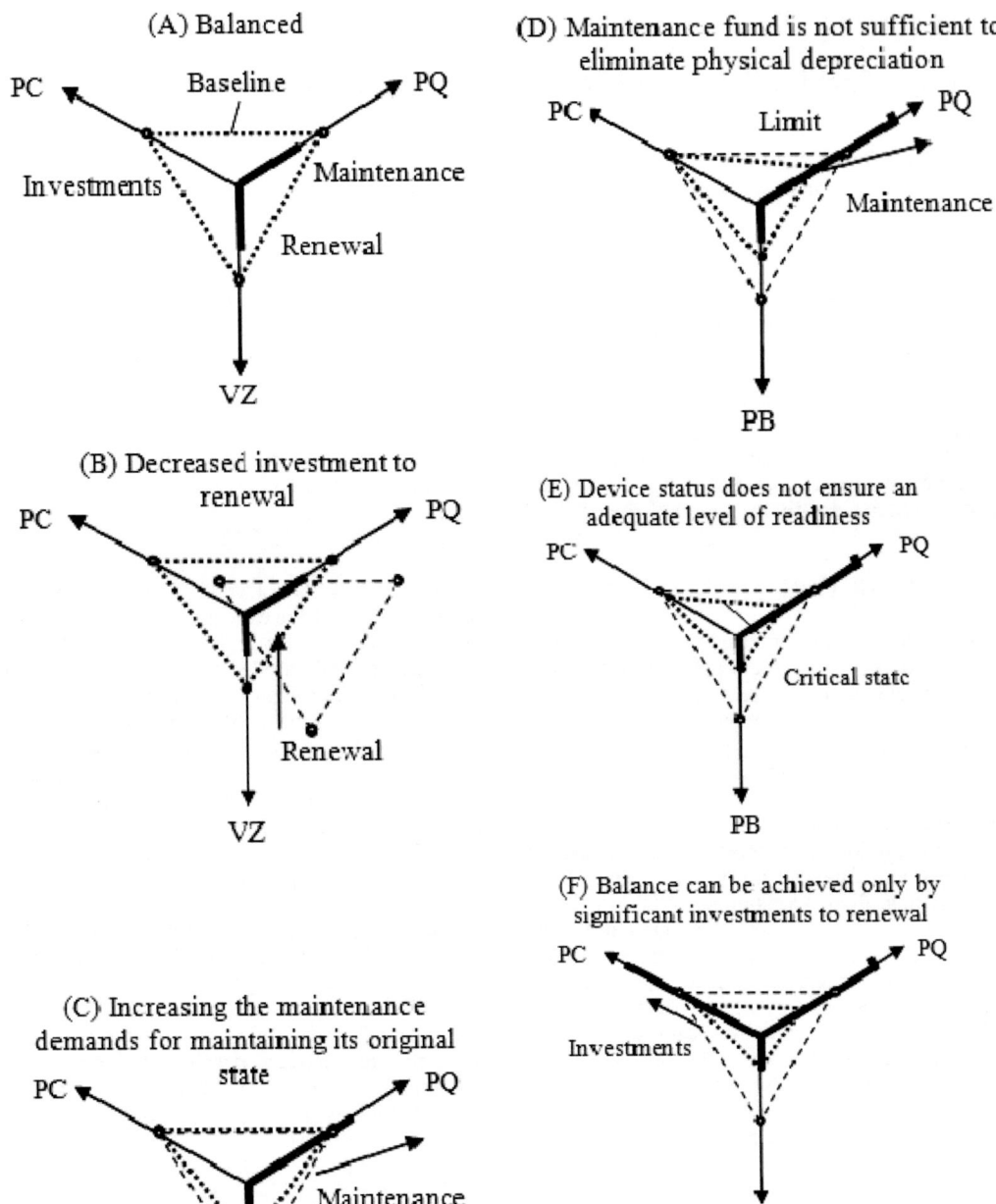

(A) Balanced

(D) Maintenance fund is not sufficient to eliminate physical depreciation

(B) Decreased investment to renewal

(E) Device status does not ensure an adequate level of readiness

(C) Increasing the maintenance demands for maintaining its original state

(F) Balance can be achieved only by significant investments to renewal

— expenses (consumption)
···· — physical quantities (production base)
PQ — production capacity
PB — production base
PQ — production quality

Figure 1.   The mechanism of loss of control over the means of production.

Resumption of production performance requires a significant amount of technical and financial resources for the recovery or renewal of production infrastructure (condition F). Unfortunately, the beneficial effects of the investment effort will not appear immediately because the maintenance expenditures are only sequentially decreasing. This inertia raises doubts among the owners and stakeholders of the company, who are reluctant to invest heavily in its entirety. Most obvious it is seen at the top level (political, economic) in which the long-term stability and strategy comes from.

2.2 Organization of maintenance process and renewal

Maintenance production infrastructure requires in most cases a partial or complete interruption of the production plant because of the implementation as "intervention (maintenance) work" as well as "support (technical and logistical) works". Rescue work is a little extra (maintenance) work at production facilities that are naturally seen as a restriction on the operation and it is required to minimize the duration.

In contrary, it is but the maintenance requirements, which is intended to have as much interventions guaranteeing a significant reduction in operating cost as possible, see Smith, Hawkins & Lean (2004). This has the effect of extending the period of intervention works which is related to related availability of the means of production and creation of additional operating cost (Figure 2).

In Figure 2 are shown the economic contradictions regarding the duration and amount of intervention works with respect to the utilization of operation. In fact, the increase in degradation causes increasing requirements for the care of the means of production, so long as we want to fully cover the increased demand. The increase in production is causing increasing requirements for the care of the means of production (intensification of maintenance work), resulting in significantly increased demand for emergency work. The number of available production base is decreasing, while demand is still growing.

2.3 Interconnection the actors of infrastructure maintenance

The two previous issues ensuring the sustainability of the means of production and optimize intervention works are interesting only if there is good communication between stakeholders (operator, infrastructure manager, supervisor). In fact and in most cases, supervisors representing the owners of the company, i.e. they co-fund the sustainable means of production and thus may affect the

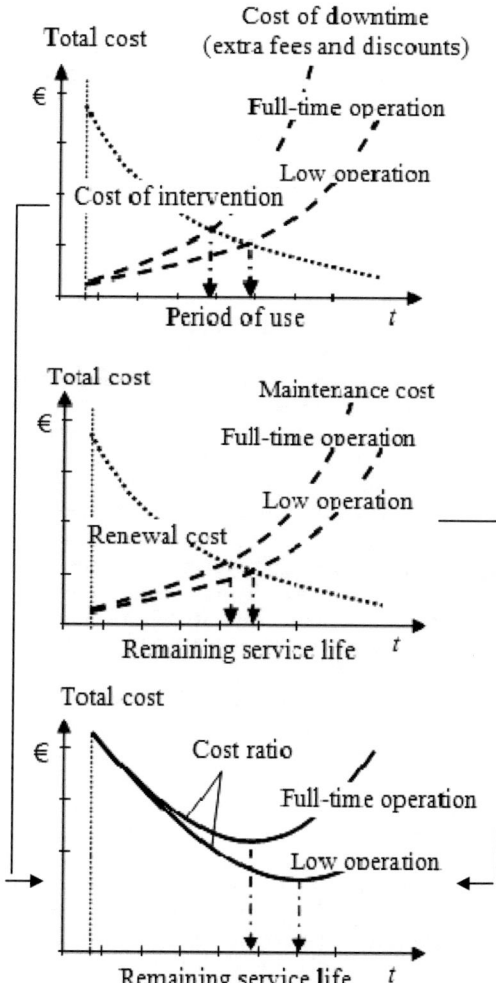

Figure 2. Economic coupling of cost and utilization.

optimizing of the efficiency of maintenance and renewal of the production base, see Jovane, Yoshikawa, Alting, Boër, Westkamper, Williams, Tseng, Seliger & Paci (2008). It must be then believed by the owners of the merits of the proposed policy and they have to realize the importance of good communication. The same approach is taken in managing maintenance work (maintenance periods); good communication between manager and infrastructure provider facilitates the consensus that results in the decision of such tactical organization of maintenance interventions which is beneficial for all parties involved. Good communication requires the existence of a particular platform that includes reliable and objective indicators representing the function of the production base and these are recognized by authority i.e. by manager and business owner. The forecasting model should then be able to show the development of these key indicators in accordance with enforced maintenance policy.

3 SUSTAINABILITY STRATEGY

3.1 Sustainability factors

Currently there are relatively few enterprises which systematically and in detail perform the monitoring and evaluating in broader context the downtime operating cost of machinery and other equipment. However, even though the most of this information is available and provided, there is no deeper interest of operation workers caused by lack of knowledge of similar processes, see Valenčík (2011). Apparently the knowledge downtime cost of operating machinery will allow workers to focus on potential areas for improving overall economic efficiency equipment. Devices that will have a lower level of downtime cost pave the way for increasing the quality and production capacity when comparing them to equipment with higher cost and this can be a serious problem.

This dispute leads to increased pressure on decision making for managers. On the one hand, the managing authorities require reducing the maintenance cost and other related activities on the production base and on the other hand, operators and consumers in turn are required to give a maximum capacity and adequate quality. It is obvious that the objective managing needs to choose the optimal tactic that will minimize the sum of the cost of maintenance and downtime cost. However, the decision requires the knowledge of the cost function for downtime (Figure 2), while maintenance costs are relatively well known. On the other hand, the cost function should be the consensus between infrastructure managers and operators, but consensus is difficult to achieve (Figure 3).

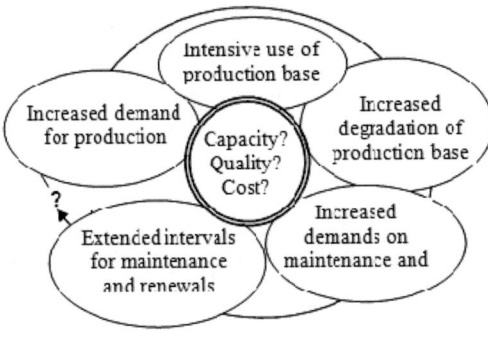

Figure 3. The mechanism of control loss over the means of production.

Cost and utilization are challenges for company managers to optimize tactical organization of intervention works (typology of maintenance work on the production infrastructure); in the context of increased utilization of the production plant, both the cost of maintenance and the cost of downtime must be involved in such decision making.

3.2 Methodology for optimizing the sustainability

Presented methodology is designed to assess the importance of maintenance strategy, based on the scenario which offers a framework for addressing the cost of maintenance and production capacity/or availability of the means of production, see Smith & Hawkins (2004). However, it is necessary to clarify the dilemma of why it is not possible to directly incorporate parameter capacity operation in methodology, rather than just be an input value? Parameter "equity production base" is almost continuous variable; parameter "production capacity" is perfectly discrete variable. Some special cases, such as investments in capacity, are subject to specific planning procedures generally leading to a limited number of possible options. For this reason, this method pushes "the amount of installed/performance" on the position of the input variable. The methodology is used repeatedly for each "capacitive variant". The optimization process evaluates various strategies of maintenance, including the key production base, and compared distinguishing criteria with the existing actual benefits. This mechanism is based on discounted life-cycle cost (Equations 1 and 2) that includes all expenses supporting priorities/growth of company. Current benefit can be expressed as follows:

$$B = -\frac{I}{(1+i)^{t_0}} + \sum_{t=1}^{t=T} \frac{A_{(t)} - R_{(t)}}{(1+i)^t} + \frac{V_r}{(1+i)^{T+t}} \qquad (1)$$

whereas:

B—discounted benefits
I—initial investment (acquisition cost)
i—discount rate (actual interest)
$A_{(t)}$—changes in surplus (consumer surplus is the difference between the amount which the consumer is willing to pay/value use for certain goods or services and actually paid value/value of the transaction and it is a benefit for the consumer from the deal) for manufacturer, purchaser, society, infrastructure managers and state (taxpayers)
$R_{(t)}$—changes in the maintenance cost and managing development of the production plant

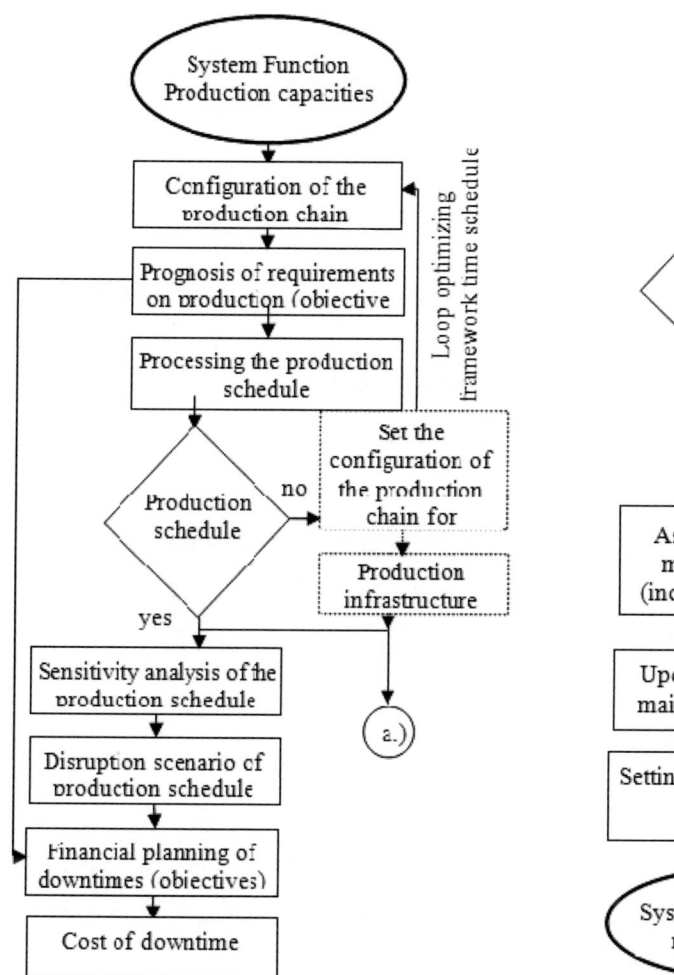

Figure 4.   Design of the optimizing procedure (steps).

V_r—residual value at the end of the planning period
t_0—initial year of investment
T—last year of the planning period

Maintenance of the means of production is divided into renewal operation $r(t)$, the repair operations and maintenance $e(t)$. Variable I represents initial investment targeted at increasing investment capacity (new machinery, workplace, etc.) to optimize and harmonize the production infrastructure. The initial view, however, remains when comparing two different technologies for the renewal of the production cell (e.g. CNC machine versus machine NC) and their impact on maintenance cost. This results here that variable I loses its meaning. In the case that we choose a sufficiently long period of

simulation, then the term representing the residual value may also be neglected, which updates formula for profit to this form:

$$B = \sum_{t=1}^{t=T} \frac{\left(A_{(t)} - r_{(t)} - e_{(t)}\right)}{(1+i)^t} \qquad (2)$$

The purpose of the methodology is therefore to clarify the benefits of discounted assessment of each maintenance strategy. The best strategy is one that brings high actual benefit. The developed methodology and models should help to assess the conditions for change in surplus $A(t)$, cost for renewal $r(t)$ and expenditure on maintenance $e(t)$, which affect the amount B. The term change of surplus/increase the efficiency $A(t)$ depends on differences in the performance of the service and it is the 1st and 2nd step of methodology proposed (Figure 4). Obviously, the cost of downtime and cost of troubleshooting are effected by factor A(t) of the equation 2 with respect to visible issues of modelling the requirements of maintenance. This is presented by distinction between the cost of regular maintenance of a preventive nature, and maintenance cost generated with correction maintenance (due to failure). The term A(t), expressed in euro per failure can be expressed by the relation:

$$A_{(t)} = u_{(t)} + ext_{(t)} + rev_{(t)} + op_{(t)} \qquad (3)$$

whereas:

$u_{(t)}$—change in utility/functionality of the customer
$ext_{(t)}$—positive and negative external aspects (pollution, production breakdowns, interruptions tax, penalties, etc.)
$rev_{(t)}$—change in revenue for the operator/business owner
$op_{(t)}$—actual expenditure on the production site

3.3 Optimization steps

Methodology for optimizing, see Valenčík (2011) sustainability/maintenance strategy and coordination of production capacity (Figure 4) has the following four steps:

1. Analysis and quantification of the sensitivity of the production program. The purpose of this step is to create a statistical sample representing the effects of the production program for the creation of infrastructure failures (statistical delay).
2. The cost of downtime, planned outages (maintenance periods) or unplanned (failure). The second step is the transformation of economic data from these samples to the corresponding calculations of discounted benefits.
3. Assessment of unit maintenance cost (effects of utilization). The third step is to integrate the variables related to capacity operation in process evaluation maintenance policies. This is done to increase the unit cost of labour through factor representing the intensity of interventions in the production plant (issue maintenance intervals) and imposes cost due to aging infrastructure.
4. Simulation of maintenance policy (evaluation of service required). The fourth step simulates generating cost (unit cost of maintenance as defined in 3) for the different maintenance strategies corresponding strategy/tactics in compliance with the performance of operations. The proposed model of coordination strategies for maintenance and production

capacity has the advantage that access to view configurations are simple and easy to use, they graphically represent their functions and these can be taken to confront specific problems. However, it should be noted that industrial practice highlights the case of turbulence that dynamically change over time and the specific case management coordination problem of maintenance strategy and production capacity will show what is needed for the selection of the most suitable method of coordination.

4 CONCLUSION

The contribution is focused on a comprehensive understanding of the maintenance process in relation to the performance of the production plant and its optimal solution. It addresses issues allowing to find a suitable path of growth and efficiency in differences in production performance, i.e. to find a balance between the demands (which means less to worry about property) and quality/value (which means extension of property/production capacity) of the structure of the production company. These issues play a key role in determining the strategy of investment capacity, i.e. mechanism for the smooth running of production but also the development of operations and growth of property and equipment manufacturing company. The solution is designed afterwards that we should avoid considering only the financial aspects of maintenance from short term perspective against the renewal aimed at longer period of time, i.e. focus on sustainability of maintenance.

ACKNOWLEDGEMENT

The contribution was prepared under the solution of grant projects VEGA No. 1/0879/13.

REFERENCES

Ben—Daya, M., Duffuaa, S.O., Raouf, A., Knezevic, J. & Ait-Kadi, D. 2009. Handbook of Maintenance Management and Engineering. Springer Dordrecht Heidelberg London New York.

Jovane F, Yoshikawa H, Alting L, Boer C, Westkamper E, Williams D, Tseng M, Seliger G. & Paci A. 2008. The Incoming Global Technological and Industrial Revolution Towards Competitive Sustainable Manufacturing. CIRP Annals Manufacturing Technology 57(2): 641–659.

Kováč, J., Valenčík, Š. & Stejskal, T. 2012. Reference physical solutions for models of production systems. In: Transfer inovácií. č. 23 (2012), s. 204–206. ISSN 1337–7094.

Kováč, J. & Mihok, J. 2013. Priemyselne inžinierstvo. Edícia vedeckej a odbornej literatúry. SjF TU v Košiciach. ISBN: 978-80-553-0806-7.

Putallaz, Y. & Rivier, R. 2003. Modelling Long Term Infrastructure Capacity Evolution and Policy Assessment Regarding Infrastructure Maintenance and Renewal. In: Conference paper STRC 2003, Session Infrastructure and Logistic. Monte Verità/Ascona, March 19–21.

Rakyta, M. & Gregor, M. 2000. Total Productive Maintenance (TPM); Zavádzanie TPM v podnikovej praxi pri zvyšovaní produktivity práce. Medzinárodná konferencia, Národné f órum údržby. Žilina.: 29–32, ISBN 80-85655-15-2.

Rakyta, M. & Gabčan, B. 2000. Skúsenosti s aplikáciou TPM na Slovensku, Certifikovaný kurz "Prom ótor pre zavádzanie TPM v podniku", odborná konferencia, Národné fórum údržby. Žilina.: 33–39, ISBN 80-85655-15-2.

R. Keith Mobley. 2014 Maintenance Engineering Handbook, Eighth Edition. McGraw-Hill Education. New York. ISBN: 9780071826617.

Smith, R. & Hawkins, B. 2004. Lean Maintenance, reduce cost, improve quality, and increase market share. Elsevier Butterworth—Heinemann, 200 Wheeler Road. Burlington, MA 01803, USA.

Tolio, T., Ceglarek, D., Elmaraghy, H., Fischer, A., Hu, S., Laperriere, L., Newman, S. & Vancza, J. 2010. SPECIES-Co-evolution of products, processes and production systems. CIRP Annals—Manufacturing Technology, 59 (2): 672–693.

Valenčík, Š. 2011. Metodika obnovy strojov. Košice EVaOL TU, 330 s.

Valenčík, Š. Aktuálne témy z údržby strojov 2. 2012. AT&T Journal. Roč. 19, č. 7.

Valenčík, Š. 2013. Aktuálne témy z údržby strojov (8). AT&T Journal. Roč. 20, č. 1.

Valenčík, Š. & Kováč, J. 2009. Modelovanie, kvantifikácia a stratégia obnovy strojov. In: Transfer inovácií. č. 14 s. 204–207. ISSN 1337–7094.

Ueda, K., Takenaka, T. & Fujita, K. 2008. Toward Value Co-Creation in Manufacturing and Servicing. CIRP Journal of Manufacturing Science and Technology 1(1): 53–58.

Production Management and Engineering Sciences – Majerník, Daneshjo & Bosák (Eds)
© 2016 Taylor & Francis Group, London, ISBN: 978-1-138-02856-2

Analysis of linear programming utilization in Slovak and foreign industrial plants

L. Veselovská, J. Závadský & Z. Závadská
Institute of Managerial Systems, Faculty of Economics, Matej Bel University, Poprad, Slovak Republic

ABSTRACT: This research paper deals with process optimization via linear programming application. It presents the results of a survey conducted on a sample set consisting of industrial plants operating both in the Slovak republic and abroad. The scientific aims of this research paper is to explore the possibilities of linear programming utilization in process optimization among Slovak and foreign industrial plants and to create foundations for a model of process optimization in selected production company using linear programming methods. The object of this research is the selected production processes and the research subject of this paper is linear programming methods. We tested three related hypotheses in this paper with the most significant benefit being the creation of an image of recent linear programming applications in practice by Slovak industrial plants, mapping its economic benefits, and proposing requirements for a model of process optimization based on a linear programming foundation.

1 INTRODUCTION

This research paper focuses on process optimization via linear programming application. Linear Programming is a widely recognized method of solving various problems in managerial practice. It is considered a powerful tool for effective achievements. The main purpose of this paper is to explore the possibilities of linear programming utilization in process optimization among Slovak and foreign industrial plants. The research results were achieved through empirical study. Our results provide insights into current applications, which stressed the resulting benefits of a company and these results thus serve as guidelines for future linear programming applications. The results of this research have various implications for process optimization, which we explored further by creating the foundations for a model algorithm based on linear programming.

Our focus is on linear programming as an optimizing method. Sarker & Newton (2008), Buresh-Openheim et al. (2011) and Baker (2011) evaluated the advantages and disadvantages of linear programming utilization and considered the application of these methods for long-term production planning to have a significant advantage as they are relatively accurate. The use of linear programming assumes the creation of the linear objective function, which describes the problem as closely as possible. The variables also enable the close modeling of the company's conditions. One of the main disadvantages of linear programming utilization is the fact that sometimes the linear function may not be the best option for modeling a processes and a situation may arise when company would have to resort to other non-linear methods for operational research. Despite this fact, there are significant advantages for using linear programming utilization in companies as their application can help companies solve many different problems as various authors have found (Vanichchinchai & Igel 2011, Floudas & Lin 2005, Gass 2010, Das 2011, Avis & Umemoto 2003).

2 MOTIVATION FOR THE STUDY

In the 21st century, industrial plants faced severe competition that put severe pressure not only on their quality requirements, but also on their production processes. It is the goal of every company's operations management to ensure the best possible outcome and gain a competitive advantage enabling them to establish a desirable market position. However, it is not a single set of managerial decisions that make this possible. A strive for excellence is a continuous process which does not only involve establishing a good market position, but it also focuses on implementing measures necessary to maintain it. One of the effective ways companies can achieve excellence is through implementing specific measures in order to achieve flexibility and cost minimization of their processes. One of the basic tools is linear programming.

The theoretical aspects of process optimization are broadly covered at various Universities all around the world and that includes application of various optimizing tools including linear programming. Still, there is a question if such methods are currently being applied in businesses and how successful they are. Thus, we consider a closer look at both Slovak reality and the situation abroad to be extremely valuable.

3 RESEARCH METHODOLOGY

The main purpose of this paper is to explore the possibilities of linear programming utilization in process optimization among Slovak and foreign industrial plants. In order to fulfill this goal we use data provided by Slovak companies and foreign companies via a survey, which was conducted in the period between March 2014 and June 2014. Our research sample file was created as a representative sample of the base file. This file consists of Slovak companies classified by the SK NACE classification as production companies. Moreover, we took into account other criteria, mainly the size of company. We focused our research on medium-sized and large-sized companies, since we assume they had a higher extent of linear programming applications. The decisive criterion was set according to the European Standard No. 96/280/EC. Foreign companies were chosen in accordance with the same criteria in order to achieve the possibilities for cross-country comparisons.

Research was carried out on a set consisting of 1,300 Slovak production companies. The companies were selected randomly and chosen respondents were addressed by email. The questionnaire was fulfilled by 236 Slovak companies, which represents an 18.15% return. In key companies, we used the method of structured interviews with the company's representatives. The overall research sample consisted of 248 Slovak companies. The creation of a sample set of foreign companies was a more complex process. Firstly, we decided which countries we would like to involve in this research. We selected the Czech Republic, Australia, New Zealand, Poland, Germany and Austria. Secondly, we selected those companies in these countries, which met the criteria of our research. Consequently, a similar questionnaire was used to gather data from these companies.

Our questionnaire consisted of 16 questions divided into 3 categories. The first set of questions was focused on exploring various applications of optimizing methods and was completed by all the companies in the survey. The second part of the questionnaire involved questions designed to gain data about linear programming utilization.

This section was fulfilled only by companies that currently use these methods or have used them sometime in the past. Lastly, we also added the socio-economic questions created in order to gain data about respondents. We asked companies to provide information about their size (the number of their employees), sector of economy, and regions where they operate.

Our sample file of Slovak companies consists of 38.31% large-sized companies with over 251 employees. Nearly two thirds of companies in our sample file have 51 to 250 employees (Table 1).

The sector structure of research sample is presented in Table 2. For specification of the production sector, we used the Statistical Classification of Economic Activities in the European Community (SK NACE).

With the use of SPSS Statistics and information about the database set, we can verify the representativeness of this sample according to size and based on the data from the Statistical bureau of the Slovak Republic we can describe its character. In 2014, there were 70,370 manufacturing companies in Slovakia. The number of medium—sized companies (based on number of employees) was 1,641 and 627 were large companies. To verify the representativeness of the sample we used the Chi-square test. We set the null hypothesis, which assumes that the sample is representative. The alternative hypothesis is an assumption of non—representativeness of the sample. From a mathematical point of view, the hypothesis is formulated as:

$$H_0: F(x) = G(x); H_1: F(x) \neq G(x)$$

Statistical testing in SPSS software is based on the following formula (1):

$$X^2 = \sum_{j=1}^{r} \frac{(n_{j-m_j})2}{m_j} \approx X^2_{(r-1)} \qquad (1)$$

where X^2 is Pearson statistics,

r is line,
n is overall frequency in the base set,
m is measured frequency.

Table 1. Structure of Slovak sample file based on the size of company.

Number of employees	Companies	
	No.	%
51–250	153	61.69
Over 251	95	38.31
Total	248	100.00

Table 2. Structure of sample file based on the sector of economy.

Sector of economy	Slovak companies		Foreign companies	
	No.	%	No.	%
Food production	31	12.50	15	15.00
Manufacture (M.) of textiles and clothing	22	8.87	9	9.00
M. of leather	1	0.40	1	1.00
Processing of wood	17	6.85	8	8.00
M. of paper	15	6.05	7	7.00
M. of chemicals	39	15.73	9	9.00
M. of pharmaceutical products	16	6.45	2	2.00
M. of rubber and plastic products	5	2.02	5	5.00
M. of metals	35	14.52	12	12.00
M. of electronic devices	12	4.84	6	6.00
M. of machinery	36	14.52	19	19.00
M. of motor vehicles, trailers and semi-trailers	11	4.44	6	6.00
M. of other transport equipment	8	3.23	1	1.00
Total	248	100.00	100	100.00

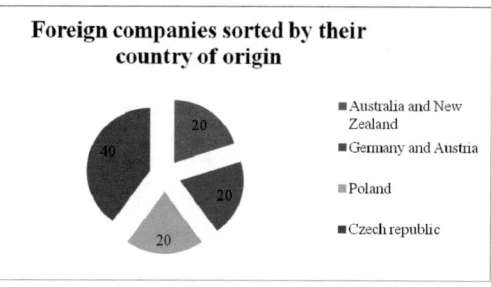

Figure 1. Foreign companies sorted by their country of origin.

Consequently, we find the critical value of $\chi 2$ distribution for $(r - 1)$ degrees of freedom and selected level of significance α from tables of critical values of the chi square. The Chi square test, however, requires the fulfillment of two conditions:

– no interval should have zero frequency;
– a maximum of 20% confidence intervals should have a frequency of less than 5 as discussed by Maloney & Byard.

The second part of our sample consisted of foreign companies with a focus on neighboring countries as well as New Zealand and Australia. This sample included 100 foreign companies. Figure 1 provides the data about these companies sorted by country.

Consequently, we also sorted these foreign companies by size. We used the same scale as was used with Slovak companies. We can therefore state that 61% of our sampled companies are medium-size and 39% are large-size.

We test one main hypothesis and two related hypotheses in this thesis. In the first hypothesis (H_1) we assume that the majority of selected production companies do not use linear programming. In hypothesis H_2 we assume that companies that use linear programming methods most frequently apply these methods for solving tasks of resource allocation. Hypothesis H_3 tests the assumption that the majority of selected companies, which use linear programming methods, consider the reduction of raw material costs to be the most important economic benefit of linear programming application.

4 RESEARCH RESULTS

In this part of the paper, we present the results of a survey conducted in order to evaluate the current state of linear programming utilization in process optimization among Slovak and foreign industrial plants. In the first part of the questionnaire, we have obtained information about the optimization of company processes in general, regardless of the used method. We have found that 68.55% of selected production companies are using optimization methods regularly.

The use of the optimization methods is equally divided between positive and negative answer in the subset of Slovak medium—sized companies. In larger companies, we can mostly identify positive answers with 93.68% of them regularly using optimization methods. Only 6 companies with more than 251 employees do not use any optimization methods (6.32%). These findings indicate that the use of optimization methods in industrial production practice is proportionally dependent on the size of the company. If a company applied optimizing methods, we further explored the various aspects of optimization. Firstly, we focused on what types of techniques these companies use with emphasis on linear programming. We found 41.76% of Slovak companies that applied optimizing tools chose linear programming—representing 28.63% of all companies in the Slovak subset. Since this aspect represents the focus of our research, we then look more closely at this aspect under different circumstances. Firstly, we analyzed utilization of linear programming based on the size of companies.

The results of this research showed us that linear programming is mostly used by large-sized companies in Slovakia, 42.11%, compared to only

Table 3. Application of linear programming structured by size of Slovak companies.

Application of LP	Yes		No	
	No.	%	No.	%
Medium-sized companies	31	20.26	122	79.74
Large-sized companies	40	42.11	55	57.89
Total	71	28.63	177	71.37

20.26% of medium—sized companies in process optimization. The difference between large-sized and medium—sized companies in usage of linear programming is rather significant (Table 3).

Moreover, we assumed that there is dependence between linear programming application and the size of companies. In order to confirm this assumption we performed the Pearson correlation test. We indeed discovered that there is a medium-strong direct dependence between these two factors. Therefore, our first significant finding is that managers of bigger production companies in Slovakia are more likely to choose linear programming as their optimizing tool.

We achieved similar results in the group of foreign companies. Based on data from the questionnaire survey, we found that 68% of surveyed foreign companies regularly optimize their processes, however only 42.65% of them (29% of all foreign companies) use linear programming as their selected optimization tool. The differentiation of the research sample by country of origin revealed that the rate of linear programming application is particularly higher in companies located in Australia and New Zealand (at 35%). Linear programming is also significantly used in German and Austrian companies (at 30%). Despite the higher rate of linear programming utilization among foreign companies, this rate is substantially lower than a 50% share.

These findings are also related to our hypothesis H_1. Since we found that only 28.63% of Slovak companies use linear programming, the hypothesis was confirmed. Moreover, we used a binomial test to serve as further evidence to this finding. Even though the rate of linear programming utilization was slightly higher in the group of foreign companies, we were unable to reach the rates necessary to disprove our hypothesis.

The following part of the questionnaire involved only companies that used/use linear programming for process optimization. Firstly, we have investigated the most important barriers of linear programming utilization in these companies. Since it was an open question, we got a variety of answers. However, we consider it very interesting that the variability of statements related with this question

was not significantly affected by country of origin. We received similar results regardless of location. The most frequent barrier was the size of the company. This barrier was especially important for medium—sized companies (51 to 250 employees). One of the most frequent answers was the inadequacy of linear programming for the needs of the company. This barrier is also related to the use of other optimization methods, especially if the company is satisfied with the achieved results of the optimization process. The lack of knowledge of linear programming by managers and subsequent distrust of these methods and achieved results were also problematic. Managers tend to use other, better-known or previous optimization method. One of the barriers of the linear programming application was the price of the software. The most problematic and hardest barrier to remove was the satisfaction of the managers with the achieved results. That is why managers pay less attention to optimization, which can consequently cause stagnation and an inevitable reduction in quality of manufactured products. These facts can cause the reduction of competitiveness as well. In practice, this fact seems to be a problem of many managers who prefer short-term positive results to sustainable future competitiveness. Inattentive long-term vision is a serious obstacle to the application of linear programming in companies both in Slovakia and abroad.

We also explored the various possible applications of linear programming. Various types of production processed were selected as provided options. Our goal was to find out which are the most common applications to the analyzed methods. Table 5 provided this information gave us an accurate image. Therefore, we also include those possibilities that were not selected. The percentage provides information about how many of all companies in the sample file choose that particular application (Table 4).

Another significant factor we wanted to explore was the linear programming applications in terms of the sector of economy in which company operates. We discovered that linear programming is mostly used by companies manufacturing chemicals and chemical products with resource allocation (27.66% Slovak companies). These companies often use linear programming also to create their production plan (23.4% Slovak companies). However, these companies also apply optimization to other processes. We marked the highest diversification of results in the group of these companies, e.g. the highest amount of different applications. Our findings also indicate that Slovak companies processing wood and manufacturing products made of wood and cork often apply linear programming to create their cutting plans. We also find that no

Table 4. Application of linear programming on different types of production processes structured by the location of company.

Optimized process	Slovak companies		Foreign companies	
	No.	%	No.	%
Resource allocation	31	43.66	13	44.83
Creating production plans	30	42.25	14	48.28
Creating cutting plans	39	54.93	18	62.07
Personnel management	7	9.86	1	3.45
Mixture preparation	8	11.27	3	10.34
Waste management	1	1.41	2	6.90
Distribution plans	14	19.72	5	17.24
Financial management	8	11.27	1	3.45
Supply-chain manage.	0	0.00	1	3.45

Table 5. Cost reduction structured by areas in the company.

Type of cost reduction	Slovak companies		Foreign companies	
	No.	%	No.	%
r. of raw material costs	41	57.75	22	75.86
r. of labour costs	23	32.39	6	20.69
r. of overheads	23	32.39	9	31.03
r. of administrative costs	0	0.00	1	3.45
r. of selling expenses	15	21.13	4	13.79

production company in our sample file applied these methods to supply chain optimization. This application can bring company many benefits and as such, we consider this area to be a gap to be fulfilled in the future. The results in the group of foreign companies were similar. The most common application of linear programming is also in the process of creating cutting plans. The biggest difference was in the rate of financial management use of linear programming application as it was significantly lower among foreign companies. These findings served as a foundation for verification of hypothesis H_2. We were unable to confirm this hypothesis despite the fact that resource allocation is a significant linear programming application. However, linear programming is not commonly applied in this process optimization, either in Slovakia or abroad.

One of the last questions included in the questionnaire was focused on analyzing the economic benefits of linear programming application in the management of industrial companies. The most general advantages are involved with various types of cost reductions. We explored the opinions of corresponding managers of companies on this subject (Table 5).

According to the data presented in this table, we can state that the most significant benefit of linear programming application is the reduction of raw material costs in both Slovak and foreign companies. Medium-sized Slovak companies consider reduction of labour costs and reduction of production overheads as important advantages of linear programming application. On the other hand, reduction of administrative costs is not perceived as a significant benefit by any Slovak company. Moreover, only one foreign company considers it important. We also looked at these benefits dependent on the sector of economy where the

company operates. According to provided data, emphasis can be placed on the fact that reduction of raw material costs is perceived as a significant benefit by companies from all economic sectors that used linear programming in optimization. These finding allowed us to confirm hypothesis H_3.

5 DISCUSSION AND CONCLUSION

This research paper focuses on application of linear programming in practice. The scientific aim of this research paper was to explore the possibilities of linear programming utilization in process optimization among Slovak and foreign industrial plants and to create a model of process optimization in selected production company using these methods. Based on the survey analysis, we can conclude that this objective was fulfilled.

We found that according to the provided data, nearly a third of medium-sized companies choose linear programming as their optimizing tool. We therefore concluded that the size of company is an important factor of linear programming application. The extent of utilization was significantly higher in the group of large-sized companies.

Moreover, we explored various aspects of practical applications of linear programming in terms of the size of company and the sector of the economy where company operates both in Slovakia and abroad. Consequently, we assessed perceived advantages of linear programming applications. Our focus included various types of cost reductions. In terms of our contribution and originality, we consider this to be the most significant aspect of this paper. The value of this article lies mainly in providing a detailed insight into the extent of linear programming applications in managerial practice of Slovak production companies and indications for possible differences in comparison to companies operating abroad. Since our sample file of foreign companies was rather small, we used data provided by them only as outline for comparison and for assessing possible differences.

These findings can serve as a foundation for further, more extended research.

Based on the information concluded from the conducted research we can assess not only the present situation of linear programming application in practice, but we can also predict basic trends for the future. Methods of linear programming are broadly covered as topics of education at universities all around the world. However, we found less than a third of Slovak production companies utilize these methods. This is merely a problem of perception of managers since they perceive their company to be "good enough". Therefore, such managers do not feel the need to seek the best optimization methods. Another both current and future problem involves perception. Many managers consider linear programming to be a set of very simple methods and in their opinion; linear programming is not suitable and instead favor other method. This problem is sometimes caused by personal characteristics of managers, especially factors concerning risk avoidance. In general, when manager do not thoroughly know linear programming, the person usually chooses other—more commonly known optimizing techniques. Academic experts clearly consider these methods suitable for practical applications whereas managers of production companies tend to favor other optimization methods. Overall, linear programming is simultaneously a very useful optimizing tool and a highly underestimated one.

Moreover, these findings served as a foundation for creating a model algorithm of selected process optimization. All of the above findings took into account the essential required basis and these guidelines provided some necessary information and helped us set:

– the objective of model algorithm—the specification of type of cost reduction that would make this model algorithm more suitable for practical needs in company and that would make its application more beneficial;
– the requirement of flexibility—implementation of various measures designed to increase flexibility of particular operations;
– the orientation of the model—since supply chain optimization is absolutely absent among linear programming applications in Slovakia, we decided not to model only the production process itself, but the whole logistics process from receiving of an order, the creation of products, and their final delivery to the customer;
– the types of secondary tasks this model algorithm needs to focus on—since linear programming is most commonly apply towards creating cutting plans and resource allocation, these problems had to be addressed (if applicable) to our model algorithm.

REFERENCES

Avis, D. & Umemoto, J. 2003. Stronger linear programming relaxations of max-cut. *Mathematical Programming* 97(3): 451–469.

Baker, K.R. (1st ed.) 2011. *Optimization modeling with Spreadsheets*. Hoboken: Wiley.

Buresh-Oppenheim, J., Davis, S. & Impagliazzo, R. 2011. A Stronger Model of Dynamic Programming Algorithms. *Algorithmica* 60(4): 938–968.

Das, K. 2011. Integrating effective flexibility measures into a strategic supply chain planning model. *European Journal of Operational Research* 211(1) 170–183.

Floudas, C.A. & Lin, X. 2005. Mixed Integer Linear Programming in Process Scheduling: Modeling, Algorithms, and Applications. *Annals of Operations Research* 139(1): 131–162.

Gass, S.I. (1st ed.) 2010. *Linear Programming: Methods and Applications*. New York: Dover Publications.

Maloney, T. & Byard, K. (1st ed.) 2013. *Quantitative Methods for Business*. Auckland: Pearson New Zealand Publishing.

Sarker, R.A. & Newton, C.S. (1st ed.) 2008. *Optimization Modelling: A Practical Approach*. Sydney: CRC Press.

Vanichchinchai, A. & Igel, B. 2011. The impact of total quality management on supply chain management and firm's supply performance. *International Journal of Production Research* 49(11): 3405–3424.

Production Management and Engineering Sciences – Majerník, Daneshjo & Bosák (Eds)
© 2016 Taylor & Francis Group, London, ISBN: 978-1-138-02856-2

Methodical procedure of fleet renewal

T. Vondráčková
The Institute of Technology and Business in České Budějovice, České Budějovice, Czech Republic

V. Voštová
CTU in Prague, Prague, Czech Republic

ABSTRACT: Global market for earth-moving machinery peaked in 2008. Since that year, construction output index decreased in the Czech Republic by 6.7%. It raised again no sooner than in 2014. The expected development in structural engineering, which includes almost all the ground works, is 5.1% for 2015. The slump in the economy has been addressed in the Czech and Slovak Republics by radical cuts, drastic restrictions on public investment with various negative impacts. For gradual recovery, it is necessary to spend investments effectively, on the basis of modern methods. Methodical process of machinery fleet renewal will be carried out by means of machine operation indicators and evaluation of economic efficiency of investment. There will be described and explained the methods of analysis of the current state and the choice of investments using multi-criteria methods and financing of the given investment.

1 INTRODUCTION

Methodical procedure of fleet renewal should depend on future orders for the company and on detailed analysis of the current situation. For every change of a machine we must create a table of available machines in the market and they must be assessed by individual indicators. These are above all indicators of machine operation, which includes the price of mechanized work, duration of use, and the purchase price with an annual machine utilization, and evaluation of economic efficiency of investment, which includes a number of indicators, whose choice depends on whether it is a short-term or long-term investment.

Choice of the investment is also assessed by technical-technological, organizational and economic assessment method of multi-criteria evaluation.

2 INDICATOR OF THE MACHINE OPERATION

Generation of profits is affected by the main variables of marketing, such as revenue from operation of the machine and operation costs of the selected type of machines with purchase price and the form of financing and with respect to the impact of the period of use and change in operating parameters, depending on the time. From these variables we can emphasize the combination of price of mechanized labour in the market with duration of use, purchase price and annual machine utilization. Other marketing variables, i.e. technical level of the equipment, technological suitability, theoretical performance, are considered as default and necessary criteria for the acquisition of machinery and its subsequent use for effective work.

Operating costs of the machine are an important indicator of the machine operation and also a criterion for comparison when purchasing new equipment. The cost of operating the machine has two basic components, fixed and variable, while the default for fixed cost tracking is annual time horizon and the default for cost tracking is an expression per unit of processed area, amount or hour of work. Along with the cost-benefit analysis in the function of machine utilisation time, it is necessary to count with annual machine utilization, which is the basis for the conversion of fixed annual costs to unit costs and variable unit costs to annual variable costs.

Fixed costs include the costs of depreciation, return on equity, combined with interest on loans or margins of financial leasing, the cost of garaging, insurance and taxes. These costs are independent on the annual use.

Variable costs include the costs of fuel and grease, repair costs, wages and operating costs and other expenses (Vondráček 2008).

3 EVALUATION OF THE ECONOMIC EFFECTIVENESS OF INVESTMENTS

Evaluation of the effectiveness of investment projects and their selection (e.g. the purchase of equipment, machinery) is carried out by the financial management. Methods for assessing the effectiveness of investments are generally divided according to whether they take into account the factor of time or not. However, the preferable methods respect the time factor. When selecting projects, it is necessary to ensure comparability. Finally, it is important to ensure real input data on capital expenditure and cash income from the investment.

The methods used for assessing the effectiveness of investment projects are understood as an effect of investment minus the income taxes. From a financial standpoint, however, the financial gain does not constitute a total cash flow of income from investments, because it does not include the income in the form of depreciation. Using different depreciation policy may decrease (increase) the profit and thus to some extent it may influence the view of the efficiency of investment projects as measured by earnings.

The methods of net present value and internal rate of return are used. The payback period is usually used as an additional aspect of the decision.

Net present value represents the absolute amount of the difference between the current value of investment revenues and the present value of expenditures spent on the investment. If the result is positive, the project is acceptable for the company. When selecting the optimal variant, we prefer the one with higher net present value.

Internal rate of return is the interest rate at which the present value of expected investment returns equals to the present value of the cost of the investment. According to the internal rate of return acceptable investment projects are those that represent a higher interest rate than the required minimum return on investment. When comparing multiple variants of investment projects the better option is the one with a higher internal rate of return.

Payback period is the time after which the investment of cash revenues returns, which it has ensured—after taxation and depreciation. The shorter the payback period, the more favourably the investment is evaluated (Hyrlišová & Klečka 2010).

4 METHOD OF MULTI-CRITERIA ASSESSMENT

This is the assessment of the technical—technological and economic—organizational criteria according to scoring.

Technical and technological criteria (requirements) for each group of earthmoving machines are based mainly on working conditions in which the machine will be used. The operators of the machine, based on their experience (e.g. a previously used machine) and anticipated trends in

their future activities, must try their best to specify technical requirements for each group of machines (range the work tool, nominal machine pressure on the underlay, the strength of the cutting forces, etc.). They must determine which requests are deemed necessary (in their absence the equipment cannot be used) and which requirements are deemed useful.

Organizational and economic assessment of other aspects of selection are very variable and in some cases represent very specific requirements depending on the "subjective" conditions in a plant, its location, market position, tradition, etc.

As an example of various aspects let us mention the following:

- efficiency of the investment,
- conditions of warranty service,
- spare parts availability (time to the repair completion),
- experience with similar machines,
- service intervention by the number of operating hours,
- the hourly rate of the fee for regular service,
- the cost of operating filling (without fuel),
- the price of spare parts (filters, etc.),
- distance of the service centre,
- the possibility (and possibly the cost) of a divestment (Vondráčková 2008).

5 METHODS OF ANALYSIS OF THE CURRENT STATE

Analysis of the current state is the first step in planning activities. Before planning it is necessary to ask which objectives have been set and compare them with reality. To understand the causes of the current state better, it is necessary to conduct a situational analysis, which focuses on internal and external factors that influenced the current situation. In our case there will be used the most frequently used analyses of the current state called SWOT analysis and Porter's model.

The SWOT method is one of the initial company state analysis models, whose specification allows setting the appropriate planned tasks and strategies selection. Strategy can be understood as a way of future business management solution that maintains or improves its competitive position, neutralize threats coming from outside, benefits from the opportunities and considers the strengths and weaknesses of the company.

Using SWOT analysis enables:

- identification of the status and trends in internal and external environment,
- identification of strengths and weaknesses and assessment of their potential impact on the realization of strategic business objectives,

- identification and evaluation of potential opportunities and threats.

The main advantage of the SWOT analysis is its simplicity, clarity, readability and illustrative nature.

Porter's model is an analysis, which aims to find a position in the business sector, which the company can best defend against competitive forces or can influence them in its favour. Knowledge of these fundamental forces of competitive pressure will clearly show the strengths and weaknesses of the company, its apparent position in the industry, clarify the areas in which changes can be the most advantageous, highlight places where industry trends offer the greatest opportunity or where threat may come from.

The company should isolate itself as much as possible from the effect of competitive forces, or use these competitive forces to its advantage (Štrach 2008).

6 METHODICAL APPROACH WHEN CHOOSING A DOZER

Criteria assessment method has two levels. First, there must be data collection on earthmoving machines—let us consider a dozer as an illustrative example. Then follows a criterion based selection of technical—technological and organizational—economic indicators. It is a selection of the most important parameters from the user's perspective, their score evaluation, and finally the selection of machines with the highest number of points accumulated.

Data collection for the group of earthmoving machinery is an essential part of the criteria assessment method when selecting a machine. It is necessary to choose the largest possible number of machines from which we choose the best three, according to the specified selection criteria. This trio will be assessed using the criteria assessment method.

Data collection uses the documentation acquired during visits to trade fairs (e.g. IBF Brno, SIMA Paris, BAUMA Munich) and it is based on promotional material, the Internet and personal contact with individual producers or distributors.

The aforementioned trade fairs are the most important fairs in the field of construction and agricultural machinery, a sector where dozers are the most frequently used for ground works. There are presented also the largest manufacturers of construction equipment. The professional community meets the public here so they can jointly get acquainted with revolutionary innovations that many manufacturers present here for the first time.

Promotional materials along with the Internet are another great source of data collection. All manufacturers now routinely present their machines in an exhaustive catalogue, and websites are now a standard information source too. It is possible to learn a lot of technical information this way. The machines can also be visually viewed in various animations.

Personal contact with manufacturers and distributors of construction equipment is an essential part of data collection, as this will provide the necessary data that cannot be found in available materials. Furthermore, there is a great opportunity to see selected machines, learn a lot of important information from real life experience, e.g. machine failure rates, maintenance, pricing etc.

Based on the above sources, data was gathered from top three manufacturers of construction equipment with a wide range of different performance classes of dozers. The manufacturers are—Caterpillar, Case—New Holland and Komatsu and dozers in the middle performance class from 80 to 110 kW.

Table 1 lists the calculated values needed for the method of multi-criteria evaluation. The relevant equations are given in the literature (Hyrlišová & Klečka 2010; Vondráčková 2008; Szabo et al. 2013). Figure 1 shows a scalar product of points and weights.

7 CONCLUSION

The paper gives a methodical description of the selection of the earthworks machine—a dozer—when the fleet undergoes renewal. The selected dozer is Komatsu D41P-6C.

The next step should be to tackle the financing of the purchase of a new machine, which can be implemented either through own resources or from long-term borrowings (Szabo et al. 2013). The own sources include depreciation and profits, external sources are investments of owners or venture capital. The most commonly used external sources are long-term investment loans, bonds and finance leasing.

It should be noted that the proposed procedure is strongly influenced by the human factor, either in a positive or negative sense. Already in the proposal of potential manufacturers and suppliers of machinery the human factor is manifested, it takes into account e.g. previous collaboration in machine supply, discounts provided by vendors for regular buying of machines etc. Also the scoring method is influenced by the subjective views of individual criteria. However, if we renewed the fleet in a lay way, i.e. without using the systematic selection methods, we would certainly not spend

Table 1. Values for multi-criteria assessment.

Technical and technological criterias required			CAT D6K		KOMATSU D41P-6C		CASE–NH D 150 STD	
The purchase price of the machine to 4,0 mil CZK			YES		YES		YES	
Blade width min. 3 m			YES		YES		YES	
Delivery term max. 12 weeks			YES		YES		YES	
Ripper			YES		YES		YES	
Service within 24 hours			YES		YES		YES	
Preventive diagnostic examinations			YES		YES		YES	

Technical and technological criterias necessary	Weight [%]	Max. points	CAT D6K Points		KOMATSU D41P-6C Points		CASE–NH D 150 STD Points	
Engine power [kW]	8	8	93	7,1	82	6,3	104	8,0
Forward speed [km.h^{-1}]	4	4	10,0	3,6	7,6	2,8	11,0	4,0
Speed reverse [km.h^{-1}]	4	4	13,3	4,0	9,4	2,8	13,0	3,9
Operating weight [kg]	5	5	12886	4,5	11520	5,0	14860	3,9
Type of blade	5	5	S	5,0	S	5,0	S	5,0
Capacity of blade [m^3]	5	5	2,9	4,7	2,9	4,7	3,1	5,0
Weight of blade [kg]	5	5	2300	4,0	2150	5,0	2670	4,0
Blade width (outer lips) [mm]	7	7	3077	4,7	3045	6,7	3180	7,0
Biggest turning angle blade [°]	5	5	25	4,2	30	5,0	28	4,6
Largest tilt angle blade [°]	5	5	8	5,0	8	5,0	5	3,1
Fuel tank capacity [l]	5	5	295	5,0	250	4,2	270	4,6
Fuelling without PHM [l]	5	5	114	2,2	154	3,0	254	5,0
Cabin FOPS-ROPS	5	5	yes	5,0	yes	5,0	yes	5,0
Outlook	5	5	stand.	5,0	stand.	5,0	stand.	5,0
Air conditioning	5	5	yes	5,0	yes	5,0	yes	5,0
Width of track shoe [mm]	5	5	560	4,0	700	5,0	550	3,9
Number of track rollers belt [p]	5	5	7	5,0	7	5,0	7	5,0
Pressure on soil [kPa]	5	5	42	2,8	24	5,0	54	2,2
Maximum depth of scarifying [mm]	5	5	360	3.6	350	3,5	495	5,0
Alternator [A]	2	2	95	2,0	60	1,3	70	1,5
Total	100	100	–	86,4	–	90,3	–	90,7

Economic—organizational criteria	Weight [%]	Max. points	CAT D6K Points		KOMATSU D41P-6C Points		CASE–NH D 150 STD Points	
Effectiveness of investments	60	60	–	–	–	–	–	–
Net present value [mil. CZK]	25	25	11,85	25	11,68	24,6	11,63	24,5
Internal rate of return [%]	25	25	4,6	24,6	4,66	25	3,38	18,1
Payback period [year]	10	10	2,93	6,3	1,89	9,8	1,86	10
Purchase price [mil. CZK]	5	5	3,9	4,6	3,7	4,8	3,6	5,0
Maintenance interval [hours]	10	10	400	10	350	8,7	400	10
Service intervention after operating hours	7	7	500	7	400	5.6	400	5,6
The hourly rate for service [CZK]	8	8	500	7,2	520	6,9	450	8
Time to services [hours	5	5	12	5	12	5	24	2.5
repayment schedule	5	5	–		–		–	
Total	100	100	–	89,7	–	90,4	–	83,7

Scalar product of points and weights	Weight [%]	Max. Points	CAT D6K Result	KOMATSU D41P-6C Result	CASE–NH D 150 STD Result
Technical and technological criteria	50	100	8,64	9,03	9,07
Economic—organizational criteria	50	100	8,97	9,04	8,37

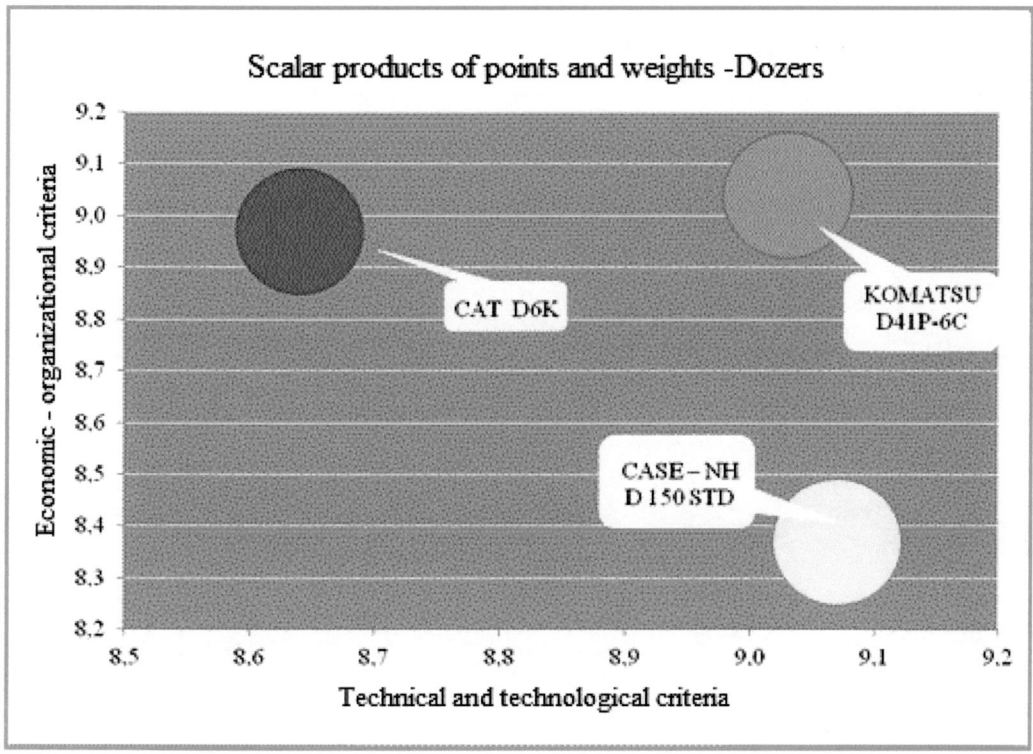

Figure 1.   Scalar product.

our investments efficiently. And this is our primary objective in the upcoming period, when an increase in construction output is expected.

REFERENCES

Hyrlišová, J. & Klečka, J., 2010. Ekonomika podniku. *Vysoká škola ekonomie a managementu, Prague.*

Štrach, P., 2008. Principy marketing. *1. vydání. Vysoká škola ekonomie a managementu, Prague.*

Szabo, S., Ferencz, V. & Pucihar, A., 2013. Trust, Innovation and Prosperity. *In: Quality Innovation Prosperity,* Technical University of Kosice, Volume IX, Issue 4(30).

Vondráček, J., 2008. Znalostní systém pro výběr zemních strojů. Disertační práce, *ČZU in Prague,* p. 120.

Vondráčková, T., 2008. Automatické řízení technologických procesů ve stavebnictví. Disertační práce, *ČZU in Prague,* p. 108.

Production Management and Engineering Sciences – Majerník, Daneshjo & Bosák (Eds)
© 2016 Taylor & Francis Group, London, ISBN: 978-1-138-02856-2

Test procedure of the window systems reaction to a shockwave load

Z. Zvaková
Faculty of Security Engineering, University of Zilina, Zilina, Slovak Republic

ABSTRACT: Explosion has various effects on its environment. Out of all possible impacts of the explosion, the shockwave, specifically overpressure in its front, has the most destructive character. The elements of the shell protection, unless there is a requirement, are not designed for such specific load and their reaction to this is unknown. There are currently no technical standards that regulate the resistance of commonly used elements of the shell protection when exposed to a shockwave. There exist technical standards which focus on the resistance of certain elements which are structurally made in the way to improve their resistance against the impacts of explosion. These technical standards are used to certify these elements into various classes of resistance to detonation. Our paper is focused on the issue of the reaction test or the resistance test of the opening fillings when hit with the shockwave of the explosion in the open area.

1 INTRODUCTION

Terrorism is a threat to the global security of the 21st century. The increase in the spread of weapons of mass destruction such as Improvised Explosive Devices (IED) is very troubling.

Explosives and their effects provide a wide range of their use against society. There are many demonstrations of such activities as well as of the means of their perpetration. Examples are vandalism, organised crime or terrorism. When we discuss the use of explosives against society, the first eventuality that occurs almost to everyone is the terrorism. The reason for this is the fact that the bombing is the most common type of a terrorist attack (Jangl, 2012). However, there are also other areas of antisocial activities where the explosives can be used.

Explosives, in case they are used against the society, are the means to kill a human or harm his health, eventually, destruction or damage of the property or objects. Explosives are also means which can be used to overcome the protection system of the premises and gain the access to a protected interest.

In our paper, we focus on the area of damage or more precisely on the destruction of the shell protection of the object—opening fillings, when they are exposed to the pressure load from the blast. Our aim is to suggest the working procedure of identifying the resistance of window systems hit by the shockwave formed after the explosion.

2 SELECTED TECHNICAL STANDARDS REGULATING THE GIVEN AREA

Currently, there are various national and international technical standards regulating the problem of the shell protection resistance of the object against the effects of the explosion. The paper provides a description of selected technical norms.

BS EN 13541—Glass in the building industry. Safety glazing. Testing and classification of the resistance to the explosion pressure. This technical standard regulates the testing and classification of the safety glazing resistance in the building industry against the explosion pressure. Classification of the resistance against the explosion pressure is based on the maximum pressure of the shockwave arising from the explosion and the acting time of the positive phase of the shockwave. The technical standard regulates the requirements that the test sample needs to meet and describes the procedure for carrying out the test, together with the description of all the equipment necessary to perform the test (support of the test sample, shock tube or other device to create required conditions and measuring equipment). The classification of the glazing resistant to the explosion pressure is divided into four classes according to the maximal gauge pressure affecting the test sample and the positive impulse. According to the classification BS EN 13 541 the duration of the positive pressure phase must be ≥20 ms (BS EN 13541).

BS EN 13123-1 Windows, doors and shutters. Explosion resistance. Requirements and classification. Part 1. The technical standard regulates requirements and classification of the explosion resistance of the windows, doors and shutters. Part one is designed for the test using a shock tube. According to this norm, there are four classes of classification following the maximal pressure (0,50–2,00 bar) and a specific positive impulse. When classifying according to BS EN 13123-1 the duration of the positive pressure phase must be ≥20 ms. The standard also provides mathematical relations valid for the maximal pressure of the shockwave, positive impulse and the duration of the positive phase, as well as the conditions, procedures and equipment necessary to perform the test (BS EN 13123-1).

BS EN 13123-2 Windows, doors and shutters. Explosion resistance. Requirements and classification. Part 2. The technical standard regulates requirements and classification of the explosion resistance of the windows, doors and shutters. Part two is designed for the test in open space. The standard specifies the criteria which must be met to achieve the classification. Based on the resistance of the tested samples, the standard defines classification groups whereby each group has a set distance of the tested sample (3–5.5 m) away from the place of explosion and the amount of explosives (3–20 kg) which should be used in the test. In the description of the explosives, the norm refers to the Annex "A" of the technical standard BS EN 13124-2 (BS EN 13123-2).

BS EN 13124-1 Windows, doors and shutters. Explosion resistance. Test method. Part 1. The technical standard describes the test method of the resistance to explosion for windows, doors and shutters. The standard is designed for the test using the shock tube (BS EN 13124-1).

BS EN 13124-2 Windows, doors and shutters. Explosion resistance. Test method. Part 2. This technical standard describes the test method of the resistance to explosion for windows, doors and shutters. The standard is designed for the test in open space. The standard contains a description of the test sample, the method of attachment, equipment necessary for the test, characteristics of the environment where the test is to be performed and the process of its performance and evaluation (BS EN 13124-2).

The explosive used in the test according to this norm is trinitrotoluene, thereinafter TNT.

The primary raw material for its production is toluene. The most advantageous in terms of transportation, storage and use is TNT in the form of flakes, therefore this is the most widely used form. TNT crystallises as light yellow crystals, its melting point in pure form is 80.7°C. TNT is insoluble in water and well soluble in organic solvents, particularly acetone. It turns brown at light which is accompanied by a small increase in sensitivity, it is neutral and does not react with metals, however, it reacts with the alkalis and oxides of the alkaline metals. Moisture has no effect on its stability and it has very low sensibility to mechanical influences (Jangl, 2012).

During the test, the explosive should be formed into a shape of ball, placed and initiated according the procedures given by this standard. Other alternative sources of explosion are allowed, but only in case the equivalent effect is proved.

On the ground of the resistance of the test samples, the standard defines classification groups whereby each group has a set distance of the tested sample (3–5.5 m) away from the place of explosion and the amount of explosives (3–20 kg) which should be used in the test.

GSA-TS01-2003—US General Services Administration Test Standard. This test standard is used to ensure the adequate measure of quality and standardisation in the field of testing window systems. It also regulates the area of so called secondary restraint mechanisms used to reduce the impacts of shards. It is applicable with tests in open space as well as with the use of shock tubes. The standard is used for certification of window systems resistant to the effects of the shockwave from the explosion.

ASTM 1642-1604—ASTM Test standard. Technical standard often used by the security forces in the USA. In its purpose and structure, this technical standard is very similar to GSA-TS01-2003.

Technical standards regulate breakthrough resistance of the opening fillings when exposed to pressure effect of the explosion, only with the aim to certify the elements for each class of explosion resistance. This raises the following problems. There is no standard that would regulate the resistance to the effect of the blast for conventionally used window systems. The marginal values of the resistance and the reaction of the commonly used window systems when hit with the shockwave are not known. Without introduction of specific requirements for increased resistance of these elements, there is no procedure, reaction test of window systems or other elements of shell protection to shockwave load. This situation is deemed as serious deficiency in the field of protection of people and property.

3 PROPOSITION OF THE WORK PROCEDURE OF THE REACTION TEST FOR WINDOW SYSTEMS HIT BY THE IMPACT OF EXPLOSION PRESSURE

Our proposal specifies the procedure to determine the reaction of window systems and their parts to the hit by a shockwave arising at detonation of explosives.

3.1 *The focus of the working procedure, description of the test sample and its fixation*

The procedure is intended for the test of resistance to the impact of the pressure generated by the use of explosive in open space. The procedure is designed for the complete window systems, including panel, frame and mounting.

The objective of the test may be the determination of the window system reaction to the load of specific pressure or the reaction to the pressure load at explosion of a specific booby trap system (specific explosive, shape of IED, its packaging, etc.).

Test sample must consist of the complete window system. The test must be complemented by drawing documentation describing the measurements and details of the test sample in scale, together with the description of all material used to create it. The documentation must specify the conditions for which the window system is designed. Fixation of the test sample and the explosive charge can be seen in Figure 1.

In case the tested window systems are intended to be used in specific climatic conditions, it is necessary to establish the same condition under which the test samples should be used, before carrying out the test. In case the tested window systems are certified as resistant to the blast effect of the explosion, it is necessary to draw from the conditions which determine the minimal values

for the class of resistance to which the window system is certified.

The side exposed to the pressure impact of the explosion must be clearly identified.

Fixture of the test window system must consist of a solid construction. The construction used for fixation:

– Must be sufficiently sturdy to prevent any deformation and subsequent damages.
– There must be no transfer of deformation on the test sample.
– Must provide the mounting of the test sample in the same way it is mounted in real conditions and thus ensures that there is no abnormal tension in the test sample and at the same time the real fixation of the sample is also a subject to the pressure load.
– Must prevent abnormal impact of the pressure from the back side of the test sample.
– Must allow an increase in the load.

3.2 *Description of the explosion charge and equipment of the test*

The charge and its properties derive from the objective of the test. Generally, TNT explosive is considered as appropriate. This charge is especially useful at identifying response to the load of specific pressure.

Because of the availability of raw materials and processes for production of ANFO explosives (ammonium nitrate + fuel), even other types of explosives are acceptable depending on particular requirements.

It is necessary to shape the charge in such way to ensure the even spread of the pressure, eventually so that the shape of the charge corresponds with the demands related to the objective of the test (reaction test of the window system to the explosion of a specific type of booby trap system, e.g. in a bottle, luggage, etc.).

The amount of explosives and their placement (distance from the test sample) are determined by the calculations based on known mathematical relations.

The placement of the charge and its initiation must be in accordance with the appropriate procedure of pyrotechnic disposal. Deployment of the pressure sensors is in Figure 2. The sensors are positioned in calculated distance around the charge, in order to ensure the evenness of the scanning of the shape of shockwave.

In the calculation, it is possible to use various mathematical relations (Henrych 1973), (Mills 1987), (Makovička 2008). Calculation processes which determine this type of loaded constructions are dealt by (Vávra 2002), (Figuli 2014), (Figul 2013).

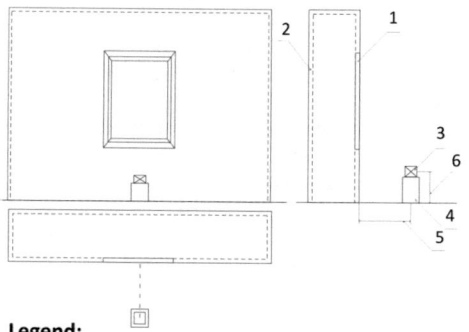

Legend:
1 The window system
2 Anchorage system of the window - back
3 The charge
4 Stand for the charge
5 Distance between the charge and windows system
6 The height of the charge

Figure 1.   The fixation of the test sample and placement of the explosion charge.

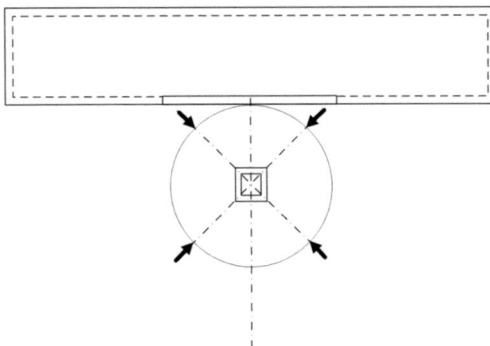

Figure 2. Deployment of the pressure sensors.

When using the relations by (Kavický 2014) it is necessary to know:

– the type of explosive,
– detonation pressure of explosive (GPa),
– density of explosive (g/cm³),
– charge tare (kg),
– explosive weight (kg),
– charge weight (kg),
– geometry coefficient of spreading shock waves.
– distance from the epicentre (m),

This calculation model is particularly suitable for ANFO explosives.

When using other approaches (Henrych 1973), (Mills 1987), (Makovička 2008) and others, it is necessary to know the explosion heat of explosives and pressure tritol equivalent (instead of explosive detonation pressure and density). These approaches are particularly useful for military explosives.

During the test, following properties must be measured and recorded:

– air temperature,
– air pressure,
– wind speed and direction,
– surface temperature of the sample,
– the highest pressure reached during the test—pressure at the front of the shockwave,
– progress of the shockwave,
– duration of the positive and negative phase of the shockwave,
– strains.

It is necessary to measure and record the above properties due to the repeatability of the tests and verification of the results. To record the mentioned properties, following instrumentation is used.

– Measuring chain to measure the shockwave with the explosion sensors, recording and evaluative equipment.
– Sound level meter.

– Pressure gauge.
– Thermometer.
– Measuring equipment for the wind speed and direction.
– Tensometers and accelerometers.

To make photo-documentation and video recording, it is necessary to use appropriate technique (high speed camera, camera).

4 CONCLUSION

Considering the global security situation development, the research in the field of breakthrough resistance (or testing the reaction of particular types of mechanical parts of the system protection of the object hit by pressure effect of the blast) is necessary. Lack of knowledge about the resistance of conventional means of shell protection hit by the explosion is an obstacle in designing security systems and ensuring the safety of people and property.

We know the values of damage for certain elements of the shell protection hit by the pressure of the blast, however, there is no knowledge of the processes obtaining this data, conditions under which the tests were conducted, description of the properties of individual elements or detailed description of their reaction—extent of the damage. The pressure values are in several cases stated in broad intervals, which significantly decreases their informative value.

The proposal of the test procedure for reaction of window systems hit by the shockwave of the explosion is a step towards unification and clarification of values determining their resistance.

The reason for the focus on the windows and window systems is that these elements together with glass make an important part of the shell protection of the object. At the same time, the opening fillings, especially windows and glass fillings are the most fragile part of the building construction. Opening fillings and facades of buildings are not constructed as support elements and unless there is a requirement, they are not designed to transmit the load which occurs in the explosion. Modern architecture takes pride on clean smooth design and airy-looking buildings which facades are very fragile and often almost entirely of glass. In recent years, a significant number of buildings with large number of people remaining inside are designed by this concept (airports, railway stations, bus stations, banks, shopping centres etc.).

In our procedure, we characterised the key points of the test aimed at detecting the response or resistance of the window systems loaded by the pressure of the explosion. We described the test sample, its mounting, charge used at the test and the instrumentation. Our proposed work procedure rises from the need of existence of such a procedure. When creating the work procedure, we used technical standards partially adapted to the given issue and own experience from the existing research activities.

ACKNOWLEDGEMENT

The article was elaborated within the project VEGA 1/0175/14

REFERENCES

BS EN 13123-1 Windows, doors and shutters. Explosion resistance. Requirements and classification. Part 1: shock tube.

BS EN 13123-2 Windows, doors and shutters. Explosion resistance. Requirements and classification. Part 2: test of load capacity.

BS EN 13124-1 Windows, doors and shutters. Explosion resistance. Test method. Part 1: shock tube

BS EN 13124-2 Windows, doors and shutters. Explosion resistance. Test method. Part 2: test of load capacity.

BS EN 1143-1 Secure storage units. Requirements, classification and methods of testing the resistance to burglary. Part 1: safes, safes for cash machines, vault doors and strong rooms.

BS EN 13541 Glass in building construction. Safety glazing. Testing and classification of the resistance to explosion pressure.

Figuli, L. & Papán, D. 2014. Dynamic analysis of blast loaded steel beam. In: *Dynamics of civil engineering and transport structures and wind engineering = DYN-WIND'2014: proceedings of the 6th international scientific conference: May 26–29, 2014 Donovaly, Slovak Republic.* Žilin.a: University of Žilina, ISBN 978-80-554-0844-6

Figuli, L., Kavický, V., Boc, K., Vidríková, D. & Jangl, Š. 2013. Analysis of blast loaded structures. In: *Science & Military.* ISSN 1336–8885.

Henrych, J. 1973. *Dynamika výbuchu a jeho užití/The dynamics of the explosion and its use*

ITOP 4-2-822: Electronic Measurement of Airblast Overpressure and Impulse Noise.

Jangl, Š. & Kavický, V. 2012. *Ochrana pred účinkami výbuchov výbušnín a nástražných výbušných systémov/ Protection against explosions of explosives and improvised explosive devices.* Žilina: Jana Kavicka—Kavicky, 294 s. ISBN: 978-80-971108-0-2.

Kavický, V., Figuli, L., Jangl, Š. & Zvaková, Z. 2014. Analysis of the field test results of ammonium nitrate: fuel oil explosives as improvised explosive device charges. In: *Structures under shock and impact XIII: [13th international conference, SUSI 2014: New Forest, United Kingdom, 3 June 2014 through 5 June 2014].—Southampton, Boston:* WITpress, ISBN 978-1-84564-796-4. ISSN 1746–4498.

Makovička, M. & Janovský, B. 2008. *Příručka protivýbuchové ochrany staveb/ Manual for blast protection of buildings* Praha: Česká technika—nakladatelství ČVUT, ISBN 978-80-01-04090-4.

Mills, C.A. 1987. *The design of concrete structure to resist exposions and weapon effects.* Edinburgh.

Vávra, P. & Vagenknect, J. 2002. *Teorie působení výbuchu/ The blast effect theory.* Pardubice: Univerzita Pardubice.

Experimental research of technologies
for remediation of tailing ponds

Production Management and Engineering Sciences – Majerník, Daneshjo & Bosák (Eds)
© 2016 Taylor & Francis Group, London, ISBN: 978-1-138-02856-2

Application of AHP in the process of sustainable packaging in company

H. Fidlerová, L. Jurík & P. Sakál
Faculty of Materials Science and Technology in Trnava, Slovak University of Technology in Bratislava, Bratislava, Slovak Republic

ABSTRACT: The paper considers using the Analytic Hierarchy Process (AHP) with selected criteria including three pillars of sustainable development in process of packaging in the manufacturing company to find out the solution. It structures the decision problem in a manner that is easy for the stakeholders to comprehend and allows them to analyze independent sub-problems by structuring into a hierarchy and using pairwise comparisons. If the AHP method is used properly, it can be helpful for managers to plan and make optimal decision. The article provides characteristics of the mentioned method; it describes the application in packaging process. The aim of project was to find out a solution for company to minimize packaging waste by purchasing a medium-sized baling press. The selection of a suitable baling press was made with AHP method using software Expert Choice.

Keywords: Multi-criteria decision-making methods; analytic hierarchy process; sustainable packaging; packaging press; case study

1 INTRODUCTION

Environment changes quickly and there many challenges for manufacturing company in Slovakia as need for cost reductions, the need for efficient allocation of funds, the stakeholders demand to enhance sustainability of logistics processes, the complicated legislation concerning environment, waste law. All this means call for changes in the way how make decisions.

The Analytic Hierarchy Process (AHP) is a multicriteria decision-making technique, developed by Thomas Saaty to support users with complex decision-making by combining their experience, judgment, and intuition with a view to selecting the best course of action from a number of alternatives.

In the paper the classical AHP is discussed to give the reader an understanding of the methodology. After familiarizing with manufacturing company and its logistics processes including packing is introduced the specific application of AHP method for selection of baling press, with defined criteria in the context of sustainable development. Finally, the authors characterize applications of AHP in various fiels in previous research and propose some outline suggestions for its application in future.

2 SUSTAINABLE LOGISTICS AS PART OF THE STRATEGY FOR SUSTAINABLE DEVELOPMENT

The first mention of the Sustainable Development (SD) was at the Conference in Paris (1968). Group of economic experts led by D. Meadows in document The Limits of the Growth in 1972, also known as the first Club of Rome report, mentioned the need for sustainable utilization of natural resources (1972).

Sustainability as a term has been used by scientist and public in the sense of human sustainability on planet Earth.

Sustainable development was defined by World Commission on Environment and Development (1987) as *"development that meets the needs of the present without compromising the ability of future generations to meet their own needs."*

Milestones in the general introduction of the conception and sustainable development were mainly the United Nations Conference on Environment and Development in Rio de Janeiro (1992)—the most important document of this conference entitled Agenda 21 is regarded as an essential prerequisite for processing sustainable development strategies at all levels, and the World Summit the sustainable development in Johannesburg in 2002. Summit declared that the aim is: "… *such a development that will ensure a balance between the three pillars: economic, environmental and social."*

The sustainability requires the reconciliation of environmental, social equity and economic demands—the "three pillars" of sustainability. Sustainability has been studied and managed over many levels of time and space and in many contexts of environmental, social and economic organization. (….Ehrefeld, 2009, Visser, 2012, Stead & Stead, 2012).

The focus ranges from the total sustainability on planet Earth for a long time to the sustainability of economic sectors, ecosystems, countries, economics, towns, neighbourhoods, individual lives, processes, individual goods and services, occupations, lifestyles, behaviour patterns and so on (Sakál et al, 2013, Zaušková A. et al, 2013).

According Stead & Stead (2012) has sustainability 3 independent coevolution features: economical, social and environmental. Sustainability is naturally complex, interdisciplinary and multidimensional. Therefore ii influence whole business environment revolution and manufacturing companies and their processes including logistics. It is important that logistics processes follow the way of sustainability as a part sustainability revolution.

Fleischmann et al. (1997) note that economic and environmental issues are often intertwined. For example, increasing disposal costs make waste reduction more economical, and environmentally conscious customers represent new market opportunities. Ideally, one would like to combine both ecological and economic advantages, as suggested by the concept of a sustainable economy.

Sustainable logistics in context corporate social responsibility means to understand and to reckon many environmental, social and culture aspects of all logistics processes of industrial enterprise (Hrdinová, G., et al, 2012).

Packaging is an important part of reverse logistics. *Packaging* provides for protection and identification of products according customer's requirements and actual legislative (Fidlerová, 2013).

Sustainable packing means using less material for every product, it should be considered additional option of recovering it and so eliminate waste material. Sustainable packaging and treating the waste form packaging process is actual because on July 2014, the European Commission published a proposal to review recycling and other waste-related targets in the EU, to encourage the transition towards a Circular Economy through the use of waste as a resource. It also aims with Extended Producer Responsibility (EPR).

3 CHARACTERISTIC OF ANALYTIC HIERARCHY PROCESS (AHP) METHOD

AHP method allows making effective decisions in complex situations, to simplify and quicken the natural process of decision-making. AHP provides a comprehensive and rational framework for structuring a decision problem, for representing and quantifying its elements, for relating those elements to overall goals, and for evaluating alternative solutions.

The author of Analytic Hierarchy Process theory, American professor Saaty, characterizes the AHP as a practical tool for supporting decision making and applied it for the quantity of practical decision problems (Saaty, 1980).

According Saaty (2008) by the usage of AHP, a decision maker knowingly and intentionally directs to increasing the quality and efficiency of all his decisions. It is used in economic, management, environmental science, administration, business, industry, health, education, etc. It is a suitable method for the evaluation in manufacturing company, where many criteria lead to the objectification of their evaluation.

The decision-making according to AHP method is based on the three principles of analytic thinking:

The principle of structuring a hierarchy. The most often alternative of viewing the hierarchy within this method is a diagram, with the aim placed at the top, the variants placed at the bottom and the criteria placed between them (Figure 1).

The principle of determining priorities. After the creation of own criteria set and hierarchy structure, all the proposed variants or criteria are compared at all levels of the evaluation, which influences the evaluation using the word meaning and numerical values. The resultant state is determined by the weight in the relative scale for variants. The basic scale of pair wise comparison of the AHP method is presented in Table 1. The value 1, 3, 5, 7, 9 represent the basic scale of the evaluation used in the

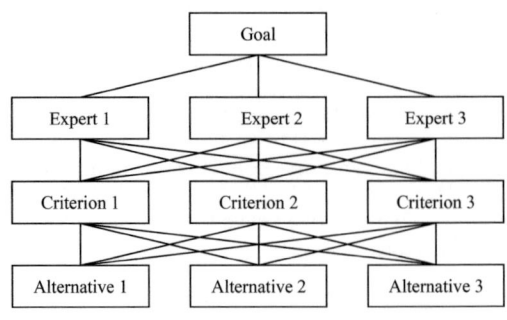

Figure 1. Hierarchical structure of AHP (Saaty, 2008).

Table 1. Measurement scale of pair wise comparison in the AHP method (Saaty, 1980).

Measurement scale	Comparison of the elements x and y
1	x is as important as y
3	x is slightly more important than y
5	x is strongly more important than y
7	x is greatly strongly more important than y
9	x is extremely important than y

AHP method. The values of 2, 4, 6 and 8 provide additional levels of discrimination (Nydick, 1992).

The principle of logical consistency. In determining relationships between objects according to the criteria, it is necessary to achieve consistency of intensity of the relationship between objects according to individual criteria. For example, if the quality of machines is compared and the machine A is rated as 2 times better quality than machine B, which is 2 times better quality than machine C, then a consistent evaluation claims that the machine is 4 times better quality than machine C (Saaty, 2000).

The general procedure of a solution includes:

1. Realization of the pairwise comparison of criteria and comparison of variants by individual criteria—the obtaining of matrices.
2. Determination of the own value (own number) of each matrix:
   a. Obtaining the characteristic polynomial
   b. Determination of the roots of the characteristic polynomial, and from them obtaining own number
3. Acquisition of the values of the own vector of matrix
4. Transformation of the own vector of matrix to normed own vector, whose components determine the weights of each criteria and weights of variants according to fulfillment of the requirements of individual criteria.
5. The final evaluation and determination of the order by means of a weighted sum.

AHP method also exists in the form of computer software Expert Choice, which takes into account qualitative, as well as quantitative information, including intuition and experience. Expert Choice is a software tool that supports decision making in the selection of variants, which are characterized by a hierarchical layout of criteria and priorities for selection.

The application of Expert Choice uses a unique method for pairing of the compared criteria within the specified priorities. The software allows to record into the model hierarchically structured criteria and their priorities for individual evaluated variants. Expert Choice combines hierarchical priorities into the overall priorities of all evaluated variants of the evaluated problem. As mentioned the problem can be solved in two ways:

1. Software Expert Choice (EC), whose procedure will be described.
2. The classical numerical method.

The solution of the problem with software EC consists from following steps:

1. The determination and inscription of the objective, criteria and variants of the decision problem.
2. The assignment of the weights to individual criteria through pair wise comparison of criteria.
3. The evaluation of the variants of solution by pair wise comparison in individual criteria.
4. The evaluation of the optimal variant of solution, eventually the sequence of individual variants.

According Sakál et al. (2012), we can perform sensitivity analysis by changing the values of the priorities and determine, what impact the change will have on the overall selection of variants.

The software Expert Choice can also be used in these areas (Sakál et al., 2012):

- Optimization of resources,
- Management of information technologies portfolio,
- Strategic planning,
- Risk assessment,
- Human resource management,
- Decision making about strategic locations.

4 APPLICATION OF AHP IN SUSTAINABLE PACKAGING IN MANUFACTURING COMPANY IN SLOVAKIA

AHP method was applied in manufacturing company with issue to enhance its logistics processes including packaging process in context of sustainable development. Managers of the company recognize that a comprehensive approach to waste management can ensure the success and competitiveness in the future.

The production of company can be characterized as on order (piece production). The packaging process is carried out according to the size and complexity of products. Each product is packaged separately, since they are produced large and oversized machines with sizes, approximately 3 m × 3 m.

The packaging process begins with placing on wooden pallets. Pallets have different sizes according to the dimensions of manufactured products. Then the product is wrapped with paperboard and plastic film from the external supplier. The packaging process is done by hand; it is caused by unusual shapes and oversize products.

The packaging of raw materials used in the transformation process is taken away directly in production. Packaging after its use becomes a waste and therefore is collected and stored in bags or containers which after filling are stored in the waste storage, where is collected the waste from transformation process. The area for waste storage is located inside the enterprise and is specially equipped and labeled for separation of waste in paper, metal, plastic and hazardous waste in bins.

Analysis in enterprise showed that more products are produced and it means also increasing production of packaging waste (Table 2). Packaging waste in company was not modified but only stored in storage and caused increasing cost for its storage, transport and recycling. Extended Producer Responsibility emphasis the waste treatment in sustainable way, what means improving the properties of waste for handling and storage and decreasing the size (bundling and molding) by adjusting the density, homogenization (mixing) etc. This can reduce for manufacturing in case study also the costs by reducing the amount of waste for storage and transport.

Therefore, it was proposed a *solution to minimize volume of packaging waste by purchasing a medium-sized baling press.* The selection of a suitable baling press was made through the application of analytic hierarchy process with use of the software tool Expert Choice.

As before mentioned AHP method is a decision-making method for determining priority of criteria when are compared more criteria and is used for a variety of decision-making areas, including research and development, project selection, evaluation of various alternatives (Nydick, 1992).

We proceed as follows, the main objective was defined, then criteria of choice and various were alternatives identified and hierarchic structure was created (Figure 2). The first step in applying the AHP method is correct definition of the objective, then the choice of alternatives and identification of criteria. *The main objective of the project was a selection of the medium–sized baling press for minimizing the packaging waste.*

Table 2. Amount of packaging waste in manufacturing company in years 2012 and 2013.

Packaging waste	2012	2013
Amount	290 kg	2070 kg

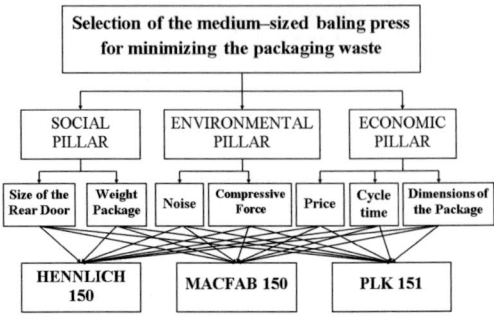

Figure 2. Hierarchical structure of decision making process. Source: (drawn by authors).

On the basis of consultation with experts are determined following variants:

1. Press HENNLICH 150
2. Press MACFAB 150
3. Press PLK 151

Variants were rated as the best technical facilities of medium-sized balers offered at Slovak market from technical catalogs for the year 2013. Selection of variants was limited by maximum accepted price for press 4 000 €.

The technical equipment HENNLICH 150 costs 3 800 € with VAT. Its thrust is 8 000 kg and noise level is 74 db. It has a vertical design and front feeding with dimensions: 470 mm × 975 mm. If the hatch is bigger, the bigger and bulkier waste can be compressed. The cycle time of processing waste is 40 seconds. It is easy to use this device and it has automated tipping package. The maximum size of the package is 1000 mm × 975 mm × 800 mm and weight is 150 kg.

The baling press MACFAB 150 is also vertical and costs 3 900 € including VAT. Compared to the press HENNLICH 150 it has similar parameters but its the rear door is larger 650 mm × 950 mm and output packet size is smaller 1000 mm × 950 mm × 800 mm. Thus, the press machine MACFAB 150 treated waste to the smaller output package.

Third variant is the baling press PLK 151, whose price is 4 200 € including VAT, so exceeds the proposed budget. Its advantage is greater thrust 10000 kg, loading hole size is 560 mm × 1093 mm and the size of the package is 1100 mm × 750 mm × 700 mm. This baling press is louder (78 db) than other two variants and its cycle time is 140 seconds.

The main criteria for the selection of the baling press defined by experts were price, compressive strength and size of package, cycle time, noise, size and weight of the rear door package.

The selected criteria can be assigned to the three pillars of sustainable development. The economic pillar includes the price (relation with evaluation of costs and benefits), cycle (purchase of one or more devices) and package size (storage costs for packages). The social pillar comprises loading aperture size (number of operators) and the weight of the package (number of operators). And the last is the environmental pillar of noise and compression force (related to noise).

Sustainable development includes:

1. Economic pillar
   − Price,
   − Cycle time,
   − Dimensions of the package.
2. Social pillar:
   − Size of the rear door,
   − Weight package.
3. Environmental pillar:
   − Noise,
   − Compressive force.

Because of limited size of paper it is not possible to describe whole process in detail, we used the software and heightening of criteria shown. Following the determination of criteria and alternatives are assigned the weights to each criterion through pair wise comparison criteria (Figure 3). The size of priorities for the individual criteria is presented in figure 4.

As before mentioned the selection of medium-sized baling press was performed numerically by means of Expert choice on the basis of pair wise comparisons. Result for selection of baling press from variants is press HENNLICH 15, as second was the MACFAB 150 and third press PLK 151 (Figure 5).

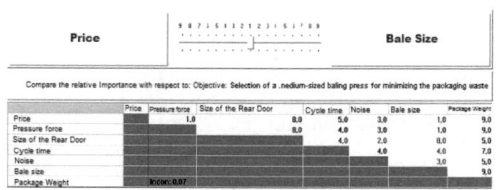

Figure 3. Pair wise comparison with Expert Choice (Pazinová, 2014).

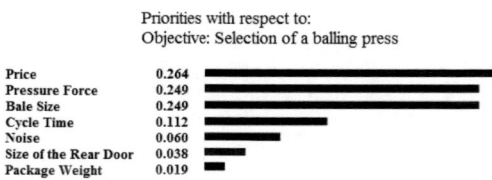

Figure 4. The size of priorities for the individual criteria. (Pazinová, 2014).

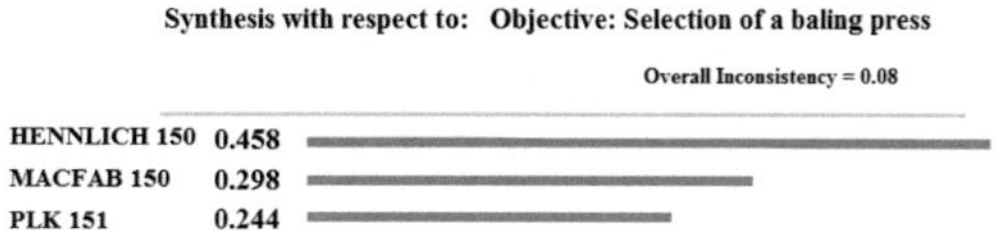

Synthesis with respect to: Objective: Selection of a baling press

Overall Inconsistency = 0.08

HENNLICH 150	0.458	
MACFAB 150	0.298	
PLK 151	0.244	

Figure 5. The result of evaluation of variants of solution. (Pazinová, 2014).

Based on results obtained with application of AHP method using software Expert Choice has been proposed as solution the purchase of the baling press HENNLICH 150. Chosen technical equipment—HENNLICH 150 is a medium press, fully automatic, with simple feeding.

According the catalogue the press HENNLICH 150 will reduce costs, reduce the volume of waste by 90%, thus ensuring space-saving, also meets the requirements of legislation for the protection of the environment and health, it is easy to work with it and no need for containers.

We expect a return on investment for the manufacturing company in the coming years. The advantages of the baling press HENNLICH 150 are that, the company in this case will not only save the cost of storage, transport and removal of untreated waste but also can be separated waste packed and sold to other companies.

5 CONCLUSIONS

AHP method is successfully used in praxis for simplification and objectification of the decision-making process, because it incorporates both the qualitative and the quantitative aspects of human thought: the qualitative to define the problem and its hierarchy and the quantitative to express judgments and preferences concisely.

AHP with its hierarchy structure and logic provides a framework for the participation of expert groups in decision making or problem solving. This makes it suitable for application in logistics process in manufacturing company as well, especially when expert from various areas of interest ought to take part in solution. (Golden, 1989, Vargas, 1990, Drieniková et al, 2011)

During our research activities and in processing the above mentioned grants we have accumulated theoretical and practical experience with the application of Analytic Hierarchy Process in other areas of management and decision making, e.g.:

1. In environmental management (Moravčík et al., 2012).
2. In risk strategic management (Naňo & Sakál, 2011)

3. In the strategy of corporate social responsibility (Drieniková et al., 2011)
4. In process evaluation competency profile of managers in industrial enterprises and university employees (Jurík & Sakál, 2014).
5. In ergonomy (Beňo et al., 2012)

In case study presented in paper took part experts from logistics, environment, production and waste management. We expect more application of AHP method in manufacturing companies in Slovakia also respecting principles of sustainable development.

ACKNOWLEDGEMENT

The paper builds on the results of APVV project No. LPP-0384-09 "Concept HCS model 3E vs. concept Corporate Social Responsibility (CSR)" and KEGA project No. 037STU-4/2012 Implementation of the subject of "Corporate Social Responsibility Entrepreneurship" into the study programme of Industrial Management in the second degree of study at STU MTF Trnava.

The paper is a part of VEGA project No. 1/0448/13, Transformation of ergonomics program into the company management structure through integration and utilization and QMS, EMS, HSMS.

REFERENCES

Beňo, R., Drieniková, K., Naňo, T. & Sakál, P. 2012. *Multicriteria assessment of the ergonomic risk probability creation by chosen groups of stakeholders with using AHP method within the context of CSR.* In Quantitative methods in economics. Multiple Criteria Decision Making XVI: Proceedings of the International Scientific Conference., Bratislava, Slovakia. Bratislava: Ekonóm, pp. 7–11.

Brundtland Commission (1987). "Report of the World Commission on Environment and Development". United Nations. [Cited: 25.3.2015].

Drieniková, K., Hrdinová, G., Naňo, T. & Sakál, P. 2011. *Case studies of using the analytic hierarchy process method in corporate social responsibility and environmental risk management*, In Materials Science and Technology, Vol. 1, 2011, pp. 1–10.

Ehrenfelf J.R. 2009. *Sustainability by design. A Subversive Strategy for Transforming our Consumer Culture.* Yale University Press.

Factsheet on Extended Producer Responsibility (EPR) for used packaging.http://www.europen-packaging.eu/news/news/80-factsheet-on-extended-producer-responsibility-epr-for-used-packaging.html [Cited: 25.3.2015].

Fidlerová, H. 2013. *Sustainable reverse logistics as unique alternative for 21st century in context of the sustainable development strategy in enterprise.* In: Acta Moraviae Vol. 5, Iss. 10, pp. 45–51, [Cited: 25.3.2015]. http://edukomplex.cz/dokumenty/acta/cisla/acta_10.pdf.

Fleischmann, M. 2001. *Reverse logistics network structures and design.* Erasmus Research Institute of Management Report Series Research In Management ERS-2001-52-LIS.

Golden, B., Wasil, E. & Harker, P. 1989. *The analytic hierarchy process: applications and studies.* Heidelberg, Springer-Verlag.

Hrablik Chovanová, H., Sakál, P., Drieniková, K. & Naňo, T. 2012. *Operačná analýza II. Operation Analysis. IInd part.* Trnava: Alumni Press. 223 p.

Hrdinová, G., Sakál, P. & Fidlerová, H. 2012. *Sustainable logistics and its role in value chain of industrial business with context of CSR and CSV.* In Carpathian Logistics Congress 2012: November 7th–9th 2012, Priessnitz Spa, Jeseník, Czech Republic. Ostrava: Tanger, 2012., http://www.clc2015.cz/files/proceedings/09/reports/1333.pdf [Cited: 25.3.2015].

Ishiyaka A. & Labib A. 2009 Analytic Hierarchy Process and Expert Choice: Benefits and Limitations, OR Insight, 22(4), pp. 201–220.

Jurík, L. & Sakál, P. *Competencies of Managers, as part of the Intellectual Capital in Industrial Enterprises.* In ECIC 2014: proceedings of the 6th European Conference on Intellectual Capital. 10–11 Apríl 2014, Trnava, Slovak Republic. 1. vyd: Academic Conferences and Publishing International Limited, 2014, pp. 368–376.

Moravčík, O., Sekera, B., Beňo, R., Sakál, P. & Šmida, Ľ. 2012. *Perspectives for Utilization of Multicriteria Decision Methods AHP/ANP to Create a National Energy Strategy in Terms of Sustainable Development.* Advanced Materials Research, Volume 616–618: Sustainable Development of Natural Resources. pp. 1585–1590.

Naňo, T. & Sakál, P. 2011. *Suggestion of the AHP (Analytic Hierarchy Process) method utilization in risk strategic management of industrial companies.* In Radioelektronika, elektrotechnika i energetika: 17. meždunarodnaja naučno-techničeskaja konferencija studentov i aspirantov, 24–25. 2. 2011, Moskva. Tom 2. Moskva: Moskovskij energetičeskij institut, pp. 302–303.

Nydick, R.L. & Hill, R.P. 1992. *Using the Analytic Hierarchy Process To Structure the Supplier Selection Procedure.* [Cited 25.3.2015]. http://www77.homepage.villanova.edu/robert.nydick/documents/Vendor%20Selection.pdf.

Pazinová, J. 2014. *Návrh využitia metódy AHP a softvéru Expertchoice pri minimalizácii odpadu v procese udržateľného balenia v podniku PSS SVIDNÍK, a.s* Diploma thesis at MTF STU Trnava, 2014. Supervisor:Ing. Helena Fidlerová, PhD.

Ramík, J. 2000. Analytický hierarchický proces (AHP) a jeho využití v malém a středním podnikání. Karviná: Slezská univerzita, 217 p.

Report on corporate social responsibility: accountable, transparent and responsible business behaviour and sustainable growth 2012/2098(INI)). http://www. europarl.europa.eu/sides/getDoc.do?pubRef = -//EP// TEXT+REPORT+A7–2013–0017+0+DOC+XML+ V0//EN [Cited: 25.3.2015].

Saaty, T.L. 1980: *The Analytic hierarchy Process: Planning, Priority Setting, Resource Allocation.* New York: McGraw-Hill Inc., 19 p.

Saaty, T.L. 2008. *Decision Making for Leaders: The Analytic Hierarchy Process for Decisions in a Complex World.* Pittsburgh, Pennsylvania: RWS Publications.

Saaty, T.L. 2010. *Principia Mathematica Decernendi: Mathematical Principles of Decision Making.* Pittsburgh, Pennsylvania: RWS Publications.

Sakál, P. et al, 2009. *Logistika výkonného podniku.* Trnava: SP SYNERGIA.

Stead, J G. & Stead, W. 2012. *Manažment pre malú planétu: prečo je dôležité meniť stratégie neobmedzeného rastu na stratégie udržateľnosti.* Bratislava: Eastone Books, 243 p.

Vargas, L. 1990. *An overview of the analytic hierarchy process and its applications.* European Journal of Operational Research 48(1), pp. 2–8.

Visser, W. 2012. *The Quest for Sustainable Business: An Epic Journey in Search of Corporate Responsibility.* Sheffield: Greenleaf.

Zaušková A., Miklenčičová, R., Madleňák, A.,Bezáková, Z. & Mendelová, D. 2013. *Environmental protection and sustainable development in the Slovak republic*, In European Journal of Science and Theology, No. 6, pp. 153–159.

Research and development of new technologies of decontamination of ash mixture tailings

M. Bosák, M. Majerník & A. Tarča
Faculty of Business Economy with Seat in Košice, The University of Economics in Bratislava, Košice, Slovak Republic

ABSTRACT: The chapter presents the results of a multiannual systematic research and development of essentially new environmental safety technology of overlapping tailing modeled in terms of Vojany thermal power plant (EVO). The newly developed and tested technology serves as a replacement for technically and economically difficult resolution of overlapping where geotextile waterproofing membrane and drainage system had been used. Nowadays the newly developed technology uses structured layers of ground and soil stabilizers (waste product of desulphurization technology) as a waterproofing material. Recultivated tailing area will ultimately be used to produce biomass (Swedish willow and selected grasses) intended for subsequent incineration of coal. Laboratory and small-scale experiments realized directly at the tailing area resulted into dimensioning parameters of new environmental safety technology of decontamination of tailings with synergistic technical, economical, environmental and safety effects.

1 INTRODUCTION

Effect of plants on the environment and is now a global problem. Fly ash and slag (dross ashes mixture) resulting from the combustion of coal is waste that burdens the environment in the wider vicinity of power plants and landfill sites are in terms of landscape stability and severe environmental-safety problems that must be effectively addressed.

Especially in the case of dumping large quantities of such wastes in an area of huge tailings ponds there is a real danger of breaking up the dam with serious Crisis consequences for the population, components of the environment and property in general. The tailings ponds are stored in very fine waste containing significant amounts of water, whose mobility if released from the pond is large, so they can migrate to long distances, particularly over the surface water flows, including transboundary impacts, impacts on the landscape areas and protected areas of European importance.

It is therefore necessary to ensure adequate management of these wastes, to ensure long-term stability and security of tailing ponds after conclusion especially preventive environmentally sound technologies (Majerník et al. 2012, Pacaiova et al. 2013).

2 TAILING PONDS

2.1 Tailing ponds in the Middle Europe

Industrial production generates some kind of waste (byproduct) toxic substance while contaminating the sites and often degrades surroundings of human habitation including air, surface and ground water.

In Slovakia there are decanting plants of various types and levels from the viewpoint of environmental safety of deposited materials. They are in various stages of their life cycles or existence (Figure 1). Inside them there are deposited mainly wastes from power plants and heating plants (slag, ash), products from ore processing (floating sludge), coal gangue etc. Generally the decanting plants, as watery building constructions are vast dangerous objects from the viewpoint of environment and therefore their safe closing or re-cultivation is an up-to-date issue not only in the Slovak but also in the European environmental safety.

Wastes can be conveniently treated before landfilling in order to minimize future emissions (Cossu et al. 2011).

Types of tailing ponds in Slovakia (Table 1–3):

25 with ash material,
20 with ore,
8 others.

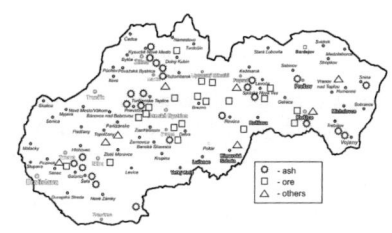

Figure 1. Registered decanting plants in Slovakia.

Table 1. The list of ash tailing ponds in Slovakia.

No.	Name	Place	District	Category
1.	Tailings pond ENO—new	Zemianske Kostoľany	Prievidza	I.
2.	Tailings pond ENO—old	Zemianske Kostoľany	Prievidza	II.
3.	Tailings pond ENO	Bystričany—Chalmová	Prievidza	II.
4.	EVO Vojany	Vojany—Drahňov	Mihalovce	II.
5.	Tailings pond KAPPA a.s.	Štúrovo- časť Obid	Nové Zámky	II.
6.	Martin—old tailings pond	Martin	Martin	II.
7.	Martin—new tailings pond	Bystrička	Martin	II.
8.	Tailings pond Poša	Poša-Nižný Hrabovec	Vranov n. T.	II.
9.	Tailings pond Snina	Snina	Snina	II.
10.	Tailings pond DUSLO	Trnovec n. Váhom,	Šaľa	II.
11.	Tailings pond Žilina	Bytčica	Žilina	II.
12.	Tailings pond Košice	Krásna nad Hornádom	Košice	III.
13.	Sereď	Dolná Streda	Galanta	III.
14.	Zvolen	Zvolen	Zvolen	III.
15.	Ash tailings pond	Horné Opatovce	Žiar n. H.	III.

Table 2. The list of ore tailing ponds in Slovakia.

No.	Name	Place	District	Category
1.	Hačava	Hačava	Rimavská Sobota	II.
2.	Hodruša Hámre	Hodruša Hámre	Žiar nad Hronom	II.
3.	Jelšava	Jelšava	Rožňava	II.
4.	Nižná Slaná	Nižná Slaná	Rožňava	II.
5.	Rudňany	Závadka	Spišská Nová Ves	II.
6.	Sedem žien	Banská Belá	Žiar n. Hronom	II.
7.	Tailings pond Slovinky	Slovinky	Spišská Nová Ves	II.
8.	Baňa Cígeľ II	Sebedražie	Prievidza	III.
9.	Dúbrava 01	Dúbrava	Liptovský Mikuláš	III.
10.	Dúbrava 02	Dúbrava	Liptovský Mikuláš	III.
11.	Dúbrava 03	Liptovský Mikuláš	Liptovský Mikuláš	III.
12.	Žiar nad Hronom	Žiar n. H.	Žiar n. H.	III.
13.	Košice—Bankov	Košice	Košice	III.
14.	Lintych	B. Štiavnica	B. Štiavnica	III.
15.	Pezinok	Pezinok	Pezinok	III.
16.	Podrečany	Podrečany	Lučenec	III.
17.	Smolník	Smolník	Spišská Nová Ves	III.
18.	Široká	Široká	Dolný Kubín	IV.
19.	Baňa Cígeľ I.	Sebedražie	Prievidza	IV.
20.	Košice Bankov	Košice	Košice	IV.
21.	Horná Ves (Kremnica)	Horná Ves	Žiar nad Hronom	IV.
22.	Hronský Beňadik	Hronský Beňadik	Nová Baňa	IV.
23.	Lubeník	Jelšava	Rožňava	IV.
24.	Pezinok	Pezinok	Pezinok	IV.
25.	Rožňava	Rožňava	Rožňava	IV.
26.	Sereď	Sereď	Galanta	IV.
27.	Špania dolina	Špania dolina	B. Bystrica	IV.

According to the summary records of water cannons, 28 tailings of I–IV category were located in Czech Republic to 1.1.2014, all listed in the following Table 4.

In Hungary, not only red mud is produced in tailings, but there also are uranium and ash tailings. Some of them are already being recultivated and prepared for liquidation. In Hungary there are 20 tailings as characterized below, Table 5.

The present Tailing ponds are still expensive and also environmentally dangerous objects. This is demonstrated by example. The accident in the Hungarian village of Ajka, where on 4th October 2010 the dam pond broke after heavy rains. Subsequently,

Table 3. The list of industrial tailing ponds in Slovakia.

No.	Name	Place	District	Category
1.	Čifáre	Čifáre	Nitra	II.
2.	Bukocel	Hencovce	Vranov n. T.	III.
3.	Dubová	Dubová	B. Bystrica	III.
4.	Tailings pond Nováky 7	Nováky	Prievidza	III.
5.	Tailings pond Handlová	Handlová	Prievidza	III.
6.	Tailings pond Sokoľany	Sokoľany-Bočiar	Košice	IV.
7.	Fámeš	Pastuchov	Hlohovec	IV.
8.	Plešivec-Gemerská Hôrka	Plešivec	Rožňava	IV.
9.	Tailings pond Veľká Ida	Veľká Ida	Košice	IV.
10.	Mokr	Veľká Ida	Košice	IV.
11.	Tailings pond Nováky 6	Nováky	Prievidza	IV.
12.	Šaľa RSTO	Šaľa	Galanta	IV.
13.	Šulekovo	Šulekovo	Trnava	IV.
14.	Veronika	Dežerice	Topoľčany	IV.

Table 4. The list of registered tailing ponds in Czech Republic.

No.	Name	Place	Category
1.	Hodějovice	České Budějovice	III.
2.	Mydlovary	České Budějovice	III.
3.	Zbrod North 1/4	Hodonín	III.
4.	Nové Chalupy	Karlovy Vary	III.
5.	Tailing ponds II.	Ostrov	III.
6.	Dolní Radechová	Náchod	III.
7.	Debrné	Trutnov	III.
8.	Tailing ponds TDK IV/3	Trutnov	III.
9.	Stráž p. Ralskem	Česká Lipa	II.
10.	Dříteč	Pardubice	III.
11.	Lhotka	Pardubice	II.
12.	Semtín č. 7	Pardubice	II.
13.	Chvaletice I.	Přelouč	III.
14.	Božkov	Plzeň	III.
15.	Panský les	Mělník	II.
16.	Tailings pond Spolana	Neratovice	III.
17.	Bytíz	Přibram	III.
18.	Rýzmburk	Vlašim	III.
19.	Ušák	Kadaň	II.
20.	SEPAP No.4	Litoměřice	III.
21.	Třískolupy	Louny	III.
22.	Barbora III.	Ústí nad Labem	III.
23.	Užín—old tailing ponds	Ústí nad Labem	III.
24.	Dolní Rožínka	Bystřice nad Pernštejnem	II.
25.	Zlatkov	Bystřice nad Pernštejnem	II.
26.	Tailing ponds Synthesia a.s.	Pardubice	IV.
27.	Tailing ponds	Mladá Boleslav	IV.
28.	Ústí—new tailing ponds	Ústí nad Labem	IV.

more than 700,000 cubic meters of red sludge flooded the neighborhood and toxic mud struck seven villages and towns. They destroyed dozens of homes and the environmental disaster has claimed up to ten human victims and over 150 injured. The table 6. shows examples of other accidents at tailings ponds fatal and environmental devastation. For this reason, the discussion about closing these ponds is very actual (Jánová & Panenka 2010).

The European Union currently allocates huge funds in development projects for member countries to prevent and remedy environmental damage, hence the restoration and rehabilitation of tailing ponds dross ashes biological mixtures.

2.2 Tailing pond of EVO Vojany

Vojany Power Plant is the biggest fossil fuel plants in Slovakia, where such fuel is used mainly semianthracite coal from Ukraine and Russia. Currently, the plant operates two facilities for disposal of waste products from coal combustion:

- Ailing ponds with dross ashes mixture.
- Dump with stabilisation material.

The pond with two cassettes of dross ash mixtures (cassette no. 1 is already closed) are stored as hydraulic transport of coal combustion products and the self-imposed dump with stabilisation material, which is a byproduct of the desulfurization of power plant technology, combustion processes.

Pond of EVO plant Vojany is water works and its operation and safety oversight within the relevant legislation. It was built in 1965 to store dross ash mixture. Located on the left bank of the river Laborec in the administrative area village Vojany and Drahňov, on the verge of PLA Latorica and is bounded on all sides by raising grass covered embankments. Pond consists of two separate approximately the same cassettes (Figs 2–3, Table 7):

- Cassette No. 1–29 ha (with dam 47,2 ha).
- Cassette No. 2 –27 ha.

Cassettes are separated by dividing dam (Fig. 4), which originally had the function of the peripheral dam cassette No. 1. This means that the area to be addressed after the final shutdown of the pond is about 56 ha.

Table 5. The list of tailing ponds in Hungary.

No.	Name	Type
1.	Pellérd	Uranium
2.	Ajka	Red mud
3.		Red mud
4.	Kurity	Red mud
5.	Mosonmagyar óvár	Red mud
6.	Neszm	Red mud
7.	Bokod	Ash
8.	Borsodn	Ash
9.	Borsodszir	Ash
10.	Estergom/Dorog	Ash
11.	Dunaújváros	Ash
12.	Gyöngyösoroszi	Ash
13.	Inota	Ash
14.	Kazincbarcika	Ash
15.	Múcsony	Ash
16.	Pécs	Ash
17.	Tatabány/Bánhida	Ash
18.	Tiszapalkonya	Ash
19.	Tiszaújváros	Ash
20.	Visonta	Ash

Table 6. Examples of tailing ponds from the world of accidents resulting in death.

Tailing pond—dam break

Town/Country	Date	Number of death	Type of pond
Zemianske Kostoľany (Slovakia)	26.5.1965	4	Ashes from heat power plant
Stava (Italy)	19.7.1985	268	Fluorite sludge
Harmony (South Africa Republic)	6.2.1994	10	Cyanide pond
Placer (Philippines)	2.9.1995	12	Sludge
Ajka (Hungary)	4.10.2010	10	Red sludge

Figure 3. Filling tailing pond.

Table 7. Base parameters of tailing pond.

Area of base:	Cassette No. 1—47,2 ha
	Cassette No. 2—48,1 ha
Overall capacity of stock ash:	Cassette No. 1—7 580 000 m³
	Cassette No. 2—5 760 000 m³
Dispozition capacity:	Cassette No. 1—full
	Cassette No. 2—850.000 m³

a)

b)

Figure 2. Place cassette no. 1 on pond in scale 1:10 000.

Figure 4. Dividing dam of tailing pond.

3 EXPERIMENT NO 1—CONTAINERS

The authors present performance of the experiments in the research and development of new remediation technologies unconventional pond dross ashes mixture by using structured layers stabilized, soil and land, to replace previous legislative solution in the form of drainage system and the overlap hydro film material.

For verification of replacement waterproofing properties of stabilizer was based experiment simulating any large-scale use of this new non-traditional, uncertified practice anywhere technology. The purpose of verification or experiments was to assess the possibility of using stabilisation material due to its potential ability to prevent solidification of penetration of rain water into the lower layers of the pond, with the risk of a subsequent accident.

Covering the energy crop is a mixture of grass varieties that are resistant to typical and local conditions as to the future consideration of a pond grown plants used as biomass for co-incineration with coal in power plants. Experimentally it is verified and the cultivation of fast growing willow Swedish and with respect to its root system was therefore used in the experiment and subsoil thickness of 500 mm (Figs 5–6, Table 8).

The laboratory experiment was set up with the following procedure:

- On the bottom of 2×11 container with size of 1000 mm × 1000 mm × 1000 mm was stratified the stabilisation material, with thickness of 0, 50, 100, 150, 200, 250, 300, 350, 400, 450, 500 mm.
- this layer was then deposited in the soil subsoil thicknesses of 300 mm for grass, 500 mm for willow, which reflect the profile subsoil rehabilitated land.
- the last build up layer of topsoil with a thickness of 200 mm is uniformly for all variants, like the subsoil, even topsoil profile describe the reclaimed area
- half of containers was located in areas with variable weather conditions and the other half for use in the local climate (Majerník et al. 2012).

Composition of grass mixtures:

- Rapid:
Barrage	30%
Bartwingo	30%
Barustic	20%
Brooklawn	20%.
- Prosoil:
Barrage	30%
Barminton	20%
Bargreen	25%
Baron	20%
Barbian	5%.

Figure 5. Containers under conditions consistent with local climate.

Figure 6. Containers in terms of controllable.

Table 8. Requirements for growing.

	Subsoil	Topsoil	Soil together
Grass	300 mm	200 mm	500 mm
Willow	500 mm	200 mm	700 mm

3.1 Stabilisation material—its analysis and usage

In the following tables we provide an analysis of hazardous substances contained in the stabilized and the Fig. 7 shows the structured layers in individual containers used in the experiment. Used stabilisation material has pH 8,45 and conductivity 38,7 (Tables 9–10).

The containers were filled with layers of stabilisation material, subsoil and topsoil layers according to their draft. The first container was filled without a layer of stabilizer, as a control container—do nothing. Individual layers after they

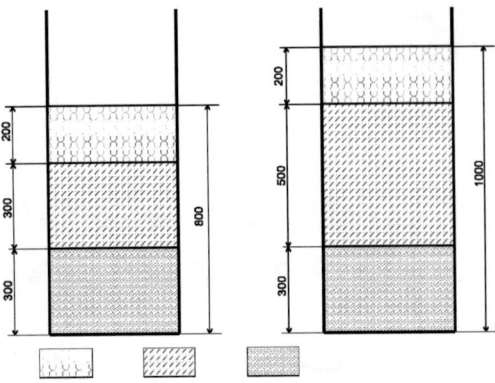

Figure 7. Diagram of the experimental variations of the experiment at 300 mm stabilizer.

Table 9. Analysis of stabilisation material based on based on Regulation of Ministry of Environment of SR No. 599/2005 Z.z.

No.	Indicator	Abbreviation	Value
1	Total organic carbon	TOC	105100
2	Benzene, toluene, ethylbenzene and xylenes	BTEX	<0,001
3	Polychlorinated biphenyls, 7 members	PCB	<0,01
4	Mineral oil (C10–C40)	NEL	<1
5	Polycyclic aromatic hydrocarbons	PAH	<0,05

Table 10. Analysis of water extract of stabilisation material based on Regulation of Ministry of Environment of SR No. 599/2005 Z.z.

No.	Indicator	Abbreviation	Value [mg]
1	Arsenic	As	0,02
2	Barium	Ba	0,58
3	Cadmium	Cd	<0,003
4	Total chromium	Cr	0,03
5	Copper	Cu	0,03
6	Mercury	Hg	0,002
7	Molybdenum	Mo	0,05
8	Nickel	Ni	<0,02
9	Lead	Pb	0,05
10	Antimony	Sb	0,01
11	Selenium	Se	<0,01
12	Zinc	Zn	0,10
13	Chlorides	Cl^-	31,7
14	Fluorides	F^-	2,2
15	Sulphurs	SO_4^{2-}	1490
16	Phenol index	FNI	<0,3
17	Dissolved organic carbon	DOC	24,3
18	Total solubles	CL	3240

were filled with compacted sufficiently to attempt what was closer to actual conditions.

3.2 Experimental method

The containers were located in areas where weather conditions are controllable and the same number of containers has been placed in areas where conditions are consistent with local climatic conditions.

Simulation of precipitation in real indoor and outdoor watering was applied to water, reflecting the maximum monthly average for last 50 years. Data on rainfall while the data drawn from the Slovak Hydrometeorological Institute, Regional Centre of Kosice from stations in Michalovce, in Milhostov, in Somotor and in Vysoká nad Uhom that are closest to the pond, Table 11.

The long-term measurements of rainfall in the area show that the average monthly values range from 40–50 mm/month, except for the rainy year 2010, when the monthly average rose to 85 mm/month.

3.3 Results and discussion

The results of the experiment can be divided into 2 groups according to different climatic conditions:

1. *Containers under conditions consistent with local climate:*
During the period November 2010–November 2011 rainfall was observed in containers on average 43.5 mm/month, which represents 1.45 mm of rainfall per day.

The natural rainfall of water do not get through the layers or in one container that stabilisation material, soil absorbed them.

2. *Containers in terms of controllable:*
The containers with variable conditions were simulated extreme daily rainfall amounts of water, min. 50 mm/day that is to say more than 7 times more than the maximum daily average precipitation. Part of the precipitation absorbed by each layer in the container and seepage water is accumulated discharge outlet in the prepared containers and continuous metering.

In carrying out experiments in order to determine the waterproofing ability of the individual layers, the

Table 11. The maximum rainfall of long-term measurements of nearby stations.

Meteo-station	Year	Month	Rainfall [mm]
Michalovce	2010	May	174
Milhostov	2010	May	219
Somotor	2010	May	226
Vysoká nad Uhom	2010	May	197
Max. average rainfall			204

daily rainfall amounts of water in controlled conditions was specified to multiple of the maximum long-term nature of rainfall in the area of the pond.

In the following Fig. 8 and 9 are shown depending on the thickness stabilizer and the amount of water tightness of the subsoil thicknesses of 300 and 500 mm.

Implementation of laboratory experiments conducted to establish the maximum daily amount of rain water at 100 mm (70-times the average daily value) are shown to be suitable alternatives to using 170 and 230 mm of stabilisation material depending on the thickness of subsoil differences for seed grass and willows. Saturated water in the subsoil and topsoil will be gradually pumped through the root system of plants planted on the surface.

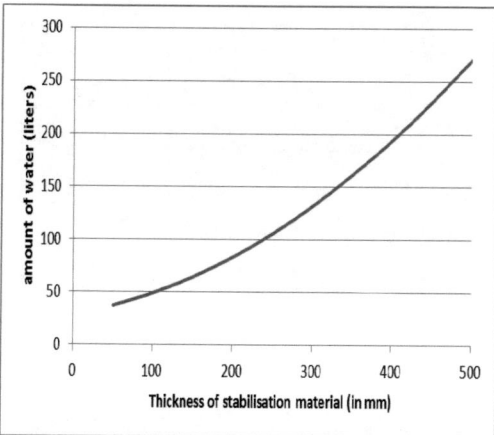

Figure 8. Dependence of the thickness of stabilizer and the amount of water tightness in thickness of the subsoil 300 mm.

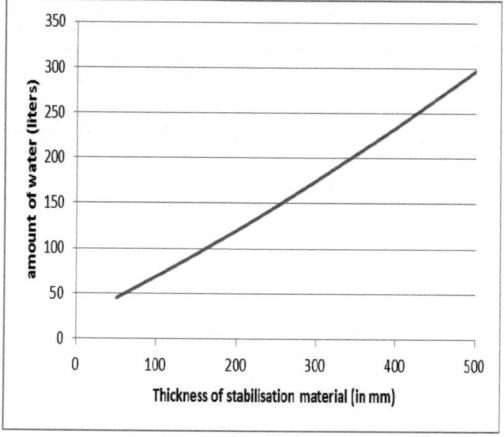

Figure 9. Dependence of the thickness of stabilizer and the amount of water tightness in thickness of the subsoil 500 mm.

4 EXPERIMENT NO 2—WILLOW CULTIVATION ON TAILINGS POND

The second phase of experiment was to implement small plot trials on the surface of tailing where four large-sized parcels of 7 meters x 20 meters in given structure have been built.

During the vegetation seasons in reality are tested the proposed alternatives of the covering layer for the biological re-cultivation of the decanting plant from the viewpoint of water permeability in natural conditions of atmospheric impacts and in the same time the possibility of willow cultivation in the closed decanting plant. The results of experimenting are supplemented with 5 years practical knowledge about the cultivation of Swedish willow on the territory of 15 ha nearby the town Kežmarok.

The average root striking of plant slips after the willow planting in 2 research stations was in the range between 66.91–89.51% and 45.67–91.35%. The planting was implemented by Slovak University of Agriculture in Nitra.

Dawson (2007) informs that in suitable conditions it is possible to achieve a root striking of more than 90 %. High root striking percentage of slips is inevitable for the optimal structure of vegetation and for optimal crop. The number of rooted pieces can be influenced by the planting way of slips during the founding of commercial plantations. Lowthe-Thomas et al. (2010) have identified better root striking in case of planting of slips horizontally to the soil surface in comparison with the classical planting vertically to the soil surface. This planting method in the same time can significantly reduce the costs of planting.

In our experiment we have achieved an average root striking of 91.76% (Table 12). From this result it is possible to conclude that the conditions for willow cultivation (structured layers of the stabilized waste, subsoil layer and arable layer) where suitably prepared and it is feasible to produce biomass directly in the power plant or in the decanting plant which is in distance of 2 km from the plant.

The optimal total quantity of precipitations in summer months should achieve 300 mm and dur-

Table 12. Number of rooted pieces of the Swedish willow in the experiment.

Plot	% potted plants
1.	88,72
2.	92,63
3.	91,14
4.	94,57
Average	91,76

a)

b)

Figure 10.   Experimental parcels on tailings pond.

a)

b)

Figure 11.   Grass mixtures rapid.

a)

b)

Figure 12.   Grass mixtures prosoil.

Figure 13.   Swedish willow on 4th parcel.

Table 13.   Structure of parcels for willow.

Parcel	Stabilisation material	Subsoil	Topsoil	Together
1st parcel	300 mm	500 mm	200 mm	1000 mm
2nd parcel	500 mm	500 mm	200 mm	1200 mm
3rd parcel	–	500 mm	200 mm	700 mm
4th parcel	–	500 mm	200 mm	700 mm

Table 14. Structure of parcels for grass.

Parcel	Stabilisation material	Subsoil	Topsoil	Together
1st parcel	300 mm	300 mm	200 mm	800 mm
2nd parcel	500 mm	300 mm	200 mm	1000 mm
3rd parcel	–	300 mm	200 mm	500 mm
4th parcel	–	300 mm	200 mm	500 mm

Table 15. Average amount of grass kg per 1 m².

Grass mixtures	2012		2013		2014	
	Green grass	Dry matter	Green grass	Dry matter	Green grass	Dry matter
Rapid	1,125	0,338	2,175	0,638	3,150	0,923
Prosoil	1,075	0,325	1,850	0,548	2,912	0,845

Table 16. Height of Swedish willow.

Season	Height of willow
2012—June	120–640 mm
2013—April	1100–1750 mm
2014—May	1900–2450 mm
2015—April	2800–3200 mm

ing the total vegetation season it should achieve 550 mm (Hall 2003). Besides the extremely rainy year 2010, the measured values for the whole vegetation season were lower than 570 mm.

The precipitation data have been gained from the 4 nearest monitoring stations to the decanting plant. The monitoring stations belong to relevant regional centre of the Slovak Hydro-Meteorological Institute.

Fast-growing tree species, specifically Sweden willow (Salix) and grass mixtures, were experimentally verified for individual variants as this composition was assumed to be suitable for these conditions. The grass mixture consists of Rapid mixture and Prosoil grass mixture (Figs. 10–13, Tables 13–16).

Since grass mixture seemed to grow unevenly due to rainfall and temperature conditions, in 2012 more of the grass has been planted. Proper mowing is being carried out in required agro-technical terms, while subsequently the substantiality of coppice in both green and dry state is being tested.

Sweden willow was planted in the spring of 2011. Unfortunately, young shoots of this willow were eaten by forest animals. As most of the shoots were damaged, it was necessary to carry out planting of new shoots of Sweden willow in spring 2012. Right after the planting it was necessary to fence these willows to avoid getting it eaten by forest animals again. In the spring of 2013 were willow root systems damaged by rodents and most of them were completely destroyed. Willow was replanted and fenced and rodent repellers were installed as well. Since the Swedish willow has been gradually replanted, various heights of willow shoots can be found on experimental fields. During the creation of appropriate conditions it is possible to achieve more than 90% of embeddedness of Sweden willow shoots. High embeddedness of shoots is essential for achieving the optimal harvest. The number of inrooted units might be influenced by the way of planting Sweden willow cuttings. Better embeddedness of shoots is being monitored during planting cuttings horizontally to the soil surface before planting the cuttings vertically to the soil surface. However, in this experiment the average embeddedness of shoots was up to 92%, thus we can conclude that conditions for growing Sweden willow were suitably prepared.

The optimal precipitation value for the entire growing season is approx. 580–600 mm of rainfall. At this level willow can produce large amounts of biomass. It can be said that according to the atmospheric precipitation of individual periods, the rainiest year was the year 2010, as shown in table 17. In the next three years the rainfall total was below the required level and did not exceed 550 mm. Consequently, Swedish willows have been drying out.

The overall process of experimental testing of the possibility of recultivating cinder/ash mixture tailings at Vojany thermal power plant through small plot trials continues. The biomass yield of mowed grass is corresponding with the expected values which increase from year to year. In terms of growing willow it would be appropriate if the annual rainfall total was around 600 mm. We might say that rainfall total for the last period was below the long-term average. Because of lower-

Table 17. The rainfall of nearby stations.

Station	2009	2010	2011	2012	2013	2014
Milhostov	585	892	526	496	529	567
Michalovce	636	929	567	635	578	625
Somotor	590	1030	486	481	522	520
Average	604	950	527	537	543	570

than-average precipitation during the year, the newly planted willow took a much lesser extent than initially expected. The growth and phenology of grass and willow, as well as the atmospheric precipitation, will be monitored within the small plot trials during the following periods.

5 EXPERIMENT NO 3—COMBUSTION OF BIOMASS

According to EU legislative from year 2008 EU has approved the energy sector goal for 2020—to provide 20% energy from renewable sources. One of the ways to achieve this goal can also be coincineration of biomass. Under the shared combustion of biomass and coal there is a reduction or better said partial elimination of the environmental impact due to low content of sulfur and nitrogen in the biomass, resulting in a reduction of CO, SO_2 and NO_x, as well as reducing emissions of heavy metals (Keoleian & Volk 2005).

Coincineration of biomass energy willows and other biomass plant with coal in the near future and in the present is currently considered one of the most promising methods for the provision of energy production in general. Despite the availability, technological and environmental benefits proclaimed by this system on a larger scale, it failed to apply in Slovakia and globally. The biggest problem appear to be the increased costs associated with the production and logistics of biomass assurance. Coincineration of biomass with coal significantly increases the clean energy ratio. It is defined as the ratio of produced electricity to the total consumption of fossil energy. When substituting a certain percentage of coal with biomass it is primarily reducing the greenhouse gas emissions from mining, transportation and combustion of coal (Heller et al. 2004). In practice, coincineration of willow chips with wood waste is being used in biomass power stations in Sweden for example it plays an important role in the local energy supply (McCormick & Kåberger 2007). Study of environmental impact of coal combustion from a power plant exists in a lot of counties and also in Poland (Kalembkiewicz & Chmielarz 2013, Kierczak & Chudy 2014).

In line with the above trends, power plant Vojany started aswell to implement in collaboration with the authors of the paper in 2009, in the form of scientific research complex project combustion of black coal with biomass in fluidized boilers, including the provision of (growing) plant biomass in the pond area surrounding the facility gaining self-made slag-ash mixture. Coincineration of biomass, mainly wood chips mixed with black coal in a 4% ratio produced its first positive results in the reduction of emissions by 40 kg per MWh produced, and operational savings associated with the consumption of limestone, creation and disposal of ash, steam and water consumption. The project's next phase was experimentally realized coincineration of biomass with a share 7% and then 15%. In terms of research focused on the potential provision of biomass showed that the surrounding of the facility has good power potential of the fast-growing energy crops - willow. In connection with this research was initiated specific purposeful cultivation of biomass on large pond that contained slag-ash mixture, about 56 ha (Bosák et al. 2013). The area of Vojany is one of most burdened areas in eastern Slovakia (Vilček et al. 2012).

5.1 The combustion process in the EVO facility before the experiments

Coincineration of biomass in EVO boilers started already in 2007. For first tests wood chips—forest biomass has been chosen. To achieve the same thermal input it was needed to deliver instead of one cubic of black coal, six cubic meters of biomass to the boiler, due to different density and heating value of wood chips and black coal calorific value, it could not be replaced by any amount of coal with biomass. Coal has a density of about 1 t/m³ and heating value of 25 MJ/kg. Wood chips have a density of 0.3 t/m³ and calorific value 10 MJ/kg.

From calculations based on the time they proved that the added mixture of wood chips containing about 4% of the heat energy mix wood chips—coal, does not negatively affect dynamic characteristics of the power plant units. Tests in 2007 also proved that it is possible smoothly combust biomass (wood chip) in a fluidized boilers in power plant Vojany. It turned out that wood chips have a positive impact on the boiler combustion mode, because of lower ignite temperature and a higher proportion of volatile matter than coal. Therefore there is more efficient combustion of irradiated fuel, which results in a decrease in the concentration of carbon monoxide in the exhaust gas (Vaszily 2012).

These partial positive results have initiated the launch of scientific-research activities aimed at complex solution for the issue of environmental energy-biomass combustion optimization processes.

5.2 Coincineration of wood chips with coal— I. stage

Replacement of a share of combusted black coal in thermal power EVO with biomass-based fuels was carried with priority to maximize the reduction

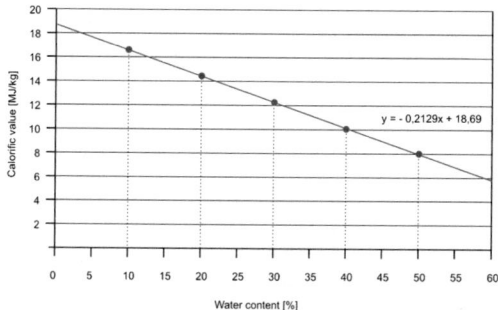

Figure 14. Dependency of the calorific value of wood chips on the water content.

of emissions especially carbon oxides sulfur by providing the required energy performance and therefore ultimately in increasing competitiveness improving economic indicators, manufacturing and energy-production companies.

The experiment was performed by the coincineration of black semi-anthracite coal wood biomass in a fluidized FK5 boilers in the examined facility.

The average heating values of black coal in the experiment ranged from 25.4 to 28.1 MJ.kg^{-1} and the wood chips ranged from 8.0 to 8.65 MJ.kg^{-1}. Course depending on the heating value of wood chips from its moisture is graphically illustrated in Figure 14.

Manufacturer guaranteed emission limits for dry combustion converted at 6% O_2 are: SO_2 400 mg.Nm^{-3}, NO_x 300 mg.Nm^{-3}, CO 250 mg.Nm^{-3}, Dust 50mg.Nm^{-3}.

Measurements within the experiment were carried out on the three different power levels block 66, 88 and 110 MW at:

a. Combustion of of black coal alone.
b. With 1.91% ratio of wood chips to total power input in the boiler.
c. With 3.91% ratio of wood chips on the total input to the boiler.

a) *The values of emissions of pollutants in the combustion of black coal alone.*

During the course of experimentation, the concentration Values Of Pollutants (VOP) in the exhausts were monitored. The process and the results are shown in Figure 15 and average values of concentration of the individual VOP are shown in Table 18.

For 1.91% share of wood chips the process and the results are shown in Figure 16 and average values of concentration of the individual VOP are shown in Table 19 and for 3.91% share of wood chips in Figure 17 and Table 20.

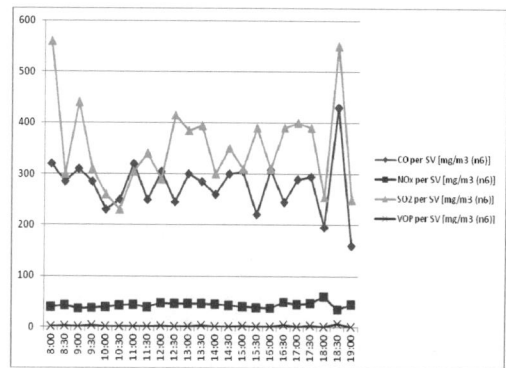

Figure 15. Time course of measured concentrations of pollutants, coal fuel.

Table 18. Average values of concentration of pollutants, fuel black coal.

Parameter	66 [MW]	88 [MW]	110 [MW]
SO_2 [mg.m^{-3} (n6)]	328,13	366,67	371,8
NO_x [mg.m^{-3} (n6)]	37,03	40,97	46,83
CO [mg.m^{-3} (n6)]	286,03	279,13	259,07
VOP [mg.m^{-3} (n6)]	5,9	5,9	6,3

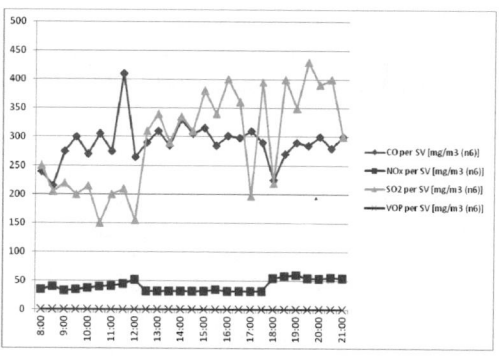

Figure 16. Time course of pollutant concentrations, with 1.91% share of wood chips.

Table 19. Average value concentration of individual pollutants with 1.91% share of wood chips.

Parameter	66 [MW]	88 [MW]	110 [MW]
SO_2 [mg.m^{-3} (n6)]	215,9	336,17	387,93
NO_x [mg.m^{-3} (n6)]	31,73	41,03	53,23
CO [mg.m^{-3} (n6)]	309,17	278,87	280,43
VOP [mg.m^{-3} (n6)]	6,1	6,3	6,3

603

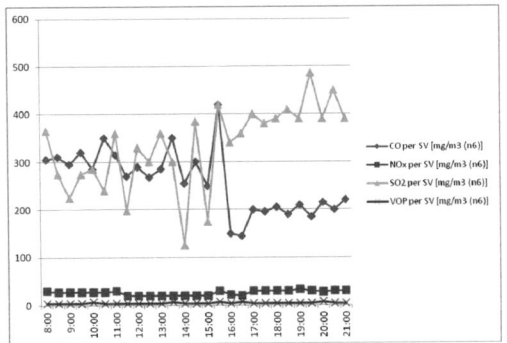

Figure 17. Time course of concentrations of pollutants, with a share 3.91% wood chips.

Table 20. Average value concentration of individual pollutants, with a share 3.91% wood chips.

Parameter	66 [MW]	88 [MW]	110 [MW]
SO_2 [mg.m^{-3} (n6)]	264,03	312,93	402,77
NO_x [mg.m^{-3} (n6)]	27,63	21,17	32,2
CO [mg.m^{-3} (n6)]	318,40	272,03	206,63
VOP [mg.m^{-3} (n6)]	6,5	6,6	6,6

b) *The total share ratio of wood chips on final boiler input 1.91%.*
c) *The total share ratio of wood chips on final boiler input 3,91%.*

From presented continuous forms of concentrations of pollutants (the individual VOP) it is significant that:

- During testing, legislative allowed emission limits for SO_2 were preserved and there has been an increase in performance and in values
- With increased performance there was a decrease in the concentration of carbon monoxide in coincineration ratio of wood chips.
- Other values of VOP are as well below the the individual emission limits.

Simultaneously the efficiency of boiler and coincineration of coal itself and the coincineration of biomass, are presented in Table 21.

Taking into account the partially different characteristics or better said quality of supplied and combusted coal and wood chip it can be stated that the coincineration of a mixture of coal and wood chip led indeed into a slight reduction in boiler efficiency, which is negligible compared to the achieved environmental-safety effects.

Based on these tests, and partial results of experimentation it has proceed to implementation I. phase in plant biomass coincineration in the power plant Vojany at the block no. 6 in July 2009.

Table 21. The effectiveness of fluidized boiler K5 in the holdings surveyed.

Electrical performance [MW]	Black coal [%]	Share of wood chip 1,91% [%]	Share of wood chip 3,91% [%]
66	93,81	93,47	92,30
88	93,55	93,32	92,91
110	93,52	92,57	93,20

a)

b)

Figure 18. Biomass—a) landfill, b) transport.

To support experiments a landfill with open capacity of 400 tons was built and customized adjusted with technology (crusher-sorter) and transport (conveyor belts) biomass (Fig. 18).

Mechanized system was used for wood chips, to transport them over conveyor belt and coal through the other one and a rated power of conveyor belts was set to properly balance mutual ratio. Use of raw wood chip (softwood and hardwood) and its share was gradually increased to 5.3%.

Share of 5.3% was not achieved by increasing the weight, but by coiniceration of better quality of the combusted wood chips with a higher calorific value than previously considered biomass which was considered in the project. This means over 11 MJ/kg compared to the projected calorific value 9.5 MJ/kg. This ratio has proved to be best possible to maintain the maximum dynamic properties of the boiler and in the execution of transporting the fuel mixture into the boiler. Subsequently, in relation to the achieved results it was determined on the implementation of II. stage of biomass project coinicineration, which consisted of the construction of independent mechanized access to the boiler especially for biomass.

Capacities of coal lines remained clear and the new path was used to transport higher volume of wood chips into the boiler.

The combustion of 1 ton of biomass eliminated approximately one ton of carbon dioxide emissions, Table 22.

5.3 Coincineration of wood pellets with coal— II. stage

Within next experiments in 2010 forest wood chips was replaced by wood pellets. Their advantage was that the bulk density was about 0.6 to 0.7 t/m³, approximately twice as much as the density of chips. Pellets additionally possessed a 1.5 times higher calorific value than wood chips. Energy parameters of wood pellets are closing up to coal properties. Their addition allowed to increase the proportion of biomass without compromising the energy operation of the power unit. Testing showed that it is possible to increase the share of biomass heat almost to 30% without any negative impact on the energy operation of the power unit.

Monitoring and measurement of individual power levels when burning a mixture of coal and biomass was determined in intervals of three hours. Stabilization of the boiler after the change of power level was 1 hour. The dependency of the calorific value of wood pellets from moisture is displayed graphically in Figure 19.

Measurements during the experiments were conducted at three individual power levels block 50, 80 and 100 MW at:

Table 22. The amount of CO_2 saved by coincineration of biomass.

Year	Wood chip share [t]	CO_2—eliminated [t]
2009	8 310	10 487
2010	21 443	27 061
2011	24 099	30 413
2012	26 917	33 969
2013	60 794	92 954
2014	48 752	84 899

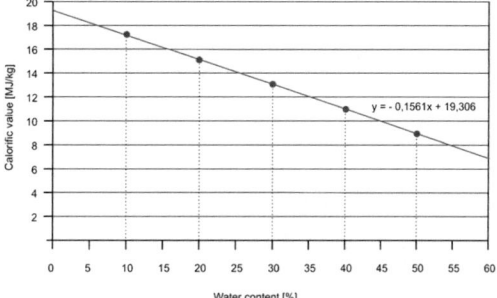

Figure 19. The dependency of calorific value of pellets on the water content.

Table 23. Characteristics of wood chips and pellets.

Type	Water content [%]	Calorific value [MJ/kg]
Wood chip	30,02	12,30
Pellets	5,50	18,45

Table 24. The average concentrations of individual pollutants, FK 6 EVO.

Parameter	Measure	Performance 50 MWe	Performance 80 MWe	Performance 100 MWe
Conc. SO_2	[mg/m³ (n6)]	284,97	241,97	239,7
Conc. NO_x	[mg/m³ (n6)]	77,53	93,73	128,7
Conc. CO	[mg/m³ (n6)]	187,0	217,9	234,5
VOP	[mg/m³ (n6)]	1,43	2,93	6,37

a. 6.64% share of biomass in total power consumption of the boiler,
b. 14.8% share of biomass in total input to the boiler.

a) *The share of total biomass boiler input 6.64%.*
Measuring and coiniceration of coal and biomass took place at the volume ratio of 1/3 mixture of wood chips and 2/3 pellets. The share of biomass in the total heat input of the boiler was 6.64%, according to the parameters of Table 23.

The average concentrations of individual pollutants are shown in Table 24.

During the experiments and measurements concentrations of pollutants in the exhaust gases were simultaneously monitored. In Figure 20. the time

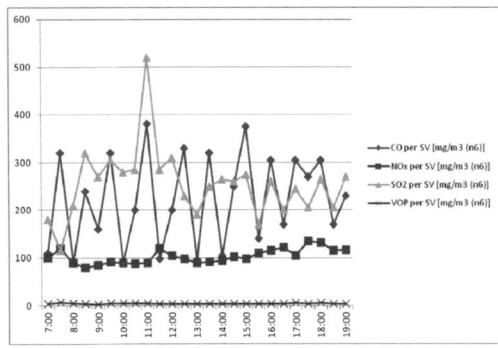

Figure 20. Time course of pollutant concentrations with a share of 6.64% of pellets.

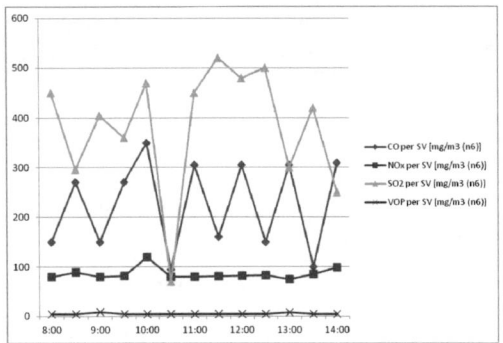

Figure 21. Time course of pollutant concentrations with a share of 14.8% of pellets.

Table 25. The average concentrations of individual pollutants, FK 6 EVO.

Parameter	Measure	Performance 50 MWe	Performance 80 MWe	Performance 100 MWe
Conc. SO_2	[mg/m³ (n6)]	348,4	446,3	363,1
Conc. NO_x	[mg/m³ (n6)]	71,15	71,10	77,35
Conc. CO	[mg/m³ (n6)]	188,3	222,7	214,6
VOP	[mg/m³ (n6)]	3,05	3,4	6,25

Table 26. Pellets properties.

Type	Water content [%]	Calorific value [MJ/kg]
Pellets	5,43	18,48

rapidly growing trees (poplar, willow) and EVO facility is a promising purchaser and will be their regular customer in the future as well. With self grown cultivation of biomass in surrounding area of the plant or in the wetlands there is potential for increasing employment in the region for people with lower qualifications.

5.4 Conclusions of the experimentation

The results of our previous experiments focused on the eco-energy optimization of biomass coiniceration with coal proved that:

- After the implementation of technological changes it is possible to improve homogenization of a mixture of coal and wood chips, in a long term combustion of this fuel mixture in thermal power SE a.s. and in general. The significant increase in the ratio of coinicerated biomass allows to build a separate route for mechanized transport of biomass from the surrounding area of the plant and pond slag-ash mixture directly into the boiler.
- Legislation required emissions of pollutants were met at 6.64% share of biomass on total power input of the fuel into the boiler, while the fuel had an average sulfur content of only 0.17%. The 14.8% share of biomass had 2-times the sulfur content in the fuel and the SO_2 emission were above the limit.
- Beneficial effect was observed by the reduction of the CO concentration with increasing quantity of biomass at lower boiler output.

course of the concentrations of pollutants in the exhaust gas is shown.

b) *Share of biomass on total boiler input 14.8%.*
During the measurements there were continuously monitored the concentrations of pollutants in the exhaust gases, Figure 21.

The average concentrations of individual pollutants are shown in Table 25.

During the experiment, the combustion of wood chips with an average calorific value of 12.30 MJ/kg and pellets with an average calorific value of 18.45 MJ/kg with the share of biomass at 6.64% and calorific value of 18.48 MJ/kg of biomass ratio of 14.8% on boiler input, Table 26.

Wood pellets in terms of calorific value and volume density are significantly suitable type of biomass, which can increase its share in the fuel mix for EVO fluidized Boilers.

The facility is obtaining its biomass currently from six local suppliers. The surrounding wetlands around the facility provide good conditions for

Production of biomass and incineration of coal in fluidized boilers, of thermal power stations SE, a.s., as well as other power plants is in terms of environmental benefits reflected in a significant reduction in emissions and production of solid waste generated by burning coal only.

The use of biomass is environmentally adequate way of power generation and has a positive impact on the environment. In cooperation with the EVO we have our own knowledge with this technology with experience for years. Daily in terms of experiment based model about tons of wood chips are coincinerated. Implementation of research results into practice, thus contributing to the European Union's commitment to continually increase the use of renewable energy of 20% by 2020.

ACKNOWLEDGEMENT

This contribution is the result of the project implementation VEGA No. 1/0936/15 Economics and Environmental Studies and experimental verification of the possibility of reclaiming tailings ash mixture in SE—EVO Vojany.

REFERENCES

Bosák, M., Majerník, M., Andrejovský, P., Hronec O. & Huttmanová, E. 2013. Applicable recycling technologies for the redevelopment of decanting plants originating from combustion processes, In *13th International Multidisciplinary Scientific Geoconference SGEM 2013: Ecology, Economics, Education and Legislation, Conference Proceedings*, Albena, Bulgaria, Volume I, pp. 79–86, ISSN 1314-2704.

Cossu, R., Lai, T. & Pivnenko, K. 2012. Waste washing pre-treatment of municipal and special waste. Journal of Hazardous Materials. Vol. 207–208, 65.

Dawson, W.M. 2007. Short Rotation Coppice Willow—Best Practice Guidelines. Renew Project, 48 p.

Hall, R.L. 2003. Short Rotation Coppice for Energy Production: Hydrological Guidelines. Report. Crown Copyright, 2003, 21 p.

Heller, M.C., Keoleian, G.A., Margaret, K., Mann, M.K. & Volk T.A. 2004. Life Cycle Energy and Environmental Benefits of Generating Electricity from Willow Biomass. In *Renewable Energy*, vol. 29, no. 7, pp.1023–1042, ISSN 0960-1481.

Jánová, V. & Panenka, P. 2010. Máme odkaliská pod kontrolou? In *Enviromagazín*, vol. 15, no 5, pp. 4–8, ISSN 1335-1877 [In Slovak].

Kalembkiewicz, J. & Chmielarz, U. 2013. Effects of Biomass Co-Combustion with Coal on Functional Specfiation and Mobility of Heavy Metals in Industrial Ash, Pol. J. Environ. Stud. 22, (3), 741.

Keoleian, G.A. & Volk, T.A. 2005. Renewable Energy from Willow Biomass Crops. Life Cycle Energy, Environmental and Economics Performance. In *Critical Reviews in Plant Sciences*, vol. 24, no. 5–6, p. 386–406, ISSN 0735-2689.

Kierczak, J. & Chudy, K. 2014. Mineralogical, Chemical, and Leaching Characteristics of Coal Combustion Bottom Ash from a Power Plant Located in Northern Poland, Pol. J. Environ. Stud. 23, (5), 1627.

Lowthe-Thomas, S.C., Slater, F.M. & Randerson, P.F. 2010. Reducing the establishments costs of short rotation willow coppice (SRC). In *Biomass and Bioenergy*, vol. 34, 2010, no. 5, pp. 677–686, ISSN 0961-9534.

Majerník, M., Bosák, M., Andrejkovič, M., Hajduová, Z. & Turisová, R. 2012, Prevention of environmental risks of tailings ponds. In 12th international multidisciplinary scientific geoconference SGEM 2012: conference proceedings, volume III, Albena, Bulgaria.—ofia: STEF92 Technology Ltd. ISSN 1314-2704. pp. 619–626.

McCormick, K. & Kåberger, T. 2007. Key barriers for bioenergy in Europe: Economic conditions, know-how and institutional capacity and supply chain coordination. In Biomass and Bioenergy, vol. 31, no.7, pp. 443–452, ISSN 0961-9534.

Pacaiova, H., Nagyova, A. & Markulik, S. 2013. BBS Program (Behavior Based Safety)—Way to Raising the Level of Health and Safety in Practice, Advances in physical ergonomics and safety, Book Series: Advances in Human Factors and Ergonomics Series, p. 70–77, ISBN 978-1-4398-7059-4.

Vaszily, M. 2012. Three years experience. Slovenská energetika, Slovenské elektrárne, a. s., p. 3–5, no. 6, vol. 37, ISSN 1335-2849, [In Slovak].

Vilček, J., Hronec, O. & Tomáš J. 2012. Risk Elements in Soils of Burdened Areas of Eastern Slovakia, Pol. J. Environ. Stud. 21, (5), 1429.

Author index